Collins
dictionary *of*

Botany

William Collins' dream of knowledge for all began with the publication of his first book in 1819. A self-educated mill worker, he not only enriched millions of lives, but also founded a flourishing publishing house. Today, staying true to this spirit, Collins books are packed with inspiration, innovation, and practical expertise. They place you at the centre of a world of possibility and give you exactly what you need to explore it.

Collins. Do more.

Collins
dictionary *of*
Botany

Jill Bailey

Collins

HarperCollins Publishers
Westerhill Road, Bishopbriggs,
Glasgow G64 2QT

www.collins.co.uk

A catalogue record for this book is available from the British Library

ISBN-13 978-0-00-721220-0
ISBN-10 0-00-721220-8

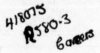

Typeset by Market House Books Ltd, Aylesbury
Printed and bound in Great Britain by Clays Ltd, St Ives plc

Preface

The Collins Dictionary of Botany encompasses in almost 6000 entries all the major fields of pure and applied plant science, including taxonomy, anatomy and morphology, physiology, biochemistry, cell biology, plant pathology, genetics, evolution, and ecology. It also covers some aspects of agricultural botany, horticulture, and microbiology. This edition is a substantial revision of the former Penguin Dictionary of Plant Sciences. In particular, we have expanded our coverage of genetics and biotechnology – reflecting important advances in these fields and the increasing use of DNA technology in horticulture and agriculture – investigative techniques, conservation, ecology, and plant anatomy. The dictionary should prove invaluable to 'A' level and undergraduate students of plant science and biology and to naturalists, geographers, and others studying or working in related fields. It is also hoped that it will be of use to anyone with a general interest in the plant world, who seeks explanations of scientific terms that are currently very much in the news.

The trivial, rather than the systematic, names of organic compounds are used throughout. However, the recommended systematic names of some of the commoner organic compounds appear as cross-reference entries in the dictionary and are also listed in the Appendix.

No attempt has been made to include named species or genera of plants, since the numbers involved make this beyond the scope of the dictionary. However, higher ranks of plant groups (phyla, classes, and the more important orders) are included, as are the subclasses and some of the large families of the flowering plants. The vernacular names of all these groups are given where appropriate. All the taxonomic entries have been revised to take account of the Five Kingdoms classification, which is now widely taught in schools and universities. However, the taxonomy of the flowering plants, fungi, and algae is currently in a state of flux and there is no single universally accepted system. Where more than one classification is in general use, we have attempted to cross-relate entries. We have based our flowering plant taxonomy on that used by Dr David Mabberley in The Plant-Book, 2nd edition, 1997, with modifications to take account of popular usage. Algal classes and orders are based on the system used in The Biology of the Algae, Philip Sze, 3rd edition, 1998. For the fungi we have tried to satisfy the several systems currently in use. Ainsworth and Bisby's Dictionary of the Fungi, 8th edition, 1998, is a useful reference for further information. We have included only a limited amount of bacterial taxonomy, as this is being substantially revised.

Jill Bailey
2006

Contributors

Consultants:
Sir John Burnett MA, DPhil, DSc
Andrew Lack BSc, PhD

Contributors:
Richard A. Blackman MSc, DIC
Jonathan Y. Clark BSc
Mary Clarke BSc
E. K. Daintith BSc
M. R. Ingle BSc, PhD, MIBiol, FLS
Sheila C. A. Jones, BSc
Sabina G. Knees, BSc, MSc
Lynn Mayers, BSc, DTA
Ian Andrew Solway BSc
Elizabeth Tootill BSc, MSc

Aa

Aapa mires ribbon-like BOGS associated with ridges across sloping ground. The bogs lie in the depressions between the ridges, and are fed by rainwater and run-off. Aapa mires are found worldwide, but especially in Scandinavia and Finland.

ABA see ABSCISIC ACID.

abaxial describing the side of leaves, petals, etc. facing away from the stem or main axis, i.e. the lower surface. Compare ADAXIAL. See also DORSAL.

abiogenesis the development of living organisms from nonliving matter as envisaged in the modern theory of the ORIGIN OF LIFE. The term is also sometimes used to mean SPONTANEOUS GENERATION.

abiotic factors the nonliving components of the environment that directly affect plant and animal life, such as water, carbon dioxide, oxygen, and light. Abiotic factors include CLIMATIC, EDAPHIC, and PHYSIOGRAPHIC FACTORS. Compare BIOTIC FACTORS.

abortive transduction see TRANSDUCTION.

abscisic acid (ABA, abscisin, dormin) a plant hormone that has numerous different, mainly inhibitory, effects on the growth and development of many species. It is a sesquiterpenoid (see TERPENOID), formula $C_{15}H_{20}O_4$ (see diagram) and, like gibberellin, may be synthesized from mevalonic acid. It is active, possibly in association with gibberellic acid, in the promotion of leaf and fruit ABSCISSION, senescence, and the control of DORMANCY. It prevents cell elongation and shoot growth and also inhibits seed germination and some tropic responses. At physiological concentrations, ABA is not toxic to plants,

Abscisic acid

which can, if necessary, remove its effects by converting it to an abscisyl glucoside by linking it to a glucose residue. A large proportion of ABA is synthesized in the chloroplasts. The rate of synthesis increases dramatically when the plant is under stress, especially from water shortage. ABA overrides the normal diurnal pattern of stomatal opening and closure and causes the stomata to close during the day. This response decreases water loss by transpiration in times of drought.

abscission the controlled shedding of a part, such as a leaf, fruit, flower, or bulbil, by a plant. The process is usually associated with a decline in auxin level within the organ to be detached. An *abscission zone* of tissue often forms at the point of separation, which is normally at the base of a petiole or pedicel (see ABSCISSION JOINT). A thin plate of cells, the *abscission* (or *separation*) *layer*, forms within the abscission zone. The pectic acid in the cell walls of the abscission layer is converted to pectin, resulting in a softening and weakening of the region. The organ is then easily dislodged from the plant by wind, heavy rain, etc. ABSCISIC ACID, which promotes leaf senescence, may also play a part in abscission. ETHYLENE (ethene) has been shown to accelerate the abscission of senescent leaves. Precocious abscission of fruits (FRUIT DROP) is a common phenomenon. See also CLADOPTOSIS.

abscission joint (struma) the articulation point on a leaf, at which the leaf will detach from the rest of the plant. These 'joints' may occur at the bases of the individual leaflets of compound leaves, or near the base of the leaf itself. The joint may be visible as a swollen area, often with a conspicuous groove around the plane of breakage, or it may not be externally obvious. See also PULVINUS.

abscission layer see ABSCISSION.

abscission zone see ABSCISSION.

absolute dating see DATING.

absolute humidity see HUMIDITY.

absorption 1 The uptake into a plant of water, solutes, or other substances by either active or passive means. Entry invariably involves movement across cell membranes. *Active absorption*, for example the uptake of a solute against an osmotic gradient, involves expenditure of energy (see ACTIVE TRANSPORT). An example of *passive absorption* is the intake of water by plant roots, which is controlled by the rate of transpiration.
2 The retention of radiant energy by the pigments of a plant. About 80% of the incident visible light and about 10% of the infrared radiation falling on a leaf is absorbed. Generally, less than 2% of this is used in photosynthesis.

absorption spectrum the pattern of bands or lines obtained by passing white light through a selectively absorbing substance into a spectroscope. It is specific for any one compound and gives a characteristic profile when plotted against wavelength. Absorption spectra have been important in the study of photosynthesis: a comparison of the absorption spectra of the various photosynthetic pigments with the ACTION SPECTRUM of photosynthesis shows which pigments are contributing absorbed light energy to the photosynthetic process. See diagram at ACTION SPECTRUM. See also COLORIMETRY.

abundance the relative quantity or number of plants in a given area. There are many different measures of abundance, some subjective, some quantitative. Quick estimates of abundance rely on the subjective assessment of frequency, the relative abundance of each species present in an area. This method has been refined into the BRAUN–BLANQUET SCALE, which combines assessment of both the COVER and the distribution pattern of the plants. More accurate quantitative assessments mainly use DENSITY (the number of individuals in a given area), cover, or the frequency of occurrence of a species in randomly placed quadrats of specific sizes. See also COVER-ABUNDANCE MEASURE, DENSITY-FREQUENCY-DOMINANCE, DOMIN SCALE.

acaulescent describing plants that have no stem or an extremely short stem, such as tufted or rosette plants.

accessory buds (supernumerary buds) the second and subsequent buds that form when more than one bud arises in the axil of a single leaf. Accessory buds that form side by side in the leaf axil are called *collateral buds*. Those that form in line with the stem axis are called *serial buds* (*superposed buds*). The order in which accessory buds subsequently develop is species-specific. They may form different organs from those produced by the main bud (for example, flowers, spines, or tendrils) or they may be used to replace a main bud damaged by frost or other injury.

accessory cell see SUBSIDIARY CELL.

accessory chromosome see B-CHROMOSOME.

accessory fruit see FRUIT, PSEUDOCARP.

accessory pigments pigments other than chlorophyll *a* found in photosynthetic cells. They include the carotenes and the xanthophylls (together known as the carotenoids), chlorophylls *b*, *c*, and *d*, and the phycobiliproteins. The latter are found only in the blue-green bacteria (see CYANOBACTERIA) and in red algae (see RHODOPHYTA). The composition of accessory pigments in algae is used as a taxonomic character.

The accessory pigments function as secondary absorbers of light in regions of the visible spectrum not covered by chlorophyll *a*. The light energy that they absorb must be transferred to chlorophyll *a* before it can be used in the photosynthetic process. As energy transfer from one molecule to another can occur only from a shorter wavelength absorbing form to a longer wavelength absorbing form, all accessory pigments have absorption maxima at shorter wavelengths than chlorophyll *a*. Energy is passed from chlorophyll *a* to the reaction centre pigments, P680 and P700 (see REACTION CENTRE.

Some accessory pigments may have a protective function, preventing photooxidation of the cell's chlorophyll at high light intensities.

accessory transfusion tissue in certain gymnosperms, e.g. *Cycas*, TRANSFUSION TISSUE that extends through the mesophyll from the midrib to the margin.

accidental species see FIDELITY.

acclimation see ACCLIMATIZATION.

acclimatization a reversible change in the morphology or physiology of an organism in response to changes in its environment. When referring to laboratory situations, the term *acclimation* is often used instead.

accumulator in plant SUCCESSION, a pioneer species whose activities enhance the build-up of nutrients in the environment.

-aceae a suffix used to denote the name of a family of plants, e.g. Liliaceae, the lily family. See also RANK.

acellular describing tissues, organs, or organisms consisting of a mass of protoplasm not divided by membranes or cell walls into discrete units. The term may be used of relatively large uninucleate organisms, such as the green alga *Acetabularia* (see CHLOROPHYTA), to distinguish them from comparatively less specialized and smaller unicellular organisms. More often acellular structures are multinucleate and result from free nuclear division (see KARYOKINESIS) with no accompanying cell wall formation. When describing multinucleate organisms, tissues, or cells the term is synonymous with COENOCYTIC. Examples of acellular multinucleate tissues are the endosperm of certain angiosperms (see ANTHOPHYTA) and the initial stages in formation of the proembryo in *Cycas* (see CYCADOPHYTA). Acellular multinucleate organisms include

siphonous green algae of the order Caulerpales and many fungi.

acellular slime moulds see MYXOMYCOTA.

acentric describing a CHROMOSOME or a fragment of a chromosome that lacks a centromere. Such a chromosome will be unable to attach to the spindle, so will be unable to migrate to the pole during anaphase of mitosis or meiosis.

acervulus (*pl.* acervuli) a small disc-shaped mass of conidiomata that erupts through the epidermis of plants infected by asexual fungi (see FUNGI ANAMORPHICI) of the order Melanconiales.

acetaldehyde (ethanal) an intermediate in the conversion of pyruvic acid to ethanol, the final stage of GLYCOLYSIS under anaerobic conditions in plants. The conversion of acetaldehyde to carbon dioxide and ethanol is energy requiring, involving oxidation of a reduced molecule of NAD. Acetaldehyde is also involved in the synthesis and breakdown of the amino acid threonine.

acetic acid (ethanoic acid) a weak organic acid, formula CH_3COOH. Acetic acid can be used as an alternative carbon source by certain algae (e.g. *Chlamydomonas mundana* and species of *Chlorella*), while green algae of the genus *Chlamydobotrys* are totally dependent on acetic acid as a carbon source. In combination with coenzyme A (see ACETYL COA) acetic acid plays a central role in aerobic energy metabolism.

acetocarmine a stain used in microscopy to colour chromosomes deep red-black. The tissue preparations are fixed in acetic acid before the acetocarmine is added.

aceto-orcein a stain used in the preparation of root-tip or anther squashes for chromosome examination. The material to be stained is placed in the aceto-orcein (which is acidified with hydrochloric acid) and heated at 60°C for 15 minutes. The material is then removed and mounted in acetic acid.

acetyl CoA (acetyl coenzyme A) a compound consisting of acetyl combined through a sulphur bridge with COENZYME A. The formation of acetyl CoA is an energy-requiring reaction, involving the conversion of ATP to AMP and pyrophosphate. Acetyl CoA plays a central role in intermediary metabolism. It is a product of the degradation of fatty acids, carbohydrates, and some amino acids. It is an essential precursor in the KREBS and GLYOXYLATE CYCLES and is the starting point for the synthesis of fatty acids, terpenes, and some amino acids.

achene any simple one-seeded indehiscent dry fruit that develops from an ovary made up of a single carpel. The CARYOPSIS, CYPSELA, and SAMARA are all types of achene. Compare NUT.

acicular needle shaped, for example the acicular crystals that, closely packed together, form a RAPHIDE, and the leaves of pine (*Pinus*) trees. *See illustration at leaf.*

acid a substance that releases hydrogen ions (protons) or H_3O^+ when dissolved in water. An acid in aqueous solution has a pH of less than 7.0. See also HYDROGEN, PH.

acid-fast bacteria see ACID-FAST STAIN.

acid-fast stain a widely applied stain in bacteriology that is used to identify bacteria that can retain a dye on washing with acid alcohol. One technique is to stain a bacterial smear with hot carbol fuchsin and, after rinsing with water, expose the smear to concentrated hydrochloric acid dissolved in water or ethanol. Following a second rinsing, the smear is counterstained with methylene blue. *Acid-fast bacteria* retain the red colour of the carbol fuchsin while other organisms appear blue. Examples of acid-fast bacteria are species of the genus *Mycobacterium*.

acid grassland GRASSLAND occurring on acidic soil, often derived from natural woodland as a result of grazing or fire. In upland areas such grasslands form distinctive communities sometimes called *grass heaths*.

acid rain rain with a high level of acidity due to pollution by oxides of nitrogen and sulphur produced by fossil fuel combustion. The oxides combine with water in the atmosphere, forming such acids as nitric and sulphuric acids. This process may be catalysed by other pollutants, such as ozone, hydrogen peroxide, and ammonia. Wind may blow the pollutants hundreds of kilometres from industrial areas, producing disastrous effects in such areas as Scandinavia,

Scotland, northern England, eastern Europe, Russia, and North America. Acid rain is highly damaging to plants, especially conifers. It renders soil toxic because the sulphuric acid forms ammonium sulphate in the soil; this liberates toxic aluminium ions and other heavy-metal ions, which may leach into waterways and drinking water supplies. Such ions inhibit metabolic processes. If stressed, certain plants release terpenes, which attract insect pests. Heavy-metal ions also damage the gills of fish. Acid rain also upsets aquatic ecosystems by lowering the pH of lakes and waterways.

acid soil a SOIL having a pH of less than 7.0. See also PEDALFER.

acid stain see STAINING.

ACP see ACYL CARRIER PROTEIN.

acquired characteristic a characteristic of a living structure or biochemical system that has been brought about by environmental factors acting during the life of the organism. For example, plants receiving fertilizer applications may grow taller and more vigorously than before and pot plants moved to a shady position may become etiolated (see ETIOLATION). It is a fundamental tenet of NEO-DARWINISM that such variations are not inherited by the succeeding generation. See LAMARCKISM.

Acrasiomycetes see ACRASIOMYCOTA.

Acrasiomycota (Acrasiomycetes, cellular slime moulds) a class of the phylum Rhizopoda in the PROTOCTISTA (formerly a phylum in its own right), also sometimes placed in the Fungi, containing organisms that feed on soil bacteria. They are found in fresh water and damp soil, on dung, and on decaying plant and fungal material. There are some 12 species in 4 genera, found in greatest abundance in woodland litter. Cellular slime moulds are made up of independently living and dividing amoebae that feed on bacteria. Under certain conditions (such as food storage) they form a sluglike aggregation of MYXAMOEBAE called a *pseudoplasmodium*. The myxamoebae have characteristic wide lobed pseudopodia, and their nuclei have a compact centrally placed nucleolus. They produce slimy masses of walled spores or *pseudospores* (encysted myxoamoebae), borne on upright stalks, by a process different from that of other moulds. The spores germinate into individual amoebae. Acrasiomycota are not known to reproduce sexually. The cellular slime moulds are often classified as an order (Acrasiales) of the MYXOMYCOTA, or as a class of the Gymnomycota (naked fungi), and were once placed in the Protozoa by zoologists.

acre a unit of area used in Britain and the USA. 1 acre = 4840 square yards or 0.447 hectare. In Anglo-Saxon times it was a strip of land measuring 660 x 66 feet (10 x 1 furlongs), but came to mean a piece of land of any shape with that area. In historical time British acres have been of different sizes in different places, ranging from 0.19 to 0.911 hectare, being the area of land that could be ploughed in one day by a wooden plough pulled by a yoke of oxen.

acrocarpous describing mosses (see BRYOPHYTA) in which the main axis is terminated by the development of the reproductive organs so subsequent growth is sympodial. In such mosses the main axis is almost always erect. Compare PLEUROCARPOUS.

acrocentric describing a chromosome in which the CENTROMERE is not centrally placed, giving two arms of different lengths. Compare METACENTRIC, TELOCENTRIC.

acronematic flagellum see WHIPLASH FLAGELLUM.

acropetal describing movement, differentiation, etc. occurring from base to apex of root or shoot. For example, the development of flowers is usually acropetal with the calyx being formed first and the gynoecium last. Movement of water through the plant is acropetal. Compare BASIPETAL.

acrostichoid condition the situation found in certain ferns, e.g. *Platycerium*, in which the sporangia are formed over the entire undersurface of a fertile frond rather than in specialized organs or sori.

acrotony a pattern of development of shoot branches in which the branches nearest to the shoot apex grow the most vigorously, so that the side shoots come to overlap the main shoot (see diagram). Compare BASITONY, MESOTONY.

actin a globular protein found in eukaryote cells, which is involved in cellular movements, motility, and the cytoskeleton. Helical strands of actin molecules (*G actin* or *globular actin*) twine around each other to form microfilaments called *F-actin*, or *fibrous actin*. Actin may make up as much as 10 to 15% of the total cell protein.

actino- a prefix meaning 'radiating'.

Actinobacteria (Actinomycetales, Actinomycetes, Actinomycota, ray fungi) a phylum of Gram-positive EUBACTERIA comprising two main classes, the true actinobacteria and the *coryneform bacteria*. They share certain features of morphology, physiology, and 16S rRNA. The coryneform bacteria are rodlike structures, but the true actinobacteria usually form branching filamentous structures resembling very small fungal mycelia, whose hyphae are rarely more than 1.5 μm in diameter. Actinobacteria are characterized by producing *actinospores*, formed from whole cells that develop thick walls to become resistant spores (compare ENDOSPORE, CONIDIUM). Most are saprobes, but some are pathogens of humans, animals, and plants. Some are important sources of antibiotics, including streptomycin and tetracycline. A few actinobacteria form lichen-like symbioses with chlorophytes, known as *actinolichens*.

actinodromous describing a form of leaf venation in which three or more primary veins originate at the base of the lamina and run out towards the margin. The leaves

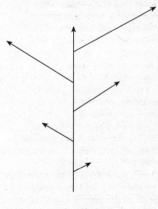

Acrotony

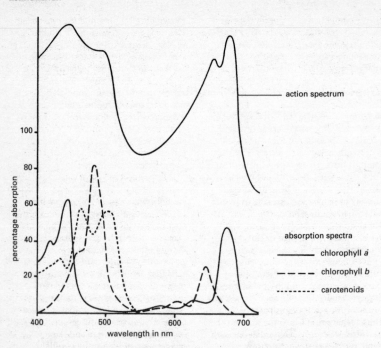

Action spectrum: a comparison of the action spectrum of photosynthesis with the absorption spectra of chlorophylls *a* and *b* and the carotenoids

of sycamore (*Acer pseudoplatanus*) are an example. In traditional terminology such venation is termed palmate or digitate. *See illustration at* venation.

actinolichen see ACTINOBACTERIA.

actinomorphy see RADIAL SYMMETRY.

Actinomycetales see ACTINOBACTERIA.

Actinomycetes see ACTINOBACTERIA.

actinomycosis any disease of humans or other animals that is caused by ACTINOBACTERIA.

Actinomycota see ACTINOBACTERIA.

actinospore see ACTINOBACTERIA.

actinostele see STELE.

action spectrum a plot of the rate of a reaction, e.g. the phototropic response or photosynthesis, at different wavelengths of light. The action spectrum for photosynthesis is obtained by measuring the photosynthetic yield for a given amount of light incident upon the plant over a range of wavelengths. For green plants the action spectrum shows that chlorophyll is the pigment responsible for photosynthesis, since peak photosynthetic activity occurs at the absorption peaks of chlorophylls *a* and *b*.

For some algae the action spectrum shows peak photosynthetic activity at the absorption peak of one of the accessory pigments. In this situation although chlorophyll is still used for transfer of light energy into the photosynthetic process, it is not the primary light-absorbing pigment.

activation energy the minimum amount of energy required to enable a specific chemical reaction to proceed spontaneously. It increases the energy levels and hence the reactivity of the participating molecules. Enzymes enhance reaction rates by lowering the activation energy.

activator 1 A protein that positively regulates the transcription of a gene. 2 A molecule or ion that associates with an enzyme or its

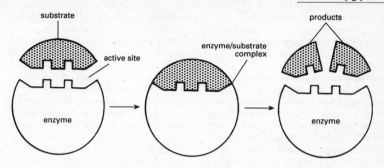

Active site: schematic representation of the induced fit hypothesis

substrate to induce a reaction. It may be a metal ion or a hormone.

active absorption see ABSORPTION.

active chamaephyte see CHAMAEPHYTE.

active layer in permafrost regions, the part of the surface layer of a soil that thaws out seasonally. This may range from a few centimetres to 3 m in thickness. See SOIL PROFILE.

active pool see BIOGEOCHEMICAL CYCLE.

active site the site at which the substrate of an enzyme is bound during catalysis. Once bound, the substrate reacts to form a product or products, which are then released from the active site. The CONFORMATION and charge distribution of the active site is carefully configured to accept only a specific substrate or class of substrates. Within the active site there are areas concerned with binding the substrate and other areas that create conditions conducive to the catalysis of substrate to products.

The theory that best explains how the active site and the substrate interact is the *induced fit hypothesis* (see diagram). This considers the active site to be flexible and to adjust its conformation in response to the presence of a substrate molecule. The resulting distortion of both substrate and enzyme puts a strain on the substrate and is one factor contributing to the increased reaction rate.

active transport the transport of substances, usually polar molecules or ions, across a membrane against a concentration gradient. Active transport is energy-requiring and is mediated by specific carrier proteins or *translocases*, which selectively bind a substrate and transport it across the membrane. The most common example of an active transport system is the sodium/potassium system, which actively pumps sodium ions out of the cell at the expense of ATP, while at the same time carrying potassium ions into the cell. The resulting concentration gradient can be very high; in the freshwater chlorophyte *Nitella clavata* the ratio of potassium inside the cell to that outside is 1065:1. See also CHEMIOSMOTIC THEORY. Compare FACILITATED DIFFUSION.

actual evapotranspiration see EVAPOTRANSPIRATION.

acumen in mosses, the upper, tapering, part of a leaf.

acuminate gradually narrowing to a point, as do certain leaves.

acute having a sharp but not extended point, as certain types of leaf.

acyl carrier protein (ACP) a low-molecular-weight conjugated protein. It forms a complex with the enzymes of fatty acid synthesis and binds the acyl group (RCO-) of the growing fatty acid molecule, thus bringing enzymes and substrate together. The prosthetic group of ACP, 4'-phosphopantetheine, is identical to the acyl-carrying portion of COENZYME A. See also FATTY ACID METABOLISM, MALONYL ACP

acylglycerol (glyceride) an ester of the alcohol GLYCEROL in which a hydroxyl group is replaced by a fatty acid radical. Acylglycerols are one of the four main types of complex LIPIDS. Often all three of the glycerol hydroxyl groups are esterified giving a TRIACYLGLYCEROL. Some diacylglycerols contain a sugar attached to the unesterified hydroxyl group. These glycosyldiacylglycerols are commonly termed GLYCOLIPIDS.

adaptation a modification to an organism, or a feature of an organism, that makes it better fitted for a particular environment. Adaptations may either be acquired during the life of the individual or, if governed by the genotype, be inherited. All organisms show adaptation to a greater or lesser extent, i.e. they have evolved from ancestors not adapted in the same way. The diversity of floral structure reflects the way different plants have become adapted to different pollinators.

adaptive enzyme see INDUCIBLE ENZYME.

adaptive peaks and valleys features on a contour map showing the ADAPTIVE VALUE of different allele combinations (genotypes) in particular environments. The peaks and valleys correspond respectively to locations where the FITNESS is strong or weak, The distribution of a given species or race will tend to coincide with the adaptive peaks for its particular genotype.

adaptive radiation the evolution of a number of different groups of organisms from a common ancestral group. It is most clearly seen in island groups, such as the various species of the rubiaceous genus *Hedyotis*, which occupy different niches in the Hawaiian Islands. See also ISLAND BIOGEOGRAPHY, SPECIATION.

adaptive value (selective value) the balance of advantages and disadvantages conferred by a particular gene or genotype for the survival and reproduction of an individual organism in a given environment. See also FITNESS, NATURAL SELECTION.

adaptive zone a TAXON and its associated ECOLOGICAL NICHE. A species or group of species that is relatively unspecialized may occupy a broad adaptive zone, exploiting and interacting with various BIOTIC and ABIOTIC FACTORS of its environment, while a more specialized taxon will have a narrow adaptive zone.

adaxial describing the side of lateral organs facing towards the stem or main axis, i.e. the upper surface. Compare ABAXIAL. See also VENTRAL.

adelphous describing an ANDROECIUM in which the filaments are fused.

adenine a nitrogenous base, more correctly described as 6-aminopurine, derived from amino acids and sugars and found in all living organisms. Adenine is a constituent of DNA, RNA, ATP, NAD, and FAD. Many CYTOKININS are derivatives of adenine and adenine itself shows cytokinin activity.

adenosine a combination of ADENINE with D-ribose, a pentose sugar. The two molecules are linked by a β-glycosidic bond and together form a NUCLEOSIDE. Adenine may alternatively combine with 2-deoxy-D-ribose (see DEOXYRIBOSE), in which case the nucleoside is called *deoxyadenosine*. The latter is the form found in DNA, while adenosine is found in RNA, ATP, ADP, and AMP. Certain derivatives of adenosine (e.g. isopentenyl adenosine or IPA) are CYTOKININS. See also NUCLEOTIDE.

adenosine diphosphate see ADP.

adenosine monophosphate see AMP.

adenosine triphosphate see ATP.

adenyl cyclase see CYCLIC AMP.

adhesorium (*pl.* adhesoria) an adhesive organelle formed from a cytoplasmic extension of a plasmodiophoran plant parasite, from which a penetrating structure develops. See PLASMODIOPHORA.

adnate describing unlike organs that are joined together, such as stamens fused with the petals. Compare CONNATE.

ADP (adenosine diphosphate) a NUCLEOSIDE with the same structure as adenosine monophosphate (see AMP) but with a second phosphate linked through a high-energy phosphate bond to the first. In PHOTOPHOSPHORYLATION ADP combines with another phosphate molecule to form ATP. This conversion is a major mechanism for the conservation of the light energy absorbed in photosynthesis. In some plants ADP acts as a sugar carrier, forming such molecules as ADP-glucose and ADP-sucrose. These ADP-sugars can then be oxidized to give energy or used in the synthesis of starch and ascorbic acid. See NUCLEOSIDE DIPHOSPHATE SUGARS.

adsorption the attachment of a gas, liquid, or solute to the surface of another substance (solid or liquid). For example, positively charged ions in soil water may become attached to the negatively charged surface of a soil particle. A soil particle, such as a particle of clay or humus, that can absorb ions and molecules is called an *adsorption complex*.

adsorption complex see ADSORPTION.

adventitious describing organs that arise in unexpected positions, such as ROOTS growing from a leaf.

adventive embryony (adventitious embryogenesis) the formation of an embryo in some position other than within an embryo sac. It is a form of APOMIXIS. Adventive embryos may develop from somatic cells of the nucellus or chalaza (APOSPORY) and may occur together with a normal zygotic embryo. One seed can contain several embryos, as is often the case in *Citrus* seeds. Alternatively an embryo may develop from the unreduced egg cell (DIPLOSPORY). Sometimes adventitious embryos may arise alongside sexually produced embryos (POLYEMBRYONY).

AE see EVAPOTRANSPIRATION.

aecidiosorus see AECIOSPORE.

aecidium (*pl.* aecidia) See AECIOSPORE.

aeciospore (aecidiospore, plasmogamospore) a dikaryotic spore formed by certain rust fungi (see UREDINALES) in a small cup-shaped sorus, the *aecium* (or *aecidium*, *aecidiosorus*). The aeciospores of *Puccinia graminis* are formed on the lower surfaces of barberry leaves in clusters of aecia, this stage of the life cycle commonly being called the cluster-cup stage. The aeciospores cannot reinfect barberry but will germinate on certain grass or cereal leaves where the subsequent infection shortly gives rise to UREDINIOSPORES. See HETEROECIOUS.

aecium (*pl.* aecia) see AECIOSPORE.

aeon see GEOLOGICAL TIME SCALE.

aerenchyma tissue with numerous large intercellular spaces. Aerenchyma is common in the cortex of the roots and stems of many aquatic plants, where it facilitates oxygenation of the roots and may increase buoyancy.

aerial mycelium see MYCELIUM.

aerial root a ROOT that arises above soil level. The term is usually applied to the tangled masses of roots developed by EPIPHYTES, which hang down in the moist air. The root

epidermis develops a sheath of dead empty cells, the VELAMEN, that helps absorb water from the atmosphere. See also PNEUMATOPHORE, CLIMBING ROOT, PROP ROOT, BUTTRESS ROOT.

aero- prefix denoting association with air or oxygen.

aerobe an organism capable of living only in the presence of air or free oxygen, the oxygen being needed for AEROBIC RESPIRATION. All plants, and most bacteria and fungi, are aerobes. In contrast to most animals they do not die instantly when deprived of oxygen but can continue to respire anaerobically for a time. Compare ANAEROBE.

aerobic respiration RESPIRATION involving the absorption of free oxygen and the complete oxidation of the organic starting materials. The oxygen is used as the terminal electron acceptor and the ultimate products are water and carbon dioxide (see diagram overleaf). Approximately 2930 kJ are released when a molecule of glucose is completely oxidized in aerobic respiration. This compares to only about 300 kJ obtained when a glucose molecule is broken down by ANAEROBIC RESPIRATION to alcohol and carbon dioxide. See also KREBS CYCLE, OXIDATIVE PHOSPHORYLATION.

aerodynamic method 1 a technique used to measure potential EVAPOTRANSPIRATION from a surface, such as a lake or area of vegetation. The calculation takes into account the influx of heat energy to the surface and the transport of water vapour away from the surface by moving air. 2 a technique used to measure PRIMARY PRODUCTIVITY in woodlands or forests, which measures differences in the carbon dioxide concentration of the air at various levels above the ground at regular time intervals, using sensors on a mast. It has the advantage that it does not involve sampling or destruction of the vegetation or its enclosure in an artificial environment. Compare CHLOROPHYLL METHOD, GAS-EXCHANGE METHOD, HARVEST METHOD.

aeropalynology see PALYNOLOGY.

aerophore see PNEUMATOPHORE.

aerotaxis a form of CHEMOTAXIS in which the cell or organism moves in response to oxygen.

aerotropism a form of CHEMOTROPISM in which oxygen is the orientating factor.

aestivation (estivation, prefloration) the arrangement of perianth parts within an immature flower bud. The definition is sometimes extended to include the arrangement of leaves in vegetative buds (*prefoliation*). The classification of aestivation types is based mainly on the extent of overlap of perianth segments and is illustrated by means of specialized FLORAL DIAGRAMS representing a horizontal slice through the bud. In such a diagram, the stem axis and position of the subtending bract are indicated, then the arrangement of the perianth parts. Bracts and bracteoles are ignored. There are four main types of aestivation: *open*, in which the edges of the sepals do not meet; *valvate*, in which the edges touch; *imbricate*, in which the edges of the sepals overlap; and *crumpled*, in which the folding is disorganized. Imbricate aestivation is further classified into *convolute (contorted)*, in which each sepal has one overlapping and one overlapped edge; *quincuncial*, in which there are five perianth segments: two outside, two inside, and the fifth half in and half out; *ascending*, in which each segment in the whorl overlaps the one posterior to it; and *descending*, in which each segment overlaps the one anterior to it. Compare VERNATION.

aetiology (etiology) the study of the causal agents of a disease.

affinity index a measure of the degree of similarity in the composition of two samples. $A = c/\sqrt{(a + b)}$, where a is the number of species in one sample that are not present in the other, b is the number of species present in the second sample but not in the first, and c is the number of species common to both samples. Compare ECOLOGICAL AND PHYTOSOCIOLOGICAL DISTANCE.

afforestation the establishment of forest on land where forests have not grown before. This may be by natural SUCCESSION or by deliberate planting.

aflatoxins toxins produced by fungi of the genus ASPERGILLUS. Aflatoxins are a cause of liver disease, and some aflatoxins are carcinogenic.

after-ripening the processes that must be undergone by certain seeds after harvest

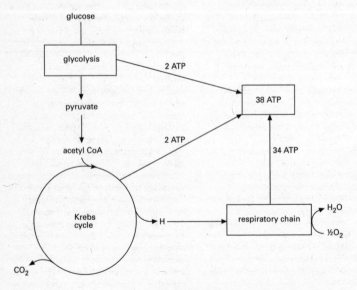

Aerobic respiration

before germination can take place.
Although the embryo appears fully mature
the seed remains dormant until the after-
ripening period is completed. It has been
suggested that after-ripening involves the
breakdown of some growth inhibitor or the
formation of a growth-promoting
substance. However since after-ripening
processes often occur in dry seeds, in which
metabolic activity is very low, it would
appear that the changes are not necessarily
metabolic in nature. An after-ripening
requirement is exhibited by many cereals
and grasses and prevents the seed from
germinating in the ear under moist
conditions.

agamospermy see APOMIXIS.

agar a mucilaginous carbohydrate that is
used in the form of a gel as a
microbiological supporting medium. It is
obtained from certain red algae (see
RHODOPHYTA), notably *Gelidium* species.
The gel is produced by dissolving agar
crystals in boiling water and then allowing
the mixture to cool. It may be sterilized by
placing in an autoclave. A nutrient agar gel
has various nutrients added to it. Agar is
widely used because it is resistant to attack
by almost all microorganisms, and, as it
does not melt until 100°C, incubation of the
selected culture can take place at high
temperatures. When agar is used in a Petri
dish it is known as an *agar plate*.

Agaricales (agarics, gill fungi) an order of
the HOLOBASIDIOMYCETES containing the
MUSHROOMS and toadstools. It contains
about 6000 species and 297 genera in 15
families. The order formerly also included
the boletes, which are now classified in an
order of their own, the BOLETALES. The
hymenium of the fleshy fruiting body is
borne on the surface of gills on the
underside of the pileus (cap). The young
fruiting body may be covered by a veil,
which breaks down as the stipe and pileus
expand, often remaining as a collar, or
volva, at the base of the stipe. Most agarics
are saprobes, and their mycelia are common
in leaf litter and decaying wood. Many form
MYCORRHIZAS. Many species, such as the
cultivated mushroom *Agaricus disporus*, are
edible. Some, such as the death cap
(*Amanita phalloides*), are highly poisonous.

Others are important as pathogens,
especially of trees; the honey fungus or
bootlace fungus, *Armillaria mellea*, of
worldwide distribution, is an example.

age-and-area hypothesis the hypothesis,
not widely accepted, that the area occupied
by a TAXON is directly proportional to its
age. This would mean that the youngest
species in a genus, for example, would
occupy the smallest area.

aggregate 1 a rock made up of fragments of
rocks or minerals. 2 a clump of soil
particles, ranging in size from microscopic
grains to small crumbs, that forms the basic
structural unit of soil. See also CRUMB
STRUCTURE.

aggregated distribution see CLUMPED
DISTRIBUTION.

aggregate fruit a fruitlike structure that has
developed from the carpels of a single
flower and is composed of a number of
separate fruits. It may be an aggregation of
achenes (as in *Anemone*), berries (as in
Actaea), drupelets (as in *Rubus*), follicles (as
in *Delphinium*), or samaras (as in
Liriodendron). The term *etaerio* is often used
to mean any aggregate fruit but its use is
sometimes restricted to a collection of
drupelets.

aggregation see CLUMPED DISTRIBUTION.

aggressin a toxin produced by a microorganisms pathogenic in humans or other
animals. which inhibits the defence
mechanisms of the host organism.

aglycone see GLYCOSIDE.

Agonomycetales (Mycelia Sterilia) an order
of fungi comprising those mycelial asexual
fungi (see FUNGI ANAMORPHICI) that do not
produce spores, although some develop
sclerotia or other nondehiscing propagules.
The order contains about 200 species and
28 genera. Some genera are important
pathogens of plants; examples include
Rhizoctonia, certain species of which cause
damping-off disease, and *Sclerotium*. In
some systems this order is placed in the
class Hyphomycetes of the Fungi
Anamorphici; in others it is put in the class
Agonomycetes of the Mitosporic Fungi.

Agrobacterium tumefaciens see CROWN GALL,
GENETIC ENGINEERING, TUMOUR-
INDUCING PRINCIPLE.

agroforestry a method of cultivation in

which forestry and arable farming are mixed together on the same land. It is especially common in rainforest regions of the tropics, as the trees shade freshly tilled soils and protect them from heavy rain and run-off, reducing soil erosion. The trees also help recycle nutrients by drawing them up from deeper soil layers to the surface layers, thus making them available to the roots of crop plants.

agrometeorology the study of the relationship between conditions in the atmosphere and on the Earth's surface, particularly in relation to agriculture.

agronomy the branch of agriculture dealing with soil management and crop production.

air bladder a swollen air-filled region of the thallus seen in many brown algae (see PHAEOPHYTA) of the order Fucales, notably *Fucus vesiculosus* (bladder wrack). The air bladders serve to increase buoyancy.

air chamber any of the air-filled cavities beneath the upper epidermis of the gametophyte thallus in many liverworts (see HEPATOPHYTA) of the order Marchantiales. Such chambers increase the surface area of the internal photosynthetic cells and so facilitate gaseous exchange. At the same time they maintain a humid atmosphere around these cells, thus reducing transpiration. Connection of the air chambers with the external atmosphere is facilitated by pores in the upper surface of the thallus.

air layering a form of vegetative propagation (see VEGETATIVE REPRODUCTION) often used to multiply greenhouse and indoor plants. A branch is stimulated to produce roots while still attached to the parent plant by making a shallow cut or removing a narrow ring of bark just below a bud. The cut is dusted with rooting powder containing synthetic auxins and kept open by inserting sphagnum moss. The whole area is then kept moist by wrapping the stem in moss and surrounding this with a plastic sleeve. Root formation is slow and may take up to two years.

air plant see EPIPHYTE.

akinete a thick-walled resting spore that is formed during unfavourable conditions by certain blue-green bacteria (see CYANOBACTERIA). Akinetes are highly resistant to temperature extremes and desiccation. On germination a HORMOGONIUM may be formed.

alanine (2-aminopropanoic acid) A simple nonpolar amino acid in which the R group is CH_3 (*see illustration at* amino acid). Alanine is formed in a TRANSAMINATION reaction in which glutamine donates an amino group to pyruvic acid to form alanine and α-ketoglutaric acid. This reaction is reversible; alanine can be deaminated to pyruvate and subsequently oxidized in the Krebs cycle. Along with aspartic acid, serine, glutamic acid, and glycine, alanine is one of the early products of photosynthesis.

albedo the ratio of the intensity of light reflected from a surface to the intensity of light received. The average albedo of the Earth's surface is about 30%, but this varies considerably with the nature of the land surface. All leaves reflect or transmit infrared radiation but different species reflect different amounts of visible radiation. Thus xerophytic plants tend to reflect more light than mesophytic plants. For grass and forest cover the albedo is 8-27%, for rock and buildings 12-18% (up to 40% for pale-coloured stone), for sand up to 40%, snow up to 90%, and cloud up to 80% (average about 55%).

albinism in plants, pale coloration due to a deficiency of chromoplasts, pigment-containing organelles that give colour to flowers and ripe fruits.

albumin any of a class of simple low-molecular-weight water-soluble proteins found in many plants; for example, in the endosperm of wheat and barley. Albumins have no prosthetic group and only one amino-acid chain. They are often found in association with globulins.

albuminous cell 1 a specialized parenchyma cell, physiologically and anatomically associated with a sieve cell in gymnosperm phloem. Albuminous cells are analogous to the companion cells of angiosperm phloem but unlike companion cells they do not usually arise from the same mother cell as the sieve element.
2 an albumin-containing cell, found in certain seeds.

albuminous seed see ENDOSPERM.

alburnum see SAPWOOD.

alcian blue a stain that enhances the preservation of cell surface materials, especially mucopolysaccharides. It is usually used during fixation with glutaraldehyde for transmission electron microscopy (see ELECTRON MICROSCOPE).

alcohol an organic compound of the general formula ROH, where R is a hydrocarbon group, for example, ethanol (C_2H_5OH).

alcohol dehydrogenase an enzyme that converts ethanol (a product of anaerobic respiration in plants) into acetaldehyde (ethanal). The ability of a plant's roots to produce alcohol dehydrogenase is a factor in determining its tolerance of waterlogging.

alcoholic fermentation a form of ANAEROBIC RESPIRATION in which glucose is broken down to form ethanol and carbon dioxide. It is carried out by yeasts and various other fungi and by certain bacteria. Fermentation takes place outside the organism and is catalysed by enzymes of the ZYMASE complex. These are either secreted by living cells or released on cell death. Fermentation usually stops because of cell poisoning when the alcohol level reaches about 15%. The process is central to the brewing, wine-making, and baking industries. Since free oxygen is not available as a hydrogen acceptor acetaldehyde is used instead. Pyruvic acid, formed by glycolysis, is broken down to acetaldehyde and carbon dioxide. The acetaldehyde is then reduced by NADH to form ethanol and NAD. The process yields about 300 kJ from each glucose molecule. This is only about 10% of the energy that would be released by complete oxidation of glucose, as in AEROBIC RESPIRATION.

aldehyde an organic compound with the general formula RCHO. Examples include acetaldehyde and formaldehyde. The aldehyde group (CHO) consists of a carbonyl group (C=O) joined to a hydrogen atom. Aldehydes are readily oxidized to acids and reduced to alcohols.

alditol see SUGAR ALCOHOL.

aldose any monosaccharide with the carbonyl (CO) group on the terminal carbon, so forming an aldehyde (CHO)

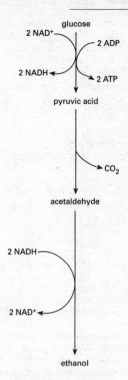

Alcoholic fermentation

group. The simplest aldose is the three-carbon sugar GLYCERALDEHYDE. Other aldoses include glucose, galactose, and mannose, which are aldohexoses, and ribose, arabinose, and xylose, which are aldopentoses (see diagram overleaf). Compare KETOSE.

-ales a suffix used to denote an order of plants, e.g. Magnoliales. See also RANK.

aleurone grain see ALEURONE LAYER.

aleurone layer (proteinaceous endosperm layer) the protein-rich outermost layer of the ENDOSPERM in seeds of the Poaceae (grasses) and Polygonaceae. The cells of the aleurone layer (*aleuroplasts*) contain *aleurone grains*, which store protein for later use by the embryo.

aleuroplast see ALEURONE LAYER.

algae an extremely diverse group of photosynthetic eukaryotes showing relatively little differentiation of tissues and

CHO
|
HCOH
|
CH_2OH
glyceraldehyde

CHO
|
HCOH
|
HCOH
|
CH_2OH
erythrose

CHO
|
HCOH
|
HCOH
|
HCOH
|
CH_2OH
ribose

CHO
|
HOCH
|
HCOH
|
HCOH
|
CH_2OH
arabinose

CHO
|
HCOH
|
HOCH
|
HCOH
|
CH_2OH
xylose

CHO
|
HCOH
|
HOCH
|
HCOH
|
HCOH
|
CH_2OH
glucose

CHO
|
HOCH
|
HOCH
|
HCOH
|
HCOH
|
CH_2OH
mannose

CHO
|
HCOH
|
HOCH
|
HOCH
|
HCOH
|
CH_2OH
galactose

Aldose: some common aldoses

organs as compared with the lower plants. There are no recognizable roots, stems, or leaves and no true vascular system. There is no layer of sterile cells surrounding the reproductive organs. Morphology ranges from unicells through colonial and filamentous forms to parenchymatous seaweeds over 50 metres long. The primary photosynthetic pigment is chlorophyll *a*. Algae are found in most habitats worldwide, but predominantly in freshwater and marine environments. Adaptation to this variety of habitats has led to the development of many distinctive biochemical traits.

The term 'algae' has no taxonomic significance; these organisms are thought to be a polyphyletic group derived from a range of primitive ancestors. In the widely used FIVE KINGDOMS CLASSIFICATION they are placed in the kingdom PROTOCTISTA. There are many distinct phyla of algae, distinguished by their pigments, food reserves, cell wall materials, number and types of undulipodia, and ultrastructural details. Formerly the algae were placed with the bacteria and fungi in the division Thallophyta. The prokaryotic blue-green bacteria (see CYANOBACTERIA) were for a long time referred to as the blue-green algae, but today the term 'alga' is restricted to eukaryotic organisms. For accounts of individual algal taxa, see CHLOROPHYTA (green algae), CRYPTOMONADA, CHRYSOMONADA (golden-brown algae), DIATOMS, DINOMASTIGOTA (dinoflagellates), EUGLENIDA (euglenas), EUSTIGMATOPHYTA (eustigs), GAMOPHYTA (conjugating green algae), HAPTOMONADA, PHAEOPHYTA (brown algae), RHODOPHYTA (red algae), XANTHOPHYTA (yellow-green algae).

algal bloom see BLOOM.

alginic acid a carbohydrate polymer consisting of D-mannuronic acid and L-glucuronic acid units (see URONIC ACIDS). It is found mainly in the cell walls and intercellular spaces of brown algae (see PHAEOPHYTA), where it functions as an ion exchange agent. Salts of alginic acid are called *algins*. When mixed with water, many form thick, viscous solutions. Alginic acid derivatives are used in the making of fireproof and disposable fabrics, as emulsifying agents, in some types of surgical suture, and as stabilizers in ice-cream and other foods.

algins see ALGINIC ACID.

algology the study of algae.

alien see EXOTIC.

aliphatic describing a compound that has an open-chain structure, e.g. an alkane. Compare AROMATIC.

Alismatidae (Alismidae) a subclass of the monocotyledons containing aquatic or semiaquatic herbaceous plants. General features of the group are unfused carpels, in contrast to other monocotyledons in which the gynoecium is usually syncarpous; trinucleate pollen, as opposed to the binucleate pollen common to most other monocotyledons; absence of a starchy endosperm; and absence of vessels in the stem and often also in the roots. There are usually two subsidiary cells associated with each stoma. Two orders are commonly recognized: the Alismatales (which includes the old orders Hydrocharitales and Najadales) and the Triuridales. The

Triuridales is sometimes placed in the subclass Liliidae. The families of the Alismatidae are mainly small, some containing only one genus (e.g. Najadaceae: *Najus*, naiads; Ruppiaceae: *Ruppia*, ditch grass), while the Scheuchzeriaceae and Butomaceae include only one species each, *Scheuchzeria palustris* (Rannoch-rush) and *Butomus umbellatus* (flowering rush). By contrast, the Potamogetonaceae (pondweeds) contains 90 species.

alkaline soil a SOIL whose ph is greater than 7.0. such soils are usually rich in calcium ions, derived from the underlying rocks. See also CALCAREOUS SOIL, PEDOCAL.

alkaloids a class of nitrogen-containing usually basic plant products, which are often poisonous. many alkaloids, e.g. morphine, codeine, nicotine, and cocaine, have been utilized in medicine and other fields. their natural functions are not well understood. some, through their bitterness and toxicity, may serve to protect the plants from herbivorous animals. others are thought to be involved in nitrogen metabolism (e.g. nicotine is involved in the absorption of nitrate through plant roots). many are thought to be simply end products of nitrogen metabolism, which are stored in leaves, fruits, and flowers before being discarded. certain families, e.g. the solanaceae, fabaceae (leguminosae), and papaveraceae, contain many alkaloidal species while monocotyledons rarely produce alkaloids. most groups of lower plants contain alkaloidal species, but alkaloids have not been found in the algae.

three classes of alkaloids have been recognized. the *true alkaloids* have a nitrogen-containing heterocyclic nucleus, examples being the ISOQUINOLINE alkaloids. The *protoalkaloids* lack a heterocyclic ring and are usually simple amines. Examples are mescaline and ephedrine. Some of the protoalkaloids may be precursors of true alkaloids. Both the true alkaloids and the protoalkaloids are derived from amino acids. The *pseudoalkaloids* are not derived from amino acids but from such compounds as terpenes, purines, and sterols. Pseudoalkaloids include theobromine and caffeine, which are both methylated

purines. See also INDOLE ALKALOIDS, PYRIDINE ALKALOIDS, PIPERIDINE ALKALOIDS, TROPANE ALKALOIDS.

allele (allelomorph) a form in which a gene may occur. Different alleles of a gene give rise to different expressions of a character. Hence alleles for 'green' and 'yellow' are alternative expressions of the gene governing the characteristic for seed colour. Alleles of a gene always occupy the same site (locus) on HOMOLOGOUS CHROMOSOMES. A diploid organism whose cells contain two identical alleles is said to be HOMOZYGOUS. One with two different alleles at a locus is said to be HETEROZYGOUS. The allele expressed in a heterozygous organism depends on the dominance relationship between the two alleles. If one is expressed to the exclusion of the other it is described as DOMINANT and the latter RECESSIVE. If the heterozygote is intermediate in appearance between the two homozygotes, the alleles are said to exhibit INCOMPLETE DOMINANCE. The term *allele frequency* is used to denote how frequently an allele occurs in a population.

allele frequency see ALLELE, GENE FREQUENCY.

allelochemical a chemical produced by a living organism to inhibit the growth of another species. See ALLELOPATHY.

allelomorph see ALLELE.

allelopathy (*adj.* allelopathic) the release of a chemical by a plant that inhibits the growth of nearby plants and thus reduces competition. For example, pines (*Pinus*) produce substances that kill any seedlings of the same species (*autotoxicity* or *autoallelopathy*) growing too close to the parent plant. Allelopathy may also be indirect, if one plant inhibits the growth of a second plant or microorganism that itself is essential to the growth of a third plant. Allelopathic substances (*allelochemicals*) have been shown to be responsible for various changes during plant successions. For example, in a field succession the pioneer weed stage is replaced by annual grasses because the weeds produce substances inhibitory to other weeds. Allelochemicals are thought to include terpenes, e.g. camphor, and phenolic compounds. See also PHYTOALEXIN.

alliance in PHYOSOCIOLOGY, a group of closely related associations.

allochthonous describing material that did not originate in its present location. For example, plant material or soil particles in a lake sediment may have been transported from other locations by rivers or floods. Compare AUTOCHTHONOUS.

allogamy (cross fertilization, exogamy) fusion of female and male gametes derived from genetically dissimilar individuals of the same species. It promotes the recombination of variable genetic material so that the total population shows greater variation than an autogamic population (see AUTOGAMY) and has greater adaptive potential. Various methods have developed in the plant kingdom to encourage or necessitate allogamy (see DICHOGAMY, DIOECIOUS, MONOECIOUS, INCOMPATIBILITY, CHASMOGAMY, HETEROSTYLY, CROSS POLLINATION). When cross fertilization is obligate, useful genetic traits may be diluted and isolated individuals cannot reproduce.

allometry the measurement and study of the rate of growth of one part of an organism in relation to the growth rate of other parts or of the whole organism or of other individuals.

allopatric (*n.* allopatry) describing a population or species that is unable to breed with a related group of organisms because of geographical separation by distance or by natural barriers, such as water, mountain ranges, deserts, etc. If two groups that have been separated in such a way develop different adaptations such that they would not be able to interbreed even if the barriers broke down then allopatric SPECIATION is said to have occurred. This process is called *allopatry*. Compare SYMPATRIC. See also GEOGRAPHICAL ISOLATION, REPRODUCTIVE ISOLATION, VICARIANCE.

allophillic see ENTOMOPHILY.

allopolyploidy a form of POLYPLOIDY that results from the combination of sets of chromosomes from two or more different species. Individuals with such a genetic makeup are called *allopolyploids*. A (diploid) interspecific hybrid is normally sterile because there is only one of each kind of chromosome per cell. Thus at meiosis no bivalents are formed and any resulting gametes are usually inviable as they contain either too many or too few chromosomes. However if some unreduced diploid gametes form and fuse then the resulting tetraploid (*allotetraploid*) will usually be fertile as its nuclei will contain pairs of homologous chromosomes. An allotetraploid may cross with a diploid to form a sterile triploid hybrid, which, if it produces unreduced triploid gametes, may give rise to a fertile *allohexaploid*. If an allohexaploid crosses with a diploid an allo-octaploid could arise in a similar fashion, and so on to higher levels of allopolyploidy. An allopolyploid is often intermediate in appearance between both parental species and cannot reproduce with either. Hence it may merit the status of a new species. Allopolyploidy has been described as *instant evolution*. Many crop plants are believed to have originated in this way. Wheat (*Triticum aestivum*), for example, is an allohexaploid. It has a chromosome number of 42 and is probably derived from the three species *T. monococcum*, *Aegilops speltoides*, and *A. squarrosa*, each with 14 chromosomes.

If at meiosis an allopolyploid forms only bivalents, i.e. it acts like a diploid, then it is termed an *amphidiploid*. *Raphanobrassica* (see ALLOTETRAPLOID) is an example. However if certain of the chromosomes from the two parent species are sufficiently similar (see HOMOEOLOGY) then multivalents may be seen at meiosis. When this occurs the allopolyploids are termed *segmental allopolyploids*. An example is the allotetraploid *Primula kewensis* derived from a cross between *P. floribunda* and *P. verticillata*. Sometimes segmental allopolyploids show reduced fertility as compared to amphidiploids and it is clear that in some allopolyploids there are mechanisms that prevent the formation of multivalents. Wheat, for example, normally behaves as an amphidiploid but if chromosome V of the B genome (the genome derived from *A. speltoides*) is missing then multivalents are formed.

allosteric enzyme a regulatory enzyme responsive to alterations in the metabolic state of a cell or tissue. Its catalytic activity is modified by the non-covalent binding of a specific metabolite (*modulator*) at a site (see ALLOSTERIC SITE) other than the active site. The most common type of allosteric enzyme is one found at the beginning of a multienzyme sequence that is inhibited specifically by the end product of the reaction sequence (FEEDBACK INHIBITION). However, the activity of an allosteric enzyme is not always decreased by the action of a modulator; some allosteric modulators increase the enzyme's activity. Allosteric enzymes may also have more than one modulator.

Reactions involving allosteric enzymes are always irreversible in the cell, as it would be impossible to regulate a reversible reaction. Allosteric enzymes are often more complex structurally than other enzymes; all known allosteric enzymes have at least two protein subunits.

allosteric site a regulatory site on an allosteric enzyme where a specific effector or modulator can reversibly bind. When bound, the modulator either activates (a positive modulator) or inhibits (a negative modulator) the enzyme by changing its shape. Some allosteric enzymes have a site for only one modulator (*monovalent enzymes*) while others have several allosteric sites (*polyvalent enzymes*).

allosyndesis pairing of chromosomes derived from different species. It is seen in some diploid hybrids and in segmental allopolyploids (see ALLOPOLYPLOIDY). See also HOMOEOLOGY.

allotetraploid an allopolyploid that has originated from an interspecific hybrid through the formation and fusion of unreduced diploid gametes. An example is *Raphanobrassica*, which is obtained by crossing *Raphanus sativus* (radish) with *Brassica oleracea* (cabbage). The F_1 is sterile but diploid gametes are sometimes formed, which on selfing give a fertile tetraploid F_2. See ALLOPOLYPLOIDY.

alluvial fan see ALLUVIAL SOIL.

alluvial soil a type of azonal SOIL formed on the flood plains of river valleys and at river mouths. New material is successively deposited on the surface when the land is subjected to flooding. As the water spreads out over the flood plain or at a river mouth,

its flow slows, and sediment (*alluvium*) is deposited, forming *alluvial fans* or deltas. In tropical regions such areas may be used as paddy fields, and in arid regions the extent of the alluvial soils may be artificially increased by irrigation methods, as along the banks of the Nile. The polders of the Netherlands are reclaimed marine alluviums.

alluvium See alluvial soil.

alpha helix (α-helix) See SECONDARY STRUCTURE.

alpha-naphthol test (Molisch's test) a standard procedure for detecting the presence of carbohydrates in solution. A small amount of alcoholic alpha-naphthol is added to the test solution in a test tube. Concentrated sulphuric acid is then poured slowly down the side of the tube. A violet ring forming at the junction of the liquids indicates a positive reaction.

alpine a major regional community (BIOME) of vegetation in high mountainous regions and on high-level plateaus. The plants generally grow in thin stony soil and are subjected to high light intensity and high wind speeds. The climate differs from that of the TUNDRA as there is daylight in winter and darkness in summer, more precipitation and wind, and a higher degree of solar radiation, but no permafrost. The number of plant species, although limited, is greater than that found in the tundra. Certain drought-tolerant species (e.g. grasses, sedges, mosses, and lichens) are common to both communities. In temperate latitudes there are many brightly coloured flowers in summer on the alpine pastures.

The vegetation changes as the altitude increases due to the associated drop in temperature. The vegetation also differs between north- and south-facing slopes. The lower limit of the alpine zone (SEE TREE LINE) varies in different mountain regions and also varies with the wetness of the locality and the mass of the mountain range. See also PAMIRS, PUNAS

alternate describing a form of leaf arrangement in which there is one leaf at each node (*see illustration at* phyllotaxis). This pattern is found in most plants. Compare OPPOSITE, WHORLED.

alternate host a plant, other than the main host, on which a pathogen or pest can live. Alternate hosts, which are often weeds, can provide a means for the pathogen to survive when its main host is not available. For example, downy mildew of beet (*Peronospora farinosa*) can overwinter in wild beet. Hawthorn (*Crataegus*) is an alternate host and important inoculum source for the fireblight bacterium (*Erwinia amylovora*), which infects pears and apples. Many RUSTS also overwinter on alternate hosts. Certain insect pests overwinter as eggs on alternate hosts, e.g. blackfly (*Aphis fabae*) migrates to the spindle tree (*Euonymus europaeus*) in the autumn. Virus diseases may be found in weed hosts near crops. The term *alternative host* may be used when the pathogen has a number of different hosts. Control of the alternate hosts can be an important way of reducing inoculum sources of some diseases. See also PARASITISM.

alternation of generations the occurrence of alternating haploid and diploid individuals in the LIFE CYCLE of an organism. In plants it usually includes the alternation of asexually reproducing individuals and sexually reproducing individuals. In bryophytes, vascular plants, many algae, and some fungi a haploid gamete-producing phase (the GAMETOPHYTE) alternates with a spore-producing diploid phase (the SPOROPHYTE). In bryophytes the dominant generation is the gametophyte, whereas in the vascular plants it is the sporophyte. If the two generations are markedly different the life cycle is termed HETEROMORPHIC, whereas if they are similar it is termed ISOMORPHIC. See also HAPLONTIC, DIPLONTIC, HAPLOBIONTIC, DIPLOBIONTIC.

alternative host see ALTERNATE HOST.

amastigote a unicellular organism that lacks UNDULIPODIA. Examples are the Diatoms.

amber fossilized resin, mainly from conifers that grew during the TERTIARY PERIOD.

ameba see AMOEBA.

ameiosis meiosis in which the nucleus divides only once, so that the number of chromosomes is not halved.

amensalism any association between two organisms in which one is harmed by the activities of the other. It includes parasitism and also the various ways in which plants protect themselves from pests and herbivores. Associations between microorganisms in which one is adversely affected are termed *antibiosis*. See also COMMENSALISM, MUTUALISM.

Amentiferae the former name for certain catkin-bearing families of angiosperms (see ANTHOPHYTA). See HAMAMELIDAE, DILLENIIDAE.

aminoacetic acid see GLYCINE.

amino acid a molecule containing both carboxylic acid and amino groups and having the general formula $RCHNH_2COOH$. The nature of the R group varies widely, from a hydrogen atom in glycine to aromatic and heterocyclic ring structures in such amino acids as tyrosine and tryptophan. Amino acids can be classified as nonpolar, polar uncharged, acidic, or basic, according to the nature of the R group (see TABLES).

Amino acids are the basic structural units of proteins with some 20 amino acids commonly occurring in proteins. In addition to these there are a few unusual amino acids that occur only in a few proteins and over 200 nonprotein amino acids that have been isolated from various plant sources. In many cases the function of these nonprotein amino acids is unclear but some are intermediates in the synthesis of common amino acids while others may have protective or storage functions.

Because they contain both acidic and basic groups, amino acids will react with both acids and bases. Consequently the charge on an amino acid varies with its pH. Each amino acid has a specific pH, known as its isoelectric point, at which the net charge on the molecule is zero. All naturally occurring amino acids, except glycine, are optically active due to the asymmetry of the α carbon atom. Most amino acids in nature are in the L form, although some D-amino acids are found in bacterial cell walls.

amino acid sequencing the determination of the amino acid sequence (primary structure) of a protein. The sequencing of homologous proteins from different species has been used as a method of determining phylogenetic relationships. It is assumed that the number of differences in the sequence may be related to the length of

amino acid	symbol	R group
alanine	Ala	$-CH_3$
valine	Val	$-CH\begin{smallmatrix}CH_3\\CH_3\end{smallmatrix}$
leucine	Leu	$-CH_2-CH\begin{smallmatrix}CH_3\\CH_3\end{smallmatrix}$
isoleucine	Ile	$-CH-CH_2-CH_3$ $\quad\ \ |$ $\quad\ CH_3$
proline	Pro	
methionine	Met	$-CH_2-CH_2-S-CH_3$
phenylalanine	Phe	$-CH_2-$ (benzene ring)
tryptophan	Trp	(indole ring structure)

amino acids with nonpolar R groups

Structures of the **amino acids** commonly found in proteins. In all except proline, where the complete structure is given, the remainder of the molecule has the structure

$$\text{HOOC}-\underset{\underset{NH_2}{|}}{\overset{\overset{H}{|}}{C}}-$$

19

amino acid	symbol	R group
glycine	Gly	—H
serine	Ser	$-CH_2-OH$
threonine	Thr	$-\overset{\overset{\displaystyle OH}{\mid}}{\underset{\underset{\displaystyle H}{\mid}}{C}}-CH_3$
cysteine	Cys	$-CH_2-SH$
asparagine	Asn	$-CH_2-C\overset{\nearrow NH_2}{\searrow_O}$
glutamine	Gln	$-CH_2-CH_2-C\overset{\nearrow NH_2}{\searrow_O}$
tyrosine	Tyr	$-CH_2-\bigcirc-OH$

amino acids with uncharged polar R groups

amino acid	symbol	R group	
aspartic acid	Asp	$-CH_2-C\overset{\nearrow O^-}{\searrow_O}$	acidic amino acids
glutamic acid	Glu	$CH_2-CH_2-C\overset{\nearrow O^-}{\searrow_O}$	
lysine	Lys	$-CH_2-CH_2-CH_2-CH_2-\overset{+}{N}H_3$	basic amino acids
arginine	Arg	$-CH_2-CH_2-CH_2-NH-C-NH_2$ $\overset{\parallel}{\underset{+}{N}H_2}$	
histidine	His	$-CH_2-C-CH$ $HN\diagdown\quad\diagup NH$ $\overset{\parallel}{\underset{\displaystyle H}{C}}{+}$	

amino acids with charged polar R groups

time since the different species diverged from a common ancestor. Phylogenetic trees have been constructed from data on the sequences of cytochrome *c* and plastocyanin. These do not always correspond with information derived from other sources. See also DNA HYBRIDIZATION, PROTEIN.

amino alcohol a compound possessing both a hydroxyl (-OH) and an amino group ($-NH_2$).

aminobutanedioic acid see ASPARTIC ACID.

aminoethanoic acid see GLYCINE.

amino group a chemical group with the formula $-NH_2$. It is characteristic of amino acids, and gives them their basic properties.

2-aminopentanedioic acid see GLUTAMIC ACID.

2-aminopropanoic acid see ALANINE.

amino sugar a monosaccharide with an amino (NH_2) group in place of one of the hydroxyl (OH) groups. The most commonly occurring amino sugars are glucosamine and galactosamine. Amino sugars are components of glycoproteins.

aminotransferase see TRANSAMINATION.

amitosis (amitotic division) nuclear division by constriction into two parts without the appearance of chromosomes. The nuclear membrane does not break down and a spindle is not formed. Amitosis occurs in some primitive unicellular organisms and has also been observed in the formation of endosperm tissue.

ammonium fixation the adsorbing (see ADSORPTION) of ammonium ions into clay minerals, which renders them unavailable to plants.

amoeba (ameba) a unicellular eukaryote that has no cell wall and moves by means of pseudopodia.

amoeboid resembling an amoeba in form and movement. The term is used to describe certain gametes, e.g. those of the Gamophyta (conjugating green algae).

AMP (adenosine monophosphate) a phosphorylated nucleoside consisting of the purine adenine and the sugar ribose phosphorylated in the 5′ position. AMP is involved in the regulation of glycolysis and gluconeogenesis, promoting the formation of fructose bisphosphate from fructose 6-phosphate (i.e. promoting glycolysis) while

inhibiting the back reaction to fructose bisphosphate. See also CYCLIC AMP.

amphi- a prefix meaning 'both' or 'on both sides'.

amphi-Atlantic species species found on both sides of the northern Atlantic, in the coastal regions of the eastern United States and western Europe. Species with this disjunct distribution are thought to be derived from the flora of a single coastal plain that at one time fringed the north Atlantic. Characteristic species include the sedge *Carex paleacea*, bog bilberry (*Vaccinium uliginosum*), and Hooker's cinquefoil (*Potentilla hookeriana*).

amphibolic see KREBS CYCLE.

amphicribal describing a concentric vascular bundle that has the phloem surrounding the xylem. Compare AMPHIVASAL.

amphidiploid see ALLOPOLYPLOIDY.

amphiesmal vesicle a membrane-bounded sac lying below the test of the DINOMASTIGOTA (dinoflagellates), thought to be involved in producing the test.

amphigastria the leaves that form in a row on the undersurface of the stem of a leafy liverwort. They are smaller than the leaves on the upper surface and are often only seen just below the apex, having been shed lower down.

amphimixis (*adj.* amphimictic) true sexual reproduction involving the fusion of two gametes. Compare APOMIXIS.

amphiphloic having phloem arranged on both sides of the xylem as, for example, in a solenostele (see STELE). Compare AMPHICRIBRAL.

amphiphloic siphonostele see STELE.

amphithecium (*pl.* amphithecia) the outer layer of the young sporophyte in bryophytes, giving rise to the capsule wall, which, in many Bryales, differentiates a PERISTOME. In *Sphagnum* and *Anthoceros*, the sporogenous tissue develops from the amphithecium, whereas in other bryophytes it develops from the endothecium.

amphitrophic describing an organism capable of photosynthesis in the presence of light and of chemosynthesis in its absence.

amphitropical species species with disjunct distribution patterns, occurring in separate geographical regions on either side of the equator. Such distributions probably arose during the Pleistocene (see QUATERNARY), when the belt of equatorial climate was much narrower.

amphitropous describing an ovule that is attached to the placenta by its centre, so that it lies parallel to the placenta. See PLACENTATION.

amphivasal describing a concentric vascular bundle in which the xylem surrounds the phloem. Compare AMPHICRIBAL.

amphoteric see ISOELECTRIC FOCUSING.

amplexicaul describing a sessile leaf in which the base of the lamina clasps the stem at the node as, for example, the upper leaves of henbit (*Lamium amplexicaule*).

amylase the HYDROLASE enzyme that catalyses the hydrolysis of the $\alpha(1-4)$ GLYCOSIDIC BONDS in starch. Amylase occurs in two forms, designated α and β.

β-amylase attacks only the nonreducing ends of a starch molecule, successively hydrolyzing alternate $\alpha(1-4)$ linkages and releasing maltose molecules. β-amylase is found in germinating seeds and is important for the production of malt in the brewing industry. Amylose is completely degraded to maltose by β-amylase but amylopectin is only partially broken down because the β-amylase cannot attack $\alpha(1-4)$ linkages beyond the first branch on each chain. A separate enzyme, $\alpha(1-6)$ glucosidase, exists to hydrolyse the $\alpha(1-6)$ branching linkages of amylopectin.

α-amylase differs from β-amylase in that it can attack $\alpha(1-4)$ bonds within the starch molecule. It thus degrades amylopectin more completely than β-amylase.

amyloid starch-like.

amylopectin a polysaccharide that, with AMYLOSE, makes up starch. It consists of glucose units linked by $\alpha(1-4)$ GLYCOSIDIC BONDS with branches formed by $\alpha(1-6)$ bonds. The molecular weight of amylopectin may be as high as 10^7. It can form either colloidal or micellar solutions with water. See AMYLASE.

amyloplast a PLASTID whose main function is to synthesize and store starch.

amylose a polysaccharide that, with AMYLOPECTIN, makes up starch. Amylose is an unbranched chain of glucose molecules linked by $\alpha(1-4)$ glycosidic bonds. The molecular weight of amylose can vary from a few thousand to over half a million. In water amylose has a micellar structure and the amylose chains form helical coils. See also AMYLASE.

amylum see STARCH.

anabiosis see CRYPTOBIOSIS.

anabolism (biosynthesis) (*adj*. anabolic) The metabolic synthesis of complex molecules from simpler ones. It requires an input of chemical energy, which is provided by ATP. The Calvin cycle is an example of an anabolic pathway. Compare CATABOLISM.

anaerobe an organism that can live in the absence of free oxygen. *Obligate* or *strict anaerobes* cannot live in the presence of free oxygen. *Facultative* or *indifferent anaerobes* can grow in the presence of oxygen but do not use it. Such organisms include the denitrifying bacteria and lactic acid bacteria. The term facultative anaerobe may also be used of such organisms as yeasts that can grow under anaerobic conditions but given free oxygen will use it to oxidize the products of anaerobic respiration. They thus grow better with free oxygen. However yeasts and similar organisms could equally well be described as facultative aerobes. Compare AEROBE.

anaerobic respiration any of various catabolic pathways by which chemical energy is obtained from organic compounds in the absence of free oxygen. The glycolytic pathway (see GLYCOLYSIS) and ALCOHOLIC FERMENTATION are the two main examples. Other types include MIXED LACTIC FERMENTATION and various other bacterial fermentations in which the end products include propionic acid, butyric acid, and acetone.

anagenesis the degree or rate of evolutionary divergence.

analogous describing organs, often similar in appearance, that carry out similar functions but have different origins. For example, phyllodes are analogous to leaf blades but are derived from petioles. Compare HOMOLOGOUS. See also CONVERGENT EVOLUTION.

anamorph (*adj*. anamorphic) The asexual ('imperfect') form (morph) of a fungus, characterized by the absence of sexual spores. Compare TELIOMORPH. See FUNGI ANAMORPHICI, STATES OF FUNGI.

anaperturate describing a pollen grain without any type of aperture.

anaphase the stage following metaphase in nuclear division (see KARYOKINESIS), during which separation of either chromatids or homologous chromosome commences. In anaphase of the first division of MEIOSIS, the homologous chromosomes of each bivalent, each with a complete centromere, become separated and move towards opposite poles of the spindle. In MITOSIS and in the second division of meiosis sister chromatids move apart towards opposite poles of the spindle. Chromatid movement is thought to be mediated by the microtubules that make up the spindle. There appear to be two distinct types of spindle fibre: long fibres running from pole to pole (*polar fibres*) and bundles of shorter fibres bound to the kinetochores in the centromere regions (*kinetochore fibres*). During separation of chromatids, the polar fibres in the equatorial region between the chromatids elongate, until the spindle eventually doubles in length, while the kinetochore fibres shorten. The exact mechanism of this lengthening and shortening is not fully understood. It involves polymerization and depolymerization of units of the protein tubulin in the constituent microtubules.

anaplerotic reactions see PYRUVIC ACID.

anastomosis in woody perennials and certain fungi, the cross-linking of branches or HYPHAE respectively.

anatomy plant structure, or the study thereof, with an emphasis on tissues and their component cells in the interior of the plant body. Compare MORPHOLOGY.

anatropous describing the form of OVULE orientation in the ovary in which the funiculus has lengthened and the ovule turned through 180° so that the micropyle is folded over and lies near the base of the funiculus (*see illustration at* ovule). This arrangement is the most common. Compare CAMPYLOTROPOUS, ORTHOTROPOUS.

ancient countryside in Britain, an area of countryside in which most of the fields,

woodlands, paths, common lands and old trees were present before AD 1700.

ancient woodland primary and secondary woodland that originated at a particular date, which in England and Wales is AD 1600, and in Scotland AD 1750. It is characterized by certain indicator species considered typical of the original natural woodland cover, such as the small-leaved lime (*Tilia cordata*), field maple (*Acer campestre*), wild cherry (*Prunus avium*), and bluebell (*Hyacinthoides non-scripta*). Compare old-growth forest.

Andreaeales see BRYOPHYTA.

andro- a prefix that denotes maleness or signifies relating to mankind.

androchory (anthropochory) the dispersal of seeds or spores by humans.

androdioecious describing plant species in which male and hermaphrodite flowers are borne on separate individuals, as in *Datisca*, *Phillyrea*, and *Mercurialis*. Compare ANDROMONOECIOUS, GYNODIOECIOUS.

androecious describing a plant that possesses only male flowers.

androecium the male component of a flower, made up of several STAMENS and sometimes also STAMINODES. The androecium usually surrounds the GYNOECIUM, although the exact arrangement or position may not be symmetrical in the more advanced forms. The androecium is described as *apostemonous* when the stamens are separate, *monadelphous* or *adelphous* when the filaments are all fused, and *syngenesious* when the anthers are fused. If the stamens form two groups, the members of each group being joined by their filaments, the androecium is *diadelphous* and if three or more groups are formed in this way, *polyadelphous*. Petalostemonous describes the condition where the filaments are joined to the petals. The androecium is represented in the FLORAL FORMULA by the letter A.

androgynophore (androphore) an extension of the receptacle between the petals and the stamens on which the androecium and gynoecium are borne. It is seen in many members of the Capparidaceae.

andromonoecious describing plant species in which male and hermaphrodite flowers are borne separately on the same individual, as in *Aesculus hippocastanum* (horse chestnut). Compare ANDRODIOECIOUS, GYNOMONOECIOUS.

androphore see ANDROGYNOPHORE.

androsporangium a sporangium in which male products of meiosis form.

androspore a specialized zoospore produced in the OEDOGONIALES. It does not participate directly in the fertilization process but swims towards and attaches itself to a female filament. Here it germinates to form a microfilament, which in turn liberates ANTHEROZOIDS from the upper disc-shaped cells (antheridia). These then fertilize the female gamete.

androsporophyll a modified leaf that bears an ANDROSPORANGIUM.

androstrobilus a cone (see STROBILUS) that bears pollen sacs or microsporangia.

anemophily (wind pollination) pollination by pollen carried on the wind. Wind-pollinated flowers often have reduced sepals and petals and often appear before the leaves. This helps ensure that the stigmas are effectively positioned for pollen interception and the stamens are free to release their pollen. The stamens often have very long filaments while the styles may be long and feathery.

The pollen of wind-pollinated plants needs to be light and smooth surfaced and is only released from the anthers on warm dry days. In catkin-bearing species it can be stored in saucer-like bracts until disturbed and transported on a windy day. Wind-transmitted pollen has a typical diameter of 20–30 μm and may be carried thousands of miles (although most pollen travels less than 1 km). Because air movement is random pollen needs to be produced in vast quantities (a hazel catkin may produce 4 000 000 pollen grains). Compare ENTOMOPHILY, HYDROPHILY.

anergized culture see HABITUATION.

aneuploidy a condition in which not all the chromosomes are present in equal numbers and hence the total number is not an exact multiple of the haploid set. It occurs when chromosomes fail to separate at meiosis (see NONDISJUNCTION), so a gamete may either lack one chromosome altogether or have an additional copy. On fertilization the resulting zygote may thus have only one

homologue of a given chromosome, and is described as a MONOSOMIC, or it may have three homologues, and is called a TRISOMIC. If both gametes lack the same chromosome then the zygote is said to be *nullisomic*. Nullisomics are often inviable. Alternatively if both gametes contain the same additional chromosome, the zygote is termed *tetrasomic*. If there are missing or additional copies of two chromosomes the zygote is described as double monosomic or double trisomic respectively. See also HOMOLOGOUS CHROMOSOMES.

Aneuploidy results in unusual segregation ratios. It has been most closely studied in the thorn apple (*Datura stramonium*) where 12 types of trisomics have been recognized (one for each of the 12 different chromosomes), each producing a different mutant phenotype.

aneuspory the production, through a modification of the meiotic process (see MEIOSIS), of an unusual number of spores (usually two) instead of the four normally formed from each spore mother cell. It is seen in the formation of megaspores in dandelions (*Taraxacum*) where, after the first meiotic division the chromosomes stay in the one cell forming a *restitution nucleus*. The second meiotic division gives rise to two cells each with an unreduced number of chromosomes. One of these develops parthenogenetically into an embryo. Crossing over and hence reassortment of the genes can occur during the first meiotic division. This accounts for some of the variation found in apomictic complexes that have arisen by aneuspory. See APOMIXIS, PARTHENOGENESIS.

angio- a prefix that denotes enclosure. For example, in angiosperms (see ANTHOPHYTA) the seeds are enclosed within carpels/fruits.

angiocarpy in fungi, development of a fruiting body in which the spore-bearing tissue is enclosed for part of the time.

angiosperms see ANTHOPHYTA.

angstrom symbol: Å. A former unit of length equal to one thousandth of a micrometre or one tenth of a nanometre, i.e. 10^{-10} metre.

aniline stain any of a group of dyes derived from aniline that are used to stain biological material. Aniline sulphate and aniline hydrochloride both stain lignin yellow. Aniline blue is often used as a counterstain with safranin. See STAINING.

anion a negatively charged ion, e.g. the hydroxyl ion (OH⁻). Compare CATION.

anisocotyly a form of ANISOPHYLLY in which cotyledons at the same node differ in shape or size.

anisogamy (*adj.* anisogamous) the production or fusion of motile gametes that differ in size. This condition is found in various algae and fungi. Compare ISOGAMY. See also OOGAMY.

anisophylly the occurrence of leaves of different shapes or sizes at the same node (*nodal anisophylly*). The term is also used to describe the regular production of leaves of more than one shape or size on the same plant. The differences may be preprogrammed early in development by differences in the leaf primordia, or they may depend on the orientation of the shoot as the leaf pair develops. Sometimes leaves on the upper and lower sides of a shoot differ, even if there is only one leaf per node (*lateral anisophylly*).

annual a plant that germinates from a seed, grows, flowers, produces seeds, and then dies within a period of less than a year. Examples are marigold (*Calendula officinalis*) and some poppies (*Papaver*). Compare EPHEMERAL, BIENNIAL, PERENNIAL.

annual ring (growth ring) the increment of SECONDARY XYLEM added to the stems or roots wood of a plant in a single year. In transverse section this often appears as one or more rings due to the seasonal variation in the diameter of xylem elements formed at the beginning and end of the growing season. Wood whose annual rings show larger vessels in the early wood, forming distinct growth rings, as in pedunculate oak (*Querus robur*), is called *ring-porous wood*. Wood whose vessels are approximately equal in diameter with no obvious growth rings is called *diffuse-porous wood*, e.g. yellow birch (*Betula lutea*). Where there is more than one growth ring in any one year, these growth rings are termed 'false annual rings'. See also DENDROCHRONOLOGY, DENDROCLIMATOLOGY.

annular thickening a type of secondary cell wall patterning in TRACHEARY ELEMENTS in which the secondary cell wall is laid down in rings (*see illustration at* tracheary elements). Annular thickening is common in tracheary elements that have not yet finished elongating, for example those in protoxylem. Compare PITTED THICKENING, RETICULATE THICKENING, SCALARIFORM THICKENING, SPIRAL THICKENING. See SECONDARY GROWTH.

annulus a band or circle of tissue, especially: **1** the ring of differentially thickened cells that encircles the sporangium of certain ferns. It aids spore dispersal by inducing tension in the sporangial wall as the water in its cells evaporates. This causes the rupture of the STOMIUM and the top of the sporangium gradually curls back. As the tension increases a certain stage is reached when the water in the cells vaporizes, so releasing the tension and causing the top of the sporangium to spring back. Such movements serve to release the spores. **2** (velum) the remnants of the ruptured partial veil that encircle the stalk of a MUSHROOM or toadstool.

anomocytic see SUBSIDIARY CELL.

anoxic describing an environment greatly lacking in oxygen.

Antarctic floristic region see BIOGEOGRAPHY, FLORISTIC REGION, PLANT GEOGRAPHY.

anther the apical portion of a STAMEN, which produces the microspores or pollen grains. An anther normally comprises four POLLEN SACS (but only two in the Malvaceae) arranged in two groups or lobes joined by the connective tissue to the filament. If the anther is attached to the filament on the dorsal surface allowing it to pivot it is called a *versatile* anther. If it is attached dorsally but there is no movement it is termed a *dorsifixed* anther and if attached at the base it is a *basifixed* anther.

The wall of the anther lobes consists of an outer epidermis below which is the ENDOTHECIUM. This surrounds an inner nutritive layer or TAPETUM. Within the cavity (*loculus*) of the lobes pollen mother cells undergo meiosis to form tetrads of pollen grains. The individual pollen sacs are joined by zones of parenchyma tissue. Compare OVULE.

anther culture the culture of excised anthers on sterile nutrient medium. If anthers are taken from a plant at a certain stage of development and cultured under appropriate conditions then embryoids (see EMBRYOGENY) and subsequently haploid plantlets may be induced to form from the pollen grains. In some species it is possible to culture isolated pollen grains. This precludes the possibility that any resulting plantlets might be derived from somatic tissue rather than a pollen grain. In culturing isolated pollen grains, anther tissue may have to be used as nurse tissue. If the ploidy level of plantlets derived from pollen grains is doubled using colchicine then completely homozygous diploid plants can be obtained. Compare OVULE CULTURE.

antheridial cell the cell from which the antheridium develops. In seed plants, it is the generative cell in the pollen grain, which divides to provide the two sperm cells in the POLLEN TUBE. Compare PROTHALLIAL CELL. See also POLLEN.

antheridiophore an upright structure consisting of a stalk and cap that bears the antheridia in certain liverworts (see HEPATOPHYTA) of the Marchantiales. The antheridia are borne in pits on the upper surface of the cap. Compare ARCHEGONIOPHORE.

antheridium the male sex organ of the lower (nonseed-bearing) plants, algae, and fungi. In the algae and fungi it is unicellular, whereas in plants it may be multicellular and surrounded by a sterile jacket. It usually produces numerous small motile gametes. Compare OOGONIUM, ARCHEGONIUM.

antherocyte (spermatocyte) a cell that differentiates into an ANTHEROZOID without further cell division. The antherozoid is usually released from the antherocyte after it has been discharged from the antheridium.

antherozoid (spermatozoid, sperm) a motile male gamete produced by lower plants and some gymnosperms, which moves by means of UNDULIPODIA. In lower plants antherozoids are released from an antheridium but in the gymnosperms they are formed in the pollen tube prior to fertilization. Most antherozoids consist of an elongated nucleus contained within a ribbon-like cell. This form enables them to penetrate the narrow neck of an archegonium. The number of undulipodia borne by an antherozoid may be used as a diagnostic character.

anthesis the period from flower opening to fruit set.

Anthocerophyta (hornworts) a phylum of nonvascular plants (see BRYOPHYTES) with a horn-shaped sporophyte whose foot is embedded in the gametophyte. There is a single class (*Anthocerotae*) and order (Anthocerotales) and some 5 genera, with about 100 living species. As in liverworts (see HEPATOPHYTA), the gametophyte is a dorsiventrally flattened green thallus. However, hornworts differ from other bryophytes in that each cell has a single large chloroplast that contains a PYRENOID, a feature normally associated with protoctists (see PROTOCTISTA) rather than plants. They also differ in having a sporophyte that grows continuously from a meristem between its foot and the sporangium. Like mosses, hornworts have a cuticle and stomata on the sporophyte, and spore dispersal is aided by ELATERS. However, hornworts do not form protonemata; the spores germinate directly into young gametophytes.

The hornwort thallus contains a mucus-filled cavity. In some genera, such as *Anthoceros*, this houses nitrogen-fixing blue-green bacteria (see CYANOBACTERIA) (in this genus, *Nostoc*), enabling the hornwort to colonize bare rock surfaces. Sexual reproduction is by swimming antherozoids, as in other bryophytes. Some species also reproduce asexually by means of GEMMAE. The hornworts are thought to have evolved independently of mosses and liverworts: no forms intermediate between hornworts and mosses have been found.

Anthocerotae see ANTHOCEROPHYTA.

anthochlor pigments a group of yellow FLAVONOID flower pigments containing the CHALCONES and AURONES. They turn red on exposure to ammonia. This reaction distinguishes them from the yellow carotenoid pigments, which do not react in this way.

anthocyanescence the development of red pigments as a symptom of disease as, for example, seen in peach leaf curl caused by the fungus *Taphrina deformans*.

anthocyanin any of a group of GLYCOSIDE pigments formed by the addition of sugars and other residues to an anthocyanidin precursor (usually pelargonidin, delphinidin, or cyanidin). A very large number of these pigments have been characterized, all of them either red, blue, or violet. They are sap soluble, and occur in the cell sap of flowers, fruits, stems, and leaves.

Anthophyta (Angiospermophyta, angiosperms, flowering plants) a phylum (formerly a class, Angiospermae) of vascular plants that bear flowers, almost all of which also produce seeds enclosed within CARPELS (fruits). The sporophyte is the dominant generation and is either herbaceous or woody. The flower comprises the reproductive axis and its associated, often brightly coloured, sepals and petals. The gametophyte, which is hidden within the flower, is reduced to the female EMBRYO SAC within the ovule and the contents of the male POLLEN grain. The pollen does not germinate directly on the ovule, as in the gymnosperms, but on a specialized extension of the carpel, the stigma. The male gametes, unlike certain gymnosperm gametes, never bear undulipodia. DOUBLE FERTILIZATION to form a zygote and a triploid endosperm nucleus is characteristic. Secondary vascular tissue is usually but not always present. The xylem contains vessels, except in certain primitive woody forms, and the phloem has distinct companion cells associated with the sieve tube elements.

Angiosperms are the most advanced, most abundant, and most widely distributed vascular plants. The group contains some 230 000 species in 76 orders containing 350 families, and is subdivided, on the basis of the number of cotyledons in the embryo, into the classes MONOCOTYLEDONAE (a single cotyledon) and the DICOTYLEDONAE (two cotyledons). The subclass MAGNOLIIDAE is often regarded as a class and is seen as ancestral to both dicotyledons and monocotyledons. Beyond these groups further subdivision into superorders and orders is based mainly on the structure of the flower and especially on the form, number, and arrangement of the stamens and carpels. Different classifications recognize various numbers of orders and the names and contents of these often differ widely between various authorities.

From fossil pollen evidence it would seem the angiosperms appeared in the late Jurassic or at the beginning of the Cretaceous. They had replaced the gymnosperms as the dominant vegetation by the second half of the Cretaceous period. This may have been due in part to the relatively rapid life cycle of angiosperms, in many of which seed set occurs days or weeks after flowering. Several groups, including the CONIFEROPHYTA, GNETOPHYTA, BENNETTITALES, and CAYTONIALES, have been postulated as angiosperm ancestors, but the origin of the angiosperms remains obscure. See also MAGNOLIOPHYTA.

anthoxanthin any of a class of yellow or cream GLYCOSIDE plant pigments normally consisting of a glucose molecule attached to a FLAVONE or xanthone molecule.

anthracnose a fungal plant disease in which the characteristic symptoms are limited lesions, NECROSIS, and HYPOPLASIA. Anthracnose diseases are generally caused by one of the MELANCONIALES (e.g. *Colletotrichum lindemuthianum*, bean anthracnose; *C. coffeanum*, coffee anthracnose or coffee-berry disease; and *Gloeosporium limetticola*, lime anthracnose). The fungi causing anthracnose diseases produce numerous spores that are spread by rain. High humidity is required for infection and the diseases are most destructive when there is some water-soaking of tissues. Anthracnose diseases also produce symptoms, including scab, leaf-spots, and blight, that are not exclusive to the group.

anthropic horizon the surface horizon of a soil (see SOIL PROFILE) that has been produced by long periods of cultivation by humans. It is a deep, dark carbon-rich layer, rich in bases and phosphates.

anion exchange capacity the capacity of a soil to hold anions, which are held mainly on the surface of colloidal particles of clay and humus. It is usually measured in milliequivalents per 100 grams (meq/100 g)

of soil. See also ADSORPTION, CATION
EXCHANGE CAPACITY, EXCHANGEABLE IONS.

antibiosis see AMENSALISM.

antibiotic a specific substance produced by a
microorganism that inhibits the growth of
another microorganism. Antibiotics are
widely used as drugs to combat bacterial
diseases. Examples are penicillin, obtained
from the mould fungus *Penicillium notatum*
and active against staphylococcal infections
and many other Gram-positive bacteria, and
streptomycin, obtained from the
actinobacterium *Streptomyces griseus* (see
ACTINOBACTERIA) and used to treat
tuberculosis. Biosynthesis of antibiotics
may be from amino acids (e.g. penicillin),
sugars (e.g. streptomycin), or from acetate
or propionate (e.g. tetracyclines).
Commercial production is usually by large
scale culture of the appropriate organism
though some simple antibiotics, e.g.
chloramphenicol, are cheaper to produce by
artificial synthesis.

Antibiotics have proved useful research
tools. Those that inhibit protein synthesis
have been used to investigate ribosome
structure and function. Most antibiotics
inhibit protein synthesis on the 70S
ribosomes of prokaryotes but not the 80S
ribosomes of eukaryotes. The susceptibility
of mitochondrial and chloroplast
ribosomes to antibiotics is taken as further
evidence that these organelles are derived
from endosymbiotic prokaryotic organisms
(see ENDOSYMBIOTIC THEORY). Some
antibiotics inhibit protein synthesis by 80S
ribosomes but not 70S ribosomes. See also
PHYTOALEXIN, ALLELOPATHY.

antibody see SEROLOGY.

antical relating to the upper surface of a
DORSIVENTRAL shoot. Compare POSTICAL.

anticlinal at right angles to the surface. The
anticlinal wall of a cell is thus arranged
perpendicular to the surface of the plant
body. An *anticlinal division* results in the
formation of anticlinal walls between
daughter cells. Such a division enables a
tissue to increase its circumference, thus
keeping pace with any increase in girth of
the organ. In cylindrical organs, such as
stems and roots, the term *radial* may be
used in place of anticlinal, especially when
describing cell walls. Compare PERICLINAL.

anticodon a sequence of three nucleotides
on transfer RNA that is complementary to a
sequence (the CODON) on messenger RNA,
to which it temporarily binds during
protein synthesis. A given molecule of
transfer RNA will possess a specific
anticodon that only complexes with one
particular amino acid. It is this absolute
correspondence between an amino acid, an
anticodon, and a codon, that enables the
type and sequence of amino acids in a
protein to be determined precisely.

antigen see SEROLOGY.

antipodal cells the haploid cells, usually
three in number, found in the EMBRYO SAC
at the opposite end to the MICROPYLE. They
are derived by mitotic divisions (see
MITOSIS) of the MEGASPORE and have no
distinct cell wall. They take no part in the
fertilization process and their function is
unknown. At fertilization they may
disintegrate or multiply and enlarge.

antiport a membrane transport protein
involved in ACTIVE TRANSPORT where the
energy released by the passive movement of
H$^+$ ions across the membrane is coupled
with the transport of another solute, such as
Na$^+$, in the opposite direction against its
electrochemical potential gradient.

antithetic see HETEROMORPHIC.

antitranspirant a compound that reduces
transpiration, either by closing the stomata
or by depositing a film over the stomata.
Antitranspirants are sprayed on some crops
and ornamentals, but their use also limits
photosynthesis and affects various other
metabolic processes.

aperturate describing a pollen grain having
one or more apertures (areas where the
exine is either thinner or absent). If the
apertures are COLPI the pollen is termed
colpate and if they are pores, *porate*.

apetalous lacking petals.

aphid any insect of the family Aphididae
(greenfly, plant lice) of the insect order
Hemiptera. Aphids are small plant bugs
that feed by sucking plant juices. Many
species are pests in their own right, such as
the blackfly (*Aphis fabae*) on broad beans.
Other species are chiefly important as
vectors of plant virus diseases, such as
cucumber mosaic virus, which is
transmitted by several species of aphid

(including *Aphis gossypii*). *Macrosiphum avenae* and *Rhopalosiphum* species both transmit barley yellow dwarf virus of cereals and grasses.

aphotic zone the part of a body of water in which light is not of sufficient intensity for photosynthesis. Compare PHOTIC ZONE.

aphototropic see PHOTOTROPISM.

Aphyllophorales (Poriales, Polyporales) an order of the BASIDIOMYCOTA (basidiomycetes) in which the fungi produce a large BASIDIOMA, which rarely contains gills. The hymenium is usually found lining tubes (pores) or on small teeth or lamellae. The fruiting body is usually tough, sometimes woody, but never fleshy. The order contains over 100 species in some 35 genera. It is usually placed in the class HOLOBASIDIOMYCETES, Euholobasidiomycetes, or Hymenomycetes, according to different classification systems. In modern classifications the order is now included within the order CANTHARELLALES. Four main groups are recognized, depending on the structure of the basidioma: the *polypore* or bracket fungi, which grow out in a bracket-like manner from both living and dead wood (e.g. dryad's saddle, *Polyporus squamosus*); the club fungi, in which the fruiting body consists of a number of finger- or clublike projections (e.g. *Clavulinopsis helvola*); fungi in which there are many tooth- or spinelike projections from the basidioma (e.g. hedgehog fungus, *Hydnum erinaceus*); and fungi in which the basidioma is more-or-less flattened (e.g. earth fan, *Telephora terrestris*). These groups were originally classified as families, but it is now realized that they are artificial groupings.

Apiaceae (Umbelliferae, umbellifers) a family of DICOTYLEDONS containing about 3540 species in about 446 genera and commonly known as the carrot family. Its members are distinguished by their characteristic inflorescence, the UMBEL. Most are herbaceous and have hollow internodes and often a characteristic odour. The leaves are compound, often highly dissected, and usually spirally arranged. The fruit is usually a schizocarp. The seeds of many species are used as spices, for example caraway (*Carum carvi*) and

coriander (*Coriandrum sativum*). Species used as herbs include parsley (*Petroselinum crispum*) and fennel (*Foeniculum vulgare*), while the carrot (*Daucus carota*) and parsnip (*Pastinaca sativa*) are important root crops.

apical control a situation in which the influence of an apical meristem on daughter branches results in axillary buds that always remain subordinate to the main bud from season to season.

apical dominance the inhibition of the development of some or all of the lateral buds by the terminal (apical) bud of a shoot. Removal of the terminal bud releases some of the lateral buds from inhibition. This implies that a substance produced at the apex, most probably AUXIN, is responsible for the inhibition, though its method of action is unclear and the experimental evidence is somewhat contradictory. Apical dominance is not so marked when nutrients are plentiful, suggesting that available nutrients are first delivered to the terminal bud and only those in excess of requirements reach the lateral buds. CYTOKININS have also been shown to promote the growth of lateral buds.

apical meristem the MERISTEM at the tip of a stem or root that gives rise to primary tissues and is responsible for increase in length rather than girth of the axis. Compare LATERAL MERISTEM. See also GROUND MERISTEM, PROCAMBIUM, PROMERISTEM, PROTODERM, HISTOGEN THEORY, TUNICA–CORPUS THEORY.

apical placentation (pendulous placentation, suspended placentation) a form of PLACENTATION, found in ovaries containing only one ovule, in which the placenta develops at the top of the ovary.

apiculate having a small broad point APICULUS at the apex.

apiculus (mucro) on a moss capsule, a short point or projection on a relatively blunt leaf tip or capsule lid, e.g. on the leaves of *Pseudoscleropodium* (feather moss).

aplanetic describing organisms in which there is no motile stage.

aplanospore a nonmotile spore as produced, for example, by fungi in the Zygomycota and by certain members of the Chlorophyta and Chrysomonada. Compare ZOOSPORE.

apocarpous describing a GYNOECIUM in which the CARPELS are free, as in buttercups (*Ranunculus*). This type of arrangement is thought to be more primitive than a SYNCARPOUS gynoecium.

apoenzyme the catalytically inactive protein portion of an enzyme that remains when the PROSTHETIC GROUP or COFACTOR has been removed. Examples of enzymes needing both apoenzyme and cofactor for catalytic activity include: alcohol dehydrogenase, which requires zinc ions; kinases, which require magnesium or manganese ions; and the cytochromes, which require ferrous, Fe(II), or ferric, Fe(III), ions. See also HOLOENZYME.

apogamy (*adj.* apogamous) the development of the sporophyte directly from the gametophyte without the formation of gametes. The resulting sporophyte therefore has the same chromosome number as the gametophyte. Although there is no nuclear alternation between the generations the morphological differences persist. The phenomenon is seen in certain ferns, fungi, and algae. If male gametes are produced they are redundant although they have been shown to be capable of functioning in certain species. For example, the male fern (*Dryopteris borreri*) produces antherozoids that can fertilize the female gametes of related ferns.

Apogamy often occurs when the gametophyte has been produced by APOSPORY. It may also be induced by ageing or chemical agents. The term may also refer to the development of an unreduced diploid cell of the embryo sac into an embryo without fertilization occurring (i.e. parthenogenesis).

apomixis any form of ASEXUAL REPRODUCTION, including vegetative reproduction. The term is very often used in a narrower sense to mean the production of seeds without fertilization occurring. In this restricted sense the term is synonymous with *agamospermy*. Seed production by apomixis may occur either by the formation of a diploid embryo (see ADVENTIVE EMBRYONY) or embryo sac by a somatic cell, or by suppression or modification of the meiotic process (see MEIOSIS) to give unreduced megaspores (see DIPLOSPORY, ANEUSPORY). Development into a mature diploid embryo can then proceed without fertilization (see PARTHENOGENESIS, PSEUDOGAMY).

Apomixis is usually associated with POLYPLOIDY. An organism that reproduces by apomixis is termed an *apomict*. Facultative apomicts, e.g. the cinquefoils (*Potentilla*), can reproduce both sexually and apomictically. In such species the incidence of apomixis may be affected by environmental factors, such as photoperiod. Obligate apomicts can only reproduce apomictically. Often these are triploids or pentaploids that cannot produce viable pollen. Plants that form apomictic complexes are notoriously difficult to classify. Some apomictic races are so constant that they have been given taxonomic status as species (for example, almost 400 species of *Rubus* have been recognized in Britain).

apomorphy in discussions of PHYLOGENY or CLADISTICS, a derived or advanced character state. Apomorphies shared by different taxa are termed *synapomorphies* (compare AUTAPOMORPHY). Only synapomorphies whose origin can be traced back to a recent common ancestor (i.e. homologous character states) are of use in constructing phylogenies or cladograms. Analogous 'synapomorphies' may arise by convergent evolution and if identified as such are ignored. Compare PLESIOMORPHY.

apophysis the slightly swollen region between the seta and the capsule in a moss sporophyte (see BRYOPHYTA). Its cells are rich in chloroplasts, there are numerous intercellular spaces, and the epidermis contains stomata, similar in structure to those of vascular plants. The apophysis is thus an active photosynthetic region and helps to nourish the developing sporogonium.

apoplast the continuum of nonprotoplasmic matter, such as cell walls and intercellular material, throughout a plant. The movement of water in nonvascular tissues is principally through the apoplast as its resistance to flow is approximately 50 times less than that of the SYMPLAST.

apospory the development of the

gametophyte from the sporophyte without meiosis and spore production, so that the gametophyte has the same number of chromosomes as the sporophyte. If fertile gametes are formed by such a gametophyte then a sporophyte with twice the original number of chromosomes is produced. A polyploid series may be built up in this way (see POLYPLOIDY). Apospory can be induced artificially in certain ferns, e.g. the lady fern (*Athyrium filix-femina*), by pinning segments of frond to damp sand. Prothalli then arise from buds on the frond.

Apospory may also be used to describe the condition in angiosperms in which a diploid embryo forms from a cell of the nucellus or chalaza and megaspore formation is bypassed (see ADVENTIVE EMBRYONY). Compare APOGAMY.

apostemonous describing an androecium in which the anthers are separate (not fused), as in *Ranunculus* species. Compare MONADELPHOUS, SYNGENESIOUS.

apothecium (*pl* apothecia) the disc- or cup-shaped ASCOMA characteristic of discomycete fungi (excepting the Tuberales). The tips of the asci are freely exposed. Most lichens contain discomycete fungi and such lichen fungi also form apothecia.

appendage a structure developed on the surface of an organ, for example a petal or sepal.

apposition the laying down of layers of CELLULOSE on the inner surface of a plant cell wall to form the secondary cell wall. The process normally occurs once extension of the cell wall is completed and it serves to strengthen the overall cell structure. Compare INTUSSUSCEPTION.

appressed decribing the arrangement of leaves or scale leaves in a bud in which two leaves at a node face each other. *See illustration at* vernation.

appressorium (*pl.* appressoria) a hyphal structure, formed by many parasitic fungi, that serves to effect penetration of the host epidermis. It is flattened and closely pressed to the outer surface of the epidermis. From the undersurface a narrow *infection hypha* or *penetration tube* is pushed through into the cell or cell spaces below.

This then expands into a HAUSTORIUM or develops into hyphae.

aquatic see HYDROPHYTE.

araban (arabinan) a polysaccharide in which the major monosaccharide subunit is the pentose sugar ARABINOSE. Arabans are found with pectic substances in mature primary cell walls and in hemicelluloses and gums.

arabinose a pentose sugar of the ALDOSE group commonly found in plants and some fungi. It often occurs in the polymerized form ARABAN. *See illustration at* aldose.

arboretum an area devoted to the cultivation of a wide selection of all the woody plants (trees, shrubs, vines, etc.) that may be grown in a particular climatic region. Arboreta are maintained both as centres of research and as educational and recreational areas. See also BOTANIC GARDEN.

arbuscular–vesicular mycorrhiza see MYCORRHIZA.

Archaea (archaebacteria) a subkingdom of the kingdom BACTERIA that contains organisms similar to those that probably thrived in the early environments on earth, such as boiling muds, volcanic craters, hot springs, saline sediments, and deep sea vents. The Archaea are characterized mainly by the RNA that comprises their ribosomes, the nucleotide sequences of which differ significantly from those of other bacteria. Since RNA is thought to evolve only slowly over geological time, such differences are considered highly significant. The ultrastructure of the Archaea ribosomes resembles that of eukaryotic ribosomes rather than eubacterial ribosomes (see EUBACTERIA). Other diagnostic features of the Archaea include cell walls that lack peptidoglycan, major lipids that are ether-linked with phytanol side chains rather than ester-linked, a single DNA-dependent RNA polymerase with more than six subunits, and ribosomes of a distinctive shape. Two phyla of Archaea are recognized: the Euryarchaeota, which contains all the methanogenic and halophilic bacteria, and the Crenarchaeota, the thermoacidophilic bacteria. Some taxonomists consider that these phyla merit the status of separate kingdoms.

Archaean eon see PRECAMBRIAN.

archaebacteria see ARCHAEA.

archegoniophore an upright structure consisting of a stalk and cap that bears the archegonia in certain liverworts (see HEPATOPHYTA). In *Marchantia* sterile raylike structures radiate from the cap giving it the appearance of a chimney sweep's brush. In *Reboulia* the archegoniophore is umbrella shaped. The archegonia are borne in groups on the lower surface of the cap, the groups being separated by involucre scales termed PERICHAETIA. The stalk of the archegoniophore does not elongate until fertilization has occurred. Thus the antherozoids can swim to the archegonia before the water film between the thallus and archegoniophore is disrupted. Compare ANTHERIDIOPHORE.

archegonium (*pl.* archegonia) the female sex organ of the Bryophyta, Lycophyta, ferns (Filicinophyta), horsetails (Sphenophyta), and most gymnosperms and of some red and brown algae (Rhodophyta and Phaeophyta). It is normally produced within the maternal tissue and is related to the development of the terrestrial habit. It is made up of a narrow neck and a swollen base (or venter) that contains the female gamete. Compare ANTHERIDIUM.

archesporium the tissue that gives rise to the spore mother cells.

arctic vegetation see TUNDRA.

Arecaceae (Palmae) a monocotyledonous family of mainly tropical trees, the palms. It contains about 2650 species in about 203 genera. The growth form of a palm is characteristic, the plant usually consisting of an unbranched trunk with a crown of spirally arranged large leaves at the apex. When branching does occur, e.g. in *Hyphaene* (doum palms), it is dichotomous. Rattans (*Calamus* and relatives) are climbers, often with very long stems. The leaves may be pinnate, as in the feather palms, or the leaflets may all arise from the tip of the midrib, as in fan palms. Palm inflorescences are large and may contain thousands of sessile flowers in a simple or branching spike. In some palms, e.g. *Metroxylon sagu* (sago palm), the inflorescence is terminal, and the plant dies after flowering. The fruits are usually one-seeded DRUPES or BERRIES, which may attain a considerable size. Many palms are of economic importance, examples being the coconut palm (*Cocos nucifera*), the date palm (*Phoenix dactylifera*), and the oil palm (*Elaeis guineensis*).

Arecidae a subclass of the monocotyledons containing both herbaceous and arborescent plants. Members of the Arecidae differ from other monocotyledons in having broad net-veined petiolate leaves. Their numerous small flowers are usually unisexual and grouped into an inflorescence subtended by a spathe. Four orders are commonly recognized: the Arecales, which contains one family, the ARECACEAE (or Palmae, palms); the Cyclanthales, which also contains one family, the Cyclanthaceae; the Arales, which contains the three families Lemnaceae (e.g. duckweeds), Araceae (e.g. the arums or aroids), and Acoraceae (e.g. sweet flag); and the Pandanales, which comprises the family Pandanaceae (the screw pines). In some classifications the order Typhales is placed in the Arecidae, while in others it is allocated to the Commelinidae.

areole (*adj.* areolar) **1** a sunken cushion representing a condensed lateral shoot from which spines, branches, and flowers arise in cacti. Areoles may occur either singly on tubercles (e.g. in *Mammillaria zeilmanniana*) or in rows along raised ridges. **2** any area outlined on a surface, e.g. a segment of leaf lamina surrounded by veins. A surface divided into areoles, e.g. lichen thalli split by cracks into hexagonal areas, is described as *areolate*.

arginine (2-amino-5-guanidopentanoic acid) A basic amino acid with the formula $H_2NC(:NH)NH(CH_2)_3CH(NH_2)COOH$ (*see illustration at* amino acid). It is important in HISTONE proteins, which are rich in this amino acid.

aril a fleshy or hairy outgrowth of a seed or fertilized ovule, commonly derived from the funiculus or hilum. It may be regarded as a modified outer integument that only becomes conspicuous following fertilization. The brightly coloured mace surrounding the seed of nutmeg (*Myristica fragrans*) is an example. In the white water lily (*Nymphaea alba*), in which seed dispersal is by water, the seed is covered by a spongy aril that helps keep the seed afloat. The red fleshy cup surrounding the seed of yew (*Taxus baccata*) is also an aril. See also CARUNCLE.

aristate describing a structure, such as a glume or lemma, bearing an awn.

arithmetic mean see MEAN.

aromatic Describing an organic molecule that contains one or more benzene rings in its structure. It has a particular type of reactivity because the benzene ring structure contains alternating single and double bonds, so that electrons become delocalized. Compare ALIPHATIC.

Arthrophyta see SPHENOPHYTA.

arthrospore see OIDIUM.

artificial classification see CLASSIFICATION.

ascocarp see ASCOMA.

ascogenous hypha any of a number of small multinucleate branches that develop from an ASCOGONIUM. Ascogenous hyphae give rise to asci by CROZIER formation.

ascogonium (*pl.* ascogonia) the large coiled multinucleate female GAMETANGIUM of certain fungi in the Ascomycota. It is fertilized either by antheridial contents or by SPERMATIA and gives rise to ascogenous hyphae.

ascolichen a lichen in which the fungal partner is an ascomycete. See ASCOMYCOTA, MYCOBIONT.

ascoma (ascocarp) (*pl.* ascomata) The fruiting body of all fungi of the ASCOMYCOTA except the Hemiascomycetae. It consists of an aggregation of hyphae surrounding the asci. The various types of ascomata are used in dividing the Ascomycota into classes. See CLEISTOTHECIUM, APOTHECIUM, PERITHECIUM, PSEUDOTHECIUM.

Ascomycetes in some classifications, a class of fungi in the Ascomycota (or Ascomycotina). Most are mainly hyphal, but some are yeast-like. In some orders an ascoma and ascogenous hyphae may be lacking; in others it is present as a cleistothecium, perithecium, or ascothecium with both paraphyses and ascogenous hyphae. The asci may have single or double walls, and ascospores may be released actively or passively. The class is often divided into two subclasses, the Hemiascomycetidae (or Protoascomycetidae; see

HEMIASCOMYCETAE) and the Euascomycetidae (see EUASCOMYCETAE). The term 'ascomycetes' is also used as a general name for the ASCOMYCOTA.

Ascomycota (Ascomycotina, ascomycetes, sac fungi) a phylum of the FUNGI (or a subdivision of the EUMYCOTA) most of whose members have simple-pored septate hyphae forming a branching mycelium, although some (the yeasts) are unicellular. Asexual reproduction is by conidia (see CONIDIUM). In sexual reproduction, conjugation takes place between hyphae of compatible mating strains or between a specialized hypha or trichogyne and conidia or microconidia. Most ascomycotes produce a multicellular ASCOMA within which asci develop; nuclear fusion and meiosis result in the formation of eight (or a multiple of eight) ascospores, which may be actively or passively discharged. The group, which shows a wide range of habitats, morphology, and life cycles, includes saprobes and parasites, especially of plants, e.g. *Fusarium*, ergot of rye (*Claviceps purpurea*), peach leaf curl (*Taphrina*), and powdery mildews (Erysiphales). The common blue and green moulds, truffles, and morels are all ascomycotes. Many ascomycotes play an important role in decomposition; others are involved symbiotically in lichens (see LECANORALES), and a few in MYCORRHIZAS. Some, such as bread mould (*Neurospora*), are economically important decomposers of food, while others, such as *Penicillium*, are sources of important antibiotics. Yeasts are involved in the fermentation process of brewing, wine-making, and bread-making; some species are pathogens of humans.

The taxonomic relationships of the Ascomycota, the largest group of the Fungi (with about 32 267 species), are very confused, and there is no general consensus on the subgroupings. About 46 orders are recognized, with over 3000 genera and about 264 families. Commonly recognized classes include the HEMIASCOMYCETAE, EUASCOMYCETAE, LOCULOASCOMYCETAE, and Laboulbeniomycetae (minute ectoparasites of insects), based on the nature and mode of development of the asci or ascospores. The Laboulbeniomycetae are alternatively considered to be an order of the Euascomycetae. See also ASCOMYCETES, DISCOMYCETES, PLECTOMYCETES, PYRENOMYCETES.

Ascomycotina see ASCOMYCOTA.

ascorbic acid (vitamin C) a lactone of a SUGAR ACID that is found in high concentrations in certain fruits and green vegetables. It acts as a cofactor in the hydroxylation of proline to hydroxyproline. However its function in this reaction can be replaced by other compounds and it is not clear why it is essential to the growth of some vertebrates. See VITAMIN.

ascospore one of usually eight haploid spores characteristically formed inside an ascus. The two nuclei of the dikaryotic ASCUS initially fuse and undergo meiosis to produce four haploid daughter nuclei. These then undergo one mitotic division resulting in eight haploid nuclei. The cytoplasm of the ascus is then isolated around each nucleus to form the ascospores. See ASCOMYCOTA.

ascus (*pl.* asci) a saclike structure in fungi of the ASCOMYCOTA in which ascospores develop. If it is cylindrical, as is usually the case, then the ascospores are violently discharged. However, in some ascomycetes (e.g. the Saccharomycetales and Eurotiales) the ascus is a globular sac and in these fungi the ascospores are released passively. The tip of the ascus may possess a cap, the operculum, or it may simply have a terminal pore. This characteristic is important in separating the operculate Pezizales from the inoperculate Leotiales. In most ascomycetes the ascus wall is a single layer (*unitunicate*) but in the Loculoascomycetae it consists of two layers (*bitunicate*), and the two layers separate as the ascospores are released.

asexual reproduction (APOMIXIS) The formation of new individuals from the parent without the fusion of gametes. This may be achieved by BUDDING, FISSION, or FRAGMENTATION in the algae, fungi, and lower plants, or by SPORE formation or VEGETATIVE REPRODUCTION in the higher plants. Individuals so formed have a genetic constitution identical to that of the parent. Compare SEXUAL REPRODUCTION. See also APOMIXIS, CLONE.

asparagine an uncharged polar amino acid with the formula $NH_2COCH_2CH(NH_2)COOH$ (*see illustration at* amino acid). Asparagine is formed by the ATP-assisted addition of ammonia to ASPARTIC ACID. The reverse of this reaction (but without ATP formation) is the route of asparagine breakdown.

Because ammonia is extremely toxic to living cells, it is used to synthesize GLUTAMINE and asparagine, in which form it can be stored for later use in amino acid synthesis.

aspartic acid (aminobutanedioic acid) an acidic amino acid with the formula $HOOCCH_2CH(NH_2)COOH$ (*see illustration at* amino acid). It is formed by a transamination, in which the amino group of glutamic acid is transferred to oxaloacetic acid so forming aspartic acid. Breakdown is by the reverse reaction, followed by further oxidation in the Krebs cycle.

Aspartate is an important precursor of nitrogenous compounds. The amino acids leucine, isoleucine, threonine, methionine, and lysine all derive from aspartic acid, and it is necessary for synthesis of purines, pyrimidines, and porphyrins. Aspartate also acts as an amino-group donor in several reactions.

Aspergillales see EUROTIALES.

aspergillosis any disease of humans or other animals that is caused by the fungus *Aspergllus*. See EUROTIALES.

asporogenic (asporogenous) not forming spores.

assart 1 to remove trees and shrubs from a forested piece of land in order to use it for arable farming or pasture, or the land that results from this process. **2** in Britain, private farmland derived from former forest, woodland, or common land.

assemblage a group of plant and animal species characteristic of a particular environment. Species assemblages are useful indicators of environmental conditions.

assemblage zone (coenozone, faunizone) a stratigraphic unit (layer of rock) characterized by a particular assemblage of plants and/or animals, and usually named after one of the distinguishing fossil species present. It does not indicate a particular interval of geological time (some zones may span much longer periods of time than others, or span different time ranges in different locations), but is rather an indicator of environmental conditions.

assimilation the incorporation of materials acquired by the digestion of food or by photosynthesis into the body structures of an organism. The incorporated material is known as the *assimilate*. In plants and algae the term is also applied to the absorption of light energy and its utilization in internal chemical reactions.

assimilator an algal filament composed of photosynthetic cells.

association a large CLIMAX community named after the dominant types of plant species, e.g. deciduous-forest association, heath association. Most associations have more than one dominant plant. See also CONSOCIATION.

association analysis a classification that relies solely on χ^2 as a measure of association between pairs of species (attributes) found at a range of sample sites ('individuals'). The species ('attribute') with the highest overall sum of χ^2 values (i.e. the closest associations) with all other species is selected as the basis for subdividing into two groups of sites (individuals), one with and one without that attribute. The process is then repeated for the next pair and so on to form a hierarchical classification. See CHI-SQUARED TEST.

association measure a measure of the degree of association between two variables as shown by qualitative or quantitative data relating to their characeristics. The methods used include correlation and regression analysis, the chi-squared test, and similar tests. See also ASSOCIATION ANALYSIS.

associes a phytosociological term for a subclimax community in a succession. See PHYTOSOCIOLOGY.

assortative mating breeding that is nonrandom. If similar phenotypes breed together it is *positive assortative mating* and may lead to INBREEDING. If dissimilar phenotypes breed together it is *negative assortative mating* or *disassortative mating* and leads to OUTBREEDING. The pollinating mechanism often has a major influence on the breeding patterns found. Thus in insect-pollinated plants, the pollinator may preferentially take pollen to flowers of a certain colour or with certain markings. In wind-pollinated plants crossing is more likely to be random but there is a tendency for plants of the same height to cross. See also SELF INCOMPATIBILITY.

aster a starlike arrangement formed by fibrils radiating from a CENTRIOLE. Asters become conspicuous at the poles of the SPINDLE as cell division commences in the cells of algae and many fungi.

Asteraceae (Compositae) one of the largest dicotyledonous families, containing about 22 750 species in some 1300 genera and commonly referred to as the daisy or sunflower family. Its members are found worldwide, except in Antarctica. They include herbs, shrubs, trees, and climbers. They are recognized by their characteristic headlike inflorescence, the CAPITULUM, which superficially appears to be a single flower. The individual flowers are called *florets* and come in two forms – *ray florets* with a strap-shaped corolla, usually situated at the periphery of the capitulum, and *disk florets* with a tubular corolla, usually in the centre of the capitulum. Many species have both types of florets in the same capitulum, but some have only ray florets, and others only disk florets. The fruit (see CYPSELA) often possesses a ring of hairs, the pappus, to aid dispersal. A few species produce drupes instead. The embryo is oily, with little or no endosperm. All composites have resin canals except most of the tribe Lactuceae, which have latex ducts. Polyfructosans, especially INULIN, are common storage compounds. Commercially important composites include the sunflower (*Helianthus annuus*), grown for the oil in its seeds, lettuce (*Lactuca sativa*), chicory (*Cichorium*), and numerous ornamental species, such as *Dahlia*, *Chrysanthemum*, and *Aster*. Many species, including dandelions (*Taraxacum*) and thistles, are weeds of arable land.

Asteridae (Sympetalae) a subclass of the DICOTYLEDONAE (dicotyledons) containing mostly herbaceous plants having flowers with fused petals. They have few stamens and usually only two carpels, and the seeds are surrounded by only one integument (*unitegmic*). The following nine orders are generally recognized: Gentianales (Contortae), including the Loganiaceae, Gentianaceae (gentians), Apocynaceae (e.g. oleanders), Asclepiadaceae (e.g. milkweeds), Strychnaceae, and Geniostomaceae; Solanales, including the Convolvulaceae (e.g. bindweeds, dodder), SOLANACEAE (e.g. potato), Polemoniaceae (e.g. phlox), Hydrophyllaceae, and Nolanaceae; Lamiales, including the Verbenaceae (e.g. verbena, teak), LAMIACEAE (Labiatae), Boraginaceae (e.g. forget-me-not), Avicenniaceae (e.g. mangroves), Stilbaceae, Lennoaceae, Cylocheilaceae, and Phrymaceae; Callitrichales, including the Callitrichaceae (e.g. water starwort), Hippuridaceae (mare's tail), and Hydrostachyaceae; Plantaginales, the Plantaginaceae (plantains); Scrophulariales, including the SCROPHULARIACEAE (e.g. foxglove), Oleaceae (e.g. olive), Buddlejaceae, Plocospermataceae, Orobanchaceae (e.g. broomrapes), Globulariaceae, Myoporaceae (e.g. boobialla), Gesneriaceae (e.g. gloxinia), Acanthaceae (e.g. black-eyed Susan), Mendonciaceae, Pedaliaceae (e.g. sesame), Bignoniaceae (e.g. Indian bean tree), and Lentibulariaceae (e.g. bladderworts); Asterales, including the Menyanthaceae (e.g. bogbean), Pentaphragmataceae, Sphenocleaceae, Campanulaceae (e.g. bellflowers), Stylidiaceae (e.g. trigger plants), Donatiaceae, Goodeniaceae, ASTERACEAE (Compositae), and Calyceraceae; Rubiales, including the RUBIACEAE (e.g. gardenia), Gelsemiaceae, and Desfontainiaceae; and Dipsacales, including the Dipsacaceae (e.g. scabious), Caprifoliaceae (e.g. honeysuckle), Adoxaceae (e.g. moschatel), Valerianaceae (e.g. valerian), and Morinaceae. The affinities of the Solanaceae are uncertain; some authorities place it in the Polemoniales or Scrophulariales.

astrosclereid (asterosclereid, star sclereid) a relatively short sclerenchyma cell (SCLEREID), differing from a BRACHYSCLEREID by its often conspicuously branched shape. Astrosclereids are usually present singly or in small groups and are often found in the mesophyll of leaves, where they act as a strengthening agent.

atactostele SEE STELE.

atavism the reappearance of a characteristic, known as a *throwback*, in an individual after several generations in which it has not been seen. This may be due to a random combination of certain alleles.

Atlantic period the period from about 6000–3000 BC that was characterized by a global warming of the climate and an increase in precipitation in north temperate regions. It was the warmest phase of the present interglacial period, and saw deciduous broad-leaved forests spread northwards. See GLACIAL.

atmometer SEE POTOMETER.

atmosphere 1 the envelope of invisible, odourless and tasteless gases that surrounds the Earth. It is a mixture consisting of almost constant proportions of nitrogen and oxygen, together with helium, hydrogen, methane, nitrous oxide and the inert (noble) gases—argon, krypton, neon, and xenon, which surrounds the earth. It also contains variable amounts of water vapour, carbon dioxide, ozone, sulphur dioxide, nitrogen dioxide, and various pollutants such as chlorofluorocarbons (CFCs) and chlorine, as well as suspended liquid and solid particles, including spores and pollen grains. Its density decreases with altitude and its composition changes above about 100 km., where incident solar radiation produces increasing numbers of charged particles.

The atmosphere has several distinct layers. The *troposphere* extends from ground level to about 15 km. Convection currents ensure mixing of air, giving a gradual vertical temperature gradient of about 6°C per kilometre. This is the layer in which water vapour and weather occur. At the top of the troposphere lies the *tropopause*, a boundary where temperature falls to about −60°C. Above this is the 50-km-thick *stratosphere*, a dry layer with a steep vertical temperature gradient, with temperature increasing with altitude. The upper part of this zone contains the OZONE LAYER, formed by the action of ultraviolet rays from the Sun acting on oxygen molecules. Beyond the edge of the stratosphere (the *stratopause*) is the *mesosphere*, where temperature declines with altitude. At its outer limit is the *mesopause*, where temperatures are around –85°C. The outer layer is the *thermosphere*, where temperature again rises with altitude. At increasing altitudes the density of the atmosphere decreases as it is affected by the solar wind (a stream of charged particles from the Sun). **2** a unit of pressure approximately equivalent to the mean atmospheric pressure at sea level or to the pressure exerted by a column of mercury 760 mm high. It is defined as 101 325 pascals (about 14.7 lbs/sq in).

atomic force microscopy (scanning force microscopy) a form of microscopy that reveals the microscopic topography of a surface. It can achieve resolutions of less than 1 angstrom. A probe linked to a cantilever is passed over the surface to be analysed. Minute intermolecular interactions between the probe and the surface cause the probe to be repelled by the surface. Its movements are exaggerated, then used to intercept a laser beam directed at a sensitive surface. In this way a three-dimensional image can be built up of such structures as chromosomes and large organic molecules. The probe can be dragged over the surface of the object, it can tap the object at intervals, or it can travel just above the surface. When the probe does not actually touch the surface, this technique can be used to study real-time changes in the surface without interfering with the process causing the changes. For example, it can reveal the changes in the structure of starch as it is digested by an enzyme.

ATP (adenosine triphosphate) the triphosphorylated form of adenosine, similar in structure to AMP but with three phosphates, linked by HIGH-ENERGY PHOSPHATE BONDS. It is the major energy-transferring molecule in all biological systems. The energy produced by PHOTOPHOSPHORYLATION and metabolic oxidation reactions is used to form ATP from ADP and inorganic phosphate. ATP then provides energy for biosynthesis, active transport of ions and metabolites, and other energy-requiring processes. It is

also essential for the synthesis of the nucleic acids DNA and RNA.

Besides its ubiquitous role as an energy-transferring molecule, ATP is involved in the regulation of sugar oxidation, being an inhibitor of the conversions of pyruvic acid to acetyl CoA and of fructose 6-phosphate to fructose bisphosphate.

ATPase (adenosine triphosphatase) an enzyme that catalyses the hydrolysis of ATP to ADP and inorganic phosphate. In OXIDATIVE PHOSPHORYLATION, this enzyme forms part of the enzyme complex ATP synthetase. Driven by the energy of the proton gradient generated by the ELECTRON TRANSPORT CHAIN, it catalyses the synthesis of ATP from ADP and inorganic phosphate, at the same time dissipating the hydrogen ion and electrical gradients.

ATP synthetase an enzyme complex that catalyses the synthesis of ATP from ADP and inorganic phosphate using energy from the proton gradient generated by the ELECTRON TRANSPORT CHAIN.

atropous see ORTHOTROPOUS.

auger a device used to sample soil. The commonest kind is the corkscrew auger, a corkscrew-shaped instrument that is twisted down into the soil, then withdrawn. The auger is usually marked at intervals to indicate the depth to which it is being inserted. Such an auger may also be used to extract a sample of wood from a tree trunk for dating purposes (see DENDROCHRONOLOGY). See also PEAT-BORER.

auricle a small earlike projection from the base of a leaf or petal. Auricles are seen at the base of the leaf blade in grasses (Poaceae). See also LIGULE.

aurones one of the two groups of ANTHOCHLOR flower pigments. They are probably formed by oxidation of CHALCONES. Aurones are found in many members of the Asteraceae, especially the genus *Coreopsis*.

Australian floristic region see BIOGEOGRAPHY, FLORISTIC REGION, PLANT GEOGRAPHY.

aut- a prefix meaning 'self' or 'individual', e.g. autecology.

autapomorphy (*adj.* autapomorphic) any derived character state (see APOMORPHY) only possessed by members of one particular TAXON. Autapomorphies distinguish a taxon from other related taxa. The number of autapomorphies possessed by one taxon as compared to another taxon assumed to be derived from the same common ancestor, provides a measure of ANAGENESIS.

autecology the study of a single species or individual and its relationship with the environment. It involves the investigation of the life cycle of the organism, recording the effects of the nonliving and living factors in the environment at each stage. Quantitative records are kept, particularly those relating to variations in numbers within and between populations. Compare SYNECOLOGY.

author in taxonomy, the person who published the first valid name of a taxon. The name of the author, or a recognized abbreviation of it, should follow the name of the taxon for it to be complete. Many plant taxa are followed by L. (for Linnaeus). If the rank of a taxon at the genus level or below is changed but the name or epithet retained then the name of the original author is also retained in brackets before the name of the person who made the change. For example, *Medicago polymorpha* var. *orbicularis* L. was raised to the species level by Allioni and is thus named *Medicago orbicularis* (L.) All. A similar procedure is followed if a taxon at the subgeneric level is transferred to another taxon but retains its epithet.

If a name proposed by one person but not published by him is published by another and ascribed to the first author then the taxon is followed by the name of the first author followed by ex (described by) and then the name of the publishing author. For example, the Corsican speedwell is *Veronica repens* Clarion ex DC. (DC. for De Candolle).

The citation of authors in this way helps other workers find the original descriptions and type specimens.

autochory the dispersal of seeds or spores by the parent organism. This is often by some kind of explosive propulsion. See also BOLOCHORY.

autochthonous describing material that originated in its present location. For example, the plant material in a peat deposit, which originally grew on that site. Compare ALLOCHTHONOUS.

autoclave an apparatus within which very high temperatures are produced by steam under pressure. It is used to sterilize laboratory equipment and is essentially a large pressure cooker.

autocolony a daughter COENOBIUM that resembles the parent colony.

autodiploid 1 a gamete containing a diploid rather than a haploid number of chromosomes. Such gametes may arise either through faulty meiosis or from TETRAPLOID tissues. Two diploid gametes may fuse to give a tetraploid zygote or such gametes may develop by PARTHENOGENESIS. 2 a diploid plant obtained by doubling the chromosomes of a haploid plant. Such plants are completely HOMOZYGOUS.

autoecious describing a rust fungus (see RUSTS) in which the various spore forms are all developed on the same host (see PARASITISM). An example is *Puccinia menthae*, mint rust. Compare HETEROECIOUS.

autogamy (self fertilization, endogamy) fusion of female and male gametes derived from the same flower. This mechanism restricts genetic variability of the population but can stabilize selected traits and ensures that isolated individuals have the opportunity to reproduce. Autogamy may be found in species where cross pollination cannot be assured, as in pioneer populations (many weeds are autogamic) or where insect vectors may be rare (as in the tundra ecosystems). Autogamy is thought to be derived from ALLOGAMY by the overcoming of self-incompatibility systems. Total autogamy is rare in the plant kingdom except in species showing CLEISTOGAMY. In most cases autogamy is facultative and may be triggered by adverse climatic conditions, such as high humidity or intense cold, or by failure to achieve cross fertilization. Autogamy may be contrasted with *geitonogamy* – fertilization that occurs between flowers of the same plant and therefore requires a vector. Genetically, the outcome is the same for both mechanisms. See also HOMOGAMY. Compare ALLOGAMY.

autogenic describing a change in a SUCCESSION due to the modification of the environment by the vegetation. This might be due to enrichment of the soil with humus, increasing of shade, or the trapping of sediment in aquatic or marine environment by plant roots.

autolysis the process of self-digestion undergone by organelles or cells when their useful life is completed. It is effected by enzymes, primarily hydrolytic in character, that are produced by the cytoplasm. The process is believed to be under the control of hormones. The products of the digestion are subsequently reabsorbed by the surrounding cells. See also AUTOPHAGY, DELIQUESCENCE.

autonomic movement a movement of a plant or plant part in response to a stimulus generated within the plant. Autonomic movements include the beating of undulipodia, cytoplasmic streaming (see CYCLOSIS), CIRCUMNUTATION, and the movement of chromosomes during nuclear division (see KARYOKINESIS). Compare PARATONIC MOVEMENT.

autopentaploid see AUTOPOLYPLOIDY.

autophagy a process utilized by certain cells to digest worn out or superfluous cell organelles. An autophagic vacuole forms around a portion of cytoplasm containing one or more unwanted organelles. Hydrolytic enzymes are secreted into the vacuole by the surrounding cytoplasm and digestion occurs followed by the subsequent reabsorption of the breakdown products. See also AUTOLYSIS.

autopolyploidy a form of POLYPLOIDY that results from the multiplication of chromosome sets from a single species. Autopolyploids may arise either through a failure in mitosis (see SYNDIPLOIDY), or in meiosis, resulting in the formation of diploid rather than haploid gametes. Autopolyploids often resemble their diploid parents, except that they may grow more slowly and flower later. Their cells are usually larger and the epidermis thicker. Autopolyploids with an uneven number of chromosome sets (*autotriploids, autopentaploids*) are often sterile because their gametes contain unbalanced numbers of chromosomes, which interfere with the pairing of HOMOLOGOUS CHROMOSOMES at metaphase of meiosis. Autotetraploids also show reduced fertility, the degree depending on whether any trivalents or univalents are formed at meiosis and also on the orientations taken up by the multivalents at metaphase. The patterns of inheritance, although they follow MENDEL'S LAWS, are very complex in autopolyploids because of the increased number of genotypes that may exist for any given locus with two alternative alleles. Thus for the two alleles A and a, a diploid can have three possible genotypes, AA, Aa, and aa, while an autotetraploid can have five, AAAA, AAAa, AAaa, Aaaa, and aaaa. In addition, while the frequencies of the gametes A and a can usually be predicted in diploids, the frequencies of the gametes AA, Aa, and aa of an autotetraploid are hard to predict because segregation can occur at either the first or second meiotic division. Compare ALLOPOLYPLOIDY.

autoradiography a technique for detecting the distribution of radioactive isotopes previously incorporated into cells. A photographic emulsion is placed over a thin piece of squashed or sectioned tissue in the dark. The developed film (*autoradiograph*) shows the location of the radioactive substance in the form of concentrated dark patches. See also ISOTOPIC TRACER.

autosome either of any pair of chromosomes that do not play a principal role in sex determination. The term is used of plant chromosomes when discussing dioecious species that have cytologically identifiable SEX CHROMOSOMES.

autospore a nonmotile spore that has the same form as its parent cell, but is smaller.

autotetraploid see AUTOPOLYPLOIDY.

autotroph an organism that needs only simple inorganic compounds to grow. Thus carbon dioxide or carbonates serve as the carbon source and can be built up into complex organic molecules using light (see PHOTOTROPH) or chemical (see CHEMOTROPH) energy. Simple inorganic nitrogen compounds are utilized in protein synthesis. Most plants and some protoctists and bacteria are autotrophic. Compare HETEROTROPH.

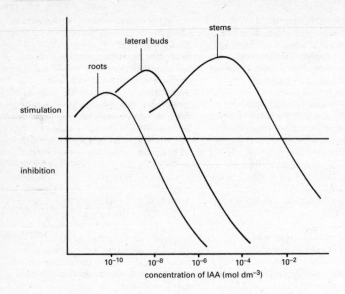

Auxin: the effects of auxin concentration on the growth of different organs

auxanometer an apparatus that is designed to measure the growth in length of plant organs. The organ is attached by a taut piece of cotton to the end of the shorter arm of a pivoted lever. Any growth results in the upward movement of the arm, which is magnified and recorded by the other longer arm of the lever as its tip moves across a calibrated scale, or as it traces the movement on a slowly rotating drum (KYMOGRAPH).

auxiliary cell a specialized cell in the gametophyte thallus of some red algae (see RHODOPHYTA), e.g. *Polysiphonia*, situated adjacent to the CARPOGONIUM. Following fertilization of the carpogonium the zygote produces filaments that grow towards and fuse with the auxiliary cell. This process appears necessary for the production of carpospores.

auxin any of a class of plant HORMONES whose principal effects are brought about by their ability to promote the elongation of shoots and roots when present in low concentrations. The most widely occurring natural auxin is INDOLE ACETIC ACID. Other naturally occurring auxins are also based on the indole ring (e.g. indole acetonitrile and indole pyruvic acid). However the indole group is not essential for auxin activity as is shown by the auxin activity of certain synthetic compounds, e.g. NAPHTHALENE ACETIC ACID (NAA) and 2,4-dichlorophenoxyacetic acid (2,4-D).

Auxins are usually synthesized in meristematic regions and transport is generally from apex to base, i.e. basipetal. The concentration of auxin is very important in determining the nature of the growth response and the optimum auxin concentration differs for different organs (see diagram). The inhibition of growth at higher auxin concentrations may be due to the auxin-promoted synthesis of ETHENE (ethylene), which inhibits cell elongation. Such inhibitory effects have been exploited in the production of herbicides based on 2,4-D and 2,4,5-T.

Auxins also promote root initiation, and certain synthetic auxins (e.g. NAA and indole butyric acid or IBA) are widely used as rooting compounds. ABSCISSION has been correlated with low levels of auxin in the organ concerned and auxins have thus been used to prevent premature fruit drop. Other phenomena in which auxins have been implicated include APICAL DOMINANCE, phototropism (see *Avena*-curvature test), and EPINASTY. Many effects are brought about by the combined action of auxin with other hormones. For example, the stimulation of cambial activity, the induction of parthenocarpy, and the enhancement of internode elongation are all more effectively promoted by a combination of auxin and GIBBERELLIN than by either substance alone. Similarly appropriate concentrations of auxin and CYTOKININ are needed in culture media to promote cell division in tissue explants (see EXPLANTATION). Depending on the relative concentration of each, root meristems (high auxin: low cytokinin) or shoot meristems (low auxin: high cytokinin) may be initiated.

auxospore a diatom cell that has been released from its siliceous test. Auxospores are usually formed following a sequence of vegetative cell divisions in which the daughter cells become progressively smaller. The decrease in size results from the nature of diatom cell division whereby the two halves of the cell separate, each half containing a daughter cell. A new half wall is then formed, the edges of which fit within those of the old half wall in a manner resembling a Petri dish. Thus the diatoms decrease in size until a certain limit is reached, at which point auxospores are released. The naked auxospores are free to expand before forming a cell wall. Auxospores may also form after sexual fusion or by APOGAMY or PARTHENOGENESIS.

auxotroph a type of BIOCHEMICAL MUTANT that can grow normally if provided with a supplement in its diet to replace the substance it has lost the ability to make. Some naturally auxotrophic fungi require exogenous vitamins; for example, *Phycomyces blakesleanus* requires thiamin. Certain mutants of the mould *Neurospora crassa* need the amino acid arginine in the culture medium to grow normally (the unmutated mould can synthesize all the amino acids it requires from ammonia). By supplying the auxotrophic mould with various precursors of arginine it can be

established which of the enzymes in the pathway to arginine is missing.

available nutrients elements, ions, or compounds in the soil solution that can be taken up by plant roots and used as nutrients by the growing plant. Many nutrients in the soil are unavailable either because they are adsorbed onto humus particles and clay minerals, or because they are insoluble.

available space theory the theory that the origin of new leaf primordia at a stem apex is governed physically by the amount of space between existing primordia. It is postulated that a new primordium only arises at a point where a certain minimum amount of space has become available. This is usually between the penultimately formed primordium and the one formed previous to that. The pattern that results from such restrictions is the same as would occur if inhibitors were released by older primordia (see REPULSION THEORY). See also PHYLLOTAXIS.

***Avena*-curvature test** a quantitative method used to correlate the degree of curvature obtained in a plant part, after the application of an external stimulus, with the amount of endogenous hormone (AUXIN) required to bring it about.

The tips of *Avena* (oat) coleoptiles are used as a standard. These are removed and placed on an agar block and auxin allowed to diffuse out for a predetermined time. The auxin-impregnated agar is then used to measure the curvature produced when pieces are placed acentrally on decapitated coleoptiles. The response is plotted against the auxin concentration in the block. Over a certain range, increases in auxin concentration result in a regular increase in the degree of bending in the coleoptile, but the linear response is lost at higher levels. Only the straight line section of the graph is useful for assays. By noting the angle of curvature produced by a given stimulus under otherwise controlled standard conditions, one can find the quantity of auxin involved in producing the response. See also TROPISM.

awn a stiff bristle-like projection, usually at the tip of an organ. The glumes and lemmas of grasses (Poaceae) commonly possess awns, as do some fruits, and less commonly leaves. Cereals and grasses having ears or spikes covered in awns are termed *awned* or *bearded*. An individual organ bearing an awn is termed *aristate*. An awn may act to bury a fruit in the soil by uncoiling in damp conditions and in so doing pushing the fruit into the ground.

axenic culture (pure culture) a culture consisting of only one type of organism or cell. Axenic cultures are derived from pure parent cultures and grown under sterile conditions. Such uncontaminated cultures are necessary in investigating the properties of a particular microorganism or cell type and in maintaining cultures of microorganisms over a long period of time.

axial describing cells derived from the FUSIFORM INITIALS, which are elongated parallel to the long axis of an organ. The *axial system* refers to all these cells collectively.

axial system see AXIAL.

axil the upper angle formed by the junction of a leaf or similar organ with the stem. Organs in the axil, such as flowers, inflorescences, meristems, and buds are termed *axillary* or lateral.

axile describing a part of a plant that is attached to the central axis.

axile placentation a form of PLACENTATION in which the placentae arise along the central axis of the ovary. It is seen in compound ovaries that are divided by septa into a number of locules, as in *Hyacinthus*.

axillary see AXIL.

axillary bud see BUD.

axoneme see UNDULIPODIUM.

azonal soil see SOIL.

azygospore see PARTHENOSPORE.

B b

baccate shaped like a berry as, for example, the arils of yew (*Taxus baccata*).

Bacillariophyceae (Diatomophyceae) an alternative name for the DIATOMS, used in classifications that consider the group to be a class of the division Chromophyta rather than a distinct phylum.

Bacillariophyta see DIATOMS.

bacillus (*pl.* bacilli) any bacterium that is rod shaped and more or less straight. Compare COCCUS, SPIRILLUM, VIBRIO.

backcross a cross between an individual and one of its parents. If the organism is short lived this may not be a practical possibility, in which case the term refers to a cross between an individual and an organism genetically identical to one of its parents. A backcross performed to ascertain the genotype of an individual is termed a TEST CROSS. See also BACKCROSSING.

backcrossing a technique used in plant breeding to introduce a desirable gene into a cultivated VARIETY. Unlike the PEDIGREE METHOD of plant breeding, the aim of a backcrossing programme is not to create entirely new varieties but to modify existing varieties.

The cultivated variety (the recurrent parent) is crossed with the donor parent, which may be quite useless agriculturally except for its possession of one particularly valued gene (e.g. for disease resistance). The progeny from this cross, which contain 50% of the donor genetic material, are screened for the character and those possessing disease resistance are crossed back to the recurrent parent. The progeny of this cross – the first backcross generation (B_1) – now contain 25% of the genetic material of the donor. The plants are again screened, the resistant plants again backcrossed to the recurrent parent, and this process is repeated until about the seventh or eighth backcross generation, by which time less than 0.25% of the donor genetic material remains. At this stage the B_7 or B_8 generation plants are selfed (crossed with each other) to produce plants homozygous for disease resistance; these may be identified by a TEST CROSS.

The process described above assumes the allele for disease resistance is dominant. If it were recessive then it is necessary to alternate backcrossing with selfing of the backcross generations. Backcrossing is more efficient with self-pollinating species but the method is not necessarily limited to self pollinators.

back mutation a MUTATION in which a mutant (non-wild-type) gene reverts to the original wild-type form.

back-scattered electron imaging in scanning electron microscopy, the use of a special detector to produce an image from the back-scattered electrons given off from below the surface of the specimen rather than the secondary electrons used in conventional imaging. The image obtained consequently includes subsurface information. See ELECTRON MICROSCOPE.

backshore the part of a beach above the level of normal high SPRING TIDES. It is affected by wave action only during exceptionally high tides or storms. It is typically colonized by HALOPHYTES and XEROPHYTES.

Bacteria (Prokaryotae, Monera) a kingdom containing all the PROKARYOTES. It contains at least 10 000 species, probably but a fraction of the true number, most of which remain undiscovered. Bacteria are distinguished by their lack of a nucleus and chromosomes. The DNA is not coated with protein; although not membrane-bounded, it is usually located in a distinct part of the cell, called the nucleoid. Enzymes for metabolic pathways, such as respiration, and photosynthetic pigments and enzymes are bound to cell membranes, but not packaged into discrete organelles. Flagella, when present, are of simple structure and are made of the protein flagellin.

Bacteria range in size from 1 to 10 μm. There is a wide range of forms, from simple cells to branching filaments and motile colonies. Multicellular forms are rare, and there is no tissue development. Bacteria also show a wide range of metabolism and include obligate and facultative aerobes and anaerobes. They obtain nutriment from a range of substrates, including many complex organic compounds and such inorganic substrates as nitrogen and sulphur compounds and hydrogen. Some are capable of tolerating extremes of temperature and salinity. Bacteria are found in every imaginable environment that can support life, and they play a vital role in the recycling of elements (see BIOGEOCHEMICAL CYCLE, CARBON CYCLE, NITROGEN CYCLE, SULPHUR BACTERIA). They make up considerable part of the body weight of other living organisms and take part in important symbioses, notably in the guts of ruminants (see also ENDOSYMBIOTIC THEORY) and in the root nodules of legumes (see NITROGEN FIXATION). They cause many diseases of plants, animals, and humans, but they are also a source of antibiotics, such as streptomycin, erythromycin, and kanamycin.

There are two major subkingdoms of the Bacteria, the ARCHAEA (archaebacteria) and the EUBACTERIA, which many systematists consider merit the status of full kingdoms. Margulis and Schwartz (1998) defend their status as subkingdoms by the argument that neither the Archaea nor the Eubacteria evolved by symbiosis, and this sets them apart from other organisms.

bacterial chromosome the name erroneously given to the naked DNA of bacteria. This does not have a linear molecule, as in eukaryotes, but a circular molecule. Unlike the CHROMOSOMES of eukaryotes, the bacterial chromosome is not complexed with proteins. It has no centromere, telomere, or other structures that are present in eukaryote chromosomes. See PROKARYOTE.

bacterial transformation the incorporation of genetic material into a bacterium from DNA in the surrounding medium. The phenomenon was discovered by F. Griffith (1928) who converted (transformed) rough-coated *Diplococcus pneumoniae* into a smooth-coated strain by mixing it with extracts from the latter under appropriate conditions. Later O. T. Avery, C. M. MacLeod, and M. McCarty (1944) showed that the substance responsible for the transformation was DNA. The latter experiment is often quoted as one of the first pieces of evidence that DNA, not protein, is the genetic material.

bactericidal describing a compound that is capable of killing bacteria. It may act by destroying part of the cell, or by disrupting a biochemical pathway.

bacteriochlorophyll any of the forms of chlorophyll found in PHOTOSYNTHETIC BACTERIA. Bacteriochlorophyll *a*, found in all bacteria, differs from the chlorophyll of plants in being a tetrahydroporphyrin rather than a dihydroporphyrin (i.e. it has more hydrogen atoms in pyrrole rings 2 and 4 of the chlorophyll molecule). It also differs in its absorption spectrum having three peaks, at 365, 605, and 770 nm. Some bacteria contain additional forms of bacteriochlorophyll (Bchl) designated Bchl *b*, Bchl *c*, and Bchl *d*. See PORPHYRIN.

bacteriocin a protein, produced by a particular strain of bacteria, which is lethal to or inhibits the activities of other bacteria, especially closely related strains.

bacteriology the study of bacteria and their use in industry, agriculture, and medicine.

bacteriophage (phage) any virus that infects bacteria. Bacteriophages are often named according to the bacteria they infect, e.g. actinophages, which infect Actinobacteria, and coliphages, which infect various strains of *Escherichia coli*. Some phages contain RNA as their genetic material, others DNA. All

RNA phages and some DNA phages bring about a *lytic response,* causing the death of the host cell by lysis. These are termed *virulent phages.* The *lytic cycle* involves: adsorption of the phage to a specific site on the bacterial cell wall; penetration of the cell by the phage; transcription and translation of the phage genetic material by the bacterium so that many new phages are formed; and lysis of the cell to release the newly synthesized particles. The genomes of some DNA phages (the *temperate phages*) become integrated in the host DNA and are replicated with it, a stable relationship known as *lysogeny.* Such a genome is termed a *prophage,* and the bacterium so infected is said to be *lysogenic.* This relationship may occasionally break down, in which case the phage may then destroy its host bacterium. Some temperate phages, e.g. the lambda virus, have been shown to transfer genes from one bacterium to another in a process termed TRANSDUCTION. Most bacteriophages infect only one particular species or strain of bacterium.

bacteriorhodopsin a purple light-harvesting pigment found in certain species of bacteria of the genus *Halobacterium,* which mediates a unique form of photosynthesis. *Halobacterium* is the only photosynthetic genus of Archaean bacteria (see ARCHAEA). Instead of using the pigment chlorophyll linked to complexes of other proteins to harness light energy and activate electrons, as in other photosynthetic organisms, it uses the single protein bacteriorhodopsin, which is situated on a specialized part of the cell membrane. The light-absorbing group in bacteriorhodopsin is retinal, the same group that is used to absorb light in the vertebrate eye. The light absorbed by this pigment is used to activate a proton (H$^+$ ion).

bacteriostatic describing an agent that prevents the growth of certain bacteria without killing them.

bacteroid a *Rhizobium* bacterium in a legume ROOT NODULE after it has undergone an increase in size (to about forty times its original size) and a change in shape to a characteristic X- or Y-shaped cell. Bacteroids lose their capacity for autonomous growth and are dependent on the host. It is in this metamorphosed state that the bacteria carry out NITROGEN FIXATION.

badlands plateau country in North and South Dakota and Nebraska, USA, which has been deeply dissected by erosion mainly due to run-off from infrequent heavy rainfall combined with lack of vegetation cover. The term is now applied more widely to similar eroded land in other parts of the world.

baeocyte see ENDOSPORE.

bakanae disease (foolish seedling disease) a serious disease of rice caused by the fungus *Gibberella fujikuroi,* in which there is excessive growth in length of the stem, believed to be caused by GIBBERELLINS produced by the fungus.

balanced polymorphism see POLYMORPHISM.

ballistospore a spore that is violently projected, e.g. the basidiospores of the Hymenomycetes.

balsam see RESIN.

Bangioideae see BANGIOPHYCIDAE.

Bangiophycidae (Bangioideae) a subclass of the red algae (RHODOPHYTA) in which cells have a single star-shaped chloroplast, growth is diffuse rather than localized to particular parts of the thallus, the life cycle is relatively simple, with just one or two phases, and there are no specialized reproductive structures. The cell walls often contain mannans and xylans. The Bangiophycidae range from unicellular forms to multicellular filaments and blades. The simplest forms reproduce by FURROWING to separate two daughter cells. More advanced forms, such as *Bangia* and *Porphyra,* have two main phases in the life cycle: a large haploid macrothallus, which may be filamentous or bladelike, alternates with a microscopic filament (which in some species is diploid), the CONCHOCELIS stage. Asexual reproduction is by spores produced either by the macrothallus or by the conchocelis. Sexual reproduction is by spermatia that fertilize simple carpogonial cells (see CARPOGONIUM), which then divide to produce diploid carpospores; these in turn produce the conchocelis phase. In some species unfertilized carpogonial cells may produce haploid carpospores, giving rise to a haploid conchocelis phase.

BAP (6-benzylaminopurine, benzyladenine) A synthetic CYTOKININ.

barachory (clitochory) the dispersal of seeds or spores as a result of their own weight.

Barfoed's reagent a mixture of copper(II) acetate and acetic acid, used to show the presence of strongly reducing sugars in solution. After boiling, monosaccharides cause the formation of a red precipitate of copper(I) oxide. Disaccharides are not such powerful reducing agents and will not show a positive reaction.

bark the tissue arising to the outside of the PHELLOGEN, i.e. the PHELLEM, when this is exposed by the sloughing off of the epidermis. At intervals the bark is interrupted by small patches of loosely packed cells (LENTICELS), which allow air to reach the underlying living tissues. The bark of different trees can be very distinctive and its characteristics are used to aid identification. In some species the same phellogen is active each year and a thick layer consisting solely of phellem is formed (e.g. oak, beech), but in most species a new phellogen arises annually in the cortex below: the bark thus consists of both phellem and dead cortex and is termed *rhytidome*. As the thickness of the bark increases, the outer layers may split in patterns characteristic of the species. Six main categories of bark pattern are recognized: smooth (e.g. *Prunus maakii*), fissured (e.g. sweet chestnut, *Castanea sativa*), cracked (e.g. sweet gum, *Liquidambar styraciflua*), scaly (e.g. London plane, *Platanus × hispanica*), dippled scaly (e.g. boldo, *Peumus boldus*), and peeling (e.g. paper birch, *Betula papyrifera*). Monocotyledons generally lack the ability to produce bark; many develop instead a tough outer layer of fibres derived from old leaf veins, as in palms.

bar of Sanio see CRASSULA.

barren a COMMUNITY of relatively sparsely distributed plants that cover less than half the ground area. Such communities are typical of some fairly level parts of the Arctic tundra, often on sandy and serpentine soils. Barrens often have few trees and are dominated by a single species, such as mountain avens (*Dryas octopetala*).

The plants are often small and stunted compared to individuals of the same species from less infertile habitats, and they often contain groups of specialized endemic species. See also PINE BARREN.

barren lands a historical name for the North American tundra.

basal body see KINETOSOME.

basal placentation a form of PLACENTATION, found in ovaries containing only one ovule, in which the placenta develops at the base of the ovary. It is seen in bistorts (*Polygonum*).

base A substance that reacts with an acid to form water and a salt; in solution, it releases ions that can combine with hydrogen ions. *Contrast* acid.

base analogue a chemical that resembles a naturally occurring PURINE or PYRIMIDINE base to the extent that it may be incorporated into a DNA molecule during replication. However such analogues are less specific in their pairing properties. For example, an analogue of adenine may pair with cytosine as well as thymine. Over two replications this would result in a GENE MUTATION, with a cytosine:guanine pair replacing an adenine:thymine pair on one of a pair of homologous chromosomes. Base analogues may also be incorporated into RNA during RNA synthesis, e.g. thiouracil and fluorouracil can replace uracil. Such changes may have various physiological effects. See 5-BROMOURACIL.

base number see GENOME.

base pairing the bonding relationship between the bases in the nucleotides of DNA and RNA. The relationship is highly specific, such that adenine (A) in one strand of DNA only pairs with thymine (T) in the complementary strand (or with uracil (U) in RNA). Similarly guanine (G) only pairs with cytosine (C). The A:T and G:C base pairs of complementary polynucleotide chains are held together by hydrogen bonds. Although these are only weak bonds, large numbers are formed between the two polynucleotide strands of DNA so that the double helix as a whole is extremely stable. The specific pairing of A:T (or A:U in RNA) and G:C means that the genetic material is accurately replicated from one generation to another. It also means that molecules of messenger RNA are accurately transcribed from the genetic material and subsequently accurately translated to protein. See also TRANSCRIPTION.

base ratio a comparison of the molar quantities of one base to another base in DNA, or of one base pair with another base pair. The observation that the amount of adenine (A) always equals the amount of thymine (T) and similarly the amount of guanine (G) equals that of cytosine (C) provided one of the first indications that adenine always pairs with thymine and guanine with cytosine in the complementary polynucleotide chains. A comparison of base pair ratios, AT:GC, or percentage GC of the total, shows a constant figure within a species but differences between species. See also GC RATIO.

base saturation the extent to which the exchange sites of the adsorption complexes in the soil are occupied by exchangeable basic cations, or by cations other than hydrogen and aluminium. It is usually expressed as a percentage of the maximum CATION-EXCHANGE CAPACITY.

base triplet hypothesis see GENETIC CODE.

basic grassland GRASSLAND occurring on alkaline soil, especially on soils developed over limestone or chalk. In Britain and northwest Europe, many basic grasslands have been grazed since Neolithic times, and today many areas of rough basic grassland are also maintained by the grazing of rabbits. These grasslands are usually rich in species, including numerous *forbs* (non-grassy herbaceous plants).

basic number see GENOME.

basic soil see ALKALINE SOIL, PEDOCAL.

basic stain see STAINING.

basidiocarp see BASIDIOMA.

basidiolichen a LICHEN in which the fungal partner is a basidiomycete. See BASIDIOMYCOTA, MYCOBIONT.

basidioma (basidiocarp; *pl.* basidiomata) the reproductive organ of all basidiomycete fungi (except the Ustilaginales and Uredinales), which produces the basidia (see BASIDIUM). It may be mushroom shaped, bracket shaped, club shaped, or a hollow sphere or cylinder. Much of the basidioma is composed of sterile

pseudoparenchymatous tissue. The basidia are borne in a fertile layer, the HYMENIUM. See BASIDIOMYCOTA.

basidiomycetes see BASIDIOMYCOTA.

Basidiomycota (Basidiomycotina, basidiomycetes) a phylum of the FUNGI (or a subdivision of the EUMYCOTA). It contains the most advanced fungi of the Eumycota, which are characterized by the production of basidia (see BASIDIUM). Other characteristic features include hyphae with CLAMP connections, typically divided by complex inflated septa with a central pore (see DOLIPORE SEPTUM). The MYCELIUM may be uninucleate or dikaryotic. In most basidiomycotes, a basidiospore germinates to form a monokaryotic septate primary mycelium; this may in some species produce oidia (see OIDIUM). Later a dikaryotic mycelium develops, either spontaneously in homothallic species or as the result of conjugation between two hyphae of opposite mating types in heterothallic species. As the mycelium grows, the nuclei divide synchronously in association with the formation of clamp connections. The dikaryotic mycelium eventually produces basidiospores, often with the formation of a simple or complex fruiting body, the BASIDIOMA. A few orders produce conidia (see CONIDIUM).

Some basidiomycotes are saprobes, others are parasites of crops and forest trees. Many form symbiotic MYCORRHIZAS. Some have persistent mycelia and over time may form FAIRY RINGS. There are some 22 000–25 000 species in 165 families and 1428 genera. Three or four classes are generally recognized: the HOLOBASIDIOMYCETES, which includes the mushrooms, puffballs, stinkhorns, and bird's-nest fungi; the TELIOMYCETES, including the rust fungi; the USTOMYCETES, including the smut fungi; and sometimes also the PHRAGMOBASIDIOMYCETES, which includes the jelly fungi. The Phragmobasidiomycetes are alternatively considered to be a subclass of the Holobasidiomycetes. An alternative system classifies the Basidiomycota as a single class, the Basidiomycetes, which is then divided into the subclasses Heterobasidiomycetidae, in which the basidium is deeply divided or septate (e.g. rusts and smuts); and Homobasidiomycetidae, in which the basidium is a single large cell. Older classification systems may use the classes HEMIBASIDIOMYCETES (rusts and smuts), HYMENOMYCETES (mushrooms, boletes, polypores), and GASTEROMYCETES (stinkhorns, puffballs, etc.).

Basidiomycotina see BASIDIOMYCOTA.

basidiospore a haploid spore, four of which are characteristically borne on a BASIDIUM.

basidium the spore-bearing structure of fungi in the BASIDIOMYCOTA. Each basidium usually bears four basidiospores on its outer surface, often on projections called *sterigmata*. The basidium is usually a single cell but in some basidiomycetes (e.g. Ustilaginales and Uredinales) it is segmented. The basidium is initially binucleate. These nuclei fuse, undergo meiosis, and form four (rarely eight) haploid daughter nuclei, which subsequently become the basidiospores. In most basidiomycetes the basidia form a fertile layer, the *hymenium*.

basifixed describing an anther that is joined to the filament at its base. See STAMEN. Compare DORSIFIXED, VERSATILE.

basifugal movement see BASIPETAL.

basipetal describing movement, differentiation, etc. occurring from apex to base of root or shoot. For example, the direction of proto- and metaxylem formation is from the apex or nodes downwards, while the transport of auxin in the stem is generally towards the base. Basipetal movement is also known as *basifugal movement*. Compare ACROPETAL.

basitony a pattern of shoot branching in which the branches furthest from the shoot apex grow most vigorously, so the shoot is wider at the bottom than at the top (see diagram). Compare ACROTONY, MESOTONY.

basket fern an epiphytic or lithophytic fern with a rosette of more or less upright fronds – the so-called basket. Such ferns tend to trap pieces of dead plant material and other debris in the basket; nutrients leached out of these by percolating rainwater are thought to be used by the fern.

basophilic describing cells or cell constituents that can be stained by a basic stain. See STAINING.

bast see PHLOEM, RAFFIA.

Baumgrenze see TREE LINE.

B-chromosome (accessory chromosome, supernumerary chromosome) any of a number of small chromosomes present in some organisms in addition to the fixed number of stable chromosomes characteristic of the species (the A-chromosomes). They are of widespread occurrence, having been reported in over 80 genera and 20 families of plants and being found increasingly in the fungi. The numbers found may differ between members of a species and even within a single individual, with gametic nuclei often having larger numbers. B-chromosomes are composed largely of HETEROCHROMATIN and in many cases appear to have little effect. However when present in large numbers they may be deleterious.

bearded 1 having many awns as, for example, the ears of the bearded fescue (*Festuca ambigua*).
2 having any other kind of stiff hair as, for example, the collection of hairs on the lower petals of the flag iris (*Iris pseudacorus*).

beard-lichens fruticose branching filamentous LICHENS of a pendant habit.

Beaumont period forty-eight hours during which the minimum temperature is 10°C and the relative humidity is 75% or more. These periods are the basis of the system used in England to forecast an outbreak of

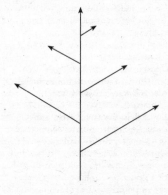

Basitony

late blight of potatoes (*Phytophthora infestans*).

Sporulation of the potato blight fungus and subsequent invasion of potato leaves is favoured by moist weather. By examining meteorological data, A. Beaumont worked out that blight could be expected ten days after a 'Beaumont period'. Careful monitoring of local meteorological data enables the advisory services to record Beaumont periods and give farmers accurate advice on when to spray their potato crops.

Beccariian body see FOOD BODY.

beet sugar see SUCROSE.

Beltian body see FOOD BODY.

Benedict's reagent see BENEDICT'S TEST.

Benedict's test a procedure used to detect the presence of reducing sugars in solution. *Benedict's reagent* is a mixture of copper(II) sulphate in solution together with a filtered mixture of sodium citrate and sodium carbonate. the reagent is added to the test solution; after boiling, a high concentration of reducing sugars in the test solution causes the formation of a red precipitate, while a lower concentration results in the formation of a yellow precipitate. it is a more sensitive test than FEHLING'S SOLUTION.

Bennettitales (Cycadeoidales) an extinct order of GYMNOSPERMS known from fossils extending throughout the mesozoic era. they resembled cycads (see CYCADOPHYTA) in habit and in leaf structure but can be distinguished from fossil cycads by certain features of the leaf epidermis, notably the possession of a SYNDETOCHEILIC stomatal complex as compared to the HAPLOCHEILIC stomata of the cycads. Like the cycads, they may have originated from the Cycadofilicales. Some had reproductive structures somewhat resembling buttercup flowers and were probably insect-pollinated.

Benson–Calvin–Bassham cycle the cycle in photosynthetic carbon dioxide FIXATION that regenerates the primary carbon dioxide acceptor. It is named after the three discoverers but is commonly abbreviated to CALVIN CYCLE.

benthic zone the habitat that comprises the sediments at the bottom of ponds, streams, lakes, and oceans. It is extremely variable, depending on the depth and turbidity of the water (and hence light penetration), its temperature, pressure, salinity, and how much oxygen is present in the sediments. In shallow water algae, diatoms, and photosynthetic bacteria form the base of the food chain, but in deeper water nutrients come mainly from the rain of detritus sinking down from the surface waters and from the decomposition of carcasses that sink down to the sediments.

benthos (*adj.* benthic) the organisms that live in the BENTHIC ZONE. The plants and algae (*phytobenthos*) that grow below low-tide mark include members of the flowering plant family Zosteraceae, such as eelgrasses (*Zostera*), certain red algae, such as dulse (*Rhodymenia palmata*), and in deeper water kelps (see LAMINARIALES). A variety of plants grow rooted in the LITTORAL ZONE of lakes and ponds, including water lilies (*Nuphar*, *Nymphaea*), pondweeds (*Potamogeton*), and water milfoils (*Myriophyllum*). There are also many micro-organisms and invertebrates living in and on the surface of the sediments, including bacteria, algae, diatoms, ciliates, amoebae, sponges, crustaceans, sea anemones, molluscs, and worms. Many benthic organisms filter tiny particles of detritus from the water, which others scavenge detritus on the sediment surface, prey on the detritus feeders, or graze on the algae, diatoms and bottom-rooted plants.

benzyladenine see BAP.

6-benzylaminopurine see BAP.

Beringia an area that encompasses the Bering Strait and adjacent areas of Alaska and Siberia. During periods of lower sea level during the late Mesozoic and Cenozoic eras, the Bering Strait was dry land and formed an important migration route for species between the Palaearctic and Nearctic BIOGEOGRAPHICAL REGIONS.

berry a many-seeded fleshy indehiscent fruit. The epicarp usually forms a tough outer skin, especially in the PEPO and HESPERIDIUM and the mesocarp becomes massive and fleshy. The epicarp and mesocarp may be highly coloured to attract the animals that act as agents of dispersal.

Examples are the tomato and grape. Compare DRUPE. *See illustration at* fruit.

betacyanin see BETALAIN.

betalain a class of nitrogen-containing pigments containing the red *betacyanins* (e.g. betanidin, the beetroot pigment) and the yellow *betaxanthins* (e.g. indicaxanthin from the fruits of *Opuntia ficus-indica*). They replace and perform the functions of the floral pigment anthocyanin in certain plants. The possession of betalains has been used as taxonomic character. Its occurrence is restricted to a group of families (nine or ten depending on the classification system used) that have all traditionally been placed in the order Caryophyllales. The occurrence of betalains in cacti is strong evidence for their inclusion in the Caryophyllales, rather than (as in some classifications) in the monotypic order Cactales (or Opuntiales). Interestingly, one family, the Caryophyllaceae, normally included in the Caryophyllales, lacks betalains.

beta-oxidation see OXIDATION.

beta-pleated sheet (β-pleated sheet) see SECONDARY STRUCTURE.

betaxanthin see BETALAIN.

Bial's reagent a mixture of orcinol in concentrated hydrochloric acid with 10% iron(III) chloride solution. It is used to detect the presence of pentose sugars in a test solution. A green colour after boiling indicates a positive reaction.

bias non-random distortion in measurements or other data due to such factors as poorly calibrated instruments, changes in a specimen due the experimental procedure or environmental conditions, or subjectivity on the part of the experimenter. Compare CONSISTENCY.

bicentric distribution the distribution pattern of a species with two distinct centres of distribution. See DISJUNCT DISTRIBUTION.

bicollateral bundle a VASCULAR BUNDLE in which the phloem occurs both internal and external to the xylem, as in the stems of some dicotyledons, e.g. members of the Solanaceae. The presence of bicollateral bundles distinguishes solanaceous plants from members of the Scrophulariaceae. See also STELE. Compare COLLATERAL BUNDLE, CONCENTRIC BUNDLE.

biennial a plant that takes one to two years to complete its life cycle. It grows vegetatively in the first year and the photosynthates are stored in PERENNATING ORGANS. The stored food is used to produce foliage leaves, flowers, and seeds the following year. The plant then dies. Some biennials almost complete their life cycle within one year; there is actually a continuous spectrum from ephemerals to long-lived perennials, such as *Agave*. Some important crops are biennials, e.g. carrot (*Daucus carota*) and parsnip (*Pastinaca sativa*). Certain garden flowers that are in fact perennials are more successfully grown as biennials, e.g. wallflower (*Cheiranthus cheiri*) and *Antirrhinum*. Compare ANNUAL, EPHEMERAL, PERENNIAL.

biflavonyls see FLAVONOIDS.

bifoliate describing a palmate leaf which consistently has only two leaflets.

bifurcate forked, forming two branches.

bigeneric describing a hybrid derived from an intergeneric cross, e.g. *Triticale* derived from a cross between *Triticum* and *Secale*.

bijugate describing a form of PHYLLOTAXIS in which the leaves are arranged in OPPOSITE pairs, but in which successive pairs are less than 90° apart. This causes a double spiral of leaves, an arrangement sometimes described as *spiral decussate*.

bilateral symmetry the arrangement of parts in an organ or organism such that it can only be split into similar halves along one given plane. Thus most leaves can only be divided into similar halves by cutting along the midrib. Bilateral symmetry in flowers is usually termed *zygomorphy*. The flowers of relatively advanced angiosperm families, e.g. Scrophulariaceae, are often zygomorphic. In a FLORAL FORMULA, zygomorphy is represented by the symbol ·|· or ↑.

biliprotein see PHYCOBILIPROTEIN.

bimodal distribution a frequency distribution that has two major peaks, not necessarily of equal height.

binary fission see FISSION.

binomial classification see BINOMIAL NOMENCLATURE.

binomial distribution (normal distribution, Gaussian distribution) the idealized symmetrical distribution of a population of statistical values centring around a mean, when departures from the mean are due to the chance occurrence of a large number of individual small independent effects. The distribution of height in humans is an example. The resulting curve is bell-shaped.

binomial nomenclature a system of naming species using a generic name and a specific epithet, established by the Swedish botanist Carl von Linné (also known by the Latinized form of his name, Carolus Linnaeus) in 1753. In this system the generic name represents the genus, and the specific epithet denotes the species. Before this the scientific names of species consisted of a short Latin description. This was too cumbersome as a name yet too short for a proper description. By separating the procedures of naming and describing, Linnaeus established the foundations of present-day nomenclature.

The generic name is a noun often based on a classical name as, for example, *Narcissus*. The name can also be honorific, commemorating the plant's discoverer, such as *Saintpaulia* (African violets), named after Baron Walter von Saint Paul-Illaire. Alternatively, the generic noun may refer to the locality in which the plant was first found, *Ligusticum* (lovage) from Liguria (northwest Italy), for example. Specific names are adjectival and often describe a particular feature of a plant, such as leaf shape or flower colour (e.g. *longifolius, albus*). Like generic names they may also be commemorative or geographical (e.g. *wilsonii, lusitanica*). See TABLE 1 IN THE APPENDIX FOR THE MEANINGS OF SOME COMMON SPECIFIC EPITHETS. Binomials are always Latinized and are usually printed in italic script, the generic name beginning with a capital letter while the specific epithet is lower case throughout (e.g. *Primula vulgaris*). When naming a species the rules of the INTERNATIONAL CODE OF BOTANICAL NOMENCLATURE must be followed.

binucleate having two nuclei per cell.

bioaccumulation the accumulation of POLLUTANTS and their toxins in the tissues of living organisms. For example, certain species or varieties of plants accumulate toxic heavy metals ions such as cadmium, nickel, lead, and zinc and can tolerate them in their tissues. A few species, called *hyperaccumulators*, can accumulate 100 times more toxins than other plants (over 10 000 mg toxic element per kilogram plant tissue). See also BIOREMEDIATION.

bioassay (biological assay) a quantitative assessment of the effect of a substance on a living organism by comparison with the effects of a similar substance of known concentration. For example, certain fungi are used to find the concentration of a particular vitamin, such as thiamin, in a substance as the growth of the fungus is directly proportional to the concentration of the vitamin. Bioassays have been used extensively to measure the effects of different auxins and gibberellins on plant growth (see AVENA-CURVATURE TEST).

bioassimilation 1 the accumulation of a substance in a habitat. 2 the incorporation of carbon, hydrogen, oxygen, nitrogen, and other elements into BIOMASS.

biochemical genetics the branch of genetics concerned with inheritance at the molecular level. It includes the study of the structure of DNA, its replication, the genetic code and transcription and translation in protein synthesis, and the regulation of gene expression.

biochemical mutant an organism possessing a MUTATION that causes a particular enzyme either not to be synthesized or to be defective in some respect. The reaction that it catalyses therefore does not take place, and, if this reaction is an intermediate step in a metabolic pathway, the subsequent steps in the pathway are also prevented. Such mutants can be induced by various methods, such as irradiation with X-rays, and have proved particularly useful in elucidating the stages of a number of pathways. See also AUXOTROPH.

biochemical oxygen demand (BOD) the standard measure of the level of organic pollution in a sample of water. The biochemical oxygen demand is the amount of oxygen used by microorganisms feeding on the organic material during a given period of time. The higher the BOD, the greater the level of pollution, as the increased levels of nutrients released by decomposition of the organic material support larger populations of micro-organisms, which require oxygen for respiration. See also CHEMICAL OXYGEN DEMAND.

biochemical pathway see BIOSYNTHETIC PATHWAY.

biochemical taxonomy see CHEMOTAXONOMY.

biochemistry the study of metabolism in prokaryotic and eukaryotic cells. As a discrete subject biochemistry has only existed for about 50 years, during which time a huge body of knowledge has accumulated. Originally biochemistry was concerned with cataloguing and investigating the biological occurrence and enzymatic reactions of a large number of organic compounds. However unifying patterns and principles have gradually emerged from the scattered body of facts and hypotheses.

biochrome a pigment produced by a living organism.

biochronology the use of biological events to measure geological time. The presence or absence of certain species of FOSSILS is used to date approximately particular rock strata. The first and last appearances of certain fossils are significant moments in geological time. See also GEOLOGICAL TIME SCALE.

bioclimatology the study of climate in relation to the environment of plants and animals, especially domestic/cultivated species and humans. It includes the effects of climate on vectors of diseases.

biodegradable describing substances that can be broken down by living organisms, usually to form harmless products.

biodiversity the number and variety of living organisms in the world, or in a specified area. Biodiversity encompasses not only ecosystems and species, but also genetic diversity within species and populations. Current estimates of the numbers of species identified to date are about 1.7 million. Estimates of species yet to be discovered range from 5 to 100 million. While species extinction is a natural part of the evolutionary process, losses today are greatly accelerated as a result of human activity. It is thought that 2–8% of the earth's species will become extinct over the next 25 years. Biodiversity is important not only for scientific research, to aid our understanding of how the natural world – and indeed our own DNA – works, but also as a source of new products (such as drugs), new varieties of crop plants, and new sources of genetic material for genetic engineering. See also CONVENTION ON BIOLOGICAL DIVERSITY.

bioenergetics the study of energy transfer in living organisms. Almost all the energy used on Earth comes directly or indirectly from the sun. The way in which it is converted in cells depends on the principles and laws of thermodynamics. The first law concerns the relationship between work, heat, and internal energy; it is equivalent to the statement that energy can be converted from one form into another, but cannot be created or destroyed. Thus, when light energy falls on a plant some is reflected, some absorbed as heat, and some converted into chemical energy in glucose by PHOTOSYNTHESIS, but the total amount of energy is constant. In fact energy transfers in living organisms tend to be inefficient. About 2% of the energy in incident radiation is converted into chemical energy. Heterotrophic organisms obtain energy from the nutrients they ingest; about 10–20% of the energy is passed on at each link in the FOOD CHAIN, which tends to have three or four links only.

The second law of thermodynamics concerns the way in which energy is transferred or converted. It states that any conversion of energy from one form into another involves some dissipation of energy as unavailable heat energy. The availability of energy in a system is determined by its entropy, which is a measure of randomness or disorder. In any process the total entropy increases (an alternative statement of the second law), and there is a constant degradation of energy in the universe into energy that is unavailable for work. The growth of organisms is characterized by decreases of entropy in the sense that disordered systems (e.g. CO_2 and O_2 gas) are converted into ordered chemical structures. However, the *total* entropy of the organism and its surroundings always increases in any change. All chemical reactions are driven by a decrease of free energy.

biogenesis the theory that new life arises only from preexisting life and never from nonliving matter. Compare SPONTANEOUS GENERATION. See also ORIGIN OF LIFE.

biogenic produced by living matter.

biogeochemical cycle (mineral cycle) the movement of chemical elements from the physical (abiotic environment) into living organisms (biotic environment) and back to the physical environment again. If the elements concerned are essential to life, such a cycle is called a *nutrient cycle*, e.g. the NITROGEN CYCLE. Physical, chemical and biological processes may be involved in regulating the cycle. In the simplest cycles, such as the PHOSPHORUS CYCLE, physical and chemical factors are the main controlling ones – the slow rate of release of phosphate from rocks by weathering limits the rate of cycling of phosphorus. In the nitrogen and carbon cycles, however, the activities of living organisms may limit the rate at which the cycle occurs. The nitrogen cycle is regulated by the rate of incorporation of nitrates into proteins in living cells and their release by decomposition (see DECOMPOSER). In the CARBON CYCLE photosynthetic organisms assimilate carbon, and carbon compounds are released by decomposition.

The part of the biogeochemical cycle in which a particular nutrient is rapidly exchanged between the biotic and abiotic components is called the *active pool* or *exchange pool*. The part of the abiotic store of nutrients that is exchanged only slowly with the active pool is called the *reservoir pool*. Normally such exchange is extremely slow, but human activities, especially those resulting in pollution, can trigger the rapid and excessive release of such nutrients, upsetting the balance in the system. For example, the nitrogen cycle is perturbed by the use of chemical fertilizers, which dissolve in rainwater and wash into rivers and lakes, causing EUTROPHICATION.

biogeographical barrier see DISPERSAL BARRIER.

biogeographical region see BIOGEOGRAPHY, FLORISTIC REGION, PLANT GEOGRAPHY, ZOOGEOGRAPHICAL REGION.

biogeography the study of the geographical distribution of plants and animals, past and present, their habitats (especially the global distribution vegetation types) and their interactions with their environment and with the human population. See also FLORISTIC REGION, PLANT GEOGRAPHY, ZOOGEOGRAPHICAL REGION.

bioinformatics the creation and maintenance of databases of biological information, ranging from genomes and proteins to species and biodiversity. Biological knowledge is expanding so rapidly that sophisticated computerized information-retrieval systems are needed to enable access to and searching of large quantities of information.

biological assay see BIOASSAY.

biological clock an internal timing mechanism possessed by some organisms and responsible for the periodic or cyclical triggering of certain physiological responses. Biological clocks are often controlled by hormones. Examples include the occurrence of flowering when a particular day length is achieved and the breaking of dormancy when conditions become suitable for seedling development. See also CIRCADIAN RHYTHM.

biological conservation see CONSERVATION.

biological control the control of pests and diseases by making use of their natural enemies or by artificially upsetting their life cycle. For example, the moth *Cactoblastis cactorum* has been used in Australia to help control the prickly pear cactus (*Opuntia vulgaris*). The caterpillars bore into the pads of the cactus, severely affecting its growth. Experiments have been conducted on controlling water weeds, such as water hyacinth (*Eichhornia*), by spraying them with disease spores. Populations of damping-off fungi in the soil can be reduced by the addition of organic supplements to encourage the growth of antagonistic saprotrophic microorganisms. Modern genetic engineering techniques are now used to genetically alter natural predators and pathogens to make them more effective. Whether such techniques can be considered as biological control is debatable. See also CHEMOSTERILANT, TRAP CROP.

biological efficiency see ECOLOGICAL EFFICIENCY.

biological oxygen demand see BIOCHEMICAL OXYGEN DEMAND.

biological species concept see SPECIES.

biology the study of living organisms, their vital processes and their interactions with their environment.

bioluminescence the emission of light from living organisms. It results either from internal chemical reactions or from the reemission of absorbed energy as radiation. In plants, luminescence is probably the result of inefficiencies during oxidation-reduction reactions. Some of the energy released excites a molecule to a high-energy state and when the molecule returns to the ground state visible light is emitted. Luminescence is exhibited by certain fungi, including some of the genus *Mycena*, and also certain bacteria and planktonic algae.

biomass (standing crop) the total weight or volume of either all the living organisms or of one species present at any one time in a community. See also PYRAMID OF BIOMASS.

biome (biotic region) any of a group of major regional terrestrial communities with its own type of climate, vegetation, and animal life. Biomes are not sharply separated but merge gradually into one another. Examples include tundra, temperate deciduous forest, and desert. See also ECOTONE.

biometrics see BIOMETRY.

biometry (biometrics) the statistical analysis of biological phenomena.

biomolecule a molecule that has been synthesized by a living organism.

bionomic strategy the features of an individual or population, such as size, fecundity, and range, which give it optimum FITNESS for its environment.

biopharming see PHARMING.

bioremediation the use of living organisms to break down waste or POLLUTANTS and help restore contaminated ecosystems to their natural state. The use of plants for this purpose is called *phytoremediation*. Some plants are capable of extracting heavy metals from contaminated soil. Others can be genetically engineered to do so. For example, *Arabidopsis* has been modified to express the enzyme mercuric ion reductase, which reduces Hg^{2+} to Hg, which is released into the atmosphere as a vapour. Plants can decontaminate water by the process of *rhizofiltration*, which uses species with extensive root systems and high rates of transpiration to draw contaminated water through their tissues. Beds of reeds (*Phragmites*) or stands of willows (*Salix*) can be used for this. Some fast-growing aquatic plants accumulate toxins; they are removed and destroyed at intervals. For organic wastes, such as oil, organic-domestic and industrial waste, and plastics, micro-organisms are used, mostly bacteria and protoctists, and a few fungi.

biorhythm a periodic physiological or behavioural change controlled by a biological clock. An example is the CIRCADIAN RHYTHM of opening and closing of stomata found in many plants.

biospecies a group of interbreeding individuals that is reproductively isolated from all other groups. See REPRODUCTIVE ISOLATION.

biosphere the zone, including the earth's surface (and surface water), the adjacent atmosphere, and the underlying crust, where life can exist. The earth is usually considered as having three spherical concentric zones: the solid LITHOSPHERE, the liquid HYDROSPHERE, and the gaseous ATMOSPHERE. The biosphere encompasses parts of all three of these zones. Its limits are not well defined; most living organisms exist in a region extending about 100 m into the atmosphere and about 150 m below the surface of the oceans. The term can also include parts of the earth and atmosphere that depend on the present or past existence of living organisms, e.g. coal deposits or atmospheric oxygen. The biosphere can be regarded as a single ECOSYSTEM and is often called the *ecosphere*.

Biosphere Reserve a specific type of NATURE RESERVE designated by UNESCO under its Man and the Biosphere Programme (MAB). Biosphere Reserves usually include a number of existing reserves. A specific type of management is imposed on the whole area, which is monitored carefully, taking into account the needs of both fauna and flora and the human inhabitants. Each Biosphere Reserve has a relatively undisturbed and highly protected 'core' area, which may possess globally significant ecological features. Buffer zones around the core may be used for human activity in such as way that they do not impinge on the core, which may serve as a reference point for comparing the effects of different types of land use. Such zones may include, for example, forests managed for timber or land used for agriculture or grazing. Still further out there may be zones of cooperation, zones of influence, and transition areas, where human activity can still affect the remainder of the Biosphere Reserve. The management of a Biosphere Reserve can integrate the interests of both nongovernmental organizations and private landowners without affecting private ownership rights.

biostratigraphic unit see BIOSTRATIGRAPHY.

biostratigraphy the use of fossil plants and animals to date and correlate the sequences of rock strata in which they occur. This can be important for establishing the chronological sequence of the strata, or for determining past environments. A zone characterized by a particular assemblage of fossils that were deposited at the same time as the sediments is termed a *biostratigraphic unit*. See also ASSEMBLAGE ZONE.

biosynthesis see ANABOLISM.

biosynthetic pathway (biochemical pathway, metabolic pathway) a series of enzymatic reactions in which more complex molecules are built from simpler ones. Biosynthetic pathways are energy requiring, this energy usually being supplied either as phosphate bond energy by ATP or as reducing power by NADPH.

Three stages of biosynthesis can be recognized. The first stage is concerned with the manufacture of simple organic molecules from inorganic molecules like carbon dioxide and water. These few simple molecules are then converted in the second stage to building blocks for the many biological macromolecules. Finally macromolecules are synthesized from their constituent subunits. The major biosynthetic pathways in plants are shown in the diagram. See ANABOLISM.

biosystematics the study of variation and relationship in populations rather than individuals. The genetic and evolutionary nature of groups is assessed by the examination of CYTOLOGY, COMPARATIVE

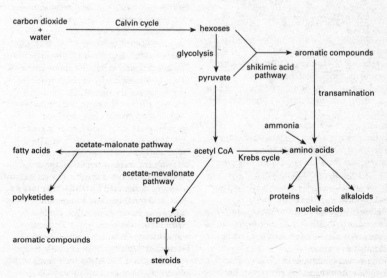

Biosynthetic pathway: the main biosynthetic pathways in plants

MORPHOLOGY, and ECOLOGY. Biosystematics is often described as the taxonomic application of GENECOLOGY or, occasionally, as experimental taxonomy. See also SYSTEMATICS, TAXONOMY.

biota the flora and fauna either of a particular region or of a given geological period.

biotechnology the use of organisms (usually microorganisms), or the enzymes they produce, in industrial processes. Biotechnology has potential applications in food production, waste recycling, alternative energy schemes, medicine, and a number of other spheres. See also GENETIC ENGINEERING, SINGLE-CELL PROTEIN, ENZYME TECHNOLOGY.

biotic association see COMMUNITY.

biotic climax see CLIMAX.

biotic factors the living components of the environment that by their activities affect the life of an organism. Compare ABIOTIC FACTORS.

biotic potential (intrinsic rate of natural increase) the maximum possible rate of reproduction of an organism or population under optimum, non-limiting conditions, i.e. conditions in which food, water, space, and other abiotic factors are not limiting and there are no predators, parasites, or diseases. Deviations from the optimum conditions constitute *environmental resistance*. The biotic potential is determined by the age at which the organism reaches sexual maturity, the frequency of reproduction, the number of offspring born at a time, and the duration of reproductive life.

biotic region see BIOME.

biotin a heat-stable COENZYME that acts as a carrier of carboxyl groups in many important metabolic reactions. An example is the oxidative decarboxylation of pyruvate to acetyl CoA, in which biotin acts as an intermediate carrier of the carboxyl group before it is released as free carbon dioxide. It is usually tightly bound to the enzyme for which it is the cofactor. Biotin is one of the B group of vitamins.

biotope a microhabitat (see HABITAT) in a large community in which several short seral stages of succession occur, forming a microsere (see SERE). An example is a cowpat dropping in a pasture.

biotopographic unit 1 a small habitat with a distinctive topography caused by the activities of an organism, e.g. a termite mound. **2** a small unit of topography that forms a distinct microenvironment for living organisms, e.g. a south-facing slope.

biotroph an organism that obtains nutrients from the living tissues of another organism. Compare AUTOTROPH, HETEROTROPH.

biotype a group of individuals belonging to the same species that have identical or almost identical genetic make-ups.

biozone 1 the total range of geological time (see GEOLOGICAL TIME SCALE) during which a specific taxon existed. **2** all the rocks deposited during the period of geological time in which a given taxon existed.

bipinnate describing a PINNATE leaf in which the leaflets themselves are further subdivided in a pinnate fashion, as seen in the sensitive plant (*Mimosa pudica*).

bird pollination see ORNITHOPHILY.

bird's-nest fungi see NIDULARIALES.

bisexual see HERMAPHRODITE.

bitunicate see ASCUS.

biuret test a test for proteins in which sodium hydroxide is added to the test solution, followed by the careful addition of drops of 1% copper(II) sulphate solution. A violet colour indicates a positive result and is caused by the presence of peptide groups (NH–CO) in the proteins or peptides, which form a purple copper(II)–peptide complex. Free amino acids will not give a positive result.

bivalent a pair of homologous chromosomes united by chiasmata (see CHIASMA). Bivalents are most clearly seen during late prophase of meiosis 1.

black earth see CHERNOZEM.

bladder any inflated organ, such as the inflated calyx of the bladder campion (*Silene vulgaris*) or the inflated fruit of the bladder nut (*Staphylea pinnata*). See also AIR BLADDER, UTRICLE.

blade see LAMINA.

blanket bog see BOG.

blastochory the dispersal of plants by means of offsets, as in the house leek (*Sempervivum tectorum*).

Blastomycetes an obsolete class comprising

anamorphic (see ANAMORPH) yeasts (imperfect yeasts or asexual yeasts). It included about 200 species in some 20 genera, and was divided into the Cryptococcales, which reproduce by budding and do not form ballistospores, and the Sporobolomycetales, which reproduce by budding and ballistospores. Most of the Cryptococcales are imperfect states (see FUNGI ANAMORPHICI) of the SACCHAROMYCETALES. The order included the parasitic genus *Candida*, which causes thrush in humans. The Sporobolomycetales (the mirror or shadow yeasts) are anamorphic states of basidiomycetes, such as the Tremellales. Many (e.g. *Sporobolomyces* and *Tilletiopsis*) are found as epiphytes on leaves. The Cryptococcales is now known to be a polyphyletic grouping. Today all these fungi are placed in different phyla of the FUNGI or in the Fungi Anamorphici.

blastospore a spore produced by budding. Blastospores are seen, for example, in the Taphrinales, in which they are formed by the budding of ascospores.

bleeding the exudation of the contents of the xylem at a cut surface. This is caused by root pressure.

blending inheritance the intermingling of the characteristics of the parents so that the offspring are intermediate in form between their parents. It is usually seen when a characteristic is controlled by many genes, as for example, height or yield. If a characteristic is governed by a single gene then blending inheritance is only seen when the alleles show INCOMPLETE DOMINANCE.

It was thought that all characteristics become blended in the offspring. Until the rediscovery of Mendel's work (see MENDELISM) this was the main criticism of Darwin's theory of evolution by natural selection (see DARWINISM), for the variation upon which it depended would always be lost by such a process. Mendel's demonstration of the particulate nature of inheritance removed this obstacle.

blepharoplast see KINETOSOME.

blight a plant disease in which leaf damage is sudden and serious. Many pathogens can cause blight-type symptoms when conditions are suitable but only a few cause such devastating symptoms on a regular basis: these include the term 'blight' in their common names. Examples include late and early blight of potatoes (caused by the oomycete *Phytophthora infestans* and the fungus *Alternaria solani* respectively), and FIREBLIGHT of pears (caused by the bacterium *Erwinia amylovora*).

bloom a visible increase in the numbers of a species, usually an algal species, in the plankton. A bloom of diatoms (see RED TIDE) is often seen in the spring, which decreases later in the year, probably as available silica is used up. Sustained algal blooms may lead to the eutrophication of a lake. In North America the term *water bloom* is used.

blue-green algae see CYANOBACTERIA.

blue-green bacteria see CYANOBACTERIA.

BOD see BIOCHEMICAL OXYGEN DEMAND.

bodos see EUGLENIDA.

body cell a fertile cell formed along with the infertile stalk cell after mitotic division of the GENERATIVE CELL in the gymnosperm pollen tube. It divides into two SPERM CELLS.

bog a region of badly drained permanently wet land that is subject to high rainfall and has a persistently moist atmosphere. These conditions result in a CLIMAX COMMUNITY with no trees. The most common plants in Britain are bog mosses (*Sphagnum*), cotton grasses (*Eriophorum*), ling (*Calluna vulgaris*), cross-leaved heath (*Erica tetralix*), bog myrtle (*Myrica gale*), rushes (*Juncus*), and sedges (*Carex*). Bogs are commonly found in upland and western areas of temperate regions. There are three main types.

Blanket bog consists of acid PEAT formed from the remains of the bog plants. It is variable in thickness and the high acidity (pH 3.0–4.5) is caused by the continuous flow of water through the peat. This washes out any bases and prevents minerals from the underlying rocks reaching the plant roots. Thus the organic acids are not neutralized and the peat is poor in nutrients (oligotrophic). Blanket bog can form over limestone.

Raised bogs develop from FENS. Rain water leaches nutrients from the upper layers of the fen peat, making it acid. This favours the growth of bog plants and prevents the growth of trees. As the bog plants become established more peat forms, especially in the central regions, resulting in the centre being raised above the periphery.

Valley bogs form in regions, such as the glens of western Scotland, receiving run-off and spring water from surrounding mountains.

Blanket bogs and raised bogs are types of *ombrogenous bog* (*ombrogenous peat*): in both types the peat-forming plant community is situated above the level of groundwater and is not in contact with the mineral soil, thus, they obtain both water and mineral nutrients from rainfall. Such bogs are highly acidic, and are dominated mainly by SPHAGNUM mosses.

bole the trunk of a tree.

Boletales an order of basidiomycete fungi of the class HOLOBASIDIOMYCETES (or Euholobasidiomycetes). The basidioma usually has a central stipe with a soft fleshy cortex and a cap in which the hymenium lines vertical downward-opening cylindrical tubes that hang down beneath the cap. The *boletes* differ from other similar fungi in that the tubes are easily separated from the flesh of the cap. The classification into families and genera is controversial, but there are over 700 species in some 70 genera and 11 families. The boletes are terrestrial fungi, usually saprobes; a few form mycorrhizas. Many are edible but a few are poisonous. See also BASIDIOMYCOTA.

bolete see BOLETALES.

bolochory (discharge dispersal) the dispersal of seeds or spores by propulsive mechanisms, as in gorse (*Ulex* spp.). See also AUTOCHORY.

bolting the premature production of flowers and seeds. It is a problem in biennial crops, such as sugar beet, which show considerably reduced yields if they 'run to seed' in the first growing season.

bomb calorimeter an apparatus that is used to measure the quantity of heat produced when a known amount of carbohydrate, fat, protein, or some mixture of these is burned. It consists of a strong cylindrical steel chamber, resistant to high pressures, surrounded by a jacket containing a known volume of water. The weighed foodstuff is placed in the chamber, oxygen is pumped in under pressure, and the food is ignited electrically. The rise in temperature of the surrounding water gives a measure of the amount of heat generated by the oxidized food.

bonsai the art of training and growing living dwarf trees in containers. Bonsai species are not natural dwarf species: the dwarfing is produced by a system of pruning roots and branches and training branches with wire. The trees range from 5 cm to about 60 cm in height. They may live for over a century and are often handed down in families. Bonsai originated in China, probably more than 1000 years ago, but has been further developed more recently by the Japanese.

borax carmine a dye used to stain nuclei in the preparation of biological material for microscopy.

Bordeaux mixture a mixture of copper sulphate and calcium hydroxide used as a fungicide. It was first used by P. M. A Millardet in France in 1883–85 to control vine downy mildew (*Plasmopara viticola*). A common mixture is 40 g copper sulphate and 40 g calcium hydroxide in 5 l water.

bordered pit a PIT possessing an extension of the secondary cell wall, i.e. a border, arching over part of the pit cavity. In many gymnosperms, bordered pits also possess a thickening of the primary wall material, termed a TORUS, in the central part of the pit membrane, the remaining unthickened part being termed the *margo*. Bordered pits mainly occur in vessel elements, tracheids, and fibres in the xylem, but may also occur in some extraxylary sclerenchyma cells. Pit pairs may be *half-bordered*, in which case one member of a pit pair is bordered and the other is not. This is sometimes the case in pits linking parenchyma cells and tracheary elements. Compare SIMPLE PIT.

border parenchyma SEE BUNDLE SHEATH.

Boreal floristic region SEE BIOGEOGRAPHY, FLORISTIC REGION, PLANT GEOGRAPHY.

boreal climate the climate associated with the boreal forest zone of Eurasia and North America, characterized by long, cold winters with temperatures below 6°C for 6 to 9 months, and short summers with temperatures averaging no more than about

10°C. Precipitation, which falls as snow in winter, is usually 380–635 mm a year. This climate is found in much of Eurasia and North America from the fringe of the tundra southwards. The southern boundary is further north in the west than in the east.

boreal forest see FOREST.

Boreal period the period in post-glacial times from about 7500 BC to 5000 BC, during which relatively dry, continental climatic conditions prevailed in temperate regions. Pollen records (see POLLEN ANALYSIS) show an increasing abundance of warmth-loving tree species. This was the last time in post-glacial history that Britain was joined to mainland Europe by a land bridge across the Dover Strait. The Boreal period was followed by the climatic optimum of the Atlantic Period.

boreal zone see CIRCUMPOLAR DISTRIBUTION.

boron symbol: B. A metalloid element, atomic number 5, atomic weight 10.81, found in very low concentrations in plant tissues. Its role is uncertain though boron deficiency quickly results in changes in cell membrane function and in cell extension. Various plant diseases, e.g. heart rot of beets and alfalfa yellows, have been attributed to boron deficiency. It has been suggested that a boron–sugar complex is involved in the movement of substrates through the phloem, though this idea is now largely rejected. Pollen germination and pollen tube growth are both greatly enhanced by addition of borate to the growth medium. Boron has been implicated in the uptake of calcium and appears necessary for the proper development of apical meristems.

botanical 1 relating to the study of plants. **2** any insecticide derived from plants, such as rotenone from roots of *Derris* and pyrethrum from *Chrysanthemum* flowers. **3** in horticulture, a cultivated plant of an unimproved wild species rather than a cultivated variety.

botanic garden an area in which a wide range of plants are grown for scientific, educational, and aesthetic purposes. Such collections aim either to include representative species of genera from all over the world or to specialize in one or more regions or types of vegetation. Large

greenhouses are often maintained to provide appropriate growing conditions for exotic plants. As well as maintaining a large and varied collection of living plants, botanic gardens often also include HERBARIA, research laboratories, and, increasingly, GENE BANKS and seed banks. In the past botanic gardens have been responsible for the introduction of many ornamental and crop plants into regions where they were previously unknown. The larger botanic gardens also finance plant-collecting expeditions to remote areas where the vegetation remains largely undescribed and uncollected. An increasing role of botanic gardens is the conservation of endangered species.

botany the study of plants, their vital processes and their interactions with their environment.

bothrosome (sagenogen, sagenogenetosome) in labyrinthulids (slime nets and thraustochytrids), an invaginated organelle at the cell surface, which links the plasma membrane with the network of intracellular membranes. It is thought to sequester calcium and aid in the production of the extracellular slime net matrix. See LABYRINTHULATA.

boundary layer the layer of liquid or gas next to a solid surface, which flows more slowly than that further away from the surface. A boundary layer of air exists at a leaf surface, its width decreasing with increasing wind speed. It is one of the factors that controls the rate of diffusion of water vapour from a leaf. Transpiration is thus greater from a leaf margin, where the boundary layer is narrower and offers less resistance, than from the centre of the lamina. The boundary layer also affects the dispersal of spores from pathogenic fungi. To escape from a leaf, the spores must be projected beyond the boundary layer. See also ESSENTIAL OIL.

BP before the present (taken to be 1950).

brachysclereid (stone cell) a relatively short, more or less isodiametric, sclerenchyma cell (SCLEREID). Brachysclereids are usually present singly or in small groups, although they occur in large numbers in some tissues, such as the fleshy part of the pear fruit.

bracket fungi see APHYLLOPHORALES.

brackish describing water that is more saline than fresh water, but less saline than sea water.

bract a leaflike organ subtending an inflorescence. Bracts are sometimes brightly coloured and petal-like, as in *Bougainvillea*. The glumes, lemmas, and paleae of grass spikes are also examples of bracts. Compare BRACTEOLE. See also INVOLUCRE.

bracteate bearing bracts.

bracteole a leaflike organ subtending a flower in an inflorescence that is itself subtended by a BRACT.

bract scale the structure that subtends the OVULIFEROUS SCALE in the female strobilus of gymnosperms.

bradytelic in CHRONISTICS, describing a species that has remained more or less unchanged for millions of years. Such species may appear primitive; however when they first appeared they may have represented the end of an increasing trend towards specialization in a particular group. Certain of the present-day gymnosperms, such as *Welwitschia*, *Ginkgo*, and *Podocarpus* are considered bradytelic. The evidence for this comes partly from fossil studies, and also from investigations of the KARYOTYPE. Such species have strikingly asymmetrical karyotypes, a feature generally considered to be advanced.

brake an area of bracken (*Pteridium*) or other fern, scrub, or brush.

branch a natural subdivision of a plant stem, especially a lateral stem or shoot arising from the main axis of the plant.

brand spore see CHLAMYDOSPORE.

Brassicaceae (Cruciferae, crucifers) a large dicotyledonous family, commonly called the cabbage family, members of which occur predominantly in north temperate latitudes. It contains about 3000 species in about 365 genera. Most crucifers are herbs with alternate leaves and four-petalled flowers borne in a raceme or corymb (see INFLORESCENCE). The petals are inserted in the shape of a cross (cruciform), hence the alternative name of the family. The fruit is also characteristic (see SILIQUA, SILICULA, LOMENTUM).

Many crucifers are important crop plants. Cultivars of *Brassica oleracea* include the cabbage, cauliflower, and Brussels sprout, while other species include the turnips, swedes, and oilseed rape. Ornamental crucifers include the wallflower (*Cheiranthus cheiri*), honesty (*Lunaria annua*), and stocks (*Matthiola*).

Braun-Blanquet scale a method of describing an area of vegetation devised by J. Braun-Blanquet in 1927. It is used to survey large areas very rapidly. Two scales are used (see TABLE). One consists of a plus sign and series of numbers from 1 to 5 denoting both the numbers of individuals and the proportion of the area covered by that species, ranging from + (sparse and covering a small area) to 5 (covering more than 75% of area). The second scale indicates how the individuals of that species are grouped and ranges from Soc. 1 (growing singly) to Soc. 5 (growing in pure populations). The information is obtained by laying down adjacent quadrats of increasing size. By combining ABUNDANCE (number) with COVER estimates, this scale helps to overcome problems due to the distribution pattern and the size of individual plants. A widely used modification is the DOMIN SCALE, which uses a larger number of divisions and is therefore more accurate. The Braun-Blanquet scale also includes a five-point scale to express the *degree of presence* of a plant. For example, 5 = constantly present in 80–100% of the areas; 1 = rare in 1–20% of the areas. See also FREQUENCY.

scale	description
+	sparse, covering small area
1	plentiful, covering small area
2	plentiful, covering at least 5% of area
3	covers 25–50% of area
4	covers 50–75% of area
5	covers over 75% of area
Soc. 1	growing singly
Soc. 2	growing in groups or tufts
Soc. 3	growing in small patches or cushions
Soc. 4	growing in small colonies, extensive patches, or as carpets
Soc. 5	growing in pure populations

Braun-Blanquet scale

breakage and reunion in genetics, the accepted model of CROSSING-OVER, in which the chromatids of homologous chromosome pairs at prophase 1 of meiosis break at points called chiasmata, the broken parts reuniting with the corresponding sections of non-sister chromatids from the same homologous pair. This results in the mutual exchange of genetic material.

breathing root see PNEUMATOPHORE.

breckland an area of south-west Norfolk and north-west Suffolk (in south-east England) with a distinctive mosaic of grass heath and heather/ling (*Calluna vulgaris*) and bracken (*Peridium aquilinum*) communities. The original forest cover was cut down by Neolithic farmers, and the area has remained deforested ever since.

breed an interbreeding group of plants or animals descended from a common ancestor but differing from the wild-type as a result of artificial selection by humans, usually for domestication (agriculture or animal husbandry), for pleasure (pets and ornamental plants), or for genetic analysis and research.

breeding system the way in which a species breeds, including the way individuals select mates (including pollination mechanisms), the genetic basis of sex determination and incompatibility systems, and other factors that affect the genetic diversity and make-up of the offspring. See also SELF INCOMPATIBILITY, SEXUAL DIMORPHISM, SEXUAL REPRODUCTION.

breeding true (pure breeding) the production of offspring whose ALLELES for certain genes are identical to those of their parents. This requires both parents to be HOMOZYGOUS for the alleles in question.

brevetoxin a neurotoxin produced by some Dinomastigota (dinoflagellates). It kills fish and causes neurotoxic shellfish poisoning.

brewer's yeast see SACCHAROMYCETALES.

brigalow scrub semi-arid scrub, dominated by *Acacia* species, found in parts of Australia.

bright-field illumination the normal method of illumination in light microscopy (see LIGHT MICROSCOPE), in which the specimen appears dark against a bright background. Several methods of setting up the condenser lenses and light source are used, of which the two most common are CRITICAL ILLUMINATION and KÖHLER ILLUMINATION. Compare DARK-GROUND ILLUMINATION.

broad the name used in East Anglia, England, for a freshwater lake, usually reed-fringed, connected to a river close to its estuary. Such lakes derive from flooded peat diggings dating back to medieval times.

brochidodromous see CAMPTODROMOUS.

bromatia in fungi cultivated by certain species of ants, swellings at the tips of the HYPHAE that serve as food for the ants. See also FOOD BODY.

5-bromouracil (5-BU) An analogue of the naturally occurring PYRIMIDINE thymine (5-methyl uracil), in which bromine replaces the methyl group on carbon five of the pyrimidine ring. During chromosome replication the analogue may be incorporated into DNA in place of thymine, but it is less specific in its pairing properties. Over a series of replications a T:A nucleotide pair could thus eventually be replaced by a C:G pair resulting in a gene mutation (see diagram overleaf). The ability of 5-bromouracil to produce gene mutations without significant chromosomal damage has made it a useful tool in genetic research. Other base analogues with similar effects include 2-aminopurine and 2,6-diaminopurine, which are analogues of adenine, and 5-bromodeoxycytidine, which is an analogue of cytosine.

brotochory see ANDROCHORY.

brown algae see PHAEOPHYTA.

brown earth a type of zonal acidic soil (PEDALFER) characteristic of broad-leaved deciduous forest (see FOREST), such as that of western Europe. The rainfall causes some leaching of lime from the A horizon, which is grey-brown in colour. The temperature is high enough for the rapid decomposition of organic matter resulting in the formation of mild humus (mull). The B horizon is dark brown because of the accumulation of various iron and other compounds. Many of these deciduous-forest lands have been cleared for cultivation. The natural fertility resulting from decomposed leaf litter is thus no longer available and both liming and

A adenine T thymine B 5-bromouracil
G guanine C cytosine
1, 2, 3 = successive replications of DNA

5-bromouracil: the figure shows
how 5-bromouracil may be
incorporated in place of thymine
and by mispairing
result in AT to GC mutation

manuring are necessary. Compare PODSOL.
See SOIL PROFILE.

Brownian motion the random motion of
microscopic particles suspended in liquids
or gases as a result of their continuous
bombardment by molecules in the
surrounding medium.

Bryales see BRYOPHYTA.

Bryata see BRYOPHYTES.

bryokinin a hormone found in certain
mosses, e.g. *Funaria*, that stimulates the
formation of buds and subsequently
leafy plants from the filamentous
protonema.

bryology the study of living BRYOPHYTES,
their vital processes, and their interactions
with their environment.

Bryophyta a phylum of nonvascular plants
(see BRYOPHYTES), generally known as
mosses. The phylum formerly also included
the liverworts and hornworts, but these are
now considered to be phyla in their own
right, the HEPATOPHYTA and
ANTHOCEROPHYTA respectively. In this
older system, the mosses constituted the
class Musci (or Bryopsida). The phylum
includes about 16 000 species in about 610
genera. Mosses are distinguished from
liverworts and hornworts by possessing
conducting cells and by the absence of
elaters. The gametophyte is the dominant
generation and exhibits two distinct
morphological stages. The first, which
arises on germination of the spore, is the
filamentous PROTONEMA, which, except for
its oblique cross walls, resembles a
heterotrichous green alga. The protonema
produces buds, from which the familiar
leafy moss plant arises. In the Sphagnales
and Andreaeales the protonema is thalloid.
The mature gametophyte, which is never
thalloid, consists of a main axis (caulid)
bearing delicate leaves (phyllids) usually
only one cell thick, although a thickened
central midrib is often seen. The leaves are
generally inserted spirally on the stem. The
stem also bears multicellular RHIZOIDS,
which distinguishes mosses from liverworts
(in which the rhizoids are unicellular). The
gametophyte may be ACROCARPOUS or
PLEUROCARPOUS. The sporophyte arises
from an apical cell and exhibits complex
spore dispersal mechanisms. The seta

elongates gradually (in contrast to liverworts, in which growth of the seta is rapid). There are no sterile elaters in the spore mass.

The Bryophyta is divided into three main orders on the basis of differences in capsule structure and in formation of the protonema. The *Bryales* (commonly termed the true mosses) is the largest order and contains about 600 genera including the advanced *Polytrichum*, which shows some internal differentiation. The *Sphagnales* (bog or peat mosses) contains a single genus, *Sphagnum*, characteristic of waterlogged acid areas. The ability of sphagnum mosses to create vast areas of peat bog arises, in addition to their low pH and nutrient tolerance, from a peculiarity in their leaf structure, which contains many dead porous cells that act as water reservoirs and release acid. The third order is the *Andreaeales*, which again contains just one genus, *Andreaea*, the members of which are known as *granite mosses*. Certain classifications elevate these orders to class status, the Bryopsida, Sphagnopsida, and Andreaeopsida. The Bryopsida is then further subdivided into some 19 orders (see DICRANALES).

The earliest fossil mosses are seen in rocks of the Carboniferous but the group generally does not have a very rich fossil record.

bryophytes (Bryata) the general name given to the nonvascular plants – the mosses, liverworts, and hornworts – which were formerly grouped together in the phylum or division Bryophyta. The three former classes Musci (mosses), Hepaticae (liverworts), and Anthocerotae (hornworts) have now been elevated to the status of phyla – the BRYOPHYTA, HEPATOPHYTA, and ANTHOCEROPHYTA, respectively. There are some 24 000–25 000 species of bryophytes. They are generally small low-growing plants, in most cases susceptible to desiccation and hence limited to damp or humid environments. Their life cycle shows a heteromorphic alternation of generations with the haploid gametophyte, which may be homothallic or heterothallic, the dominant generation. The ephemeral sporophyte is partly or completely parasitic

on the gametophyte and consists solely of a stalk bearing the spore capsule. All bryophytes show some tissue differentiation, but they lack the highly differentiated xylem and phloem tissues of the vascular plants. They lack roots, having only slender rhizoids.

There are some similarities between mosses and the Chlorophyta (green algae), especially between chlorophyte filaments and the moss PROTONEMA, which suggest that bryophytes evolved from certain green algae. As in green algae, their chloroplasts contain chlorophylls *a* and *b* and such carotenoids as β-carotene, the main storage compound is starch, they have certain cell wall constituents in common, and their antherozoids have two forwardly-directed undulipodia. However, the moss gametophyte bears multicellular gametangia (archegonia or antheridia), which are surrounded by layers of sterile tissue, and the sporophyte produces aerial spores. Water is still needed for dispersal of antherozoids and for fertilization. Liverworts are also thought to have evolved from green algae, but independently of mosses. The origins of hornworts are unknown. Despite their primitive characteristics, bryophytes are the dominant vegetation in certain areas, notably the bogs of temperate latitudes.

Bryopsida see BRYOPHYTA.

Bryopsidophyceae an obsolete class of the CHLOROPHYTA containing certain SIPHONOUS GREEN ALGAE now included in the class ULVOPHYCEAE.

bud 1 an undeveloped condensed region of a shoot consisting of a short stem terminated by a meristem and, in foliage buds, numerous leaf primordia, leaf buttresses, and young rolled or folded leaves. Flower buds contain the immature flower. Buds are found at the apex of a shoot (*apical* or *terminal buds*) and in the axils of leaves (*axillary* or *lateral buds*). In some species accessory buds develop in addition to the axillary bud in a leaf axil. Often axillary buds remain dormant unless the apical bud is injured or removed (see APICAL DOMINANCE). Adventitious buds may arise anywhere on the plant. Suckers, for example, develop from adventitious buds on the roots. The dormant winter buds of deciduous trees and shrubs possess protective, often resinous, *bud scales* or *cataphylls* to resist desiccation. Transpiration may be further reduced by a covering of fine hairs. Many trees can be identified in winter by the form and colour of their resting buds. The arrangement of leaves in the bud and the pattern of leaf folding or rolling (vernation) are useful diagnostically.

2 the protrusion formed from a unicellular organism during the asexual reproductive process of BUDDING.

budding 1 a form of asexual reproduction in which a new individual develops as an outgrowth of a mature organism. Yeast cells exhibit budding and sometimes the new cells so produced can themselves give rise to further buds before the chain of individuals separates. Compare FISSION, FRAGMENTATION.

2 (bud grafting) a type of GRAFTING in which a bud (the *scion*) of the desired variety, together with a small piece of bark, is removed and inserted into a slit made in the bark of the chosen rootstock (the *stock*), and secured with raffia or tape. The rootstock is usually cut back and any buds removed. The technique is widely used to propagate rosebushes. Bush roses are produced by inserting buds as near to the ground as possible while standard roses are produced by inserting buds higher up on the stem of the rootstock.

bud grafting see BUDDING.

bud scale see BUD.

buffer a chemical solution that counteracts small changes in pH when acids or alkalis are added to it. Buffers play an important role in cells and tissues, which usually function best at or near neutrality (pH 7) since changes in pH adversely affect metabolic processes. Examples of buffers in cells are phosphates, borates, and bicarbonates (hydrogen carbonates).

Buffers are used in microscopy to reduce the production of artefacts by maintaining the specimen at its original pH during fixation. Sorenson's phosphate buffer and sodium cacodylate are two of the commonest buffers for electron microscope preparations of plant material.

building phase see HUMMOCK AND HOLLOW CYCLE.

bulb a fleshy underground PERENNATING ORGAN formed by many monocotyledons. It is a highly modified shoot, the bulk of which is made up of colourless swollen SCALE leaves or leaf bases. The central apical bud contains the immature foliage leaves, the future flower, and rudimentary adventitious roots at its base. It is surrounded by numerous layers of fleshy scales, which may be complete modified leaves or the leaf bases of previous years' foliage leaves. The first sign of growth is the rapid elongation of the adventitious roots. When these have established themselves in the soil the apical bud sprouts and the foliage leaves and inflorescence emerge, growing at the expense of the food reserves in the scale leaves. In some bulbs, e.g. tulip, the inflorescence develops from the apex of the bud and further apical growth is consequently prevented. Photosynthates from the foliage leaves are passed down to one or more lateral buds in the axils of the scale leaves. This is an example of SYMPODIAL BRANCHING. In other bulbs, e.g. daffodil, the inflorescence develops in the axil of one of the foliage leaves and the apex of the bud remains within the bulb. It persists from year to year, each year giving rise to foliage leaves and a lateral inflorescence. Propagation may be effected by the expansion of a lateral bud in the axil of the outermost scale leaf. Bulbs such as daffodil show MONOPODIAL BRANCHING rather than sympodial growth. The food reserves of the bulb may be starch (as in tulip) or sugars (as in onion). Compare CORM.

bulbil a small bulb that develops from an aerial bud. Bulbils are easily detached and function as a means of VEGETATIVE REPRODUCTION. They may form from lateral buds, as in the lesser celandine (*Ranunculus ficaria*), or develop in place of flowers, as in many species of *Allium*. Certain forms of APOMIXIS give rise to bulbils, as seen in the lesser bulbous saxifrage (*Saxifraga cernua*). The term is also applied to various outgrowths formed by lower plants that become detached and develop into new plants. For example, the fern *Asplenium* *bulbiferum* produces bulbils on the upper surface of its fronds and *Lycopodium selago* has bulbils in the axils of the uppermost leaves.

bulliform cell any of the large cells that occur in longitudinal rows in the leaf epidermis of certain grasses. They are believed to be involved in the unrolling of young leaves and in the rolling and unrolling movements of the leaf in response to water status.

bundle cap one or more layers of sclerenchyma or thickened parenchyma cells seen at the xylem and/or phloem poles of a vascular bundle.

bundle sheath a region surrounding the small vascular bundles in the leaves of vascular plants. It usually consists of parenchyma cells but may sometimes consist largely of sclerenchyma. In angiosperms, the bundle sheath is a type of ENDODERMIS, which may be differentiated as a starch sheath. The cells of the bundle sheath have their long axes parallel to the direction of the vascular bundle. In dicotyledons, the parenchymatous bundle sheath is sometimes termed *border parenchyma*. Bundle sheath extensions, in which the cells of the bundle sheath extend to one or both epidermal layers, are present in many monocotyledons. They are believed to be concerned with conduction or support. See also KRANZ STRUCTURE.

bunt a covered SMUT disease of wheat, sometimes called *stinking smut* because of the smell of rotten fish given off by the spore masses. The disease is caused by *Tilletia caries*. Dwarf bunt of wheat is caused by *T. contraversa*. Infected grains are transformed into balls of bunt spores. At threshing the spores are released and contaminate healthy seed as well as the combine harvester and grain-handling machinery. Contaminated seed in turn gives rise to more infected plants.

bur (burr) a type of PSEUDOCARP in which the fruit is surrounded by a persistent barbed involucre as, for example, in burdocks (*Arctium*) and cocklebur (*Xanthium*). Often the term refers to any barbed fruits, e.g. those of cleavers (*Galium aparine*) and enchanter's nightshade (*Circaea lutetiana*) both of which have hooks borne directly on the pericarp. Burs cling to passing animals and are thus widely dispersed.

Burgundy mixture a copper fungicide, similar to but stronger than BORDEAUX MIXTURE and used especially against rusts. It contains 60 g sodium carbonate and 50 g copper sulphate in 5 l water.

bush 1 a low, densely branched shrub, or a dense thicket of shrubs that appears to be a single plant. **2** a area of wilderness or uncleared land, usually covered in scrub or forest, with very few, if any, human habitations.

bush veld see VELD.

butanedioic acid see SUCCINIC ACID.

buttress root a form of PROP ROOT that is asymmetrically thickened to give a planklike outgrowth on the upper side, providing extra support for the tree. Buttress roots are common in many tropical trees, e.g. certain figs (*Ficus*).

Cc

caatinga a type of thorn FOREST found in the dry interior of northeastern Brazil. The trees are often stunted but drought-resistant and remain leafless for long periods to conserve water. The dominant species are leguminous trees (family Fabaceae), especially of the genera *Caesalpinia*, *Mimosa*, and *Zizyphus*, euphorbias (Euphorbiaceae), and members of the Bombacaeae, and cacti. In many places the vegetation is rather sparse, but where water is abundant enough, there may be dense thickets up to 10 m high. Unlike typical savanna woodlands, these forests have very little grass.

Cactaceae a large family of dicotyledonous xerophytic plants, the cacti. It includes over 1650 species in about 130 genera; some authorities, however, recognize many more, over 300 genera being listed in some classifications. The cacti are limited in distribution to the drier regions of South and Central America and central and southern North America and (*Rhipsalis*) to southern Africa. Some have become naturalized elsewhere, e.g. prickly pears (*Opuntia*) in Australia and the Mediterranean.

Cacti are succulents and, being leafless (the leaves are reduced to spines), are a particularly distinctive group of plants, popular as pot plants. Photosynthesis is carried out by the green swollen stems, which show various advanced characteristics in the arrangement of vascular tissues and many other features. By comparison the flowers, which are borne singly on the stem, are comparatively primitive, with large numbers of stamens, petals, sepals, and bracts arranged spirally. Leaves, when present, are also spirally arranged. The fruit is a berry. In addition to the water-storing photosynthetic stems and reduced leaf and surface area, adaptations to dry conditions include a shallow spreading root system to take advantage of brief showers, and CRASSULACEAN ACID METABOLISM, which enables the plant to undergo carbon dioxide uptake and the dark reactions of photosynthesis at night, so the stomata can remain closed by day. This process generates organic acids, BETALAINS, and often also alkaloids. The spines deter animals seeking to obtain moisture by eating the plant and also help to condense dew and retain air. Many cacti have ribbed stems, which allow for expansion or contraction as water is taken in or lost, in much the same way as a concertina changes volume.

Very few cacti are of economic use, except as ornamentals, though prickly pears are grown for their fruits in California, Mexico, and some Mediterranean countries, and cactus hedges are sometimes planted for their deterrent spines. See also XEROPHYTE.

cactus (*pl.* cacti) a member of the family CACTACEAE. The term is sometimes used incorrectly to refer to other succulent prickly plants, such as species of *Agave*.

caducous describing parts that fall off easily or at an early stage, such as protective nonphotosynthetic stipules.

Caenozoic see CENOZOIC.

caespitose having a tufted form of growth, as have many grasses, e.g. *Deschampsia caespitosa* (tufted hair-grass).

Cainozoic see CENOZOIC.

Calamitales see SPHENOPHYTA.

calcareous 1 consisting of or containing calcium carbonate. 2 growing on limestone or chalk or in soil rich in calcium carbonate.

calcareous soil an ALKALINE SOIL with a pH greater than 6.0, which contains enough free calcium carbonate to effervesce visibly, releasing carbon dioxide gas, when treated with cold 0.1N hydrochloric acid. Such soils are rich in calcium ions and often in other nutrients, too. They are typically found over limestone or chalk bedrock, or on calcareous sand near coasts. They are commonest where rainfall is light. See also RENDZINA, TERRA ROSSA.

calcicole describing a plant that flourishes in lime-rich soil. The presence of such a plant is so characteristic of such conditions that it can be used as an indicator. An example of a calcicole plant is the common spotted orchid (*Dactylorhiza fuchsii*). Compare CALCIFUGE.

calcifuge describing a plant that grows in lime-deficient soils, e.g. sandy or peaty soils. Tormentil (*Potentilla erecta*) is an example of a calcifuge plant. Compare CALCICOLE.

calcimorphic soil see SOIL.

calcisol see RENDZINA.

calcium symbol: Ca. A soft metal element, atomic number 20, atomic weight 40.08. It is essential to plant growth due to its role, in the form of calcium pectate (see PECTIC ACID), in the development of the middle lamella. Calcium ions increase the rigidity of the cell wall by cross-linking the carboxyl groups of adjacent chains of pectic acid. Calcium is also found as crystals of calcium oxalate in many plant cells. Calcium deficiency is first apparent in the younger parts of the plant because there is little or no translocation of calcium from mature tissues to the growing points.

Calcium is easily leached from soils resulting in an acid soil. This may be countered by adding calcium carbonate, calcium hydroxide (slaked lime), or calcium sulphate (gypsum) to the soil. These compounds also help improve soil structure (see LIMING).

callose a structural polysaccharide found in the sieve plates of the phloem of higher plants, consisting of glucose residues linked through $\beta(1-3)$ glycosidic bonds. It is deposited steadily throughout the growing season eventually causing blockage of the pores of mature sieve tubes. The functions of blocked tubes are taken over by new phloem formed from the cambium. In some species the blockage is reversible, the callose being hydrolysed in the spring. Callose may be laid down very rapidly in response to injury of the phloem. Callose is also deposited around a growing pollen tube in the style and, in an incompatible pollination, blocks the tube's progress.

Callose was the first structural plant polysaccharide to be synthesized *in vitro*, using an enzyme isolated from bean (*Phaseolus aureus*) seedlings.

callus 1 a mass of parenchyma cells that forms at a wounded surface, for example where a branch has been cut off a tree. The callus tissue is produced by the cambium and initially forms a ring of thickening (*wound wood*) around the wound. Eventually it may completely cover the exposed wood. Callus tissue is also formed at the base of cuttings before roots are produced. Similarly callus tissue is important when propagating plants by leaf cuttings. For example, when a *Begonia* leaf is pegged down on the soil surface and the veins are cut, it is the callus tissue at the wound sites that gives rise to the new plantlets. Grafts (see GRAFTING) also depend on the formation of callus for a successful graft union. The regenerative capacity of callus tissue is utilized considerably in tissue culture. Callus cells are grown *in vitro* on suitable nutrient media and under correct conditions new plantlets will develop. See also CANKER.

2 an accumulation of the carbohydrate CALLOSE on phloem sieve plates.

calmodulin a Ca^{2+}-binding protein that mediates many calcium-regulated processes in cells. When it binds to Ca^{2+} ions, calmodulin changes shape and binds to other proteins, such as kinases and phosphatases, which are involved in the activation or inactivation of signalling pathways.

calorific value the number of units of heat liberated when a unit weight of a substance is burned completely in oxygen to produce carbon dioxide, oxygen, sulphur dioxide, nitrogen, water and ash. It is an expression of the amount of chemical energy stored in the substance. The calorific value of organic matter is usually expressed as joules per gram dry weight, but the old units of calories or kilocalories per gram are still in use.

Calvin cycle (Benson–Calvin–Bassham cycle) the sequence of reactions, making up the dark, or light-independent, reactions of photosynthesis, in which carbon dioxide is reduced to carbohydrate using ATP and NADPH derived from the light-dependent reactions (see PHOTOPHOSPHORYLATION). The Calvin cycle takes place in the chloroplast stroma and begins with the CARBOXYLATION and cleavage of ribulose 1,5-bisphosphate (RuBP) to form two molecules of glycerate 3-phosphate. The RuBP is then regenerated by a complex series of reactions involving 3-, 4-, 5-, 6-, and 7-carbon sugar phosphates. During this process glucose is formed by reversal of the GLYCOLYSIS reactions (see diagram). The overall series of reactions can be written as:

$$6 \text{ RuBP} + 6 \text{ CO}_2 + 18 \text{ ATP} + 12 \text{ NADPH} +$$
$$12 \text{ H}^+ + 12 \text{ H}_2\text{O} \rightarrow 6 \text{ RuBP} + \text{glucose} + 18 \text{ P}_i +$$
$$18 \text{ ADP} + 12 \text{ NADP}^+.$$

The glucose is subsequently converted to starch, cellulose, and other polysaccharides.

Calvin and his associates worked out this cycle of reactions by illuminating green algae (see CHLOROPHYTA) in the presence of radioactive carbon dioxide for a couple of seconds and then immersing the cells in boiling water to prevent further reaction. They then found which metabolites first became radioactively labelled using chromatography. See also HATCH–SLACK PATHWAY.

calyciform shaped like a calyx or cup.

calyptra (pileorhiza, root cap) A hood-shaped cap of parenchyma cells at the tip of the root that protects the root meristem from abrasion as it grows through the soil. It is derived from a separate meristem, the CALYPTROGEN, and may be augmented by periclinal divisions of the cells at the surface of the root.

calyptrogen the layer of meristematic cells at the apex of a root tip that continually cuts off new cells to its outer edge. These replace the cells of the root cap (calyptra) that are worn away as the root pushes through the soil.

calyx (pl. calyces) the collective term for the sepals, constituting the outer whorl of the PERIANTH. It usually encloses and protects the other floral parts during their development. A calyx tube is formed when the lateral margins of the individual sepals fuse, as seen in the primrose (Primula vulgaris). Compare COROLLA.

CAM see CRASSULACEAN ACID METABOLISM.

cambium (pl. cambia) a LATERAL MERISTEM found in vascular plants that exhibit SECONDARY GROWTH. It gives rise to secondary tissues mostly by periclinal divisions of initials. There are two cambia, the vascular cambium and the PHELLOGEN (cork cambium).

Cambrian the earliest period of geological time in the Palaeozoic era from about 570 to 500 million years ago. Cambrian rocks contain abundant fossils of the marine organisms of that period, i.e. algae and such invertebrate animals as trilobites and brachiopods. Colonization of the land had not yet occurred. It is believed that the climate was uniformly warm. See GEOLOGICAL TIME SCALE.

cAMP see CYCLIC AMP.

campanulate bell-shaped.

CAM plant any of a group of succulent plants that employ CRASSULACEAN ACID METABOLISM (CAM) for fixing atmospheric carbon dioxide (see FIXATION).

campo cerrado see CAMPOS.

campos tropical savanna GRASSLAND with scattered broad-leaved trees, found in Brazil. It is usually found on very poor soils. Periodic fires prevent it reverting to woodland or scrub. There are several types, including campo cerrado, with a reasonable tree cover; campo sujo, with only scattered trees and shrubs; and cerradão, where the vegetation forms an almost closed semi-evergeen woodland.

campo sujo see CAMPOS.

camptodromous describing a form of leaf venation in which there is a single primary vein and the secondary veins arising from it curve upwards towards the leaf margin. The secondary veins may join together in a series of conspicuous marginal loops

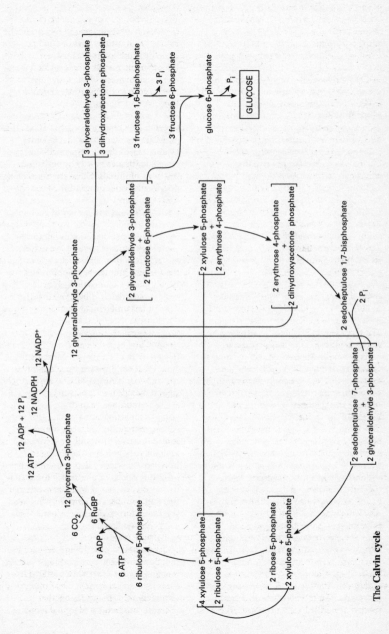

The **Calvin** cycle

(*brochidodromous* venation, e.g. in wild cherry, *Prunus avium*) or may remain separate (*eucamptodromous* venation, e.g. in downy rose, *Rosa tomentosa*). *See illustration at* venation.

campylodromous describing a form of leaf venation in which several primary veins originate at the base of the lamina and run in downward-curving arches towards the apex, as seen in black bryony (*Tamus communis*). *See illustration at* venation.

campylotropous describing a form of ovule orientation in which the ovule develops horizontally and the funiculus appears to be attached half way between the chalaza and the micropyle (*see illustration at* ovule). This arrangement is not very common but may be found in the Malvaceae and Caryophyllaceae. Compare ANATROPOUS, ORTHOTROPOUS.

Canada balsam a yellowish resin obtained from the fir *Abies balsamifera* that is used as a MOUNTING medium for microscope slides. It has similar optical properties to glass.

cane sugar see SUCROSE.

canker a plant disease in which there is a well defined area of NECROSIS of the cortical tissue, which becomes surrounded by layers of CALLUS tissue. Cankers occur mainly on woody stems but may also appear on herbaceous plants. For example, in apple canker, caused by *Nectria galligena*, the fungus penetrates the apple stem through a wound or leaf scar. As the fungus grows into the stem the tree produces callus tissue to seal off the infected zone. Further penetration by the fungus results in the production of more callus tissue, or wound wood, forming the characteristic canker.

canopy the uppermost layer of vegetation in a forest, woodland, or shrub community, formed from the branches of the tallest plants. The degree to which the canopy is *closed* (the branches have no gaps between them) affects the structure and nature of the vegetation below, controlling both light intensity and light spectrum, as well as air movement, temperature, and rainfall penetration. Compare GROUND LAYER.

Cantharellales an order of basidiomycete fungi with a large basidioma, which may be funnel-shaped, tubular, or stalked and pileate. The hymenophore may be smooth or folded into gills. Most of these fungi are terrestrial saprobes. The order contains some 677 species in 52 genera and 12 families. It is usually placed in the class HOLOBASIDIOMYCETES or Euholobasidiomycetes. The order includes the edible chantarelle (*Cantharellus cibarius*) and horn of plenty (*Craterellus cornucopioides*), and the club fungi. See BASIDIOMYCOTA. See also APHYLLOPHORALES.

cap cell a cell, seen in filaments of green algae (see CHLOROPHYTA) in the order Oedogoniales, that has one to several 'caps' of cell wall material at its anterior end. These result from a distinctive form of cell division that involves the formation of rings of new cell wall material prior to nuclear division. *See illustration at* Oedogoniales.

Cape floristic region see BIOGEOGRAPHY, FLORISTIC REGION, PLANT GEOGRAPHY.

capillarity (capillary action) the rise or fall of liquids within narrow tubes as a result of the surface tension of the liquid. Capillarity is also responsible for the formation of drops, bubbles, etc., and for the retention of water droplets on plants following dew formation, mist, rain, and guttation. The supply of water to all parts of a plant is due, in large part, to capillarity. Water will rise against gravity providing that the forces between the surfaces of the water and the solid it is in contact with are greater than the cohesive forces within the body of water.

capillary action see CAPILLARITY.

capillary moisture (capillary water) the moisture that is left in the soil after water has drained off under the influence of gravity. It is held on to the surface of soil particles by surface tension, and also forms tiny bodies of water in the pores between particles. Capillary water can move through soil by capillarity; some, but not all, is available to plants.

capillary water see CAPILLARY MOISTURE.

capillitium in the Myxomycota (slime moulds) a mass of sterile threads in the spore-bearing structures.

capitiform shaped like a head.

capitulum an inflorescence consisting of a

head of small closely packed stalkless flowers or florets arising at the same level on a flattened axis. The whole is surrounded or subtended by an INVOLUCRE of bracts and simulates, in appearance and function, a single large flower. The capitulum is typical of members of the family Asteraceae. Capitula are often made up of two distinct types of floret: *disc florets*, in which the corolla tube terminates in five short teeth; and *ray florets*, in which the tube is extended into a conspicuous strap. When both types of floret are present the disc florets form the centre of the capitulum and the ray florets are arranged around the edge, giving a daisy-like flower. Some composites have ray florets only, e.g. dandelions (*Taraxacum*) and chicory (*Cichorium intybus*), while others have only disc florets, e.g. thistles (*Carduus*, *Cirsium*) and groundsel (*Senecio vulgaris*). *See illustration at* inflorescence.

capsid the protein coat of a virus, which surrounds and protects its nucleic acid. The nature of the capsid determines the virus's host range and the efficiency of its infectivity. Viruses can be identified by the serological (see SEROLOGY) reaction of the capsid proteins.

capsomere any of the identical polypeptide subunits making up the CAPSID protein of a virus.

capsule 1 any dry dehiscent fruit derived from two or more many-seeded fused carpels (*see illustration at* fruit). Capsular fruits are classified by the nature of dehiscence and the number of carpels in each fruit. For example, if dehiscence is along the dorsal suture the capsule is *loculicidal*, as in willow herbs (*Epilobium*), and if along the septa, *septicidal*, as in St John's worts (*Hypericum*). If dehiscence is through pores, it is *poricidal*, as in poppies (*Papaver*). If the seeds remain attached to a central column, dehiscence is *septifragal*, as in the orchid genus *Epidendrum*.
2 the structure containing the spores of the SPOROPHYTE generation in mosses, liverworts, hornworts, and ferns, the wall of which may be uni- or multicellular. The capsule is borne on the end of a stalk and may rupture by a variety of mechanisms.
3 a transparent layer of gelatinous or

mucilaginous material that envelops some bacterial cells. The polysaccharides of the capsule are often characteristic of a species. Polypeptides may also be present. In some species the capsules hold together a number of bacteria, forming a structure termed a *zoogloea*.

carbohydrate (saccharide) a class of compounds with the general formulae $C_x(H_2O)_y$. Carbohydrates are aldehyde or ketone derivatives of polyhydroxyl alcohols. The simplest carbohydrates are the MONOSACCHARIDES or simple sugars and commonly contain five or six carbons. OLIGOSACCHARIDES are more complex carbohydrates containing between two and ten monosaccharide units. POLYSACCHARIDES contain many monosaccharide units in long linear or branched chains; they function as energy storage molecules and as structural elements in the plant body.

carbon symbol: C. An amorphous or crystalline element, atomic number 6, atomic weight 12.01, essential to life as it forms the backbone of most organic molecules. It is particularly suited for this role since it readily forms covalent bonds with other carbon atoms and with such elements as hydrogen, oxygen, nitrogen, and sulphur. Thus many different types of functional groups can be incorporated into organic molecules. Being a light element, these bonds are particularly stable as the strength of a covalent bond increases with decreasing atomic weight. In addition, many different three-dimensional structures can be formed by carbon–carbon bonding due to the tetrahedral arrangement of electron pairs around singly bonded carbon.

All organic carbon is ultimately derived from atmospheric CARBON DIOXIDE fixed during PHOTOSYNTHESIS (see CARBON CYCLE, FIXATION).

carbon cycle the circulation of carbon between living organisms and the environment. Atmospheric carbon dioxide is fixed as carbohydrates during photosynthesis and animals obtain carbohydrates by eating green plants. Carbon dioxide is returned to the atmosphere by respiration and by the burning of fossil fuels such as coal, oil, and peat. See also FIXATION.

carbon–14 dating see RADIOMETRIC DATING.

carbon dioxide a heavy gas, formula CO_2, that makes up 0.033% by volume of the atmosphere. It is the basic carbon source of all life, being converted to carbohydrates by PHOTOSYNTHESIS. The buffering capacity of the oceans, which contain large amounts of dissolved CO_2, helps to counteract changes in atmospheric CO_2 due to the increased burning of fossil fuels and DEFORESTATION (see also GREENHOUSE EFFECT). CO_2 was much more abundant in the primitive atmosphere, but disappeared as a result of bioassimilation. The extreme solubility and rapid diffusion of CO_2 in water is a major factor in the maintenance of aquatic life.

The rate of photosynthesis can be increased by raising CO_2 levels. This has been put to practical use in the cultivation of some glasshouse crops. However yield increases are not as great as might be expected since increased CO_2 concentrations also increase stomatal closure.

carbon dioxide method a method of measuring primary productivity by monitoring the changing carbon dioxide concentration of the atmosphere surrounding a plant or plants, as a measure of the balance between carbon dioxide uptake in photosynthesis and its release in respiration. Techniques used may include conductivity measurements, infrared gas analysis (see INFRARED GAS ANALYSER), or the use of radiocarbon tracers. See *also* gas-exchange method, oxygen method.

Carboniferous the second period of the Upper Palaeozoic era from about 345 to 280 million years ago. During this period the climate was generally warm and humid and there were great forests and swamps, dominated by arborescent lycopods (see LYCOPHYTA) and sphenophytes (see SPHENOPHYTA), from which the coal measures were formed. In the US the term Carboniferous is not widely used because rocks of this age in North America can be separated into two distinct types. American geologists refer to rocks of the Lower Carboniferous (345–310 million years ago) as *Mississippian* and of the Upper

Carboniferous (310–280 million years ago) as *Pennsylvanian*. In North America only the Pennsylvanian rocks contain coal. Carboniferous fossil remains include mosses and liverworts, and herbaceous lycopod fossils similar to present-day *Lycopodium* and *Selaginella* are abundant. Some of the arborescent members of theLycophyta, as represented by *Lepidodendron* (see LEPIDODENDRALES), appear to have had a rudimentary ovule. The Sphenophyta are represented by the now extinct arborescent Calamitales (giant horsetails) and the extinct herbaceous Sphenophyllales. *Equisetites hemingwayi* in the Upper Carboniferous is one of the few herbaceous forms resembling the present-day *Equisetum*. The ferns were well represented as were the GYMNOSPERMS with fossils of the Cycadofilicales (seed ferns), Cordaitales, and Coniferophyta all being found. See GEOLOGICAL TIME SCALE.

carboxydismutase see RIBULOSE BISPHOSPHATE CARBOXYLASE.

carboxyl the chemical group –COOH (or CO_2H), present in carboxylic acids.

carboxylase any enzyme that catalyses the transfer or incorporation of carbon dioxide into a substrate molecule. CARBOXYLATION reactions usually require energy in the form of ATP or a related compound; a coenzyme, usually biotin, is also a part of most carboxylases. An example is pyruvate carboxylase, which catalyses the formation of oxaloacetate from pyruvic acid. As with many carboxylases, this is an allosteric enzyme, its activity being increased by the binding of acetyl CoA. See also RIBULOSE BISPHOSPHATE CARBOXYLASE, PHOSPHOENOLPYRUVATE CARBOXYLASE.

carboxylation the transfer or incorporation of carbon dioxide or a carboxyl group into a molecule. This is the essential first step in carbon FIXATION. See also CARBOXYLASE.

carboxylic acid any organic acid having one or more –COOH groups. The intermediates of the Krebs cycle are important di- and tricarboxylic acids, while acetic acid is a monocarboxylic acid. Carboxylic acids in metabolic pathways are often phosphate esters or thioesters of the acid, e.g. acetyl CoA is a thioester of acetic acid.

Some higher plant cells contain high

concentrations of certain carboxylic acids, for example citrus fruits are rich in citric acid, and oxalic acid is present in high concentrations in rhubarb and tobacco leaves. In the Crassulaceae and some other succulents and semisucculents there is a large diurnal fluctuation in carboxylic acid levels. This is due to incorporation of carbon dioxide into malic acid in the dark (see CRASSULACEAN ACID METABOLISM).

carboxysome an organelle found inside plastids, which is thought to contain the carbon dioxide-fixing enzyme RIBULOSE BISPHOSPHATE CARBOXYLASE. See also FIXATION.

carcerulus (*pl.* carceruli) a type of CAPSULE (fruit) that breaks up on maturity into one-seeded segments or nutlets (see NUT), as in the Lamiaceae.

cardiac glycoside any of a group of steroid GLYCOSIDES characterized by their stimulative effects on the heart. Cardiac glycosides resemble SAPONINS in being diterpenes and by foaming in solution. They differ however in having a lactone ring attached at carbon 17 of the steroid nucleus. An important commercial source is the foxglove (*Digitalis purpurea*), which yields digitonin. See also TERPENOID.

carina see KEEL.

carinal canal (protoxylem lacuna) a longitudinal channel in the stem internode of *Equisetum* (horsetail) and some of its fossil relatives, positioned radially opposite a raised stem ridge (*carina*). Each carinal canal is arranged on the inner side of a vascular bundle. The carinal canals apparently result from breakdown of PROTOXYLEM and are believed to function in water conduction. Compare VALLECULAR CANAL.

carnivorous plant see INSECTIVOROUS PLANT.

carotene (carotin) any of a class of orange, red, or yellow pigments all of which are hydrocarbons containing eight ISOPRENE SUBUNITS and hence have the formula $C_{40}H_{56}$. The most commonly occurring carotenes are α- and β-carotene (see also VITAMIN A). Another carotene, *lycopene*, gives tomato fruits their red colour. Carotenes are ACCESSORY PIGMENTS in most photosynthetic cells, contributing to the process of gathering light energy. However they are relatively inefficient as light absorbers and are thought to have a second function of protecting the other photosynthetic pigments from PHOTOOXIDATION by ultraviolet light. XANTHOPHYLLS are oxygenated carotenes. See CAROTENOIDS.

carotenoids yellow, orange, or red fat-soluble pigments found in all photosynthesizing cells, where they act as ACCESSORY PIGMENTS in photosynthesis. Their ABSORPTION SPECTRUM suggests they may also be involved in the phototropic response. They are also found in other organs, e.g. roots, notably carrot roots, and petals. There are two groups of carotenoids; the CAROTENES, which are hydrocarbons, and the XANTHOPHYLLS, which are oxygenated derivatives of the carotenes.

During leaf senescence the carotenoids do not break down as quickly as the chlorophylls and their colours are temporarily revealed.

carotenol see XANTHOPHYLL.

carotin see CAROTENE.

carpel the structure that bears and encloses the ovules in flowering plants. It normally comprises the OVARY, STYLE, and STIGMA. The typical enclosed carpel of the angiosperms has probably evolved from an open carpel bearing MEGASPORANGIA on the margins, and is thought by many to be homologous with the MEGASPOROPHYLL of club mosses, horsetails, and ferns and the OVULIFEROUS SCALE of gymnosperms.

In more primitive angiosperms there are often large numbers of unfused carpels in the gynoecium. In advanced forms the number of carpels is reduced and they tend to be fused and inserted asymmetrically.

carpogonium (*pl.* carpogonia) the female sex organ in the red algae (Rhodophyta), which has a swollen basal portion containing the egg and a number of other cells, and an elongated terminal projection or TRICHOGYNE, which attracts the male gamete. It develops by the expansion of a single cell.

carposporangium a SPORANGIUM that develops directly from the zygote in situ in some of the red algae (see RHODOPHYTA), and which is dependent on the parental tissue for nutrition.

carpospore a spore formed following division of the zygote in red algae (RHODOPHYTA). It may be haploid if formed following meiosis as, for example, in *Porphyra*. On germination it then gives rise to a haploid gametophyte or occasionally a short-lived filamentous juvenile form termed the *chantransia stage*. The carpospores of the more advanced red algae, e.g. *Polysiphonia*, are diploid and give rise to a diploid plant similar in form to the haploid generation. See also ALTERNATION OF GENERATIONS.

carposporophyte in some red algae (see RHODOPHYTA), a carpospore-producing diploid (sporophyte) phase that develops on the female gametophyte. The carposporophyte may develop directly from the fertilized CARPOGONIUM, or the zygote nucleus may be transferred to an auxiliary cell in another branch, which then produces the carposporophyte. The carposporophyte is made up of diploid branching filaments (*gonimoblast filaments*) derived from the zygote. These filaments may be intermingled with those of the gametophyte, or they may form a distinctive structure, the CYSTOCARP. See also ALTERNATION OF GENERATIONS.

carr see FEN.

carrageenan (carragheen) a mucopolysaccharide, derived from galactose, present in the walls of some red algae (RHODOPHYTA), especially Irish moss (*Chondrus crispus*) and *Gigartina* species. It has various commercial uses, acting as a stabilizer in paints, in sauces and creams, and in pharmaceuticals. The name is also sometimes used for the seaweed itself.

carragheen see CARRAGEENAN.

carrying capacity the maximum number of individuals of a particular species that can live on a specific area of land. This number is usually limited by food (and/or water) resources.

caruncle (*adj.* carunculate) a horny outgrowth near the hilum of a seed that is formed from the integuments. The warty outgrowth covering the hilum and micropyle of the seed of castor oil (*Ricinus communis*) is a caruncle. Carunculate seeds

are also found in other genera of the Euphorbiaceae, e.g. *Euphorbia* and *Jatropha*. Compare ARIL.

caryogamy see KARYOGAMY.

Caryophyllidae (Centrospermae) a subclass of the dicotyledons containing mostly herbaceous plants usually with bisexual flowers. The seeds often contain perisperm, and BETALAIN alkaloids are usually present. Three orders are usually recognized: the Caryophyllales; the Polygonales, containing the single family Polygonaceae (including buckwheat and sorrel); and the Plumbaginales, with the single family Plumbaginaceae (including sea lavender, thrift, and statice). Some taxonomists also recognize the Batales, comprising the single genus *Batis* (saltworts), usually placed in the Caryophyllales; and the Theligonales, the sole genus of which, *Theligonum*, has been variously placed in the Caryophyllales or Haloragales (subclass Rosidae) and, more recently, in the Rubiales (subclass Asteridae). The Caryophyllales is divided into some 15 families, the more important of which are the CACTACEAE (cactus family), the Aizoaceae (including the mesembryanthemums), the Caryophyllaceae (carnation family), and the Chenopodiaceae (beet family).

caryopsis (grain) a fruit that resembles an ACHENE except that the seed wall fuses with the carpel wall during embryo development. The caryopsis is typical of cereals and grasses.

cascade a sequence of biochemical reactions, each triggered by the previous one. Cascades play an important role in signalling pathways: extracellular signals, such as hormones, bind with receptors on the plasma membrane, triggering changes in membrane proteins which then activate proteins inside the cell and set off a chain of reactions to bring about a response. This may be, for example, the activation or inactivation of a gene in the nucleus, or the induction of secretion.

Casparian strip (Casparian band) a band of suberized and/or lignified wall material in the radial (anticlinal) and transverse walls of cells of the ENDODERMIS. It ensures that water and solutes pass through the living protoplast of the endodermal cells, rather than through the cell walls, thus facilitating selective filtering of the sap solutes and control of the rate of flow.

catabolism (*adj.* catabolic) the metabolic breakdown of complex molecules to simpler ones. For example, respiration involves various catabolic reactions that degrade sugars to carbon dioxide and water and release energy, which is stored as ATP. Compare ANABOLISM.

catalase a haem-containing enzyme that catalyses the breakdown of hydrogen peroxide to water and oxygen and the oxidation of substrates by hydrogen peroxide. The fastest-acting enzyme known, it is found in small organelles in the cytoplasm called PEROXISOMES.

catalyst (*adj.* catalytic) a compound that enhances the rate of a chemical reaction without itself being used up in the process. It achieves this by combining transiently with the reactants to form a transitional complex having a lower ACTIVATION ENERGY than that of the uncatalysed reaction. Biological catalysts are known as enzymes.

cataphyll 1 see BUD.
2 see SCALE LEAF.

catastrophe a major and widespread change in the environment that causes the death of many organisms. An example is the impact of a large asteroid, which upset global climate and is thought to have contributed to the mass extinction at the end of the CRETACEOUS PERIOD. Some catastrophes recur from time to time but at such long intervals that species are unable to adapt to them by natural selection, e.g. large volcanic eruptions. By eliminating many species and changing environments, catastrophes can be triggers for bursts of evolution. See ADAPTIVE RADIATION, CATASTROPHISM.

catastrophism the theory that the differences in the fossils found in different layers (strata) of rock are the products of repeated catastrophes followed by new creations. A modification of this in the 20th century was *neocatastrophism*, which claims that geological history represents a series of cycles of mountain building, inundations by the sea, and other major environmental perturbations, associated with the evolution and extinction of species.

catena a sequence of different soil types that vary with the topography of the land, being repeated in areas of similar topography. The term is usually applied to soils of the same age that are derived from the same parent material. It is not used for soils derived from glacial deposits. The variations in SOIL PROFILES are related to the differing environments at different positions on the slope.

cation a positively charged ion. Compare ANION.

cation exchange capacity (CEC) the capacity of a soil to hold cations, which are held mainly on the surface of colloidal particles of clay and humus. CEC is usually measured in milliequivalents per 100 grams (meq/100 g) of soil. It is related to the soil's ability to store plant nutrients: the plants obtain nutrients by exchanging one cation for another where their root colloids meet the soil colloids in the soil solution. Clay minerals are efficient cation exchangers, storing cations in spaces in their crystal lattices and allowing their substitution by other cations of similar size. CEC increases with increasing pH, and with increasing proportions of clay and/or humus in the soil. The efficiency of CEC depends on the relative concentrations of the cations, their valences and diffusion rates. See also EXCHANGEABLE IONS, ADSORPTION.

catkin an inflorescence modified for wind pollination (see ANEMOPHILY). It is made up of numerous sessile usually unisexual flowers. The calyx and corolla are normally reduced or absent, which allows maximum air circulation around the flower; the catkin itself develops in an exposed position on the plant. Male catkins shed vast amounts of light dry pollen. Female catkins usually have long hairy styles and stigmas to enhance pollen interception. Catkins are formed by many tree species, e.g. the male flowers of the Betulaceae (birches, hazels, alders, hornbeams) are always aggregated into catkins. *See illustration at* inflorescence.

caudex 1 the trunk of a palm or tree fern.
2 the persistent swollen stem base of certain herbaceous perennials.

Caulerpales (Siphonales) an order of SIPHONOUS GREEN ALGAE of the class ULVOPHYCEAE. The Caulerpales have a thallus made up of a single multinucleate cell, often giant-sized (*Caulerpa* may be over a metre long). Membranes dividing the thallus develop only during the formation of reproductive structures or the sealing-off of wounds. The Caulerpales includes algae of the orders Derbesiales and Codiales recognized by some classification systems. Members of the Caulerpales include *Caulerpa*, *Halimeda*, *Derbesia*, *Bryopsis*, and *Codium*. Growth forms vary from branching filaments to clusters of tubular structures, sometimes with distinct palisade-like photosynthetic regions (*Codium*). Rhizoids may be present for anchorage. Some species secrete calcium carbonate and magnesium carbonate to protect them against grazing. In some genera, both LEUCOPLASTS and chloroplasts are present. A variety of life cycles is found; *Derbesia* and *Bryopsis* have ALTERNATION OF GENERATIONS, while many other genera are DIPLONTIC. Members of the Caulerpales are found in the benthic communities of oceans in many parts of the world and in fresh water. Some species are weedlike pests of commercial shellfish beds; a few are used as food. The calcareous alga *Halimeda* is an important producer of reef sand.

caulescent in the process of becoming stalked. The young or basal leaves of some plants show a gradual progression from sessile to stalked with age or distance up the stem.

caulid the 'stem' of a moss or liverwort.

cauliflory (*adj.* cauliflorous) the production of flowers on tissue that has been secondarily thickened, such as the branches or trunks, rather than at apical meristems. Flowers arising in this way normally develop from suppressed side shoots, but more infrequently they may be formed from the phellem. This trait is found most frequently in species that make up the lower canopy of tropical forests, e.g. cocoa (*Theobroma cacao*). It is believed in some cases to be associated with pollination or fruit dispersal by bats.

cauline pertaining to the stem, especially applied to leaves arising on the upper part of the stem, as compared to the RADICAL leaves.

cavitation the formation of vapour-filled cavities in a flowing liquid due to a decrease in pressure. It is the main cause of disruption of water columns in the xylem when a plant is suffering water stress.

Caytoniales an extinct order of gymnosperms known from fossils of the Jurassic period. Their habit is uncertain but it is known they had compound leaves consisting of three to six leaflets with reticulate venation. These arose from the tip of the rachis giving a fanlike or palmate leaf. The female reproductive axis bore short pinnae each of which ended in a round hollow structure or CUPULE containing the ovules. The pinnae of the male reproductive organs bore SYNANGIA longitudinally divided into four pollen sacs, which appear to have been very similar to the anthers of flowering plants.

C_2 cycle see PHOTORESPIRATION.

C-14 dating see RADIOMETRIC DATING.

cDNA see COMPLEMENTARY DNA.

CEC see CATION EXCHANGE CAPACITY.

cell the fundamental unit of a living organism and the basis of its structure and physiology. A living plant or algal cell comprises CELL WALL and PROTOPLAST. Bacterial cells are prokaryotic, i.e. they have no nucleus (see PROKARYOTE). The cells of all other organisms are eukaryotic and have a well-defined nucleus containing the chromosomes enclosed in a nuclear membrane (see EUKARYOTE). In many plant tissues, e.g. sclerenchyma, cells are in a nonliving state, consisting of a cell wall only or cell wall thickened with additional nonliving material such as LIGNIN.

cell biology see CYTOLOGY.

cell culture see TISSUE CULTURE.

cell cycle the sequence of events in a cell resulting in the formation of two complete daughter cells. In addition to the observable changes during MITOSIS, biochemical processes occur in the nucleus and cytoplasm. CYTOKINESIS is followed by a growth phase (G_1 phase) during which there is a high rate of RNA formation and protein synthesis. A lag phase of variable duration is followed by a period of DNA replication and histone synthesis (*S* phase) as the chromosomes divide into chromatids. Finally there is a short growth period (G_2 phase) before prophase of mitosis begins.

The length of each period varies considerably. The cycle is completed when the cytoplasm divides into two daughter cells (CYTOKINESIS). In actively dividing cells in culture the full cycle takes twenty four hours while in living tissue the time taken to double the number of cells can be as short as eight hours.

cell division the formation of two daughter cells from a parent cell. PROKARYOTE cell division is relatively simple: the NUCLEOID attaches itself to a part of the CELL MEMBRANE that is expanding and divides in two, the two parts moving apart as the membrane grows between them and divides the cell in two. In EUKARYOTES, the division of the nucleus (KARYOKINESIS) is followed by the division of the cytoplasm (CYTOKINESIS). In VEGETATIVE REPRODUCTION and somatic cell division, two daughter cells are formed, each with a nucleus containing the same number of chromosomes as the mother cell (see MITOSIS). After reduction division (see MEIOSIS) the nuclei of the daughter cells contain half the number of chromosomes of the mother cell.

cell extension the increase in cell length and volume that occurs a certain distance behind the apex following cell division. The process involves both the uptake of large quantities of water into the cell vacuole by osmosis (*vacuolation*) and the synthesis of new cytoplasmic and cell wall materials. Cell extension accounts for a far greater proportion of overall plant growth than cell division.

cell fractionation see DIFFERENTIAL CENTRIFUGATION.

cell fusion the *in vitro* fusion of nuclei and cytoplasm from different somatic cells to create a HYBRID cell. The cells involved are usually derived from tissue cultures, sometimes of different species. The process has been used particularly to gain information about the position of genes on chromosomes, and in comparative studies of GENOMES from different organisms.

cell lineage the sequence of cells, produced by cell division and cytological changes, that results in the formation of differentiated cells (see DIFFERENTIATION) with a specific function. For example, the germ cells in flowering plants are produced by the sequence of changes, commencing with undifferentiated cells in the anthers and nucellus respectively, that produce the microspores (pollen grains) and megaspores (embryo sacs) and eventually the male and female gametes.

cell membrane (membrane) any membrane surrounding either the PROTOPLAST or any of the organelles within the protoplasm. The term is frequently used as a synonym for the plasma membrane. All cell membranes have essentially the same structure (see PLASMA MEMBRANE) and their selective permeability controls the passage of molecules and ions within the cell, particularly into and out of organelles. This ensures that metabolic processes are segregated without preventing links between metabolic pathways.

cellobiase see CELLOBIOSE.

cellobiose a reducing disaccharide composed of two glucose units linked by a $\beta(1-4)$ GLYCOSIDIC BOND. It is formed by the incomplete hydrolysis of cellulose by the enzyme CELLULASE. Complete hydrolysis to glucose units is achieved by the enzyme *cellobiase* (β-glycosidase). See also REDUCING SUGAR.

cell plate an opaque colloidal layer established by membrane-bound cavities in the equatorial region of the SPINDLE at the TELOPHASE stage of mitosis. It separates the cytoplasm into two daughter protoplasts, each with a nucleus. It is formed by the coalescence of vesicles containing PECTIC SUBSTANCES, which contribute to the formation of a MIDDLE LAMELLA between the newly formed cells.

cell sap the contents of a plant cell vacuole.

cell theory the theory that all organisms are made up of cells, and that growth and development is the result of the division and differentiation of cells. It was formulated by Schleiden and Schwann in 1839.

cellular slime moulds see ACRASIOMYCOTA.

cellulase a hydrolytic enzyme found in some bacteria, fungi, and seedlings that attacks alternate $\beta(1-4)$ linkages in CELLULOSE, degrading it to the disaccharide cellobiose, which is hydrolysed to its constituent glucose units by the enzyme cellobiase.

Cellulase is one of the very few enzymes capable of attacking the β(1–4) GLYCOSIDIC BOND; cellulose is not degraded by other glycolytic enzymes such as amylase. Cellulase produced by bacteria and fungi living in the gut of ruminant animals, enables such animals to derive nutrients from the fibrous component of their diet.

cellulose a polysaccharide composed solely of glucose units linked by β(1–4) GLYCOSIDIC BONDS. It is the most abundant structural polysaccharide in the plant kingdom and probably the most abundant of all compounds found in living organisms. In cell walls it is present in highly organized MICROFIBRILS.

cellulose synthetase SEE CELL WALL.

cellulytic capable of breaking open (lysing) cells.

cell wall the external covering of the cells of plants, algae and many other protoctists, fungi, and bacteria. The plant cell wall is composed mainly of cellulose molecules organized into MICROFIBRILS. Small quantities of proteins have also been identified, e.g. enzymes such as sucrase, phosphatase, and ATPase, whose functions are related to the uptake of nutrients and their passage from cell to cell. Enzymes responsible for cellulose synthesis are produced in the Golgi apparatus and reach the cell surface, where they become functional enzyme complexes (known as *cellulose synthetase*) within the membranes of vesicles. Microtubules beneath the plasma membrane are thought to have a role in the organization of cellulose into microfibrils. Following cell division the *primary cell wall* is laid down by the deposition of microfibrils on the MIDDLE LAMELLA. The orientation of the microfibrils differs depending on whether the cells are destined to be parenchymatous or more specialized tissue. The *secondary cell wall* normally consists of three layers laid down after cell extension is complete. The microfibrils are closely packed and aligned in the same direction in each layer, but in different directions in successive layers. In maturing fibres, the microfibrils are mainly parallel to the long axis, while in developing xylem vessels they are laid down in rings or in helical strands. The hydrated nature of the microfibrils renders the wall elastic and permeable to water and solutes, including the soluble respiratory gases. The pressure of cell contents against the walls causes stretching and confers turgidity on plant tissues (see TURGOR). This is a major factor in the provision of mechanical support in the nonwoody tissues of plants. In the cells of sclerenchyma and xylem tissues, lignin is deposited within the layers of the secondary walls and the protoplasts of these cells eventually disintegrate. Lignin confers considerable strength and these tissues form the wood of plants. Other substances, including suberin and callose, may also be deposited. The walls of spores are often impregnated with SPOROPOLLENIN.

Most algae have cellulose cell walls, but in a few species XYLANS or MANNANS are present instead of cellulose. Silica is an important constituent of certain algal cell walls (e.g. diatoms) and is also found in some grasses and sedges. Certain calcareous algae secrete a hard layer of calcium carbonate (and sometimes also magnesium carbonate) on the cell wall for protection against herbivores. The cell walls of fungi differ in that they usually contain chitin (those of a few species contain xylans or mannans), and the walls of some fungal spores contain melanin, a pigment that protects against damage from ultraviolet radiation. Bacterial cell walls consist mainly of mucopeptide substances (see PEPTIDOGLYCAN).

Cenozoic (Cainozoic, Caenozoic) the present era of geological time beginning about 65 million years ago. It has seen four periods of extensive land elevation separated by periods when the oceans expanded. The Andes, Rockies, Himalayas, and Alps were formed in the Cenozoic. Traditionally it has been divided into the Tertiary and Quaternary periods. Certain authorities now recognize the Palaeogene, equivalent to the earlier epochs (Palaeocene, Eocene, and Oligocene) of the Tertiary, and the Neogene for the later Tertiary epochs (Miocene and Pliocene) and the Quaternary. See GEOLOGICAL TIME SCALE.

central cell in angiosperms, the binucleate cell formed by the two POLAR NUCLEI in the centre of the EMBRYO SAC of the ovule, which contributes two of the three nuclei that give rise to the endosperm tissue. See also DOUBLE FERTILIZATION.

Centrales (centric diatoms) an order of the DIATOMS that have radially arranged markings on the test. The cells have a large central vacuole, surrounded by a layer of cytoplasm containing many disc-shaped chloroplasts. CHRYSOLAMINARIN granules may be present, but the more common food reserves are lipids. The sperm have a single mastigonemate UNDULIPODIUM (tinsel flagellum). Usually a single egg is produced, which may be released into the water or retained inside the test until fertilization. Some species form dormant cysts when conditions are hostile to growth. The Centrales are considered to be more primitive than the pennate diatoms (PENNALES).

centrarch describing a PROTOSTELE in which the protoxylem is at the centre of the axis. Compare ENDARCH, EXARCH, MESARCH.

centre of diversity see GENE CENTRE.

centre of origin the geographical area in which a particular group of organisms is thought to have originated. It is often considered to coincide with the GENE CENTRE (centre of diversity) of the group.

centric bundle see CONCENTRIC BUNDLE.

centric diatoms see CENTRALES, DIATOMS.

centrifugal developing from the centre outwards so the oldest structures are in the middle and the youngest at the perimeter. An example is the differentiation of secondary xylem. Compare CENTRIPETAL. See also ENDARCH.

centrifuge an apparatus designed for rapid separation of solids in suspension by rotating them in a container at very high speeds. The organelles of ruptured cells can be separated in this way. An *ultracentrifuge* rotates at much higher speeds (up to a million times per second) and is used to separate colloidal particles such as large protein molecules. As the sedimentation rate (see SEDIMENTATION COEFFICIENT) depends on the particle size, the size of particles can be estimated. The ultracentrifuge operates under refrigeration in a vacuum. See also DENSITY-GRADIENT CENTRIFUGATION, DIFFERENTIAL CENTRIFUGATION.

centriole a cylindrical structure some 500 nm long and 150 nm in diameter composed of nine longitudinal fibrils. Centrioles are commonly found in pairs at right angles to each other in animal cells and have been identified in the cells of some algae and other protoctists. They appear to act as centres for the polymerization of the microtubules that make up the mitotic and meiotic spindles. At the start of cell division fibrils appear, radiating from the centrioles and forming ASTERS. These form the poles from which the SPINDLE fibres arise. The basal bodies (see KINETOSOME) of UNDULIPODIA have an almost identical structure. Centrioles are not found in plant cells, which nevertheless produce spindles during cell division. See also CENTROSOME.

centripetal developing from the outside inwards, so the oldest structures are at the edge and the youngest are in the middle. An example is the differentiation of xylem in club mosses (*Lycopodium*). Compare CENTRIFUGAL. See also EXARCH.

centromere the region within a chromosome where DNA is present, as part of the structure, but has no genetic significance. When chromosomes are contracted in cell division, the centromere appears as a narrowed region because the DNA is not coiled. Its position on a chromosome is a distinctive feature enabling HOMOLOGOUS CHROMOSOMES to be identified. Centromere positions are described as median, terminal, subterminal, etc. (see also ACROCENTRIC, METACENTRIC, TELOCENTRIC) and are useful taxonomic characters. KINETOCHORES are visible structures situated in the centromere region and these attach chromosomes to the equatorial region of the SPINDLE during cell division. During nuclear division centromeres hold the chromatids together at the equatorial plate until they are pulled apart at anaphase.

centrosome an organelle that controls the assembly of the microtubules that comprise the CYTOSKELETON, and hence controls cell division, shape, and movement. The centrosome has no specific shape. Microtubules radiate from it, influencing the distribution of other components of the cytoskeleton, such as ACTIN filaments. During cell division the centrosome organizes the SPINDLE. In animal cells and some algal cells, the centrosomes contain the CENTRIOLES. In these cells, the microtubules of the spindle grow outwards from the area surrounding the centrioles, not from the centrioles themselves. During interphase, the microtubules extend from the centrosomes throughout the cytoplasm. Since microtubules also direct the transport of molecules and the positioning of other cell organelles, the centrosomes may indirectly also control other aspects of cellular organization. The centrosomes also play an important role in organizing the cytoskeleton during the differentiation of cells as an organism develops, and are known to direct polarity in the cells of some embryos.

Centrospermae see CARYOPHYLLIDAE.

cephalodium in a lichen thallus, a tissue containing blue-green bacteria (see CYANOBACTERIA).

cerradão see CAMPOS.

certation competition in growth rate between pollen tubes of different genotypes, which results in unequal chances of achieving FERTILIZATION.

CFCs see CHLOROFLUOROCARBONS.

Chaetophorales an order of filamentous green algae (Chlorophyta) of the class CHLOROPHYCEAE that are differentiated into prostrate basal branches, which anchor the alga to the substrate, and more widely spaced upright branching filaments, which are the main site of photosynthesis. The cells are connected by plasmodesmata. Branching of the filaments may increase when nutrient levels are low, and the tips of the branches may develop long tapering colourless hairs. Examples include *Stigeoclonium, Fritschiella*, and *Draparnaldia*.

chalaza the basal region of the ovule where the nucellus and integuments fuse. It may or may not coincide with the position of the funiculus depending on the mode of ovule orientation. See also CHALAZOGAMY.

chalazogamy a method of FERTILIZATION in seed plants in which the pollen tube does not reach the nucellus through the micropyle (*porogamy*) but grows or digests its way through the placenta and CHALAZA

instead. This may be observed in certain trees and shrubs, such as beech, elm, and hazel.

chalcones one of the two groups of ANTHOCHLOR flower pigments. Chalcones are less variable than other types of FLAVONOID pigments with only some 20 kinds being known. In the synthesis of the various flavonoids from acetates and cinnamic acid the chalcones are the first compounds to be formed along the biosynthetic pathway. This has led some to consider the possession of chalcones a primitive characteristic as compared to the possession of AURONES, which are oxidized derivatives of chalcones. The yellow colours of gorse, dahlia, and carnation flowers are due to chalcones.

chalk grassland a form of species-rich GRASSLAND that forms as a subclimax (see CLIMAX) community on calcareous soils as a result of grazing, usually by rabbits or sheep, as in the downs of southern England. Here it is characterized by such species as salad burnet (*Sanguisorba minor*), upright brome (*Bromus erectus*), and meadow oat grass (*Helictotrichon pratense*). If grazing ceases, the succession will progress to scrub and woodland.

chalybeate describing natural waters that contain iron.

chamaephyte a low-growing plant whose perennating buds are borne either at or near (within 0.25 m of) the soil surface. Chamaephytes include small bushes and herbaceous perennials and are commonly found in cold or semiarid climates. The buds are often protected by snow in cold climates. Three types of chamaeophyte are recognized. In *passive chamaeophytes* the aerial stems fall over as they die back. and buds develop on horizontal stems close to the ground; this is the normal growth form of *active chamaephytes*; in *suffruticose chamaephytes* the aerial shoots die back only partially and the buds arise from the lower, persistent parts of the stems; *cushion chamaephytes* are telescoped suffruticose chamaephytes. See also RAUNKIAER SYSTEM OF CLASSIFICATION, PERENNATING ORGAN.

chañaral a type of thorny scrub with abundant chañar (*Gourliaea decorticans*), found in south-central South America. See also FOREST.

chantransia stage a filamentous juvenile stage of certain red algae (Rhodophyta) of the subclass FLORIDEOPHYCIDAE. See CARPOSPORE.

chaparral see MAQUIS.

character displacement see SYMPATRY.

characteristic species see KENNARTEN SPECIES.

characters see TAXONOMIC CHARACTERS.

Charales (stoneworts) an order of green algae (Chlorophyta) of the class CHAROPHYCEAE. The stoneworts are macroscopic plantlike organisms that are more complex in structure and development than other members of the Chlorophyta. They are commonly found in freshwater ponds attached to the bottom by rhizoids and are strongly associated with unpolluted water. They have a distinct main axis with whorls of branches arising from the nodes and superficially resemble certain higher plants. The thallus often becomes encrusted with calcium carbonate, hence the vernacular name. Sexual reproduction is oogamous and complex. The gametes are produced at the nodes of the plant in complex multicellular structures, the sperm in *globules*, the eggs in *nucules*. The zygote becomes dormant and overwinters. During germination it undergoes meiosis to produce a filamentous haploid PROTONEMA, which later gives rise to the main thallus, a situation found among algae only in the Charales. This and the disclike form of the plastids suggests that, together with the closely related COLEOCHAETALES, the Charales are more similar to the plants than are other orders of the Chlorophyta.

Charophyceae a class of green algae (Chlorophyta) that have an open mitotic spindle: during mitosis, the nuclear envelope breaks down. Cell division involves the formation of a cell plate or phragmoplast, as in plants. Cells with undulipodia have an unusual multilayered microtubular structure associated with the KINETOSOMES of the undulipodia. The basal bodies are parallel to the undulipodia, which arise from one side of the cell just below the anterior end. The Charophyceae

may be unicellular or multicellular, ranging from solitary cells to complex branching filaments. The main stage of the life cycle is haploid, and the zygote often forms a dormant resting stage, undergoing meiosis immediately upon germination.

The Charophyceae are considered to be closely related to the BRYOPHYTES and vascular plants. The class includes the orders Charales (stoneworts), including the filamentous *Chara*, which may reach 50 cm in length; the COLEOCHAETALES; and the KLEBSORMIDIALES. Some systems include the conjugating filamentous green algae and desmids as an order, Zygnematales, of this class, but the FIVE KINGDOMS CLASSIFICATION places these organisms in a separate phylum, GAMOPHYTA.

Charophyta (stoneworts) in some classifications, a division of algae containing a single order, the CHARALES. Today, the Charales are more commonly placed in the class CHAROPHYCEAE of the phylum CHLOROPHYTA (green algae).

chase 1 a lane between two woodlands. **2** in Britain, a royal forest that has passed into private ownership.

chasmogamy the production of flowers that open to expose the reproductive organs. This allows CROSS POLLINATION but does not preclude SELF POLLINATION. Compare CLEISTOGAMY.

chelate see CHELATION.

chelation the process of forming a stable chemical complex (a *chelate*) with a substance, usually a metal ion, that involves the formation of at least two chemical bonds. A chelate is usually an organic ring compound, with the added ion completing the closure of the ring. An example is chlorophyll, in which the porphyrin ring is bound to iron. Chelating agents are used to remove toxic ions from water. Some plants form their own chelates, usually derived from malate or citrate, to enable them to tolerate toxic ions. The chelates are deposited in the vacuole, where they cannot interfere with cell processes.

chemical fossil any of various chemicals, such as alkanes and PORPHYRINS, found in certain rocks and thought to indicate the former presence of life. Such compounds are often the only evidence for the existence of living organisms in rocks of the Precambrian era.

chemical oxygen demand (COD) the amount of oxygen consumed during the oxidation of the organic and oxidizable inorganic matter in a sample of water, usualy expressed as milligrams of oxygen per litre of water. The COD is an indicator of the quality of water or effluent, and is used to monitor chemically polluted water and industrial waste. The COD is determined by incubating a known volume of water with a known quantity of a chemical reagent (e.g. the oxidizing agent potassium dichromate) at about 150°C until oxidation is complete. The amount of reagent reduced is determined using colorimetry or spectrophotometry. Compare BIOCHEMICAL OXYGEN DEMAND, BIOLOGICAL OXYGEN DEMAND.

chemiosmotic theory a theory developed by the British biochemist Peter Mitchell in 1961 to explain the mechanism that couples electron transport to ATP synthesis during respiration and photosynthesis. The basis of the process is a membrane that is impermeable to the movements of protons (hydrogen ions) from outside to inside (in mitochondria this is the inner membrane). This is thought to be achieved by the distribution of electron carriers in the membrane. Electron flow along an ELECTRON TRANSPORT CHAIN causes protons to be drawn from the internal matrix to the space outside the membrane (the so-called *proton pump*). This causes a build-up of protons and creates a pH gradient across the membrane. To maintain this gradient, energy from electron transfer along the electron transport chain is used to prevent the protons returning across the membrane. This gradient provides a source of potential energy that can be harnessed for various purposes. In respiration and photosynthesis protons are allowed to move back across the membrane through specific sites where the enzyme ATPASE can use this energy to drive ATP synthesis.

chemoautotroph an organism that synthesizes organic molecules using energy derived from the oxidation of inorganic compounds present in the environment. Examples are bacteria of the genera

Nitrosomonas, which obtain energy by oxidizing ammonia to nitrite, and *Nitrobacter*, which oxidize nitrite to nitrate. Compare CHEMOHETEROTROPH. See also AUTOTROPH, CHEMOTROPH.

chemoheterotroph an organism that obtains the energy required for the synthesis of organic molecules from the oxidation of organic compounds. All heterotrophs, excepting the photoheterotrophs, are chemoheterotrophs.

chemolithotroph an organism that obtains the energy required for the synthesis of organic molecules from the oxidation of inorganic compounds or elements.

chemo-organotroph an organism that requires organic compounds as a source of both energy and carbon.

chemoreceptor an organ or other body structure or cell that responds to a particular chemical stimulant.

chemosterilant any chemical used to sterilize pests and hence reduce the number of offspring in the subsequent generation. A common technique is to regularly release large numbers of laboratory-reared sterile males into the wild population, which compete with normal males. The method has been used successfully to control various fruit flies.

chemosynthesis the synthesis of organic materials by organisms (chemotrophs) using chemical energy derived from chemical reactions. Compare PHOTOSYNTHESIS.

chemotaxis (*adj.* chemotactic) a taxis in response to a chemical stimulus. For example, bacteria commonly show *positive chemotaxis*, swimming towards regions containing higher concentrations of food substances such as peptone and lactose. *Negative chemotaxis* may be shown in response to toxic substances.

chemotaxonomy (biochemical taxonomy) the application of the principles and procedures of chemical analysis and the results obtained to the classification of plants. Although the science has diverse and ancient origins (e.g. the medicinal and ethnic interest in the chemical constituents of plants) it has made significant growth only in recent decades. Plant product analysis has been simplified by the development of chromatography and electrophoresis, which with other new techniques, make routine analysis of a representative selection of material a possibility that was not once available. Many different compounds are potentially of taxonomic value and three main groups are recognized; primary and secondary metabolites and SEMANTIDES. Primary metabolites are those involved in the essential METABOLIC PATHWAYS of the plant. As such they are virtually ubiquitous through the plant kingdom and are rarely useful taxonomically. Secondary metabolites are those substances accumulated by plants that have traditionally been regarded as waste products, e.g. terpenoids, alkaloids, phenolics, etc. Often distinct discontinuities are found in the distribution of secondary metabolites, which have been used to delimit taxa.

chemotroph an organism that derives the energy necessary for the synthesis of organic compounds from the oxidation of either organic or inorganic compounds (as in chemoheterotrophs and chemoautotrophs respectively). Light energy plays no part in the process. Compare PHOTOTROPH.

chemotropism a TROPISM in response to a chemical stimulus. Thus, when a pollen tube grows through an agar culture towards a piece of implanted pistil, it exhibits two forms of chemotropism, namely negative aerotropism in growing away from oxygen and positive chemotropism in growing towards those chemicals exuding from the pistil.

chernozem (black earth) a type of zonal alkaline soil (a pedocal) found in regions with light summer rainfall, as in the pampas of South America and the prairies of North America. The top of the A horizon is very thick, rich in humus, and black, and the brown lime-rich layer beneath merges into, and is often indistinct from, the B horizon (see SOIL PROFILE). The slight leaching due to the spring and summer rains is counteracted by surface evaporation in the hot summer resulting in the accumulation of lime in the A horizon.

Such soils are very fertile over a long period of time. See SOIL PROFILE.

chestnut-brown soil a type of zonal alkaline soil (a PEDOCAL) that forms in areas of dry grassland receiving about 340–360 mm of precipitation per annum. The A horizon is much thinner than in a CHERNOZEM soil because the supply of organic matter is limited.

chiasma (*pl.* chiasmata) a point at which chromatids from different homologous chromosomes in a BIVALENT are seen to be in close contact as separation of the paired chromosomes commences during the first meiotic PROPHASE (see MEIOSIS). Genetic evidence shows that at these points non-SISTER CHROMATIDS break and rejoin in such a way that portions are exchanged, resulting in the recombination of maternal and paternal alleles on the homologous chromosomes. Chiasma formation is also important for the distribution of chromosomes into the nuclei of gametes, as it holds homologues together until the start of anaphase. If chiasmata fail to form, homologues do not pair effectively at first prophase and they are unable to move to the poles of the spindle. See CROSSING OVER.

chiasma interference the effect a crossover occurring at one point has on the chances of a second crossover occurring in the same pair of chromatids. If the term is used without qualification, it normally implies the chances of a second crossover are reduced (positive interference). Negative interference, where the chances of a second crossover are increased, is rarer.

chimaera a plant or plant part that consists of two or more genetically different types of cell. In a *graft hybrid* the distinct cell types are derived from two different species. Such chimaeras are formed when a bud grows from the region of a graft union (see GRAFTING) and contains tissues from both scion and stock (see PERICLINAL CHIMAERA). Chimaeras can also arise by a mutation in a meristematic region. This may give rise either to a sector of different cells (see SECTORIAL CHIMAERA) or to a complete layer of different cells, as might occur if the mutation is in the top layer of the tunica, which subsequently gives rise to the epidermis.

Chimaeras consisting of cells of different chromosome number may occur in cytologically unstable tissues. This may result from wounding. It is also seen in trisomic plants in which some cells lose the extra chromosome to revert to the normal chromosome number. In such cases both normal shoots and shoots showing the morphological changes associated with trisomy can occur.

chi-squared test (X^2 test) A test usually used to find how well observed frequencies of results correspond with expected frequencies. It is used when the observations fall into discrete classes, such as red-, pink-, or white-flowered plants. X^2 is found by calculating

(observed–expected)^2/expected

for each class and adding these together. The sum is then checked in a table of X^2 values for different degrees of freedom at various PROBABILITY levels.

chitin a polymer containing chains of acetylglucosamine units, having the empirical formula $C_{15}H_{26}O_{10}N_2$. Microfibrils of chitin make up the hyphal wall of most fungi.

chitosan a partly deacylated form of CHITIN found in the cell walls of ZYGOMYCETES.

Chlamydomonadales SEE VOLVOCALES.

chlamydospore a thick-walled resting spore formed by certain fungi and fungus-like protoctists. The term refers both to thickened segments of hyphae (*gemmae*), as found in the Saprolegniaceae and Mucoraceae, and to the dikaryotic *brand spores* of the Ustilaginales that develop in sori on infected plants.

chlorazol black a temporary acid stain that colours walls grey or black and is especially suitable for demonstrating pitting. It is also used for chromosome counts, in which case the tissue is first fixed in acetic acid. After STAINING, a drop of acetocarmine is added and the chromosomes stain a deep reddish-black colour. The stain gives a clear-cut picture and is therefore useful for photomicrography. See MICROGRAPH.

chlorenchyma photosynthetic tissue containing numerous chloroplasts and often having relatively large intercellular spaces. The MESOPHYLL of leaves is composed of chlorenchyma.

chlorinated hydrocarbon any hydrocarbon with any or all of the hydrogen atoms replaced by chlorine atoms. Such compounds are often used as insecticides, examples being aldrin, dieldrin, and DDT.

chlorine symbol: Cl. A reactive nonmetallic element, a greenish-yellow gas at room temperature, belonging to the halogens (group 17, formerly group VIIa) of the periodic table, atomic number 17, atomic weight 35.5, important as a MICRONUTRIENT for plant growth. The chloride ion (Cl$^-$) can readily pass through cell membranes, and is involved in the regulation of osmotic and ionic balance in the plant and suspected of involvement in the light reaction of photosynthesis. Chlorine deficiency leads to wilting, and young leaves turn shiny and blue-green, later becoming bronze and chlorotic (see CHLOROSIS).

chlorinity a measure of the chlorine content of sea-water. It is the number of grams of chlorine in 1 kilogram of sea water. Since chlorine is a major component of salt, the chlorinity is also a measure of the salinity of the water: the salinity is equivalent to 1.80655 times the chlorinity.

Chlorococcales an order of green algae (CHLOROPHYTA) in the class CHLOROPHYCEAE, containing unicellular (e.g. *Chlorella*), colonial (e.g. *Scenedesmus*), coenobial (e.g. *Pediastrum*), and coenocytic coenobial (e.g. *Hydrodictyon*) algae that lack undulipodia. Motility is restricted to the gametes and zoospores. Vegetative cells lack undulipodia and associated structures. Sexual reproduction is isogamous or anisogamous but never, so far as is known, oogamous. Asexual reproduction is by motile zoospores or nonmotile APLANOSPORES (autospores). Most Chlorococcales live in freshwater plankton or in moist terrestrial habitats. Evidence from ribosomal RNA suggests that the order is not a natural one, but is of polyphyletic origin.

chlorofluorocarbons (CFCs) a group of compounds derived from hydrocarbons in which some hydrogen atoms have been replaced by chlorine or fluorine atoms. They are chemically inert, and used to be widely used as coolant liquids in refrigerators, as propellants in aerosols, and as blowing agents in the manufacture of plastics. They are implicated in the destruction of the ozone layer in the stratosphere (see ATMOSPHERE), where the high light intensities cause them to dissociate, releasing chlorine atoms that become involved in the catalytic destruction of ozone. They are also greenhouse gases, with an effect on GLOBAL WARMING 3000–13 000 times greater than that of carbon dioxide. See GREENHOUSE EFFECT.

Chlorophyceae a class of green algae (Chlorophyta). In some classifications it contains all the orders of the Chlorophyta except the CHARALES, but in most it comprises just the orders VOLVOCALES, TETRASPORALES, CHLOROCOCCALES, MICROSPORALES, CHAETOPHORALES, and OEDOGONIALES. The class is distinguished by having a cruciate arrangement of microtubules in the cells with UNDULIPODIA, and a PHYCOPLAST, which separates cells during cell division. The Chlorophyceae range from single cells (with or without undulipodia) and colonies to filaments. The main vegetative phase is haploid, with the zygote becoming dormant before germinating and immediately undergoing meiosis. Most species are found in freshwater or terrestrial habitats.

chlorophyll method a method of estimating the PRIMARY PRODUCTIVITY of an ecosystem by using the chlorophyll content of the community in a given area as an index of productivity. It is especially useful for aquatic ecosystems.

chlorophylls the main class of photosynthetic pigments. They absorb red and blue-violet light and thus reflect green light, so giving plants their characteristic green colour. Chlorophylls are involved in the light reactions of photosynthesis and are located in the chloroplast. The molecule consists of a planar PORPHYRIN 'head' and a long phytol 'tail' (see diagram). A molecule of magnesium is present in the centre of

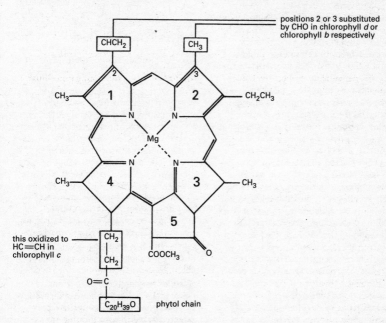

Chlorophyll: the structure of chlorophyll a

the porphyrin ring. In vivo the chlorophyll molecule is bound to membrane proteins via the phytol chain. Its function in photosynthesis is to absorb light energy and initiate electron transport. There are four groups of chlorophylls: *a*, *b*, *c*, and *d*. Chlorophyll *a* is found in all autotrophic plants and algae, chlorophyll *b* in the Chlorophyta and in land plants, and chlorophylls *c* and *d* only in certain algae. See also BACTERIOCHLOROPHYLL.

Chlorophyta (chlorophytes, green algae) a large phylum of the PROTOCTISTA (in some classifications a division of the Plantae) comprising green-pigmented algae found in marine, freshwater, and moist terrestrial habitats. Their main photosynthetic pigments are chlorophylls *a* and *b*, with associated carotenes and xanthophylls, their main carbohydrate reserve is starch, they usually have cellulose cell walls, their motile cells have smooth UNDULIPODIA, and the chloroplasts comprise stacks of two to six thylakoids inside a double envelope. In these respects chlorophytes resemble bryophytes and higher plants more than any other algal phylum does. Chlorophytes show a wide range of form and of sexual and asexual reproductive methods. In structure they range from unicellular and colonial cells to filamentous forms. In some chlorophytes mitosis involves the complete breakdown of the nuclear membrane, while in others a PHYCOPLAST or CELL PLATE is formed. In the most primitive species, the main phase of the life cycle is haploid, and the zygote is the only diploid stage. Most species show alternation of generations, which may be ISOMORPHIC or HETEROMORPHIC. A few have a dominant diploid phase, with the gametes being the only haploid stage. Four classes are commonly recognized, mainly on the basis of ultrastructural differences, especially in undulipodium and kinetosome structure and microtubular organization: the PRASINOPHYCEAE, CHLOROPHYCEAE, ULVOPHYCEAE, and CHAROPHYCEAE. Some classifications divide the Chlorophyta into just two classes, Chlorophyceae and Charophyceae, while others elevate the Charophyceae to divisional rank (Charophyta), dividing the remaining chlorophytes into four classes: Chlorophyceae, Oedogoniophyceae, Bryopsidophyceae, and Zygnemaphyceae.

chloroplast a green PLASTID with an internal membrane system incorporating the pigment molecules, including chlorophylls *a* and *b*, that are essential to photosynthesis. Chloroplasts are present in most of the cells of autotrophic plants and algae that are exposed to light. In algae they are frequently large and may have a complex shape, e.g. the spiral chloroplasts of *Spirogyra*. Many algal cells contain only one chloroplast. In bryophytes and all higher plants they are generally lens-shaped and present in large numbers (up to 100 per cell in some palisade cells). They vary in size, average dimensions being 5 μm by 2.3 μm, and develop from the minute PROPLASTIDS of meristematic cells.

The chloroplast is surrounded by a double membrane with no pores, the *chloroplast envelope*. The matrix (*stroma*) of the chloroplast is a complex hydrophilic proteinaceous sol, containing particles that vary in size and distribution. Some are the temporarily stored products of photosynthesis, e.g. starch grains, PLASTOGLOBULI, various crystals, and other storage compounds in some species. Others are aggregates of RIBULOSE BISPHOSPHATE CARBOXYLASE, ribosomes, and POLYSOMES, and circular fibrils of DNA. Chloroplast ribosomes are smaller than cytoplasmic ribosomes. In this respect, and in their reactions to antibiotics, chloroplast ribosomes resemble those of prokaryotes (see ENDOSYMBIOTIC THEORY).

The internal chloroplast membranes form a complex system of granal and intergranal lamellae (see diagram). GRANA are formed from two or three up to approximately a hundred dislike flattened vesicles (*thylakoids*) stacked on top of each other. They are orientated in a variety of directions relative to the long axis of the chloroplast and their number and size varies with different species. The chloroplasts in the bundle sheath cells of C_4 PLANTS do not contain grana. The intergranal lamellae are flexible interconnecting channels, continuous with and linking together the channels of individual grana.

The precise arrangement of pigment molecules within the membranes is not clear but the thylakoid membranes have the densest concentration. Spherical structures some 17.5 nm and 11.0 nm in diameter have been identified on the thylakoid surface. These are believed to be embedded in the lipid layers with their edges protruding from the membrane surface. These particles have been shown to be essential to the light reactions of photosynthesis. They consist of lipids and proteins, including various photosynthetic pigments and electron carriers. The pigment in the smaller particles is the P_{700} form of chlorophyll a associated with photosystem I, and that of the larger particles is the P_{680} form associated with photosystem II (see PHOTOSYSTEMS I AND II). Formerly termed *quantasomes*, these structures are now more commonly known as *photosynthetic units*. Superficial particles, loosely attached to the granal membranes, have also been identified. These consist of molecules of a calcium-dependent ATPase. The light reactions of photosynthesis thus occur within the granal layers. Water molecules are split, oxygen is evolved, and ATP and NADPH are formed. The dark (light-independent) reactions occur in the stroma, glycerate 3-phosphate being reduced to glyceraldehyde 3-phosphate by NADPH. This is the basis for the biochemical reactions that ultimately produce all the organic molecules of the organism. See PHOTOSYNTHESIS. See also CHROMOPLAST.

chloroplast DNA (cpDNA) the DNA found in chloroplasts. It occurs in a circular form, rather like bacterial DNA, and codes for some, but not all, of the proteins needed by the chloroplast.

chloroplast envelope see CHLOROPLAST.

chlorosis (*adj.* chlorotic) a condition of green plants in which they become unhealthy and pale or yellow in colour. The yellowing is due to a fall in chlorophyll levels, which may be caused by a number of factors. Deficiency of iron, magnesium, or copper is

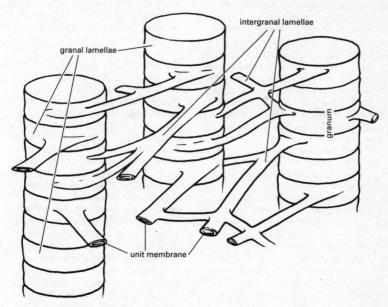

intergranal lamellae

granal lamellae

granum

unit membrane

Chloroplast: the arrangement of granal and intergranal lamellae in a chloroplast

a common cause (see DEFICIENCY DISEASES). Poor light conditions, and certain plant diseases, notably virus infections, also cause chlorosis. See also YELLOWS.

chloroxybacteria see CYANOBACTERIA.

chlor-zinc-iodide see SCHULTZE'S SOLUTION.

Chordariales an order of brown, often crustose, algae (PHAEOPHYTA). The branching sporophyte thallus is usually pseudoparenchymatous, often differentiated into a central medulla and an outer photosynthetic cortex. The order shows heteromorphic ALTERNATION OF GENERATIONS. Most members of the order are marine, some of them epiphytic on other seaweeds.

-chory a suffix that denotes seed-dispersal.

chromatid a thread formed by the longitudinal division of a chromosome. During the S phase of the CELL CYCLE, the DNA of the chromosomes replicates and the protein components of chromatin are synthesized. As chromatin becomes condensed in MITOSIS, the chromosomes can be seen to be divided along their length, with the exception of the centromere, into two sister chromatids. In MEIOSIS chromatids do not become apparent until the diplotene stage of the first prophase. Each is structurally a complete chromosome. During anaphase of mitosis and the second division of meiosis, when the centromeres divide, chromatids are separated to become the chromosomes of the daughter nuclei.

chromatid interference the non-random involvment of non-sister chromatids of a tetrad in successive crossovers during meiosis. It causes a deviation from the expected 1:2:1 ratio for the resulting frequencies for 2-, 3- and 4-strand double crossovers.

chromatin the complex of proteins, DNA, and small amounts of RNA of which CHROMOSOMES are composed. HISTONES are the principal nuclear proteins, although nonhistone acidic proteins are also present. Chromatin stains deep red when treated with Feulgen reagent (see FEULGEN'S TEST). See also EUCHROMATIN, HETEROCHROMATIN.

chromatin loop a region of relatively unwound DNA, about 200 nm long, extending outwards in a loop from the chromosome. It is thought that in this uncoiled condition TRANSCRIPTION can occur. If so, chromatin loops correspond to genes actively involved in protein synthesis.

chromatogram see CHROMATOGRAPHY.

chromatography a technique used to separate and identify the components of mixtures of similar compounds, such as different amino acids or chlorophyll pigments, by selective adsorption. The mixture, dissolved in a liquid or gas *mobile phase*, is passed through another solid or liquid medium, the *stationary phase*. This may be, for example, a column of charged resin (see ION-EXCHANGE CHROMATOGRAPHY) or a piece of filter paper (see PAPER CHROMATOGRAPHY). Some of the compounds in the mixture will pass more slowly through the stationary phase than others, as they are more strongly adsorbed by the solid, or more soluble in the liquid. Consequently, the mixture will separate out. The column, strip of paper, or other material containing the separated components is termed a *chromatogram*. In some techniques, the compounds are collected as they emerge from the column, or their progress along the column is measured against a known standard. Colourless components of the mixture can be identified by electronic detection, ninhydrin developer, or radioactive labelling. See also ELUTION, GAS–LIQUID CHROMATOGRAPHY, THIN-LAYER CHROMATOGRAPHY.

chromatophore a pigmented lamellar or vesicular structure that can be isolated from disrupted photosynthetic bacteria. These organisms have no chloroplasts. Their plasma membrane may be projected in folds into the cytoplasm (see LOMASOME) forming lamellae that have, therefore, double unit-membrane structure. The pigments and most of the enzymes required for the light-induced electron transport and phosphorylation processes of photosynthesis are located in the plasma membrane and lamellae. The principal photosynthetic pigment in the blue-green bacteria (see CYANOBACTERIA) is chlorophyll *a*; in purple bacteria, bacteriochlorophyll *a* or *d*; and in green bacteria, bacteriochlorophyll *b* or *c*.

Chromista in some classifications, a kingdom comprising certain fungus-like organisms (see OOMYCOTA, HYPHOCHYTRIOMYCOTA, LABYRINTHULATA), together with a range of algae including the Dinomastigota, Chrysomonada, Haptomonada, Euglenida, Cryptomonada, Xanthophyta, Diatoms, and Phaeophyta. In the FIVE KINGDOMS CLASSIFICATION, all these organisms are currently placed in the kingdom PROTOCTISTA; in other classifications the algae are placed in the kingdom Plantae, and the fungus-like members in the division PARAMYCOTA or MASTIGOMYCETIA of the kingdom Fungi. The Chromista are characterized by being basically phototrophic, their cell walls contain a structural cellulose-like molecule and β-glucan, their chloroplasts are surrounded by a periplastid membrane (thought to be evidence of their endosymbiotic origin), and undulipodia (when present) include at least one mastigonemate UNDULIPODIUM.

chromomere a deeply staining band of chromatin often showing up as a beadlike structure on chromosomes during prophase of mitosis and meiosis when the chromosome is relatively uncoiled. Chromomeres vary in size and shape, homologous chromosomes having identical patterns along their lengths. It is thought that the chromatin is more tightly coiled or folded in the region of the chromomeres. Stains specific for DNA indicate that there is more DNA in these regions. Specific genes have been associated with specific bands, but it is not known if more than one gene is located in each chromomere.

chromonema a thin thread of CHROMATIN seen in the interphase nucleus. Individual chromosomes are not distinguishable at this stage, when the chromatin remains relatively uncondensed and tangled. CHROMOMERES are often visible along the chromonema.

Chromophyta in some classifications, a division of algae that have chlorophylls *a* and *c* and whose motile cells are heterokont (having one smooth undulipodium and one hairy undulipodium). The chloroplasts have thylakoids arranged in groups of three. The scope of the Chromophyta varies in different classification systems, but may include the CHRYSOMONADA, PHAEOPHYTA, DINOMASTIGOTA, CRYPTOMONADA, EUSTIGMATOPHYTA, and XANTHOPHYTA, or just the Chrysomonada and Phaeophyta. It is thought to be closely related to the Oomycota, which also have heterokontous motile cells.

chromoplast a coloured PLASTID. The CHLOROPLAST is a specialized form of chromoplast. Other chromoplasts, in which carotenoid pigments predominate, give the characteristic colours to red, yellow, and orange flower petals, stamens, and fruits.

chromosome a threadlike structure, several of which can be seen in the nuclei of eukaryotic cells during cell division. They are composed of CHROMATIN and hence carry the genetic information. The chromatin becomes condensed prior to cell division, which renders the chromosomes visible under the light microscope. Following replication in late interphase, the chromosome is divided along its length into two identical strands, the CHROMATIDS, joined at some point by the CENTROMERE. The chromosome thus has four arms, the lengths of which depend on the position of the centromere. The number, size, and shape of the chromosomes as seen at prophase are generally characteristic of a species (see KARYOTYPE). During interphase the chromosomes uncoil into long narrow threads of DNA about 2 nm in diameter. These bear beadlike structures, the NUCLEOSOMES, which are highly organized aggregations of histones and DNA. The complex coiling and recoiling of the DNA and histones that occurs prior to cell division considerably reduces the length of the chromosome and increases the diameter to between 10 and 30 nm. During cell division, the chromosomes become attached to the mitotic or meiotic SPINDLE at a region of the centromere called the KINETOCHORE. See also BACTERIAL CHROMOSOME, HOMOLOGOUS CHROMOSOMES, B-CHROMOSOME, CHROMOSOME NUMBER.

chromosome aberration see CHROMOSOME MUTATION.

chromosome inversion see INVERSION.

chromosome mapping the assigning of

genes to chromosomes and the calculation of their relative sequence and distance from each other on particular chromosomes. In initially assigning a gene to a particular chromosome, the inheritance of particular chromosome mutations may be studied, coupled with microscopic techniques. Linkage studies will indicate whether genes are on the same or different chromosomes and, if the former, their relative order and distance. Absolute distance cannot be determined by linkage studies: the distance between two genes on a genetic map (the *map distance*) is based on the cross-over percentage.

Chromosome mapping indicates that genes are arranged in linear order along chromosomes. In contrast to this, circular genetic maps are often obtained in prokaryotic (nonchromosomal) organisms and viruses.

When linked genes on homologous chromosomes undergo crossing over, the formation of chiasmata leads to the production of recombinants among the offspring, in which paternal and maternal alleles are intermingled. From these, the *recombination frequency* is calculated using the formula:

(number of recombinants/total number of offspring) × 100

This *crossover value* (*COV, crossover frequency*), a percentage, is a measure of the number of crossovers that have occurred during gamete formation. It represents a hypothetical distance along the chromosome: a crossover value of 5% would equate to a distance of 5 *map units*. Since crossovers are more likely to occur between genes widely separated on the same chromosome than between those close together, these values can be used to determine the order of genes on the chromosome. See LINKAGE MAP.

chromosome mutation (chromosome aberration) an alteration in the number, order, or arrangement of genes caused by structural changes to a chromosome. Sometimes chromosome mutation may involve a large number of genes and the aberration may be visible by microscopy. There are four classes of chromosome mutation (see DELETION, DUPLICATION, INVERSION, TRANSLOCATION).

chromosome number the number of chromosomes possessed by a species, which may be given as either the haploid (n) or diploid ($2n$) number. Chromosome number is important taxonomically since it is usually constant within a species. There are, however, many exceptions especially in species exhibiting reproductive abnormalities, e.g. APOMIXIS. Among flowering plants the average haploid chromosome number is about 16, though this varies from $n = 2$ in the composite *Haplopappus gracilis* to high numbers in certain polyploid series. For example, in the allopolyploid *Spartina anglia* (saltmarsh grass) $n = 60$. It is thought that the original haploid chromosome number in angiosperms was in the region 7–9 and that numbers higher than $n = 14$ are polyploid in origin. Some ferns have extraordinarily high chromosome numbers; adder's-tongue ferns (*Ophioglossum* spp.) have the highest chromosome numbers known – in *O. reticulatum* $2n = 1260$.

chromosome polymorphism the occurrence in the same population of one or more chromosomes in two or more alternative structural forms.

chromosome theory of heredity the theory proposed by W. S. Sutton in 1902 that Mendel's laws of inheritance can be explained by assuming that genes are located at specific sites (see LOCUS) on chromosomes.

chronistics the study of the time taken for evolutionary events to occur. If phylogenetic trees are drawn with the vertical axis representing the time scale then a diagram is obtained showing the chronistic relationship between organisms. Organisms that, from fossil evidence, appear to have evolved rapidly are termed *tachytelic* while those that appear to have changed slowly are termed *bradytelic*. *Horotelic* is used of organisms showing moderate rates of evolution. Lines that are evolving slowly or rapidly at one time may be horotelic at another period during their evolution.

chronospecies (evolutionary species) a species that is reproductively isolated (see

REPRODUCTIVE ISOLATION) from its nearest relatives because it occurs in a different period of geological time (see GEOLOGICAL TIME SCALE). The evolution of species over geological time may be regarded as a succession of chronospecies or as a single species that is continuously changing. Since the definition of a chronospecies is time-based, and there is no way of knowing if it could interbreed with its relatives, it is not equivalent to a biological species (see SPECIES).

Chroococcales an order of CYANOBACTERIA comprising unicellular and colonial coccoid •bacteria that reproduce by binary fission. The palmelloid colonial forms arise when daughter cells become bound together in a mucilaginous sheath after cell division. Heterocysts and hormogonia are lacking. Common genera include *Gloeocapsa* and *Chroococcus*. The genus *Entophysalis* is responsible for the building of STROMATOLITES.

chrysolaminarin (leucosin) a polysaccharide, made up of $\beta(1–3)$-linked polymers of glucose, that forms the main photosynthetic food reserve of the golden-brown algae (CHRYSOMONADA) and DIATOMS. It is often stored in membranous intracellular vesicles.

Chrysomonada (Chrysophyta, golden-brown algae) a phylum of the PROTOCTISTA (in some classifications a division of the Plantae) comprising predominantly unicellular algae that possess UNDULIPODIA, but also containing some colonial and complex filamentous forms. Some motile species form shells of silica, often beautifully sculpted. The plastids (*chrysoplasts*) contain golden yellow pigments, mainly FUCOXANTHIN, and also chlorophyll *c*. There are usually two golden-brown plastids in each cell, and food is stored as the polysaccharide CHRYSOLAMINARIN or as fat. At some stage in the life cycle the Chrysomonada are heterokont – they have two anteriorly attached undulipodia of different size, one of which is a mastigonemate (hairy) undulipodium. Asexual reproduction is often by heterokont zoospores called swarmers. Sexual reproduction has not been observed this phylum. Some

freshwater species produce statocysts, dormant overwintering cells encased in a silicified membrane inside the cell wall and containing granules of chrysolaminarin. During germination the amoeboid chrysomonad emerges through a pore at the top of the statocyst, which is often surrounded by a conical collar. Chrysomonads are common in fresh temperate waters; a few are marine. See CHRYSOPHYCEAE, RAPHIDOPHYCEAE, SYNUROPHYCEAE.

Chrysophyceae a class of golden-brown algae (see CHRYSOMONADA), in some classifications placed in the division Chromophyta. The class includes a wide range of forms and both motile and nonmotile cells. The cells are usually covered in scales or LORICAE. Dormant stages called statocysts, whose walls have siliceous linings, are also found. The main photosynthetic pigment, FUCOXANTHIN, gives the algae their golden-brown colour. The chloroplasts also contain chlorophylls c_1 and c_2.

Chrysophyta see CHRYSOMONADA.

chrysoplast a photosynthetic plastid characteristic of members of the phylum CHRYSOMONADA.

Chytridiales an order of fungi of the class Chytridiomycetes (see CHYTRIDIOMYCOTA) containing unicellular or coenocytic organisms distinguished by certain ultrastructural features. They are considered the most primitive members of the EUMYCOTA (subdivision Chytridiomycotina). In the Five Kingdoms classification system they are placed in the kingdom Protoctista (phylum Chytridiomycota, class Chytridia). Chytrids number about 566 species in some 77 genera and 5 families. They are usually aquatic, although some species, such as *Synchytrium endobioticum* (which causes potato wart disease), are found in damp soil. Most chytrids are saprobes or parasites of algae, plants, and other fungi. Some species produce a branching rhizoid-like mycelium (*rhizomycelium*) from the posterior end of the main thallus. Chytrids produce motile cells, each with a single large lipid globule and a single smooth posterior UNDULIPODIUM. These cells may behave asexually as ZOOSPORES or sexually as gametes, fusing in pairs.

Chytridiomycetes see CHYTRIDIOMYCOTA.

Chytridiomycota (chytrids) in some classification systems, a phylum of primitive fungi of the EUMYCOTA. There is a single class (Chytridiomycetes or Rumpomycetes) with some 793 species in about 112 genera and 18 families. Five orders are usually recognized: CHYTRIDIALES (unicellular), Blastocladiales (with branching mycelia and often complex life cycles, sexual reproduction by planogametes), Monoblepharidales (with specialized antheridia and oogonia), Neocallimastigales (anaerobic saprobes living in the guts of ruminants; zoospores without mitochondria), and Spizellomycetales (zoospores with distinctive ultrastructure and often more than one lipid globule). In form they range from single motile cells, through simple thalli with rhizoids, to septate hyphal forms. The walls contain chitin, the ribosomes are usually confined to a well-developed nuclear cap region, and the mitochondria are closely associated with the nucleus. The zoospores have one or more posterior smooth undulipodia, a single anterior lipid globule, and a *rumposome* (an organelle with a honeycomb-like surface situated close to the cell wall).

In the Five Kingdoms classification, the Chytridiomycota is regarded as a phylum of the kingdom PROTOCTISTA. This system recognizes four classes, the Chytridia, Blastocladia, Monoblepharida, and Harpochytridia, distinguished by the ultrastructure of their KINETOSOMES.

Chytridiomycotina in some classifications, a subdivision of the EUMYCOTA containing the class Chytridiomycetes (see also CHYTRIDIOMYCOTA). Members have chitinous walls, a structure ranging from a simple thallus with rhizoids to aseptate hyphae, zoospores with one or more posterior smooth undulipodia (whiplash flagella). Sexual reproduction is unknown in many species.

chytrid see CHYTRIDIOMYCOTA.

cicatrix in a cereal inflorescence, a scar left at the base of the lemma by a discarded floret.

These are commonly seen in, for example, oat (*Avena sativa*).

ciguatera poisoning caused by eating fish contaminated with *ciguatoxin* and other toxins from certain tropical dinomastigotes (dinoflagellates). See DINOMASTIGOTA.

ciguatoxin see CIGUATERA.

ciliary body the basal body of a cilium. See UNDULIPODIUM, KINETOSOME.

ciliate describing a structure, such as a leaf margin, fringed with fine hairs. *See illustration at leaf.*

cilium (*pl.* cilia) See UNDULIPODIUM.

cincinnus see INFLORESCENCE.

cingulum (girdle) in the test of a dinoflagellate, the plate of the equatorial groove. See DINOMASTIGOTA.

circadian rhythm (diurnal rhythm) an ENDOGENOUS RHYTHM in which physiological responses occur at 24-hourly intervals. For example, some plants exhibit a characteristic change in leaf position on a daily cycle. The opening and closing of stomata and flowers and the frequency of cell divisions also exhibit a circadian rhythm. Such rhythms will persist for several days if the plant is kept in continual darkness. It also appears that, in some species, the photoperiodic response is governed by circadian rhythms (see PHOTOPHILE, PHOTOPERIODISM).

circinate describing a type of individual leaf folding in which the leaf is rolled in on itself from the apex to the base, so that the apex is in the middle of the coil (*see illustration at ptyxis*). It is seen in the leaf primordia of most ferns (except the Ophioglossales) and in certain of the cycads and extinct seed ferns.

circumantarctic distribution the distribution of organisms around the South Pole. This usually includes species with circumpolar and circumaustral distributions, e.g. the water fern *Blechnum penna-marina alipna*.

circumarctic distribution the distribution of organisms around the North Pole. this usually includes species with circumpolar and circumboreal distributions, e.g. *Saxifraga oppositifolia*.

circumaustral distribution the distribution of organisms around the high latitudes of the southern hemisphere, adjacent to the

Antarctic zone, e.g. *Ranunculus acaulis*, which occurs from New Zealand to Chile, the Falkland Islands and Tasmania.

circumboreal distribution the distribution of organisms around the high latitudes of the northern hemisphere, adjacent to the Arctic zone, e.g. *Empetrum hermaphroditum*, which occurs from North America to Iceland, Europe, and Siberia. Many circumboreal species also extend into the Arctic (see CIRCUMARCTIC).

circumnutation (nutation) the circular or elliptical movement of the stem tips of many plants. Due to a shift in the region of most active cell division in the stem, the growing tip swings in a characteristic manner, either clockwise or anticlockwise depending on the species. If during these movements (which seem to be dictated both by internal stimuli and by gravity) contact is made with a solid object, the plant begins to twine around it. The importance of gravity to the response can be demonstrated by inverting a plant that has already entwined a support. The tip and the last two or three coils straighten out and become erect before once again coiling upwards around the support in the characteristic direction.

circumpolar distribution the distribution of organisms around the North and South Poles. The term should be used only for those with circumarctic or circumantarctic distributions.

cis **arrangement (coupling)** the situation in which an individual is heterozygous for two linked GENES (either two units of function or two mutations in the same functional unit), and one chromosome carries both the recessive alleles. The homologous chromosome thus carries both the dominant alleles and will mask the effects of the recessives in the phenotype. Compare *TRANS* arrangement. See also CIS–TRANS test.

cis-**octadec-9-enoic acid** see OLEIC ACID.

cisternae interconnecting intracellular compartments consisting of tubular or flattened channels and vesicles that extend throughout the cytoplasm. The ENDOPLASMIC RETICULUM and GOLGI APPARATUS are made up of membranes formed into cisternae.

cis–*trans* **test** a method for deciding whether two mutations having similar effects occur in the same or different genes (defined as units of function). If the mutations occur in the same gene, only heterozygotes with the *cis* arrangement will show the normal phenotype because the gene in the homologous chromosome is dominant wild-type and codes for normal protein. In the *trans* arrangement both copies of the gene contain a mutation and thus there is no wild-type copy of the gene and hence no normal protein. If the mutations occur in different genes, specifying different polypeptides, all heterozygotes will be normal in appearance because the homologous chromosomes carry wild-type copies of the genes.

cistron a gene defined as a unit of function, i.e. a length of DNA that codes for a single polypeptide. The end of one cistron and the beginning of the next can be determined by the CIS–TRANS test.

CITES see CONVENTION ON INTERNATIONAL TRADE IN ENDANGERED SPECIES OF WILD FAUNA AND FLORA.

citric acid (2-hydroxypropane-1,2,3-tricarboxylic acid) a six-carbon tricarboxylic acid with the formula $HOOCCH_2C(OH)(COOH)CH_2COOH$. Citric acid is the first intermediate in the KREBS CYCLE, being formed by condensation of acetyl CoA and oxaloacetate. The reaction is the rate-limiting step in the Krebs cycle. Citrate is itself an allosteric modulator (see ALLOSTERIC ENZYME) for several reactions, e.g. the formation of fructose 1,6-bisphosphate in glycolysis is inhibited by citrate. High concentrations of citrate are found in the fruits of some plants, notably lemons. See CARBOXYLIC ACID.

citric acid cycle see KREBS CYCLE.

cladistic describing genetic relationships due to recentness of common ancestry.

cladistics a method of classification in which the relationships between organisms are represented by a diagram, or CLADOGRAM, based on selected shared characteristics. It is basically a phylogenetic system since the selected characteristics are those that are assumed to have been derived from a common ancestor. However some practitioners of cladistics, the 'transformed cladists' believe that cladograms are simply summaries of patterns of shared characteristics that provide a logical basis for classifications but do not have to be linked with and explained by the evolutionary process.

Cladistics differs from traditional phylogenetic approaches in the methods used to determine evolutionary relationships and in its definition of a true natural group. Cladistics assumes the closeness of relationship depends on the recentness of common ancestry, which itself is decided by the number and distribution of shared 'derived' character states (*synapomorphies*) in the organisms in question. Synapomorphies are those character states that can be traced back to the same character states in a recent common ancestor. They do not include shared derived characters that have arisen by convergent evolution. 'Primitive' character states (*symplesiomorphies*), inherited from more remote common ancestors, are ignored as misleading because some may have been inherited by one descendant but not by another, closely related, descendant. The same character can be both derived and primitive, depending at what level in the taxonomic hierarchy one is working. Thus a derived character at the class level will probably be a primitive character at the species level. In constructing a cladogram, branches are based upon the simplest explanation (often called the most parsimonious) of the distribution of shared derived characters.

Cladistics also differs from some traditional phylogenetic methods in regarding the only true natural groups as those that contain *all* the descendants of a common ancestor (see MONOPHYLETIC, PARAPHYLETIC). Cladists also believe that those groups that are each other's closest relative – sister groups – should be given equal taxonomic rank.

cladode (phylloclade) a specialized stem structure resembling and performing the functions of a leaf. Examples are the stem joints or pads in *Opuntia* (prickly pears) and the 'leaves' of *Ruscus aculeatus* (butcher's broom). In these genera the true leaves are reduced to spines or scales respectively. The

formation of cladodes is an adaptation to arid conditions. A cladode may be distinguished from a leaf by the presence of buds on its surface. Compare PHYLLODE.

cladogenesis the branchings of a CLADOGRAM.

cladogram a type of branching diagram (*dendrogram*) representing the relationships between organisms as determined by cladistic methods. A cladogram may be regarded as showing the phylogenetic relationships of organisms, in which case it is effectively an evolutionary tree, or it may be seen as simply representing a pattern based on shared characteristics. In the first case, the points of branching are thought of as actual or hypothetical ancestral species, while in the second case (the position held by 'transformed' cladists), the points simply represent shared characteristics. See CLADISTICS.

Cladophorales an order of SIPHONOUS GREEN ALGAE of the class ULVOPHYCEAE (Bryopsidophyceae in older classifications). These are multicellular coenocytic algae with saclike or filamentous thalli, often with a holdfast made up of rhizoid-like branches. Asexual reproduction by FRAGMENTATION or diploid zoospores is common. Most species show ISOMORPHIC alternation of generations. Apical cells of the sporophyte release haploid zoospores, which give rise to the gametophyte generation. Isogametes produced by apical cells of the haploid thallus then fuse to form zygotes, which develop directly into new sporophytes. In some species the gametophyte generation is absent.

cladoptosis the shedding of all or part of a branch or shoot. This is usually due to the formation of an ABSCISSION ZONE. Alternatively, the branch may die down to its point of attachment to the axis, then rot or break off; the wound thus produced is then sealed by wound tissue or, as in *Eucalyptus* species, by gum. In some species, whole shoots or branches are shed in a distinct pattern as part of the plant's development, and the process makes an important contribution to the plant's overall architecture.

Cladoxylales see FILICINOPHYTA.

clamp connection a characteristic form of

growth seen in dividing dikaryotic hyphae of basidiomycete fungi (see illustration). Before cell division occurs the two nuclei divide simultaneously and at the same time a backward-pointing side branch develops from the leading cell. One of the daughter nuclei moves into this while another daughter nucleus of different genetic constitution moves down the main filament away from the hyphal tip. Two cross walls are then formed, one across the main filament, so separating off the other two nuclei in the hyphal tip, and the other across the side branch. The wall between the branch and the subapical cell then dissolves so the nucleus in the side branch can pass through to join the remaining daughter nucleus. This process has apparently developed to avoid the random movement of daughter nuclei, which could lead to the production of homokaryotic cells. See also CROZIER.

class a major category in the taxonomic hierarchy, comprising groups of similar orders. The Latin names of plant classes end in -ae (e.g. Monocotyledonae) or in -opsida (e.g. Gnetopsida). Classes may be subdivided into subclasses, with Latin names ending in -idae (e.g. Dilleniidae, Rosidae).

classical genetics (Mendelian genetics) the branch of genetics concerned with the study of inheritance through observation of the progeny obtained in breeding experiments. See MENDEL'S LAWS, MENDELISM.

classification the process of establishing and delimiting taxa within the hierarchy of classes. The resulting system of classification should either help in the rapid identification of organisms or should express their natural interrelationships (preferably both). Four kinds of classification can be recognized: artificial, natural, phenetic, and phylogenetic. *Artificial classifications* are based on a few predetermined characters usually selected to aid identification and consequently may not reflect the true relationships of the plants. The sexual system of Linnaeus based on the numbers of stamens and carpels is an example (see also KEY). *Natural*

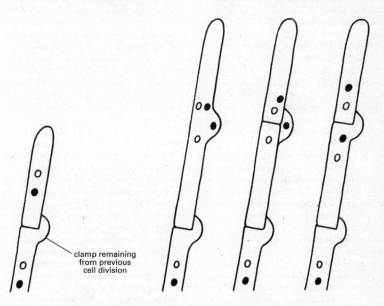

clamp remaining from previous cell division

Clamp connection: stages in the formation of a clamp connection

classifications are based on many characters and have a predictive value. Thus if one species of a particular genus is found to possess a certain character that was not used when the genus was originally defined, it is nevertheless quite likely that other species within the genus will also possess the character. *Phenetic classifications* are based on similarities between as many characters as possible and allow for characters to be weighted if desired. *Phylogenetic classifications* are also broadly based, but have the additional intention of reflecting the evolutionary history and relationships of a group of taxa. Most modern systems aim to be phylogenetic. Compare CLADISTICS.

Clavices purpurea see ERGOT.

Clavicipitales see HYPOCREALES.

claviform club-shaped.

clay mineral particles with a diameter below 0.002 mm. They consist of aluminium and silica arranged to form a layered crystal structure. See SOIL TEXTURE.

clear-cutting the felling of timber that involves the removal of all the trees from the site, resulting in increased runoff and leaching of nutrients, especially nitrates.

clear-felling the felling of all the trees in a forest.

clearing a stage in the preparation of permanent slides for microscopical examination in which the stained dehydrated cells or tissue sections are immersed in a clearing agent to remove the alcohol that was used to dehydrate the preparation. The process is essential if the dehydrating agent and the mounting or embedding medium are not miscible. Examples of clearing agents include xylene (xylol), toluene, and 1,1,1,-trichloroethane and various essential oils, e.g. clove oil and thyme oil. The term is sometimes used to mean the dissolving away of cell contents before staining so that the tissue distribution may be better observed. Sodium hypochlorite is commonly used for this.

cleavage polyembryony the splitting of an embryo into several identical parts each of which is capable of developing into a mature embryo. It is seen in some gymnosperms and is particularly common in *Pinus*, where the zygote divides to form four nuclei each of which divides to give a chain of cells or suspensor. At the tip of each suspensor is a group of embryo initials. Usually, however, only one embryo develops to maturity.

cleistocarp see CLEISTOTHECIUM.

cleistogamy (*adj.* cleistogamous) the production of flowers that do not open to expose the reproductive organs, so preventing cross pollination. Some species are obligately cleistogamous but others may be cleistogamous only under certain climatic conditions. For example, most *Viola* species produce cleistogamous flowers in summer, when temperatures are high, whereas earlier flowers are chasmogamous. In some grasses only the lower florets are cleistogamous. The evolutionary origin of this mechanism is not known. Compare CHASMOGAMY. See also AUTOGAMY.

cleistothecium (cleistocarp) (*pl.* cleistothecia) The type of ASCOMA characteristic of fungi in the PLECTOMYCETES. It is a globose structure with no specialized opening to the exterior.

climacteric the rise in respiration rate observed in certain fruits during ripening. It can be artificially induced by exposure to low concentrations of ethylene (ethene), a gas emitted by ripening fruit. Ripening can therefore be stimulated by placing unripe fruit alongside ripe fruit. Conditions that prevent the onset of the climacteric (low oxygen and high carbon dioxide concentrations) can be employed during storage to maintain the quality of stored fruit.

climatic climax see CLIMAX.

climatic factors those aspects of the weather, such as temperature, rainfall, light, humidity, and air movement, that influence the life of organisms. See also BIOTIC FACTORS, EDAPHIC FACTORS.

climatic optimum the period during which the highest prevailing temperature since the last Ice Age were recorded. In most parts of the world this was between 4000 and 8000 years ago. In Britain it is represented by the ATLANTIC PERIOD.

climax (climax community) the final stable community that results after a series of changes (succession) in the vegetation and animal life in a particular area. However, there are often factors that halt or slow down succession and prevent the establishment of a climax community. These factors include changes in climate, natural hazards, disease, animal interference, fire, and human activity. Often intensive grazing prevents the growth of shrubs and trees. The equilibrium achieved in this way is termed a *biotic climax, disclimax, plagioclimax,* or *subclimax vegetation.* However, all communities are dynamic in nature, so the concept has only limited value. See also CLISERE.

climax adaptation number a phytosociological hierarchy in which species are assigned a value from 1 to 10 according to their *importance values* (the sum of a species' relative abundance, relative density, and relative frequency, expressed as percentages). The importance value may range from 0 to 300.

climax community see CLIMAX.

climax theory the theory that a vegetation succession will eventually reach a state of equilibrium with the environment. There are two versions of the theory. *Monoclimax,* proposed by F. E. Clements in 1904 and 1916, predicts a specific equilibrium point determined by climate – a *climatic climax. Polyclimax,* proposed by A. G. Tansley in 1916 and 1920, predicts a range of equilibrium communities determined by edaphic conditions, fire, and other factors in addition to climate. Polyclimax is the most widely accepted theory, as environmental conditions are rarely stable for long periods of time.

climbing ferns ferns of the genus *Lygodium* (family Schizaeaceae),which climb up other plants by means of the long, thick flexible midrib (rachis) of each frond, which twines around other plants. Individual fronds may be up to 10 metres long.

climbing root any of the short ADVENTITIOUS roots that develop from the stems of certain climbing plants, e.g. ivy (*Hedera helix*), and serve to attach the plant to its support. Climbing roots are negatively phototropic and thus grow into darkened tissues in bark or crevices in walls. Their function is enhanced by the secretion of mucilaginous substances from the root tip.

climograph a two-dimensional graphical representation of the relationship between two climatic variables, e.g. temperature and rainfall. Usually the months of the year are placed along the horizontal axis of the graph, and the climatic variables on the two vertical axes. The values for each variable may be joined sequentially by lines. Because the trends in the two variables are superimposed on the same graph, it is easy to see the relationship between the two, and to compare the climates of different locations.

clinal speciation the type of ALLOPATRIC speciation that occurs when a geographic barrier disrupts a CLINE, separating members of a species into two distinct populations that can no longer interbreed. See also GEOGRAPHICAL ISOLATION.

clinandrium see ORCHIDACEAE.

cline a gradual change in a character over the range of a species. The term may also be applied to a change in the relative proportions of two or more distinct forms. For example, the leaves of lords and ladies (*Arum maculatum*) may or may not be spotted. The frequency of the spotted form increases towards the southern part of its range in Britain. See also ECOCLINE, ECOTYPE.

clinostat see KLINOSTAT.

clint see LIMESTONE PAVEMENT.

clisere a succession of different CLIMAX COMMUNITIES in a particular area brought about by changes in the climate.

clitochory See BARACHORY.

clonal dispersal in a modular organism, the growth or movement away from each other of the component modules. For example, whorls of leaves on a stem become separated as the internodes elongate. Clonal dispersal can be a form of vegetative reproduction, as in strawberry (*Fragaria ananassa*), where large numbers of plantlets grow at intervals from buds on runners, and become independent plants when the intervening section of runner rots away. These plants are genetically identical, i.e. they are clones of their parent plant. As the plant produces ever younger modules, some species can survive for many years and cover large areas, e.g. some clones of bracken (*Pteridium aquilinium*) are over 1400 years old; a single clone of trembling aspen (*Populus tremuloides*) has been known to cover 14 hectares.

clone 1 a population of genetically identical cells or individuals. Such a population is obtained by mitotic division (see MITOSIS) or asexual reproduction and is widespread in plants, e.g. reeds (*Phragmites* spp.) and strawberries (*Fragaria* spp.). The term *strain* is sometimes used synonymously. See also PURE LINE.

2 see GENE CLONING.

closed canopy see CANOPY.

closed population a POPULATION in which there are barriers to gene flow.

closed spindle a mitotic SPINDLE that is enclosed by the nuclear envelope during mitosis.

cloud forest a tropical montane FOREST, usually above 1000 m altitude, in regions of heavy rainfall that are shrouded in cloud for much of the day. The trees are often low-growing, but the moist atmosphere supports a lush growth of epiphytes, especially climbing ferns, filmy ferns, clubmosses (Lycophyta), mosses, and lichens. The forest clearings typically have very large ferns and herbs, such as begonias.

club fungi (coral fungi) basidiomycete fungi of the family Clavariaceae, order Cantharellales, with club-shaped basidiomata. Most are ground- or wood-dwelling saprobes; a few are associated with lichens. See BASIDIOMYCOTA.

club mosses (clubmosses) see LYCOPHYTA.

club root (finger-and-toe disease) a disease of the Brassicaceae (crucifers) caused by *Plasmodiophora brassicae* in which the roots become swollen and malformed. Above ground the symptoms, which may not appear for a while, are wilting and later stunting and yellowing of infected plants. Infection occurs via the root hairs and once the cambium is infected club roots develop rapidly. Spores can remain viable in the soil for several years making control difficult. Liming of the soil is a traditional control method. See also PLASMODIOPHORA.

cluster cup the cup-shaped AECIUM formed on the leaves of plants infected by certain rust fungi (see UREDINALES).

CoA see COENZYME A.

coacervate (protobiont) a complex of large organic compounds derived from colloidal hydrophilic complexes of protein molecules that became surrounded by envelopes of water molecules, and eventually by envelopes of lipid molecules. It is suggested in certain theories on the ORIGIN OF LIFE that such aggregates were the precursors of living organisms. The lipid molecules would have acted rather like a primitive cell membrane, conferring stability on the coacervate and determining which substances could enter or leave it. The internal composition of the coacervate would depend on the local surrounding medium. Metal ions could have been incorporated to form enzymes. Growth of the coacervate by intake of materials might result in fragmentation and the production of more identical coacervates. When polymers resembling RNA and DNA developed and became associated with coacervates, they could store and transfer information, a major step in the evolution of life. The result would be a primitive heterotrophic self-replicating organism feeding on the surrounding medium.

coadaptation the evolution and persistence of characteristics that enhance the relationship between two species. For example, the adaptation of flowers to attract their pollinators, and the adaptations of the pollinators to specialize in pollinating particular species of flowers (such as evolving long mouthparts or beaks). See COEVOLUTION, MUTUALISM.

coal a carbon-rich deposit derived from the fossilized remains of plants. Most of today's coal was formed during the CARBONIFEROUS PERIOD some 300 millions of years ago, when vast swamp forests flourished in tropical or subtropical climates. There are two main kinds of coal. *Woody (humic) coal* is derived from decomposed plant materials such as peat, which under anaerobic conditions and great pressure with the gradual loss of moisture and volatile components and concomitant increase in the proportion of carbon gradually becomes lignite, then subbituminous coal, semianthracite, and finally anthracite. *Sapropelic coal* is derived from more finely divided plant material, spores, and algae.

See DECOMPOSITION, FOSSIL FUEL, GREENHOUSE EFFECT.

coating the application of a thin film of heavy metal, commonly gold, to the surface of a specimen during preparation for scanning electron microscopy (see ELECTRON MICROSCOPE). Coating is particularly useful in preparation of biological specimens because it results in the production of many more secondary electrons and the improved conductivity helps prevent a charge building up in the specimen and causing image deterioration. Coating units are of several types; some use fine filaments as the source of coating metal while others, known as sputter coaters, have a large solid block of metal that is ionized by a beam of ions.

cobalt see MICRONUTRIENT.

coccoid spherical in shape. The term is usually applied to bacteria or algae.

coccolith see HAPTOMONADA.

coccolithophorids see HAPTOMONADA.

coccus (*pl.* cocci) any spherical or ellipsoidal bacterium. Cocci rarely possess flagella. Compare BACILLUS, SPIRILLUM, VIBRIO.

COD see CHEMICAL OXYGEN DEMAND.

Codiales see CAULERPALES.

codiolum stage a unicellular diploid photosynthetic stage in the life cycle of some green algae of the order ULOTRICHALES, formed by the enlargement of a zygote. This stage was formerly thought to be a separate genus, *Codiolum*. It later divides by meiosis to produce haploid ZOOSPORES that develop into new filaments.

codominance the phenomenon in which both alleles in a heterozygote are expressed in the phenotype. Thus in certain plants that possess a series of alleles governing incompatibility (see *s* alleles), an individual plant that has two different alleles both relatively high in the dominance series may be incompatible with any plant possessing either of these dominant alleles (i.e. both alleles are exhibited by the phenotype). In other cases the one allele may be completely dominant to the other and such a plant would be able to cross with another plant possessing the same recessive allele. This illustrates that codominance relationships do not necessarily exist for all

the alleles of a gene. Compare INCOMPLETE DOMINANCE.

codon a sequence of three nucleotides (TRIPLET) that codes for one amino acid in a polypeptide. A sequence of codons will thus determine the type and sequence of amino acids in a polypeptide. There are about 20 different kinds of amino acids in proteins. Most amino acids are coded for by more than one codon: the GENETIC CODE is thus said to be *degenerate*. The term codon is given both to the sequence of triplets on DNA and also to the sequence on messenger RNA. However these are not identical but complementary copies of each other.

coefficient of coincidence in genetics, the observed number of double recombinants (formed by double crossover events) divided by the expected number (see CROSSING OVER).

Coelomycetes a class of mycelial asexual fungi (see FUNGI ANAMORPHICI) in which the conidia are borne in various types of cavity lined by fungal or fungal and host hyphae. Various types of conidiomata are found, including pycnidia, acervuli (as in the MELANCONIALES), or pycnothyria (as in the PYCNOTHYRIALES). This nonphylogenetic class is placed in the Fungi Anamorphici (Mitosporic Fungi). It contains over 9000 species in some 700 genera, most of which are saprobes or parasites on plants, lichens, fungi, and vertebrates.

coeno- a prefix denoting more or many.

coenobium (*pl.* coenobia) a colony of cells that behaves as a single integrated unit, the number of component cells remaining constant. A coenobium often gives rise to internal daughter colonies. An example is the colonial green alga *Volvox*.

coenocarpium (*pl.* coenocarpia) a MULTIPLE FRUIT that incorporates the ovaries, floral parts, and receptacles of many flowers. It has a fleshy axis. The pineapple (*Ananas comosus*) is an example. Compare SOROSIS.

coenocline a gradient of communities, usually along a topographical or environmental gradient, that shows a sequence of changes in the relative importance of different species in populations.

coenocyte a multinucleate mass of protoplasm enclosed by a cell wall. See COENOCYTIC.

coenocytic describing a cell containing many nuclei or a thallus that lacks separating walls between nuclei. Such a structure is often threadlike. Examples are the hyphae of many fungi and the thalli of certain algae in the Xanthophyta and Caulerpales. See ACELLULAR.

Coenopteridales see FILICINOPHYTA.

coenosorus a continuous line of sporangia. It is seen in certain ferns, e.g. *Pteridium*, in which discrete sori are not produced.

coenozone SEE ASSEMBLAGE ZONE.

coenzyme an organic molecule that acts as an enzyme COFACTOR. Coenzymes usually function as intermediate carrier molecules, transferring or removing functional groups, atoms, or electrons. The coenzyme is only loosely bound to the enzyme and acts like a second substrate of the enzyme. Examples of important coenzymes are NAD and NADP, coenzyme A, and biotin. Compare PROSTHETIC GROUP.

coenzyme I see NAD.

coenzyme II see NADP.

coenzyme A a heat-stable coenzyme that acts as a carrier of acyl groups in many metabolic processes, such as fatty-acid and pyruvate oxidation. Coenzyme A is a sulphur-containing molecule formed from ATP, PANTOTHENIC ACID, and cysteine. The –SH group of coenzyme A can react with acetate to form ACETYL COA in the reversible reaction:

$$CoA–SH + CH_3COOH \rightarrow CoA–SCOCH_3$$

Acetyl CoA can then react with an acyl acceptor to form an acylated product and release free coenzyme A. An example is the reaction of acetyl CoA and oxaloacetate to form citrate.

coenzyme Q (ubiquinone) a lipid-soluble electron-carrying coenzyme that participates in the transport of electrons from NAD to oxygen in the RESPIRATORY CHAIN. It is a reversibly reducible quinone. PLASTOQUINONE, a closely related compound, performs a similar function but in photosynthetic electron transport.

coevolution the complementary evolution of two or more ecologically interdependent species, e.g. a flowering plant and its insect

pollinators. The evolution of an adaptation in one species, triggers the evolution of a complementary adaptation (see COADAPTATION) in the other(s). For example, the evolution of tube-like flowers favours the evolution of insects with long mouthparts to reach nectar at the base of the flower. The insect benefits from having less competition from other insects, and the plant benefits because the insect is more likely to visit similar flowers and thus achieve pollination. See also MUTUALISM, POLLINATION.

cofactor a nonprotein component of some enzymes that is necessary for catalytic activity. Cofactors may be either metal ions or COENZYMES. They are usually heat stable, unlike most enzyme proteins, which are denatured by heat. Many enzymes require metal ions as cofactors. Examples are ferredoxin, which requires ferrous, Fe(II), and ferric, Fe(III), ions, and alcohol dehydrogenase, which requires zinc ions. In some enzymes the metal ion is the primary catalytic centre, while in others it may act to bind enzyme and substrate together, or to stabilize the enzyme in its catalytically active conformation. See also PROSTHETIC GROUP, APOENZYME, HOLOENZYME.

cohesion theory the currently accepted theory explaining the ascent of sap in the xylem. It postulates that the reduction in water potential in the leaf (due to an increase in the rate of evaporation) causes water to flow through the plant to replace the lost water. The column of water in the xylem can withstand considerable tension before breaking because of the very strong cohesive forces between the water molecules. See also CAVITATION.

cohort 1 a group of individuals that began life at the same time in the same population, and are therefore of the same age. **2** in plant taxonomy, an obsolete term for a group of related families.

colchicine an alkaloid drug, extracted from the meadow saffron (*Colchicum autumnale*), that reacts reversibly with the protein TUBULIN and prevents its polymerization into microtubules. Consequently microtubule-assisted processes, such as the movement of organelles, are inhibited. Colchicine is often applied to dividing cells when making a chromosome preparation. It inhibits spindle formation and thus the condensed chromosomes are scattered rather than aligned on the spindle. This greatly facilitates their examination. If colchicine is applied to cells undergoing meiosis then the failure of the homologous chromosomes to separate results in the formation of diploid gametes. Colchicine is thus an important tool for the artificial induction of POLYPLOIDY and in other experimental studies, such as investigations of pollen development. It may also be used to render sterile hybrids fertile by doubling their chromosome number so the chromosomes have a homologue to pair with at meiosis. In humans it is carcinogenic.

cold acclimation see COLD TOLERANCE.
cold forest see FOREST.
cold hardening see COLD TOLERANCE.
cold tolerance (cold acclimation, cold hardening, frost tolerance, hardening) the development of the ability to tolerate very low temperatures, especially those below freezing, in advance of the cold temperatures actually occurring. The occurrence of low, but not damagingly low, temperatures before the onset of winter stimulates many plants to accumulate free amino acids (especially proline) or other organic compounds, such as oligosaccharides, in their cells. This helps to prevent damage to cell membranes if the temperature falls still further.

Coleochaetales an order of green algae (CHLOROPHYTA) of the class CHAROPHYCEAE that have HETEROTRICHOUS filaments; members are sometimes included in the Chaetophorales. The more advanced members of the order, such as *Coleochaete*, are often suggested to be the line of algae from which land plants evolved. Advanced features include the development of parenchymatous thalli with localized regions of growth, formation of a cell plate during division, oogamy, multicellular antheridia, retention of the zygote on the gametophyte thallus and its enclosure in a protective layer of cells, and the production of more than four spores from each zygote.

coleoptile the sheathlike structure around the EPICOTYL in the seeds of grasses. It

protects the erect plumule during its growth to the soil surface. In most grasses the coleoptile is very sensitive to light and has thus been used to investigate light-directed tropisms. Compare COLEORHIZA.

coleorhiza (*pl.* coleorhizae) the protective sheathlike structure around the radicle in the seeds of grasses. Compare COLEOPTILE.

coliform bacteria (faecal coliforms) gram-negative (see GRAM STAIN) rod-shaped bacteria that commonly live in the intestines of humans and other warm-blooded animals, e.g. *Escherichia coli*, *Salmonella*. They derive their energy aerobically or by anaerobic fermentation of sugars with the production of acid or gas. Some are present in soil and water or pathogenic in plants. Their presence in water indicates that the water is contaminated with faeces and may well contain other pathogens. The *coliform index* is a rating of water purity based on a count of the faecal bacteria present.

coliform index see COLIFORM BACTERIA.

colinearity the correspondence between the sequence of bases on DNA, excluding introns, and the sequence of amino acids in the protein it codes for.

coliphage any DNA bacteriophage that infects the bacterium *Escherichia coli*. Examples are the intensively studied T series of phages (T1 through to T7), which have been developed on a particular strain of *E. coli*.

collateral buds see ACCESSORY BUDS.

collateral bundle a VASCULAR BUNDLE in which the phloem occurs on only one side of the xylem. In the stem the xylem is usually internal to the phloem. This type of bundle is common in both angiosperms and gymnosperms. See also STELE. Compare BICOLLATERAL BUNDLE, CONCENTRIC BUNDLE.

collateral protostele see STELE.

collenchyma a tissue composed of relatively elongated cells with thickened nonlignified primary cell walls. Collenchyma is sometimes considered to be PARENCHYMA with exceptionally thick cellulose cell walls, specialized as a supporting and strengthening tissue in organs lacking (or almost lacking) secondary growth. Examples are young herbaceous stems and leaves. The thickened cell walls of collenchyma cells are sometimes used as a supply of cellulose for the other tissues in times of shortage. The wall thickenings may be mainly on the tangential walls (lamellar collenchyma), in the corners of the cells (angular collenchyma), or adjacent to the intercellular spaces (lacunar collenchyma). Compare SCLERENCHYMA.

colloid a substance that consists of particles dispersed throughout another substance that are too small to be viewed with the light microscope, but too large to pass through a selectively permeable membrane (size range 10^{-7}–10^{-3} mm in diameter). The particles remain suspended as a result of their small size and electrical charge. Because they have the same charge, they repel each other and cannot clump together into larger particles, which would settle out.

colony a type of growth form in which many cells of similar form and function are more or less loosely grouped together. Colonies of cells are characteristic of many genera of algae, e.g. *Volvox* and *Gonium*. The colony is often surrounded by mucilaginous substances. See also COENOBIUM.

colpus (*pl.* colpi) an oblong to elliptic germinal aperture in a pollen grain, at least twice as long as it is broad. Pollen with such apertures is termed *colpate*. Compare PORE.

columella a small column, especially: **1** the central portion of the root cap of certain roots in which the cells are arranged in longitudinal lines.
2 the sterile tissue in the centre of the sporangium in mosses, liverworts, and fungi.
3 a rodlike radial element in a pollen or spore wall.

columnar root see PROP ROOT.

Commelinidae a subclass of the monocotyledons containing mainly herbaceous plants, most of which have bisexual flowers with a more or less reduced perianth. It contains the following orders: Commelinales, which comprises four families, including the Commelinaceae (spiderwort family) and the Xyridaceae (yellow-eyed grasses); Eriocaulales; Restionales; Cyperales, which includes the families POACEAE (Gramineae, grass family) and CYPERACEAE (sedges); Juncales, which

includes the Juncaceae (rushes); Hydatellales; and Typhales, which includes the Sparganiaceae (bur-reeds) and Typhaceae (reedmace, bulrush, and cattails). Some taxonomists also include the Bromeliales, comprising the single family Bromeliaceae (bromeliads, pineapples); and the Zingiberales, which includes (among others) the families Musaceae (banana family) and Zingiberaceae (containing various aromatic plants, e.g. ginger, *Zingiber officinale*), but these two orders are more commonly placed in the subclass ZINGIBERIDAE, although the Zingiberales is sometimes placed in the LILIIDAE. The Typhales are sometimes placed in the ARECIDAE and the Poaceae in a separate order, Poales.

commensalism a relationship between two living organisms in which one of the participants, (the *commensal*) benefits but the other neither benefits nor loses. For example, an epiphyte, by germinating on the upper branches of a tree, benefits from the increased supply of light but the tree simply acts as a support and is not affected either beneficially or adversely. See also AMENSALISM, MUTUALISM.

community (biotic association) a group of plants and animals living together and interacting with one another in the same environmental conditions. Communities range in size from, for example, small woodland communities to large expanses of temperate deciduous forest or temperate grassland.

community biomass see BIOMASS.

community ecology the study of ecology with particular emphasis on the living (biotic) components of the ecosystem. It focuses on interactions within communities, such as partitioning of resources, and the changes in communities during succession, using techniques such as classsification and ordination. See also PHYTOSOCIOLOGY.

companion cells a specialized PARENCHYMA cell, characterized by a dense cytoplasm, numerous mitochondria, and conspicuous nucleus, associated with an adjacent SIEVE TUBE ELEMENT in the phloem of angiosperms. Companion cells are analogous to the ALBUMINOUS CELLS of gymnosperms. A sieve tube element and its associated companion cell arise from a single division of the same mother cell. The protoplast of the companion cell is connected with the enucleate protoplast of the sieve tube element by means of PLASMODESMATA and is believed to regulate the rate of flow along the SIEVE TUBE.

comparative biochemistry the recognition of discontinuities in biochemical variation and their subsequent use in the construction of classifications or correlation with existing classifications. See also CHEMOTAXONOMY.

comparative morphology the study of differences in morphology, their use in constructing classification systems, and their significance in phylogeny.

compass plant a plant whose leaf edges are permanently aligned due north and south, e.g. the compass plant of the prairies, *Silphium laciniatum*. This alignment allows the plant to take maximum advantage of the morning and evening sunlight coming from the east and west respectively, but prevents the powerful midday sun from shining directly onto its leaf blades.

compensation depth the depth of a body of water at which the rate of carbon fixation by photosynthesis equals the rate of carbon loss through respiration.

compensation point the lowest steady carbon dioxide concentration achievable in a closed system containing a photosynthesizing plant. When this minimum level is reached, the uptake of carbon dioxide during photosynthesis is exactly balanced by its respiratory release, indicating that the rate of synthesis of organic material is equal to the rate of breakdown by respiration. Low compensation points are indicative of photosynthetic efficiency as the plant is then using the maximum amount of available carbon dioxide (see C_3 PLANT, C_4 PLANT). Following a period of darkness a plant will take a certain amount of time, termed the *compensation period*, to reach its compensation point. Shade plants often have shorter compensation periods than sun plants as they can make better use of dim light.

The point at which photosynthesis no longer increases with increasing light intensity is called the *light compensation point*. C_4 plants have a much higher light saturation point than C_3 plants.

competence a characteristic exhibited by embryonic cells whereby they can go on to develop into any one of several different types of cell. Competence is therefore a feature that they possess before their ultimate cellular type is determined. See also TOTIPOTENCY.

competition the situation that arises when two or more organisms of the same or different species need the same limited resources. Organisms compete with one another for light, water, nutrients, etc. Competition may be resolved in various ways, namely: certain organisms or species being eliminated completely; migration of the less successful competitors; use of the resources at different times; physiological changes occurring; or the organisms involved living together, but not as successfully as they would at lower densities (i.e. the reproductive rate falls off). When members of different species compete for most or all of the same resources, one species may flourish and the other may decline. This is known as the *competitive exclusion principle* or *Gause principle*. See also THIESSEN POLYGONS.

competitive exclusion principle see COMPETITION.

competitive inhibition a form of enzyme inhibition in which the inhibitor competes with the substrate for the enzyme active site. The inhibitor can bind to the active site but does not react to form products. An example is the inhibition of the enzyme succinate dehydrogenase by certain compounds resembling succinate, e.g. malonate and oxalate. Studies of competitive inhibitors have provided valuable information about the way in which substrates are bound to the active site. Compare NONCOMPETITIVE INHIBITION.

complanate in mosses, describing shoots that are flattened dorsiventrally so that the leaves lie in a single plane, as in *Neckera* (flat feather moss).

complementarity the phenomenon in which mutant ALLELES at different gene loci work together to produce a WILD-TYPE phenotype. Two forms are recognized: *dominant complementarity*, in which the dominant alleles of two or more genes are required for the given phenotype; and *recessive complementarity*, in which the trait is displayed only if the recessive alleles of both genes are present. See also COMPLEMENTARY GENES, COMPLEMENTATION.

complementary DNA (cDNA) in genetic engineering, a form of DNA synthesized from a messenger RNA (mRNA) template by the use of a REVERSE TRANSCRIPTASE enzyme. Unlike the original DNA, cDNA lacks intron and promoter (see OPERON) sequences. It is used in cloning to obtain gene sequences from mRNA taken from the tissue to be cloned. Labelled single-stranded cDNA is also used as a GENE PROBE when searching for gene sequences common to different tissues or species. See also GENE CLONING.

complementary genes (reciprocal genes) nonallelic genes that cooperate in the expression of a single characteristic. The genes may be linked or unlinked. In some cases a standard Mendelian ratio is maintained, but there may be unexpected phenotypes. Thus in the inheritance of flower colour in sweet peas, purple-flowered offspring may be obtained when two white-flowered plants are crossed. This is because flower colour is affected by two genes and if either is a double recessive then no colour develops. In other cases, no unusual phenotypes appear but there may be some apparent modification of the Mendelian ratio in the F_2, such as 9:3:4 (= 9:3:[3+1]), or 9:7 (= 9:[3+3+1]).

complementation the phenomenon whereby two recessive mutant strains can supply each other's deficiencies so that growth of both can occur. Thus adenine-requiring (ad^-) strains or histidine-requiring (his^-) strains of the fungus *Neurospora* will grow in the presence of the other in minimal media (lacking both adenine and histidine) but not separately.

complementation test see CIS–TRANS effect.

complete dominance the expression of the dominant allele in a heterozygote, so that the phenotype is exclusively that corresponding to the dominant allele. Compare CODOMINANCE, INCOMPLETE DOMINANCE.

complex gene see SUPER GENE.

Compositae see ASTERACEAE.

composite fruit see MULTIPLE FRUIT.

compost rotted plant material, made from garden rubbish and kitchen vegetable waste and used as an organic fertilizer. The decomposition of compost can be hastened by adding a compost activator, such as nitrochalk.

compound a chemical combination of atoms of different elements to form a substance in which the atoms are combined in a fixed ratio specific to that substance. For example H_2O has two hydrogen atoms to each oxygen atom present, and no other elements are present. The constituents can be separated only by a chemical reaction, not by any physical means.

compression wood see REACTION WOOD.

concentric bundle (centric bundle) a VASCULAR BUNDLE in which one vascular tissue surrounds the other so they appear to be arranged concentrically when viewed in transverse section. Compare BICOLLATERAL BUNDLE, COLLATERAL BUNDLE, MERISTELE. See also AMPHICRIBAL, AMPHIVASAL.

conceptacle a flask-shaped cavity in the thallus of some brown algae (Phaeophyta), e.g. *Fucus*, in which gametangia are formed. A female conceptacle is lined with unbranched sterile hairs or paraphyses, and the oogonia develop from short stalks arising from the chamber wall. The male conceptacle is filled with branched hairs that bear antheridia. The conceptacle opens via an ostiole.

conchocelis stage in certain red algae (RHODOPHYTA) of the class BANGIOPHYCIDAE, a diploid (or sometimes haploid) phase of the life cycle in which the thallus consists of a microscopic branching filament, produced by the carpospore, which bores into the shells of molluscs and crustaceans. These filaments may reproduce asexually by spores or they may produce a different kind of spore, a *conchospore*, that germinates into the larger haploid thallus. Until its connection with the Bangiophycidae was recognized, this filamentous phase was thought to be a distinct genus (*Conchocelis*) of the class Florideophycidae.

concrescence a joining of two or more distinct organs by the growth of tissue between them.

condensation a chemical reaction in which two molecules join together with the elimination of a simpler molecule or group. The formation of GLYCOSIDIC BONDS between glucose molecules to form starch is an example, with water being eliminated. Compare HYDROLYSIS.

conditioned media see NURSE TISSUE.

conducting tissue 1 see VASCULAR SYSTEM. 2 see TRANSMITTING TISSUE.

conduplicate 1 describing a kind of individual leaf folding in which the leaf is folded towards the adaxial surface. *See illustration at* ptyxis.
2 describing an arrangement of leaves in a bud in which each leaf is folded in a U-shape around the next youngest leaf. *See illustration at* vernation.

cone see STROBILUS.

cone scale the part of a cone (see STROBILUS) that bears an ovule or a microspore.

confocal microscopy a form of scanning optical microscopy that involves the use of two microscopes. The first microscope focuses an illuminated aperture onto the specimen, while the second microscope, on the opposite side of the specimen, also focuses on the specimen but receives the light transmitted through it. This process produces images of a single plane in the specimen, an 'optical section'. It can be repeated in different planes to build up a three-dimensional view of the specimen.

conformation the spatial arrangement of atoms in molecules. The conformation of a molecule determines its shape and the distribution of chemical groups and electrostatic charges on its surface, and hence its reactivity. In complex proteins, such as enzymes, small perturbations in conformation may be sufficient to prevent reactions taking place (see ALLOSTERIC ENZYME). Various types of bonds (covalent, hydrogen, ionic, and disulphide bridges) affect the conformation of proteins. Different levels of complexity of protein structure are recognized (see PRIMARY STRUCTURE, SECONDARY STRUCTURE, TERTIARY STRUCTURE, QUATERNARY STRUCTURE).

Congo red a scarlet benzidine dye often used to stain fungal hyphae.

conidioma (*pl.*conidiomata) a conidia-bearing structure made up of several specialized hyphae. See CONIDIUM.

conidiophore a hypha bearing conidia-producing cells. See CONIDIUM.

conidiospore see CONIDIUM.

conidium (conidiospore) (*pl.* conidia) An asexual spore produced by many fungi, especially ascomycetes and Fungi Anamorphici, on specialized erect hyphae termed *conidiophores*. The conidia are usually formed in chains, which radiate from the globose or branched tip of the conidiophore. See also PHIALOSPORE.

Coniferales (Pinales) an order of the phylum CONIFEROPHYTA or, in some classifications, of the division Pinophyta (subdivision Pinicae) or of the subdivision Coniferopsida. This is the largest and most widely distributed order of gymnosperms, containing about 550 species in about 50 genera, mostly evergreen trees. It is sometimes extended to include the TAXALES. Six families are recognized: the Pinaceae (pines); Taxodiaceae (swamp cypresses), in some classifications extended to include the Cupressaceae (cypresses and junipers); Podocarpaceae (podocarps); Cephalotaxaceae (cow's-tail pines); and Araucariaceae (including the Paraná pine and monkey puzzle tree). The Cupressaceae and Taxodiaceae are sometimes included as subfamilies in the Pinaceae.

These conifers are particularly abundant in the high latitudes of the northern hemisphere, where they form the climax vegetation. Typically they show a pyramidal growth pattern and bear simple leaves, often needles or scales. Most conifers are monoecious but bear the male and female reproductive organs in separate compact cones (see STROBILUS) on different parts of the tree.

Trees of this order are commercially important as a source of timber for the paper-making, building, and furniture industries. They develop a less dense wood than angiosperm trees and are thus commonly called softwoods. In cross section distinct pores are normally absent in the wood (consequently termed nonporous wood), in contrast to the RING-POROUS or DIFFUSE-POROUS WOOD of angiosperms.

Coniferophyta (conifers) a phylum of cone-bearing GYMNOSPERMS. There are about 550 species in some 50 genera and 9 families. Most conifers are trees, but a few are shrubs or prostrate plants. The leaves are simple, usually needle-shaped or scalelike, often with a thick waxy cuticle. Resin canals run through the leaves and are especially well developed in species of arid habitats. Many species are evergreen but some, such as the larches, are deciduous. Most conifers are monoecious, bearing separate male and female cones (see STROBILUS) on the same plant. Male and female cones are quite distinct: the male cones are smaller, and bear small sporangia (MICROSPORANGIA), while the female cones are larger, and bear MEGASPORANGIA. The conifers show a reduced form of ALTERNATION OF GENERATIONS. The main plant is the diploid sporophyte, and the gametophyte generation does not lead an independent existence. The microsporangia produce microspores in the form of pollen grains, which in some species have air floats to aid wind dispersal. Each pollen grain contains vegetative cells equivalent to a minute male gametophyte. The megasporangium contains a megaspore mother cell, which divides to produce a multicellular female gametophyte, which produces an egg. The megasporangium is enclosed by an ovule and opens to the outside through a micropyle. Pollen grains that land on the micropyle are trapped by a sticky fluid and grow pollen tubes to transport the male nuclei to the egg. The zygote develops into a seed, which remains embedded in the megasporangium on the underside of an ovuliferous scale until dispersal. Most conifer seeds are wind-dispersed, but some are released from the parent cone only by fire.

Conifers are more advanced than the cycads and ginkgo in that they produce pollen grains; thus antherozoids are replaced by the two sperm nuclei, which are delivered to the egg by the pollen tube. It is thought that this feature may have contributed to the success of the conifers as compared with other gymnosperms. *See also* Coniferales, Taxales.

Coniferopsida in some classification systems, a subdivison of the Gymnospermae (see GYMNOSPERMS) made up of the four orders: Cordaitales (now extinct), Ginkgoales (now the phylum Ginkgophyta), Coniferales, and Taxales.

coniocarp a lichen with a MAZAEDIUM fruiting body.

Conjugales (Conjugatophyceae) in old classifications, an order (or class) of the CHLOROPHYTA the members of which undergo the characteristic reproductive process of CONJUGATION. They were later renamed Zygnematales (Zygnemaphyceae), and in the Five Kingdoms system they are classified as GAMOPHYTA.

conjugated protein a protein that contains other components besides polypeptide chains. Conjugated proteins are classified according to the nature of the nonpeptide portion of the molecule. Glycoproteins are found in plant and bacterial cell walls, while cell and organelle membranes contain lipoproteins. Many enzymes contain metal ions and some dehydrogenases are FLAVOPROTEINS. Ribosomes and viruses are nucleoproteins.

conjugating green algae see GAMOPHYTA.

conjugation 1 in eukaryotes, such as conjugating green algae (GAMOPHYTA) and certain fungi (e.g. *Rhizopus*), a form of sexual reproduction in which isogametes (see ISOGAMY) fuse and the contents of one or both cells migrate to fuse with the contents of the other to form a zygote (see diagram). **2** in PROKARYOTES, a process by which genetic information is transferred from one bacterium to another across a bridgelike structure.

conjunctive symbiosis see SYMBIOSIS.

connate describing similar organs that are joined together, such as petals fused to form a tube. The term is often used of paired leaves at a node, the bases of which have become fused to completely encircle the stem. Compare ADNATE.

connecting fibre in cells with undulipodia, a protein fibre that connects the

KINETOSOMES; it forms part of the cytoskeleton.

connective the parenchymatous tissue that joins the pollen sacs in the ANTHER. It is a continuation of the filament and contains the vascular strand supplying each lobe.

connivent converging.

conservation (biological conservation) The active management of wild animals, plants, and other organisms in such a way as to ensure their survival as a natural resource for future generations, i.e. to preserve habitat and species diversity (see BIODIVERSITY) and the genetic diversity within species. This may take the form of preserving specific habitats or it may involve special programmes to ensure the survival of particular species.

Conservationists may choose to allow the natural succession of plant communities to take its course (noninterventionist) or they may take measures to arrest the succession at a particular stage (interventionist). Examples of the latter are the cutting of reedbeds to ensure some open water remains for floating plants, fish, and water birds and the culling of such animals as deer to allow regeneration of vegetation. Since the genetic diversity of living organisms depends on the survival of their habitats, conservation also includes the management of BIOGEOCHEMICAL CYCLES to conserve the abiotic resources. Modern conservation recognizes that people, too, are part of natural ecosystems, and often a compromise must be reached to allow long-term sustainable use of natural resources/sustainable yield from

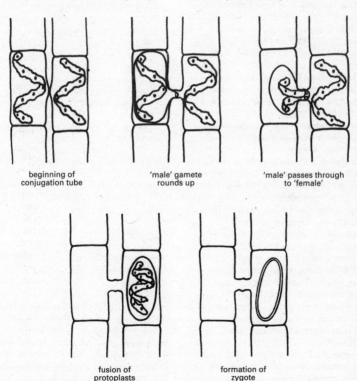

beginning of
conjugation tube

'male' gamete
rounds up

'male' passes through
to 'female'

fusion of
protoplasts

formation of
zygote

Conjugation: stages in *Spirogyra*

ecosystems. In most countries, conservation of certain areas and species is enshrined in law. See also BIODIVERSITY, BIOSPHERE RESERVE, CONVENTION ON INTERNATIONAL TRADE IN ENDANGERED SPECIES OF WILD FAUNA AND FLORA, CONVENTION ON WETLANDS OF INTERNATIONAL IMPORTANCE, NATIONAL PARK, NATURE RESERVE, PLANT CONSERVATION SUBCOMMITTEE, WORLD CONSERVATION UNION, WORLD CONSERVATION MONITORING CENTRE, WORLDWIDE FUND FOR NATURE.

consistency (consistence) the absence of bias in experimental procedures or data analysis.

consociation a comparatively small climax community named after one dominant species, e.g. oakwood, beechwood, etc. Consociations can be grouped together to form an ASSOCIATION. For example, in England the deciduous-forest association consists of consociations, i.e. smaller woods, of oak, ash, and beech.

conspecific describing individuals that belong to the same species.

constant species (constant) a species that is found in a particular community or association but is also found outside it, as compared to a faithful species (see FIDELITY), which is almost exclusively confined to a particular community or association. See PHYTOSOCIOLOGY.

constellation diagram a representation of species affinities that uses the chi-squared (χ^2) values as a measure of the degree of association between the species. The reciprocal of the χ^2 value for each pair of species is plotted on the diagram. Species that are highly associated have large χ^2 values, and are positioned closely together. See ORDINATION.

constitutive enzyme an enzyme that is always present in nearly constant amounts in a given cell. A constitutive enzyme is part of the permanent and basic machinery of a cell. The enzymes of GLYCOLYSIS are examples of constitutive enzymes in nearly all organisms. Compare INDUCED ENZYME.

consumer an organism that feeds on other living organisms, i.e. animals and parasitic and insectivorous plants. In a FOOD CHAIN, herbivores that eat green plants are *primary consumers*, and carnivores that eat

herbivores are *secondary consumers*. Compare DECOMPOSER, PRODUCER.

contact describing an insecticide or herbicide that kills an insect or plant on contact rather than relying on ingestion or absorption. Examples of contact insecticides are DDT and pyrethrin. Contact herbicides include paraquat and diquat. Compare STOMACH INSECTICIDE, SYSTEMIC.

contagious distribution see DISPERSION.

continental drift the theory that the existing continents moved to their present position following the break up of an ancient landmass. The idea was first proposed by A. Wegener in 1912 but was discarded by the geophysicists of the time. However evidence from biogeographic and oceanographic studies and the emergence of plate tectonics theory has established its validity. It is thought that the original landmass, *Pangaea*, broke up in the Mesozoic into *Gondwana* (or Gondwanaland) and *Laurasia*. The subsequent break-up of Gondwana resulted in the formation of the landmasses of the southern hemisphere, i.e. Africa, India, South America, Australia, Antarctica, and New Zealand, while the break-up of Laurasia gave rise to North America and Eurasia. The theory helps to explain the present disjunct distributions of certain plant and animal groups. See also VICARIANCE.

contingency table a table of data showing the values for two or more variables, which can be used for statistical analysis of association between these properties using a CHI-SQUARED TEST. Contingency tables are used to compare different methods of classification of species or individuals using values for various attributes used in the classification.

continuous distribution the pattern of values of a variable that shows a continuous spectrum of values, e.g. the heights of different individuals in a species.

continuous variation (quantitative variation) the type of variation shown by a characteristic that is exhibited to a greater or lesser extent by all members of a population or species. Often the characteristic has an average value, about which individuals in the population are distributed. An example would be the size of ears from a cereal crop. Continuous variation is common in characteristics that are strongly affected by environmental influences, and in characteristics determined by many genes, each with a small effect. Compare DISCONTINUOUS VARIATION.

continuum a gradual change in the species composition of overlapping populations in a COMMUNITY that spans an environmental gradient such as rainfall or altitude. The concept was originally developed at a site in southern Wisconsin, where a moisture gradient from dry to moist sites resulted in a gradual change in dominant woodland species from oak and aspen to basswood, ironwood, and sugar maple. The change in composition can be quantified by assigning each population an abitrary value (a *continuum index*) according to how closely their species composition resembles that of either end of the continuum. A continuum may be regarded as a seral succession, populations with the highest values corresponding most closely to the mature CLIMAX community. Alternatively, populations with lower values may be considered to be local climax communities adapted to particular edaphic or topographic conditions, rather than individual populations that form part of a larger community.

contractile root a specialized thickened root that serves to pull down a corm, bulb, rhizome, etc. to an appropriate level in the soil. Once the tips of the contractile roots are firmly anchored in the soil the upper region contracts, owing to changes in the shape of the cortical cells. Prominent contractile roots are seen, for example, in *Crocus* and *Allium* species.

contractile vacuole a cavity in the cytoplasm of many freshwater Protoctista that enlarges as it fills with water and contracts as water is expelled from it. The external medium is hypotonic with respect to the algal cell contents. To help prevent osmotic lysis, energy from ATP is used to pump water into the contractile vacuoles, the pulsations being visible with a light microscope. Water is released through the cell membrane or,

in the case of *Euglena*, into the gullet. Such vacuoles are particularly important in those organisms that do not have a rigid cell wall to help limit water intake.

Convention on Biological Diversity an international environmental convention signed at the *Rio Summit* organized by the United Nations in 1992, which came into effect at the end of 1993. Its aim is the conservation of BIODIVERSITY, the sustainable use of its components, and the fair and equitable sharing of the benefits arising out of the utilization of genetic resources. It is concerned with all aspects of biological diversity, from genetic resources and species to ecosystems, and recognizes that the conservation of biological diversity is an integral part of the development process. Its focus is on sustainable development rather than simply conservation of the natural world. It details the obligations and rights of the consenting parties, with respect to scientific, technical, and technological cooperation, and makes provision for a financial mechanism and an established body to give scientific, technical, and technological advice. A major feature of the convention is that it takes account of the need to share costs and benefits between developed and developing countries. In response to the Convention, the United Nations Environment Programme (UNEP) commissioned the *Global Biodiversity Assessment*, with contributions from over 80 countries, to put together and review a scientific analysis of all current issues, theories, and views regarding biodiversity from a global perspective. This has revealed the rapid rate at which the environment is being changed as a result of human activity, with the loss of species and genetic variability and the modification of natural ecosystems. Fauna and Flora

Convention on International Trade in Endangered Species of Wild Fauna and Flora (CITES) an agreement signed by some 80 countries in 1973, prohibiting commercial trade in ENDANGERED SPECIES of wild animals and plants and in products derived from them. 375 species were initially covered by this convention, and a further 239 species could be traded only on possession of a permit granted by both the importing and the exporting countries. More species have been added to the list since.

Convention on Wetlands of International Importance (Ramsar Convention) an international convention designed to coordinate international action for the conservation of WETLANDS of international importance. The Convention acknowledges the interdependence of humans and their environment and the roles of wetlands as regulators of the hydrological regime and as habitats for wild fauna and flora, as well as their economic, cultural, scientific, and recreational value. It aims to restrain the increasing loss of wetlands to drainage, cultivation, and human settlement. States that are members of the United Nations are entitled to sign the Convention; signatures may or may not be followed by ratification with the United Nations Educational, Scientific and Cultural Organization (UNESCO). The original Convention was established at a meeting in Ramsar, Iran, in February 1971 and has since been amended in 1982 and 1987. As a result of the Convention, a List of Wetlands of International Importance, commonly referred to as *Ramsar sites*, was drawn up and is continually updated. The IUCN keeps a register of the nature and management information relating to these sites.

convergent evolution (convergence) the development of ANALOGOUS structures in unrelated groups of organisms as a result of living in similar habitats or sharing other features, e.g. similar pollinators. An example is the development of the insectivorous habit in the families of pitcher plants Sarraceniaceae and Nepenthaceae. See also PARALLEL EVOLUTION.

convolute 1 describing an arrangement of leaves or scale leaves in a bud in which each leaf overlaps another on one side and is itself overlapped on the other side. *See illustration at* vernation.

2 see SUPERVOLUTE.

3 see AESTIVATION.

coordinated growth see SYMPLAST.

copper symbol: Cu. A metal element, atomic number 29, atomic weight 63.55, essential in trace amounts for growth. It is found in the enzyme cytochrome *c* oxidase, which is thought to catalyse the transfer of electrons from cytochrome a_3 to oxygen in respiration. In this reaction the two copper atoms of the molecule undergo transition from the cupric, Cu(II), to the cuprous, Cu(I) state. Copper also plays a role in photosynthetic electron transport, being found in the protein PLASTOCYANIN.

Copper deficiency leads to chlorosis, which suggests the element may also play a role in the synthesis of chlorophyll.

copper fungicide an inorganic FUNGICIDE containing copper. Some of the earliest known fungicides were copper compounds including the BORDEAUX and BURGUNDY MIXTURES. Other copper-containing fungicides are often based on copper oxychloride or copper oxide and may be formulated with other types of fungicide.

coppicing a form of woodland management in which trees are cut back to ground level regularly (every 10–15 years) to encourage growth of numerous adventitious shoots (*coppice shoots*) from the base. The resulting thicket is termed a copse or coppice. The young shoots are harvested for firewood, charcoal burning, fencing, etc. Many British woodlands have been managed in this way. Often selected trees may be left to mature, giving a coppice-with-standards.

coprophilous describing organisms that live on or in dung as, for example, fungi of the Pilobolaceae.

coral bleaching the loss of symbiotic ZOOXANTHELLAE from corals, which results in the corals losing their colour.

coral fungi see CLUB FUNGI.

coralloid roots in cycads, specialized secondary roots growing at or above the soil surface, which have a layer of greenish tissue containing blue-green bacteria just below their surface. The bacteria, usually species of *Anabaena*, which are usually filamentous, assume a coccoid shape inside the roots. The cycads probably benefit from the nitrogen-fixing ability of the bacteria.

Cordaitales an extinct order of gymnosperms, thought to have originated some time before the Carboniferous period and to have persisted into the Triassic. They were arborescent and in some of their vegetative structures resembled certain modern conifers, notably the monkey-puzzle (*Araucaria araucana*).

cordate heart shaped, such as the leaves of the sweet violet (*Viola odorata*) or the lemmas of quaking grasses (*Briza*). See *illustration at* leaf.

coremium a sheaf-like tuft or bunch of hyphae or conidiophores.

co-repressor a metabolite that becomes conjugated to a REPRESSOR molecule and thus binds to the operator gene in an OPERON to prevent the synthesis of an enzyme.

coriaceous having a leathery texture.

cork see PHELLEM.

cork cambium see PHELLOGEN.

corm a short swollen underground stem that serves as an organ of perennation and of vegetative propagation in such genera as *Crocus* and *Gladiolus*. The foliage leaves and flowers form from one or more axillary buds and grow at the expense of the food reserves in the corm. At the end of the season the corm is reduced to a withered mass and one or more new corms form above it at the base of each flowering stem. The protective brown scales surrounding the corm are the remains of the previous season's leaf bases. Compare BULB, RHIZOME. See also PERENNATING ORGAN.

corolla the collective term for the petals, constituting the inner whorl of the PERIANTH. A *corolla tube* is formed when the margins of the individual petals are completely or partially fused. Compare CALYX.

corona 1 a crownlike outgrowth from a corolla tube. It is especially prominent in daffodils (*Narcissus*).
2 a crownlike structure at the tip of the oogonium in certain chlorophytes of the class Ulvophyceae (e.g. *Acetabularia*) and Charophyceae (e.g. *Chara*).

corpus 1 see TUNICA–CORPUS THEORY.
2 the main body of a bladdered pollen grain, as in many conifers, e.g. *Pinus*.

correlative inhibition the suppression of the growth of certain plant parts by a compound, such as a food substance or a hormone, produced in another area of the plant. An example is the production of a substance by the root system that suppresses the development of roots on the stem. If the influence of the root is removed, a detached petiole or stem may subsequently produce a profusion of adventitious roots.

cortex 1 the tissue (including the endodermis) between the epidermis and the STELE in the stem or root. Although consisting mainly of ground parenchyma, the cortex often contains some collenchyma and sclerenchyma cells. Sometimes part of the cortex is photosynthetic chlorenchyma, especially in xerophytes. 2 in fungi, a fairly thick outer covering.

cortication in certain algae, such as *Chara* (stonewort, Charophyta) and the red alga *Batrachospermum* (frogspawn alga, Rhodophyta), the growth of short specialized branches from the node of a filament to form a layer over the adjacent internodal cells.

corticolous growing on the bark of trees and shrubs.

corymb an inflorescence in which the flowers are formed on lateral stalks of different lengths, the longest at the base, resulting in a flat-topped cluster of flowers. This form of floral arrangement, found frequently in the Brassicaceae, may aid insect pollination by giving easy access to individual flowers and in providing a flat landing platform, as in candytufts (*Iberis*). *See illustration at* inflorescence.

coryneform bacteria see ACTINOBACTERIA.

cosmid a type of PLASMID used in cloning DNA. A cosmid is able to incorporate a fragment of DNA up to 45 kilobases in length. This enables large fragments of eukaryotic DNA to be propagated as plasmids (cosmids) in the cells of such bacteria as *Escherichia coli*. Cosmids are also used in chromosome analysis. The order along a chromosome of fragments of DNA produced by restriction endonuclease enzymes can be determined by creating cosmids with a complementary base code linked to a fluorescent dye, and hybridizing them with the unbroken chromosomal DNA. This will reveal the position of the template DNA on the chromosome.

cosmopolitan describing species that have a worldwide distribution and are not

restricted to specific areas. Compare
ENDEMIC.

costa 1 a longitudinal rib, as seen in many
cacti.

2 see NERVE.

3 see COSTAPALMATE.

costapalmate describing a leaf shape
intermediate between palmate and pinnate;
palmately arranged leaflets are borne on a
very short rachis (*costa*). The fishtail palms
(*Caryota*) provide an example.

cothallus see THALLUS.

cotyledon (seed leaf) the first leaf or leaves
of the embryo in seed plants. In
angiosperms the number of cotyledons is
an important taxonomic character used to
separate the two classes Monocotyledonae
and Dicotyledonae. In endospermic seeds
the cotyledons may be thin and
membranous, as in the castor oil seed, but
in nonendospermic seeds, e.g. pea, they take
over the food-storing functions of the
endosperm. Depending on the pattern of
germination the cotyledons may remain in
the seed (see HYPOGEAL) or emerge to
become the first photosynthetic organs (see
EPIGEAL). Gymnosperms may have several
cotyledons.

coumarin a toxic white crystalline substance
with the formula $C_9H_6O_2$, found in certain
plants, particularly in the testa or fruit
where it inhibits germination. Only when it
has been removed or destroyed, for example
by exposure to light, will the seed
germinate. It has a characteristic smell of
new-mown hay, since it is found in the
vegetative parts of sweet vernal grass
(*Anthoxanthum odoratum*), common in hay
meadows. It has been manufactured
artificially for use in the perfume industry.

counterstaining see STAINING.

coupling see CIS arrangement.

cover the proportion of the ground
occupied by individual plants of a given
species, as determined by a perpendicular
projection onto the ground of their aerial
parts. It is usually expressed as a percentage
of the total area, and may be used as a
rough guide to plant ABUNDANCE (see also
DOMIN SCALE. It may be estimated visually
(often with the aid of a wire grid) or by
taking a number of sample points, using a
frame of pins (point quadrats) and

determining which species (if any) are
covering the ground at those points. The
points are lowered until they touch each
species below them. Where there is overlap
of the leaves of different species, this
estimate results in cover exceeding 100%.
The exact measurement will depend on the
diameter of the pin. In dense vegetation
visual estimates may therefore be more
meaningful than the pin method.

covered smuts see SMUTS.

cover-sociability scale a method of
recording vegetation, devised by
Braun–Blanquet, which includes number
and cover of a species as well as the spatial
arrangement of the individuals. See
BRAUN–BLANQUET SCALE.

cpDNA see CHLOROPLAST DNA.

C_3 plant any plant that produces, as the first
step in photosynthesis, GLYCERATE 3-
PHOSPHATE, which contains three carbon
atoms. Most plants of temperate regions are
C_3 plants. They exhibit PHOTORESPIRATION
and are relatively inefficient
photosynthetically compared to C_4 PLANTS.
They generally have lower carbon dioxide
fixation rates and higher COMPENSATION
POINTS than C_4 plants.

C_4 plant any plant that produces, as the first
step in photosynthesis, OXALOACETIC ACID,
which contains four carbon atoms (see
HATCH–SLACK PATHWAY). Over 100 species
of C_4 plants have been identified, most of
which are tropical. Examples include maize,
sugar cane, sorghum, Bermuda grass, and
many desert plants. C_4 plants require 30
molecules of ATP and 24 molecules of water
to synthesize a molecule of glucose (C_3
PLANTS need only 18 molecules of ATP and
12 molecules of water). However C_4 plants
produce more glucose for a given leaf area
than C_3 plants and consequently grow more
quickly. They can also continue to
photosynthesize at high light intensities
and low carbon dioxide concentrations (see
COMPENSATION POINT) and, most
significantly, do not exhibit
PHOTORESPIRATION. Compare
CRASSULACEAN ACID METABOLISM. See also
KRANZ STRUCTURE.

craspedromous describing a form of leaf
venation in which there is a single primary
vein running down the centre of the leaf

with secondary veins branching off this in an essentially parallel fashion. In traditional terminology such venation is described as *penni-parallel*. An example is the leaf of the sweet chestnut (*Castanea sativa*). *See illustration at* venation.

crassula (bar of Sanio) one of a pair of barlike thickenings made up of primary cell wall and intercellular material found adjacent to the BORDERED PITS of most gymnosperm tracheids.

crassulacean acid metabolism (CAM) a form of photosynthesis first described in the family Crassulaceae and since found in many other succulent plants. CAM plants keep their stomata closed during the day to reduce water loss by transpiration. Carbon dioxide can therefore only enter at night when, instead of combining with ribulose bisphosphate (as in conventional C_3 PLANTS) it combines with the three-carbon compound phosphoenolpyruvate (PEP) to give the four-carbon oxaloacetic acid. This is then converted to malic acid, which can be stored in the cell vacuoles until daylight, when it is transferred to the cytoplasm. Here it is broken down to release carbon dioxide, which is then fixed (see FIXATION) in the normal manner. This adaptation allows such plants to flourish in arid habitats, as they no longer need to keep their stomata open by day to take in carbon dioxide, but their growth rate is slow. CAM can be induced in certain C_3 plants by water shortage. See also HATCH–SLACK PATHWAY.

creationism (special creation) a view that opposes evolutionary theory and envisages the vast variety of living organisms, both existing and fossilized, as having been specially designed by a Creator. The attempted construction of phylogenetic pathways based on the idea that living forms have evolved from ancestral forms is interpreted by creationists as evidence of a 'Great Design'. Creationism is difficult to disprove by experiment. Some creationists believe in the theory of catastrophism, in which it is thought that there have been a number of creations at different times, each having been destroyed by some kind of natural catastrophe, such as a flood. See INTELLIGENT DESIGN.

cremocarp a type of SCHIZOCARP, derived

from two fused carpels, that divides into two one-seeded units at maturity. It is typical of the Apiaceae (umbellifers).

Crenarchaeota see ARCHAEA.

crenate describing a leaf margin that has rounded projections (*See illustration at* leaf). A leaf margin with very small such projections is termed *crenulate*.

crenulate see CRENATE.

Cretaceous the final period of the Mesozoic era between about 145 and 65 million years ago during which much of the present-day land surface was covered by shallow seas. Cretaceous rocks largely consist of chalk formed from fossilized calcareous plates (coccoliths) of marine plankton. It is thought that forms similar to present-day members of the Filicales probably evolved towards the end of the period. The Bennettitales and Caytoniales (both gymnosperm orders) died out and the gymnosperms generally declined in importance, although forms similar to modern species, such as pines, yews, firs, and giant redwoods, arose. Most significant was the emergence of the angiosperms, which became the dominant vegetation, forming large areas of broad-leaved forest.

crista (*pl.* cristae) a shelflike structure in a MITOCHONDRION formed by infolding of the inner membrane. The cristae greatly increase the surface area of the inner mitochondrial membrane so providing a large area for the reactions of the electron transfer system to occur.

critical day length the period of daylight, specific in length for any given species, that appears to initiate flowering in long-day plants or inhibit flowering in short-day plants. In actual fact long-day plants will not flower if the dark period exceeds a certain maximum and conversely short-day plants will not flower unless the dark period exceeds a certain minimum. These periods are termed *critical dark periods* and must be continuous to have effect (see NIGHT-BREAK EFFECT). See also PHOTOPERIODISM.

critical illumination a form of BRIGHT-FIELD ILLUMINATION used with light microscopes in which the light from a uniform source is concentrated by the condenser lens and then focused directly onto the plane of the

specimen. Compare KÖHLER
ILLUMINATION.

critical-point drying a technique for
removing the water from biological
specimens without causing shrinkage or
collapse. The specimen is dehydrated by
suspending it in acetone or alcohol and
driving off the solvent under controlled
conditions of temperature and pressure.
Compare FREEZE DRYING.

cross 1 the process of cross-fertilization.
2 an organism resulting from cross-
fertilization.

cross-breeding see BREEDING.

cross dating see DENDROCHRONOLOGY.

cross fertilization see ALLOGAMY.

cross field in the wood of conifers, the area
where the walls of ray cells abut onto those
of axial tracheids, as seen in radial
longitudinal section. This region is usually
pitted and the form of these cross-field pits
is often characteristic of a genus or group
of genera.

crossing over the exchange of
corresponding segments between
chromatids of homologous chromosomes.
The process is inferred genetically from the
recombination of linked genes (see
LINKAGE) and cytologically from the
formation of chiasmata. Normally crossing
over is reciprocal (equal) but unusual
segregation products in ordered tetrads
indicate that it can very occasionally be
nonreciprocal (unequal crossing over).
Work with imperfect (asexual) fungi (see
FUNGI ANAMORPHICI) indicates that
crossing over can also occur during mitosis.
Chromosome reassortment and crossing
over together constitute the main source of
genetic variation in sexually reproducing
organisms. Crossing over is suppressed by
chromosome inversions. See also
CHROMOSOME MAPPING, INDEPENDENT
ASSORTMENT, PARASEXUAL RECOMBINATION,
TETRAD ANALYSIS.

crossover an exchange of genetic material
between nonsister chromatids in a pair of
HOMOLOGOUS CHROMOSOMES during
meiosis. This results in the recombination
of linked genes (see LINKAGE), and the
mingling of maternal and paternal alleles
in the daughter cells. The frequency with

which crossovers occur is used in
CHROMOSOME MAPPING.

crossover frequency see CHROMOSOME
MAPPING.

crossover region the segment of a
chromosome situated between two marker
genes, for example, in gene mapping
studies that analyse the production of
recombinants following meiosis.

crossover suppressor see INVERSION.

crossover value see CHROMOSOME MAPPING.

cross pollination the transfer of pollen from
the anthers of one individual to the stigma
of another individual of the same species.
This is achieved by intermediary agents
such as insects (see ENTOMOPHILY), wind
(see ANEMOPHILY), or water (see
HYDROPHILY). Birds (see ORNITHOPHILY)
and bats often act as pollinators in the
tropics. Cross pollination does not
necessarily preclude self pollination,
although some mechanisms, such as the
closing of the stigma lobes after pollination
with compatible pollen (as in *Mimulus*), may
reduce the possibility of it occurring. Plants
may be cross-pollinated by hand by using a
small paint brush to transfer pollen from
one individual to the stigma of another
individual of different genotype. See also
COMPATIBILITY, SELF POLLINATION,
OUTBREEDING, S ALLELES, SELF-
INCOMPTABILITY.

cross protection the inoculation of a young
plant with a mild strain of a virus to protect
it from other virulent strains. This
procedure has been used in the control of
tobacco mosaic virus and citrus tristeza
virus.

crown gall a gall disease caused by the soil-
borne bacterium *Agrobacterium tumefaciens*.
Galls are produced above and below ground
on a wide range of host plants – particularly
fruit trees. It is not generally a serious
disease and even large galls on fruit trees
seem to have little economic effect. The
disease is more serious on nursery stock.
Careful inspection of fruit-tree nursery
stock and the use of resistant rootstocks are
helpful control measures.

The galls are of interest to physiologists
because they display HABITUATION and are
self-sufficient for auxin when grown in

culture. See also GENETIC ENGINEERING, TUMOUR-INDUCING PRINCIPLE.

crozier the hooked dikaryotic tip of an ASCOGENOUS HYPHA. Its two nuclei divide simultaneously and the resulting four daughter nuclei are partitioned by two septa into three cells. The middle cell contains two genetically different nuclei and the apical and basal cells either side contain one nucleus each. The apical cell bends round to fuse with the basal cell, thus reestablishing the dikaryotic condition in both cells resulting from the cell division. The middle cell, now at the tip of the ascogenous hypha, develops into an ascus. The cell below may extend and go on to form another crozier. Supposed similarities between this process and the CLAMP CONNECTIONS of certain basidiomycetes have led to the suggestion that basidiomycetes evolved from ascomycetes.

cruciate cross-shaped.

Cruciferae (crucifers) see BRASSICACEAE.

crumb structure the texture of a SOIL, expressed in terms of the size of the soil particles in the surface horizons and the degree to which they tend to stick together. This is often assessed simply by rubbing the soil with the fingers. Sandy soils have large particles with a loose crumb structure and good drainage and aeration, while clay soils have smaller particles that tend to stick together to give a dense crumb structure and poor drainage and aeration.

crustose (crustaceous) describing lichens that have a crustlike thallus closely appressed to and virtually inseparable from the substratum. An example is *Lecanora*.

cryophilic describing organisms that grow best at low temperatures.

cryptobiosis (anabiosis) The ability to survive very cold (even freezing) or dry conditions in a state of suspended animation, a form of dormancy. This is shown especially by some microbial spores.

cryptochrome a protein used by plants to monitor light. Cryptochromes are stimulated by light in the blue-green and ultraviolet wavelengths (500–320 nm). They are involved in the plant's sensitivity to day/night changes, daylength, light intensity, and the direction of incident light. Cryptochromes appear to work by activating chalcone synthase, an enzyme involved in the control of many genes and in the synthesis of various pigments. Cryptochrome 2 is involved in triggering the switch from vegetative growth to flower development in response to changing daylength (see PHOTOPERIODISM).

Cryptococcales see BLASTOMYCETES.

cryptogam in early classifications, a plant whose method of reproduction is not immediately apparent, i.e. a plant in which the reproductive structures are not borne in conspicuous flowers or cones (see STROBILUS). Cryptogams thus included the algae, fungi, bryophytes, and most pteridophytes. Compare PHANEROGAM.

Cryptomonada (Cryptophyta) a phylum of motile algal protoctists that are dorsiventrally flattened and have an oblique

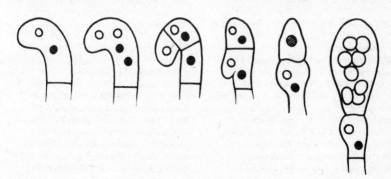

Crozier: stages in crozier formation

furrow ('gullet') with two anterior undulipodia. Some cryptomonads are pigmented and photosynthetic; they contain green pigments (including chlorophylls *a* and c_2) and blue and red pigments and store food as starch in a PYRENOID outside the plastid. They generally lack β-carotene, but contain α-carotene and xanthins. Others are colourless and heterotrophic, preying on bacteria and small protoctists. These species often have trichocysts lining the gullet that produce poisons to kill their prey, which is then ingested through the gullet. Asexual reproduction is by longitudinal fission; sexual reproduction is unknown.

Cryptophyta see CRYPTOMONADA.

cryptophyte a plant with its perennating buds hidden either below ground (see GEOPHYTE, HELOPHYTE) or in water (see HYDROPHYTE). Cryptophytes are more common in arctic and temperate regions than in tropical regions. See also PERENNATING ORGAN, RAUNKIAER SYSTEM OF CLASSIFICATION.

crystal a nonliving cytoplasmic inclusion, often of calcium oxalate or less frequently of calcium carbonate. Crystals occur in a variety of forms (see DRUSE, RAPHIDE).

crystallochory the dispersal of seeds or spores by glaciers.

culm the jointed stem of members of the Poaceae and Cyperaceae. In grasses it is usually hollow or more rarely filled with pith. In sedges it is usually solid and triangular in cross section.

cultivar any variety or strain produced by horticultural or agricultural techniques and not normally found in natural populations; a *cultivated variety*.

culture medium (nutrient medium) a mixture of nutrients used to support the growth of microorganisms or of the cells, tissues, or organs of animals or plants. Such media may be solid, if mixed with a gelling agent (e.g. agar), or liquid. Whenever possible media used in experimental work should be chemically defined, i.e. the identity and concentration of their constituents should be known. Sometimes however a culture can only be grown when complex undefined additional substances, e.g. coconut milk, are added. A culture medium must contain all the macro- and micronutrients essential to growth. A carbohydrate source is necessary and is usually supplied as sucrose. Carbohydrates are necessary even for plant tissue cultures since photosynthesis either does not occur or is insufficient in culture. Plant tissue cultures also require hormones normally an auxin or cytokinin or very often both in a particular ratio. Other factors, e.g. gibberellins or vitamins, are often necessary. Ideal media for plant tissue culture are usually also suitable for the growth of microorganisms. Culture must therefore be carried out under sterile conditions to avoid contamination.

cuneate describing leaves that are wedge shaped with the point of the wedge forming the base of the lamina.

cup fungi see DISCOMYCETES, PEZIZALES, LEOTIALES.

cup lichen the common name for lichens of the genus *Cladonia*, in which the fruiting bodies (podetia) are shaped like tiny cups or goblets.

cupule any of various cup-shaped structures, especially: 1. The structure that partially or completely encloses the fruits of trees in the Fagaceae, such as the cup encircling the bottom of an acorn and the spiny husk around sweet chestnuts (*Castanea sativa*). It is formed from fused extensions of the pedicel and is believed by some to provide a link between angiosperms and seed ferns.

curare the dried aqueous extract of a vine (especially *Strychnos toxifera* or *Chondodendron tometosum*) that contain toxic alkaloids used by some native South Americans as an arrow poison. The alkaloids compete with acetylcholine for receptors on the post-synaptic membrane of the synapse, blocking nerve transmission.

curved ptyxis a form of individual leaf folding (see PTYXIS) in which the unfurling leaf is slightly curved, but not rolled. Compare INVOLUTE, REVOLUTE.

cushion chamaephyte see CHAMAEPHYTE.

cushion plant a plant that forms a compact low hummock close to the ground as an adaptation to cold and dry or windy situations, for example, thrift (*Armeria* *maritima*). Cushion plants usually have small, thick leaves to minimize water loss, or hairy leaves or thick cuticles to reduce transpiration. They are common in arctic–alpine habitats (see ARCTIC–ALPINE SPECIES), and in some locations their low growth habit allows them to be covered by an insulating blanket of snow in winter.

cuspidate suddenly narrowing to a point. The term is especially used of leaves.

cuticle the continuous layer of CUTIN secreted by the EPIDERMIS and covering the aerial parts of the plant body, broken only by stomata and lenticels. The cuticle affords some physical protection, but its main function is to prevent excessive water loss.

cutin a mixture of complex macromolecules forming the waxy cuticle that covers the aerial parts of most higher plants. Cutin contains long-chain fatty acids and polyhydroxy derivatives of these fatty acids (e.g. 10,16-dihydroxypalmitic acid). These molecules are cross-linked through ester bonds in a more or less random fashion to form a large interlinking matrix, relatively impervious to water and gases.

cutting a detached portion of a living plant, such as a bud or leaf, that can produce a new daughter plant if grown in soil or in a suitable culture medium. Taking cuttings is a common horticultural practice for plant propagation. Often the cut end of a stem cutting is dipped into a rooting powder, which contains a natural or synthetic auxin that stimulates the rapid development of adventitious roots and the establishment of a new daughter plant.

cyanelle an endosymbiotic cyanobacterium in a eukaryote cell, in which it functions as a chloroplast.

Cyanobacteria A phylum of Gram-negative oxygen-producing photosynthetic bacteria that contains the *blue-green bacteria* and the chloroxybacteria. The blue-green bacteria, with over 1000 genera, were formerly classified as algae (blue-green algae, Cyanophyta). Their main photosynthetic pigments are chlorophyll *a* and PHYCOBILIPROTEINS (phycocyanin and phycoerythrin) and the main carbohydrate reserve is starch. Unlike most other Eubacteria, they contain organelle-like photosynthetic structures in the form of thylakoids reminiscent of the chloroplasts of photosynthetic algae. The phycobiliproteins form spherical structures called *phycobilisomes* on the outer surfaces of these thylakoids. The blue-green bacteria are a very ancient group, their fossils dating back over 3.5 billion years. The chloroplasts of red and green algae (which are thought to have given rise to the plants) are believed to have evolved from symbiotic blue-green bacteria. The cell wall of blue-green bacteria is made of peptidoglycan and is surrounded by a mucilaginous sheath. The blue-green bacteria range from aggregations of single cells (e.g. *Gloeocapsa*) to colonial forms (e.g. *Microcystis*) and filamentous forms (e.g. *Oscillatoria*). Many filamentous forms (e.g. *Nostoc*) can fix nitrogen in specialized cells called *heterocysts*. Some nitrogen-fixing species are important sources of nitrates in rice paddies. Some blue-green bacteria produce *akinetes* (resting cells), which germinate to produce new filaments. Blue-green bacteria are common members of the plankton, sometimes forming toxic blooms that are fatal to many vertebrates. Many species use gas vacuoles to help them float. They thrive in low nutrient waters in the tropics and warm temperate oceans (see also STROMATOLITES). They also form mats on the surface of damp soil. Some are endosymbionts. The chief form of reproduction is simple fission by ingrowth of the cell membrane and wall. Filamentous forms reproduce by fragmentation. Such fragments of filaments are called *hormogonia*.

The second, much smaller, group of Cyanobacteria are the green *chloroxybacteria* (prochlorophytes). Once thought to be green algae, they are coccoid in form and nonmotile, resembling the chloroplasts of green plants. Unlike other photosynthetic bacteria but like plants, they contain both chlorophyll *a* and chlorophyll *b* and carotenoids. In some, the main storage carbohydrate is $\alpha(1-4)$ glucan, a substance typical of eukaryotes, but some (*Prochloron*) have a starchlike form of amylose similar to that found in green algae and plants. Their cell walls, however, contain muramic acid, typical of bacteria, and their DNA is in the form of fibrils dispersed through the cell.

Most chloroxybacteria are associated with marine sea squirts in the tropics and subtropics, usually as surface dwellers or internal symbionts.

cyanocobalamin (vitamin B₁₂) a complex molecule consisting of a corrin ring (similar to a porphyrin ring) and a ribonucleotide. The corrin ring contains a cobalt atom, which is bonded to a cyanide group and to the ribonucleotide. Cyanocobalamin is not synthesized by plants or animals and commercial production is from bacterial cultures, e.g. of *Streptomyces olivaceus*. It is not needed by plants but animals need it for the normal development of erythrocytes.

cyanogenesis the production of cyanide GLYCOSIDES by certain plants, including cherry laurel (*Prunus laurocerasus*), birdsfoot trefoil (*Lotus corniculatus*), and white clover (*Trifolium repens*). The glycosides, found largely in water solution in the vacuole, yield a sugar and hydrogen cyanide on hydrolysis. Their function is uncertain but they may serve to detoxify harmful by-products of necessary metabolic reactions. They may also act as an emergency sugar store and, because of their poisonous nature, can play a role in deterring invasion or damage by disease and pest organisms.

cyanophage a virus (bacteriophage) that infects blue-green bacteria.

cyanophycin a food reserve characteristic of the blue-green bacteria.

Cyanophyta see CYANOBACTERIA.

cyathium a specialized cup-shaped type of inflorescence found in members of the Euphorbiaceae. It resembles a single flower and consists of a female flower in the centre surrounded by many simple male flowers enclosed within an involucre. The involucre also bears a number of petaloid glands along its rim.

Cycadales (cycads) a group of tropical and subtropical GYMNOSPERMS that constitute the sole living order of the phylum CYCADOPHYTA.

Cycadatae see CYCADOPHYTA.

Cycadeoidales see BENNETTITALES.

Cycadicae see PINOPHYTA.

Cycadofilicales (Pteridospermales, seed ferns) an extinct order of gymnosperms known from fossils of the Devonian to Triassic periods. Although they had fernlike fronds, similar to those of the fern order Marattiales, their stems show secondary thickening (see SECONDARY GROWTH) and more significantly there is evidence of seed formation. The ovules were not borne in strobili but developed along the margins or on the surface of megasporophylls, similar to the foliage leaves. Their seeds were characteristically radially symmetrical, a feature shared by the CYCADALES, BENNETTITALES, and CAYTONIALES and thought to indicate that these orders possibly originated from the Cycadofilicales.

Cycadophyta (cycads) a phylum of cone-bearing gymnosperms with palmlike compound leaves, dioecious reproductive organs, and special CORALLOID ROOTS that have symbiotic blue-green bacteria (see CYANOBACTERIA) in their tissues. The stems are thick and usually unbranched, and the leaves are clustered in a rosette at the apex of the stem. The trunks have a scaly appearance, due to a thick covering of the sclerenchymatous bases of shed leaves, which provide most of the mechanical support of the stem. Scale leaves are also present on the upper part of the stem and apex. Cycads are very slow-growing and vary considerably in height between species; some have a large proportion of the stem underground and reach a height of only a few metres, while others attain heights up to 15 m. Many cycads have deep taproots and can tolerate relatively arid environments; the secondary coralloid roots grow at or above the soil surface. The pollen is dispersed by wind or beetles. In the latter case, the insects are attracted by sugary exudates. Cycads bear exceptionally large female cones (see STROBILI) and ovules. Pollen landing on an ovule of the female cone is trapped in a pollination droplet and drawn into the ovule, where it grows a pollen tube to transport the motile sperm to the egg. Cycad sperm is the largest known: up to 300 μm long, it has a spiral band of undulipodia. The embryo has two cotyledons, which supply it with food.

There are some 145 living species of cycads, in a single class, Cycadopsida or Cycadatae, and order, Cycadales, with 8 genera and 4 families, confined to tropical and subtropical regions. Cycads first appeared in the Permian and show affinities to the extinct Cycadofilicales. The phylum is thought to be polyphyletic in origin, with closer affinities to the angiosperms than to modern conifers (Coniferophyta). Many species are cultivated as ornamentals. The starchy nature of the stem (and of the seeds) has led to the use of certain cycads, the sago palms, as a food source, although they require careful treatment as they can be toxic.

Cycadopsida see CYCADOPHYTA.

cycads see CYCADOPHYTA.

cyclic AMP (3,5-adenosine monophosphate, cAMP) An organic compound produced from ATP by the action of the enzyme adenyl cyclase. It acts as a secondary messenger in the cell: the binding of a hormone to a receptor site on the cell membrane causes the release of cyclic AMP, which travels to the part of the cell required to make the appropriate response.

cyclic photophosphorylation in photosynthesis, the flow of electrons from PHOTOSYSTEM I, when excited to a higher energy level by absorbed light energy, along the ferredoxin/cytochrome *b/f*/plastocyanin electron transport chain and back to photosystem I to absorb more light energy. Energy released during cyclic photophosphorylation powers ATP synthesis, but does not produce NADPH. See PHOTOSYNTHESIS.

cyclins regulatory proteins that help control progress through the various stages of the cell cycle by activating cyclin-dependent protein kinases.

cyclosis (cytoplasmic streaming) the movement of CYTOPLASM within cells. The viscosity of cytoplasm can vary as a result of changes in the nature of protein molecules, particularly ACTIN. In the fluid form it flows freely and can be observed carrying organelles and other inclusions in the cell. Cyclosis is very variable in plant cells. For example, in leaves of *Elodea* (waterweeds) it is more active in cells in the midrib region than in cells near the margin. The rate of flow increases with rise in temperature but is normally less than 0.1 mm per second. It is probable that contractile forces generated

by F-actin microfilaments in the cytoplasm contribute to the movement. The streaming of cytoplasm probably assists the distribution of metabolites between cells via PLASMODESMATA, augmenting diffusion, which is a slower process. It is also believed to contribute to the passage of material through sieve tubes and is more pronounced in young phloem sieve tubes than in the older elements. Cyclosis is not seen in prokaryotic cells.

cygneous shaped like a swan's neck. The term is used to describe the setae of some mosses, e.g. *Grimmia pulvinata* (grey cushion moss).

cyme (cymose inflorescence) see INFLORESCENCE.

Cyperaceae a large family of monocotyledonous plants, the sedges, numbering some 4350 species in about 98 genera. It is cosmopolitan in distribution but members are concentrated in the wetter regions of temperate and cold latitudes. Sedges superficially resemble grasses but may often be distinguished vegetatively by their solid three-angled stems (not present in all species), by the absence of ligules, and by the leaf sheaths, which are usually closed as compared to the open sheathing bases of grasses. The plants arise from swollen underground stems, which in some species, e.g. Chinese water chestnut (*Eleocharis tuberosus*), are edible. The flowers are small, having a reduced perianth, and arranged in spikes, each flower being subtended by a single glume. Many are monoecious, including nearly all members of the largest genus, *Carex*. The fruit is an achene.

A few species are of limited commercial importance, being used for making mats, baskets, hats, and paper. Others are used as ornamentals in water gardens.

cyphella a clear-cut breathing pore on the lower surface of a bryophyte thallus.

cypsela a fruit similar to an ACHENE except that it develops from an inferior ovary and thus also includes noncarpellary tissue. It is typical of the Asteraceae, where the fruit is surrounded by hairs derived from the calyx (the PAPPUS). This type of fruit is traditionally considered a PSEUDOCARP.

cyst any thick-walled resting spore. See also AKINETE, ZYGOSPORE.

cysteine a polar sulphur-containing amino acid with the formula HSCH$_2$CH(NH$_2$)COOH (*see illustration at* amino acid). It is synthesized from the amino acids methionine and serine. Breakdown is to pyruvate and may occur by several routes.

Cysteine is involved in the synthesis of the cofactor biotin. Sulphur-substituted derivatives of cysteine are found in the storage organs of many plants and the sulphide thus stored can be utilized in periods of rapid growth. See also CYSTINE.

cystine a derivative of the amino acid CYSTEINE, formed by the oxidative linkage of two cysteine molecules through a DISULPHIDE BRIDGE. The formation of sulphide bridges is important in the maintenance of the tertiary structure of many globular proteins. Cysteine residues in different parts of a protein chain may cross-link so causing folding of the chain. Cystine is also necessary for the synthesis of thiamin pyrophosphate, a coenzyme in many decarboxylation reactions.

cystocarp a swollen urn-shaped fruiting body found in some red algae (Rhodophyta), e.g. *Polysiphonia*. It forms following fertilization of a CARPOGONIUM and contains the developing carpospores. The surrounding pseudoparenchymatous tissues develop from cells of the filament below the carpogonium.

cystolith a large intracellular structure formed by the deposition of lime on an ingrowth of the cell wall. An enlarged epidermal cell containing such a structure is termed a *lithocyst*. Species in which cystoliths are formed include *Ficus elastica* (rubber plant) and *Urtica dioica* (stinging nettle).

cytidine (cytosine nucleoside) a nucleoside formed by the joining of cytosine to D-ribose (see RIBOSE) with a β-glycosidic bond.

cytochrome any iron–porphyrin-containing electron-carrying protein forming part of the electron transport chains of respiration and photosynthesis. The PORPHYRIN prosthetic group of cytochromes contains a chelated Fe ion, which changes oxidation state from ferric, Fe(III), to ferrous, Fe(II), when the cytochrome is reduced. Most cytochromes are tightly membrane bound.

Cytochrome *aa*$_3$ (also called cytochrome *c* oxidase or the *respiratory enzyme*) is responsible for the final step in the respiratory chain in which water is formed from oxygen and hydrogen ions.

cytochrome oxidase see CYTOCHROME.

cytogamy the fusion of gametes during fertilization. Compare KARYOGAMY.

cytogenetics the application of techniques used in CYTOLOGY to the problems of inheritance, particularly the study of chromosome structure and its relation to the phenotype.

cytokinesis the process of cytoplasmic division as distinct from nuclear division (karyokinesis). In primitive algae, as telophase proceeds, ingrowth of the cell membrane causes a constricting furrow to develop around the cell equator, eventually pinching it in two. This growth of the cell membrane is controlled by a contractile ring of ACTIN. In most algae and in plants, a CELL PLATE is organized in the cytoplasm in the equatorial region of the spindle, dividing the cell into two daughter cells.

cytokinin (kinin) any of a group of plant hormones, derivatives of adenine, whose primary effect is to stimulate cell division. Cytokinins were discovered during work on tissue culture media when it was found that cells of tobacco pith explants could be stimulated to divide by adding the purine adenine to the medium. Subsequently various adenine derivatives, e.g. kinetin, were found to have even greater effects on cell division. However these effects are not seen in the absence of auxin. Moreover, by changing the proportion of cytokinin to auxin, different types of meristematic activity may be induced (see AUXIN). Cytokinins are probably active in most aspects of plant growth and development. Their most obvious effects include the delay of senescence, the induction of flowering in certain species, and the breaking of dormancy in axillary buds and some seeds.

All natural cytokinins are derivatives of the base adenine. Some synthetic cytokinins are substituted phenylureas. It has been suggested they may act by regulating nucleic acid activity, particularly that of transfer RNA (see IPA). Endogenous cytokinins are found in the greatest concentrations in embryos and developing fruits, e.g. in the 'milk' of the coconut. Before fruit set, it is probable that most cytokinin synthesis is in the root. Abnormally high levels of cytokinin are associated with certain plant diseases, e.g. witches broom and crown gall. See KINETIN, ZEATIN.

cytology (cell biology) the study of the structure, organization, and functioning of cellular material.

cytomictic channels see CYTOMIXIS.

cytomixis the apparent migration of cytoplasm and nuclear material from one pollen mother cell to another through pores (cytomictic channels) in the cell walls. It occurs prior to and during the prophase stage of meiosis. The reason for this is unknown though it may be a response to the nutritional problems of developing reproductive cells.

cytopharynx see GULLET.

cytoplasm (hyaloplasm) (adj. cytoplasmic) The part of the protoplasm outside the nucleus in which the CISTERNAE and membrane-bound organelles lie. The soluble phase of cytoplasm contains the principal components of the cell's metabolic pathways, i.e. ions, dissolved gases, and most of the enzymes and substrate molecules for metabolic processes. The cytoplasmic matrix also contains storage products, e.g. lipid droplets, aleurone grains, and starch grains. Complex networks and linear arrangements of fine actin microfilaments and microtubules are commonly present. The microfilaments, 0.5–7.0 nm in length, are composed of helical chains of F-actin, and when supplied with ATP they contract. Those at the periphery of the cell are involved in movements of the components of the plasma membrane, e.g. in endocytosis. Cytoplasm shows transitions between viscous (plasmagel) and fluid (plasmasol) phases. This is displayed in CYCLOSIS and in amoeboid movement shown by some unicells. The F-actin molecules of the microfilaments, particularly at the periphery of cells, are unstable, readily breaking up into soluble

components so that the cytoplasm becomes fluid. See also CYTOSKELETON.

cytoplasmic inheritance (extrachromosomal inheritance) the determination of a characteristic by genes in the cytoplasm (plasmagenes) rather than in the chromosomes. The expression of a cytoplasmically determined character is therefore not related to the behaviour or movement of chromosomes, and consequently such characters fail to segregate in Mendelian ratios. They are normally transmitted through the female gamete, which contributes most of the cytoplasm to the zygote, the male gametes only contributing a nucleus. Genes inherited in this manner are found on DNA present in small quantities in the chloroplasts, mitochondria, and sometimes the cytosol itself. An example of a characteristic inherited in this fashion is a form of male sterility in maize (Zea mays).

cytoplasmic streaming see CYCLOSIS.

cytosine (4-amino-2-oxypyrimidine) A nitrogenous base derived from amino acids and sugars and found in all living organisms. Cytosine is found in one of the NUCLEOTIDES present in DNA and RNA and pairs specifically with the purine base guanine. Substantial amounts of cytosine in DNA and RNA are found in the modified forms of 5-methylcytosine and 5-hydroxymethylcytosine. The affinity of these substances for guanine is similar to that of cytosine but the significance of their occurrence is uncertain. Recent evidence suggests that the regions of DNA with a large number of methylated cytosines are inactive. Thus gene activation may be brought about, at least in part, by selective demethylation.

cytoskeleton a complex of hollow protein MICROTUBULES and ACTIN filaments, about 24 nm in diameter, with walls about 5 nm thick, that helps to determine the shape of cells during development and maintain the form of differentiated cells. It is often found concentrated in a plane just below the cell membrane, especially in areas where there is activity, such as cell plate formation and mitotic or meiotic spindle formation. It is also closely involved in the alignment of cellulose fibres as the cell wall is laid down. The cytoskeleton is a dynamic system: microtubules and actin filaments can assemble and disassemble rapidly as their subunits in the cytoplasm polymerize and depolymerize. Accessory proteins link the filaments to one another and to cell organelles and membranes, and can interact with them to produce movement, using energy from ATP.

cytosol the soluble fraction of cytoplasm after particulate matter has been removed by centrifugation. See CELL FRACTIONATION, CENTRIFUGE.

cytosome see GULLET.

cytotaxonomy the use of chromosome studies (i.e. number, structure, and behaviour) in taxonomic work. Chromosome number is probably the most frequently quoted feature of the karyotype used by taxonomists. Chromosome counts are usually made on sporophyte tissue at mitosis and are therefore the diploid number ($2n$). When dealing with a polyploid series, the base number (i.e. the number of chromosomes present in the original haploid genome) may be given. The position of the centromere is a reliable feature of chromosome structure and consequently makes a good taxonomic character. More detailed studies of meiotic behaviour can reveal, for example, the heterozygosity of some inversions. Such a feature may be consistent for a particular taxon, thus providing additional taxonomic evidence. Cytological data is sometimes thought to be of special significance, and may be ascribed more weight (see WEIGHTING) than other taxonomic evidence. GENE SEQUENCING is of increasing importance in determining phylogenetic affinities.

CZI see SCHULTZE'S SOLUTION.

D*d*

2,4-D (2,4-dichlorophenoxyacetic acid) A synthetic AUXIN of the PHENOXYACETIC ACID type with two substituted chlorine atoms. It is the auxin most widely used commercially as a selective weedkiller. See also 2,4,5-T, MCPA.

damping-off a disease of seedlings encouraged by cold wet soil and crowded conditions. In pre-emergence damping-off the germinating seed rots before the plumule breaks through the soil surface. In post-emergence damping-off the seedlings rot at soil level and then collapse and die. Damping-off is caused by numerous soil-inhabiting fungi and oomycetes, including *Fusarium*, *Phytophthora*, *Pythium*, and *Rhizoctonia* (*Corticium*). Correct planting conditions, seed dressings, and the use of sterile soil in seed trays are the usual preventative methods. See also SEEDLING BLIGHT.

dark-ground illumination (dark-field illumination) a microscopical technique used to examine living cells or microorganisms. These are not suitable for examining under a normal light microscope (BRIGHT-FIELD ILLUMINATION) as they are usually transparent. Thus little detail can be seen against a light background. In dark-ground illumination direct light is prevented from forming an image by illuminating the specimen from the side so only diffracted light from the specimen passes through into the objective. The image of the specimen then appears luminous against a completely dark background. This technique exploits the differences in refractive index between organelles and the surrounded protoplasm that causes the organelle boundaries to reflect more light.

dark reactions (light-independent reactions) the sequence of light-independent reactions that utilize the energy (in the form of ATP) and reducing power (in the form of NADPH), produced during the LIGHT REACTIONS of PHOTOSYNTHESIS, to reduce carbon dioxide. In eukaryotes, this process usually occurs in the chloroplast stroma and it can take one of two forms, depending on whether the subject is a c_3 PLANT or a c_4 PLANT. The details of the fixation of carbon dioxide differ in the two types of plants but the end result in both cases is the production of carbohydrates via the CALVIN CYCLE. See also CRASSULACEAN ACID METABOLISM.

darwin a measure of the rate of evolution, expessed as units of change per unit time.

Darwinian fitness see ADAPTIVE VALUE AND SELECTION.

Darwinism the theory, put forward by Charles Darwin in 1859, that species evolve by NATURAL SELECTION. Darwin noted that although organisms produce more than enough offspring to replace themselves, the numbers of a species tend to remain constant. He concluded that there is a struggle for existence and organisms compete with one another for food, space, etc. He also noted the considerable variation exhibited by members of the same species and concluded that only those best adapted to the environment survive. As many of the variations are hereditary, those that survive because of favourable variations will transmit these to their offspring. Those with unsuitable characteristics will not survive and such characteristics will be lost. This process of selective birth and selective death or 'survival of the fittest' as it came to be called Darwin termed natural selection. Finally, because the environment changes, natural selection continuously acting on a large number of variations could result in the formation of new species, the origin of species. Darwin could not explain how variation arose and made no distinction between continuous and discontinuous variation. He was later driven to accept Lamarck's theory of the inheritance of acquired characteristics (see PANGENESIS). The origin and maintenance of variation was later explained by work in genetics (see NEO-DARWINISM). See also WEISMANNISM. Compare LAMARCKISM.

Dasycladales an order of SIPHONOUS GREEN ALGAE of the class ULVOPHYCEAE that comprise very large unicells, sometimes several centimetres long, e.g. *Acetabularia*. Some species are uninucleate; others become coenocytic early in development. The thallus has a central erect axis, anchored by rhizoids, and whorls of lateral branches. The cell walls are composed mainly of MANNANS, often coated with a calcareous layer. There is no significant alternation of generations, the diploid thallus undergoing meiosis just before gametes are produced and the new thallus arising directly from the zygote. Uninucleate species have a large nucleus, which may be diploid or polyploid, at the base of the thallus. This divides to produce several haploid nuclei just before reproduction. These migrate to lateral branches (gametangia) that form cysts, each containing a single haploid nucleus, in some species protected by a cellulose cell wall. The cysts are shed into the water, where the gametes are released. Gametes of different mating types fuse.

dating the determination of the age of rocks, minerals, organic materials, and archaeological remains. There are two approaches. *Relative dating* assesses the age of a specimen in relation to other specimens. For example, there is a relative time scale based on the correlations between fossil and stratigraphic data. Annual rings offer another means of correlation (see DENDROCHRONOLOGY). *Absolute dating* is the assessment of the actual age of a specimen by using a measure of time. For example, organic materials may be dated by RADIOCARBON DATING. However this method can only date specimens from the later part of the Quaternary period. VARVE DATING is also limited to this period. Age determinations of older rocks rely on RADIOMETRIC DATING.

daughter cells the cells resulting from the division of a single cell.

day degrees the number of degrees by which the average daily temperature of a particular site differs from a standard temperature, such as the minimum temperature required for growth or flowering. Day degrees are useful whem making comparisons between growing seasons, for example in different locations or in different years. See also MONTH DEGREES.

day length see CRITICAL DAY LENGTH, PHOTOPERIODISM.

day-neutral plant a plant that is capable of flowering regardless of the amount of light it receives each day. The onset of flowering is therefore controlled by other factors. Compare LONG-DAY PLANT, SHORT-DAY PLANT. See also PHOTOPERIODISM.

deamination the removal of an amino group from a compound. Breakdown of amino acids is by oxidative deamination. Glutamate, which is formed from many amino acids by TRANSAMINATION, is deaminated by the enzyme glutamate dehydrogenase, yielding α-ketoglutarate, ammonium ions, and reduced NAD or NADP.

decarboxylase any enzyme that catalyses the removal of carbon dioxide from a substrate (*decarboxylation*). An example is the enzyme pyruvate decarboxylase, which catalyses the formation of acetaldehyde (ethanal) from pyruvate, a reaction in the alcoholic fermentation of glucose. Decarboxylases have a tightly bound coenzyme, often biotin or thiamin pyrophosphate, which acts as an intermediate carboxyl carrier. See also CARBOXYLASE.

deciduous describing woody PERENNIAL plants that shed their leaves before the winter or dry season. Leaf fall is an adaptation that reduces water loss by transpiration when little or no water is available to the plant roots (see PHYSIOLOGICAL DROUGHT). Various environmental factors, such as daylength, temperature, and light intensity, influence the onset of leaf ABSCISSION. Compare EVERGREEN.

deciduous forest see FOREST.

decidous summer forest the commonest kind of broad-leaved FOREST in middle latitudes of the northern hemisphere, but not found in the southern hemisphere. Most of the trees are deciduous, shedding their leaves at the onset of winter. This is thought to be an adaptation to conserve water in winter, when the soil is frozen or low temperatures make uptake by the roots extremely slow.

decomposer an organism that feeds on dead organic material, breaking it down into simpler substances and bringing about decay. In this way, organic material is recycled as the products of decomposition can be used by plants (PRODUCERS). Examples of decomposers are some bacteria and fungi. Compare CONSUMER. See also FOOD CHAIN.

decomposition the breakdown of dead organic remains into simpler inorganic substances of lower energy content. The organic material may be animal carcasses, fallen trees, leaves, flowers, root caps, faeces, and so on. Decomposition involves first larger animals that break down the dead material into smaller pieces, then smaller organisms such as bacteria, fungi, and DETRITIVORES complete the process, respiring the final products of decomposition and releasing carbon dioxide, water, and inorganic nutrients back into the environment to be recycled (see BIOGEOCHEMICAL CYCLE). Under anaerobic conditions particular organisms, especially certain bacteria and fungi, respire anaerobically (see FERMENTATION), releasing alcohol and organic acids. This increases the acidity of their surroundings, influencing the succession of DECOMPOSERS. In deep anoxic ocean sediments the decomposers include bacteria specializing in denitrification and sulphate reduction, and methanogens. See

also DETRITUS AGRICULTURE, DETRITUS FOOD CHAIN, SAPROBE.

decumbent describing a stem that lies along the ground. Decumbent stems often have an upturned tip.

decurrent describing leaf bases that extend down the stem beyond their point of insertion, forming a wing on the stem. The leaves of the common mullein (*Verbascum thapsus*) are examples. The term is also used of the gills of basidiomycete fungi that run down the stem, as seen, for example, in the genus *Clitocybe*.

decussate see OPPOSITE.

dedifferentiation the loss of the specialized features and reversal to a meristematic state of a differentiated cell. Dedifferentiation often occurs immediately prior to further reorganizational changes as in the production of a secondary meristem. The process may also be brought about by wounding or stimulation by hormones. It is thought to involve the removal of inhibitors that normally prevent the expression of the complete genetic complement. See TOTIPOTENCY. See also CALLUS.

deficiency see DELETION.

deficiency disease a disease caused by the lack of an essential nutrient element (see MACRONUTRIENT). Mineral deficiencies can often be recognized by characteristic symptoms. For example, nitrogen deficiency results in stunting of the plant and CHLOROSIS of older leaves. Iron deficiency results in interveinal chlorosis, particularly of younger leaves, giving a striped appearance in cereal and grass leaves. Boron deficiency causes the disease of heart rot in sugar beet and mangolds (*Beta vulgaris*). See also YELLOWS.

definite growth (determinate growth) a form of growth where a maximum size is reached beyond which there is no further increase, i.e. growth is limited. It is seen in annual and biennial plants and in many plant organs, e.g. internodes, which do not continue to lengthen throughout the life of the plant. The term is also used of cymose INFLORESCENCES, in which growth is terminated by the production of a terminal flower bud. Compare INDEFINITE GROWTH.

definite inflorescence see INFLORESCENCE.

definitive nucleus the diploid nucleus

found in the centre of the EMBRYO SAC after the fusion of the two haploid POLAR NUCLEI. A male gamete released from the pollen tube may fuse with this to form the triploid primary endosperm nucleus from which the endosperm develops. See also DOUBLE FERTILIZATION.

deflexed describing a structure that is bent sharply downwards.

defoliation the removal of leaves from plants. The term is applied particularly to the use of herbicides called *defoliants*, which cause leaf abscission. It is also used to describe the effects of outbreaks of pests such as the caterpillars of certain moths, which can almost totally defoliate trees.

deforestation the clearing of FORESTS. The term is used especially in connection with the clearing of forests for human settlement and agriculture and logging for timber. Deforestation is major cause of loss of natural habitats and of biodiversity and in the worst cases may also lead to floods, soil erosion, and increased incidence of drought in the surrounding areas. Some forests, especially in the tropics, lie on poor soils that contain inadequate nutrients for crops. Most of the nutrients in these forests are contained within the vegetation and are recycled between the trees and the upper layers of the soil. Once this vegetation is lost, the remaining nutrients are quickly washed out of the thin soil. Outside the tropics, with the exception of large areas of Russia, most deforestation occurred long ago – in the Mediterranean about 8000 years ago, in China 4000 years ago, and in Europe well over 500 years ago. Most of North America's deciduous forests had been cleared by the end of the 19th century, as had the forests of Australia and New Zealand. The 20th century saw a great acceleration of deforestation in the tropics, exacerbated by years of low rainfall associated with the El Niño climatic phenomena, when wildfires and fires deliberately set for legal or illegal clearances ran riot, destroying large areas of vegetation in the Far East and South America. Forests are the largest reservoirs of carbon in the world (see CARBON CYCLE), containing some 80% of the carbon above ground in terrestrial habitats, and about 33% of the carbon below ground. Much of the carbon is locked up in the tissues of plants. Forests at high latitudes are particularly important, and further carbon is also contained in peat deposits below ground. Deforestation has contributed to the rise in carbon dioxide in the atmosphere. This is likely to increase still further as large tracts of forest in Russia (which holds about 25% of the world's forest carbon) and Amazonia (which contains 20%) are logged or burned for timber or human settlement. As well as carbon dioxide and oxygen from photosynthesis, forests also release carbon monoxide, methane, and nitrous oxide. More than half of the carbon monoxide given off by forests is released during burning. Forests are also important in the water cycle. Rainforests in particular recycle rainfall, absorbing it as it falls and reducing runoff, then releasing moisture over a period of time through evaporation and transpiration. In such regions as the Amazon basin, it is estimated that forests supply the moisture for half the rainfall of the whole region. Local, and possibly even global, changes in climate may result from large-scale deforestation. On a smaller scale, the deforestation of catchment areas and watersheds can exacerbate incidents of flooding lower down the river basins, as the 'sponge' effect of the forests is lost and rain runs off straight into rivers and streams.

degenerate see CODON, GENETIC CODE.

degenerate phase see HUMMOCK AND HOLLOW CYCLE.

degree of freedom in an analysis of experimental results, a comparison between data made independently of any other comparisons. The number of degrees of freedom (N) in an analysis is therefore one less than the number of observations.

degree of presence see BRAUN-BLANQUET SCALE.

dehiscence (*adj.* dehiscent) the splitting open along predetermined lines of certain plant organs, such as anthers, spore capsules, and fruits, to release their contents. Dehiscence is often caused by the gradual drying out of the enclosing walls. In addition, variations in the degree of wall thickening may create tensions that result in a violent opening of the structure. See also ANNULUS.

dehydration a stage in the preparation of permanent slides for microscopical examination, in which water is removed from the specimen. The material is dehydrated slowly by being placed in a series of successively more concentrated ethanol solutions culminating with absolute alcohol. Plant cells must be dehydrated far more gradually than animal cells as there is a tendency for the protoplasm to shrink from the cell wall. Dehydration is followed by CLEARING.

dehydrogenase any OXIDOREDUCTASE enzyme that catalyses the transfer of electrons from a reduced substrate or electron donor to an electron-accepting COENZYME; the transfer usually involves hydrogen. The equation for a dehydrogenase reaction is generally: reduced substrate + oxidized coenzyme → oxidized product + reduced coenzyme. The reduced coenzyme is usually reoxidized in the RESPIRATORY CHAIN. Dehydrogenases are classified according to the nature of the coenzyme, for example, those requiring the pyridine dinucleotide coenzymes NAD and NADP are termed pyridine-linked dehydrogenases, while those requiring FAD or FMN are termed flavin-linked dehydrogenases or flavoproteins.

Delaunay triangulations see THIESSEN POLYGONS.

deletion (deficiency) a CHROMOSOME MUTATION involving the loss of part of a chromosome. Since a number of genes are lost as a result, deletions are usually harmful and often lethal if the corresponding segment of the homologous chromosome contains any recessive deleterious mutations. Compare ANEUPLOIDY.

deliquescence a gradual dissolving of tissue by AUTOLYSIS. It is seen, for example, in *Coprinus* (ink-caps), in which the gills change into an inky black fluid after the basidiospores have been discharged.

deltate shaped like a delta.

deme a subpopulation of a SPECIES. A deme is a distinct interbreeding group of individuals with genetic or cytological characteristics that distinguish them from other groups, from which they are also spatially separated, though this may be by a very short distance. Within the deme there is an equal chance of all possible male and female pairings, but exchange of genetic material with other demes is rare. Typically different demes lie adjacent to each other, with no dispersal barriers between them. See POPULATION. Compare SUBSPECIES.

demethylation *See* cytosine.

denaturation change in the structure of a globular protein or a DNA molecule following exposure to temperatures above 60–70°C, to pH outside the normal physiological range, or to certain chemicals. Denaturation of proteins involves the uncoiling of the polypeptide chain so that its TERTIARY STRUCTURE is lost, resulting in loss of biological activity. The process is sometimes reversible, the protein regaining its characteristic coiled structure (native form) and its activity. This is termed *renaturation*. Denaturation of DNA involves the uncoiling of the double helix, forming single-stranded DNA. The double helix molecule will quickly reform if denaturation has not proceeded too far (i.e. if a small part of the molecule has retained the double helical form). Renaturation (or annealing) is slow if the strands have completely separated.

dendrochronology the use of the annual rings of trees to date historical events. The dating of archaeological sites depends largely on *cross dating*. This involves taking small cores from old living trees and comparing them with the rings of timbers at the site. The year the timbers were felled is determined by finding the point at which ring patterns of the living tree correspond to those of the archaeological specimens. Bristlecone pines (*Pinus longaeva*) are often used as they can live for over 4000 years. If sufficiently aged living trees cannot be found in a particular region then chronologies may be built up successively further back in time by matching the ring patterns of a number of wood samples whose ages overlap.

The average tree-ring chronology for a particular location is called the *master chronology*. It forms a baseline against which new ring series can be compared and dated.

Where the annual ring evidence from dead wood in a particular area does not overlap with chronologies from living trees in the same area, it cannot be dated accurately, and is termed a *floating chronology*.

Annual rings also provide a record of past environmental conditions (see DENDROCLIMATOLOGY). Thus ring width is normally positively correlated with water abundance. The presence and level of certain pollutants, e.g. lead, may also be recorded in the rings. It is necessary in such work to ensure that only those tree species that produce one growth ring a year are used. Certain trees, e.g. many juniper species, produce multiple rings each year, while in others it may not be possible to distinguish one season's growth from the next. See also DENDROECOLOGY.

dendroclimatology the study and reconstruction of past climates using information recorded in the annual rings of trees (see DENDROCHRONOLOGY). The amount of radial growth of a tree varies with the weather during a particular year, especially the seasonal changes in water availability. This affects the date at which growth begins in the that year, the rate of growth, and the diameter of the xylem vessels. The number of cells in each ring, the diameter of their lumens, and differences in the thickness and composition of the cell walls provide information about the climatic conditions at the time they were formed. Annual rings tend to be thicker in years of plentiful rainfall. The normal pattern of increasing cell wall thickness and decreasing cell diameter as the growing season progresses is disrupted by drought, which may lead to an extra ring (*false ring*). Climatic stress may lead to rings that do not show this gradation, or to incomplete or missing annual rings, or distortion in various locations on the circumference or at different heights in the tree. By analysing today's responses to climate, it is possible to infer past climates from details of the annual rings in old wood.

dendroecology a branch of dendrochronology concerned with the relationship between the annual ring pattern and the abiotic and biotic factors that affect them.

dendrogram a branching usually 'rooted' diagram that reflects the relationships of a group of taxa. The taxonomic hierarchy can be represented in a dendrogram, with kingdom at the base and subforms as the terminal branches. Specific types of dendrogram can be recognized. *Phenograms* represent the degree of phenetic similarity

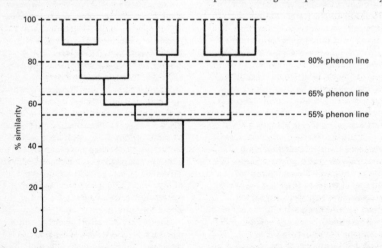

Dendrogram: showing phenetic similarity; the phenon lines delimit groups of the same rank

and are based solely on phenetic data. *Phenon lines* drawn at right angles to the phenogram (see diagram), show values of similarity as percentages (the greater the value the higher the level of similarity). Clusters of similar groups are termed *phenons* and phenons may be assigned rank according to the level on the phenogram at which they branch off. Thus in the diagram the four clusters above the 80% phenon line could be assigned the rank of genus while the two clusters at the 55% similarity level might be assigned the rank of subfamily. Phylogenetic trees are also dendrograms, but unlike phenograms, the vertical axes represent time (though not necessarily to scale) or relative advancement. See also CLADOGRAM.

dendrograph an instrument that continuously records the circumference of a tree trunk. It is used, for example, to measure daily fluctuations in girth due to differences in water content.

dendroid tree-shaped.

dendrometer an instrument that measures precisely the circumference of tree trunks and/or branches. It usually consists of a thin metal band with a vernier scale.

denitrification the loss of nitrate from the soil through the action of various *denitrifying bacteria* (e.g. species of *Clostridium*, *Pseudomonas*, *Micrococcus*, and *Thiobacillus*), which use nitrate as the terminal electron acceptor during anaerobic respiration. Molecular nitrogen, nitrous oxide, or ammonia may be given off. See NITROGEN CYCLE.

denitrifying bacteria see DENITRIFICATION.

density the number of individuals in a specific area. Density is usually measured by counting the number of individuals in a series of randomly distributed quadrats, summing the results, then calculating the average number of individuals per unit area. This gives an accurate measure of ABUNDANCE and allows different areas and different species to be compared. When assessing a species' role in a community, the *relative density* is sometimes used. This is expressed by:

(number of individuals of the species/total number of individuals of all species) × 100

See also THIESSEN POLYGONS.

density dependence the regulation of population size by mechansims that are themselves controlled by the density of the population. For example, the rate of increase in numbers of individuals in the population depends on the number already present in the population. Such mechanisms involve environmental factors. A population in a new environment may show a typical *S-shaped growth curve*, in which growth is slow at first, then increases rapidly as increasing numbers of the population begin to reproduce, then declines to a fairly stable peak, which represents the carrying capacity of the environment for that species. This is achieved because the birth or growth rate in the population decreases or the death rate increases as population density increases.

density–frequency–dominance (DFD measure) an outdated measure of abundance used in early American ordination schemes. It expresses the *relative abundance* of a species in terms of its *importance value*, the sum of its *relative density*, *relative frequency*, and *relative dominance*. The importance value may range from 0 to 300. The relative density is the number of individuals of a given species in a given area expressed as a percentage of all the species present; the relative frequency is the frequency of a given species expressed as a percentage of the sum of the frequency values for all the species present; and relative dominance is the basal area of a given species expressed as a perentage of the total basal area of all the species present. Compare DENSITY INDEPENDENCE, RELATIVE-IMPORTANCE VALUE.

density-gradient centrifugation a technique for separating and isolating pure samples of cell organelles, i.e. cell fractionation. The cells are first broken up in a homogenizer to release the contents and the homogenate is filtered to remove cell wall fragments. It is then poured on top of prepared salt or sucrose solutions of different concentrations that have been layered according to their density in a glass tube. Provided an appropriate centrifugal force is applied, the cell inclusions will band together at the regions in the density gradient that correspond to their own density and can then be separated off by various methods. Compare DIFFERENTIAL CENTRIFUGATION.

density independence the tendency for the rate of increase of a population to remain unaffected by density-dependent factors as the population density increases. Such factors only start to limit growth at a point when the population undergoes a dramatic crash, producing a characteristic *J-shaped growth curve*. Initial growth is exponential, but growth stops abruptly due to factors such as the end of the breeding season or the onset of winter limiting growth. Such growth curves are typical of an organism in a new environment. Such populations typically show 'boom and bust' cycles, e.g. algal blooms. The actual rate of change of population growth depends on the population size and the biotic potential: $dN/dt = rN$, where N is the number of individuals in the population, t is time, and r is a constant representing the intrinsic rate of increase (biotic potential) of the organism. Compare DENSITY DEPENDENCE.

dentate describing a leaf margin that is toothed, with outwardly-pointing notches. Leaf margins finely toothed in this way are termed *denticulate*. *See illustration at* leaf.

denticulate see DENTATE.

deoxyribonucleic acid see DNA.

deoxyribose a pentose sugar and a component of the nucleotides present in DNA, i.e. adenosine, cytosine, guanidine, and thymidine phosphates (AMP, CMP, GMP, and TMP). Although not conforming to the general formula of monosaccharides $(C_x(H_2O)_y)$, deoxyribose $(C_5H_{10}O_4)$ is in other respects a typical five-carbon sugar.

deoxy sugar a sugar in which a hydroxyl group (OH) has been replaced by hydrogen (H), with the loss of oxygen, for example, deoxyribose.

deplasmolysis the entrance of water into a plasmolysed plant cell (see PLASMOLYSIS), which results in the the plasma membrane returning to make contact with the cell wall.

depside a lipid-based group of substances found in lichens. Depsides are often brightly coloured.

Derbesiales see CAULERPALES.

derived character a character that has changed from the form in which it appeared in an ancestor.

dermal relating to the epidermis or periderm.

dermatogen see HISTOGEN THEORY.

dermatophyte a fungus that lives on skin as a parasite.

derris a relatively safe insecticide derived from plants of the tropical genus *Derris* (family Fabaceae).

desert a major regional community (BIOME) characterized by low rainfall and consequently supporting little or no vegetation. The term 'true desert' is used of regions completely devoid of higher plant life. Cold deserts include the TUNDRA and regions permanently covered with ice and snow. Hot deserts have a mean annual rainfall of under 250 mm while hot semideserts have a mean annual rainfall of under 400 mm. The rain falls in brief heavy showers and varies in amount from year to year. The plants are few in number and are usually short and sparsely distributed. The permanent perennial plants consist of succulent or xerophytic trees, shrubs, and herbs. There are also EPHEMERAL plants whose seeds lie dormant until a brief rainstorm prompts them to germinate, flower, and set seed in a short space of time.

desertification the process of desert formation, which may be due to climate change, but is often due to the removal of vegetation and subsequent soil erosion due to overgrazing, overcultivation, or the overextraction of water for irrigation, or for domestic or industrial use. Once it begins, desertification develops its own momentum and is very difficult to stop or reverse: the water table falls and the groundwater and topsoil (see SOIL PROFILE) become increasingly saline as water evaporates from the ground surface, leaving behind the salt that had been dissolved in it. The decrease in surface water increases the rate of soil erosion, and both salinization and loss of topsoil increase the rate of vegetation loss. The loss of vegetation increases the albedo of the land surface, which in turn leads to an increasingly dry climate. An example of desertification is the advance of the Sahara

Desert into former steppe and scrub woodland in the Sahel region of Africa as a result of severe drought exacerbated by overgrazing, overcultivation, and the removal of trees and shrubs for firewood.

desiccation loss of water, drying-out.

Desmarestiales an order of brown algae (Phaeophyta) in which the thallus is complex and pseudoparenchymatous, with a branching axis that supports blades or wiry branches. Vegetative growth takes place by means of a meristem at the base of a hair (*trichothallic meristem*) at the tip of each axis. Sexual reproduction is by oogamy.

desmids see GAMOPHYTA.

desmokont a type of dinoflagellate that has two anterior undulipodia. See DINOMASTIGOTA.

determinate growth see DEFINITE GROWTH.

determinism see EVOLUTIONARY DETERMINISM.

detrital pathway see DETRITUS FOOD CHAIN.

detritivore (detritus feeder) a HETEROTROPH that feeds on detritus (dead organic material). The detritus may be of plant and/or animal origin, and has usually been produced by DECOMPOSERS. It is also the habitat for other organisms, such as nematodes, very small insects, and microorganisms, which will also be taken in and consumed by the detritivore. In aquatic and marine habitats, many detritivores are filter-feeders, extracting minute particles of detritus from the water. Animals that feed on larger dead items, such as carcasses or dung, are usually called *scavengers* as distinct from detritivores.

detritus organic material produced by DECOMPOSERS.

detritus agriculture the deliberate production of detritus as a source of food, using microorganisms to convert it to a more palatable substance, e.g. silage production.

detritus feeder see DETRITIVORE.

detritus food chain (detrital pathway) a food chain in which the dead remains of primary producers, e.g. leaf litter, are acted on by DECOMPOSERS and DETRITIVORES, which in turn are eaten by primary consumers, such as earthworms, which in turn are eaten by secondary consumers, such as birds, and so on up the food chain. See also FOOD CHAIN.

Deuteromycota see FUNGI ANAMORPHICI.

Deuteromycotina see FUNGI ANAMORPHICI.

developmental spiral see GENETIC SPIRAL.

Devensian the last glacial stage of the recent Ice Age in Britain, which lasted from about BP 70 000 to BP 10 000 years. During the late Devensian there were unusually high rates of extinction of large mammals, considered by some to be the result of human activity, though this is debated. The Devensian is roughly synchronous with the Weichselian Glaciation in northern Europe, the Würm Glaciation in the Alps, and the Wisconsinian Stage in North America. See also DRYAS, GLACIAL PERIOD, ICE AGE, LATE GLACIAL, POST-GLACIAL.

deviation symbol: *d*. The departure of a value or observation from what was expected. See also MEAN DEVIATION, STANDARD DEVIATION.

Devonian the first period of the Upper Palaeozoic era from about 395 to 345 million years ago in which the first extensive invasion of the land by plants and animals took place. A possible bryophyte fossil, *Sporognites*, with a well-developed capsule has been found in the Lower Devonian, but most fossil bryophytes appear later. Many well preserved fossils of the Psilophytales (see PSILOPHYTA) have been found in the RHYNIE CHERT of the Devonian in Scotland. Lycophytes (Lycophyta) and sphenophytes (Sphenophyta) are also found. The ferns (Filicinophyta) arose in Devonian times possibly from a form resembling *Protopteridium*, an intermediate between the psilophytes and the Filicinophyta. The gymnosperms probably arose in the late Devonian, being represented by the extinct Cycadofilicales, which include the earliest seed plants *Archaeopteris* and *Archaeosperma*, and also the extinct Cordaitales. See GEOLOGICAL TIME SCALE.

dextran a POLYSACCHARIDE made up of branching chains of D-glucose units, which is used by bacteria and yeasts as a storage compound.

dextrin a polysaccharide SUGAR produced when starch is hydrolysed by amylase enzymes. Some dextrins are used as adhesives.

dextrose the former name for GLUCOSE.

DFD see DENSITY–FREQUENCY–DOMINANCE.

diadelphous see ANDROECIUM.

diageotropism see DIATROPISM.

diagnosis the statement, in Latin, that describes how a new taxon differs from its closest relatives.

diagravitropism see DIATROPISM.

diakinesis see PROPHASE.

diallele cross a breeding experiment in which each of a number of males is mated to each of a number of females.

diallelic describing an individual in which two different alleles exist at a particular gene locus (even if the individual is polyploid).

dialysis a process in which large molecules, such as proteins and starch, and small molecules, such as amino acids, glucose, and salts, are separated by selective diffusion through a selectively permeable membrane. For example, a mixture of starch and glucose molecules in solution can be separated by placing them in a selectively permeable dialysis tube made, for example, of cellophane, which is immersed in a container of distilled water. The large starch molecules remain in the dialysis tube and the small glucose molecules pass out through the membrane and can be detected in the water in the container.

diaminopimelic acid an amino acid found in bacteria. It is involved in the biosynthesis of lysine in such organisms. It is also often found in the peptide side chains of the acetylmuramic acid molecules that partly make up the mucopeptides (see PEPTIDOGLYCAN) of bacterial cell walls.

diaphototropism see DIATROPISM.

diarch describing a root with two strands of xylem. See also STELE.

diaspore a dispersal unit, such as a seed or fruit.

diastase the name originally given to the active preparation from malt extract that breaks down starch to sugar. It includes the enzyme β-amylase. See AMYLASE.

diatomaceous earth (kieselguhr) a sedimentary deposit made up of fossil diatoms, which has many industrial uses, the most important of which are as a filtering medium for clarifying sugar and syrups; as a filler in paper, paint, brick, ceramics and plastics; as insulation for boilers and furnaces; as a mild abrasive agent in metal polishes and toothpaste; and as a component of dynamite.

Diatoms (Bacillariophyta) a phylum of the kingdom PROTOCTISTA consisting of unicellular or colonial unicellular organisms lacking undulipodia and having delicately sculpted silica-impregnated cell walls (tests) divided into two overlapping valves. There are two classes, the radially symmetrical or *centric diatoms* (the CENTRALES) and the bilaterally symmetrical or *pennate diatoms* (the PENNALES). The cell walls are extremely resistant to decay and form deep deposits of diatomaceous earth or kieselguhr in lake and ocean beds. There are some 10 000 species of diatoms in at least 250 genera, most of them aquatic or marine but some living in soil. Their fossil record dates back to the CRETACEOUS PERIOD. They form the basis of many marine and freshwater food chains. Most are photosynthetic, but a few are saprobes. As in the CHRYSOMONADA (golden-brown algae), with which they are sometimes classed, the plastids contain chlorophylls *a* and *c*, β-carotene, and the xanthophylls FUCOXANTHIN, lutein, and diotoxanthin, and the main food reserve is CHRYSOLAMINARIN. However, they differ radically from the Chrysomonada in cell structure, cell division, and life cycle. They are primarily diploid, forming haploid gametes just before fertilization. Asexual reproduction is by FISSION and AUXOSPORE formation.

diatropism a TROPISM in which the plant part is aligned at right angles to the direction of the stimulus acting on it. A diatropic response to gravity (*diagravitropism* or *diageotropism*) is seen in the horizontal growth of rhizomes and stolons. A diatropic response to light (diaphototropism) is often seen in the positioning of leaf blades at right angles to the incident light. Compare ORTHOTROPISM, PLAGIOTROPISM.

diazotroph see NITROGEN FIXATION.

dibiontic see DIPLOBIONTIC.

dicarboxylic acid see CARBOXYLIC ACID.

dicaryon see DIKARYON.

dichasial cyme see INFLORESCENCE.

dichasium (*pl.* dichasia) See INFLORESCENCE.

2,4-dichlorophenoxyacetic acid see 2,4-D.

dichogamy (*adj.* dichogamous) the maturation of anthers and stigmas at different times in the same flower so that pollen reception and pollen presentation do not coincide. This reduces the chances of self fertilization (see AUTOGAMY) and enhances outbreeding. Compare HOMOGAMY. See also PROTANDRY, PROTOGYNY.

dichotomous (*n.* dichotomy) describing the system of branching where each division is into two equal parts. It occurs in many lower plants and algae: for example, the branching of the thallus in *Fucus* and the branching of the stem in *Lycopodium* and *Selaginella*.

dicliny (*adj.* diclinous) the separation of male and female reproductive parts into different flowers. Diclinous plants may either have female and male unisexual flowers on the same individual (monoecy) or on different individuals (dioecy). The term diclinous may also refer to the flowers themselves, in which case it is equivalent to the term unisexual. Compare HERMAPHRODITE.

Dicotyledonae (Magnoliopsida, dicotyledons) (*adj.* dicotyledonous) A class of the ANTHOPHYTA (angiosperms) in which the embryos usually have two cotyledons. Other features that distinguish the Dicotyledonae from the class Monocotyledonae include: having flower parts inserted in fours or fives; having a persistent primary root that develops into a taproot; and having the vascular bundles in a ring (see EUSTELE). Dicotyledons usually possess a cambium and may thus be either woody or herbaceous. There are some 193 700 species in over 10 500 genera and 321 families, ranging from the primitive Magnoliaceae to the advanced Asteraceae. In recent classifications the class Dicotyledonae is divided into some six subclasses (see MAGNOLIIDAE, CARYOPHYLLIDAE, HAMAMELIDAE, DILLENIIDAE, ROSIDAE, ASTERIDAE). In certain other classifications the Dilleniidae, Rosidae, and sometimes also the Asteridae are grouped together. Compare MONOCOTYLEDONAE.

Dicranales one of the most diverse and widely distributed orders of mosses (see BRYOPHYTA) with some 13 families. The plants usually have long, narrow leaves with a single nerve, erect gametophores and sporophores at the tips of stems. The peristome teeth are deeply forked. The order includes the genera *Campylopus*, *Ceratodon*, *Dicranum*, *Dicranella*, and *Leucobryum*.

Dictyosiphonales an order of brown algae (PHAEOPHYTA) containing algae with large parenchymatous sporophyte thalli that are not differentiated into holdfast, stipe, and lamina. The thalli may be solid, tubular, branched, saclike, or flat and foliose. Examples include *Stictyosiphon* and *Punctaria*.

dictyosome see GOLGI APPARATUS.

dictyostele see STELE, EUSTELE.

Dictyotales an order of brown algae (PHAEOPHYTA) that have parenchymatous thalli growing from an apical meristem. Reproduction is oogamous, with isomorphic alternation of generations. The sporophyte produces sporangia, which release aplanospores. The Dictyotales are commonest in tropical oceans. Some, such as *Dictyota*, are like branching ribbons, while others, e.g. *Padina*, form fan-shaped thalli, in some species encrusted with calcium carbonate. *Padina pavonia* (peacock's tail) is common on the shores of southern England.

differential centrifugation a technique for separating and isolating pure samples of cell organelles, i.e. cell fractionation. The cells are first broken up in a homogenizer to release the contents and the homogenate is filtered to remove cell wall fragments. It is then centrifuged and decanted several times in sequence, increasing the time and speed of centrifugation at each successive stage. Nuclei, membranes, chloroplasts, mitochondria, vesicles, endoplasmic reticulum, and ribosomes are collected in that order, the largest organelles separating at the slower speeds. Compare DENSITY-GRADIENT CENTRIFUGATION.

differentially permeable membrane see OSMOSIS, SELECTIVELY PERMEABLE MEMBRANE.

differentiation 1 the process by which amorphous cells, produced by meristematic cell division, undergo various physiological and morphological changes during maturation in order to become specialized for a particular function. Differentiation thus produces the specific cell types that make up the separate tissues composing a plant body. At a higher level, differentiation results in the formation of the various plant organs and the division into shoot and root and into vegetative and reproductive structures. Compare GROWTH. See also DEDIFFERENTIATION.

2 see STAINING.

diffuse-porous wood see ANNUAL RING.

diffusion the movement of molecules or ions in a fluid from areas of high concentration to areas of low concentration. In a closed system this will continue until the solution or gas mixture is evenly mixed. Diffusion is responsible for such processes as TRANSPIRATION and the uptake of carbon dioxide. The rate of diffusion depends on concentration differences and on the nature of the pathway the molecules have to take. Thus transpiration is greater when the air is dry and when the stomata are fully open.

diffusion pressure deficit (DPD, suction pressure) in older terminology, the net force or pressure that causes water to enter a plant cell. Its magnitude is the difference between the OSMOTIC PRESSURE (Π) and the TURGOR PRESSURE (TP), i.e. DPD = Π – TP. See WATER POTENTIAL.

digitate see PALMATE.

dihybrid an organism that is heterozygous for two particular genes, as in AaBb. Dihybrids are most conveniently produced from crossing parents that are homozygous for the genes in question, e.g. AABB × aabb or AAbb × aaBB. A TEST CROSS normally yields four kinds of offspring: A_B_ (both dominant characters expressed); A_bb, and aaB_ (only one dominant character expressed); and aabb (no dominant characters expressed). If the genes A/a and B/b are not linked, equal numbers of all kinds of offspring will be produced. If they are linked, two classes, the parentals, will be more common, and two classes, the recombinants (cross overs), will be rarer. Selfing a dihybrid normally yields a 9:3:3:1 ratio (see DIHYBRID RATIO) in the case of independent assortment but a much more complex result in the case of linkage. See also MONOHYBRID, TRIHYBRID.

dihybrid cross an experimental cross between two individuals, each of which is homozygous for different alleles of each of two genes. For each gene (trait), the first generation (F_1) offspring inherit a pair of nonidentical alleles (i.e. they are heterozygous for both genes).

dihybrid ratio a 9:3:3:1 ratio of phenotypes among the offspring of a single cross. Such a ratio can only be obtained providing that all the following conditions are met: two genes at different loci are involved, e.g. gene A/a and gene B/b; the immediate parents of these offspring are heterozygous for both genes, i.e. AaBb; these heterozygotes are selfed or crossed with one another, i.e. AaBb × AaBb; the genes are unlinked, i.e. INDEPENDENT ASSORTMENT; there is no interaction between the genes, e.g. no EPISTASIS; A is dominant to a, and B is dominant to b, i.e. no CODOMINANCE, etc.; fertilization is random, and all the gametes, zygotes, and offspring have an equal chance of survival.

If there is reason to expect a 9:3:3:1 ratio in a particular cross and this result is not obtained, one or more of the above conditions is not being fulfilled. For example, the ratio may be modified to 9:3:4 (= 9:3:[3+1]) in some cases of gene interaction (epistasis).

A dihybrid ratio is an example of a simple *Mendelian ratio*, being an expansion of the monohybrid 3:1 ratio, viz: $(3:1)^2$.

dikaryon (dicaryon) (*adj.* dikaryotic) A mycelium in which each segment contains two usually genetically distinct nuclei. Dikaryosis is common in basidiomycete fungi and results when two monokaryotic mycelia of different mating strains fuse. See HETEROKARYOSIS.

Dilleniidae a subclass of the dicotyledons containing both woody and herbaceous plants. The flowers may be unisexual or bisexual and the stamens develop in a centrifugal manner. The leaves are usually simple and spirally arranged. There are many primitive anatomical features. Some 14 orders are recognized, including: the Buxales, including the Buxaceae (boxes); the Dilleniales, including the Dilleniaceae (dillenias and hibbertias) and the Paeonaceae (paeonies); the Violales, including the Violaceae (violets, pansies), Salicaceae (willows), Passifloraceae (passion flowers), Caricaceae (papaya), Cucurbitaceae (pumpkins, gourds), and Begoniaceae (begonias); the Nepenthales, including the Droseraceae (sundews) and Nepenthaceae (pitcher plants); Malvales, including the Bixaceae (annatto), Dipterocarpaceae (dipterocarps), Tiliaceae (limes, basswoods), Sterculiaceae (cocoa, kola), Bombacaceae (baobab, balsa, durian), Malvaceae (mallows, cotton, hollyhocks), and Geraniaceae (geraniums); the Urticales, including the Ulmaceae (elms, hackberries), Moraceae (figs, hemp, mulberries), Urticaceae (stinging nettles), and Cannabaceae (cannabis, hop); the Euphorbiales, including the Euphorbiaceae (spurges, rubber) and Thymeliaceae (daphnes); the Theales, including the Theaceae (tea, camellias), Ebenaceae (ebonies, persimmons), Aquifoliaceae (hollies), and Sarraceniaceae (pitcher plants); the Lecythidiales, including only the Lecythidiaceae (Brazil nuts, cannonball tree); the Sapotales, including the Sapotaceae (gutta-percha, sapodilla); the Guttiferales, including the Clusiaceae (mangosteen, mammey apple); the Primulales, including the Myrsinaceae and the Primulaceae (primroses); the Ericales, including the Ericaceae (heaths) and Empetraceae (crowberries); and the Capparidales, including the Resedaceae (mignonette), Capparidaceae (Capparaceae) (capers), Brassicaceas (cabbages, mustards), Bataceae (Batidaceae) (saltwort), and Tropaeolaceae (nasturtiums). Other orders sometimes recognized include: the Paeoniales (here included in the Dilleniales); the Passiflorales, Cucurbitales, and Tamarales (here included in the Violales); and the Thymeleales (here included in the Euphorbiales). The Urticales is sometimes included in the Hamamelidae.

dimension analysis in studies of productivity, the detailed measurement of plant dimensions to determine relationships between external dimensions and dry-matter production. This enables future assessments of productivity to be made without the need for destructive sampling. The method is particularly important in studies of forest productivity. See PRIMARY PRODUCTIVITY.

dimer a protein made up of two polypeptide chains, for example, glucose-phosphate isomerase. When the amino acid sequence of the two subunits is identical, the protein is said to be *homomeric*; if the subunits are different, it is *heteromeric*.

dimerous (2-merous) Describing flowers in which the parts of each whorl are inserted in twos, as in the enchanter's nightshades (*Circaea*).

dimictic lake a lake in which there are two periods of vertical mixing of the water, in spring and autumn. Compare MEROMICTIC LAKE, MONOMICTIC LAKE.

dimorphic fungi fungi that can exist either as mycelia (hyphae) or as single, yeast-like cells, depending on environmental conditions.

dimorphic lichen a LICHEN whose thallus has characteristics of both foliose and fruticose lichens. The primary thallus is foliose, but bears an erect body of fruticose lichens, the secondary thallus.

dimorphism (*adj.* dimorphic) the existence of two distinct forms. The term may be applied to organelles (e.g. bundle sheath chloroplasts and mesophyll chloroplasts), appendages (e.g. sun and shade leaves or juvenile and adult leaves), stages of a life cycle (gametophyte and sporophyte), individuals (e.g. males and females, i.e. *sexual dimorphism*), etc. See also POLYMORPHISM.

dinitro compounds a group of chemicals including dinocap, which is effective in controlling powdery mildews, and dinitro-orthocresol (DNOC), used as a fungicide, herbicide, and insecticide.

dinitrogenase see NITROGENASE.

2,4-dinitrophenol SEE UNCOUPLING AGENT.

dinoflagellates see DINOMASTIGOTA.

dinokaryotic see DINOMASTIGOTA.

Dinomastigota (Dinoflagellata, Dinophyta, Pyrrophyta, dinoflagellates, dinomastigotes) a phylum of the kingdom Protoctista (in some classifications a division of the Plantae) consisting predominantly of unicellular marine algae with two undulipodia. There is no cell wall, but there may be a series of cellulose plates (the test or theca) under the cell membrane. In some species there are TRICHOCYSTS associated with pores in the theca. Unlike the typical eukaryote nucleus, the dinomastigote nucleus has chromosomes that are attached to the nuclear envelope and remain condensed throughout the cell cycle. There are no associated histone proteins. This condition is often described as *dinokaryotic* (or *mesokaryotic* – intermediate between prokaryotic and eukaryotic). The nuclear envelope remains intact during nuclear division and the spindle forms outside it. Microtubules enter the nucleus through channels in the envelope; some also attach to the nuclear envelope beside the chromosomes.

In a typical dinomastigote the test is divided into semicircular halves by a transverse groove. The plate comprising the transverse groove is termed the *cingulum* or *girdle*. The part of the dinomastigote test above the transverse groove is termed the *epicone*, and the part below is the *hypocone*; the hypcone has a longitudinal groove in the theca. There are typically two anterior laterally inserted undulipodia, one lying in each groove; the transverse undulipodium may be coiled over the cell surface.

Some dinomastigotes are photosynthetic and some are heterotrophic. Many are involved in symbiotic associations with marine coelenterates (such as sea anemones and corals), clams, and flatworms. Photosynthetic symbiotic species are often called *zooxanthellae*. Photosynthetic species are often dark brown owing to the presence of the pigment PERIDININ. The plastids usually also contain chlorophylls *a* and *c*$_2$, and sometimes also *c*$_1$, carotenes, and xanthins (sometimes including fucoxanthin). The plastids have a triple envelope, suggesting an endosymbiotic origin. The main photosynthetic storage product is starch, sometimes in PYRENOIDS. A distinctive dinomastigote organelle is the *pusule*, a series of vesicles near the base of the undulipodia, which appears to be involved in osmoregulation and in the uptake and secretion of macromolecules. Most dinomastigotes are haploid, but a few heterotrophic species are diploid. They may reproduce asexually by binary fission or zoospore formation, or sexually by the

production of gametes. The zygote develops into a thick-walled resting cyst, which undergoes meiosis immediately upon germination.

Dinomastigotes are found in the phytoplankton (see PLANKTON) in marine and freshwater environments worldwide. In nutrient-rich waters they may multiply rapidly to form RED TIDES. Some species are bioluminescent, while others release toxins that poison fish; the eating of contaminated fish or shellfish can lead to food poisoning. A few species are benthic (see BENTHOS).

Dinophyta see DINOMASTIGOTA.

dinucleotide see NUCLEOTIDE.

dioecious describing plants in which the female and male reproductive organs are separated on different individuals. Dioecy makes cross fertilization obligatory and ensures genetic variability in the population but this is done at the cost of lower seed-setting efficiency and also prevents isolated individuals reproducing. The sex of a dioecious plant may sometimes be determined by SEX CHROMOSOMES, as in red campion (*Silene dioica*), but sex expression is flexible in some plants, with occasional flowers of the opposite sex produced on a plant. In a few species, including bog myrtle (*Myrica gale*), individuals may change sex in different years. Compare HERMAPHRODITE, MONOECIOUS. See also DICLINY.

diphosphopyridine nucleotide (DPN) see NADP.

diplanetic describing organisms that successively produce two different types of zoospore (planospore), as do certain species of *Saprolegnia* (see SAPROLEGNIALES). Compare MONOPLANETIC.

diplobiontic describing a life cycle in which two types of vegetative plant are formed, one haploid and the other diploid (i.e. a life cycle in which there is an ALTERNATION OF GENERATIONS). The life cycles of ferns are diplobiontic. Compare HAPLOBIONTIC.

diploid a nucleus or individual having twice the haploid number of chromosomes in the nuclei of its somatic cells. In many lower plants and about 50% of flowering plants, diploidy is established at fertilization by the union of two haploid gametes, each containing a single set of chromosomes from its respective parent. The symbol 2*n* is used to denote the diploid number. Diploid plants usually have between 12 and 26 chromosomes in their nuclei. If they contain more than this then it is likely they are polyploid in origin. Compare HAPLOID. See also POLYPLOID.

diplontic describing a life cycle in which the diploid phase predominates and where the haploid stage is limited to the gametes. Such life cycles are seen in the diatoms and members of the Fucales. Compare HAPLONTIC. *See illustration at* life cycle.

diplophase the part of the life cycle in which the nuclei are diploid. Compare HAPLOPHASE.

diploplontic describing a life cycle that has no ALTERNATION OF GENERATIONS; a diploid organism produces gametes by meiosis.

diplospory a form of APOMIXIS in which the embryo forms directly from the megaspore mother cell. It is found, for example, in mountain everlastings (*Antennaria*).

diplotene see PROPHASE.

directional selection a form of selection in which the selection pressure on the range of PHENOTYPES for a particular characteristic in the population moves the mean phenotype towards one extreme of phenotype, so the frequency of certain alleles changes in a constant direction. This is often a response to a progressive change in environmental conditions. In artificial selection it is used to shift the population's phenotype in a direction desired by the breeder, for example towards plants that come into flower earlier in the year. Compare DISRUPTIVE SELECTION, STABILIZING SELECTION.

directed speciation a trend of speciation in which species do not form a continuum of gradually adapting PHENOTYPES across an environmental gradient, but rather a succession of species, each having evolved from the previous one or from a common ancestor. An example is Jacob's ladder (*Polemonium*) in which an ancestral form produced *P.carneum* in the cool, moist coastal region, then *P. caeruleum* evolved in the coniferous forest zone, *P. californicum* in the conifer/subalpine zone, followed by *P. eximum* on the mountain peaks.

disaccharide a carbohydrate in which two

MONOSACCHARIDE units are joined by a GLYCOSIDIC BOND. The commonest disaccharide in plants is SUCROSE. Other common disaccharides include MALTOSE and CELLOBIOSE. Less common disaccharides include the nonreducing trehalose, composed of two glucose units linked by an $\alpha(1-1)$ glycosidic bond. This is the main free sugar in *Selaginella* species.

Some disaccharides have specific functions, such as the translocation function of sucrose, while others (e.g. cellobiose) are breakdown products of larger molecules.

disassortative mating see ASSORTATIVE MATING.

disc floret see CAPITULUM.

discharge dispersal see BOLOCHORY.

disclimax see CLIMAX.

Discomycetes (cup fungi) a former class of the ASCOMYCOTA or Ascomycotina containing those ascomycetes that produce an APOTHECIUM. It included the orders RHYTISMATALES (formerly the Phacidiales), LEOTIALES (formerly the Helotiales), Ostropales, PEZIZALES, and Elaphomycetales, and the majority of the fungi that form lichens (see LECANORALES). The class is no longer used in modern classifications as it is considered not to be a natural grouping, but the term 'discomycete' is still used colloquially.

discontinuous distribution see DISJUNCT DISTRIBUTION.

discontinuous species see DISJUNCT SPECIES.

discontinuous variation (discrete variation, qualitative variation) the expression of a characteristic in perceptibly different ways by different individuals in a population. Examples include the tall and short forms of garden pea used by Gregor Mendel. Discontinuous variation is most obvious in characteristics determined by only one or two genes whose expression is not greatly affected by environmental factors. Such variation may be due either to recurrent mutation or to the existence of a balanced POLYMORPHISM. It is sometimes called *discrete variation*, and the population can be divided into discrete groups on the basis of expression of this characteristic. A character in which variation is discontinuous is termed a *qualitative* character. Compare continuous variation.

discrete variation see DISCONTINUOUS VARIATION.

disease resistance see RESISTANCE.

disjunct distribution (discontinuous distribution) the distribution pattern of a SPECIES that is not continuous within its range, i.e. it has at least two distinct centres of distribution. For example, magnolias grow in southeast Asia, eastern North America, and Central America, but nowhere in between. A disjunct distribution is caused by shrinking of the range of a formerly more widespread species as a result of changing environmental conditions, for example, cooling or warming of the climate. Some disjunct species may formerly have been more widespread, but climatic changes may later have restricted them to certain *refugia* (isolated areas) of more favourable microclimate. Examples are species of northern Europe and North America that were widespread during the recent glacial period but are now confined to isolated mountain tops, where they escape competition from more warmth-loving plants. Other species have had their former range fragmented by continental drift and consequent climate change. See also LUSITANIAN FLORAL ELEMENT.

disjunction the separation of HOMOLOGOUS CHROMSOMES at ANAPHASE of mitosis and meiosis, and their movement towards the poles of the nuclear spindle. Compare NONDISJUNCTION.

disjunctive symbiosis see SYMBIOSIS.

disk floret see ASTERACEAE.

disomy the condition in which there are two representatives of a particular chromosome in a cell, i.e. the diploid condition. Compare NULLISOMY, POLYSOMY, TETRASOMY, TRISOMY. See also HAPLOID, NONDISJUNCTION.

dispersal the tendency for seeds, fruits, spores, or other dispersal units to travel away from the parent plant, and to germinate in another place. A seed or fruit may show passive dispersal, simply falling under the influence of gravity and rolling away); the dispersal units may actually be clones spread on runners, rhizomes, or tubers and later separated from the parent; or dispersal may involve an agent such as wind (e.g. the winged fruits of sycamore, *Acer pseudoplatanus* and the winged seeds of pine, *Pinus*), water (e.g. coconuts), raindrops (e.g. the splash cups of some liverworts and lichens), or animals (e.g. burs and other hooked seeds) that catch in the coats of animals, and fleshy fruits that attract animals to eat them, passing out the intact seeds in their faeces. The pattern of dispersal and the distances travelled are important considerations in the design of nature reserves, and in attempts to regenerate damaged or destroyed habitats.

dispersal barrier (ecological barrier) an area of unsuitable habitat separating two areas of suitable habitat, large enough that individuals of a particular species are unable to cross it to colonize suitable habitats on the other side and to interbreed with other populations of the same species there. Examples include agricultural land separating two areas of woodland, or desert separating greener areas along river banks. See also GEOGRAPHICAL ISOLATION.

dispersalist biogeography the approach to BIOGEOGRAPHY that assumes that new species or groups of species arise in a centre of origin, and spread out from there in stages. Compare VICARIANCE BIOGEOGRAPHY. See *also* gene centre.

dispersion 1 in statistics, the pattern in which the values for one or more characteristics (variables) in different individuals in a population are distributed about the mean value.

2 the pattern of spatial distribution of individuals in their environment, either relative to a particular location or to each other. In *random dispersion* there is an equal chance of finding an individual at any place at a given time. This occurs only where individuals are not affected by the presence of other individuals or by patchiness in their environment. In *regular dispersion* the individuals are non-randomly distributed and tend to be more regularly spaced. This may be due to competition between individuals of the same population for resources, such as water or light, or where roots produce exudates that inhibit the roots of other individuals from growing too

close to them. An example is the creosote bush (*Larrea divaricata*) in North American deserts. In *aggregated dispersion* individuals occur in groups, This may be due to the patchy nature of their environment, or to a pattern of reproduction in which the offspring stay close to the parent, either because they arise from asexual reproduction, e.g. house leek, or because the seeds, fruits, or other dispersal units are released close to the parent, for example, mangroves.

dispersion coefficient a measure of the spread of data about a mean value or other standard reference point, for example, the standard deviation. See also DISPERSION.

disphotic zone see APHOTIC ZONE.

disruptive selection see SELECTION.

disseminule any part of a plant that may give rise to a new plant.

dissepiment see SEPTUM.

distal denoting the region of an organ that is furthest away from the organ's point of attachment. Compare PROXIMAL.

distely see stele.

distichous describing a form of ALTERNATE leaf arrangement in which successive leaves arise on opposite sides of the stem so that two vertical rows of leaves result. It is seen, for example, in grasses. The term also describes a form of OPPOSITE leaf arrangement in which the pairs of leaves all arise in the same plane, again giving two rows of leaves. Compare MONOSTICHOUS, TRISTICHOUS.

disulphide bridge (disulphide bond) in proteins, a bond formed between the sulphydryl (–SH) groups of neighbouring CYSTEINE molecules. Such bonds or 'bridges' can hold together different parts of a polypeptide or different polypeptides in a protein; they are important in determining the CONFORMATION of the protein.

dithiocarbamate fungicide any fungicide based on the organic sulphur compound dithiocarbamic acid. Such compounds include the commonly used fungicides ziram, thiram, zineb, and maneb.

diurnal 1 occurring in the daytime.
2 occurring daily, e.g. the opening and closing of many flowers. See also CIRCADIAN RHYTHM.

diurnal curve method a method of assessing gross primary or community productivity in aquatic ecosystems by measuring their oxygen production. Dissolved oxygen levels are recorded over a 24-hour period to assess the balance between oxygen production and consumption.

diurnal rhythm see CIRCADIAN RHYTHM.

divarication a distinctive wide-angled form of branching found in certain shrubs, which results in a three-dimensional interlacing of the branches. Sometimes the effect is exaggerated by the shoot bending at each node to form a zigzag axis.

divergent evolution (divergence) the development of different forms from a single basic structure as a result of different selective pressures acting on that structure. These changes may be associated with a new function that is added to or replaces the original function. For example, the general form of the angiosperm stem is aerial and elongated. Its function is to support and space out the leaves and flowers as well as conducting food and water. However certain xerophytic plants with reduced leaves have expanded leaflike stems to increase photosynthetic capacity. Other plants have short swollen stems that serve as underground perennating organs. See ADAPTIVE RADIATION, HOMOLOGOUS.

diversity (species diversity) the number of species in an area or community. This is usually combined with an assessment of their relative abundance. See also DIVERSITY INDEX.

diversity index a mathematical expression of the species diversity of a particular area or community that also takes into account the relative abundance of different species, e.g. the Shannon–Wiener index. Diversity indices are used to compare the structure and stability of different communities and to assess the effects of environmental factors on them.

division the second highest category in the taxonomic hierarchy. Kingdoms are made up of divisions, which are composed of classes. It is the equivalent of a phylum, the term used in the FIVE KINGDOMS CLASSIFICATION and in most animal classifications. The modern tendency is to use the term phylum rather than division, but the term division has tended to persist in plant and fungus classification. The number of divisions varies with the system of classification. The Latin names of plant divisions terminate in -phyta, e.g. Bryophyta, and those of fungi end in -mycota, e.g. Ascomycota.

divisive method a system of hierarchical classification that progressively breaks down taxa into smaller and smaller units.

DNA (deoxyribonucleic acid) the substance of which genes are made and that determines the inherited characteristics. In eukaryotes, it is mostly confined to the chromosomes, where it is found in association with basic proteins called HISTONES. Bacterial, mitochondrial, and chloroplast DNA is naked and forms circles or loops. Typically a DNA molecule consists of two polynucleotide chains, forming a right-handed helix with one coil positioned more or less underneath the other (the double helix, see diagram). There are only four types of NUCLEOTIDES in the polynucleotide chains, although each chain can be thousands of nucleotides long. Each nucleotide is constructed from DEOXYRIBOSE sugar, esterified to phosphate and a base. The base may be adenine or guanine (the purine bases), or cytosine or thymine (the pyrimidine bases). The sequence of nucleotides in a polynucleotide chain ultimately specifies the sequence of amino acids in proteins, i.e. forms the basis of the GENETIC CODE. The phosphate groups are on the outside of the helix. Each polynucleotide chain is established and held together by the formation of further ester bonds between the phosphate group of one nucleotide and the deoxyribose sugar of an adjacent nucleotide in the same chain. For this reason, each polynucleotide is sometimes described as having a 'sugar-phosphate' backbone (see diagram). The two polynucleotide chains in a double helix are antiparallel, i.e. in one chain the sugar-phosphate backbone is composed of 3′–5′ PHOSPHODIESTER BONDS, while in the other, it is composed of 5′–3′ phosphodiester bonds. The bases are on the inside of the helix, and each forms hydrogen bonds with a complementary base on the other helix. By the formation of these hydrogen bonds, the two helices are held together. The base

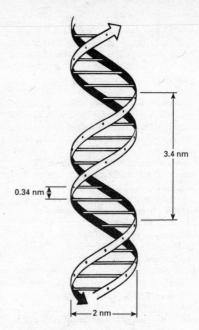

DNA: dimensions of the double helix

pairing is highly specific, adenine:thymine and guanine:cytosine being the only permissible combinations.

Evidence that DNA is the genetic material is substantial. By means of SEMICONSERVATIVE REPLICATION, DNA produces copies of itself with great accuracy, so that the genes (polynucleotide sequences) are normally passed on unaltered from one generation to the next.

DNA may be contrasted with RNA, the other main type of nucleic acid found in cells. The latter plays an important role in TRANSCRIPTION (copying) and TRANSLATION (decoding) the genetic message during protein synthesis. In viruses the genetic material is much more variable and can be double-stranded DNA (e.g. T-even bacteriophage), single-stranded DNA (e.g. φ174 bacteriophage), double-stranded RNA (e.g. φ6 bacteriophage), or single-stranded RNA (e.g. tobacco mosaic virus).

DNA fingerprinting see GENETIC FINGERPRINTING.

Segment of the **DNA** molecule showing the sugar-phosphate backbones running in opposite directions and hydrogen bonding between complementary base pairs

phosphodiester bond linking deoxyribose units of two nucleotides

T thymine A adenine
C cytosine G guanine ----- hydrogen bond

DNA hybridization a laboratory technique for measuring the similarity of the DNA of two species. DNA from each species is denatured by heat into single strands and a mixture of the DNA strands is incubated allowing them to recombine into hybrid DNA, a process called annealing. The amount of annealing is directly proportional to the similarity of the DNA strands and can be measured by heating the hybrid DNA and recording the temperature at which the strands separate. DNA hybridization studies in the grasses showed wheat and rye to be closely related and both to be close to barley but not to oats. This supports the traditional classification of wheat, rye, and barley in the tribe Triticeae and oats in the tribe Aveneae. The same technique is also used to hybridize RNA. Nucleic acid hybridization has many applications. One of the most important is the use of gene probes designed to bind to certain complementary base sequences in a mixture of different DNA or RNA fragments.

DNA ligase an enzyme that joins together short sections of newly synthesized polynucleotide chains (*Okazaki fragments*) during DNA REPLICATION, using energy from ATP. It is also used in the *in vitro* synthesis of DNA molecules.

DNA polymerase an enzyme that joins together free DNA nucleotides to form a complementary strand of DNA during DNA REPLICATION, using energy from ATP. See POLYMERASE.

DNA repair the natural repair of mutations that occur due to mistakes in DNA replication or nuclear division (see KARYOKINESIS), or to outside agents (mutagens) such as ionizing radiation or chemicals. Not all mistakes are repaired, but many are. Specific genes code for the enzymes involved in repair of different kinds of damage. For example, prokaryotes have *uvr* genes that aid in repair of damage done by ultraviolet light. In the process of *excision repair*, endonucleases cut out a section of one strand of the DNA molecule that includes the mutation, then DNA polymerase catalyses the joining of new bases to complete the strand correctly, and DNA ligase seals the ends of the new

section. However, if the mutation occurs during DNA replication, there is no second strand to act as a template for this kind of repair. In this case, *recombinant repair* occurs, where the damaged strand swaps the damaged section with one of the strands of an intact DNA molecule not undergoing replication. This second DNA molecule then undergoes excision repair. During replication, the DNA polymerase that adds nucleotides to the growing new DNA strand actually proofreads it for correct base pairing; if it finds the wrong base has been added, it acts as an exonuclease to remove the incorrect nucleotide, rather like using the delete key when word processing. If a mismatched base pairing escapes detection at this stage, it causes a distortion in the resulting DNA double helix, which may be recognized and repaired by excision repair. The enzyme determines the correct strand to change because newly formed DNA strands lack methyl groups, which are added shortly after replication. Ultraviolet light often causes a specific type of mutation, in which adjacent bases become linked vertically rather than horizontally (like rungs on a ladder). Many organisms, including PROKARYOTES and some plants and animals (but not humans) contain enzymes that can directly reverse this change using blue light. There are other complex mechanisms of repair that are used less frequently. Enzyme-mediated DNA repair is sometimes called *repair synthesis*. A measure of the success of DNA repair comes from bacteria, where the actual mutation rate during DNA replication is about one in 10^5 nucleotides added, but natural repair reduces this to one in 10^7.

DNA replication the production of new DNA molecules. The process involves the splitting of the original DNA molecule and the assembly of new complementary DNA strands alongside each old strand, using the parent strands as templates (see SEMICONSERVATIVE REPLICATION). A HELICASE enzyme unzips the DNA molecule, breaking the weak hydrogen bonds between the two strands and separating them. DNA POLYMERASE adds complementary free nucleotides (attracted by BASE PAIRING) to the strand running in the 5→3 direction. DNA polymerase moving in the opposite direction along the other strand produces only short segments of DNA (*Okazaki fragments*). These are joined together by DNA LIGASE.

doliform barrel-shaped or jar-shaped.

dolipore septum a type of septum found in many basidiomycete fungi, in which there is a narrow pore surrounded by a thickened rim and enclosed on both sides by caps of endoplasmic reticulum, often perforated by pores.

Dollo's law a law that states that evolution is irreversible. It probably applies only in the case of particularly specialized organisms, in which the mutations that would allow the organism to survive in its extremely narrow niche are very limited. In these cases, there is a steady directional selection pressure resulting in a self-perpetuating evolutionary trend (orthoselection).

domain 1 a distinct part of the TERTIARY STRUCTURE of a protein that performs a particular function; for example, a regulatory domain, or a domain that binds a particular cofactor or substrate.
2 see KINGDOM.

domatium (*pl.* domatia) A cavity inside a plant, usually in a stem, leaf, or root, that houses ants or mites. Domatia are formed by the plants even if the animals are absent; they are often associated with the production of FOOD BODIES or extrafloral nectaries. Some are simply hollows in internodes, petioles, or stipular spines (e.g. whistling thorns), others (as in *Myrmecodia*) are more elaborate systems of cavities. The benefits to the plant from this arrangement are not always clear, but in many plants the ants defend the plants against grazing animals or competition from neighbouring plants. In other cases it is thought that the remains of dead ants inside the plant may provide a supplementary source of nitrogenous compounds. The ants gain shelter and often an additional food source in the form of food bodies. The term domatia is sometimes also applied to hydathodes and substomatal cavities that are colonized by bacteria. These are found especially in the Rubiaceae.

domestication the modification (especially by artificial selection) and cross-breeding of naturally occurring species of plants and animals by humans. For example, modern wheat has been created over thousands of years by crossing different species of wild wheats, and by artificial selection for such features as better yields, or easier mechanical harvesting. See also BREEDING SYSTEM, CROSS POLLINATION, SELECTION.

dominant 1 describing an allele that masks the expression of a different (RECESSIVE) allele at the same locus. Thus the phenotype of the heterozygote resembles that of a plant homozygous for the two dominant alleles. The dominant allele is usually the normal and more common form in natural populations, i.e. the wild-type allele. Not all characteristics are governed by simple dominant/recessive relationships. See also CODOMINANCE, INCOMPLETE DOMINANCE.
2 describing the most abundant plant species in a community, e.g. in the UK *Quercus robur* (pedunculate oak) is dominant in most deciduous woodlands on nonacid clays and loams.
3 in forestry, a tree whose crown is more than half exposed to full illumination.

Domin scale a system for measuring the ABUNDANCE of plants in a particular area. It allows for a reasonably accurate assessment of plan COVER. The Domin scale is derived from the BRAUN–BLANQUET SCALE, but has more divisions (see TABLE).

dormancy (quiescence) an inactive phase often exhibited by seeds, spores, and buds, during which growth and developmental processes are deferred. It may be a means of surviving adverse environmental conditions, as in the formation of dormant perennating buds and the dormant resting spores of certain bacteria, fungi, and algae. However seeds often exhibit dormancy despite prevailing favourable conditions, in which case it may serve to give the seed time to mature fully. Dormancy may be triggered by changes in light, temperature, and moisture conditions. In some cases the organism enters dormancy in advance of the onset of adverse conditions. This is called *predictive dormancy*. In higher plants dormancy is controlled by plant hormones, particularly ABSCISIC ACID (formerly referred to as *dormin*), an increase in which tends to suppress growth and induce dormancy. The breaking of dormancy is often associated with an increase in the concentration of GIBBERELLINS and/or CYTOKININS. Seed dormancy may be broken in various ways, including the gradual degradation of an impervious testa or the progressive destruction of growth inhibiting substances such as coumarin, sometimes found in the seed coat. Seed or bud dormancy may also be broken by a specific light treatment. See also AFTER-RIPENING, COLD TOLERANCE, PERENNATING ORGAN.

dormin see ABSCISIC ACID.

dorsal In thallose plants, the upper surface away from the substrate. Compare VENTRAL.

dorsifixed describing an anther that is joined to the filament for some distance along its dorsal edge. Compare BASIFIXED, VERSATILE.

dorsiventral (n. dorsiventrality) describing a structure that has distinct upper and lower surfaces.

double fertilization the process, characteristic of flowering plants, in which two male gametes enter the EMBRYO SAC and both participate in fertilization. One male gamete fuses with the female gamete or egg nucleus to form a ZYGOTE, which develops into the embryo. The other male gamete fuses with either the POLAR NUCLEI or the DEFINITIVE NUCLEUS to form a triploid primary endosperm nucleus, which will give rise to the endosperm.

A form of double fertilization is also seen in some species of the gymnosperm genus *Ephedra*, in which one male gamete fuses with the egg cell and the other fuses with the VENTRAL CANAL CELL. However the product of the second fertilization does not undergo further development.

double flower a flower having more than the usual number of petals usually due to the transformation by a mutation of stamens into petals. Extremely double flowers, in which the carpels have also become petals, are termed *flore pleno*. An example is *Ranunculus aconitifolius* 'flore pleno' (fair maids of France). The transformation of reproductive organs into petals is called *petalody*.

double helix a molecule composed of two

similar polymer chains coiled in the same direction about the same axis, as in DNA. Actin, a fibrous protein found in most eukaryotic cells, also takes the form of a double helix. Double helices are, however, relatively rare configurations. Much more common is the single helix (α-helix; see SECONDARY STRUCTURE) characteristic of globular proteins.

double recessive a diploid individual homozygous for (containing two copies of) the same recessive allele of a gene, as indicated by the expression of the recessive allele in the phenotype.

double staining see STAINING.

downland GRASSLAND in lowland Britain that has been created and maintained by grazing. It is most common on chalk and limestone hills, where the usual grazers are sheep and rabbits. Thse grasslands are often rich in species of plants, butterflies and other invertebrates. Low-growing species such as dwarf thistle (*Cirsium acaule*), wild thyme (*Thymus praecox*), and bastard-toadflax (*Thesium humifusum*) are typical on rabbit-grazed downland.

downy mildew a plant disease caused by oomycetes of the family Peronosporaceae in the order PERONOSPORALES. The pathogen penetrates the tissues more deeply than POWDERY MILDEWS and produces a white or grey mealy growth on the leaves and stems of the host in humid conditions. This surface growth is composed of sporangiophores, which grow out from the stomata. Examples include downy mildew of vine (*Plasmopara viticola*), downy mildew of onion (*Peronospora destructor*), and downy mildew of lettuce (*Bremia lactucae*).

DPD see DIFFUSION PRESSURE DEFICIT.

DPN see NADP.

drip tip the tip of the leaf of many tropical plants, which is drawn out into a long narrow point that facilitates the run-off of rainwater from the leaf surface.

dropper (sinker) a slender rootlike structure that bears a bud produced in the axil of a corm or bulb. Later rotting away of the 'root' allows the bud to develop into an independent plant. The rootlike structure may be derived from the axis between the parent axis and the first leaf of the bud, or it may arise as an adventitious root adjacent to the bud and fused to it. Large droppers are usually called tubers.

drought a prolonged period during which water loss by evapotranspiration exceeds precipitation, so that soil moisture is reduced, and levels of groundwater and streamflow fall. Certain environments, such as the North American grasslands, experience a regular *drought cycle* of repeated temporary periods of drought (in these grasslands at intervals of about 22 years).

drought cycle see DROUGHT.

drupe a fleshy indehiscent fruit in which the seed or seeds are surrounded by a hardened sclerenchymatous endocarp, as in wild cherry (*Prunus avium*) and holly (*Ilex aquifolium*). The endocarp may replace the testa in its protective role and may also play a part in the dormancy mechanism. *See illustration at* fruit.

drupelet any of the individual small drupes that make up some fruits, such as blackberry (*Rubus fruticosus*).

druse (sphaeroraphide) a globular mass of needle-like crystals found either attached to the cell wall or free in the cytoplasm.

Dryas a part of the sequence of climatic changes between the last ice advance (the Devensian glaciation) and the warmer present (Flandrian) interglacial period. It is characterized by deposits rich in the pollen of mountain avens (*Dryas octopetala*), a typical tundra species. The Younger Dryas was an interval of abrupt cooling across Europe that began about 11 000 years ago.

dryfall the settling out of suspended particles from the atmosphere in dry weather. The particles are a source of nutrients for living organisms.

dry-matter production a measure of productivity in terms of the dry weight of organic matter procuced per unit area per unit time.

dry rot a fungal disease in which there is disintegration of the tissues and the affected cells crumble into a powdery mass. Dry rot of stored potatoes is caused by the fungus *Fusarium solani* var. *caeruleum* and that of timber by *Serpula lacrimans*.

dry season a period of very little or no precipitation that occurs at a particular time of year every year. In the Mediterranean region, on the west coasts of continents, and in subtropical climates it occurs in summer, when evapotranspiration is further reduced by the high temperatures. In the tropics the dry season occurs in winter, except very close to the Equator, where there may be two dry seasons a year as the equatorial rain belts migrate north and south of the Equator.

dune see SAND DUNE.

dune heath see SAND DUNE.

duplex double, or having two distinct parts. The term is especially used in describing the Watson–Crick model of the DNA double helix.

duplication 1 the occurrence of two or more copies of the same gene on a chromosome.

2 the occurrence of two or more copies of the same segment of chromosome sequentially on a chromosome.

duramen see HEARTWOOD.

duricrust a weathered surface layer of SOIL that consists of a hardened mass of silica (SiO_2), alumina (Al_2O_3), and iron oxide (Fe_2O_3), and often other oxides as well. Duricrusts are found particularly where forest cover has been removed, resulting in the lowering of the water table, or where the climate has become more arid. Duricrusts in some parts of Africa that currently have reasonable rainfall are attributed to past periods of aridity. They may also represent former hard pan soil horizons (see PODSOL) that have been exposed by soil erosion. The term is also used for crusts of calcium carbonate, gypsum, and salt.

Durvillaeales an order of kelplike brown algae (PHAEOPHYTA) whose members have a diffuse growth pattern and a free-living haploid phase.

Dutch elm disease a serious disease of elms (*Ulmus* spp.) caused by the fungus *Ceratocystis ulmi* (an ascomycete). The fungus is spread from tree to tree by elm-bark beetles (*Scolytus scoytu*). It spreads through the xylem vessels, and the plant responds by producing tyloses, bladder-like ingrowths of the parenchyma cells that penetrate infected xylem vessels through their pits to seal them off, thus starving the tree of water and minerals. This results in yellowing and curling of the leaves, wilting, and rapid death firstly of certain branches, but soon of the whole tree. An especially virulent strain of Dutch elm disease was introduced to Britain in the 1970s in infected timber imported from Canada, which killed most of the elms in England. This changed the appearance of the English countryside, which had been dominated by hedgerow elms and by elms grown in fields as shade for livestock.

dwarfism stunted growth, often due to a fault in genetic composition. The condition often involves the alteration of the proportions of the various body parts. In plants, dwarfism is frequently caused by a GIBBERELLIN deficiency, as in the case of the pea. The application of exogenous gibberellin usually causes internode elongation and the production of a plant of normal size.

dynein see UNDULIPODIUM.

dysphotic zone the part of the PHOTIC zone that lies below the compensation depth. Light penetration is so low that oxygen production by photosynthesis cannot exceed oxygen uptake by respiration.

dystrophic describing water that is brownish in colour owing to its high content of organic material (humus). Compare EUTROPHIC, OLIGOTROPHIC.

E*e*

E see ELECTRODE POTENTIAL.

-eae a suffix used in plant taxonomy to denote a tribe. See also RANK.

early wood (spring wood) the inner portion of a GROWTH RING, comprising wood produced in spring when growth is relatively rapid. The XYLEM vessels are wider in diameter than in late wood, and the wood is therefore less dense. Compare LATE WOOD.

earth stars see LYCOPERDALES.

ebractate lacking bracts.

ecesis the establishment of the first stage or SERE in a SUCCESSION.

echinulate covered in small spines.

ecocline a CLINE that has been shown to be due to a specific environmental factor. For example, a gradual increase in heavy-metal tolerance can be correlated with increasing metal concentration in the soil.

ecological amplitude the range of TOLERANCE of a species to a specific range of environmental factors. When plotted graphically it forms a bell-shaped curve. Species with a narrow ecological amplitude are restricted to a narrow niche (see ECOLOGICAL NICHE), so are often used as indicator species.

ecological and phytosociological distance a measure of the degree of dissimilarity in the composition of two samples. For example, $c \rightarrow a + b$, where a is the number of species in one sample that are not present in the other, b is the number of species present in the second sample but not in the first, and c is the number of species common to both samples. Compare AFFINITY INDEX.

ecological backlash the situation where the unpredicted consequences of a modification of the environment, such as the building of a dam or the straightening of a river, are so detrimental that they negate the expected benefits of the project.

ecological barrier see DISPERSAL BARRIER.

ecological efficiency (biological efficiency) the efficiency of energy transfer between successive trophic levels (see FOOD CHAIN). It is usually expressed as the percentage of the energy of the BIOMASS of a particular trophic level that was produced by the donor trophic level (i.e. by the organisms it consumed). Compare LINDEMANN'S EFFICIENCY.

ecological energetics the study of energy transformations and energy transfer within ecosystems.

ecological factor see LIMITING FACTOR.

ecological homeostasis the tendency of a plant population to adapt to environmental conditions, i.e. to be in balance with their environment.

ecological indicator see INDICATOR SPECIES.

ecological isolation the separation of groups of organisms, especially of populations or subpopulations, as a result of changes in their environment. Isolation limits or prevents gene flow between populations or smaller breeding groups, and over time this results in changes in the allele frequencies of the different groups as the different locations impose different SELECTION PRESSURES. Ecological isolation is one of the main processes leading to speciation, as groups may eventually become so different that they are unable to reproduce. See GENETIC DRIFT, NATURAL

SELECTION. See also GEOGRAPHIC ISOLATION, REPRODUCTIVE ISOLATION.

ecological niche (niche) the place and role occupied by an organism in a community, determined by its nutritional requirements, habit, etc. Different species may occupy a similar niche in different areas, for example, the grass species of the Australian grasslands, though different from those of the North American grasslands, occupy the same niche. Also, the same species may occupy a different niche in different areas. Generally the breadth of the niche varies depending on the adaptability of the species, the more adaptable species occupying a wider niche than the less adaptable or more specialized species. Niches for plants are generally poorly defined, as many basic requirements are similar. Differences may be defined by biotic interactions, e.g. mycorrhizas or diseases.

ecological pyramid see PYRAMID OF BIOMASS.

ecological system see ECOSYSTEM.

ecology the study of the relationships between living organisms and the living (biotic) and nonliving (abiotic) factors in the environment.

ecospecies see ECOTYPE.

ecosphere see BIOSPHERE.

ecosystem (ecological system) a unit comprising a COMMUNITY of living organisms and their environment. There is a continuous flow of energy and matter through the system. Ecosystems may be small or large, and simple or complex in structure. They range from small freshwater ponds or pools to the earth itself.

ecotone a transition zone or region separating two BIOMES. For example, in central Asia an area of wooded steppe, the grove belt, separates the coniferous forest from the temperate grassland.

ecotype a distinct population of organisms within a species that has adapted genetically to its local habitat. For example, some organisms may be able to tolerate different conditions of temperature or light intensity from other members of the same species. This may result in changes in their morphology or physiology. However, they are able to reproduce with other ecotypes of the same species and produce fertile offspring. These ecotypes may be sufficiently distinct to be given subspecies names, in which case they may be termed *ecospecies*. Compare CLINE.

ectexine see EXINE.

ectocarp see EXOCARP.

Ectocarpales an order of filamentous and pseudoparenchymatous brown algae (PHAEOPHYTA) with ISOMORPHIC alternation of generations and diffuse growth (apical growth in a few species). The group includes the simplest of the Phaeophyta. The thallus is differentiated into a branching basal area, spreading over the substrate to provide anchorage, and erect, more openly branching, filaments. In a few species (e.g. *Ralfsia*) the erect filaments coalesce to give a crustlike appearance. Chloroplast morphology varies greatly between species, ranging from simple platelike chloroplasts to ribbon-shaped, discoid, and irregularly shaped forms.

ectohydric describing a moss that absorbs water over its whole surface, including the leaves. Compare ENDOHYDRIC.

ectomycorrhiza see MYCORRHIZA.

ectoparasite see PARASITISM.

ectophloic siphonostele see STELE.

ectophytic mycorrhiza see MYCORRHIZA.

ectoplasm in the cells of plants and some protoctists (see PROCOTISTA), an outer, gel-like layer of the cytoplasm, containing closely packed layers of microtubules that lies immediately beneath the plasma membrane.

ectotrophic mycorrhiza see MYCORRHIZA.

edaphic factors (soil factors) the physical, chemical, and biological properties of SOIL that influence the life of organisms. The main edaphic factors include water content, organic content, texture, and pH. See also ABIOTIC FACTORS, BIOTIC FACTORS, CLIMATIC FACTORS.

edge effect 1 sampling errors that occur at the edges of sampling plots, due to items overlapping the periphery, or to items at the edge of a plot experiencing different environmental conditions from those in the centre of the plot. The effect is most pronounced in small sampling plots. **2** the tendency for an ecotone (transitional zone between two communities) to contain a greater variety of species and a greater density of populations than the communities on either side of it. **3** the tendency for water vapour to evaporate at a faster rate around the edge of a pore than in the middle. This has implications for the rate of transpiration through the stomata.

Edman degradation SEE PROTEIN SEQUENCING.

EDTA (ethylenediamine tetra-acetic acid) a compound that reversibly binds iron, magnesium, and other positive ions, i.e. that acts as a chelating agent (see CHELATION). Used in culture media complexed with iron it slowly releases iron into the medium as required. It may also act as a noncompetitive inhibitor of certain enzymes that require metal ions as cofactors. See also SEQUESTROL.

effector (inducer) a molecule that can combine with a repressor in an OPERON, preventing it from acting, and thus allowing the production of messenger RNA coding for a particular protein. See also JACOB–MONOD MODEL.

egg a large nonmotile female gamete, such as an OOSPHERE or a MEGASPORE.

egg apparatus the three haploid nuclei at the micropyle end of the EMBRYO SAC in most flowering plants. The central nucleus is the female gamete, and the nuclei on either side are called the SYNERGIDAE.

EIA SEE ENVIRONMENTAL IMPACT ASSESSMENT.

ejectosome in the CRYPTOMONADA, an organelle associated with the pellicle that contains tightly coiled ribbons, which can be discharged to poison prey in the same manner as TRICHOCYSTS.

ektexine SEE EXINE.

elaioplast a LEUCOPLAST in which oil is stored.

elaiosome (oil body) an oil-secreting structure found on fruits or seeds that attract ants as dispersal agents. In seeds it may be derived from a modified CARUNCLE or STROPHIOLE.

elater any of the numerous elongate helically thickened cells intermixed with the spores in the capsule of most liverwort and hornwort sporophytes. As the mature capsule dries out, tension builds up in the thickening of the elater. This tension is released when the capsule dehisces, dispersing the spores over a distance. See also HAPTERON.

electrode potential (E) **(reduction potential)** a quantitative measure of the tendency of an element to form ions in solution, i.e. the ease with which it is oxidized or reduced. It is defined in relation to a hydrogen half-cell, which is connected to a half-cell containing an aqueous solution of the oxidized and reduced forms of the test element by means of a salt bridge. The electrode potential represents the voltage needed to prevent the flow of electrons to or from the test half-cell.

electromagnetic radiation energy that flows through space or through a material medium at the speed of light in the form of fluctuating electrical and magnetic fields that make up oscillating waves, ranging from relatively low-energy radio waves and microwaves to visible light, ultraviolet light, X-rays, and gamma rays. Electromagnetic radiation interacts with charged particles in atoms, molecules, and larger pieces of matter. The energy and intensity of the radiation is related to the wavelength of the electromagnetic wave. In terms of quantum theory, electromagnetic radiation may be defined as the flow of photons (light quanta) through space. Photons are packets of energy of value $h\nu$ that travel at the speed of light, where h is Planck's constant, and ν is the frequency.

electron carrier a molecule capable of accepting and donating electrons and/or protons in an ELECTRON TRANSPORT CHAIN.

electron density the contrast between parts of a specimen observed by transmission electron microscopy. Electron dense structures deflect most of the electron beam and appear dark whereas electron lucent features do not deflect many electrons and appear bright. Such differences are often accentuated by the use of electron stains that render parts of the specimen electron dense that would otherwise appear electron lucent.

electron micrograph SEE MICROGRAPH.

electron microscope an instrument that uses electromagnetic lenses to focus a parallel beam of electrons and produce an image by differential electron scattering.

The RESOLVING POWER of an electron microscope is far greater than that of a LIGHT MICROSCOPE because the wavelength of an electron beam is 0.005 nm as compared to about 600 nm for visible light. In theory this should mean that objects only 0.0025 nm apart could be distinguished. However in practice aberrations in the lenses (electromagnetic fields), deterioration of the specimen during observation, and other technical difficulties reduce resolution to about 1 nm. This is still some 300 times better than that using light.

Two principal types of electron microscope are used. In the *transmission electron microscope* (TEM) the electron beam, which is usually produced from a tungsten filament, passes through a condenser lens system and is focused onto the specimen, which is normally an ultrathin section held on a fine copper grid. The image of the specimen, focused by the objective lens and magnified by the projector lens, is then projected onto a fluorescent screen or photographic plate. The entire microscope column is under high vacuum to prevent the beam of electrons from being scattered by the molecules in air. The casing of the microscope must therefore be completely dry and the specimen dehydrated. The magnification is altered and the focusing adjusted by varying the current through the electromagnetic lenses.

In the *scanning electron microscope* (SEM), the surface of the specimen is observed and a three-dimensional appearance is obtained. The beam of incident electrons ejects secondary electrons and back-scattered electrons from the specimen, which is usually treated by COATING. These reflected electrons are collected by a scintillator and an image is built up on a high-resolution cathode-ray tube as the electron beam passes over the specimen in a sequence of scanning movements. The resolution obtained in the scanning electron microscope is lower than in the transmission electron microscope but is constantly being improved upon and instruments capable of resolving 2 nm are now available.

Some electron microscopes are capable of operating in several different modes, either as scanning or transmission electron microscopes or in a scanning transmission electron microscope (STEM) mode. Microscopes combining optical and transmission electron microscopy are also available for making high and low magnification observations of the same specimen.

Since the electron beam produces X-rays from the specimen, attachments are also available that permit the composition of the specimen, or part of it, to be determined by comparison of the radiation produced.

One of the main limitations of the electron microscope apart from its expense is that living material cannot be viewed. The specimen has to be fixed, dehydrated, etc., before observation. Hydrated specimens can, however, be examined by LOW-TEMPERATURE SCANNING ELECTRON MICROSCOPY and one direction of current research is aimed at producing a special specimen chamber that could be operated at atmospheric pressure and isolated from the high vacuum system of the microscope. See also ATOMIC FORCE MICROSCOPY, COATING, STAINING, FREEZE DRYING, FREEZE FRACTURING, REPLICA PLATING, SHADOWING.

electron stain see STAINING.

electron transport chain a series of membrane-linked oxidation–reduction reactions in which electrons are transferred from an initial electron donor through a series of intermediates to a final electron acceptor (usually oxygen). The electron transport systems of respiration and photosynthesis are the major chemical energy sources in aerobic organisms.

A number of other electron transport systems are known besides those of respiration and photosynthesis. Heterotrophic bacteria have a RESPIRATORY CHAIN similar to that in eukaryotes and a microsomal electron transport system is responsible for desaturation of fatty acids.

electrophoresis a technique used to separate mixtures of solute molecules or colloidal particles in solution by placing them in an electric field. Molecules with a net positive charge (cations) move towards the cathode (negative electrode) and those with a net negative charge (anions) move towards the anode (positive electrode). The speed of movement of the molecules depends on their net charge, which in turn depends on the pH of the medium. The size of the molecule and the strength of the voltage applied also affect the speed of movement. Often the solvent is a gel of, for example, starch, agar, or polyacrylamide. These prevent the passage of small molecules, which become caught within the molecules of the gel (see GEL FILTRATION), so enabling clearer separation of the larger molecules. Electrophoresis is often used to separate the components of protein mixtures and may be used in conjunction with paper chromatography. The technique has been used extensively in studies of genetic variation, since the products of different ALLELES may have different charges, so will separate on the gel. This has revealed many details about plant population structure, breeding systems, and relationships. See also IMMUNOELECTROPHORESIS.

electrovalent bond see IONIC BOND.

elfin forest (Krummholz) the dwarfed and deformed trees that are found in the zone between the TIMBER LINE and the TREE LINE. They tend to grow along the ground and are thus better able to withstand the strong winds in such regions. The zone of elfin forest is termed the *Kampfzone.*

elicitor a molecule that initiates a cell signalling pathway resulting in the activation of protective genes leading to the production of defensive chemicals such as PHYTOALEXINS in response to infection by a pathogen.

El Niño (El Niño-Southern Oscillation Event, ENSO) a warm ocean current that flows southwards along the Pacific coast of tropical South America at intervals of approximately seven years. It is associated with a change in the atmospheric circulation called the *Southern Oscillation.* During an El Niño event, the easterly trade winds prevailing in the southern midsummer become weaker than usual, which allows the Equatorial countercurrent to strengthen. Warm surface ocean water, which is normally driven westwards towards Indonesia by the trade winds, instead flows westwards to overlie the cold

water of the northward-flowing Peru Current. In a severe El Niño event the normal upwelling of nutrient-rich cold water along the Pacific coast of South America is prevented, causing declines in fish stocks and widespread death of plankton. El Niño not only causes increased rainfall in the desert areas fringing the Pacific coast of South America, but also affects the climate across the Pacific region and causes droughts in parts of Africa. See also LA NIÑA.

elongation the process of cell enlargement that usually follows mitotic cell division. The main event is the uptake of water by osmosis accompanied by the enzyme-mediated stretching of the cell wall until turgor is reached. The process is effected by hormones, especially AUXINS. Differential elongation of cells on different sides of the root or shoot results in curvature. See also TROPISM.

elongation factor see TRANSLATION.

Eltonian pyramid see PYRAMID OF BIOMASS.

eluate see ELUTION.

eluent see ELUTION.

elution the removal of an adsorbed substance in a CHROMATOGRAPHY column or an ion-exchange column by means of a solvent (the *eluent*), forming a solution called the *eluate*, which drains out of the column. When several substances are adsorbed on the chromatography column, a process called graded elution is used, in which the composition of the eluent is changed in a stepwise manner, beginning with a nonpolar solvent and gradually changing to a more polar solvent. This ensures the separation of components as they are washed from the column.

eluviation the removal in solution or suspension of components from the surface soil horizons, and their (partial) deposition in lower soil horizons. The term LEACHING is more commonly used for removal in solution, and the term eluviation is mainly used to refer to removal of substances in suspension. See also SOIL PROFILE.

emarginate describing a leaf, petal, or sepal that is indented at its tip. The notched petals of many cranesbills (e.g. *Geranium versicolor* and *G. pyrenaicum*) are examples.

emasculation the removal of the anthers of

a flower to prevent either self-pollination or the pollination of surrounding plants.

Embden–Meyerhof–Parnas pathway see GLYCOLYSIS.

embedding a process in microscopical preparation in which delicate tissue is impregnated by, and embedded in, a solid supporting medium, such as paraffin wax, to enable thin sections of the material to be cut with a microtome. Tougher embedding materials (e.g. epoxy resins) are used in preparations for the electron microscope.

embryo the young plant individual after fertilization or parthenogenesis when the PROEMBRYO has differentiated into embryo and suspensor. Embryo cells are typically thin walled with dense cytoplasm and maintain mitotic activity. Cells exhibit polarity from early on, and in higher plants a plumule, radicle, and cotyledons may be identified. In seed-bearing plants the embryo is protected by integuments, which later form the seed coat. Further embryo development in the seed is usually controlled until environmental conditions are appropriate. Compare ZYGOTE. See EMBRYOGENY.

Embryobionta see EMBRYOPHYTA.

embryo culture the growth of isolated plant embryos on suitable media in vitro. The technique is useful in plant breeding as it enables certain hybrids to be raised that would abort if left on the plant, either because of endosperm breakdown or endosperm incompatibility. See also OVULE CULTURE.

embryo dune see SAND DUNES.

embryogenesis see EMBRYOGENY.

embryogeny (embryogenesis, embryony) the development of an embryo, normally from a fertilized egg cell. The first divisions of the zygote are at right angles to the long axis of the archegonium or embryo sac. In angiosperms and in some gymnosperms and pteridophytes these initial divisions give rise to a chain of cells, the SUSPENSOR. The embryo proper develops from a large cell at the tip of the suspensor at the end furthest from the micropyle. The pattern of embryo development differs markedly between groups of plants but normally globular, heart-shaped, and torpedo-shaped stages can be recognized. Embryogeny often

lags behind the development of other parts of the seed so the embryo may grow at the expense of the previously formed endosperm. Certain cells of the embryo may be polyploid if DNA synthesis outstrips nuclear division.

The term embryogenesis is also applied to the development of embryos from diploid somatic cells in suspension cultures. This phenomenon has been described in a number of angiosperm species and demonstrates the TOTIPOTENCY of cells when removed from the inhibiting influences of the plant body. Such nonzygotic embryos are often termed *embryoids* to distinguish them from zygotic embryos. However they closely resemble zygotic embryos in their development and if transferred to suitable culture conditions may give rise to normal plants. Embryoids can also be induced to form from pollen grains and subsequently may develop into haploid plants.

See also ADVENTIVE EMBRYONY, ENDOSCOPIC EMBRYOGENY, EXOSCOPIC EMBRYOGENY, OVULE CULTURE.

embryoid see EMBRYOGENY.

embryology the study of the development of embryos, usually from the moment of fertilization.

embryony see EMBRYOGENY.

Embryophyta (Embryobionta) in old classifications, one of two subkingdoms of plants (the other being the THALLOPHYTA). It contained the bryophytes and vascular plants. These plants were grouped together because they all have an embryo stage that is dependent on the parent plant for a greater or lesser period.

embryo sac a large oval cell in the nucellus of the ovule of flowering plants within which fertilization occurs. Initially it is the megaspore mother cell, which divides by meiosis, normally forming four megaspores. Typically three of these abort and the remaining one divides by mitosis to give the haploid cells of the embryo sac. This form of development is termed *monosporic*. Usually there are three mitotic divisions to give an eight-nucleate embryo sac. The nuclei are arranged as shown in the illustration. However in some plants, e.g. *Oenothera*, there are only two mitotic divisions giving a four-nucleate embryo sac (lacking the three antipodal cells and one of the polar nuclei).

In *bisporic development*, two megaspores contribute to the formation of the mature embryo sac. For example, in *Allium* the cells of the embryo sac develop from one of the two cells formed after the first division of meiosis (the other product of the reduction division aborting at this stage). Three divisions of this cell give rise firstly to two megaspores (a dyad) and then to an eight-nucleate embryo sac.

In some plants all four megaspores may continue development (*tetrasporic development*). In such cases there may be one mitotic division to give an eight-nucleate embryo sac, e.g. *Adoxa*, or two mitotic divisions, giving a sixteen-nucleate embryo sac, e.g. *Drusa*. The arrangement of cells in embryo sacs derived by tetrasporic development may show a variety of forms.

The mature embryo sac represents the female gametophyte, the egg cell being the gamete. See also DOUBLE FERTILIZATION.

emergence a structure of epidermal and subepidermal origin that develops on a shoot or root but not at the location of a leaf or shoot primordium. Emergences may take the form of spines, prickles, or thorns, or they may be more specialized, as in the case of FOOD BODIES. They are more substantial structures than hairs or trichomes.

emergence marsh the uppermost zone of a SALT MARSH, from the normal mean high-water level to the mean high-water level of spring tides. It experiences fewer than 360 submergences a year, usually with less than one hour of submergence during the day time. The minimum period of continuous exposure to the air may exceed 10 days.

emergent 1 an individual tree or group of trees that grows significantly higher than the continuous canopy of tropical rainforest. 2 an aquatic plant in which most of the vegetative parts are above water.

Emerson effect (enhancement effect) the observation (made by Robert Emerson in 1957) that photosynthesis, which proceeds very slowly using light of 700 nm wavelength, can be greatly increased when chloroplasts are also illuminated with light of shorter wavelength (650 nm). This was a surprising observation as it was then thought that light absorbed by the chlorophylls and other pigments was all passed on to a small percentage of chlorophyll *a* molecules (the energy trap) absorbing at 700 nm. This and later work indicated a second energy trap absorbing at 680 nm.

enation 1 an outgrowth, usually from a leaf. Enations may occur in response to virus infection, e.g. cotton leaf curl virus. 2 a small lateral outgrowth of the stem that is produced by certain primitive vascular plants and may be an early stage in MICROPHYLL evolution.

endangered species a species of plant, animal, or other living organism that is in danger of extinction. See also CONVENTION ON INTERNATIONAL TRADE IN ENDANGERED SPECIES OF WILD FAUNA AND FLORA, RED DATA BOOK.

endarch describing xylem maturation in which the older cells (protoxylem) are nearer the centre of the axis than the younger cells. Development is thus centrifugal. Compare EXARCH, MESARCH.

endemic describing a species that grows in a specific area and has a restricted distribution. Some species (broad endemics) are restricted to a particular large region. Other species (narrow endemics) are confined to a much smaller area, such as a few square kilometres, and tend to be very specialized. Examples of broad endemics are the sugar maple (*Acer saccharum*) in the eastern United States and the cacao tree (*Theobroma cacao*) in the Amazon basin. An example of a narrow endemic is *Darcycarpus viellardii* in the island of New Caledonia. Some endemics (palaeoendemics) represent the relicts of once widespread species. For example, *Lyonothamnus floribundus*, which today is only found on the California Islands, grew throughout California in the Tertiary. Compare COSMOPOLITAN. See also DISJUNCT DISTRIBUTION.

endexine see EXINE.

endobiont see ENDOPHYTE.

endobiotic growing inside a living organism.

endocarp the innermost layer of the PERICARP of an angiosperm fruit, internal to the mesocarp and exocarp and external to the seed(s). Sometimes the endocarp consists of a stony layer, as in the fruit of the peach (*Prunus persica*).

endocytosis the entry of particles or fluid into cells by methods other than diffusion or active transport across the plasma membrane. In *phagocytosis*, exhibited by some unicellular holozoic algae, protrusions of the outer regions of the cell, formed by flowing movements of the cytoplasm, surround food particles, enclosing them in a membrane-bound food vacuole. Lysosomes become associated with the food vacuoles, the intervening membranes break down, and hydrolytic enzymes are released to digest the particles.

Pinocytosis is a process by which submicroscopic particles (macromolecules or molecular aggregates) or droplets of extracellular fluid enter cells. The particles adhere to the plasma membrane (absorptive endocytosis) and an invagination forms. This becomes pinched off so taking the particles into the cytoplasm in a membrane-bound VESICLE – a *pinocytotic* or *endocytotic vesicle*. Alternatively extracellular fluid enter a pitlike invagination that becomes pinched off and enters the cytoplasm (fluid endocytosis). Water is absorbed from the vesicles as they move through the cytoplasm. Eventually the membrane breaks down, releasing the contents into the cytoplasm. Compare EXOCYTOSIS.

endocytotic vesicle see ENDOCYTOSIS.

endodermis a layer of cells at the boundary of the CORTEX and stele, usually regarded as the innermost layer of the cortex. It is generally clearly seen in all roots and in pteridophyte and certain dicotyledon stems. A CASPARIAN STRIP is usually present. When the endodermis contains numerous starch grains, it is known as a *starch sheath*.

endogamy see AUTOGAMY.

endogenous describing any process, substance, or organ that arises from within an organism. Examples are ENDOGENOUS RHYTHMS, the origin of lateral roots from the pericycle, and the capacity of certain tissue cultures to manufacture their own hormones. Compare EXOGENOUS.

endogenous rhythms sequential physical or biochemical processes that occur in a plant or plant part in response to internal stimuli. An example is the apparently autonomic component of the nyctinastic movements of certain plants, the leaflets of which continue to follow a periodic folding and unfolding sequence even if kept in continuous darkness (see NYCTINASTY). See also CIRCADIAN RHYTHM, BIOLOGICAL CLOCK.

endogenous root see ROOT.

endohydric describing a moss that absorbs water through the rhizoids at the base of the shoots. Compare ECTOHYDRIC.

endolithic living inside rocks.

endomitosis a sequence of changes in the nucleus resulting in division of the chromosomes, as in MITOSIS, but no separation of the chromatids into daughter nuclei. The resulting nucleus is therefore POLYPLOID. The process can be induced in isolated tissues by treatment with colchicine, which prevents spindle formation so the CENTROMERES of the daughter chromosomes are unable to move apart into separate nuclei. It may occur as an error in part of a plant, producing, for example, a tetraploid branch on a diploid plant. It occurs as a normal feature in some tissues of higher plants, e.g. the phloem cells of some leguminous plants are polyploid. This type of polyploidy, where some of the cells of a plant have more than the normal complement of chromosomes for the species, is known as *endopolyploidy*. If endomitosis occurs in cells in the germ line or during the second division of meiosis then unreduced gametes may result.

Endomycetales see SACCHAROMYCETALES.

endomycorrhiza see MYCORRHIZA.

endonuclease an enzyme that catalyzes the hydrolysis of bonds inside polynucleotides such as RNA and DNA, producing shorter chains of nucleotides. See also RESTRICTION ENDONUCLEASE.

endoparasite see PARASITISM.

endopeptidase see PEPTIDASE.

endophyte (endobiont) a bacterium, protoctist, or fungus that lives inside the tissue of a plant or other photosynthetic organism. Compare EPIBIONT.

endophytic mycorrhiza see MYCORRHIZA.

endoplasm in the cells of plants and some

protoctists (see PROTOCTISTA), the layer of cytoplasm situated below the ECTOPLASM, which contains the main organelles.

endoplasmic reticulum (ER) a complex interconnecting system of flattened membrane-surrounded channels and vesicles (cisternae) that spreads throughout the cytoplasm of eukaryotic cells. The membranes appear to be continuous with the PLASMA MEMBRANE at the outer surface and with the TONOPLAST and NUCLEAR MEMBRANE, and have the same general structure. Some cisternae have a granular appearance in electron micrographs due to ribosomes that are attached to the cytoplasm side of the enclosing membranes. They are thus called *rough endoplasmic reticulum*. Others are devoid of ribosomes and are called *smooth endoplasmic reticulum*. Polypeptides can be identified in the channels of the rough endoplasmic reticulum. The function of the endoplasmic reticulum is the segregation of newly synthesized products and their further processing and intracellular transport. See also GOLGI APPARATUS.

endopolyploidy see ENDOMITOSIS.

endoribonuclease see RIBONUCLEASE.

endoscopic embryogeny EMBRYOGENY in which the embryo develops from the inner of the two cells that result from the first division of the zygote. The outer cell often forms the suspensor. Endoscopic embryogeny is seen in the Lycophyta, most of the Filicinophyta, and all the gymnosperms and angiosperms. Compare EXOSCOPIC EMBRYOGENY.

endosperm the storage tissue in the seeds of most angiosperms (but no other seed plants), derived from the fusion of one male gamete with two female polar nuclei. The endosperm is a compact triploid tissue, lacking intercellular spaces and storing starch, hemicelluloses, proteins, oils, and fats. Seeds containing an endosperm or PERISPERM at maturity are termed *albuminous*, whereas those lacking such tissue are termed *exalbuminous*. See also ALEURONE LAYER.

endospore 1 a spore produced inside a bacterium or fungus, which is highly resistant to heat and desiccation. In endospore-forming bacteria, each parent cell produces a single spore, which is released when the parent cell shrivels up and may be dispersed by wind or water. Endospores may survive for many years without water or nutrients until conditions become favourable for resuming growth. 2 in fungi, the inner wall of a spore.

endosporic gametophyte a gametophyte that develops within a spore. For example, the female gametophyte of *Selaginella* is contained within the megaspore and at maturity only a portion, bearing the archegonia, is exposed through the spore wall. Such gametophytes are better able to withstand dry conditions and this pattern of development may be seen as a step in the evolution of the seed habit.

endosymbiont a symbiont that lives inside the cells of its host.

endosymbiotic theory the proposal that over the course of evolution complex cells have arisen by the symbiotic association of simpler types of cells. The chloroplasts of eukaryotic cells are thought to be derived from symbiotic photosynthetic bacteria, and the mitochondria from other aerobic bacteria. These cell organelles possess a double outer membrane and have their own DNA and ribosomes. In some groups of algae, there appear to be plastids derived from more than one type of symbiont. See SYMBIOSIS.

endothecium 1 (fibrous layer) the subepidermal layer of the ANTHER wall in angiosperms. The cells of the endothecium often develop thickenings on the anticlinal and inner tangential walls as the anther matures, which are believed to aid in anther dehiscence.

2 the inner layer of cells in the young sporophyte of bryophytes giving rise to the columella and (except in *Sphagnum*) sporogenous tissue. Compare AMPHITHECIUM.

endotoxin see TOXIN.

endotrophic mycorrhiza see ENDOMYCORRHIZA.

endozoochory see ZOOCHORY.

end-product inhibition see NEGATIVE FEEDBACK.

energy flow the transfer of energy between the different trophic levels of a food chain or food web.

energy transformation the changing of one kind of energy into another. For example, during photosynthesis light energy is transformed into chemical energy, while in respiration chemical energy is transformed into heat energy.

enhancement effect see EMERSON EFFECT.

enhancer a region on a DNA molecule where a regulator protein binds. Such a region may be some distance from the promoter and the start site for transcription of the regulated gene. See also OPERON.

ensiform sword-shaped.

ENSO see EL NIÑO.

enterobacteria bacteria of the family Enterobacteriaceae, a large family of rod-shaped, Gram-negative (see GRAM STAIN) bacteria that includes many parasites and pathogens of plants, animals, and other organisms. They are CHEMO-ORGANOTROPHS, and can can live in anaerobic conditions, such as are found inside their hosts. Many also live as saprobes in soil and water.

enterotoxin a toxin produced by certain bacteria, which acts on the lining of the intestines in animals, causing complaints such as gastroenteritis and diarrhoea.

entire describing a leaf, petal, or sepal margin that has a smooth undivided outline. *See illustration at* leaf.

entomochory dispersal of seeds or spores by insects.

entomophily (insect pollination) POLLINATION by pollen carried on insects. Insect-pollinated plants need to be able to attract the pollinating agent and provide a suitable landing space for it and then deposit pollen onto and collect pollen off the visitor. Insects may be attracted to a flower through the provision of food, by pseudo-mating signals (for example, some orchids mimic the shape, colour, and odour of the female insect), or by provision of brood or shelter sites. Pollen itself may be provided as the food source as it is protein rich, but it needs to be produced in quantities sufficient to offset the loss. More often nectar is offered as a high-energy food. This is usually secreted from a nectary so placed relative to the reproductive parts that pollen collection and deposition is ensured. Fats, oils, and water are also used to attract pollinating insects. Some plants deceive insects into visiting although no food is offered.

Adequate landing sites may be provided by increase in individual petal size, by increase in size of all the petals, or by the clustering of flowers into a compact inflorescence, such as an umbel. Recognition of the plant by the pollinating insect is achieved through secondary attractants, such as brightly coloured petals or insect-like movements.

Special arrangements of the reproductive parts can be seen in many insect-pollinated species. Flowers that can be pollinated by a number of different insects are termed *allophilic* or promiscuous, while those that can only be pollinated by one specific agent are termed *euphilic* (see also MUTUALISM). Some flowers may deposit pollen all over the agent but more specialized flowers deposit pollen only on certain areas of the vector. The pollen grains tend to be heavily sculptured and sticky in order to adhere to the agent's body.

Compare ANEMOPHILY, HYDROPHILY.

Entomophthorales an order of fungi of the ZYGOMYCOTA (zygomycetes) in which the spores are forcibly discharged. Usually the whole sporangium is ejected. Most are parasites on animals, especially insects; others live on dung.

entropy 1 a measure of disorder – of the degree to which matter and energy is dispersed to form a uniform, inert system, as opposed to being organized into complex molecules and structures. 2 a measure of the unavailable energy in a thermodynamic system.

enucleate 1 describing a cell that lacks a nucleus.

2 to remove a nucleus from a cell.

enumeration data data in which the material under observation falls into discrete classes.

environment the conditions in which an organism lives. The external environment includes the nonliving physical and chemical factors such as light, nutrients, etc., together with the effects of other living organisms. Genetically similar plants growing in different environments may appear different (see PLASTICITY). Such intraspecific variation is termed environmental variation and is not inherited. An organism also has an internal environment, which is the result of its own metabolism. See ABIOTIC FACTORS, BIOTIC FACTORS.

Environmental Impact Assessment (EIA) a study that attempts to predict the effects of proposed industrial developments, new housing, engineering projects, legislation, etc. on the natural and manmade environment, and hence on human health and wellbeing. This may include both positive and negative effects.

environmental resistance a sum of deviations from the optimum, non-limiting (biotic and abiotic conditions, i.e. those conditions in which an organism or population can attain its maximum possible rate of reproduction (see BIOTIC POTENTIAL). Such factors include the availability of food, water, and space, the numbers of predators, parasites and diseases, pollution, and intraspecific and interspecific competition.

environmental science the study of environments. the term may be used to encompass both biotic and abiotic environments (see ABIOTIC FACTORS, BIOTIC FACTORS), as well as cultural and social factors, or it may be restricted to the physical environment.

environmental stochasticity random variation in the abiotic environment.

environmental variance the proportion of the variation in the PHENOTYPES in a population that is due to differences in the environment experienced by individuals in the population. Compare GENETIC VARIANCE.

enzyme a protein molecule specialized to catalyse biological reactions. The extremely high specificity and activity of enzymes enables the living cell to function at physiological temperatures and pH values. Enzymes work by lowering the ACTIVATION ENERGY of the reaction. Without enzymes, nearly all metabolic processes would require high temperatures and extreme pH values, or would produce excessive amounts of heat.

The catalytic activity of an enzyme is due to its possession of an ACTIVE SITE. The three-dimensional conformation and the charge distribution of the active site are critical; in some enzymes these are entirely maintained by the TERTIARY STRUCTURE of the protein, but in other enzymes COFACTORS or COENZYMES are required for catalytic activity.

Enzymes are classified according to their function by an internationally recognized system. There are six classes of enzyme, the OXIDOREDUCTASES, the TRANSFERASES, the HYDROLASES, the LYASES, the ISOMERASES, and the LIGASES. Each class has a code number, and each is subdivided into subclasses and sub-subclasses. Thus any enzyme has a common or recommended name, a systematic name indicating the reaction that it catalyses, and a four-digit code number.

See also SUBSTRATE.

enzyme technology the use of isolated purified enzymes as catalysts in industrial processes. The enzymes used are normally extracellular enzymes with no requirements for complex cofactors. Examples are proteases, amylases, cellulases, and lipases extracted from bacterial or yeast cultures. Cell-free enzymes have certain distinct advantages over the use of microorganisms, a major one being that only the desired enzymatic reaction will occur and substrate is not wasted in the formation of unwanted by-products or microbial biomass.

Eocene see TERTIARY.

eon see GEOLOGICAL TIME SCALE.

eosin an acid stain that colours cytoplasm pink and cellulose red. It is often used with a counterstain, such as methylene blue or haematoxylin. Wright's stain, Leishman's stain, and Giemsa stains are all mixtures of eosin and methylene blue mixed with alcohol.

Ephedrales see GNETOPHYTA.

ephemeral (ephemerophyte) a plant with a short life cycle that may be completed many times in one growing season. Examples are groundsel (*Senecio vulgaris*) and shepherd's purse (*Capsella bursa-pastoris*). Some desert plants also have short life cycles that can be completed in a short period following rain. Many ephemeral plants have seeds that can remain dormant for long periods. Compare ANNUAL, BIENNIAL, PERENNIAL.

ephemerophyte see EPHEMERAL.

epi- prefix denoting 'on the outside'. For example, the epicarp is the outer layer of a fruit wall.

epibasal cell the outer of the two cells formed by the first division of the zygote. It gives rise to the embryo in plants showing EXOSCOPIC EMBRYOGENY and to the foot in those showing ENDOSCOPIC EMBRYOGENY. Compare HYPOBASAL CELL.

epibiont an organism that lives on the surface of another organism. Such organisms are described as *epibiotic*.

epibiontic describing endemic taxa (see TAXON) that have a very restricted distribution today, but which were once more widely distributed.

epiblem see EPIDERMIS.

epicalyx a calyx-like extra whorl of floral appendages, positioned below the calyx. The individual segments resemble sepals and are termed *episepals*. An epicalyx is found in the strawberry flower (*Fragaria vesca*) and in the tree mallow (*Lavatera arborea*). Compare INVOLUCRE.

epicarp see EXOCARP.

epicautical root in parasitic or hemiparasitic plants, a runner-like structure that spreads over the outer surface of the host, putting out sucker-like discs at intervals, from which haustoria penetrate the host.

epicone the part of a dinoflagellate cell (see DINOMASTIGOTA) anterior to the CINGULUM (girdle).

epicormic branching the development of twigs or clusters of twigs at discrete points on the trunk or branch of a tree, often marked by swellings. The buds giving rise to these twigs are of two types. Adventitious buds may arise deep in the existing trunk tissue (endogenous origin) by reactivation of a meristem. 'Preventitious' buds develop from normal axillary buds on the trunk, often multiplying by producing further axillary buds in the axils of their own scale leaves.

epicotyl the apical end of the axis of an embryo above the cotyledon(s), which gives rise to the stem and associated organs. Compare HYPOCOTYL, PLUMULE.

epidemiology the study of factors causing and influencing widespread outbreaks of disease.

epidermis (*adj.* epidermal) the outermost cells of the primary plant body, usually consisting of a single layer but sometimes several layers thick (multiple or multiseriate epidermis). In either case the cells differentiate from the PROTODERM. In stems and roots exhibiting secondary growth, the epidermis is usually replaced by the PERIDERM. The terms *epiblem* or *rhizodermis* are sometimes used in place of epidermis for the outermost layer of cells in the root. The main function of the epidermis is to protect the underlying tissues from excessive water loss and, to some extent, from physical injury and attack from pathogens.

epigeal describing seed germination in which there is considerable elongation of the HYPOCOTYL so the cotyledons are raised above the surface of the ground to form the first leaves (seed leaves) of the plant. Epigeal germination is seen in many plants and prominently in trees, e.g. sycamore (*Acer pseudoplatanus*). In some species, e.g. onion, the cotyledon is raised above ground level by the rapid growth of the cotyledon itself, rather than by extension of the hypocotyl. Compare HYPOGEAL.

epigenesis (*adj.* epigenetic) the theory that an organism develops from a fertilized egg by a gradual series of interdependent physiological and physical changes, brought about by the genes, that result in an increase in the complexity of its separate parts. It refutes the earlier *preformation theory*, which stated that either the male or female gamete contained a miniature version of the adult and that only an increase in size followed thereafter.

epigenetics the study of the mechanisms by which genes express themselves in the PHENOTYPE.

epigyny an arrangement of floral parts in which the stamens, sepals, and petals are inserted above the ovary, giving an *inferior ovary*. Epigynous flowers are seen in many members of the Rosaceae, e.g. quinces (*Chaenomeles*). The receptacle is urn-shaped and completely encloses the ovary, the other floral parts arising from the top of the receptacle. A type of pseudocarp, the POME, commonly forms from such flowers. Compare HYPOGYNY, PERIGYNY.

epilimnion in a stratified lake, the upper layer of warm water. See STRATIFICATION.

epilithic attached to or growing on rock surfaces.

epimatium a specialized type of OVULIFEROUS SCALE that bears and completely encloses a single inverted ovule. It is found in the monkey-puzzle tree (*Araucaria araucana*) and in certain of the Podocarpaceae.

epimerase an enzyme that catalyses the transfer of a hydroxyl group from one position to another within a molecule. Epimerases are a form of ISOMERASE and are important biologically in the interconversion of sugars.

epinasty a NASTIC MOVEMENT in which the resultant bending of the plant part is downwards, due to increased growth on the upper side of an organ. This occurs in the opening of many flowers when the bracts and perianth parts curl downwards to expose the sexual organs. Epinasty can also occur in leaves where the petiole bends so that the leaf points to the ground rather than upwards. Compare HYPONASTY.

epipelic growing on mud.

epipetalous growing on, and united with, the petals. For example, in some flowers the stamens are attached to the petals.

epiphragm a membrane that seals the mouth of a moss spore capsule before the spores are released.

epiphyllous see EPIPHYLLY.

epiphylly the growth of a plant, often a moss, on the leaf of another plant. Such a plant is called an *epiphyll* or is described as *epiphyllous* or *folicolous*. See also EPIPHYTE.

epiphyte (air plant) (*adj.* epiphytic) A plant that has no roots in the soil and lives above the ground surface, supported by another plant or object. It obtains its nutrients from the air, rain water, and from organic debris on its support. Many orchids and ferns are epiphytes and numerous species are found growing in the canopies of tropical rainforests. Their aerial roots form a tangled network that catches falling leaves and other organic material, which provide a source of mineral salts for the plant. In addition, the mesh of roots and organic debris acts as a sponge to collect and hold water. Mosses and lichens, growing on the bark of trees, are examples of epiphytes in temperate regions and some bromeliads, such as Spanish moss (*Tillandsia usneoides*), are examples of xerophytic epiphytes. Epiphytes growing on the leaves of another plant are termed *epiphyllous* while those growing on rock outcrops are called LITHOPHYTES. Epiphytes are also a group of plants in the RAUNKIAER SYSTEM OF CLASSIFICATION.

epiphytotic a widespread outbreak of disease among plants.

episepal see EPICALYX.

episepalous growing on, and united with, the sepals.

episodic evolution the view of EVOLUTION as characterized by periodic extinctions followed by periods of rapid evolution, as indicated by fossil evidence. See also PUNCTUATED EQUILIBRIUM.

episome see PLASMID.

epistasis (*adj.* epistatic) a form of gene interaction in which one gene affects the expression of a second gene. it usually arises because the two genes affect sequential steps in the same metabolic pathway. it is most easily detected as a modification of the MENDELIAN RATIO in the F_2 GENERATION. The phenomenon was first reported in the inheritance of flower colour. For example, suppose there are two enzymes, A′ and B′, produced by the normal alleles of two unlinked genes, A and B, but not by their recessive alleles, a and b. The substrate of A′ is a white pigment and the product of B′ is a purple pigment. In addition the product of A′, a red pigment, is also the substrate of B′. Thus the sequence of reactions white pigment to red pigment to purple pigment can only take place if there is at least one normal allele at both gene loci. The genotype aa_ would give a white phenotype (no enzyme A′; gene B is immaterial), genotype A_bb would give a red phenotype (enzyme A′ but not B′), and genotype A_B_ would give a purple phenotype (both A′ and B′). Instead of the F_2 forming a 9:3:3:1 ratio, a 9:3:4 ratio (= 9:3:[3+1]) would be obtained, since aaB_ would be indistinguishable from aabb. In this case, gene A is the *epistatic gene*, and gene B the *hypostatic gene*. Other variants are

9:7, 13:3, and 12:3:1 ratios. See also COMPLEMENTARY GENES.

epitheca see FRUSTULE.

epithecium the layer of tissue covering the hymenium in an ASCOMA. It is formed by the branching ends of the paraphyses above the asci.

epithelium the lining of either a resin canal in gymnosperms or, more rarely, a gum duct in dicotyledons. When present the epithelium usually consists of the cells pulled apart during the often schizogenous formation of the canal. These epithelial cells often have a secretory function.

epixylous growing on wood.

epizoochory the dispersal of fruits or seeds by attachment to an animal. For example, the hooked fruits of goosegrass (*Galium aparine*) are dispersed in this way.

epoch see GEOLOGICAL TIME SCALE.

equatorial plate the arrangement of the chromosomes during metaphase of MITOSIS and MEIOSIS, in which they come to lie in a single plane across the centre of the spindle.

equilibrium species a species whose main survival strategy is its ability to outcompete other species rather than its reproductive rate or dispersal suucess. Such a strategy is typical of successful species in stable environments. Equilibrium species are also found in unstable environments, where they survive unfavourable conditions by entering a state of dormancy or aestivation, reducing metabolic consumption of energy to a minimum, rather than relying on producing large numbers of seeds or spores that will germinate when conditions improve.

Equisetales the sole extant order of the SPHENOPHYTA.

equitant describing an arrangement of leaves or scale leaves in a bud (vernation) in which one CONDUPLICATE leaf completely overlaps another. *See illustration at* vernation.

ER see ENDOPLASMIC RETICULUM.

era see GEOLOGICAL TIME SCALE.

ergastic matter nonprotoplasmic substances that are produced as by-products of protoplasmic activity. It includes crystals, starch grains, tannins, and oil droplets. Many ergastic substances are regarded as waste matter, some are storage products, while the functions of other materials are unknown.

ergot a fungal disease of many cereal crops, caused by the ascomycete *Claviceps purpurea*. The hard black sclerotia (see SCLEROTIUM) of the fungus replace the grains of the cereal. They contain alkaloids that can cause severe poisoning or even death if the infected crop is eaten by humans or other animals.

Ericaceae a large family of dicotyledonous, usually shrubby plants, commonly called the heath or heather family, numbering about 2400 species in some 107 genera. It is virtually cosmopolitan in distribution, though members are poorly represented in Australia. Ericaceous plants have simple, often evergreen, leaves that lack stipules and may be reduced to needles, e.g. bell heather (*Erica cinerea*), or rolled, e.g. cowberry (*Vaccinium vitis-idaea*). The flowers are usually actinomorphic and bisexual and may be borne singly or in racemes. They are often pendulous and bell-shaped. The petals, normally four or five, are usually fused at the base. Many species are associated with acidic sandy or peaty soils, often as dominant species in the vegetation. Many form mycorrhizal associations, the so-called 'ericoid mycorrhizas' (see MYCORRHIZA).

Over half the described species are in the two genera *Rhododendron* (about 850 species) and *Erica* (heaths; about 750 species). Both show unusual distributions, species of *Erica* being concentrated in South Africa (the Cape heaths number some 450 species) while *Rhododendron* species are concentrated in the Himalayan foothills and New Guinea. Over half the *Rhododendron* species are cultivated as ornamentals (the azaleas are included in this genus). Many heaths and heathers (*Calluna*) are also grown as ornamentals. The berries of certain species, notably bilberry (*Vaccinium myrtilus*) and cranberry (*Vaccinium oxycoccus*), are valued as food in many areas. See also DILLENIIDAE.

erumpent bursting out.

Erysiphales an order of the ASCOMYCOTA containing some 437 species in 21 genera and a single family, Erysiphaceae. They have bitunicate asci, and some authorities consider them to be closely related to the Dothideales. Commonly called POWDERY MILDEWS, because the superficial hyphae produce masses of powdery conidiospores, they cause many common plant diseases.

erythrose a four-carbon aldose sugar (*see illustration at* aldose). In the phosphorylated form, erythrose 4-phosphate, it is an intermediate in the CALVIN CYCLE. In the SHIKIMIC ACID PATHWAY, the first reaction step is the joining of erythrose 4-phosphate with phosphoenolpyruvate to form a seven-carbon keto sugar acid.

esculent edible (by humans).

essential amino acid see AMINO ACID.

essential element see MACRONUTRIENT.

essential fatty acid see FATTY ACID.

essential oil (ethereal oil) any of the volatile oils secreted by aromatic plants that give them their characteristic taste or odour. Most essential oils are TERPENOIDS, while some are derivatives of benzene. Many xerophytic plants produce essential oils as a means of reducing their transpiration rates. In hot conditions the evaporating oils contribute to the density of the BOUNDARY LAYER at the leaf surface. The higher the density of the boundary layer, the greater resistance it offers to diffusion of water vapour. Some essential oils repel insects while others, e.g. those of *Myoporum deserti*, are toxic to grazing animals. Many allelopathic substances (allelochemicals) are essential oils. See also ALLELOPATHY.

established cell line cells derived from a particular initial cell culture, which can be subcultured indefinitely *in vitro*.

ester a compound formed by the CONDENSATION reaction between an acid (especially a carboxylic acid) and an alcohol, where R is a hydrocarbon group:

$$RCOOH + HOR_1 \rightarrow RCOOR_1 + H_2O.$$

estivation see AESTIVATION.

etaerio see AGGREGATE FRUIT.

-etalia see ZURICH–MONTPELLIER SCHOOL OF PHYTOSOCIOLOGY.

-etea see ZURICH–MONTPELLIER SCHOOL OF PHYTOSOCIOLOGY.

ethanal see ACETALDEHYDE.

ethanedioic acid see OXALIC ACID.

ethanoic acid see ACETIC ACID.

ethene (ethylene) a gaseous hydrocarbon, formula C_2H_4, that is produced in small quantities by many plants and that can be considered to act as a plant hormone. The production of ethene is frequently stimulated by auxins and the ethene so produced acts, by a feedback loop, to inhibit auxin synthesis. Ethene production also increases after wounding and exposure to disease. Ethene may enhance an auxin response but it normally inhibits longitudinal growth and causes radial expansion of tissues. It can promote flowering in certain species, e.g. pineapple (*Ananas comosus*), and speeds up the ripening of fruit, an effect that has been put to commercial use in the citrus industry. Other effects mediated by ethene include: the induction of epinasty; the induction of root hairs; the stimulation of seed germination in certain species; the promotion of leaf abscission; and the inhibition of auxin transport.

The route of ethene synthesis in plants is uncertain though the amino acid methionine has been suggested as a precursor.

ethereal oil see ESSENTIAL OIL.

ethnobotany the study of the use of plants by humans.

ethylene see ETHENE).

ethylene chlorohydrin (ethylene chlorhydrin) a chlorinated derivative of ethylene glycol, formula CH_2ClCH_2OH, used to break bud dormancy.

ethylenediamine tetra-acetic acid see EDTA.

etiolation the condition, seen when plants are growing with an inadequate light supply, in which there is abnormal chlorophyll development. It causes the plant to look pale and there is excessive elongation of the internodes and poor development of the leaves. It is an adaptation to hasten the plant's location of a light source. When plants are growing from underground stems, etiolation is manifested by extreme elongation of the petioles. This phenomenon is exploited in the growth of certain vegetables, such as rhubarb and celery, which may be kept in darkness to either blanch (whiten) the leaf stalks or to force them on early. See also ETIOPLAST.

etiology see AETIOLOGY.

etioplast a partially differentiated PLASTID, 1.5–2.0 μm in diameter, found in the cells of plants that have been kept in darkness (etiolated plants). A three-dimensional lattice, the PROLAMELLAR BODY, occupies almost the whole of the central region. Etioplasts contain protochlorophyll, which is converted to chlorophyll on exposure to light. Compare PROPLASTID.

-etosum SEE ZURICH–MONTPELLIER SCHOOL OF PHYTOSOCIOLOGY.

-etum SEE ZURICH–MONTPELLIER SCHOOL OF PHYTOSOCIOLOGY.

eu- prefix denoting 'new'.

Euascomycetae (Euascomycetes) a class of ASCOMYCOTA in which the asci usually develop from ascogenous hyphae enclosed in a true ascoma. The asci are unitunicate: their inner and outer walls do not separate during spore dispersal. The class includes the orders Elaphomycetales, Halosphaeriales, Medeolariales, Erysiphales, and Arthoniales. It encompasses the older groupings of Pyrenomycetes, Discomycetes, and Laboulbeniomycetes. The class is not recognized by all modern classification systems, and is often incorporated into the class ASCOMYCETES (as the subclass Hemiascomycetidae).

Eubacteria a major subkingdom of the BACTERIA. The fundamental distinction between the Eubacteria and the other subkingdom, the ARCHAEA, is based on analysis of their ribosomal RNA. Most known bacteria belong to the Eubacteria, including all phototrophic and multicellular bacteria. Some classifications recognize three main divisions of Eubacteria on the basis of their cell walls: the Firmicutes, which are Gram-positive nonphotosynthetic bacteria with thick PEPTIDOGLYCAN walls that lack an outer lipoprotein layer; the Gracilicutes, which are all Gram-negative and have relatively thin cell walls; and the Tenericutes, which lack a rigid cell wall and cannot synthesize peptidoglycan. The Firmicutes includes the Endospora (endospore-forming bacteria and related groups), Pirellulae (bacteria with proteinaceous walls), ACTINOBACTERIA, Deinococci (radiation- and heat-resistant Gram-positive bacteria), and Thermotogae (thermophilic fermenting bacteria). The Gracilicutes includes the Proteobacteria (purple bacteria), Spirochaetae, CYANOBACTERIA (blue-green bacteria and chloroxybacteria), Saprospirae (fermenting gliding bacteria), Chloroflexa (green nonsulphur bacteria), and Chlorobia (green SULPHUR BACTERIA). The Tenericutes comprises just the Aphragmabacteria (MYCOPLASMAS).

eucamptodromous SEE CAMPTODROMOUS.

eucarpic describing the mode of development in certain fungi and algae in which reproductive structures form on specific parts of the thallus, the remainder of the thallus continuing to grow and function vegetatively.

eucaryote SEE EUKARYOTE.

euchromatin an expanded region of chromatin in the interphase nucleus, which stains lightly with basic dyes. Metabolically active nuclei show large amounts of euchromatin. Investigations using autoradiography support the view that messenger RNA is transcribed in the euchromatic regions, but the amount of euchromatin is related to the number of genes being transcribed rather than the amount of mRNA produced. Compare HETEROCHROMATIN.

Eufungi an obsolete name for the kingdom FUNGI.

Euglenida (euglenids) a class of algal protoctists of the phylum Discomitochondria (unicellular organisms and their colonial relatives that are characterized by having mitochondria with discoid cristae and no meiotic sexual fertilization cycles). The Euglenida, formerly the Euglenophyta, are a very variable group, mostly motile unicellular organisms. They include both pigmented (autotrophic or facultatively heterotrophic) and unpigmented (obligately heterotrophic) genera. Euglenids are thought to have evolved from free-living motile unicells called *bodos*, which have two undulipodia. The phototrophic forms are believed to have acquired their plastids during evolution by ingesting plastids from their prey, probably partly digested chlorophytes (see CHLOROPHYTA). The chloroplasts have three enveloping membranes (indicating an endosymbiotic origin), and the thylakoids are arranged in groups of three. Like chlorophyte plastids, euglenid plastids contain chlorophylls *a* and *b* and various other pigments, but they also have various carotenoids not found in the Chlorophyta. Unlike chlorophytes, euglenids do not store starch; the main carbohydrate reserve is a polymer of glucose called paramylum (paramylon), which occurs as granules in the cytoplasm. There is no cell wall, but a pellicle made up of proteinaceous strips is associated with the inner surface of the cell membrane. In some euglenids this is flexible, as the strips can slide relative to each other, enabling the organism to ingest food by phagocytosis (see ENDOCYTOSIS). Most euglenids have one or two anteriorly placed undulipodia, which arise from a small depression, the reservoir. Associated with the reservoir is an eyespot consisting of pigmented granules containing FLAVIN, which are sensitive to the direction of light. The main undulipodium is hairy and is used for locomotion. The cells rotate as they swim. As in the dinomastigotes, the chromosomes remain condensed and the nuclear envelope remains intact during mitosis and there is no metaphase plate. Asexual reproduction is by longitudinal division of the cell, sometimes followed by the formation of resistant dormant cysts. Sexual reproduction is unknown. Euglenids occur in both freshwater and marine habitats, especially in nutrient-rich waters, where they sometimes form BLOOMS. Some species are parasitic.

Euglenophyta a former phylum (or division in some classifications) of very variable, predominantly unicellular, motile algae that are now classified as a class, EUGLENIDA, of the protoctist phylum Discomitochondria.

euhaline see BRACKISH.

Eukarya in the FIVE KINGDOMS CLASSIFICATION, a superkingdom that contains the kingdoms PROTOCTISTA, Animalia, FUNGI, and PLANTAE. Members of the Eukarya are distinguished by having eukaryotic cells (see EUKARYOTE). They exhibit Mendelian genetics and do not have unidirectional gene transfer by means of naked DNA. The cells have tubulin–actin cytoskeletons. See also KINGDOM.

eukaryote (eucaryote) an organism whose cells have a nucleus (compare PROKARYOTE). The genetic material of eukaryotes consists of CHROMATIN and is divided into a number of chromosomes, which are located in the nucleus. Eukaryotic cells usually divide either by mitosis or meiosis while division in prokaryotes is amitotic. In the Five Kingdoms classification all eukaryotic organisms are placed in the superkingdom EUKARYA, to emphasize the difference between these and the prokaryotes (bacteria), superkingdom PROKARYA.

eulittoral zone see LITTORAL.

Eumycota in certain classification systems, a division or kingdom of 'true' fungi, which produce hyphae (or more rarely nonmotile unicells) whose walls contain chitin and β-glucans and/or mannans. The main storage polysaccharide is glycogen, and the dominant sterol is usually ergosterol. The spores are usually nonmotile, and sexual reproduction is widespread. The Eumycota comprise the five subdivisions (or divisions) CHYTRIDIOMYCOTINA, Zygomycotina, Ascomycotina, Basidiomycotina, and FUNGI ANAMORPHICI (Mitosporic Fungi, Deuteromycotina). The equivalent groupings in the Five Kingdoms system are the three phyla Zygomycota, Ascomycota, and Basidiomycota, in the kingdom Fungi, together with the Chytridiomycota, which is placed in the kingdom Protoctista. See also FUNGI, PARAMYCOTA.

euphilic see ENTOMOPHILY.

Euphorbiaceae a large family of dicotyledonous plants, mainly tropical in distribution, containing over 8000 species in some 300 genera. It is commonly known as the spurge family. The stems and leaves often have cells or tubes of latex. Most members have simple alternately arranged leaves and unisexual flowers. In the tribe Euphorbieae the small flowers are grouped into an inflorescence, which, due to the presence of various petal-like structures (e.g. highly coloured bracts and glandular appendages), resembles a single flower. The 'flowers' of poinsettia (*Euphorbia pulcherrima*) are examples. The specialized inflorescence of *Euphorbia* species is termed a CYATHIUM. The fruit is often a capsular schizocarp, or sometimes a drupe, samara, or berry, and in many genera the seeds are carunculate.

The family contains many commercially important species, including *Hevea brasiliensis*, the main source of natural rubber, *Manihot esculenta* (cassava), and *Ricinus communis* (castor oil), and many spiny succulents, particularly in Africa.

euphotic zone SEE PHOTIC ZONE.

euploidy the condition in which the number of chromosomes in a nucleus is an exact multiple of the haploid number. This is the most common and stable condition. Compare ANEUPLOIDY.

Eurotiales (Plectascales, Ascosphaerales, Aspergillales) an order of the ASCOMYCOTA containing some 232 species (52 genera) of various moulds whose ascomata (when present) are small, usually solitary, and form CLEISTOTHECIA. Asci are produced by CROZIER formation and there is no tissue between them. They are usually club-shaped or saclike, sometimes forming chains. The ANAMORPHS are usually thick-walled conidia. Most members of the Eurotiales are saprobes, living in the soil or on rotting plant material. Many are of medical and industrial importance (e.g. *Penicillium, Aspergillus*). See also PLECTOMYCETES.

eury- prefix denoting wide, broad.

Euryarchaeota SEE ARCHAEA.

eurychoric describing organisms with a wide geographic distribution that spans a range of climates.

euryhaline describing organisms that can tolerate a wide range of salinity (and hence osmotic pressure) in their environment.

euryhydric describing organisms that can tolerate a wide range of moisture conditions.

euryoecious describing organisms that can live in a wide range of habitats.

eurythermal describing organisms that can tolerate a wide range of temperatures.

eurosporangiate SEE EUSPORANGIUM.

eusporangium (*adj* eusporangiate) a type of sporangium that is derived from several initials. There are several layers of cells in its wall and it usually produces a large number of spores. The eusporangium is typical of ferns in the Marattiales and Ophioglossales. Compare LEPTOSPORANGIUM.

eustele see stele.

Eustigmatophyta (Eustigmatophyceae, eustigs) a phylum of algal protoctists containing the chlorophylls *a*, c_1, c_2 and *e*, carotenoids, and usually violaxanthin, but lacking chlorophyll *b* and fucoxanthin. Most are unicells with a single chloroplast and no pyrenoid; a few are multicellular. Most have a single anterior hairy undulipodium with a T-shaped swelling near its base. There is an adjacent eyespot. Some species also have a second, smooth, undulipodium directed posteriorly and associated with the plastid. Asexual reproduction is by BINARY FISSION; sexual reproduction is unknown. Most eustigs live in fresh or marine habitats or on soil. On account of their yellow-green pigments, the eustigs used to be grouped with the Xanthophyta. Some modern classifications place them in a distinct class, Eustigmatophyceae, within the division Chromophyta.

eutrophic rich in minerals and bases. The term is usually used of nutrient-rich lakes and ponds but it may be applied in other contexts, e.g. to fen peats that contain relatively high proportions of minerals and bases. Eutrophic lakes and ponds support a prolific growth of algae and aquatic plants. When these organisms die they accumulate as a thick layer on the lake bottom. The organisms decomposing this material diminish the dissolved oxygen in the water so fish are scarce or absent. *Eutrophication* may unwittingly be accelerated by humans when, for example, sewage effluents enter a lake, or water rich in dissolved fertilizers seeps in from surrounding fields. In such instances the amount of oxygen used to decompose the plant remains may be so great that the fish die from oxygen deprivation. Compare DYSTROPHIC, OLIGOTROPHIC.

eutrophication SEE EUTROPHIC.

evagination the release of the contents of membranous vesicles to the surrounding environment.

evaporation the conversion of liquids to the vapour state by the expenditure of heat energy. Losses of water due to evaporation at plant surfaces can be serious and many species use various means, such as waxy or thick cuticles, to minimize the problem. However evaporation does play an important role in keeping the leaf cool. Evaporation of water from the stomata gives rise to the TRANSPIRATION STREAM.

evaporimeter an instrument for measuring evaporation.

evapotranspiration the combined loss of water by transpiration and by evaporation from a water surface or soil surface. *Actual evapotranspiration* (AE) is the evaporation that occurs if the total amount of water is limited, while *potential evapotranspiration* (PE) is the amount of evapotranspiration that would occur if water supply were unlimited. PE is calculated taking into account local factors such as the mean monthly temperature and daylength.

Everglades an area (about 10 000 sq km) of low, flat, marshy land in southern Florida through which water moves slowly towards the coast. There are large areas of sedges (see CYPERACEAE), especially saw-grass or saw-sedge, *Cladium effusum*, and on the coast, mangrove swamps. There are substantial accumulations of peat. On slightly higher ground there are many palms and pines. Distinct patches of tropical rainforest occur, locally called *hammocks*.

evergreen mixed forest a FOREST in which the dominant trees are evergreen broad-leaved species and conifers (see CONIFEROPHYTA). Such forests are widespread in New Zealand, Tasmania, Chile, and southern Africa, but in the Mediterranean region and eastern Asia the once-extensive evergreen mixed forests have largely been cleared.

evergreen a woody perennial plant that retains its foliage throughout the year by continuously shedding and replacing a few leaves at a time. Many evergreens are tropical or equatorial plants that are not subject to periods of winter cold or dry seasons. Evergreens growing in temperate and cold latitudes show various adaptations to overcome water loss by transpiration. Compare DECIDUOUS.

evolute turned back, unfolded.

evolution the process by which genetic changes have taken place in populations of animals and plants over successive generations in response to environmental changes (see NATURAL SELECTION). Evolution has resulted in the formation of new species and, usually, an increase in complexity. Evidence for evolution comes from palaeontology, biogeography, genetics, and comparative anatomy and physiology. Compare CREATIONISM. See DARWINISM.

evolutionary determinism the theory that events in nature are directly caused by preceding events or by natural laws. Evolutionary determinism is the non-random change in allele frequencies caused by deterministic processes. See also ORTHOSELECTION.

evolutionary rate the amount of evolutionary change that occurs in a given unit of time, e.g. the number of new genera appearing per million years. See also CHRONOSPECIES, DARWIN, MOLECULAR CLOCK.

evolutionary species see CHRONOSPECIES.

evolutionary tree a branching diagram showing the phylogenetic relationships between the different taxa. When a new taxon evolves from an existing one, it forms a branch from the line representing the parent taxon.

evolutionary trend a steady change in phenotype over many generations in a particular direction (usually assumed to be of adaptive value). This may be in a particular characteristic such as flower size or shape, or it may encompass the direction of speciation in general. See also EVOLUTIONARY DETERMINISM, ORTHOGENESIS, ORTHOSELECTION, DIRECTIONAL SELECTION.

ex- prefix denoting outside or outer.

exalbuminous seed a seed that lacks an ENDOSPERM at maturity. In such seeds the cotyledons absorb the food reserves from the endosperm and act as storage organs.

exarch describing XYLEM maturation in which the older cells (protoxylem) are further from the centre of the axis than the younger cells. Development is thus centripetal. Compare ENDARCH, MESARCH.

exchangeable ions ions that are adsorbed onto sites of opposite charge on the surface of clay and humus colloids in the soil (see

ADSORPTION COMPLEX). Exchangeable ions can subsitute for each other on this surface. Many, such as magnesium and phosphate, are important sources of plant nutrients. See also ADSORPTION, ANION EXCHANGE CAPACITY, CATION EXCHANGE CAPACITY, EXCHANGE CAPACITY.

exchange capacity the total ionic charge of the soil adsorption complex that is capable of adsorbing anions or cations. Clay minerals have particularly high exchange capacities.

exchange pool see BIOGEOCHEMICAL CYCLE.

excipulum in an APOTHECIUM, the cup-shaped layer of sterile tissue that contains the hymenium.

exclusion principle the principle that two species with identical ecological requirements (i.e. identical ecological niches) cannot coexist in the same location.

exclusive species in phytosociology, a species that occurs almost exclusively in a particular community. Such as species is assigned to optimum-fidelity class 5. See FIDELITY.

excretion the elimination of waste products of metabolism from the body or from sites of metabolic activity. In plants many waste products accumulate in the vacuoles, whose membranes prevent their unwanted interference in cell reactions. In small protoctists and fungi, waste products often diffuse across cell membranes and cell walls into the external medium.

excurrent describing a leaf in which the midrib extends beyond the end of the flat lamina, sometimes forming a HAIR POINT or AWN.

exiguous extremely scanty, small, and narrow.

exine the outer part of the wall of a pollen grain external to the INTINE. Several alternative systems exist for recognizing different layers in the exine (see diagram). In some morphological systems it is divided into an outer *sexine* and an inner *nexine*. Other systems based on chemical differences recognize an outer *ektexine* (*ectexine*) and an inner *endexine*.

Various projections of the exine surface may confer a characteristic sculpturing to the pollen grain. This sculpturing, together with information on pollen size and shape and the positioning of the germinal apertures, can be so distinctive that individual genera and even species can be recognized.

The exine is composed of a carotenoid polymer called SPOROPOLLENIN, which is highly resistant to decay. This has enabled palynologists to study fossil pollen and deduce the composition of earlier floras. Such *pollen analysis* studies have also thrown light on the evolution of plants.

exo- prefix denoting outside, outer, or giving out.

exocarp (epicarp, ectocarp) the outermost layer of the PERICARP of an angiosperm fruit, external to the mesocarp. It is usually only a thin layer, as in the plum, where it forms the outer skin.

exocytosis the removal or secretion of

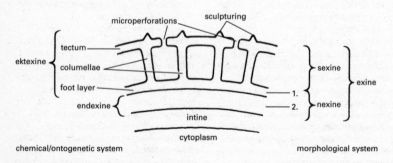

Exine: section through a pollen grain wall showing details of the exine

material from a cell by means of internal vesicles that transport the material to the plasma membrane, then fuse with it to release their contents to the outside. Compare ENDOCYTOSIS.

exodermis a layer of cells, often lying immediately beneath the epidermis or velamen of the root, that is essentially a specialized HYPODERMIS. Exodermal cells often resemble those of the ENDODERMIS and usually possess a suberized lamella (see SUBERIN) on the inside of the primary cell wall.

exoenzyme an enzyme that is discharged from a cell to the surrounding environment.

exogamy SEE ALLOGAMY.

exogenous describing any process or substance arising from outside the organism, or organs arising in the peripheral layers of the body. Examples are the external supplies of nutrients and hormones needed to maintain growth in cultures, and the origin of branches from the stem. Compare ENDOGENOUS.

exon a polynucleotide sequence in a structural gene that codes for a protein, i.e. a section between two INTRONS. This terminology is not applicable to bacteria, which do not have introns.

exonuclease an enzyme that catalyzes the hydrolysis of bonds at the ends of polynucleotides such as RNA and DNA, thus removing the terminal nucleotides. Compare ENDONUCLEASE. See also RESTRICTION ENDONUCLEASE, RIBONUCLEASE.

exoribonuclease SEE RIBONUCLEASE.

exopeptidase SEE PEPTIDASE.

exoscopic embryogeny EMBRYOGENY in which, of the two cells resulting from the first division of the egg cell, the outer gives rise to the embryo and the inner to the foot. It is seen in the bryophytes, Psilophyta, and Sphenophyta. Compare ENDOSCOPIC EMBRYOGENY.

exoskeleton a hard supportive external covering, such as the deposition of calcium carbonate on certain algae in the CHARALES and RHODOPHYTA. Such encrustations are involved in the formation of coral reefs.

exospore in certain orders of CYANOBACTERIA, a spore that is borne externally and is not necessarily resistant to heat and desiccation.

exosporic describing an organism in which spore formation occurs on the outside of the spore-producing organ. For example, in basidiomycetes the basidiospores are borne at the tips of outgrowths of the BASIDIUM called sterigmata.

exotic (alien) describing an organism that has originated from another region and is not native to the area in question. The term is used especially of tropical plants grown in greenhouses or indoors in temperate regions. An exotic plant that has become adapted to its new environment and can grow successfully without help from man is termed *naturalized*. An example in the UK is the monkey flower (*Mimulus guttatus*), which originated from North America. Many naturalized plants are garden escapes.

exotoxin see TOXIN.

exozoochory see ZOOCHORY.

experimental error an error that arises because experimental samples show more variation than was assumed. This may be due to the materials or methods used rather than to intrinsic differences. Experimental error needs to be monitored so that it can be taken into account when statistical methods of analysis are applied to the data.

explantation the removal of parts of living organisms (explants) for culture in a suitable artificial medium. This may be done to study the growth and development of tissues and organs and how they respond to different nutrient conditions. Explants are often only small segments of tissue, such as leaf discs or meristems. Their placement into or onto the culture medium is termed *inoculation* by analogy with microbiological techniques. The regenerative capacity of plant explants had led to their use as a means of plant propagation.

explicative describing a kind of folding of an individual leaf (see PTYXIS) in which the leaves are folded (rather than curled; compare REVOLUTE) towards the abaxial surface.

explosive evolution a sudden diversification of a group of organisms over a relatively short period of time. See also ADAPTIVE RADIATION.

exponential growth a type of population growth in which the rate of increase in the number of members is proportional to the number present. Exponential growth involves an increasing *rate* of growth: the more individuals there are, the faster the population increases. A graph of number against time in such growth has a J-shaped growth curve. In practice, exponential growth does not continue indefinitely. Other factors (e.g. shortage of food) become important and slow down the growth rate. The resulting growth curve is S-shaped (sigmoid). See also LOGARITHMIC PHASE, LOGISTIC EQUATION.

expressivity the degree to which a particular genotype is expressed in the phenotype.

exserted describes a structure that is protruding, for example, stamens that project beyond the corolla.

exstipulate lacking stipules.

extant describing a TAXON that contains some members that are living at the present time. Compare EXTINCT.

extensin see GLYCOPROTEIN.

external fertilization see FERTILIZATION.

extinct describing a TAXON that has no members living at the present time. Compare EXTANT.

extinction the permanent elimination of a taxon.

extracellular taking place or located outside the cell.

extrachromosomal DNA in EUKARYOTES, DNA that is found outside the nucleus of the cell and which replicates independently of the DNA in the chromosomes. It is found in cytoplasmic organelles such as mitochondria, chloroplasts, and plastids. See CYTOPLASMIC INHERITANCE.

extrachromosomal inheritance see CYTOPLASMIC INHERITANCE.

extrafloral nectary see NECTARY.

extremophile a microorganism that lives in extreme environments, such as highly saline or acidic water or hot springs. See HALOPHILE, HYPERTHERMOPHILE, METHANOGEN, PSYCHROPHILE, THERMOPHILE.

extremozyme any of a range of enzymes found in EXTREMOPHILES, which remains active in extreme conditions of temperature, salinity, acidity, or alkalinity at which other enzymes would be denatured or inactive. Such enzymes have a wide range of industrial uses, for example in biological detergents.

extrorse describing anthers that release their pollen to the outside of the flower so promoting cross pollination. Compare INTRORSE.

eyespot (stigma) a pigmented structure in motile unicellular algae and in motile reproductive bodies of some multicellular algae. It consists of one or more concentrically arranged curved plates composed of orange-red CAROTENOID granules. The eyespot is situated close to the bases of the undulipodia. It functions as a light-absorbing shield, which, depending on the position of the organism relative to the light source, prevents light from reaching photoreceptive areas at the bases of the undulipodia. This enables the organism to detect the direction of light and respond by movements of the undulipodia.

FAA see FORMALIN-ACETIC-ALCOHOL.

Fabaceae (Leguminosae) a large family of dicotyledonous plants, commonly called the pea family, containing about 18 000 species in about 642 genera. Legumes usually have pinnately compound leaves and root nodules containing nitrogen-fixing bacteria of the genus *Rhizobium* (see NITROGEN FIXATION). The inflorescence is usually a raceme and the individual flowers have five fused sepals and five petals often arranged in a shape fancifully resembling a butterfly (see also STANDARD, KEEL, WING), hence the name of the largest subfamily, Papilionoideae. This and the other two subfamilies, the Mimosoideae and Caesalpinioideae, are sometimes classified as families (Papilionaceae, Mimosaceae, and Caesalpiniaceae). The fruit is typically a LEGUME. Many of the Papilionoideae are important food crops, e.g. *Phaseolus* and *Vicia* (various kinds of bean), *Pisum sativum* (pea), *Lens culinaris* (lentil), and *Arachis hypogea* (peanut). Others, such as *Trifolium* (clovers) and *Medicago sativa* (lucerne), are used for forage. Ornamentals include *Wisteria* and *Lupinus*.

facilitated diffusion the passage of solutes into cells along a concentration gradient, facilitated by interactions with specific carriers, usually proteins. Unlike ACTIVE TRANSPORT, facilitated diffusion cannot move solutes against a concentration gradient.

facilitation in a SUCCESSION of communities, the way in which the activities of earlier species changes the environment in a way that facilitates the establishment of later species. For example, the presence of reeds at the edge of a lake traps sediments, eventually building up the mud to a level where less aquatic species can establish themselves.

factor (Mendelian factor) the inherited component that is responsible for the determination of a characteristic. In modern terminology, chromosomal GENE or cistron may be used instead.

factorial experiment an experiment in which a number of different factors are investigated simultaneously. The material is divided into a number of groups, such that each combination of treatments can be tried separately on at least one group and there are enough groups for every combination to be applied. See also RANDOMIZED BLOCK, LATIN SQUARE.

facultative possessing the ability to utilize certain circumstances or environmental conditions but not being dependent upon them. For example, a facultative parasite can live either as a parasite or as a saprobe. Compare OBLIGATE.

FAD (flavin adenine dinucleotide) a riboflavin-derived COENZYME that acts as a PROSTHETIC GROUP to several DEHYDROGENASE enzymes, accepting electrons from the substrate and thus being reduced to $FADH_2$. The most important FAD-linked enzymes are succinate dehydrogenase, which oxidizes succinate to fumarate, and dihydrolipoyl dehydrogenase, a component of the pyruvate dehydrogenase complex, which oxidizes pyruvate to ACETYL COA. See also FLAVOPROTEIN.

faecal coliforms see COLIFORM BACTERIA.

Fagaceae a family of hardwood trees and

shrubs containing about 700 species but only 8 genera. These include *Quercus* (oaks), *Fagus* (beeches), *Castanea* (chestnuts), and *Nothofagus* (southern beeches), which together make up a large proportion of the broad-leaved forests of the world. In terms of biomass, this is one of the most abundant families of flowering plants. Representatives are not, however, found in the equatorial rainforests of Africa and South America. The leaves are simple, and the flowers unisexual, usually borne in catkins or spikes. The fruit is a nut, which is partly or completely enclosed within a CUPULE. Certain species of chestnut, especially sweet chestnut (*Castanea sativa*), are grown for their fruits, and beech nuts and acorns are food for livestock and wild pigs. Timber, however, is the most economically important product of this family and, on a far smaller scale, certain species of oak yield cork (*Quercus suber*) and tannins. Many species are grown as ornamentals.

fairy rings rings of the mycelia and fruiting bodies of fungi, usually basidiomycetes (some 60 species are known to do this), most often found in grassland. They are formed by a mycelium that grows at its outer edges, leaving inside a zone of decaying mycelium and exhausted soil. Some rings grow over 30 cm a year and may reach a diameter of more than 200 m. Rings can persist for many years (rings up to 600 years old are recorded). The rings produced by some fungi often damage the vegetation ahead of the ring while leading to increased growth of vegetation behind due to the release of nutrients during fungal decay.

faithful species see FIDELITY.

falcate sickle-shaped. The term is usually used of leaves.

falcato-secund curved to one side. The term is used particularly to describe moss leaves.

fall any of the three large drooping outer petals of an *Iris* flower. In some species (e.g. *I. germanica*) they are bearded. See also STANDARD.

fallow describing land that is allowed to be uncultivated for a period of time. It may allow land to recover from nutrient depletion, or to accumulate moisture.

Fallow land may be sprayed to reduce weeds.

false dichotomy the forking of a stem axis into two almost equal halves that is due not to the organization of the apical meristem into two centres of growth, but to the growth of axillary buds, the apical meristem having aborted or formed an inflorescence or other structure.

false fruit see PSEUDOCARP.

false rings see DENDROCLIMATOLOGY.

false vivipary see VIVIPARY.

family a major category in the taxonomic hierarchy, comprising groups of similar genera. Families are thought by some to represent the highest natural grouping. The Latin names of families usually terminate in -aceae, e.g. Ranunculaceae, Papaveraceae, etc. However, there are eight exceptions, (Compositae, Cruciferae, Gramineae, Guttiferae, Labiatae, Leguminosae, Palmae, and Umbelliferae) whose names have been conserved by the International Code of Botanical Nomenclature. Alternative names have been proposed and are now often used in preference in many works (including this dictionary). These are Asteraceae, Brassicaceae, Poaceae, Hypericaceae, Lamiaceae, Fabaceae, Arecaceae, and Apiaceae, respectively. Groups of similar families are placed in ORDERS. Large families may be split into TRIBES. See also NATURAL ORDER.

farinose having a powdery covering.

far-red light electromagnetic radiation of approximately 740 nm wavelength. It has been shown that far-red light inhibits many physiological processes in plants, e.g. the germination of lettuce seeds. It also counteracts the effect of RED LIGHT. For example, if an etiolated plant is exposed to short periods of red light this results in more normal growth. However if the periods of red light are followed by periods of far-red light the plant remains etiolated (see ETIOLATION). It is believed that far-red light is absorbed by a pigment, PHYTOCHROME, which mediates such responses.

fasciation an abnormal form of growth in which a shoot becomes enlarged and flattened, giving the appearance of several shoots fused together. The apical meristem has become wide and flat instead of dome-shaped. It is often seen in dandelions (*Taraxacum*). It may be caused by mechanical injury, by fungi or mite attack, or by infection with a bacterium of the genus *Phytomonas*, especially *P. fasciens*. In some plants, e.g. cock's comb (*Celosia argentea cristata*), it is due to a mutation.

fascicle a tuft of leaves or branches that all arise from the same point.

fascicular cambium (intrafascicular cambium) the part of the VASCULAR CAMBIUM that originates within the VASCULAR BUNDLES (fascicles) between the xylem and phloem. Compare INTERFASCICULAR CAMBIUM.

fast green a permanent stain used to stain cellulose and cytoplasm. It is often used as a counterstain with safranin and, when mixed with acid fuchsin, stains pollen tubes.

fastigiate describing a tree in which the branches grow almost vertically, such as the Lombardy poplar (*Populus nigra italica*).

fat a triacylglycerol that is solid at room temperature. The predominant fatty acid in such compounds is palmitic acid. The term is often used in a wider sense in much the same way as the term LIPID.

fatty acid a long-chain aliphatic CARBOXYLIC ACID. Fatty acids are to a large extent responsible for the physical properties of complex LIPIDS. They may be saturated, monounsaturated (monoenoic), or polyunsaturated (polyenoic). Some of the less common ones may also contain polar or cyclic groups. The nature of the hydrocarbon chain greatly alters the solubility and reactivity of the fatty acid.

The majority of fatty acids have an even number of carbon atoms, although some odd-carbon fatty acids are found. The commonest fatty acids in plants are OLEIC ACID (C_{18}, monounsaturated) and PALMITIC ACID (C_{16}, saturated), but in specialized tissues, such as the chloroplast, LINOLENIC ACID (C_{18}, triunsaturated) is predominant. Other important fatty acids are STEARIC, LINOLEIC, myristic, lauric, and palmitoleic acids. Linoleic and linolenic acid can be synthesized by plants but not by animals. As precursors of prostaglandins they are essential in animal diets and are termed *essential fatty acids*. See also TRIACYLGLYCEROL.

fatty acid metabolism the breakdown or synthesis of FATTY ACIDS. In plants fatty acids can be broken down to yield carbon dioxide, water, and energy but in many lipid-metabolizing tissues, especially germinating oil-rich seeds, fatty acids are metabolized to sugars. The major route for fatty acid breakdown is the β-oxidation cycle, which degrades fatty acids to ACETYL COA. The acetyl CoA can then either undergo further oxidation or can be converted into sugar via the GLYOXYLATE CYCLE.

Fatty acid synthesis is a membrane-associated process. The enzymes responsible for fatty acid synthesis aggregate in SYNTHETASE complexes, which are found associated with the various membrane-bound organelles of the cell. Three different types of synthetase complex have been identified, each with a different specificity. See also MALONYL ACP.

fauna the animal species present in a particular geographical region or geological time period.

faunizone see ASSEMBLAGE ZONE.

fecundity the production of large numbers of offspring.

feedback inhibition (end-product inhibition) a type of inhibition in which the end product of a multienzyme sequence inhibits the activity of an enzyme at or near the beginning of the pathway. An example is the inhibition by isoleucine of the enzyme threonine dehydratase, the first enzyme in the biosynthetic pathway from threonine to isoleucine. The opposite of feedback inhibition, *feedforward stimulation*, occurs when the first substrate of a reaction sequence stimulates subsequent reactions in the sequence. Variations of simple feedback inhibition are found in branching pathways.

feedforward stimulation see FEEDBACK INHIBITION.

Fehling's solution a mixture of two solutions used to detect the presence of REDUCING SUGARS and aldehydes in solution. Fehling's A, which is a copper(II) sulphate solution, is added to the test solution followed by an equal amount of

Fehling's B (a mixture of sodium potassium tartrate and sodium hydroxide solutions). On boiling, a brick red precipitate of copper(I) oxide indicates a positive result. See also BENEDICT'S TEST.

fell an open mountainside with short vegetation.

fell field in the TUNDRA, an area of frost-shattered rock debris interspersed with fine particles. It supports a mixed community of plants that often cover less than half the ground area. Alternating freezing and thawing of the soil often results in the stony material forming a characteristic pattern (so-called *patterned ground*).

female describing either the reproductive parts or a whole organism that bears the MEGASPORE-producing apparatus. After fertilization the female may nurture the developing embryo. More strictly, the term applies to the gametophyte that produces archegonia. Compare MALE. See also CARPEL, ARCHEGONIUM, EGG.

fen a flat region of land that has developed from open stretches of base-rich water that have gradually silted up and passed through the HYDROSERE or HALOSERE stage of vegetation. There is a build up of basic peat but the persistence of wet marshy conditions means that large trees cannot grow, resulting in a subclimax (see CLIMAX) community. If the level of peat is raised sufficiently then certain woody plants that can tolerate wet conditions, such as alders (*Alnus*), willows (*Salix*), and guelder rose (*Viburnum opulus*), may become established. These fen woodlands are known as *carr*. Drained fenland, as for example that of the eastern English counties of Lincolnshire, Cambridgeshire, and Norfolk, is agriculturally very productive.

fenestrated describing a plant part that has many small perforations or transparent areas. For example, the flowers of some *Aristolochia* species have transparent windows at the base of the flower tube to lure insects deep inside, where they remain trapped until pollination has been achieved. The leaves of some species of *Monstera* (cheese-plants) have many large holes near the margin when young, which often increase in size until they merge with the leaf margin, giving it a divided edge.

feral describing formerly domesticated plants or animals and their descendants that have escaped into the wild.

fermentation the anaerobic breakdown of glucose and other organic fuels to obtain energy. Fermentations are thought to be the oldest energy-yielding catabolic pathways (see CATABOLISM).

Various types of fermentation are known, which differ in their substrate and their final product or products. The best known fermentation is GLYCOLYSIS, in which glucose is broken down to pyruvic acid. ALCOHOLIC FERMENTATION in yeast yields ethanol and carbon dioxide as end products. Other fermentations in microorganisms and bacteria lead to such end products as butyric acid and acetone.

fermenter an apparatus that is designed to contain, control, and monitor a biological process and collect the product of that process. For example, ALCOHOLIC FERMENTATION, in which alcohol is produced by yeast cells, can be carried out on an industrial scale in a suitably designed fermenter. Industrial fermenters are used in brewing, effluent disposal, yeast production for bakeries, and in the production of important organic substances such as antibiotics, amino acids, SINGLE-CELL PROTEIN, etc.

ferns see FILICINOPHYTA.

ferredoxin one of the components of the ELECTRON TRANSPORT CHAIN in chloroplasts. Ferredoxins are nonhaem iron–sulphur proteins, i.e. they utilize an iron-containing reaction centre but the structure is not that of a haem, as in cytochromes, but iron and acid-labile sulphur are present instead. During photosynthesis in chloroplasts ferredoxin mediates the transfer of electrons between photosystem I and the ultimate electron acceptor, NADP. It accepts an electron, via a ferredoxin reducing substance, from the excited P700 chlorophyll *a* molecule, and transfers it, through a reaction catalysed by ferredoxin-NADP reductase, to NADP. Bacteria and animals also contain iron–sulphur proteins, some free in solution and some membrane bound.

fertilization (syngamy) an essential part of SEXUAL REPRODUCTION involving the fusion of haploid male and female gametes to form a diploid zygote, from which a new individual develops. *External fertilization* occurs in lower plants, where gametes are released and exist independently of the parent before fusion occurs. In this process the gametes are totally dependent on the availability of water. *Internal fertilization* occurs in higher plants, where fusion takes place within the tissues of the female parts, which serve to protect and nurture the developing zygote. Internal fertilization, in its more advanced forms, removes the dependence on water and has enabled such plants to exploit drier habitats.

fertilizer any substance that is applied to land to raise soil fertility and increase plant growth. Fertilizers include natural products such as farmyard manure, guano, and bonemeal, as well as proprietary chemicals. Organic fertilizers tend to have longer lasting effects than inorganic fertilizers but are generally more expensive and slower acting. However they have the advantage of improving soil structure. The three main constituents of fertilizers are usually nitrates, phosphates, and potash (potassium oxide). The relative amounts of these in a compound fertilizer are expressed by the N:P:K ratio. See also COMPOST, MACRONUTRIENT, MICRONUTRIENT.

Feulgen's test a test used to detect DNA in nuclei, especially during cell division. A section of tissue is first placed in dilute hydrochloric acid for ten minutes at 60°C to hydrolyse DNA, removing the purine bases and exposing the aldehyde groups of deoxyribose. When the tissue is soaked in SCHIFF'S REAGENT the location of the DNA is shown by the development of a magenta colour.

F factor see PLASMID.

F_1 generation the first generation of offspring (first *F*ilial generation) from a cross between two individuals homozygous for contrasting alleles. In this case the F_1 must necessarily be heterozygous for the gene under investigation and the phenotypes are consequently similar. The F_1 generation thus shows far less variation than subsequent generations (see F_2 GENERATION) in which segregation becomes

apparent. The term F_1 is also commonly, though incorrectly, used to mean the offspring produced from unspecified parents. See also F_1 HYBRID.

F_2 generation the offspring (the second *F*ilial generation) produced by selfing the F_1 GENERATION, or allowing them to breed among themselves. Provided that a characteristic is governed by simple Mendelian genes, it is in the F_2 that MENDELIAN RATIOS, such as 3:1, 9:3:3:1, become apparent (see MONOHYBRID, DIHYBRID).

F_1 hybrid a crop variety that has been produced by crossing two selected parental pure lines. F_1 hybrids are favoured because they can combine the qualities of the parental lines and because they usually show hybrid vigour. However they do not breed true and seed must consequently be produced each year. In such seed production selfing within the parental lines, which would give nonhybrid seed, must be prevented. This may be achieved by emasculation and hand pollination but this is only commercially feasible for particularly valuable crops. Selfing can be prevented by ensuring that the parental lines both contain (different) reliable incompatibility alleles, i.e. *S* alleles high in the dominance series. This method is employed in producing F_1 hybrid varieties of certain brassica vegetables. Alternatively one parental line may be male sterile; however use of male sterility halves the yield of hybrid seed since seed set is only possible on one parent line. F_1 hybrid varieties of maize may be produced this way.

Fibonacci angle see PHYLLOTAXIS.

fibre a relatively long sclerenchyma cell, often with inconspicuous simple pits, that is usually differentiated directly from meristematic cells. Fibres that occur in the xylem are called *xylary fibres* and include the FIBRE-TRACHEIDS. The inner layers of the secondary cell wall of certain xylem fibres, notably in tension wood, may be gelatinous and swell with the uptake of water. Of the *extraxylary fibres*, those occurring in the phloem are often referred to as *bast fibres*.

Fibres are of great economic importance in the textile industry and for the making of rope and baskets. Commercially, fibres are divided into 'soft' and 'hard' categories. The 'hard' fibres come from monocotyledons; examples are those of the leaves of *Musa textilis* (abaca) and *Agave sisalana* (sisal). Examples of soft fibres are the pappus of *Gossypium* (cotton) and the bast fibres of *Linum usitatissimum* (flax), *Corchorus capsularis* (jute), and *Cannabis sativa* (hemp).

fibre-tracheid a xylary FIBRE that resembles a TRACHEID in its possession of bordered pits, although these are usually less conspicuous than those of tracheids. Fibre-tracheids appear to exhibit characters intermediate between those of fibres and tracheids and are believed to represent evidence for a phylogenetic relationship between the two.

fibril a small fibre or other thread-like structure.

fibrilla a fibrous filament.

fibrous layer see ENDOTHECIUM.

fibrous protein see PROTEIN.

fibrous root system the type of root system that develops when the RADICLE or primary root does not persist as the main root but is either replaced by ADVENTITIOUS (seminal) roots, as occurs in grasses, or becomes highly branched. Fibrous root systems are typical of monocotyledons.

fiddlehead the name commonly given in North America to a young unfurling fern frond, whose curled-over shape resembles the head of a fiddle.

fidelity in PHYTOSOCIOLOGY, the degree to which the distribution of a species is confined to a particular association, i.e. the degree to which it is a *faithful species*. In the Braun–Blanquet school of phytosociology, five levels of fidelity are recognized: Class 1 – *accidental species* occur only rarely in the community in question, either as chance migrants from another community or as relics from a previous community on the same site; Class 2 – *indifferent species* are not rare in the community, but are also just as common in other communities. Class 3 – *preferential species* are especially abundant and vigorous in one particular community, but are also reasonably abundant in several other communities. Class 4 – *selective species* are found most commonly in a particular community, but also occur occasionally in

other communities. Class 5 – *exclusive species* (*companion species*), the highest level of fidelity, are confined almost exclusively to a particular community. See also ASSOCIATION, KENNARTEN SPECIES.

field capacity the amount of water held in the soil by CAPILLARITY after excess or gravitational water has run off or percolated down to the water table. At field capacity the water potential of the soil is high. The percentage of water held at field capacity varies with soil texture, clay having a high field capacity and sand a low field capacity. See also PERMANENT WILTING POINT.

field layer the herb and shrub layer of a plant community.

field resistance see RESISTANCE.

field theory see REPULSION THEORY.

filament 1 the stalk of a STAMEN, bearing the anther at its apex.

2 a long strand composed of numerous similar cells joined end to end. The thalli of many algae, e.g. *Spirogyra*, are filamentous. Certain plants, e.g. the mosses and the Filicales, have a short-lived phase of filamentous growth following germination of the spore (see PROTONEMA). The nonphotosynthetic filaments forming the mycelia of fungi are termed HYPHAE.

Filicales (Polypodiales) the largest order of the FILICINOPHYTA, containing about 9000 species. The term was once used to encompass all the ferns, but now refers only to ferns with LEPTOSPORANGIA: its members differ from those of the MARATTIALES and OPHIOGLOSSALES in the development of the sporangium, which originates from one initial cell, while in other orders it develops from a group of initials. Although secondary vascular tissue is not clearly developed in any of the Filicinophyta, the tree ferns attain heights of 20 m. In these species bands of sclerenchyma strengthen the stem. The lamina of the fern frond usually shows differentiation into palisade and spongy mesophyll, but in the filmy ferns the leaf is only one cell thick.

The larger homosporous families of the Filicales include: the Osmundaceae (royal ferns), with some 22 species in 3 genera; the Gleicheniaceae, with some 125 species in 5 genera; the Grammitidaceae, with some 450 species in 4 genera; the Polypodiacae, with about 700 species in 52 genera; the Schizaeaceae, including the climbing ferns, with about 180 species in 4 genera; the Pteridaceae (Adiantaceae), including the maidenhair ferns, with about 825 species in 35 genera; the Cyatheaceae, with about 620 species in the genus *Cyathea*; and the Aspleniaceae, including the spleenworts and lady ferns, with about 720 species in the genus *Asplenium*. The heterosporous families all contain aquatic ferns and are often classified into separate orders: the Marsileales, comprising the Marsileaceae (about 70 species in 3 genera); and the Salviniales, comprising the Salviniacae (10 species in the genus *Salvinia*) and the Azollaceae (6 species in the genus *Azolla*).

Filicinae see FILICINOPHYTA.

Filicinophyta (Pterophyta) a phylum of mainly terrestrial vascular plants commonly called ferns, formerly classified as a class (Filicinae or Polypodiopsida) of the PTERIDOPHYTA or PTEROPSIDA. They do not produce seeds, and are distinguished from lower phyla of plants by possessing MEGAPHYLLS, leaf- or frondlike structures derived from lateral branches of the main axis, which form the familiar fern FRONDS. There are some 12 000 species, mainly tropical and herbaceous. Except for the heterosporous water ferns of the Salviniales and Marsileales, all ferns are homosporous and normally bear the spores on the abaxial surface of the frond in distinct groups of sporangia called sori (see SORUS). In many species groups of sori are covered by a protective sheet of tissue, the INDUSIUM, which eventually dries and shrivels to allow spore release and wind dispersal. The main fern plant is the sporophyte generation. The free-living gametophyte requires humid conditions to survive and is a vulnerable part of the life cycle. It often resembles a heart-shaped liverwort and bears rhizoids, but filamentous and subterranean gametophytes also occur. In some species the gametophytes bear only antheridia or archegonia; in others they bear both. Fertilization is also a water-demanding part of the life cycle, as the sperm need to swim by means of undulipodia to the ova. This confines the

ferns to habitats that are moist for at least part of the time.

The Filicinophyta is divided into two groups (classes or subclasses according to the classification system used): the Leptosporangiatae, in which the sporangium develops from a single cell (see LEPTOSPORANGIUM), and the Eusporangiatae, in which the sporangium develops from a number of cells (see EUSPORANGIUM). The FILICALES belong to the Leptosporangiatae, and the OPHIOGLOSSALES and MARATTIALES to the Eusporangiatae. The Filicinophyta has a rich fossil record going back to the Devonian, and made up a large part of the Carboniferous flora. Many forms became extinct during the Palaeozoic, and two extinct orders are recognized, the Cladoxylales and the Coenopteridales, both of which appear to have been predominantly homosporous.

filiform having a threadlike form.

filmy ferns ferns of the family Hymenophyllaceae, characterized by delicate, translucent fronds just one cell thick, with no stomata. They can take up water directly through the leaf surface, so need a moist atmosphere for survival. Ranging in size from less than 2.5 cm to 0.6 m, they tend to grow up the bases of tree trunks and on tree ferns in humid tropical forests. They are also found in humid, shady habitats in temperate regions.

fimbria see PILUS.

fimbriate having a margin that is fringed, usually with hairs.

fimicolous growing in dung or manure.

fine structure see ULTRASTRUCTURE.

finger-and-toe disease see CLUB ROOT.

fireblight a serious disease of pears and other trees of the family Rosaceae caused by the bacterium *Erwinia amylovora*. The disease attacks blossoms and leaves in the spring and, as the disease progresses, individual branches or whole trees look as if they have been scorched by fire. The host forms CANKERS in response to infection. The bacteria overwinter in the cankers and are spread by insects and rainsplash to infect flowers in the spring. Hawthorn (*Crataegus*), an ALTERNATE HOST, can also be a source of inoculum.

Firmicutes a division of EUBACTERIA that includes all the Gram-positive bacteria.

FISH see FLUORESCENCE IN SITU HYBRIDIZATION.

fission a method of ASEXUAL REPRODUCTION in which the nucleus and then the cytoplasm splits into equal parts to form new individuals genetically identical to the parent. It is seen in many microorganisms, such as bacteria and the fission yeasts (*Schizosaccharomyces*). The new cells may often sporulate in adverse conditions. The parent usually divides into two equal parts (*binary fission*). Compare BUDDING, FRAGMENTATION.

fitness the relative ability of an organism to produce large numbers of viable offspring that survive to a reproductive age and themselves contribute to the gene pool of the next generation. Fitness in this sense is not synonymous with the common usage of the word, i.e. healthy, as it does not always follow that the physically fittest are the fittest genetically.

Five Kingdoms classification a classification system that recognizes at least 96 phyla in 5 kingdoms. It makes a fundamental distinction between eukaryotic and prokaryotic organisms, separating them into two superkingdoms, the EUKARYA and PROKARYA, respectively. In this system, the Prokarya contains a single kingdom, the BACTERIA (formerly Prokaryotae; equivalent to the older kingdom Monera), which is divided into two subkingdoms, the ARCHAEA and the EUBACTERIA. The superkingdom Eukarya contains the remaining four kingdoms. The Animalia (animals) are defined as multicellular heterotrophic diploid organisms that develop from two different haploid gametes, an egg and a sperm. The diploid zygote develops into a hollow ball of cells called a blastula. The plant kingdom (Plantae) excludes the algae, which were included in older definitions of the plant kingdom. Plants are distinguished from animals by their lack of a blastula, and from both animals and fungi by their alternation of haploid and diploid generations and their possession of plastids and photosynthetic pigments. They are distinguished from algae in developing from diploid multicellular embryos that are nourished or protected by nonreproductive maternal tissue. Those plants which do not have sexual reproduction, but reproduce only vegetatively, are descended from ancestors that were once sexual.

The FUNGI are elevated to kingdom status, being defined as heterotrophic eukaryotes that form spores and lack undulipodia at all stages of the life cycle. They do not form embryos, and most produce asexual spores. Their cell walls are made of chitin, and they feed by secreting enzymes and absorbing the soluble products of digestion. The fifth kingdom, PROTOCTISTA, is not a natural grouping but includes all those organisms that do not fit into the other four kingdoms. It thus includes the protozoa, algae, and various fungus-like microorganisms, such as the chytrids (Chytridiomycota), oomycetes (Oomycota), slime nets (Labyrinthulata), and slime moulds (Acrasiomycota and Myxomycota). Most protoctists have nuclei and mitochondria and possess the characteristic eukaryotic undulipodia at some stage in their life cycle.

fixation 1 an initial stage in the preparation of cells or tissues for microscopical examination, in which the material is killed and preserved to prevent distortion, decay, and self-digestion (autolysis). Various chemicals (*fixatives*) can be used to penetrate the cells and preserve their structure either by denaturing the protein (e.g. picric acid, ethanol) or by tanning (e.g. acetic acid, formaldehyde). Plant specimens are often prepared for electron microscopy by fixing with Luft's fixative, a 1% solution of potassium permanganate. Common botanical fixatives include Navashin's solution (chromic-acetic-formaldehyde) and Carnoy's solution, an alcoholic fixative that rapidly penetrates the tissues and is especially suitable for hard materials such as seeds.

2 the attainment of a frequency of 100% by an allele in a population due to the complete loss, either by chance or by natural selection, of all other allelic forms of that gene. The likelihood of fixation occurring is greater in small populations.

3 The incorporation of carbon, nitrogen, phosphate, or other inorganic materials into organic compounds by living organisms. For example, carbon is incorporated into new organic compounds during photosynthesis. See also NITROGEN FIXATION.

fixative see FIXATION.

flaccid describing the wilted condition that results when a plant is deprived of water. The cells have lost their TURGOR and the plant therefore lacks rigidity. See WILTING.

flagellar apparatus the structure consisting of UNDULIPODIA together with the bodies associated with them, such as KINETOSOMES (basal bodies) and associated microtubules of the CYTOSKELETON.

flagellate 1 describing a unicellular organism or a cell that possesses UNDULIPODIA or FLAGELLA.
2 a flagellate organism (especially a eukaryote) or cell.

flagellin see FLAGELLUM.

flagellum (*pl.* flagella) a threadlike projection arising from the surface of motile bacteria (the term is currently restricted to prokaryotes, equivalent structures in eukaryotes now being called undulipodia). Flagella have no membrane or axoneme (see UNDULIPODIUM). They are composed of units of the protein *flagellin*, which resembles the protein myosin in contractile muscle fibres. These units are assembled in 11 helical spirals to form a wave-shaped hollow cylinder, some 10–20 nm in diameter. The flagellum does not beat, but rotates about its base on what appear to be ring-shaped bearings, thus propelling the cell in a corkscrew fashion.

Flandrian (Holocene) the present interglacial period, which succeeded the DEVENSIAN glacial period. Once considered to be the post-glacial period, it is now thought possible that glacial conditions may return in the future. The highest temperatures reached during the the Flandrian occured in the ATLANTIC PERIOD, about 6000 BP. See also ICE AGE.

flavanones a group of FLAVONOID compounds the members of which lack a double bond between carbons two and three of the C_3 group in the centre of the flavonoid nucleus. An example is naringenin, which is responsible for the bitterness of lemons.

flavescent yellow, or yellowing.

flavin any of a group of light-sensitive yellow plant pigments with an absorption peak around 370 nm that are active in certain plant growth responses to light. Flavins are also involved in phototropism and are also thought to transfer light energy from blue to red wavelengths, so influencing the function of phytochrome. RIBOFLAVIN is an example. See also FAD, FMN.

flavin adenine dinucleotide see FAD.

flavin mononucleotide see FMN.

flavones a group of flavonoid pigments the members of which contain an unaltered FLAVONOID nucleus. It includes flavone, found in certain *Primula* species, and the more widespread apigenin and luteolin. *Isoflavones* are isomers of flavones in which the B group of the flavonoid nucleus is attached to the third rather than second carbon of the central C_3 group. Isoflavones are particularly common in the Fabaceae. Some isoflavones have oestrogen-like properties, which has led to sterility in sheep grazing on certain legume fodder crops, notably subterranean clover (*Trifolium subterraneum*).

flavonoids a group of plant compounds all of which contain a 2-phenylbenzopyran nucleus (see diagram). They include the FLAVONES, ANTHOCYANINS, FLAVANONES, CHALCONES and AURONES, and FLAVONOLS. Flavonoids usually occur in combination with sugars as glycosides. They have been isolated from bryophytes and vascular

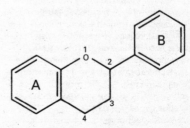

Flavonoid: the flavonoid nucleus (2-phenyl-benzopyran)

plants but not from protoctists, fungi, or bacteria. Certain types of flavonoid may be characteristic of a group of plants, the best example being the widespread occurrence of *biflavonyls* (dimers of the flavone apigenin) in the gymnosperms. Biflavonyls are scarce or absent among other plant groups. Flavonoids have been more often used than any other group of plant substances in chemotaxonomic studies.

Flavonoids have traditionally been thought of as waste products and their functions largely remain obscure. Many act as flower pigments while some have been implicated in the control of IAA activity. For example, flavonols, such as quercitin, with two hydroxyls in the B ring (forming a catechol group) inhibit IAA oxidase while flavonols with one hydroxyl (e.g. kaempferol) promote the enzyme's activity. Other flavonoids serve as PHYTOALEXINS. An example is pisatin, which accumulates in pea (*Pisum sativum*) tissues in response to invasion by many fungal species. Certain isoflavones have oestrogen-like activity while others, e.g. rotenone, resemble the saponins and are extremely toxic to certain forms of animal life.

flavonols a group of FLAVONOID compounds, the members of which have a hydroxyl group on the third carbon of the C_3 group in the centre of the flavonoid nucleus. Common flavonols include quercitin and kaempferol.

flavoprotein an enzyme containing a flavin molecule (e.g. FMN, FAD) as a PROSTHETIC GROUP. Flavoproteins are usually DEHYDROGENASES, e.g. succinate dehydrogenase, which is a Krebs cycle enzyme. The general equation for a flavoprotein reaction is:

flavoprotein + reduced substrate → reduced flavoprotein + oxidized substrate.

Like NAD, reduced flavoproteins can be reoxidized in the RESPIRATORY CHAIN. Flavoproteins are also components of the electron transport system of photosynthesis and the microsomal electron transport system (see ELECTRON TRANSPORT CHAIN).

flexuous hypha see HYPHA.

flimmer flagellum see UNDULIPODIUM.

floating chronology see DENDROCHRONOLOGY.

floccose describing a surface that appears woolly, fluffy, or cotton-like.

flocculation see LIMING.

flora all the plant species in a given area, or an assemblage of fossil plants present in a particular period of geological time, or in a certain area in a particular period of geological time.

floral diagram a stylized representation of the structure of a flower in which the whorls of floral parts are shown as a series of concentric circles. All floral segments arising at the same level are placed, in their correct relative positions, in the same circle. When appropriate, fusion of parts is also indicated. The ovary is represented in cross section in the centre of the whorls. The bottom of the diagram represents the anterior side of the flower, i.e. the side furthest from the axis of the inflorescence. Any bracts subtending the flower are inserted at the base of the diagram, while the inflorescence axis is shown as a circle at the top (posterior) side of the diagram. Floral diagrams are usually presented in conjunction with a longitudinal section through the centre of the flower (see illustration).

floral element (floristic element) 1 a species, genus, or other taxon characteristic of a particular flora or ASSOCIATION. Floral elements may also be described in terms of morphological or other features of the plants, for example, cryptogams are the main floral element of the Antarctic flora. 2 a part of a flower, e.g. petal, sepal, stamen, etc.

floral formula a method of recording floral structure by a series of symbols, letters, and numbers. The formula begins either with the sign ⊕, representing actinomorphy, or with ·|· or ↑, representing zygomorphy. This is followed by a series of letters in the order K (for calyx), C (for corolla), A (for androecium) and G (for gynoecium). The number of parts (sepals, petals, stamens, and carpels respectively) in each whorl is given by a number following the appropriate letter. If this number exceeds 12 the symbol ∞ (infinity) is used. If the parts are fused then the number is bracketed. If the parts are in distinct whorls or groups this is indicated by splitting the number

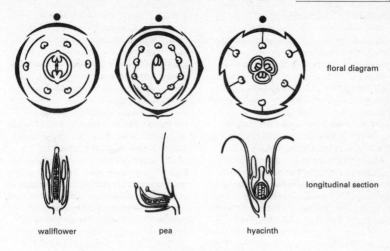

floral diagram

longitudinal section

wallflower   pea   hyacinth

Floral diagrams

appropriately. Thus in the garden pea (*Pisum sativum*) there are ten stamens, nine of which are fused and the tenth separate. This is represented A(9)+1. If two whorls are united then the representative symbols are joined by a single bracket. Thus in the primrose (*Primula vulgaris*) the stamens are inserted on the corolla tube. This is represented

$$C\overset{\frown}{(5)}A5$$

If the sepals and petals are indistinguishable then K and C are replaced by P representing perianth. The position of the ovary is indicated by a line, which in a superior ovary is placed beneath the number of carpels and in an inferior ovary above. Examples of floral formulae are ⊕K4C4A4+2G(2) for the wallflower (*Cheiranthus cheiri*) and ⊕P3+3A3+3G(3) for the tulip (*Tulipa* species). Floral formulae are often accompanied by FLORAL DIAGRAMS.

floral kingdom SEE DOUBLE FLOWER.

florepleno SEE FLORISTIC REGION.

floristic kingdom SEE FLORISTIC REGION.

floristic region (floristic kingdom, floral kingdom) any of a number of phytogeographical regions into which the world is divided on the basis of the distribution of plant species. The regions are divided into provinces. Each region contains unique families and genera, and contains one or more centres of diversity (gene centres). There are several systems and names for these regions, but the ones most commonly used are the *Boreal*, covering the whole north temperate zone; the *Neotropical*, covering tropical Central and South America; the *Palaeotropical*, covering tropical Africa and Asia; the *Australian*; the *Cape* Region, covering a small area around the Cape of Good Hope; and the *Antarctic*, which includes New Zealand and temperate South America as well as Antarctica. Oceanic islands, which often have very distinctive floras, form minor provinces. The distinctions between adjacent regions often blur near their boundaries. See BIOGEOGRAPHICAL REGION, PLANT GEOGRAPHY. See also ZOOGEOGRAPHICAL REGION.

floret a small flower, such as the disc or ray florets making up the CAPITULUM in plants of the Asteraceae. In grasses the lemma, palea, and the flower that they enclose comprise the floret.

Florideae SEE FLORIDEOPHYCIDAE.

floridean starch the assimilation product of the Rhodophyta (red algae). It has a structure similar to amylopectin.

Florideophycidae (Florideae) one of the two major subclasses of the red algae (Rhodophyta). Floridean algae are usually branching filaments or pseudoparenchymatous thalli; they show apical growth and have cellulose cell walls with distinct pit connections between adjacent cells, and a characteristic grooved plug in each pit. There is a wide range of growth forms, from the open filaments of such species as *Polysiphonia*, the encrusting calcified thalli of *Hildenbrandia*, and the calcified filaments of *Corallina*, to the soft tubelike branches of *Dumontia* and the flattened fronds of *Chondrus*. Floridean algae usually have many small disc-shaped chloroplasts. The life cycle is complex, with two or three phases and well-differentiated reproductive structures. A typical three-phase life cycle includes a free-living, often unisexual, gametophyte phase, which produces *spermatia* and carpogonia. The carpogonium, a flask-shaped equivalent of an oogonium, is embedded in the gametophyte thallus; its tip forms an extension called a *trichogyne*. The nonmotile spermatia are carried by the water to the trichogyne, and the spermatium nucleus travels down to fuse with the ovum nucleus. A fertilized carpogonium forms a diploid *carposporophyte* attached to the female gametophyte. It is composed of branching *gonimoblast filaments*, which release carpospores into the water. Carpospores give rise to the diploid tetrasporophyte generation, which produces tetraspores by meiosis; these germinate into new gametophytes. Detailed taxonomy of the subclass is based on the structure of the pit plugs and on details of carposporophyte development. See also ALTERNATION OF GENERATIONS.

florigen a hypothetical plant hormone that is said to cause flower initiation in many species. Florigen has yet to be isolated from plants although there is much circumstantial evidence for its existence. Thus although flowering often depends on the photoperiod experienced by the plant it is the leaves and not the growing apex that perceive photoperiod (see PHOTOPERIODISM). Thus the stimulus must be transported from the leaves to the apex.

The flowering stimulus can be transferred, by grafting, from one plant to another. If leaves from short-day plants grown under short days are grafted onto long-day plants grown under short days then the stimulus from the grafted leaves will promote flowering in the long-day plant. Such work shows that florigen is similar in long- and short-day plants. The rate of florigen movement has been measured by removing the leaves from a plant at varying intervals after exposure to an appropriate photoperiod and noting whether flowering still occurs. In some plants the rate appears to be slower than that expected if transport were via the phloem. See also VERNALIN.

floristic element see FLORAL ELEMENT.

floristics the branch of botany concerned with identifying and listing all the plant species present in a particular area. The check list so produced gives the *floristic composition* of the region. Once this has been established further work may be undertaken to describe the ABUNDANCE and distribution of the various species and to consider the affinities of the regional flora.

flower the reproductive unit of the angiosperms. It consists of microsporophylls (stamens) and megasporophylls (carpels) concentrated on a terminal apex of limited growth. The unit comprises a central axis or receptacle, and usually nonessential floral parts (the sepals and petals) surrounding the essential floral parts, the stamens and carpels. The arrangement of floral parts within the flower is usually related to the method of pollination and seed dispersal. The essential floral parts are involved in seed production, and the female parts persist after fertilization to form the fruit. The nonessential parts may be incorporated into the fruit after fertilization but more usually senesce and disintegrate. See also INFLORESCENCE.

flowering glume see LEMMA.

flowering plants see ANTHOPHYTA.

fluid-mosaic model see PLASMA MEMBRANE.

fluorescence in situ hybridization (FISH) a technique in which fluorescent probe molecules are used to bind ('hybridize') with specific sequences of DNA. Different colour probes may be used to distinguish different sequences of DNA. Some FISH probes bind to the centromeres or telomeres of specific chromosomes, allowing the chromosomes to be distinguished from each other. FISH probes are important in genetic analysis for detecting and locating mutant genes, euploidy, deletions, and other chromosomal aberrations and for identifying the source of DNA in hybrid cells genetically engineered from two different species. See also CHROMOSOME MUTATION, FLUORESCENCE MICROSCOPY.

fluorescence microscopy a form of microscopy in which fluorescent compounds (*fluorescent probes*) are added to the object under investigation. These compounds bind to specific parts of the cell or to particular substances within it. For example, they may bind to specific parts of certain chromosomes, or to sites of calcium ion release, or to particular protein receptor molecules. Different coloured probes can be used to highlight more than one structure at a time. See also FLUORESCENCE IN SITU HYBRIDIZATION, CONFOCAL MICROSCOPY.

fluorescent probe see FLUORESCENCE MICROSCOPY.

flushing the process by which soluble substances in the lower layers of the soil are washed upwards and deposited on or near the soil surface. It occurs, for example, in water-meadows, fenland, and where there are springs. Compare LEACHING.

FMN (flavin mononucleotide) a riboflavin-derived COENZYME, similar to FAD, that acts as a PROSTHETIC GROUP to several DEHYDROGENASES. NADH dehydrogenase, the enzyme that catalyses transfer of electrons from NADH to COENZYME Q in the respiratory chain, is an FMN-containing enzyme. See also FLAVOPROTEIN.

folic acid a member of the B group of VITAMINS first isolated from spinach leaves. In its activated form tetrahydrofolic acid it acts as a COENZYME in reactions involving the transfer of hydroxymethyl, formyl, and methyl groups. Such transfers are important in the synthesis of the purines and the pyrimidine thymine.

folicolous see EPIPHYLLY.

Folin–Ciocalteau reaction a quantitative colorimetric method for determining the presence of proteins. The test solution is reacted with a phosphomolybdotungstic acid reagent. If a protein is present the tyrosine it contains will react to produce a blue colour.

foliose describing organisms that bear leaflike structures, used especially of leafy lichens (e.g. *Peltigera*) and sometimes of leafy liverworts (e.g. *Diplophyllum*).

follicle a dry dehiscent many-seeded fruit derived from one carpel, which on ripening splits down one side only, usually the ventral suture, to expose the seeds, as in *Delphinium.*

food body a structure developed on the surface of a leaf, stem, or seed that produces edible substances, such as proteins or oil. Anatomically such structures are usually derived from trichomes or emergences. Various types of food body have been given specific names. They include simple food cells in *domatia* on the petioles of species of *Piper, Beccariian bodies* on the leaves and stipules of *Macaranga* species, *Beltian bodies* on the leaflets of *Acacia* species, the ELAIOSOMES on the seeds of many plants, *Müllerian bodies* on a specific swelling at the base of the petioles of *Cecropia* species, and *pearl bodies* on the leaves and stems of *Ochroma* species. Often these food bodies contribute to a mutualistic relationship between the plant and certain insects, particularly ants. For example, the Müllerian bodies of *Cecropia* provide glycogen for ants, which live in their hollow stems; in return the ants drive away herbivores and the seedlings of other plants, such as vines, which might compete with the host plant for light or minerals. See also BROMATIA.

food cell see FOOD BODY.

food chain a series of organisms, each successive group of which feeds on the group immediately previous in the chain, and is in turn eaten by the succeeding group. Green plants or other autotrophic organisms (PRODUCERS) are normally the first step in a food chain, herbivores (primary CONSUMERS) the second step, with carnivores (secondary consumers) making up the remaining two or three stages. The different stages are termed *trophic levels* and organisms that are removed from the beginning of the chain by the same number of steps are said to occupy the same trophic level. The chain is effectively a series of energy transfers, with considerable amounts of energy being lost at each transfer (see BIOENERGETICS). A food chain consequently rarely has more than four or five stages since there is not enough energy in the system to maintain more. A consequence of this is that, for a given consumer level, considerably more individuals can be maintained if an earlier consumer stage is omitted. For example, a given amount of grain will support many more people if it is eaten as grain, rather than being first converted to meat by feeding to livestock.

Food chains rarely exist in isolation but are interconnected to form a *food web*. Thus any of the links in one food chain may, through the activity of decomposers, saprobes, or parasites, lead into a number of different food chains. In addition, most plants and herbivores have a number of different predators.

food web see FOOD CHAIN.

foolish seedling disease see BAKANAE DISEASE.

foot the basal portion of an embryo, sporophyte, or spore-producing body, which is embedded in the parental tissue. It serves as an anchor and to absorb nutrients.

foot rot a fungal disease of plants in which there is rotting of the root system and the bottom part of the stem. Foot rot of tomato is caused by *Phytophthora cryptogea* and *P. parasitica*. Stem tissues around soil level appear brown or blackish and the root system is found to be rotten.

forb see BASIC GRASSLAND.

forcing bringing a plant into flower earlier in the year than it would normally flower, by using such means as controlling day length or supplying extra heat. It may be used to provide plants for sale at particular seasons, e.g. Christmas, or to synchronize the flowering dates of plants for use in breeding programmes.

foredune see SAND DUNE.

foreshore the part of the shore between the normal high- and low-water marks.

forest any major community (BIOME) in which the dominant plants are trees. There are three main types of forest: *cold forests*,

temperate forests, and *tropical forests*. The cold forests include the *northern coniferous forest* (*boreal forest* or *taiga*), where there are relatively few types of tree, mainly fir, pine, and spruce with some deciduous birches and larches. Boreal forest is found in North America and northern Eurasia. The *mountain coniferous forest* and *pine forest* of the southeast United States are also boreal forests.

Temperate forests have fairly evenly distributed rainfall with moderate temperatures, but there is a marked seasonal change. The vegetation consists of broad-leaved deciduous trees, an adaptation to the lack of available water during winter (see PHYSIOLOGICAL DROUGHT). There are usually several different species of trees and the term *mixed deciduous forest* is often used. *Warm temperate* and *cool temperate rainforests* also exist, e.g. in south China and New Zealand respectively. *Broad-leaved evergreen forest* is characteristic of a Mediterranean-type climate with hot dry summers and mild wet winters.

Tropical forests include tropical rainforest, monsoon forest, and thorn forest. *Rainforests* have regular heavy rain, constantly high temperatures, and therefore prolific plant growth. Rainforest regions, such as the Amazon basin and the Malay–Indonesian region, contain thousands of different species. Essentially similar forests, known as *riverine forests*, occur along river banks in drier regions. *Monsoon forests* have a period of drought, which may last for several months. They do not have such a great variety of species as tropical rainforest and have a more open canopy and very dense undergrowth. Many of the trees shed their leaves in the dry season. They are found in India, Burma, and Indo-China. *Thorn forest* is a transitional type of vegetation with some similarities to SAVANNA and semidesert vegetation. It grades into tropical forests and is found in Central and South America, Australia, and Africa.

A distinction is made between *primary forest*, which is believed never to have been affected by any human activity, and *secondary forest*, which has developed on the site of a former forest that has been cleared.

Secondary forest consists mainly of pioneer species, and represents an earlier stage of succession. See also DEFORESTATION.

forestry 1 the study of trees and timber production. 2 the growing and management of forests (both natural and manmade) and plantations for commercial timber production. It includes artifical selection and other forms of genetic modification for specific uses or environments.

form (forma) the lowest rank normally used by practising taxonomists for sporadic distinct variants that sometimes occur in populations. These may be relatively minor genetic variants (i.e. based on one or a few linked characters) but their effects can be conspicuous. The most obviously recognizable forms are those in which flower colour is modified (e.g. white-flowered individuals occurring in a population of purple-flowered plants). The category form may relate directly to species but it is more commonly used to distinguish variants of subspecies and varieties.

formalin a mixture of about 8% methanol, 40% formaldehyde and 52% water. It is a powerful reducing agent, and is used preserve biological specimens. It is also used as a fungicide and general disinfectant. See also FIXATION.

formalin-acetic-alcohol (FAA) a fixative containing 17 parts of 70% alcohol to 2 parts 40% formaldehyde to 1 part glacial acetic acid. See also FIXATION.

forma specialis (*pl.* formae speciales) See PHYSIOLOGICAL RACE.

formation (plant formation) a unit of vegetation classified by the form and structure of the plant community rather than by its component species. The formation has different meanings and hierarchical status in different traditions of PHYTOSOCIOLOGY. A *formation type* or *formation class* is roughly equivalent to a climatic climax (see CLIMAX) vegetation type or a major world biome, with a fairly uniform appearance and characteristic life forms, e.g. tropical rainforest or savanna. Compare ASSOCIATION, ASSEMBLAGE.

form genus 1 any genus in which anamorphic states of fungi are placed.

Similarly a *form species* is a species of anamorphic (asexual) fungus (see FUNGI ANAMORPHICI). When the connection between an ANAMORPH and a TELIOMORPH is established then the name of the latter takes precedence, although both names are still valid. See also STATES OF FUNGI. 2 a group of fossil forms whose affinities are unknown.

form species see FORM GENUS.

fossil fuel a fuel derived from the fossilized remains of ancient organisms, for example, coal, peat, petroleum, and natural gas. Such fuels were formed by the mainly anaerobic decomposition or partial mineralization of organic remains under great pressure. Such processes take millions of years, so these fuels are effectively non-renewable. Burning of fossil fuels releases into the atmosphere carbon dioxide that was fixed (see FIXATION) millions of years ago, thus contributing to the GREENHOUSE EFFECT and global warming. See non-renewable resource.

fossils the remains of organisms from past geological ages preserved in sedimentary rocks either as actual structures or as impressions, casts, or moulds. Plant fossils are rarely as well preserved as animal fossils because their tissues do not normally contain calcified structures. They are usually therefore completely decomposed before the processes of fossilization, including carbonization and petrification, act to preserve them. The remains of macroscopic structures, such as branches, leaves, fruits, and seeds, are termed *megafossils* while those of pollen and spores are called *microfossils*. If a fossil cannot be assigned to any genera containing extant species then its genus is termed an *organ genus*. Similarly if it cannot be assigned to a family it is placed in a FORM GENUS.

The study of fossils has helped in the construction of phylogenetic classification schemes and has also thrown light on how some of the complex structures of extant plants have evolved. See also BIOCHRONOLOGY, PALAEONTOLOGY.

founder effect the proposition that a small pioneer community that is establishing itself in genetic isolation from the main population will only possess a small and possibly nonrandom selection of genes

from the parental GENE POOL.
Consequently, the community that develops may take a different evolutionary path from the parental population. Relative uniformity among the community's members and the frequent occurrence of distinct or unusual characteristics is seen as evidence in support of the principle. See also GEOGRAPHICAL ISOLATION, REPRODUCTIVE ISOLATION.

founder population the first generation of a breeding population.

founder principle the proposition that a small pioneer COMMUNITY that is establishing itself in genetic isolation from the main population will only possess a small and possibly nonrandom selection of genes from the parental GENE POOL. Consequently, the community that develops may take a different evolutionary path from the parental population. Relative uniformity among the community's members and the frequent occurrence of distinct or unusual characteristics is seen as evidence in support of the principle.

fox-fire a form of BIOLUMINESCENCE emitted by moist rotting wood and by certain fungal fruiting bodies.

fraction 1 protein see RIBULOSE BISPHOSPHATE CARBOXYLASE.

fragmentation a form of ASEXUAL REPRODUCTION in which the parent splits into several pieces, each of which may develop into a new individual. It is seen in many filamentous algae, e.g. *Ulothrix* and *Spirogyra*. It is also a feature of some aquatic plants, e.g. waterweeds (*Elodea*), and of the gametophytes of certain bryophytes. Compare BUDDING, FISSION.

frame-shift mutation see GENE MUTATION.

free-central placentation a form of PLACENTATION in which the placentae develop on a central dome or column of tissue. It is seen in unilocular compound ovaries, such as those of the Primulaceae and Caryophyllaceae.

free space the part of the protoplasm where there is relatively free exchange of inorganic ions between the protoplasm and extracellular fluids. This occurs because the outer membrane of the protoplasm is permeable to most inorganic ions. The membrane surrounding the cell vacuole is not so permeable and though ions may enter by ACTIVE TRANSPORT they cannot pass out again; certain salts thus tend to accumulate in the vacuole. The composition of the free space and the vacuole may therefore be quite different.

freeze drying a technique for the removal of water from a specimen by sublimation of ice under vacuum. It is used as a method of drying in cases where the action of heat would cause damage, e.g. the concentration of labile solutes such as enzymes and the drying of fragile specimens for electron microscopy and storage (e.g. herbarium specimens). It is also used when the presence of liquid water would cause distortion of tissue by surface tension. Rapid freezing prevents the formation of ice crystals that would otherwise cause disruption.

freeze etching SEE FREEZE FRACTURING.

freeze fracturing a technique used in electron microscopy (see ELECTRON MICROSCOPE) in which the specimen is frozen quickly at very low temperatures and then fractured along lines of weakness. The new surface so formed is coated with a thin layer of carbon or platinum, which is then floated off to form a replica of the surface. *Freeze etching* is a modification of the technique in which the water vapour is allowed to evaporate from the fracture surface before the replica is made. This takes longer but allows higher resolution of the specimen.

frequency a measure of the chance of finding a species in any QUADRAT in a given area. To determine frequency, the presence or absence of the species in a series of randomly placed quadrats is totalled, the frequency being expressed as a percentage of quadrats in which the species occurs. The value of frequency varies with the quadrat size, the size and number of shoots of individuals of the species, and the patchiness of distribution of the individuals.

frequency distribution the distribution of the values of a variable, usually espressed as a graph, histogram, or table (*frequency table*) with the number of individuals on the horizontal axis and the value of the variable on the vertical axis. See also BICENTRIC

DISTRIBUTION, BIMODAL DISTRIBUTION, BINOMIAL DISTRIBUTION, UNICENTRIC DISTRIBUTION.

frequency-dependent selection the situation where a genotype or phenotype is at a selective advantage when it is rare in the population but at a disadvantage when it is common.

frequency table see FREQUENCY DISTRIBUTION.

fresh water water containing little or no chlorine, usually defined (Venice system) as having less than 0.03% chlorine or less.

friable easily crumbled. The term is usually used in relation to soil.

fringing forest see GALLERY FOREST.

frond a large leaf or leaflike structure. The term is most frequently used of the dissected leaves of ferns, cycads, and palms. More rarely it is applied to the thalli of foliose lichens and certain macroscopic algae.

frost resistance the ability, possessed by many temperate and arctic plants, to withstand subzero temperatures. The resistance is technically *frost tolerance* as the plant cannot avoid the cold conditions and resultant freezing of water in the intercellular spaces. This causes a lowering of the WATER POTENTIAL in the intercellular spaces so water moves out of the protoplast to regions of ice crystallization. Frost damage is thus often a result of severe cellular dehydration. However this dehydration increases the solute concentration of the cell and thus makes ice formation within the cell less likely. Dormant organs and seeds tend to show more frost resistance than actively growing tissue. The mechanism of frost resistance is uncertain but membrane elasticity and reduction in cell size appear to play a part. See also HARDENING.

frost tolerance see FROST RESISTANCE.

fructification any seed- or spore-bearing structure, especially one that is easily visible as, for example, the aerial fruiting bodies of many higher fungi.

fructosan see FRUCTOSE.

fructose (fruit sugar) a ketohexose sugar (*see illustration at* ketose). In solution fructose forms a FURANOSE RING to give fructofuranose. Fructose is the commonest

keto sugar and is a component of the disaccharide sucrose. Fructose is, with glucose, one of the few monosaccharides to occur free in plant tissues, e.g. fruit juices. Polymers of fructose are termed *fructosans*, the most important of which is inulin.

fructose 1,6-bisphosphate a phosphorylated derivative of fructose, situated at the main control point of GLYCOLYSIS. In glycolysis fructose 1,6-bisphosphate is formed from fructose 6-phosphate by the ALLOSTERIC ENZYME phosphofructokinase, which is inhibited by ATP and citrate, and stimulated by ADP and AMP. This reaction is the rate-limiting step of glycolysis. The reverse reaction, which is a step in glucose synthesis, also involves an allosteric enzyme, fructose bisphosphatase. Fructose 1,6-bisphosphate is also an intermediate in the PENTOSE PHOSPHATE PATHWAY.

fruit 1 the structure that develops from the ovary wall (PERICARP) as the enclosed seed or seeds mature. A fruit may be classified as succulent or dry depending on whether or not the middle layer of the pericarp (the MESOCARP) develops into a fleshy covering. It may be further classified as dehiscent (see DEHISCENCE) or INDEHISCENT according to whether or not the fruit wall splits open to release the seed. Fruits that develop from the gynoecium of a single flower are termed simple or true fruits (see illustration overleaf for the main types of true fruits). If they are derived from a single ovary they are termed *monocarpellary* while those that incorporate a number of fused ovaries are termed *polycarpellary*. An AGGREGATE FRUIT may develop from an APOCARPOUS gynoecium. The fruit may incorporate tissues other than the gynoecium (see PSEUDOCARP) and some fruits may develop from a complete inflorescence (see MULTIPLE FRUIT). In some cases a fruit may develop even though the ovule has not been fertilized (see PARTHENOCARPY).
2 loosely, any of various fleshy structures that may be associated with a gymnosperm seed, such as the succulent ARIL of yew (*Taxus baccata*) or the fleshy OVULIFEROUS SCALES of some members of the Cupressaceae, such as junipers (*Juniperus*).

fruit drop the premature ABSCISSION of fruit

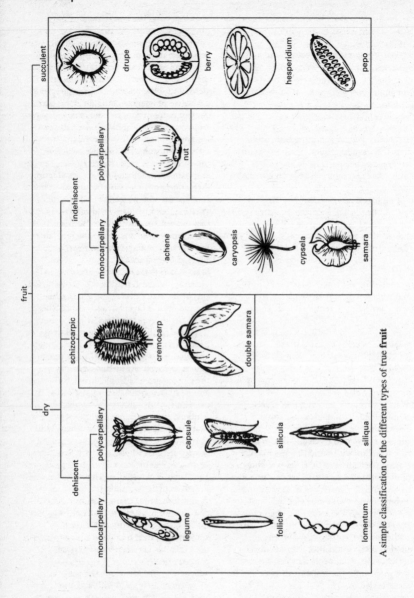

A simple classification of the different types of true **fruit**

before it is fully ripe. It is a normal process and in many fruits certain peak periods of fruit drop can be identified. For example, apple fruits are lost immediately following pollination (*postblossom drop*), when the embryos are developing rapidly (*June drop*), and during ripening (*preharvest drop*). Like LEAF FALL, fruit drop is associated with low

AUXIN levels and auxin sprays have been used to prevent excessive fruit drop.

fruiting body a spore-bearing structure that shows differentiaton of tissues, e.g. the ascomata of ascomycetes and the basidiomata of basidiomycetes.

fruit sugar SEE FRUCTOSE.

frustule the silica test ('shell') of a diatom, which is in two parts, one part (the *epitheca*) overlapping the other, rather like a lid on a box. The epitheca is the oldest part of the frustule: when the diatom divides, the two parts of the frustule separate, and each daughter cell grows a new, smaller 'base' part. See also DIATOMS.

fruticose describing organisms that are erect and branching, especially the erect or pendant branching lichens (e.g. *Usnea*).

fuc-, fuco- a prefix denoting seaweeds. For example, *fucin* and *fucoidin* are colloids found only in the walls of brown algae (Phaeophyta). FUCOXANTHIN is a carotenoid pigment found in the Diatoms, Phaeophyta, and Chrysomonada. *Fucosterol* is the main type of sterol found in seaweeds.

Fucales an order of marine brown algae (PHAEOPHYTA) whose thalli are parenchymatous, with growth from an apical cell. Some, such as the wracks and *Pelvetia*, live on rocks in the intertidal zone, others are benthic, while a few, such as some sargassum weeds, are floating. Sargassum weed forms extensive rafts in the part of the Atlantic Ocean called the Sargasso Sea; these are home to a unique floating community of invertebrates and small fish, which are camouflaged to resemble the sargassum weed. The life cycle of the Fucales is considered advanced, the main thallus being diploid, with no free-living gametophyte. The thallus is differentiated into a cortex and a medulla, and often also a midrib, and shows dichotomous branching. The surface is dotted with small cavities called *cryptostomates*, whose openings are surrounded by tufts of sterile hairs, which may have a role in nutrient uptake. Many fucoids, such as *Fucus vesiculosus*, *Ascophyllum*, and *Sargassum*, have gas bladders to keep the fronds floating when submerged, so increasing the surface area available for photosynthesis, nutrient absorption and gas exchange.

The gametes are produced at the tips of branches in special receptacles, which in some species are swollen with mucilage. Just under the surface of a receptacle are tiny cavities (CONCEPTACLES) that produce the gametes. Male and female gametes may be produced in the same conceptacle or on different parent thalli, according to the species. Packets of sperm and eggs are released into the sea, where the gametes are released. Sperm are chemically attracted to the eggs. The zygote secretes an adhesive and attaches to the substrate. In some species the whole receptacle is detached, which helps to disperse the gametes over a larger area. Floating species, such as *Sargassum natans* and *S. fluitans*, reproduce vegetatively by fragmentation and do not produce receptacles.

fuchsin (magenta) dark green crystals of rosaniline hydrochloride, which dissolve in water to form a purple-red solution used as a dye and to stain certain cell components.

fucoxanthin a CAROTENOID pigment that plays an important role in photosynthesis in the DIATOMS, CHRYSOMONADA, HAPTOMONADA, PHAEOPHYTA, and some DINOMASTIGOTA, contributing to the brownish colour of the chloroplasts in these phyla. Fucoxanthin strongly absorbs blue light, which penetrates to greater depths than the longer wavelengths as red light. See also XANTHOPHYLL.

fugacious short-lived. The term usually refers to plant parts other than floral organs.

fugitive species (opportunist species) a species that is able to rapidly colonize an area, but which is unable to survive COMPETITION from later invaders. Fugitive species are characteristic of unstable environments such as deserts, newly cleared land, and temporary ponds. They usually have efficient methods of long-distance dispersal, such as wind-borne seeds or spores, and short life cycles. See also EPHEMERAL, PIONEER SPECIES, SUCCESSION.

fuliginous soot-coloured, dusky.

fulvous tawny, of a dull brownish yellow colour.

fumaric acid (*trans*-butanedioic acid) An

intermediate in the KREBS CYCLE formed by the dehydrogenation of SUCCINIC ACID with the concomitant production of $FADH_2$ from FAD. It has the formula $COOH(CH)_2COOH$. In the next step of the cycle fumaric acid reacts with water to form malic acid. Fumaric acid is also involved in amino acid metabolism. It is formed by the degradation of phenylalanine and tyrosine. In amino acid synthesis it is thought ammonia may react with fumaric acid to form aspartic acid.

fumigant a volatile chemical used to kill pathogens in the soil or in glasshouses. Fumigants are usually applied to the soil by injection. Chemicals used as fumigants include methyl bromide, chloropicrin, formaldehyde, metham, ethene, dibromide, and carbon disulphide.

fundamental niche the ECOLOGICAL NICHE that an organism could potentially occupy. The actual niche occupied will depend on environmental pressure and pressures from other species.

fungal sheath in a MYCORRHIZA, the fungal tissue that surrounds the root of the host plant.

Fungi (Mycota) a kingdom of saprobic, symbiotic, or parasitic eukaryotic organisms containing some 100 000 recognized species divided among about 6800 genera. However, it has been estimated that the actual number of species may be at least 2 million. The name 'fungi' has also been used as a general term, lacking any systematic meaning. As defined in the FIVE KINGDOMS CLASSIFICATION system, members of the kingdom Fungi form spores and lack undulipodia at all stages of the life cycle. Certain primitive fungi, such as the chytrids (see CHYTRIDIOMYCOTA), that possess motile stages are placed in another kingdom, the PROTOCTISTA. Other classification systems include these groups within the Fungi, which are instead defined by a range of criteria, as follows. They are heterotrophic, usually obtaining food by the secretion of extracellular enzymes, followed by absorption of the products of digestion. Fungal cell walls characteristically contain chitin as the principal structural carbohydrate. The basic fungus body is

made up of slender tubelike filaments called HYPHAE, which grow from the apex and may or may not be incompletely divided at intervals by septa. Even where septa are present, they may not completely separate adjacent sections of the hypha. Thus each 'cell' of the hypha may contain more than one haploid nucleus (see also DIKARYOTIC, HETEROKARYOSIS). The main fungus body is usually a mass of branching hyphae called a MYCELIUM, which shows indeterminate growth.

Reproductive structures are built from an aggregation of hyphae. Mushrooms and brackets are the largest. Asexual reproduction is by various types of spores, which germinate into new hyphae. In sexual reproduction nuclei may fuse and immediately undergo meiosis (haploid life cycle), or they may remain as discrete synchronously dividing pairs of haploid nuclei for some time before eventually fusing to form a diploid zygote cell (haploid-dikaryotic life cycle). The zygote immediately undergoes meiosis to restore the haploid state by the production of haploid spores, which germinate directly into hyphae. In HETEROTHALLIC fungi the conjugant nuclei are of different mating strains derived from different parents; in HOMOTHALLIC forms the nuclei are identical.

In the FIVE KINGDOMS CLASSIFICATION, the Fungi comprise three phyla: ZYGOMYCOTA, ASCOMYCOTA, and BASIDIOMYCOTA. Some primitive fungus-like organisms that used to be included in the Fungi are here assigned to discrete phyla of the kingdom PROTOCTISTA. These include the LABYRINTHULATA (slime nets), ACRASIOMYCOTA (cellular slime moulds), MYXOMYCOTA (plasmodial slime moulds), PLASMODIOPHORA, HYPHOCHYTRIOMYCOTA, CHYTRIDIOMYCOTA, and OOMYCOTA. The Five Kingdoms system formerly recognized two further phyla, the Deuteromycota (fungi that lack sexual stages) and Mycophycophyta (fungi in lichen associations). Since members of both these groups have evolved from the Ascomycota or Basidiomycota, they are now placed in their parent phyla. In other modern classification systems two major divisions may be recognized: the PARAMYCOTA, which includes the Oomycetes and Hyphochytriomycetes; and the EUMYCOTA or 'true fungi', whose component subdivisions are the Chytridiomycotina, Zygomycotina, Ascomycotina, Basidiomycotina, and Fungi Anamorphici (Mitosporic Fungi). In some classifications the Oomycota, Hyphochytriomycota, and Labyrinthulata are placed in the kingdom Chromista.

Fungi are extremely important DECOMPOSERS, common in soil, on rotting material almost everywhere, and in fresh water. Many, especially parasitic forms, cause disease in both plants and animals. Others form symbiotic relationships with a wide range of autotrophs: MYCORRHIZAS are found on most forest trees and on many plants, and LICHENS are symbioses between fungi and algae or blue-green bacteria. Some fungi, such as *Penicillium*, are important sources of antibiotics, and many fungi, including the yeasts, are of industrial importance, being used in fermentation processes.

Fungi Anamorphici (Mitosporic Fungi, Fungi Imperfecti) in certain modern classification systems, a polyphyletic group classified as a subdivision of the EUMYCOTA, comprising fungi that lack a sexual phase. It was formerly known as the Fungi Imperfecti, Deuteromycotina, or (in the Five Kingdoms classification) Deuteromycota. These fungi typically have conidia formed by mitosis; they lack ascospores, basidiospores, teliospores, and other similar spores and any meiotic structures. There are some 15 000 species in about 2600 genera. Most are ANAMORPHS of Ascomycota, a few of Basidiomycota. Two or three classes are often recognized: the HYPHOMYCETES, which do not produce conidiomata; the COELOMYCETES, which produce conidiomata; and sometimes also the Agonomycetes (see AGONOMYCETALES), based on the structure of the spore-producing bodies. See also STATES OF FUNGI.

fungicide a chemical that kills fungi. Fungicides are used to prevent or control fungal diseases. They are applied to seeds as a dust or slurry and to standing crops usually as a spray. For control of postharvest diseases fruits may be dipped or sprayed, or packed in fungicide-impregnated materials. Fungicides can also be applied to the soil for control of damping-off and root diseases.

Fungicides, like other pesticides, usually have three names – the chemical name of the active ingredient, the approved common name, and the trade name. For example, methyl N-[1-(butylcarbamoyl)-2-benzimidazole] carbamate is the chemical name of benomyl, which has the trade name Benlate.

Fungicides fall into two major categories: the inorganic fungicides including COPPER FUNGICIDES and SULPHUR DUST; and the organic fungicides, a group comprising the more modern fungicides, including SYSTEMIC fungicides. See also QUINONE FUNGICIDE, ORGANOMERCURIAL FUNGICIDE, DITHIOCARBAMATE FUNGICIDE, DINITRO COMPOUNDS, FUMIGANT, SEED DRESSING.

fungicole an organism that grows on fungi.

Fungi Imperfecti see FUNGI ANAMORPHICI.

funicle see FUNICULUS.

funiculus (funicle) the stalk attaching the ovule, and later the seed, to the PLACENTA or ovary wall in angiosperms. It serves as an anchor and provides a vascular supply to the ovule and seed. See also RAPHE.

furanose ring a ring containing four carbon atoms and an oxygen atom. Such rings are formed by ketose sugars with five or more carbon atoms. When such sugars are in solution the ketone group on the second carbon of the sugar reacts with the hydroxyl group on the fifth carbon atom. Fructose existing in this form is termed fructofuranose. Aldose sugars can also form furanose rings. However aldoses also form PYRANOSE RINGS, which are far more stable and so predominate in solutions of aldose sugars.

furfuraceous covered with small scales.

6-furfurylaminopurine see KINETIN.

furrowing a form of cell division in which the plasma membrane invaginates to pinch the cell into two daughter cells.

fuscous dark-coloured, brownish grey or black.

fusiform elongated and tapering at each end, as does a fusiform initial.

fusiform initials more or less elongate INITIALS in the vascular cambium that give rise to the components of the axial (longitudinal) system of the secondary xylem and secondary phloem (e.g. vessel elements, tracheids, and sieve tube elements). They also give rise to the RAY INITIALS.

future-natural see NATURAL.

Gg

Gaia hypothesis the theory, proposed by the British scientist James Lovelock in 1979, that the whole planet, including its living (biotic) and nonliving (abiotic) components, functions as a single homeostatic system (see HOMOEOSTASIS). According to Lovelock, the activities of living organisms in response to environmental changes in turn influence the physical and geochemical cycles (including those of climate) of the planet so it continues to sustain life. See also BIOGEOCHEMICAL CYCLE.

galactan a polysaccharide in which the major monosaccharide subunit is GALACTOSE, though other sugars, notably ARABINOSE, are often also present. Galactans are structural polysaccharides, often found among the PECTIC SUBSTANCES of cell walls.

galactolipid a LIPID that contains GALACTOSE.

galactose an aldohexose sugar commonly found in plants (*see illustration at* aldose). It normally does not exist in the free state but as polymers (see GALACTAN). It is also a constituent of the oligosaccharides raffinose and stachyose. Oxidation of galactose yields galacturonic acid, which on polymerization forms pectic acid.

galacturonic acid SEE URONIC ACIDS.

Galápagos Islands a group of islands in the Pacific Ocean about 970 km from the west coast of South America, and part of Ecuador. They were visited by Charles Darwin in 1835, and the thousands of endemic species of plants and animals he found there were a significant factor in the development of his theory of evolution by natural selection (see DARWINISM).

gall an abnormal localized swelling or outgrowth produced by a plant as a result of attack by a parasite. Galls are caused by bacteria, fungi, nematodes, insects, or mites or by a combination of these agents. CROWN GALL is a common gall produced as a result of bacterial infection. The root knots produced by nematodes of the genus *Meloidogyne* are another example. Galls caused by insects are very diverse and include the oak apple caused by the gall wasp *Neuroterus lenticularis* and the pincushion galls on roses caused by the gall wasp *Rhodites rosae*.

gallery forest (fringing forest) a FOREST that grows along the banks of a river in an area otherwise devoid of trees. Such forests are typical of the tropics and subtropics, where gallery forests extend from RAINFORESTS into the adjacent grasslands. They are generally similar to rainforests, but less luxuriant.

gametangiophore an upright structure that bears the female gametes (see ARCHEGONIOPHORE) or male gametes (see ANTHERIDIOPHORE) in certain liverworts (Hepatophyta). Gametangiophores are extensions of the thallus and are most elaborate in the genus *Marchantia*.

gametangium (*pl.* gametangia) a cell or organ in which gametes are formed. The term is mostly used of the single-celled gametangia of certain algae and fungi in which there is no differentiation into male and female. However antheridia, oogonia, archegonia, etc. may also be described as gametangia.

gamete a cell or nucleus that may participate in sexual fusion to form a

ZYGOTE. It is normally haploid and thus on fusion of two gametes a diploid zygote is formed. In virtually all plants (exceptions are those with a DIPLONTIC life cycle) meiosis is separated from gametogenesis by the development of a somatic gametophyte generation (see ALTERNATION OF GENERATIONS). In the primitive algae and fungi the gametes are often naked and isogamous (see ISOGAMY). In more advanced forms there is a trend through ANISOGAMY to OOGAMY and specialization of the gametes, so that they become better protected and less dependent on water for survival and dispersal. See also SEXUAL REPRODUCTION.

gametic equilibrium see LINKAGE EQUILIBRIUM.

gametic meiosis meiosis that occurs immediately before gametogenesis.

gametocyte a cell that divides by meiosis to produce gametes. See MEGASPORE MOTHER CELL, MICROSPORE MOTHER CELL.

gametogenesis 1 the formation of gametes. Gametes are usually formed by mitosis by the gametophyte generation, though in some groups (see DIPLONTIC) gametogenesis corresponds with meiosis. Gametes may form in any cell of the gametophyte, as occurs in the green alga *Ulva*, or they may be confined to specialized organs, such as antheridia and archegonia. Compare SPOROGENESIS.

2 the growth and development of the GAMETOPHYTE generation.

gametophore a modified branch that bears the sex organs (gametangia). The term is used especially with reference to the BRYOPHYTA.

gametophyte (*adj.* gametophytic) the gamete-producing, usually haploid generation in the life cycle of a plant or alga. The gametophyte arises from haploid spores produced as a result of meiosis in the SPOROPHYTE, or diploid, generation. In the bryophytes the gametophyte constitutes a major part of the life cycle and the sporophyte is either partially or totally dependent on it for anchorage and nutrition. In the vascular plants the gametophyte becomes progressively less prominent so that in the angiosperms it is represented by the pollen tube and its

nuclei and the embryo sac. In lower plants the gametophyte is unprotected and tends to be highly susceptible to dehydration. It therefore tends to be confined to damp habitats. See also ALTERNATION OF GENERATIONS, SPOROPHYTE.

gamone any substance released by a gamete that serves to attract another gamete. For example, in certain ferns malic acid is believed to be released from the archegonia and to attract the male gametes.

gamopetalous (sympetalous) having petals that are fused along their margins to the base, forming a corolla tube. Most plants possessing such flowers are grouped together in the ASTERIDAE (Sympetalae).

Gamophyta (conjugating green algae) a phylum of algal protoctists, with chlorophylls *a* and *b*, that do not have undulipodia at any stage of the life cycle. Asexual reproduction is by FISSION or FRAGMENTATION. Sexual reproduction involves the production of amoeboid gametes that fuse to form a zygote; this usually develops into a resistant ZYGOSPORE, which eventually undergoes meiosis to produce haploid cells and germinates into a new thallus. On the basis of their photosynthetic pigments, the Gamophyta are often retained within the CHLOROPHYTA, being regarded as a distinct class, Zygnemaphyceae, or an order (Zygnematales) of the class CHAROPHYCEAE.

There are two main groups of Gamophyta, the conjugating filamentous green algae and the desmids. Conjugating filaments, such as *Spirogyra* and *Zygnema*, are common in freshwater habitats, where they form floating masses of filaments, sometimes anchored by rhizoids from basal cells. Most have a single large nucleus in each cell. Genera are often distinguished by the form of their chloroplasts, which range from disclike (*Mougeotia*) to spiral (*Spirogyra*, *Zygnema*) and stellate forms. A PHRAGMOPLAST forms during cell division, which is accompanied by cell-plate formation and/or FURROWING. Sexual reproduction is by conjugation, in which cells of adjacent filaments develop tubelike extensions that grow towards each other and fuse, forming a conjugation tube. The protoplasts of these cells round up to form amoeboid gametes. Gametes from one filament migrate into the other and fuse with its gametes. The zygotes develop into zygospores.

The second group of Gamophyta comprises the *desmids*, a group of mostly single-celled algae. The *saccoderm desmids* are simple cells that undergo conjugation, and on a molecular basis are closely related to the filamentous forms. The cells of the *placoderm desmids*, such as *Closterium* and *Cosmarium*, are often deeply divided into *semicells*, with a narrow bridge (the isthmus) at the centre housing the nucleus. Each semicell contains its own chloroplast. The cell wall is often impregnated with silica and intricately sculpted, enveloped in a mucilaginous sheath. Asexual reproduction is by fission, while sexual conjugation involves the pairing of cells and the splitting of each cell at the isthmus, followed by the shrinkage and liberation of the protoplasts prior to fusion and zygospore production. Desmids are also common in freshwater habitats.

gamosepalous having sepals that are fused along their margins to form a tubular calyx.

gap analysis a technique for selecting ecosystems in need of CONSERVATION. The numbers of particular rare or ENDANGERED SPECIES are mapped across a geographical area, then the maps are overlaid onto a single map, which also shows the location of reserves and other protected areas. This shows up the areas where there are considerable numbers of such species outside the existing reserves. Compare HOT SPOT.

gap phase in forestry, the pioneer stage of SUCCESSION when trees start to colonize a newly cleared site.

garigue (garrigue) a type of scrub woodland characteristic of limestone areas with low rainfall and thin poor dry soils. The low-growing vegetation consists of aromatic, often spiny, species, such as sage (*Salvia officinalis*), thymes (*Thymus*), and lavender (*Lavandula vera*). Many bulbous plants grow between the shrubs and there are numerous species that are also present in MAQUIS. This type of vegetation is widespread in Mediterranean countries.

gas chromatography a chromatographic technique in which the mixture of substances to be analysed is vaporized and carried along a column containing a non-volatile liquid (*gas–liquid chromatography*, *GLC*) or a solid absorbent (*gas–solid chromatography*, *GSC*) by an inert carrier gas such as nitrogen. The carrier gas is termed the mobile phase, and the liquid or solid is the stationary phase. The components of the mixture separate out at different times according to their gas–liquid or gas–solid partition coefficients (see PARTITION CHROMATOGRAPHY). Those that are most soluble in the gas emerge from the column first and those most soluble in the liquid or solid emerge last. The nature and amount of the different gases emerging from the column is registered using a gas detector. Fatty acids and other substances, e.g. sterols and hydrocarbons, that are volatile at reasonably low temperatures are separated in this way. See also CHROMATOGRAPHY.

gaseous exchange the DIFFUSION of gases into and out of the cells of an organism. For exchange to occur, the gases usually have to pass through a liquid boundary layer. The rate of diffusion of gases dissolved in water is much slower than diffusion through air, an important consideration in calculations of gaseous exchange. During photosynthesis carbon dioxide is taken in and oxygen given off; the reverse occurs in respiration. The passage of water vapour is usually in one direction only – out of the plant.

gas-exchange method a method of measuring PRIMARY PRODUCTIVITY by measuring the rate of uptake of carbon dioxide (during photosynthesis) and the rate of release of oxygen (during respiration) by the plant or plants being studied. This is usually done by gas analysis. See also CARBON DIOXIDE METHOD, INFRA-RED GAS ANALYSER, OXYGEN METHOD.

gas–liquid chromatography (GLC) *See* gas chromatography.

gasoline SEE PETROLEUM.

gas-solid chromatography see GAS CHROMATOGRAPHY.

Gasteromycetes an obsolete class of the BASIDIOMYCOTA containing fungi that have well-developed basidia and do not forcibly discharge their spores, which are usually totally enclosed within the basidioma. It contained some 1169 species in about 164 genera. A polyphyletic grouping, it is no longer considered a natural taxonomic unit; members are now included in the HOLOBASIDIOMYCETES.

gas vacuole (gas vesicle) a structure made up of numerous gas-filled vesicles, many of which may be found in the cells of certain aquatic bacteria, including the blue-green bacteria (see CYANOBACTERIA). The density of the cell depends on how much gas is contained within the vesicles. The position of the organism in a column of water can thus be regulated by these structures. The vesicles are unusual in being bound by a protein rather than phosphoglyceride membrane.

Gause principle see COMPETITION.

Gaussian distribution see BINOMIAL DISTRIBUTION.

GC ratio (G+C ratio) the ratio of the number of guanine and cytosine base pairs (see BASE PAIRING) in a piece of DNA to the total number of base pairs. The greater the differences in GC ratios between two species, the less likely they are to be phylogenetically related. The GC ratio of the rRNA in the 16S ribosomes of bacteria is a key factor in their classification into phyla in the Five Kingdoms classification.

geitonogamy see AUTOGAMY.

gelatinous fungi see TREMELLALES.

gelatinous lichen a lichen in which the PHYCOBIONT is a cyanobacterium (see CYANOBACTERIA).

gel filtration a form of CHROMATOGRAPHY in which the mixture to be separated is washed through a column packed with beads of an inert gel. The polymer forming the gel contains internal pores, into which molecules below a certain size can penetrate. In general, large molecules, which cannot penetrate the pores, move quickly down the column; smaller molecules, which enter the gel, travel more slowly. The technique exploits differences in size (i.e. molecular weight) and measurements of the rate of travel can give estimates of molecular weight. It is used particularly to separate protein mixtures, but can also be applied to cell nuclei, viruses, and other large molecules. It is

sometimes referred to as *molecular-exclusion chromatography*.

gemma 1 a specialized multicellular unit of VEGETATIVE REPRODUCTION found in certain mosses and liverworts, and in *Psilotum*. It may take various forms, from being disc- or platelike to being filamentous or heart shaped. When separated from the parent it develops into a new individual identical to the parent. *Gemma cups* may be formed where several gemmae are produced on a protective receptacle, as seen in the moss *Tetraphis pellucida*. This process is called *gemmation*. Clusters of gemmae may also be seen at leaf tips or in the leaf axils, on specialized stalks or pseudopodia, and on rhizoids. Reproduction by gemmation occurs more frequently than spore formation and in some species it is the only form of reproduction seen.

2 see CHLAMYDOSPORE.

gemmation see GEMMA.

gemmiferous producing GEMMAE in asexual reproduction.

gene the unit of inheritance. Defined no more precisely than this, the term 'gene' corresponds with Mendel's 'factors' responsible for the manifestation of a particular characteristic.

A particular gene may be further defined as a segment of DNA located at a specific point (*locus*) on one of the chromosomes and capable of existing in alternative forms or ALLELES. Thus there would be a gene for 'seed colour' in peas, with alternative alleles of 'green' and 'yellow'. In a diploid cell, alleles of a gene are paired, one on each of a pair of homologous chromosomes. If contrasting alleles are present in the same organism, e.g. one for 'green' and one for 'yellow', then the phenotype of the organism will depend upon the dominance relationship (see DOMINANT) between the alleles.

A gene may also be defined as a piece of DNA, several hundred nucleotides long, that acts as a functional unit (see CISTRON) controlling the synthesis of a single polypeptide or a messenger RNA molecule whose GENETIC CODE is contained within the gene. In modern usage the term gene is most commonly applied to functional units of DNA. Thus STRUCTURAL GENES determine protein structure, and regulator, promoter, and operator genes control the regulation of protein production. One or more structural genes may be associated with promoter, regulator, and operator genes that control their expression to form a larger functional unit, the OPERON. However, in addition to determining characteristics, inherited information also mutates and recombines. A gene defined as a unit of mutation (*muton*) may be a single nucleotide, compared to the hundreds of nucleotides of a structural gene. The unit of recombination (*recon*) lies between these two extremes. Genes are extremely stable, so spontaneous mutations occur very rarely. However, mutation is the ultimate source of all genetic variation, i.e. the reason why the 'seed colour' gene exists in different forms, such as green and yellow. Once mutation has created alternative alleles, a further source of genetic variation is by genetic RECOMBINATION, the rearrangement of whole chromosomes or part of chromosomes during meiosis.

Although the majority of DNA of eukaryotic cells, and hence the genes, is confined to the chromosomes, some DNA is present in such organelles as mitochondria and chloroplasts (see ENDOSYMBIOTIC THEORY). The pattern of inheritance shown by cytoplasmic genes differs from that of chromosomal (Mendelian) genes and they are referred to as *plasmagenes* in order to distinguish them from the latter.

gene amplification the production of multiple copies of a GENE.

gene bank 1 an institution where plant material, in danger either of becoming extinct in the wild or of being lost from cultivation, is stored in a viable condition. In many such centres the emphasis is on maintaining collections of plants of demonstrated or potential use to humans, e.g. crop varieties containing useful genes that have been superseded by improved CULTIVARS. Normally seed material is dried down to about 4% moisture and stored at around 0°C preferably in hermetically sealed containers. For many seeds such conditions will maintain viability for 10–20 years and longer. The seeds of some species,

including many tropical plants, cannot be dried without killing them and yet have a limited lifespan in the hydrated state. Such material must be maintained in the growing condition. The problems of space and maintenance that this entails may in some instances be overcome by tissue-culture methods. All material must periodically be sown to check viability and multiply seedstock. Pollen may also be stored in gene banks though generally the longevity of pollen is less than that of seeds. See also GENETIC EROSION, GENETIC RESOURCES.

2 see GENE LIBRARY.

gene centre (centre of diversity) an area that shows considerable genetic diversity of certain crop plants and their relatives. Some gene centres, the primary centres, are believed to correspond to the region where a particular crop originated. For example, numerous forms of wheat exist in the Middle East and, since many of the wild relatives of wheat are also found in this area, it is believed the crop was first domesticated there. Other centres, the secondary centres, do not contain wild relatives and are believed simply to be areas where the crops have been cultivated for a considerable period. Ethiopia is an example of a secondary centre.

Gene centres are typically found in mountainous regions. This may be because human communities are more isolated and climatic conditions more variable in such areas – both conditions that would lead to more rapid divergence of crop populations. The recognition of gene centres is important in conservation and plant breeding work as such areas are important reservoirs of natural genetic variability. See also GENETIC EROSION.

gene cloning the production of multiple exact copies of a gene or group of genes. The DNA containing the gene(s) is cut into fragments by RESTRICTION ENZYMES, and the fragments inserted into plasmids or bacteriophages, which act as vectors and transfer the DNA segments to host cells, often of the bacterium *Escherichia coli*. Transfer to the host cells may also be achieved directly, by stimulation of the bacteria with a weak electrical current, but this is an inefficient method. Inside the host cell the DNA to be cloned undergoes replication as the host replicates. DNA probes can be used to screen the resulting bacterial colonies for the required DNA segments. See also GENETIC ENGINEERING, GENE LIBRARY.

genecology the study of variations in GENE FREQUENCY within a species in relation to changes in the environment.

gene conversion a process during the correction of mistakes in the DNA (MUTATIONS) whereby one member of a GENE FAMILY acts as a model for the correction of mistakes in other members, which are repaired to match it. In the case of a new mutation, if this acts as the model, then the mutation spreads through the GENOME to other members of the same gene family. If another member serves as the model, then the mutation is prevented from spreading.

gene diversity a measure of the genetic diversity in a population, defined as the mean expected HETEROZYGOSITY per locus.

gene duplication a process by which a gene is copied twice, so that there are two copies side by side on the same chromosome. This can aid evolution, since the second copy may sometimes be mutated further without affecting the function of the gene, since the original copy still codes for the production of the required proteins.

gene family a group of similar or identical genes that are usually situated alongside each other on the same chromosome, since they have arisen by gene duplication. Some members of the family may work together; others may act as PSEUDOGENES.

gene flow the movement of genes between populations of the same species by interbreeding. Compare INTROGRESSION.

gene-for-gene coevolution a form of COEVOLUTION in which the degree to which a species resists a pathogen or a pathogen succeeds in infecting a host species depends on the presence of certain alleles of particular genes in both species – a gene conferring resistance in the host, and a gene determining virulence in the pathogen. Such associations are found between crop plants and pathogens (such as fungi, viruses, bacteria), and between plants and nematodes or insects.

gene frequency the proportion of an ALLELE in a population compared with other alleles of that gene. Thus if a gene A has two alleles, a and A, and the frequency of a = 0.2 (20%) then the frequency of A must be 0.8 (80%) since A + a = 1 (100%). Although the term 'gene frequency' is commonly used by population geneticists, it would be more consistent with standard definitions of the word 'gene' to use the term *allele frequency*. See also HARDY–WEINBERG LAW.

gene library (gene bank) a collection of cloned DNA fragments (see GENE CLONING) that constitutes the entire GENOME of an organism. The gene library consists of bacterial cells or bacteriophage vectors, either frozen or in culture, which contain the component fragments of the genome. GENE PROBES are used to identify specific segments of the genome, which may then be used for GENETIC ENGINEERING or other forms of experimentation.

gene machine a device for the automated synthesis of oligonucleotides.

gene mutation (point mutation) an alteration to a single gene resulting from a change in the number, type, or sequence of bases specifying the amino acids in a polypeptide. Such MUTATIONS are not visible by microscopy. Gene mutations result in new ALLELES of a gene. If the mutation results in the loss or gain of one or two bases, it is a *frame-shift mutation*, altering the 'reading frame' of translation, so that the wrong triplets will be read. The genetic code is read in units of three bases. Imagine placing a frame over each unit as it is read– a *reading frame* – it is important to start it on the right base, or the entire sequence from that point on will be read wrong. For example, a single base deletion (G) changes AGC/TAT/C to ACT/ATC. Such a mutation is likely to have a much more deleterious effect than a simple substitution, which affects only one amino acid in the polypeptide. See also CHROMOSOME MUTATION, MUTAGEN.

gene pool the sum total and variety of all the genes and their alleles present in a breeding population or species at one time. See also GENE CENTRE, GENETIC EROSION.

gene probe a single-stranded fragment of DNA or RNA that will bind by base pairing to a specific complementary base sequence on a nucleic acid. The probe is labelled with a radioisotope or a fluorescent dye to enable its identification when later separated. Probes may be of varying lengths and styles. They are valuable in determining the sequence of bases on a length of DNA or RNA by reconstruction from fragments generated by RESTRICTION ENZYMES, a process called *restriction mapping*. Gene probes are also used to find gene and chromosome mutations. See also FLUORESCENCE MICROSCOPY, FLUORESCENCE IN SITU HYBRIDIZATION.

generation time the interval between the commencement of consecutive cell divisions. In meristematic tissue or in a colony of unicells a period of time elapses after cell division while the daughter cells enlarge and establish their complete cellular organization (see CELL CYCLE) before they divide.

generative cell one of the cells found in the POLLEN TUBE of seed plants (see also TUBE CELL, VEGETATIVE NUCLEUS). In gymnosperms it gives rise to the BODY CELL and stalk cell whereas in angiosperms it gives rise directly to the two male gametes or GENERATIVE NUCLEI.

generative hypha see HYPHA.

generative nuclei in angiosperms, the two male gametes that are formed by division of the GENERATIVE CELL. They migrate down the POLLEN TUBE behind the VEGETATIVE NUCLEUS. When the pollen tube enters the EMBRYO SAC the tip of the tube breaks down to release the generative nuclei. On entering the nucellus and embryo sac they participate in the fertilization process. One fuses with the egg nucleus to form the zygote and the other usually fuses with the POLAR NUCLEI or definitive nucleus to form the PRIMARY ENDOSPERM NUCLEUS.

generic name see BINOMIAL NOMENCLATURE.

gene sequencing the process of determining the nucleotide sequence of a gene. The most widely used method is the *Sanger method*. This involves synthesizing a new DNA strand using as a template single-stranded DNA (such as a fragment produced by a RESTRICTION ENZYME or one from a GENE LIBRARY) from the gene under investigation. Modified nucleotides called dideoxynucleotides are added to halt synthesis of the new strand at any desired base (they are incorporated into the new strand but are unable to form further bonds). For example, dideoxyGTP will cause it to terminate at a guanosine. The strand to be sequenced is mixed with a solution containing free nucleotides of the four types, a small amount of one kind of dideoxynucleotide, a short radioactively labelled DNA sequence that acts as a primer, and DNA POLYMERASE. Replica experiments are set up with each of the four dideoxynucleotides. The resulting mixture contains strands of varying lengths with a radioactive label at one end (the start) and the specific nucleotide (matching the dideoxynucleotide) at the other. The fragments are separated by ELECTROPHORESIS and an autoradiograph prepared. The nucleotide sequence can be worked out from the positions of each of the four nucleotides.

gene splicing the stage in TRANSCRIPTION in eukaryotic cells in which the INTRONS (noncoding segments) are removed from the initial messenger RNA transcript and the remaining EXONS are spliced together to form the mRNA molecule that leaves the nucleus to take part in TRANSLATION.

genetic code the sequence of nucleotides along the DNA of an organism within which is incorporated the information necessary for protein synthesis. By controlling the type and amount of protein manufactured the genetic code controls the growth, development, and characteristics of an organism. The observation that DNA could specify the incorporation of twenty different types of amino acids into protein but itself consisted of only four bases implied that more than one base must code for a single amino acid. The shortest sequence of any four bases that could specify all the types of amino acids found in protein is three, since $4^3 = 64$ possible combinations (see TABLE). (Pairs of any four bases could only code for $4^2 = 16$ amino acids, which is not enough.) Thus arose the *base triplet* or *triplet code hypothesis*. It was further proposed that the triplets were sequential, not overlapping. This would account for the almost infinite variation in the sequence of amino acids in a protein. If there was a tendency for some amino acids to follow others in a polypeptide, then this might indicate an overlapping code, i.e. one where the last one or two bases of one triplet acted as the first one or two in the next triplet. However, this is not the case. Since there are 64 possible base combinations and only 20 amino acids coded for, it is not surprising that some amino acids are coded for by more than one triplet, i.e. the code is *degenerate*. For some amino acids, only the first two bases of the triplet seem to be important; substitution of another base for the third base would still leave the triplet coding for the same amino acid. Other triplets code for the starting or stopping of transcription or translation, and yet others (*nonsense codons*) have no known function.

The theory that the genetic code consists of base triplets received considerable support from *frame-shift experiments*. These demonstrated that mutations resulting from the addition or loss of three base pairs close together produced protein resembling that of normal individuals. However, mutations resulting from the addition or loss of one, two, or four pairs produced significantly different protein. When three bases were added or lost only a short segment of protein was affected, whereas if other numbers were added or lost, the code for the whole protein was thrown out of sequence.

genetic distance a measure of the genetic difference between individuals or

first base	second base				third base
	U	C	A	G	
U	UUU Phe	UCU Ser	UAU Tyr	UGU Cys	U
	UUC Phe	UCC Ser	UAC Tyr	UGC Cys	C
	UUA Leu	UCA Ser	UAA stop codon	UGA stop codon	A
	UUG Leu	UCG Ser	UAG stop codon	UGG Trp	G
C	CUU Leu	CCU Pro	CAU His	CGU Arg	U
	CUC Leu	CCC Pro	CAC His	CGC Arg	C
	CUA Leu	CCA Pro	CAA Gln	CGA Arg	A
	CUG Leu	CCG Pro	CAG Gln	CGG Arg	G
A	AUU Ile	ACU Thr	AAU Asn	AGU Ser	U
	AUC Ile	ACC Thr	AAC Asn	AGC Ser	C
	AUA Ile	ACA Thr	AAA Lys	AGA Arg	A
	AUG Met (start codon)	ACG Thr	AAG Lys	CUG Arg	G
G	GUU Val	GCU Ala	GAU Asp	GGU Gly	U
	GUC Val	GCC Ala	GAC Asp	GGC Gly	C
	GUA Val	GCA Ala	GAA Glu	GGA Gly	A
	GUG Val	GCG Ala	GAG Glu	GGG Gly	G

The genetic code

populations. The genetic distance between two populations that have the same allele frequencies at a particular LOCUS, when based solely on that locus, is zero. Where more than one allele is involved, the genetic distances for these alleles are averaged. Genetic distance is usually based on DNA sequences, allele frequencies, or phenotypic traits.

genetic drift (Sewall Wright effect) a change in GENE FREQUENCY that is due purely to chance, as opposed to natural selection. Genetic drift is more likely to occur in very small populations, since in such populations mating is likely to be nonrandom compared to matings in larger populations. See also FIXATION, FOUNDER EFFECT.

genetic engineering (genetic manipulation) the isolation of useful genes from a donor organism or tissue and their incorporation into an organism that does not normally possess them. The term is often extended to include the alteration of an organism's existing DNA (genes), and the incorporation of artificially synthesized genes. Genes may sometimes be isolated from living tissues by isolating the RNA produced when the gene is active, then using the enzyme reverse transcriptase to convert it back to DNA. The foreign DNA is introduced into the host by means of a vector (carrier), usually a PLASMID, bacteriophage, or other kind of virus (for DNA segments up to 24 000 base pairs or 24 kb long) or, for larger pieces of DNA, COSMIDS (up to 45 kb) or YEAST ARTIFICIAL CHROMOSOMES (YACs, up to 50 kb). Once there, it is replicated and passed to daughter cells at cell division together with the rest of the host's DNA. Techniques have also been developed for introducing foreign DNA into the cells of plant tissue cultures, which can then be grown into whole plants. Genes for herbicide resistance enable crop plants to be sprayed with herbicides to remove weeds. Other useful genes include those that produce compounds that render crops unpalatable or even fatal to insect pests and those that delay maturation and rotting of fruit.

The technique most commonly used in genetic engineering is called *recombinant DNA technology*. It uses RESTRICTION ENZYMES (restriction endonucleases), which cut DNA at specific base sequences. Different kinds of restriction enzymes cut DNA at different base sequences, and the same restriction enzymes are used to cut the DNA of the vector. Some restriction enzymes cut the DNA in such a way that single-stranded ends are left, which will base-pair with complementary single strands on the vector DNA. In the presence of DNA ligase, the cut ends of the foreign DNA and vector DNA are joined together and the vector inserted into a bacterium, which will now produce the new protein coded for by the foreign gene. In order to recognize bacterial colonies that have been successfully transformed in this way, a second marker gene is usually added together with the foreign gene. This is often a gene for antibiotic resistance. By growing the bacteria on a medium containing the antibiotic, only those colonies that contain the foreign gene will survive. See also GENE CLONING, GENE LIBRARY.

genetic equilibrium (linkage equilibrium) the situation in which the frequencies of two alleles at the same locus are in equilibrium and remain the same through many generations. The tendency for a population to reach an equilbrium resulting in a stable genetic composition is called *genetic homoeostasis*.

genetic erosion the loss of genetic variation and the consequent narrowing of the genetic base of cultivated plants through the introduction of new improved varieties of crops that can be grown over wide areas. The continued existence of locally adapted primitive varieties is threatened by their large-scale replacement with adaptable high-yielding varieties. The situation is particularly serious in GENE CENTRES, where valuable genes for such characters as disease resistance may be lost. See also GENETIC RESOURCES.

genetic fingerprinting (DNA fingerprinting) a technique that uses RESTRICTION ENZYMES to fragment the DNA of an organsim; the resulting fragments are then mixed with GENE probes and analysed for the pattern of repetition of particular nucleotide sequences. Such a pattern is characteristic of the individual rather than just the species. While genetic fingerprinting can be used to identify humans and animals (to ascertain stock lines for animal breeding, for instance), it can also be used to determine local varieties of crops that may be adapted for specific local circumstances, in order to check that commercially offered stocks of seeds are from the local variety they are claimed to be from, and also to reveal relationships between different populations of a plant species. See also DNA HYBRIDIZATION.

genetic homoeostasis see GENETIC EQUILIBRIUM.

genetic load the proportion of disadvantageous genes that are present in and sustained by a population. Genetic load has three components: *mutational load*, *segregational load*, and *substitutional load*. The first arises because most mutations are harmful. The second may occur in cases of heterozygous advantage, since selfing a fit heterozygote Aa generates deleterious AA and aa offspring. The third, also called environmental load or frequency-dependent load, occurs during transient POLYMORPHISM.

genetic manipulation see GENETIC ENGINEERING.

genetic map a map of the order of specific genes or sequences of DNA on a chromosome. Genetic mapping uses RESTRICTION ENZYMES and DNA probes to work out the order of nucleotides or groups of nucleotides. A rough idea of the relative positions of genes on a chromosome can be obtained by analysing the frequency of recombination between the alleles of linked genes (see CHROMOSOME MAPPING, LINKAGE).

genetic resources the diversity and availability of alleles in both natural and artificially maintained stocks of organisms. For a number of years concern has been expressed about the need to maintain a diverse range of genotypes in domesticated plants and animals and their close relatives in the wild. A broad genetic base increases the capacity for variation and successful adaptation, such as resistance to pathogens or hostile environments. Both national and international programmes exist to maintain genetic resources of agriculturally important organisms. This had led to the establishment of a number of GENE BANKS or seed banks where potentially valuable seeds are evaluated, multiplied, and stored under ideal conditions. GENE LIBRARIES are an extension of this conservation process. See also GENETIC EROSION.

genetics the study of inheritance. The subject is now so large that it impinges on all branches of biology. It also has wide applications in agriculture (e.g. plant and animal breeding), medicine (e.g. blood typing), and industry (e.g. synthesis of antibiotics, hormones, etc. by genetic engineering). See also BIOCHEMICAL GENETICS, CLASSICAL GENETICS, CYTOGENETICS, POPULATION GENETICS.

genetic spiral (developmental spiral) a hypothetical spiral formed by drawing a line through and joining the centres of successive leaf PRIMORDIA at a shoot apex. The form of the spiral depends on the angle of divergence and the distance between successive primordia. The first is often 137.5° and the second depends on the rate of vertical growth of the apex. It is often found that this distance increases geometrically. Compare PARASTICHY. See PHYLLOTAXIS.

genetic system the way in which the genetic material is organized in species or other taxa and the way in which it is transmitted from generation to generation.

genetic variance the proportion of the variance in the PHENOTYPE that results from the different GENOTYPES present in the individuals of a population. The total phenotypic variance shown by individuals in a population is the sum of the genetic variance and the environmental variance.

geniculate resembling a knee or able to bend like a knee, for example the awns of some grass species.

genome a complete haploid chromosome set. The term is used in discussions of POLYPLOIDY. A diploid organism contains two homologous genomes unless it is an interspecific hybrid, in which case it contains two different genomes. An autopolyploid (see AUTOPOLYPLOIDY) contains three or more homologous genomes while an allopolyploid (see

ALLOPOLYPLOIDY) has two or more pairs of different genomes. The number of chromosomes in a genome (termed the *base* or *basic number*) is usually between six and thirteen and is represented by the symbol *x*. In diploid organisms the number of chromosomes in the genome is equivalent to the haploid number *n*, and so $2n = 2x$. However in tetraploid organisms $2n = 4x$, in hexaploid organisms $2n = 6x$, etc. For example, allohexaploid wheat, *Triticum aestivum*, contains six genomes designated AA, BB, and DD each of which contains seven chromosomes, i.e. $x = 7$. Thus $n = 3x = 21$ and $2n = 6x = 42$.

genotype the genetic constitution of an organism as opposed to its physical appearance (PHENOTYPE). The latter is not necessarily a full expression of the genotype and will depend on the dominance and epistatic relationships between genes and on the environment.

genotype frequency the percentage of individuals in a population that possess a specific genotype. It is calculated using the HARDY–WEINBERG LAW.

genus (*pl.* genera) an important rank in the taxonomic hierarchy, which is subordinate to FAMILY, but above the rank of SPECIES. It is a group of obviously homogeneous species. The generic name forms the first part of the binomial (the second being the specific epithet), e.g. *Quercus robur* (see BINOMIAL NOMENCLATURE). It is usually a singular noun and is written in Latin, usually in italics, with a capital initial letter, but lacks a uniform ending. Collections of similar genera are grouped into families. Large genera, e.g. *Rhododendron*, may be further subdivided into sections and series, with the possibility of designating the additional subordinate ranks of subgenus, subsection, and subseries. See RANK.

geocarpic describing plants that produce their fruits below ground, e.g. peanut (*Arachis hypogaea*).

geographical barrier see DISPERSAL BARRIER.

geographical isolation the separation of two populations of the same species by a physical barrier, such as a large river or sea or a mountain range. The different environmental (selective) pressures in the two locations may lead over many generations to divergent gene pools in the two populations (see ADAPTIVE RADIATION), and ultimately to sexual incompatibility and the evolution of two separate species.

geological time scale a table showing the sequence of the geological periods and the lengths of time they are assumed to have occupied (see illustration). It has been constructed by studying rock strata, where these have been exposed by excavations or mining or where rivers have cut deeply into the earth's crust. It may be assumed, providing there is no evidence of large earth movements, that the lower the rock layer, the older it is and the more ancient are the fossils it contains. With a knowledge of rates of erosion and deposition the intervals occupied by the different periods can be very roughly estimated by measuring the relative thicknesses of the strata. However considerably more accurate dating can now be provided by measuring the rates of decay of radioactive materials in the rocks (see RADIOMETRIC DATING). The results of such work are summarized in the table, which also shows the types of plants that have been found and their very approximate relative abundances. The earth is estimated to be about 4600 million years old and life is believed to have originated about 3000–3500 million years ago.

Complete agreement on the terminology used in such tables has not been reached. However, generally the major divisions are *eras*, divided into *periods*, which are then subdivided into *epochs*. The different periods are recognized on the basis of changes in fossil composition and the occurrence of major geological events, such as episodes of mountain building or major changes in the level of the seas. The earliest era, the PRECAMBRIAN, has few fossils. The succeeding PALAEOZOIC, MESOZOIC, and CENOZOIC eras have abundant fossils. Certain authorities use the term *eon* (*aeon*) for the very longest periods of time comprising several eras grouped together. The term eon may also be used to denote one billion years.

geometric series a quantitative measure of the relationship between species in a community expressed in terms of their RELATIVE IMPORTANCE VALUES, such that at

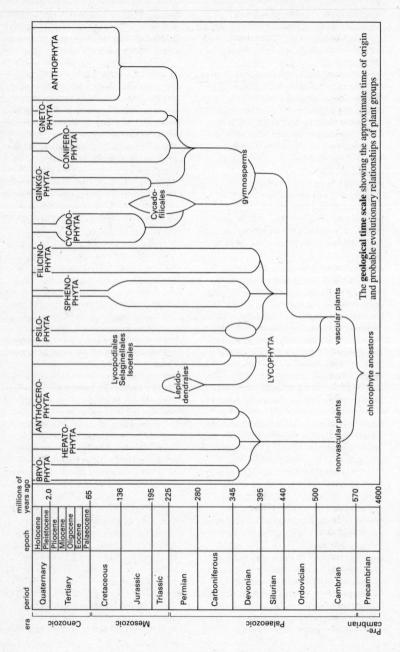

The **geological time scale** showing the approximate time of origin and probable evolutionary relationships of plant groups

each level of the hierarchy, the less important species occupies the same proportion of the remaining space as the previous one (for example, if the most important species occupies 60% of the space, the next species will occupy 60% of the remaining space, and so on, resulting in a geometric progression of importance values). Geometric species are most typical of the early (pioneer) stage of a SUCCESSION.

geophyte a plant with its perennating buds situated below ground on a rhizome, tuber, bulb, or corm. See CRYPTOPHYTE, PERENNATING ORGAN.

geotaxis see GRAVITAXIS.

geotropism see GRAVITROPISM.

germ cell any cell in the GERM LINE that eventually produces gametes.

germinal aperture (germinal pore, germ pore) a thin-walled area on a spore or pollen-grain wall through which the germ tube may emerge on germination. In the pollen grain the germinal aperture is an area where the exine is either thinner or absent. Pollen grains may be identified to some extent by the shape and number of the germinal apertures. See COLPUS, PORE.

germinal pore see GERMINAL APERTURE.

germination the physiological and physical changes undergone by a reproductive body, such as a seed, pollen grain, spore, or zygote, immediately prior to and including the first visible indications of growth. The process will not occur unless both internal and external conditions are favourable. In seeds the various factors causing DORMANCY must be overcome and dehydrated seeds must have a supply of water. Temperature and light can also affect germination. Following imbibition of water by a dry seed, the respiration rate increases markedly as food reserves are broken down and protein synthesis commences. The radicle is normally the first organ to emerge through the testa, followed by the plumule. See also EPIGEAL, HYPOGEAL.

germ line the sequence of cells whose descendants produce the gametes.

germ plasm genetic material, especially that contained within the reproductive (germ) cells. See WEISMANNISM.

germ pore see GERMINAL APERTURE.

germ tube the filament that emerges on

germination of a spore. The structure that emerges from a pollen grain is termed a POLLEN TUBE.

giant ferns see MARATTIALES.

gibberellic acid (GA₃) the first of the GIBBERELLINS to be isolated and characterized. It was obtained from the fungus *Gibberella fujikuroi*, which causes a disease of rice seedlings characterized by excessive elongation of shoots and leaves. The basic structure of gibberellic acid and most other compounds with gibberellin-like activity is the gibbane carbon skeleton (see diagram). Gibberellic acid is a TERPENOID and is synthesized from MEVALONIC ACID. Other GIBBERELLINS are synthesized by the same pathway and there is probably much interconversion in the plant between the different gibberellins.

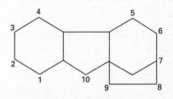

Gibberellic acid: the gibbane skeleton

Gibberellic acid is the gibberellin most widely used in experiments.

gibberellin any of a group of plant HORMONES first discovered through their ability to cause greatly increased stem elongation in intact plants. Gibberellins have subsequently been shown to affect numerous aspects of plant growth and development. Most are chemically related to GIBBERELLIC ACID. A few compounds with gibberellin-like activity, e.g. helminthosporol from the fungus *Helminthosporium sativum*, have a quite different chemical structure. Gibberellin can overcome certain forms of genetic dwarfism and dwarf varieties are often used in bioassays for gibberellin. Gibberellin-induced stem extension is due to the effect of gibberellin on cell expansion, which it increases by influencing cell-wall expansibility. This however does not occur in the absence of AUXIN, as illustrated by the fact that gibberellin does not cause extension of excised internodes, cut off from their auxin supply.

Gibberellins have also been found to break dormancy in buds and seeds that normally have a light or chilling requirement. They can also partly or completely replace the photoperiod (see PHOTOPERIODISM) or cold requirement necessary to some species for flowering. Gibberellin levels are high in young leaves and, if applied to ageing leaves, can delay senescence. Levels are also high in developing seeds and fruits. Gibberellin applications can induce PARTHENOCARPY and this has been put to commercial use in the production of seedless varieties of fruit. In barley seeds gibberellin has been found to stimulate the synthesis of the enzyme α-amylase. It does this by making available the messenger RNA responsible for α-amylase synthesis. This effect has proved of use in the brewing industry where α-amylase activity is essential for the production of malt. Gibberellin can even promote the enzyme's activity in inviable seed.

Gibberellins interact with other hormones in various ways. There is evidence that ABSCISIC ACID reduces gibberellin levels, hence the antagonistic effects of these substances. Gibberellin is believed to interact with auxin in the control of sex expression in dioecious plants. Gynoecious plants usually have low levels of gibberellin and high levels of auxin. Application of gibberellin can induce the formation of male flowers. This effect has been put to use in the cucumber industry where certain hybrid varieties are naturally gynoecious. Since pollen production is necessary for fruit set some of the plants are sprayed with gibberellin to produce the necessary male flowers.

gibbous swollen or bulging on one side. The calyx of the narrow-leaved vetch (*Vicia angustifolia*) is an example of a gibbous structure.

gigantism abnormal overgrowth due to an increase in the number of cells (*hyperplasia*), often in response to an infection such as club root, or due to an increase in the size

of cells (*hypertrophy*), for example witches broom, cankers, and galls, which are also responses to infection by pathogens,

gills the radially arranged lamellae that hang from the undersurface of the cap in certain fungi of the Basidiomycota. The gills, which are covered by the HYMENIUM, are usually wedge-shaped in longitudinal section and attached at the wider end of the wedge. This is believed to help prevent the spores becoming caught in the gills if the fruiting body is tilted slightly from the vertical. All gill-bearing fungi are traditionally placed in the family Agaricaceae, but this family is now sometimes split into a number of more homogeneous families.

Ginkgoales see GINKGOPHYTA.

Ginkgophyta a phylum of gymnosperms with a single living class (Ginkgoopsida) and order (Ginkgoales) and only one living species, the maidenhair tree (*Ginkgo biloba*) from China, which is known only in cultivation. The phylum was much more widespread in the Mesozoic era, and fossil remains have been found in North America. The group may have been derived from the CORDAITALES. The maidenhair tree is a tall deciduous tree with fan-shaped leaves, usually with a distinct notch opposite the petiole. The leaf venation is exceptional among extant gymnosperms in being dichotomously branching. The ginkgo is dioecious: male trees bear catkin-like microstrobili, which produce haploid microspores for wind dispersal; female trees produce fleshy ovules on peduncles in the axils of leaves or scale leaves. These contain two or three archegonia. Pollen grains are sucked through the micropyle by a POLLINATION drop. They produce a pollen tube that releases two large antheridia bearing spiral bands of undulipodia. The embryo has two cotyledons, and the integument of the ovule develops into a fleshy covering for the seed, which is further protected by a stony inner layer. Seeds are eaten in parts of Asia, but in Europe females are rarely planted because of the unpleasant smell of the seed covering. Maidenhair trees are very long-lived, some specimens being over 1000 years old.

girdling cutting a ring around a stem or

trunk so that the PHLOEM is severed and downward transport of substances is no longer possible. This eventually kills the plant as much of it no longer receives essential nutrients manufactured by photosynthesis.

glabrous describing a surface that is devoid of hairs or other projections.

glacial a period during which the climate cools, and glaciers and ice sheets advance or remain stationary, covering large areas of the Earth's surface. An ice age normally consists of a series of glacial periods (e.g. the Devensian) interspersed with warmer *interglacials* (e.g. the Flandrian). The term is also used for the ice age itself. The terms glacial and interglacial are used particularly in relation to the Pleistocene period. Evidence for these fluctuations comes from pollen analysis, which reveals the vegetation succession that occurred as the climate changed. See ICE AGE.

glaciation the formation and expansion of glaciers and ice sheets and their effects on the land. As the glaciers advance, plants and other organisms are forced to retreat to lower latitudes or face extinction. When the ice finally retreats, the newly exposed landscape is almost devoid of soil, although often covered in boulders, sand, gravel, and clay deposited by the melting ice. See also ICE AGE, GLACIAL, REFUGIUM.

gladiate sword-like.

gland one or a group of cells whose main function is to secrete a specific chemical substance or substances. A gland may retain its secretions, secrete them into a special reservoir or canal, or discharge them to the outside. Examples are the HYDATHODES of certain leaves, which secrete water; NECTARIES; the salt-excreting glands of some desert plants and halophytes; and the glands of some insectivorous plants, which secrete sticky substances and digestive juices. See also RESIN CANAL, LATICIFER, OIL GLAND, TRICHOME.

glaucophyte a eukaryotic cell that contains symbiotic blue-green bacteria (see CYANOBACTERIA) termed *cyanelles*.

glaucous describing surfaces having a waxy greyish-blue bloom, such as the leaves of rape or swede (*Brassica napus*).

GLC see GAS–LIQUID CHROMATOGRAPHY.

gleba the mass of spore-producing tissue that forms in the fruiting bodies of such fungi as truffles, puffballs, earthstars, and stinkhorns. It is enclosed by the PERIDIUM.

glei *See* gley.

gley (glei) a waterlogged (hydromorphic) intrazonal SOIL lacking in oxygen. In such conditions there is little decomposition of organic matter by bacteria, resulting in the accumulation of mor or raw humus. Beneath it is a gley horizon of blue-grey clay including ferrous compounds. Localized areas of oxidized rust-coloured ferric compounds may occur giving the soil a mottled appearance. Gleyzation may be seen in bog, meadow, and tundra soils.

gliadin a storage protein found in the caryopses of wheat. See PROLAMINE.

gliding bacteria see MYXOBACTERIA.

Global Biodiversity Assessment see CONVENTION ON BIOLOGICAL DIVERSITY.

global warming see GREENHOUSE EFFECT.

globose spherical.

globular protein see PROTEIN.

globules see CHARALES.

globulin a globular PROTEIN that is more or less insoluble in pure water, but which dissolves in dilute salt solutions. Globulins serve as storage proteins in seeds, especially those of members of the Fabaceae (legumes).

glochid a short barbed hair, many of which arise in tufts from the AREOLES of some cacti. It is a type of GLOCHIDIUM. The presence of glochids is one character used to separate the subfamily Opuntioideae, whose members possess glochids, from the other two subfamilies of the Cactaceae, Pereskioideae and Cactoideae, in which glochids are absent.

glochidium (*pl.* glochidia) any hairlike projection with a hooked tip. In certain water ferns, e.g. *Azolla*, glochidia arise from the surface of the MASSULAE and serve to attach the microsporangia to the megasporangium prior to fertilization. See also GLOCHID.

Glomales see ZYGOMYCOTA, VESICULAR–ARBUSCULAR MYCORRHIZA.

glucan any polysaccharide made up exclusively of glucose subunits. Examples of glucans are starch and cellulose.

gluconeogenesis the formation of glucose

from various precursors, such as pyruvate, certain amino acids (glucogenic amino acids), and intermediates of the KREBS CYCLE. Most of the stages in the pathway are reversals of the reactions involved in GLYCOLYSIS, with two exceptions. Phosphoenolpyruvate cannot be formed from pyruvate by reversal of the pyruvate kinase reaction of glycolysis. An alternative sequence of reactions involving the formation of oxaloacetate and malate and the input of a molecule each of ATP and GTP is performed to bypass this step. Similarly fructose 1,6-bisphosphate cannot be converted to fructose 6-phosphate by reversal of the glycolytic reaction catalysed by phosphofructokinase. The reaction is instead catalysed by the enzyme fructose bisphosphatase.

In plants glucose is formed predominantly by photosynthesis. Amino acids may be converted to glucose by the Krebs and GLYOXYLATE CYCLES though, unlike animals, no distinction is made between glucogenic and nonglucogenic amino acids, all being suitable for gluconeogenesis by this pathway. Plants can also bring about the net synthesis of glucose from fatty acids via the succinate produced by the glyoxylate cycle. This process is utilized by germinating fatty seeds.

glucosamine (2-amino-2-deoxy-D-glucose) a derivative of glucose in which the carbon at position 2 is replaced by an amino group. It is a common component of GLYCOLIPIDS and POLYSACCHARIDES, including chitin. See also AMINO SUGAR.

glucosan see GLUCAN.

glucose (dextrose, grape sugar) an aldohexose sugar (*see illustration at* aldose). In solution it forms a six-membered PYRANOSE RING. Glucose is the major fuel source of nearly all organisms and the basic substrate from which starch, cellulose, sucrose, and other carbohydrates are synthesized. Glucose is synthesized by several routes; de novo synthesis from carbon dioxide is via the CALVIN CYCLE, but synthesis from fatty acids via ACETYL COA and the GLYOXYLATE CYCLE is also a major route in some cells (e.g. germinating oil-bearing seeds). It can be oxidized through GLYCOLYSIS and the KREBS CYCLE. Synthetic pathways from glucose involve activation, by phosphorylation or by formation of a nucleotide diphosphate sugar.

glucose 1-phosphate a phosphorylated derivative of glucose, formed by isomerization from glucose 6-phosphate. Glucose 1-phosphate is an intermediate in the synthesis of UDP-glucose, from which starch and cellulose are formed. It is also the predominant breakdown product of starch and cellulose.

glucose 6-phosphate a phosphorylated derivative of glucose. It can be formed by the action of hexokinase on glucose or by isomerization from glucose 1-phosphate. Glucose 6-phosphate is an important activated form of glucose. In GLYCOLYSIS and in the PENTOSE PHOSPHATE PATHWAY the first step is activation of glucose to glucose 6-phosphate. It is also an intermediate in sucrose formation.

glucosidase see AMYLASE.

glucoside see GLYCOSIDE.

glucuronic acid see URONIC ACIDS.

glume 1 one of a pair of BRACTS subtending each spikelet in the inflorescence of grasses. Compare PALEA, LEMMA. *See illustration at* poaceae.

2 the bract subtending a flower in the inflorescence of reeds and sedges.

glutamic acid (2-aminopentanedioic acid) An acidic amino acid with the formula $HOOC(CH_2)_2CH(NH_2)COOH$ (*see illustration at* amino acid). Glutamic acid plays a central role in the cell's nitrogen metabolism. Along with GLUTAMINE it is the primary product of nitrogen assimilation, being formed from α-ketoglutaric acid by reductive addition of ammonia. Some amino acids are formed directly from glutamate (e.g. proline, ornithine), while others receive their amino group from glutamate in a TRANSAMINATION reaction. Glutamate is also a precursor in the synthesis of purines, pyrimidines, and porphyrins.

In amino acid CATABOLISM, glutamate is formed in TRANSAMINATION reactions with other amino acids. The carbon skeleton of these amino acids can then be further broken down, while the glutamate can undergo OXIDATIVE PHOSPHORYLATION to reform α-ketoglutarate.

glutamine a polar uncharged amino acid with the formula $NH_2CO(CH_2)_2CH(NH_2)COOH$ (*see illustration at* amino acid). Glutamine is formed by the ATP-assisted addition of ammonia to glutamic acid. Breakdown is by a reversal of this reaction but without concomitant formation of ATP. Like asparagine, glutamine is important in the neutralization and storage of free ammonia. It is also an amino group donor, particularly in purine and pyrimidine synthesis.

glutelin any of a group of plant proteins that are soluble in dilute acids and alkalis but insoluble in neutral salt solutions and in water and alcohol. *Glutenin* from wheat and *oryzenin* from rice are examples. Glutenin is the binding agent in flour pastes and dough. Compare PROLAMINE.

gluten the mixture of gliadin (see PROLAMINE) and glutenin (see GLUTELIN) that remains when the starch of wheat grains has been removed.

glutenin see GLUTELIN.

glycan see POLYSACCHARIDE.

glyceraldehyde the simplest of the monosaccharides containing an aldehyde group. It has the formula $CHOCH(OH)CH_2OH$ (*see illustration at* aldose). In its phosphorylated form, glyceraldehyde 3-phosphate (formula $CHOCH(OH)CH_2OPO_3{}^{2-}$), it is important in many metabolic pathways. In GLYCOLYSIS glyceraldehyde 3-phosphate is formed, together with dihydroxyacetone phosphate, by the cleavage of fructose 1,6-bisphosphate. Glyceraldehyde 3-phosphate is also one of the compounds formed in the CALVIN CYCLE.

glycerate 3-phosphate the phosphorylated derivative of the three-carbon glyceric acid, formerly called phosphoglyceric acid (PGA). Glycerate 3-phosphate is the first product of the dark reactions of photosynthesis, two molecules being formed by the carboxylation and cleavage of ribulose bisphosphate. It is also an intermediate in glycolysis, being formed from glyceraldehyde 3-phosphate, two molecules of ATP being formed in the process. It is

then converted to phosphoenolpyruvate. Glycerate 3-phosphate is the precursor of the amino acid serine.

glyceride see ACYLGLYCEROL.

glycerol (glycerine, 1,2,3-propanetriol) a trihydroxy alcohol with the formula $CH_2OHCHOHCH_2OH$. Glycerol is a basic component of nearly all complex LIPIDS. TRIACYLGLYCEROLS are esters of glycerol and fatty acids, while PHOSPHOGLYCERIDES and GLYCOLIPIDS are esters of glycerol derivatives and fatty acids. Phosphorylated derivates of glycerol are intermediates in GLYCOLYSIS and other areas of metabolism. Glycerol is often used as a mounting medium in microscopy. See also TRIOSE PHOSPHATE.

glycerolipid a LIPID derived from glycerol 3-phosphate by the addition of fatty acid chains and a 'head' group. They include TRIACYLGLYCEROLS, which are important storage compounds in seeds. Triacylglycerols can be broken down by the action of lipases to release fatty acids, which then undergo β-oxidation to release energy or produce sugars.

glycerophosphatide see PHOSPHOGLYCERIDE.

glycine (aminoacetic acid, aminoethanoic acid) the simplest amino acid. It has the formula NH_2CH_2COOH (*see illustration at* amino acid). There are several pathways of glycine biosynthesis. It can be made directly from ammonia, carbon dioxide, and a methyl group donated by the methylating agent N^5,N^{10} methylenetetrahydrofolic acid (mTHFA). In another reaction involving mTHFA, glycine can be formed from serine. A third pathway, from glycerate 3-phosphate involves transamination of glyoxylate. Glycine can be broken down either by reversal of the first or second synthetic pathway, or by deamination to glyoxylate and hence into the Krebs cycle via malic acid.

Glycine is one of the basic components of PORPHYRINS, along with SUCCINIC ACID. It is also necessary in purine synthesis.

glycogen a storage polysaccharide, similar in structure to STARCH, important in animals but not found in plants. It is however present in certain bacteria and fungi. It is a branched glucan, like amylopectin, but the branching is more frequent, occurring every 8–12 glucose residues. The enzyme that catalyses glycogen synthesis in bacteria is similar to starch synthetase in that it requires ADP-glucose as a substrate. Breakdown of glycogen, like starch breakdown, is catalysed by AMYLASE.

glycolic acid (hydroxyacetic acid, hydroxyethanoic acid) the substrate that is oxidized during PHOTORESPIRATION in C_3 plants. It is formed by the oxidation of RIBULOSE BISPHOSPHATE to glycerate 3-phosphate and phosphoglycolic acid. Hydrolysis of the phosphoglycolic acid yields free glycolic acid. During photorespiration the enzyme glycolate oxidase catalyses the oxidation, by molecular oxygen, of glycolic acid to glyoxylic acid. This reaction, which also produces hydrogen peroxide, takes place in the PEROXISOMES. The glyoxylic acid is then transformed into glycine. In the mitochondria two molecules of glycine react together to form serine and carbon dioxide. It is this carbon dioxide that is released in photorespiration.

glycolipid (glycosyldiacylglycerol) any of a group of ACYLGLYCEROLS containing a carbohydrate group, commonly a mono-, di-, or trisaccharide (see POLYSACCHARIDE) or an amino sugar. Glycolipids are the major lipid constituents of chloroplasts. No TRIACYLGLYCEROLS are found in chloroplasts and PHOSPHOGLYCERIDES are present only in small quantities. The commonest glycolipids are monogalactosyl- and digalactosyldiacylglycerols. They are synthesized from diacylglycerols and UDP (see NUCLEOSIDE DIPHOSPHATE SUGARS) derivatives of GALACTOSE.

glycolysis (Embden–Meyerhof–Parnas pathway) (*adj.* glycolytic) The metabolic pathway by which a molecule of glucose is anaerobically degraded to two molecules of pyruvate (see diagram). In the glycolysis of one molecule of glucose, two molecules of ATP are used in the phosphorylation reaction at the beginning of the pathway and four molecules of ATP are formed later, giving a net yield of two ATP molecules. The enzymes of glycolysis are located in the cytosol, the most important being

phosphofructokinase, which is the major regulatory enzyme in glycolysis. It is inhibited by ATP and citrate and stimulated by ADP and AMP. Glycolysis is an example of anaerobic fermentation.

Under anaerobic conditions in plants, fungi, and some bacteria, the pyruvate is decarboxylated to ethanal, then reduced by $NADH_2$ to form ethanol, in the process of ALCOHOLIC FERMENTATION. In some bacteria it is simply reduced to lactate (see LACTIC ACID), in the process of lactate fermentation. Neither of these processes produces any more ATP. Under aerobic conditions, pyruvate enters the KREBS CYCLE, where is its further oxidized (by dehydrogenation reactions) and decarboxylated to carbon dioxide and water, with the release of a further 36 ATP molecules.

glycoprotein a macromolecule consisting of a protein backbone along which short oligosaccharide branches are attached at intervals. The most important plant glycoprotein is extensin, a structural component of cell walls. The protein backbone of extensin is unusual in that it contains a large number of HYDROXYPROLINE residues (about 30% of the total amino acid residues). The hydroxyproline serves to provide attachment points for the carbohydrate chains – in extensin these side chains are tetrasaccharides of ARABINOSE.

glycosidase *See* hydrolase.

glycoside a compound formed by the reaction of a pyranose sugar (see PYRANOSE RING) with a nonsugar molecule (an aliphatic or aromatic hydrocarbon) termed the *aglycone*. The aglycone replaces the hydrogen in the hydroxyl group of carbon atom one of the PYRANOSE RING. Glucose is the sugar component of many glycosides, such compounds being called *glucosides*. Some rare sugars are only found in glycosides, e.g. digitalose, which has been detected only in certain *Digitalis* glycosides. Major classes of glycosides include the ANTHOXANTHIN glycosides, important as plant pigments, the steroid glycosides (see SAPONIN, CARDIAC GLYCOSIDE), and the cyanogenic glycosides, which release hydrogen cyanide on hydrolysis. An

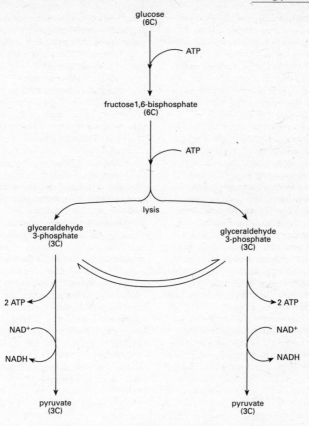

Glycolysis

example of the last group is the glucoside amygdalin, which is obtained from certain members of the Rosaceae, e.g. almond (*Prunus amygdalus*) and peach (*Prunus persica*). Cyanogenic glycosides may act to deter grazing animals.

glycosidic bond a chemical bond formed between monosaccharides to form polysaccharides and disaccharides. The bond is usually formed between the carbon-1 on one sugar and the carbon-4 on the other (1–4 glycosidic bond) by the removal of a hydrogen and hydroxyl group by a condensation reaction; in sucrose, however, it is formed between C-1 of glucose and C-2 of fructose (see illustration). If the

resulting bond lies below the plane of the glucose (or other sugar) ring the bond is called an α-glycosidic bond, if above this

Glycosidic bond: the glycosidic bond in a sucrose molecule

plane a β-glycosidic bond. β(1–4) glycosidic bonds are found in cellulose, and α-glycosidic bonds in starch.

glycosylation the addition of a carbohydrate to another molecule; for example, in the production of glycoproteins and glycolipids. The glycosylation process occurs in the rough endoplasmic reticulum and Golgi apparatus. See also TRANSGLYCOSYLATION.

glycosyl group see TRANSGLYCOSYLATION.

glyoxylate cycle (glyoxalate cycle) a cyclic series of reactions involving KREBS CYCLE intermediates in which one molecule of succinate (see SUCCINIC ACID) is formed from two molecules of acetyl CoA (see diagram). The reactions that differ from those of the Krebs cycle are indicated by broken arrows in the diagram. These reactions are catalysed by isocitrate lyase and malate synthase respectively. The glyoxylate cycle avoids the carbon dioxide forming steps of the Krebs cycle and thus allows net synthesis of carbohydrates from fatty acids via succinate (see GLUCONEOGENESIS), since acetyl CoA is formed by the β-oxidation of fats. The enzymes of the glyoxylate cycle are active in germinating oil-bearing seeds and in other fat-metabolizing plant tissues. The cycle also takes place in microorganisms but not in higher animals.

glyoxysome a MICROBODY that contains the five enzymes of the GLYOXYLATE CYCLE. Acetyl residues that are derived from the fatty acids of stored fats are converted to carbohydrates via succinic acid released in the glyoxylate cycle. Glyoxysomes are therefore found in those cells of higher plants where fats are utilized in this way, e.g. in the endosperm or cotyledons of fat-storing seeds. They also contain an enzyme (catalase) that catalyses the decomposition of hydrogen peroxide to water and oxygen.

glysophosine a synthetic plant growth regulator used to ripen sugar cane.

Gnetales see GNETOPHYTA.

Gnetophyta a phylum of cone-bearing desert gymnosperms whose cones (see STROBILUS) are compound and lack resin canals. It includes woody shrubs, vines, and large-leafed trees. There is a single class (Gnetopsida) containing three genera: *Ephedra* (joint firs) with about 40 species, *Gnetum* with about 30 species, and *Welwitschia*, containing the single species *W. mirabilis*. Certain species of *Ephedra* are important as sources of the alkaloid drug ephedrine.

Gnetophytes have some characteristics

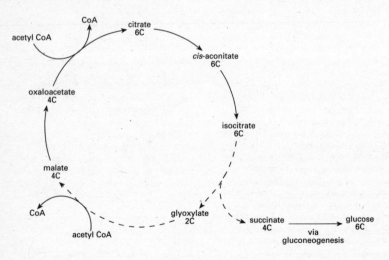

Glyoxylate cycle

reminiscent of the angiosperms. For example, their tracheids are arranged in columns and have highly perforated end walls similar to angiosperm vessels, while in *Gnetum* the sieve cells are closely associated with parenchymatous cells, suggestive of companion cells. The leaves of *Gnetum* have a broad oval lamina with reticular venation, very similar to certain angiosperm leaves. They also have compound male and female strobili with superficially sepal- or petal-like structures, while the female gametophytes of *Gnetum* and *Welwitschia* lack archegonia. In some species of *Ephedra* there is a form of double fertilization, (or more often multiple fertilization with three or four fertilizations) in which one sperm nucleus fuses with the egg cell and another with the ventral canal cell. However, only the zygote undergoes further development. The sperm are nonmotile, but the seeds are naked, as in gymnosperms.

The taxonomic position of the Gnetophyta is controversial. They were once considered to be a link between the Coniferophyta and Anthophyta, but are now thought to be more closely related to the angiosperms. Pollen studies show that *Welwitschia* evolved some 300 million yeas ago, probably from cone-bearing plants ancestral to modern conifers. Some taxonomists allocate gnetophytes to a single order, Gnetales, while others recognize three distinct orders: Gnetales, Ephedrales, and Welwitschiales. Classification systems that place all the gymnosperms into a single division, PINOPHYTA, put the gnetophytes in the subdivision Gneticae.

golden algae see CHRYSOMONADA.

golden-brown algae see CHRYSOMONADA.

golden-yellow algae see CHRYSOMONADA.

Golgi apparatus (Golgi body, dictyosome) an ORGANELLE formed from flattened membrane-bound CISTERNAE and a variable number of associated spherical vesicles. In most cells, between three and twelve saucer-like cisternae are stacked together with their concave surfaces directed towards the cell surface. The expanded 'rims' of the cisternae give rise to the surrounding spherical vesicles, which contain the products of biochemical activity within the

Golgi apparatus. If these vesicles move to the cell surface to release their contents, their membranes become incorporated into the PLASMA MEMBRANE. The Golgi apparatus is thus important in cell membrane synthesis.

Within the Golgi cisternae, synthesis of glycoproteins, which commences on the POLYSOMES of the ROUGH ENDOPLASMIC RETICULUM, is completed. These are moved within vesicles to the cell surface where they may be discharged from the cell or become incorporated into the cell membrane. Complex polysaccharides are also formed in the cisternae and are similarly distributed in vesicles to the cell surface. Cellulose-synthesizing enzymes have been identified in the membranes of these vesicles, but they only become functional when incorporated in the plasma membrane. It is probable that hydrolytic enzymes in the Golgi cisternae are organized into membrane-bound LYSOSOMES. In some unicellular algae (Haptomonada) scales that have been observed with the electron microscope on the cell surface originate in the Golgi apparatus.

Gondwana (Gondwanaland) see CONTINENTAL DRIFT.

gonidium (*pl.* gonidia) **1** any of the algal cells in a lichen.

2 a cell in a *Volvox* colony that gives rise sexually to a daughter colony.

gonimoblast filament see CARPOSPOROPHYTE, FLORIDEOPHYCIDAE.

goodness of fit the degree to which observed data agree with (fit) the values predicted by a model. Statistical methods for measuring goodness of fit include the CHI-SQUARED TEST.

G₁ phase see CELL CYCLE.

G₂ phase see CELL CYCLE.

GPP see PRIMARY PRODUCTIVITY.

Gracilicutes a division of EUBACTERIA that contains all the Gram-negative bacteria (see GRAM STAIN), which have thin peptidoglycan walls.

gradate sorus see SORUS.

gradualism see PHYLETIC GRADUALISM.

graft hybrid see CHIMAERA.

grafting a horticultural method of plant propagation in which a segment (the SCION) of the plant to be propagated is inserted onto another plant (the STOCK) in such a way that their vascular tissues combine, forming a *graft union*, so allowing growth of the grafted segment. The technique, which is mainly used on woody species, relies on the natural regenerative capacities of plants following wounding. In addition to the cambia of scion and stock growing together to form a continuous column, prolific development of callus around the graft area ensures a firm union. Grafting is only successful between closely related species. Thus apples are usually grafted on to different varieties of apple rootstock. Pears, however, are usually grafted onto quince rootstocks. See also CHIMAERA.

grain see CARYOPSIS.

Gramineae see POACEAE.

Gram-negative bacteria see GRAM STAIN.

Gram-positive bacteria see GRAM STAIN.

Gram reaction see GRAM STAIN.

Gram stain a stain used in bacteriology to distinguish between two physiologically distinct types of bacteria. The procedure involves staining the organisms with a basic dye, such as gentian violet or crystal violet, and adding a mordant, such as iodine or picric acid. In *Gram-positive bacteria* these form a complex that cannot be removed by such decolorizing agents as acetone or alcohol. However *Gram-negative bacteria* lose the violet colour and on counterstaining with a red dye, such as carbol fuchsin or neutral red, will take up the red colour. Gram-positive organisms, on counterstaining (see STAINING), retain the original violet colour and can thus easily be distinguished from the red Gram-negative bacteria. It is believed these differences may be related to cell wall structure. Thus Gram-negative cell walls contain far more lipid than those of Gram-positive cells, and a wider range of amino acids. However it is not clear whether it is the materials of the wall that act as the substrate for the differential staining reaction or whether they act by affecting the permeability of the wall.

Gram-positive and Gram-negative bacteria differ in many other ways apart from their staining reactions. For example, Gram-positive bacteria are often more

exacting in their nutritional needs and more susceptible to antibiotics but more resistant to PLASMOLYSIS. Examples of Gram-positive bacteria are the ACTINOBACTERIA. Gram-negative bacteria include the CYANOBACTERIA.

grana (*sing.* granum) distinct stacks of LAMELLAE seen within CHLOROPLASTS. They contain the pigments, electron transfer compounds, and enzymes essential to the light-dependent reactions of photosynthesis.

granite mosses see BRYOPHYTA.

grape sugar see GLUCOSE.

grasses see POACEAE.

grass heath see ACIDIC GRASSLAND.

grassland a major regional community (BIOME) in which grasses (POACEAE) are the dominant vegetation. Usually the rainfall is insufficient for trees to grow (about 250–500 mm annually in temperate regions and 750–1500 mm in the tropics). Grasslands are common in continental interiors where the rain falls mainly in spring and summer. Most grasses are perennials but some are annuals or biennials. There are the low-growing closely packed turf-forming grasses of temperate regions and the tufted or tussock grasses, which grow in separate clumps and occur widely in temperate, tropical, and tundra regions.

Temperate grasslands do contain other herbaceous plants though trees are found only along rivers and streams. They include the North American PRAIRIES, the STEPPES of southwest Russia, the grasslands of Manchuria and Mongolia, and the South American PAMPAS. See also CHALK GRASSLAND.

Tropical grasslands, known as *savanna*, have tall coarse tufted grasses and there are often scattered trees. In Africa baobab (*Adansonia digitata*) and *Acacia* and *Euphorbia* species are common. The climate alternates between cool dry winters and hot summers with heavy rains. Savanna is found in large areas of South America, East and South Africa, Southeast Asia, and northern Australia.

graticule see MICROMETER.

graveolent strongly scented, often unpleasantly.

gravitational water water in the soil that is free to move under the influence of gravity. See FIELD CAPACITY. Compare HYGROSCOPIC WATER, CAPILLARY MOISTURE.

gravitaxis (geotaxis) a change in the direction of locomotion (e.g. in motile cells and protoctists) in response to gravity.

gravitropism (geotropism) a TROPISM exhibited in response to gravity. Thus, when placed in a horizontal position with all-round illumination, shoots grow upwards and roots downwards, the former being negatively and the latter positively gravitropic. A klinostat, consisting of a drum to which plant seedlings can be attached, revolves so as to cancel out the effects of gravity. Using it, one can demonstrate the lack of root and shoot curvature when no net gravitational forces are operating. Like PHOTOTROPISM, the removal of the shoot tip destroys the gravitropic response. Certain parts, such as tertiary roots, are naturally agravitropic. See also STARCH-STATOLITH HYPOTHESIS.

grazing food chain a food chain in which the primary CONSUMERS are grazing herbivores.

green algae see CHLOROPHYTA.

greenhouse effect the process by which radiation from the sun that is re-emitted as long-wave infrared radiation by the earth is trapped by water vapour, carbon dioxide, methane, chlorofluorocarbons (CFCs), and certain other so-called *greenhouse gases*, which leads to the warming of the atmosphere and consequently the warming of the earth's surface (*global warming*). This effect is increasing as the concentrations of greenhouse gases in the atmosphere increase, owing mainly to the emission of carbon dioxide from the burning of fossil fuels. The likely effects of increased global warming are destabilization of climate and weather patterns and disruption of agriculture. It could also result in melting of the polar ice caps, leading to a rise in sea levels.

greenhouse gas see GREENHOUSE EFFECT.

green manure a fast-growing inexpensive crop that is sown towards the end of the growing season with the intention of ploughing or digging it in a short while later when it is still green. This increases

the amount of organic matter in the soil. Legumes (see FABACEAE), such as clovers (*Trifolium*) and lupins (*Lupinus*), also increase the nitrogen content of the soil, are often used.

green revolution a dramatic increase in crop productivity that occurred during the second half of the 20th century as a result of advances in genetics and machinery, and in fertilizers, pesticides, and other chemicals.

green sulphur bacteria see SULPHUR BACTERIA, PHOTOSYNTHETIC BACTERIA.

grid analysis in the analysis of vegetation pattern, the use of a grid of contiguous QUADRATS rather than randomly placed quadrats. This method allows for the grouping of adjacent quadrats into blocks of increasing sizes, which has a significant effect on the detection of A pattern. The size of quadrat at which non-random distribution of species occurs often provides a clue as to the cause of CLUMPING.

grike see LIMESTONE PAVEMENT.

Grime's habitat classification a scheme for classifying habitats devised by ecologist J. P. Grime, which classifies habitats in terms of the concentration of resources and level of stability they offer plants and their effects on plant life strategies. Habitats with good levels of resources that are seldom disturbed tend to favour large, long-lived plants whose shoots and roots occupy a considerable space, and which are good competitors. Habitats that are lacking in resources or are cold or dry favour long-lived, slow-growing plants that can tolerate long periods of stress, reproducing only when conditions are favourable. Habitats that are highly disturbed favour fast-growing ephemeral plants that produce large numbers of seeds or spores, e.g. weeds.

gross primary productivity see PRIMARY PRODUCTIVITY.

ground layer the lowest layer of a plant COMMUNITY, usually made up of low-growing forms such as mosses, lichens, fungi, and low-growing herbs with rosettes of leaves or prostrate stems. See WATER TABLE.

ground meristem the central part of the APICAL MERISTEM, the derivatives of which give rise to the ground tissue (ground PARENCHYMA) and associated tissues. Compare PROCAMBIUM, PROTODERM.

ground tissue see PARENCHYMA.

groundwater underground water contained in the pore spaces in the soil and underlying bedrock in the zone that is saturated with water.

group translocation a mechanism for transporting sugars across a cell membrane that involves the modification of the sugar during transport. For example, glucose is activated by an additon of a high-energy phosphate group and transported as glucose-6-phosphate. Membrane proteins are thought to be involved in the process. It is not ACTIVE TRANSPORT, since there is no adverse concentration gradient during transport, but it is an energy-saving mechanism. Group translocation occurs especially in bacteria under anaerobic conditions, when energy is limiting.

grove 1 a small wood without underwood, usually less than 8 hectares in area. 2 a small planting of fruit or nut trees. 3 a patch of timber trees with no underwood within a larger forest of woodland that does have underwood.

growing season the part of the year during which conditions allow germination, growth, and reproduction of indigenous vegetation and cultivated crops. Definitions vary in different countries. In Britain it is the period when the mean temperature exceeds 60°C. but the temperature is variously taken to be the ground or air temperature, and the daily, weekly, or monthly means. The required temperature varies between different crops. For example, wheat needs a temperature of at least 5°C to germinate, while rice requires about 20°C. Where temperatures remain close to the threshold, growth and maturation are likely to be slow. However, in high latitudes the extra duration of daylight in summer helps to compensate for the lower temperatures. In the United States the growing season is defined as the period between the last killing frost of spring and the first killing frost of autumn. Most crops require a frost-free growing season of at least 90 days. The growing season varies not only with latitude, but also with altitude, and with

proximity to warm ocean currents or warm winds that raise the temperature of adjacent land masses. In the tropics, conditions my be favourable to growth all year round, with no distinct growing season.

growth the sum total of the various physiological processes that combine to cause an increase in the dry weight of an organism and an irreversible increase in size. In most plants, growth is accomplished by the assimilation and FIXATION of inorganic substances from the surrounding environment. In contrast to animal growth, plant growth is usually confined to MERISTEMS, where cell division occurs, and to the regions adjacent to these where CELL EXTENSION takes place. See also DEFINITE GROWTH, INDEFINITE GROWTH.

growth correlation the relationship that exists between the different growth rates of various parts of a plant body. Growth rates depend on the balance of hormones in the region and the competition between parts for nutrients.

growth factor any substance that affects the growth rate of a plant or plant part. The term covers food reserves and minerals as well as the various hormones and GROWTH INHIBITORs found within plant systems.

growth form the MORPHOLOGY of a plant, especially in relation to its physiological adaptations to its environment.

growth inhibitor any substance that retards the growth of a plant or plant part. Almost any substance will inhibit growth if present at high enough concentrations but two common hormonal inhibitors are ABSCISIC ACID and ETHENE (ethylene), which both have effect at very low concentrations, similar to the effective concentrations of auxins, gibberellins, and cytokinins. Other inhibitors, which must be present in much higher concentrations to take effect, include certain of the phenolics, quinones, terpenes (see TERPENOIDS), fatty acids, and amino acids.

growth promoter a hormone that stimulates growth by increasing cell division (e.g. cytokinin) or cell elongation (e.g. gibberellin).

growth retardant a synthetic substance that inhibits or slows down the growth of plants. Such compounds are used, for example, to prevent the sprouting of such vegetables as onions and to restrict the height of cereal crops to facilitate harvesting.

growth ring see ANNUAL RING.

growth substance See HORMONE.

GSC see GAS CHROMATOGRAPHY.

GTP (guanosine triphosphate) a nucleoside triphosphate containing the base guanine. It is formed from GDP (guanosine diphosphate) and phosphate during the deacylation of succinyl CoA to succinate in the Krebs cycle. GTP plays a role in the activation of fatty acids, in GLUCONEOGENESIS, and in the initiation and elongation of polypeptide chains during protein synthesis.

guanine a nitrogenous base, more correctly described as 2-amino-6-oxypurine, derived from amino acids and sugars and found in all living organisms. Guanine combines with pentose sugar phosphates to form one of the nucleotides making up RNA and DNA. In addition, hydrolysis of the NUCLEOSIDE guanosine triphosphate (GTP) releases energy to drive some energetically unfavourable reactions. See also NUCLEOTIDE.

guanosine triphosphate see GTP.

guard cells the pair of specialized crescent-shaped epidermal cells immediately surrounding the stomatal pore and forming the STOMA. The opening and closing of the stoma are controlled by changes in TURGIDITY of the guard cells, facilitated by the pronounced thickening of the cell walls adjacent to the stomatal pore. Changes in osmotic potential necessary to effect these changes are believed to be due to the accumulation of potassium ions and associated anions, such as malate (see MALIC ACID) and other organic acids, in the guard cells during the day in response to light. Compare SUBSIDIARY CELL.

gullet an invagination at the anterior end of the cell of euglenids (see EUGLENIDA). It has an expanded basal region or reservoir, a narrow passage rising to the surface, the *cytopharynx*, and an opening to the exterior, the *cytosome*. In the reservoir, in a region where there is no pellicle, particles from the surrounding medium can be taken into the cytoplasm by ENDOCYTOSIS.

gum any substance that swells in water to form gels or sticky solutions. Structurally gums are mostly complex, highly branched polysaccharides, although a few gums with relatively simple structures are known.

Three main classes of gum are recognized. Acidic polysaccharides of glucuronic or galacturonic acids (see URONIC ACIDS) form glassy, hard gums; these are often produced by plants as a result of injury. Examples include GUM ARABIC and tragacanth. In algae another class of acidic gums are found in which the acidity is due either to sulphate acid ester groups or to uronic acids. Examples are AGAR, ALGINIC ACID, and CARRAGEENAN. The third class contains gums that are obtained from seeds, examples being the seeds of the carob tree (*Ceratonia siliqua*). Natural gums are increasingly being replaced by synthetic substitutes in the manufacture of adhesives.

gum arabic a polysaccharide, found in the cell walls of certain plants, made up of D-galactose and D-glucuronic acid (see URONIC ACIDS), and arabinose and rhamnose. It is obtained commercially from *Acacia* and is used in glues and pastes and as a mounting medium in microscopy.

gummosis a symptom of certain plant diseases in which there is an abundant formation of gum. Gummosis often occurs on trees but may occur on herbaceous plants, e.g. infection of young cucumber fruits with the cucumber scab fungus (*Cladosporium cucumerinum*) results in secretion of gum at the edge of the lesions.

guttation the exudation of liquid water onto a plant surface. It occurs under conditions of high humidity when the saturated atmosphere prevents TRANSPIRATION. The increase in ROOT PRESSURE forces water out of special HYDATHODES. The secreted water may contain calcium salts, which dry as a white crust at the leaf margins. Morning 'dew' on grass is often the product of guttation, as the lower temperatures at night provide ideal conditions for the process to occur.

Gymnomycota see MYXOMYCOTA.

Gymnospermae see GYMNOSPERMS.

gymnosperms those seed plants that differ from the angiosperms (ANTHOPHYTA) in having naked seeds with no enclosing carpellary structure. Double fertilization does not occur and thus no true endosperm is developed. The gametophyte generation is not as reduced as in angiosperms, but neither is it autotrophic. The female gametophyte typically consists of at least 500 cells and distinct archegonia are produced in all genera except *Welwitschia* and *Gnetum*.

The term 'gymnosperm' no longer has any taxonomic significance; plants formerly assigned to the class Gymnospermae have now been reclassified. There are four major living groups, which in the FIVE KINGDOMS CLASSIFICATION are assigned phylum status: CYCADOPHYTA (cycads), GINKGOPHYTA (maidenhair tree), CONIFEROPHYTA (conifers and yews), and GNETOPHYTA (comprising the genera *Welwitschia*, *Ephedra*, and *Gnetum*). They probably arose independently of the angiosperms from the now-extinct seed ferns (CYCADOFILICALES). Extinct orders of gymnosperms include the Cycadofilicales, CAYTONIALES, BENNETTITALES, CORDAITALES, and PENTOXYLALES. Gymnosperms have a particularly rich fossil record and were abundant in Carboniferous times, contributing largely to the formation of coal deposits. They originated during the Devonian period and were the dominant vegetation during the Jurassic and early Cretaceous. Towards the end of the Cretaceous their dominant position was taken over by the angiosperms. See also PINOPHYTA.

gynaecium see GYNOECIUM.

gynandrous describing stamens that are inserted on the gynoecium.

gynodioecious describing plants that bear female and hermaphrodite flowers on separate individuals, as in thymes (*Thymus*) and oregano (*Origanum vulgare*). Compare ANDRODIOECIOUS, GYNOMONOECIOUS.

gynoecium (gynaecium) the female component of the angiosperm flower, which may be made up of one or more CARPELS. The gynoecium usually occupies the most central part of the floral axis, although in more advanced forms it may not be symmetrically placed. If the gynoecium contains only one carpel it is termed *monocarpellary, unicarpellous,* or *stylodious*. If there are two or more separate carpels it is *apocarpous* and if the carpels are fused *syncarpous*. The gynoecium is represented by the letter G in the FLORAL FORMULA. Compare ANDROECIUM. See OVARY.

gynomonoecious describing plants that bear female and hermaphrodite flowers on the same individual. For example, many members of the Asteraceae have female ray florets and hermaphrodite disc florets. Compare GYNODIOECIOUS, ANDROMONOECIOUS.

gynophore an extension of the receptacle between the androecium and gynoecium that bears the ovary. Gynophores are found in many members of the Capparidaceae, e.g. the spider flower, *Cleome spinosa*. See also ANDROGYNOPHORE.

gynosporophyll in seed plants, a leaf bearing MEGASPORANGIA.

gynostemium the column formed by the fusion of the ANDROECIUM and GYNOECIUM in certain genera, e.g. *Stylidium*.

gynostrobilus a female cone bearing MEGASPORANGIA, ovules, or seeds.

gypsophilous describing organisms that tolerate and thive on limestone soils.

gyrose marked with wavy or spiral lines or ridges.

gyttja a rapidly accumulating muddy lake deposit derived from the oxidized remains of colloidal organic material that was formerly in suspension. Pollen analysis of the pollen in gyttja may provide a record of the past climatic conditions of the lake.

H*h*

habitat the area in which an organism or group of organisms lives. It includes the climatic, topographic, BIOTIC, and, in the case of terrestrial habitats, EDAPHIC FACTORS of the area. There may be considerable variations in conditions within a habitat and also seasonal variations. Examples of habitats are seashore, woodland, pond, etc. *Microhabitats* are much smaller areas, for example, a bird's nest, a cow pat, or a log.

habituation the process of becoming accustomed to a new environment. The term is applied specifically to the ability of certain long-established CALLUS cultures to synthesize AUXIN and hence become independent of exogenous supplies. Such habituated cultures are also called *anergized cultures*.

Hadean eon see PRECAMBRIAN.

haem (heme) an iron-containing porphyrin that acts as the prosthetic group in cytochromes.

haematochrome a red pigment that accumulates in certain terrestrial green algae (see CHLOROPHYTA), e.g. *Trentepohlia*, under drought conditions.

haematoxylin a blue dye obtained from the leguminous tree logwood (*Haematoxylon campechianum*) that stains nuclei and cellulose cell walls blue. Haematoxylin itself is not a stain until it has been oxidized to haematin. It also usually requires a mordant, such as iron alum or an aluminium salt. Different haematoxylin solutions can be prepared, e.g. Delafield's haematoxylin, which can be used as a counterstain with safranin.

hair a TRICHOME. The terms trichome and hair are often regarded as synonymous, but sometimes the term hair is restricted to relatively simple nonglandular trichomes. See also ROOT HAIR.

hair point a leaf tip that tapers sharply to form a colourless hair, usually formed by the midrib.

halo- prefix denoting something that relates to salt or salinity.

halocline in a body of water, a zone of rapid change in salinity with depth.

halomorphic soil see SOIL.

halophile (*adj.* halophilic) an EXTREMOPHILE bacterium of the domain Archaea that thrives in an extremely saline environment. See also THERMOPHILE.

halophyte a plant that is adapted to live in soil containing a high concentration of salt. Such plants are abundant in SALT MARSHES and mud flats. Halophytes must obtain water from soil water with a higher osmotic pressure than normal soil water. To achieve this the root cells of some halophytes have a very high concentration of salts and so are able to take up water by osmosis. SUCCULENT halophytes also store water for use when the salt concentration of the soil water rises further as a result of evaporation at low tide. An example of a succulent halophyte is sea rocket (*Cakile maritima*). There are also certain halophytic grasses that grow so abundantly they have played a major part in land reclamation. The most successful is the C_4 plant *Spartina anglia*, which was deliberately introduced in the Netherlands in 1924 specifically for the purpose of land reclamation. See also HALOSERE.

halosere a pioneer plant community that

develops in a low-lying region of land originally beneath the sea and subsequently silted up with mud, silt, and shingle. The plants must be adapted to survive inundation by the tide twice a day. A major pioneer plant is glasswort (*Salicornia europaea*), which gradually gives way to salt-marsh grasses (*Puccinellia*) or sea aster (*Aster tripolium*). When the salinity decreases other species, such as thrift (*Armeria maritima*) and seablites (*Suaeda*), grow. The region becomes stabilized to form mud flats and SALT MARSHES. See also SERE.

Hamamelidae (Amentiferae) a subclass of the dicotyledons containing mainly woody plants, often with small unisexual apetalous flowers frequently borne in catkins. The number of orders recognized varies between about 7 and 15. Usually recognized are the Trochodendrales, including the Cercidophyllaceae; the Hamamelidales, including the Platanaceae (e.g. plane trees) and Hamamelidaceae (e.g. witch hazels); the Fagales, including the FAGACEAE (e.g. beeches, oaks, and chestnuts) and Betulaceae (e.g. birches, alders); the Juglandales, including the Juglandaceae (e.g. walnuts) and Myricaceae (e.g. sweet gale); and possibly the Casuarinales (e.g. casuarina) and Eucommiales. Additional orders include the Urticales, Balanopales (often included in the Buxales), Didymelales (also included in the Buxales), and Barbeyales (included in the Urticales), all here placed in the DILLENIIDAE.

hamate describing leaves that are hooked or strongly curved, often due to an extension of the midrib, as in the moss *Cratoneuron filicinum*.

hammock see EVERGLADES.

hapaxanthic see SEMELPAROUS.

haplobiontic (haplobiont) describing a life cycle in which either the sporophyte or gametophyte generation is lacking. For example, in the chlorophyte *Chlamydomonas* the vegetative cell is haploid and the diploid condition is represented only by the zygospore, which gives rise to haploid zoospores immediately on germination (see HAPLONTIC). In diatoms the vegetative cell is diploid and arises directly by the fusion of gametes formed by the sporophyte generation (see DIPLONTIC). See

ALTERNATION OF GENERATIONS. Compare DIPLOBIONTIC.

haplocheilic describing a gymnosperm stomatal apparatus in which the subsidiary cells are not derived from the same initial as the guard cells, as occurs in cycads (see CYCADOPHYTA). Compare SYNDETOCHEILIC. See also PERIGENOUS.

haplodiplontic see DIPLOBIONTIC.

haploid a nucleus or individual containing only one representative of each chromosome of the chromosome complement. The haploid condition, denoted by the symbol n, is established by meiotic division of a diploid nucleus. In plants meiosis establishes a haploid generation, the GAMETOPHYTE. Sooner or later the haploid gametophyte produces gametes by mitosis. In algae the situation is extremely variable. Some, such as *Fucus*, follow the pattern more characteristic of animals, where meiosis results directly in the formation of gametes. In other cases, such as *Spirogyra*, the only form of the alga is haploid, and the diploid stage is restricted to a single-celled zygote.

In flowering plants the haploid gametophyte generation is reduced to the pollen tube in the male and the embryo sac in the female. However haploid plants can be obtained by culturing pollen grains (see POLLEN CULTURE) under suitable conditions. Haploid plants may also be obtained when a zygote formed from an interspecific cross sheds all the chromosomes of one parent as it undergoes development. This phenomenon has been demonstrated when barley (*Hordeum vulgare*) is fertilized with pollen from the wild barley *H. bulbosum*. Haploid plants have great potential in plant breeding as it is possible, by doubling the chromosomes of a haploid plant, to obtain a completely homozygous plant. This may be impossible by other means, especially with self-sterile plants. See also ALTERNATION OF GENERATIONS.

haplontic (haplont) describing a life cycle in which the haploid phase predominates and the diploid stage is limited to the zygote. Such life cycles may be found in the filamentous Chlorophyta. See ALTERNATION

OF GENERATIONS. Compare DIPLONTIC. See also HAPLOBIONTIC.

haplophase the part of the life cycle in which the nuclei are haploid. Compare DIPLOPHASE.

haplostele see stele.

haplotype a group of closely linked alleles that tend to be inherited together. See LINKAGE.

hapteron (*pl.* haptera) **1** see HOLDFAST. **2** the outer wall of the spore in species of *Equisetum* (horsetails). It differentiates into an elongated X-shaped structure, the arms of which are tightly coiled around the spore. Following release of the spores from the sporangium the haptera dry out and uncoil, this movement assisting in the dispersal of the spores. They are thus similar in function to ELATERS. **3** an attachment disc produced by the EPICAUTICAL ROOTS of a parasitic plant. HAUSTORIA grow out from the base of the hapteron and penetrate the host tissues. **4** a holdfast formed by the adventitious roots of certain plants of the Podostemaceae.

Haptomonada (Haptophyta) a phylum of the PROTOCTISTA (in some classifications a division of the Plantae) containing mainly unicellular algae with two undulipodia, most of which have a HAPTONEMA between the two undulipodia, which are of the smooth type and usually located the anterior end. There is a Golgi body near the base of the undulipodia. The haptomonads have golden plastids containing chlorophylls *a*, c_1, and c_2 and fucoxanthin and the main carbohydrate reserve is *chrysolaminarin*, stored in PYRENOIDS. The naked cells are covered in tiny scales, formed from carbohydrates in the Golgi vesicles then extruded in layers onto the cell surface. One very numerous group, the *coccolithophorids*, have calcified scales (*coccoliths*) formed of calcite (calcium carbonate) crystals. The shape and sculpting of these scales, visible only with an electron microscope, is diagnostic of the species. Haptomonads are common members of the marine plankton and also occur in fresh water. Blooms of certain species, such as *Prymnesium*, can be toxic to fish. Other species are fed to farmed mussels and

oysters. Cretaceous chalks contain the remains of vast numbers of coccolithophorids, whose species serve to distinguish rock layers of different ages.

Asexual reproduction is by fission. Reports of sexual reproduction in this group are scarce. Many species have more than one phase in the life cycle, and motile stages may alternate with nonmotile bottom-dwelling stages. Some haptomonads, such as *Phaeocystis*, may form large colonies of nonmotile cells embedded in mucilage, a PALMELLOID stage. Many haptomonads, such as *Chrysochromulina*, are mixotrophs, able to live either by photosynthesis or by assimilation of organic material: suspended particles of organic matter are captured by the haptonema and taken in by phagocytosis (see ENDOCYTOSIS).

haptonasty a nongrowth NASTIC MOVEMENT in which a plant part moves in response to a tactile stimulus. An example is the rapid and progressive collapse of *Mimosa pudica* (sensitive plant) leaflets, often throughout the whole plant, after those of one particular branch are touched. See also PULVINUS.

haptonema a threadlike structure that arises between the two undulipodia in algae belonging to the HAPTOMONADA. It is not involved in cell movement, but may be used for temporary anchorage. The length varies and in some cases it is coiled when the cell is motile. Derived from a KINETOSOME-like structure, it contains six doublets of microtubules arranged around a hollow centre or a single central microtubule and surrounded by three concentric membranes.

Haptophyta see HAPTOMONADA.

haptotropism (thigmotropism) a TROPISM in response to an external contact stimulus. The movements are most clearly exhibited by the TENDRILS of certain species. In passion flowers (*Passiflora*), the tendril is coiled spirally with the lower side facing outwards. The tendril later straightens and the sensitive tip undergoes CIRCUMNUTATION. If it touches a solid body, it bends towards the stimulated side, a process occupying perhaps several minutes. The bending results in fresh parts of the tendril making contact with the stimulus and so the process is continuous, the tendril gradually encircling the support.

hardening the gradual exposure of a plant to increasingly lower temperatures to increase its resistance to frost. This is usually achieved by placing plants in a cold frame and gradually increasing the ventilation. Seedlings raised in greenhouses are often hardened off in this manner before planting out.

hard pan see PODSOL.

hard seed any seed with a tough impervious outer coat that will not allow the entry of water. IMBIBITION and germination therefore cannot occur until the seed coat is ruptured, either by scarification or microbial action. The sweet pea (*Lathyrus odoratus*) has hard seeds.

hardwood see SECONDARY XYLEM.

Hardy–Weinberg law the law stating that, provided certain conditions are met, the gene (allele) frequencies in a population of organisms will remain constant and be distributed as p^2, $2pq$, and q^2 for the genotypes AA, Aa, and aa respectively where p = the frequency of the dominant allele and q = the frequency of the recessive allele, such that $p + q = 1$ (i.e. A and a are the only alleles). The law only holds providing that: the population is large (theoretically infinite); the population has been produced by random breeding; there is no natural selection for or against any particular genotype; there is no differential migration into or from the population; and there is no mutation.

Despite these conditions, Hardy and Weinberg's law is the basic theorem of population genetics. From it can be calculated the frequencies of A and a, even though a significant proportion of the a alleles are masked in heterozygotes. Thus the frequency of a $(q) = \sqrt{}$(frequency of homozygous recessives), and the frequency of A$(p) = 1 - q$. If a population does not fit the distribution $p^2 + 2pq + q^2$ then one or more of the conditions stated above are not being fulfilled. The usual reason is natural selection against a particular phenotype. The theorem can be extended to enable the effects of natural selection on gene frequencies to be calculated. In effect, this provides a yardstick by which the rate of evolution can be measured and quantitatively defined.

Hartig net see MYCORRHIZA.

harvest method a method of measuring PRIMARY PRODUCTIVITY that involves harvesting a sample area at intervals throughout the growing season, then estimating the dry weight or calorific value of the harvested material. It is used particularly for annual crops, new grasslands, and certain heath and ponds where there are few primary consumers. Normally, nondestructive methods are preferred. Compare AERODYNAMIC METHOD, CARBON DIOXIDE METHOD, CHLOROPHYLL METHOD, DIURNAL CURVE METHOD, GAS-EXCHANGE METHOD, OXYGEN METHOD.

hastate describing a leaf shaped like a spear, with three lobes, one pointing forwards and two pointing sideways either side of the petiole. The leaves of the copse buckwheat or bindweed (*Polygonum dumetorum*) are an example. *See illustration at leaf.*

hastula in some PALMATE and COSTAPALMATE leaves, a ridge of tissue at the junction between the lamina and petiole.

Hatch–Slack pathway an alternative form of carbon dioxide (CO_2) FIXATION found in C_4 PLANTS (see diagram). In such plants the first product of CO_2 fixation is not the three-carbon glycerate 3-phosphate (see CALVIN CYCLE) but the four-carbon oxaloacetate. This is formed by the carboxylation of phosphoenolpyruvate (PEP) by the enzyme PEP carboxylase. The oxaloacetate is then either reduced to malate or transaminated to form aspartate. These reactions occur in the cells of the mesophyll. The malate or aspartate is then transported to bundle-sheath cells situated around the leaf veins (see KRANZ STRUCTURE) and decarboxylated to form CO_2 and pyruvate. The CO_2 so released reacts with ribulose 1,5-bisphosphate to form two molecules of glycerate 3-phosphate. The normal Calvin sequence of reactions then commences. The pyruvate is returned to the mesophyll cells where it is converted to PEP with the concomitant formation of a molecule of AMP from ATP.

This step, which uses up two high-energy phosphate bonds, is the reason why, overall, C_4 plants require 30 molecules of ATP for each molecule of glucose synthesized whereas C_3 plants only require 18.

PEP carboxylase has a far higher affinity for CO_2 than rubisco and C_4 plants are consequently more efficient at fixing CO_2 than C_3 plants. This accounts for their lower COMPENSATION POINTS. See also PHOTORESPIRATION.

haulm a stem, used especially for members of the grass family (Poaceae).

haustorium an organ produced by a parasite to absorb nutrients from its host. Most commonly the term refers to the specialized hyphal extensions formed by obligate fungal parasites, which protrude into individual host cells. A successful fungal parasite does not kill the host cells and manages to insert haustoria without rupturing the plasma membrane. If the host cells are killed, the host is said to be hypersensitive and the infection cannot develop. The suckers of parasitic angiosperms, such as dodders (*Cuscuta*), are also called haustoria. These penetrate the vascular tissue of the host plants.

The term haustorial is also used of nonparasitic organs that absorb nutrients from surrounding tissues, such as the pollen tubes of seed plants and the foot of a sporophyte plant.

heartwood (duramen) the central part of the SECONDARY XYLEM in some woody plants containing nonfunctional tracheary elements, often blocked by TYLOSES and infiltrated with organic compounds such as resins, tannins, gums, and many aromatic substances and pigments. The heartwood is derived from the SAPWOOD as it deteriorates due to age and damage. Heartwood is more

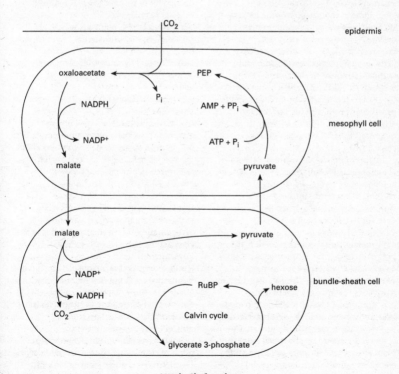

Hatch–Slack pathway

highly prized than sapwood for the making of furniture, due to its greater density, darker colour, and greater resistance to decay.

heath (heathland) a region of land with poor sandy well-drained soil (often not particularly exposed), dominated by dwarf shrubs of the heath family, Ericaceae; in Europe the dominant species is usually common heather or ling (*Calluna vulgaris*). Heaths are normally subclimax communities (see CLIMAX) maintained by a combination of grazing and cutting and, in places, burning. Other members of the Ericaceae that occur in Europe include bell heather (*Erica cinerea*), cross-leaved heath (*E. tetralix*), and bilberry (*Vaccinium myrtillus*). Members of the legume family, Fabaceae, are also well represented, particularly the gorses (*Ulex* spp.). In Britain the habitat is mainly southeastern in distribution and has come under serious threat from development and forestry. Some heaths, such as those in Dorset, contain many nationally rare species, such as the Dorset heath (*Erica ciliaris*) and marsh gentian (*Gentiana pneumonanthe*), and some rare animals (Dartford warbler, sand lizard).

The term 'heath' may also be applied to MOORLAND, dominated by *Calluna vulgaris*, or to chalk heath, a peculiarity of East Anglia's Breckland, in which heather is mixed with plants typical of CHALK GRASSLAND.

In other parts of the world the term is loosely applied to any area of usually sandy soil dominated by dwarf shrubs of the Ericaceae or related families (e.g. the Epacridaceae in South Australia).

heath forest see KERANGA.

heathland see HEATH.

heavy metals metals that have a density greater than 5g/cm^3, e.g. copper and zinc.

heavy-metal shadowing see SHADOWING.

heavy-metal tolerance the ability of certain plants to grow in areas where the concentration of heavy metals, e.g. copper, lead, and zinc, is so high as to prevent the growth of most plants. Tolerance may be achieved by excluding the metal from the plant altogether but more usually results from detoxification, when the metal is converted to a harmless form. Some *Agrostis*

species exhibit copper tolerance, restricting the metal to the roots and thus protecting the more sensitive shoots.

hectare (ha) a metric unit of area, representing an area of land 100 metres square (10 000 sq m). 1 hectare = 2.47 acres.

helical thickening see SPIRAL THICKENING.

helicase any one of a class of enzymes that catalyse the unwinding of double-stranded nucleic acids, such as DNA, with the aid of energy from ATP. At the start of DNA REPLICATION, a helicase binds to a single-stranded part of the DNA and to ATP and moves along the DNA, separating the strands.

heliciform coiled, like the shell of a snail.

heliophyte a plant characteristic of sunny habitats. Compare SCIOPHYTE.

heliosis see SOLARIZATION.

heliotropism another word for PHOTOTROPISM, though implying movement in response to sunlight rather than artificial light.

helophyte a marsh plant with its perennating buds situated in the mud at the bottom of a lake or pond. Examples are bulrushes (*Typha*) and water plantains (*Alisma*). See CRYPTOPHYTE. See PERENNATING ORGAN.

Helotiales see LEOTIALES.

heme see HAEM.

Hemiascomycetae (Hemiascomycetes) a class of the ASCOMYCOTA whose members do not develop ascomata. It contains about 330 species in some 60 genera. Such fungi are unicellular or produce a poorly developed mycelium. The group includes the SACCHAROMYCETALES, TAPHRINALES, and Protomycetales. The class is not recognized by all modern classification systems, and is often incorporated into the class ASCOMYCETES (as the subclass Euascomycetidae).

Hemibasidiomycetes an obsolete group of basidiomycete fungi that included the Uredinales (now placed in the class TELIOMYCETES) and the Ustilaginales (now placed in the class USTOMYCETES).

hemicellulose any of a variety of polysaccharides found in plant cell walls often in close association with cellulose. Hemicelluloses differ from cellulose in being composed of pentose sugars (arabinose or xylose) or hexose sugars other than glucose (e.g. mannose and galactose). They also differ in that, as well as having a structural function, like cellulose, in cell walls, they can be broken down by enzymes and so act as a nutrient reserve. Cellulose by contrast is metabolically inactive once it has been incorporated into the cell wall.

hemicryptophyte a plant with perennating buds situated at or just below the soil surface. Hemicryptophytes are usually herbaceous perennials and are commonly found in cold moist climates. ROSETTE PLANTS and partial rosette plants, which have a basal rosette of leaves but also other leaves further up the stem, are hemicryptophytes. Hemicryptophytes in which the lowest leaves on the stem are scale-like or smaller than the others, so providing extra protection for the bud, are called *protohemicryptophytes*. See PERENNATING ORGAN. See also RAUNKIAER SYSTEM OF CLASSIFICATION.

hemiparasite (partial parasite) a plant, such as mistletoe, that carries out its own photosynthesis but depends on another plant for its water and mineral salts.

hemizygous the situation in which the normal diploid nucleus contains only one copy of a particular gene (or chromosome). Thus the unpaired section of the X chromosome in the heterogametic sex (XY) is described as hemizygous.

hep see HIP.

Hepaticae see HEPATOPHYTA.

Hepatophyta a phylum of nonvascular plants (see BRYOPHYTES) containing the thallose and leafy liverworts, which number about 6000 species. They were formerly classified as a class, Hepaticae or Marchantiopsida, of the phylum (or division) BRYOPHYTA, which now includes only the mosses. The Hepatophyta differ from the mosses in showing marked dorsiventrality in the gametophyte. The antheridia and archegonia may be borne on the surface of the thallus or on fleshy stalks (see GAMETANGIOPHORE). The capsule of the sporophyte, which contains sterile ELATERS as well as spores, matures before the seta lengthens, while in mosses the reverse occurs. The capsule does not contain a central pillar of sterile cells (columella) as is found in the mosses and in the hornworts (see ANTHOCEROPHYTA), and it does not have a lid. When ripe, it usually splits into four valves. The Hepatophyta contains some five orders, of which the Jungermanniales and MARCHANTIALES are the largest. A typical thallose liverwort is *Pellia*, which consists of a flattened dichotomously branching and frequently deeply lobed thallus. The ventral surface has numerous unicellular rhizoids growing from the area around the midrib. The leafy liverworts generally have three rows of leaves arising from a prostrate stem, although usually only the two dorsal rows are fully developed.

heptamerous having seven parts.

herb (*adj.* herbaceous) a small, non-woody seed-producing plant whose aerial parts die back at the end of the growing season, or which dies completely at that time. Compare ANNUAL, EPHEMERAL, SHRUB, TREE.

Herbaceae see LIGNOSAE.

herbaceous perennial see PERENNIAL.

herbage 1 the living plants on which domestic livestock feed. **2** herbaceous vegetation grown as a crop. **3** the payment made to an owner of pasture for permission to graze livestock on that land.

herbarium (hortus siccus) (*pl.* herbaria) A collection of dried pressed plants, mounted on sheets of thin card, accompanied by data labels and stored in pest-proof wooden or metal cabinets. Smaller organs, including pollen grains, are perfectly preserved in this way, and many other features of the plant (e.g. anatomy, morphology, and chemistry) may be retained virtually unaltered. Details of floral structure can be observed by boiling or soaking in a wetting agent. The data labels, besides giving the plant's name, usually also include the data and place of collection and the collector's name. Notes on habitat, local names, and local uses of the plant (for food, medicine, etc.) may be invaluable in later searches for new sources of drugs, etc.

Specimens are normally arranged according to a particular taxonomic system (e.g. that of Bentham and Hooker in many UK herbaria) but occasionally material is filed in alphabetical order of family, genus, and species. Often specimens are further segregated into geographical regions within their family or generic groups. Herbaria may be small local collections, containing, for example, a county flora, or large international assemblages as at the Royal Botanic Gardens, Kew, where approximately five million specimens are housed. The larger herbaria are centres for taxonomic research and usually provide a plant identification service to other institutions and to the public.

herbicide (weedkiller) any chemical that, when applied to a plant, will either destroy it or seriously inhibit its growth. Herbicides exist in many forms but may be subdivided into two basic categories: the CONTACT and the SYSTEMIC herbicides. See also SELECTIVE.

herbivore an animal that eats plants. The term is used especially of herbivorous mammals such as cattle, sheep, and rabbits.

heredity the phenomenon by which offspring resemble their parents and the laws that govern this.

heritability the proportion of the total variance of an observable characteristic that may be accounted for by genetic factors. Knowledge of the heritability of a trait is of particular value to agriculturalists. For example, if a variable characteristic like yield shows low heritability in a particular crop, this suggests that most of the variation is due to environmental factors and thus selection of high-yielding plants over a number of generations would not improve the yield.

hermaphrodite (bisexual, monoclinous) having both male and female reproductive parts in the same flower. This condition is thought to be derived from a primitive unisexual floral arrangement involving wind pollination (see ANEMOPHILY) and to have developed simultaneously with animal pollination. The hermaphrodite arrangement lends itself to self pollination and inbreeding but many species have intricate self incompatibility systems. Compare DICLINY, DIOECIOUS, MONOECIOUS.

hesperidium (*pl.* hesperidia) a type of BERRY that has a leathery epicarp, such as a citrus fruit. Fluid-filled TRICHOMES fill the locule of each carpel to form the characteristic segments.

hetero- prefix denoting different, of different kinds.

heteroallelic describing ALLELES of a gene that have mutations at different sites within the same gene. Recombination between such heteroalleles may produce a functional cistron. Compare HOMOALLELIC.

Heterobasidiomycetidae see BASIDIOMYCOTA.

heteroblastic development a progressive change in the form and size of successive organs. In many grasses, for example, successive leaf blades are progressively longer and only after a maximum length is achieved will the plant begin to flower. In many plants with compound leaves it may be seen that the younger leaves are simpler than the later leaves. For example, in hemp (*Cannabis sativa*) successive leaves have a greater number of lobes and the margins become more serrated (see diagram). This trend to greater complexity is often reversed after flowering.

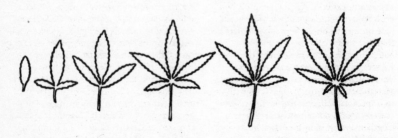

Heteroblastic development: in leaves of *Cannabis sativa*

heterocaryon see HETEROKARYON.

heterochromatin a condensed region of CHROMATIN in the interphase nucleus that stains heavily with basic dyes. There are large amounts of heterochromatin in inactive nuclei. Compare EUCHROMATIN.

heterochromosome a chromosome composed mainly of heterochromatin.

heterocyst any of the large cells that occur at intervals in the filaments of certain species of blue-green bacteria (see CYANOBACTERIA). They do not contain chlorophyll but do possess large amounts of DNA. Narrow pores at one or both poles are also a distinguishing feature. Heterocysts are involved in NITROGEN FIXATION.

heteroecious (heteroxenous) describing a rust fungus (see RUSTS) in which the various spore forms are developed on two different and usually unrelated hosts. An example is *Puccinia graminis*, which causes black stem-rust of grasses and cereals and also infects barberry (*Berberis*). The UREDINIOSPORES, TELIOSPORES, and BASIDIOSPORES develop on the grass host and the pycnospores (see PYCNIDIUM) and AECIOSPORES develop on barberry. Compare AUTOECIOUS.

heterogamous (*adj.* heterogametic) describing the situation in which two different kinds of flowers are borne on the same plant, for example the disk florets and ray florets of some members of the Asteraceae. See also INFLORESCENCE.

heterogamy the fusion of gametes that differ in size or form.

heterogeneous RNA see MESSENGER RNA.

heterokaryon see HETEROKARYOSIS.

heterokaryosis (*adj.* heterokaryotic) the occurrence of normally two nuclei of different genotypes in a fungal cell or mycelium. *Heterokaryons* tend to grow better than their constituent *homokaryons*. If one type of nucleus possesses a mutation such that it cannot synthesize a particular compound, then this deficiency is overcome by the other type of nucleus, a phenomenon termed COMPLEMENTATION. Heterokaryosis also allows recombination in imperfect and homothallic fungi (see PARASEXUAL RECOMBINATION).

heterokont describing an organism having two different types of undulipodia, as have the motile stages of algae in the XANTHOPHYTA, hence the former name of the taxon, Heterokontae. Compare ISOKONT.

Heterokontae see XANTHOPHYTA.

heterolactic fermentation see MIXED LACTIC FERMENTATION.

heteromerous (*adj.* heteromeric) describing lichens in which the algal cells are restricted to a layer situated between an upper PLECTENCHYMA cortex and a lower loosely woven medulla of fungal tissue. It is the more usual tissue arrangement. Compare HOMOIOMEROUS.

heteromorphic (antithetic) (*n* heteromorphism) Describing a life cycle in which the alternating generations are morphologically and physiologically distinct, as in certain algae, the bryophytes, and the ferns. See ALTERNATION OF GENERATIONS. Compare ISOMORPHIC.

heterophasic describing a life cycle that has different phases.

heterophylly the possession of two or more leaf types, often differing widely in morphology and function. For example, certain species of *Lycopodium* and *Selaginella* have two rows of expanded lateral leaves and one or two rows of smaller adaxial or abaxial leaves. Many aquatic and semiaquatic plants have dissected submerged leaves and entire floating or aerial leaves (see illustration overleaf). The submerged leaves are adapted to reduce resistance to water flow while the floating leaves have a broad lamina to maintain buoyancy. Compare HETEROBLASTIC DEVELOPMENT.

heteroplasty the occurrence of different kinds of PLASTIDS in the same cell.

heteropolysaccharide see POLYSACCHARIDE.

heterosis see HYBRID VIGOUR.

heterospory the production of more than one type of spore by a species. Usually large MEGASPORES containing food reserves and small MICROSPORES are formed. The megaspore eventually produces female gametes and the microspore male gametes. Heterospory is seen in certain members of the Lycophyta, e.g. *Selaginella*, in the water ferns, and in all the seed plants. Compare DIPLANETIC.

heterostyly the existence of two or more different arrangements of the reproductive

submerged | aerial or floating

water buttercup (*Ranunculus aquatilis*)

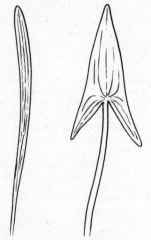

arrowhead (*Sagittaria sagittifolia*)

Heterophylly: in two aquatic species

parts of a flower in one species. These differences are often related to the position assumed by the pollinating agent when visiting the flower. For example, in *Primula* two different arrangements are found (see illustration), one where the style is long and the stamens inserted at the base of the corolla on short filaments (*pin*), and the other where the style is short and the stamens are inserted in an almost sessile fashion on the neck of the corolla (*thrum*).

The different arrangements were thought to ensure cross pollination between the different types. Recent evidence does not support this hypothesis although the differences do appear to reinforce existing physiological barriers to self fertilization. See INCOMPATIBILITY, ALLOGAMY.

heterothallic having morphologically identical but physiologically distinct thalli within a species between which fertilization can take place but among which fertilization does not occur. Thus gametes released from the same plant do not fuse and self fertilization is prevented. Heterothallism is thought to be the most primitive form of sexuality and is found in certain protoctists and fungi, e.g. the green alga *Ulothrix* and the fungus *Mucor mucedo*. It is also seen in certain plants that produce morphologically identical but self sterile gametophytes. Compare HOMOTHALLIC.

heterotrichous describing a filament made up of both erect and prostrate parts, as found in green algae (Chlorophyta) of the order CHAETOPHORALES and in brown algae (Phaeophyta) of the ECTOCARPALES.

heterotroph (*adj.* heterotrophic) an organism that depends for its nourishment on organic matter already produced by other organisms. All animals and fungi are heterotrophs. Parasitic plants and many bacteria also exhibit heterotrophism. Compare AUTOTROPH. See also PHOTOHETEROTROPH.

heterozygosity the presence of different alleles at a particular gene locus. The frequency of individuals that are heterozygous at a particular gene locus gives a measure of the genetic variation in a population, while the proportion of gene loci with heterozygous alleles is a measure of the genetic variation of an individual. See also HETEROZYGOUS.

heterozygous describing an individual that has been formed either from gametes possessing contrasting alleles of a single gene or from gametes differing in the arrangement of genes (see INVERSION). If only one of the alleles is expressed in the phenotype, then that allele is described as dominant, and the masked allele as recessive. Heterozygosity may be inferred from the failure of an individual to breed true for the characteristic under investigation. Compare HOMOZYGOUS.

heterozygous advantage see POLYMORPHISM.

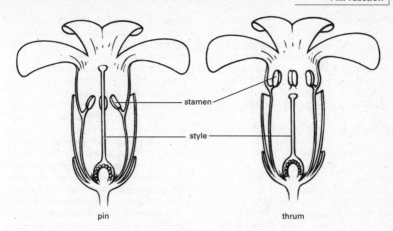

stamen

style

pin

thrum

Heterostyly: in *Primula*

hexadecanoic acid see PALMITIC ACID.

hexamerous having six parts.

hexaphosphoinositol see PHYTIC ACID.

hexaploid having six sets of chromosomes
($2n$). See POLYPLOIDY.

hexokinase an enzyme that accelerates the
phosphorylation of hexose sugars, e.g. the
formation of glucose-6-phosphate from
glucose and ATP.

hexose any sugar with six carbon atoms.
Most of the common monosaccharides are
hexoses, e.g. glucose, fructose, galactose, and
mannose. See also ALDOSE, KETOSE.

hexose monophosphate shunt see PENTOSE
PHOSPHATE PATHWAY.

hibernal in the winter.

hierarchical classification a system in which
individuals are grouped into an ascending
series of successively larger and broader
categories, so that lower groups are always
subordinate to, and included in, those that
are higher in the hierarchy. Within the
hierarchy, classification is based on the
affinities of the component units. The order
in which the 12 principal ranks are used is
governed by the INTERNATIONAL CODE OF
BOTANICAL NOMENCLATURE and, although
all available ranks in the hierarchy need not
be used, their sequential order must not be
altered. In ascending order the ranks are:
form, variety, species, series, section, genus,
tribe, family, order, class, division, and

kingdom. Additional subcategories can also
be designated (e.g. subseries and subclass)
thus further extending the total number of
ranks to 24.

high-energy phosphate bond a phosphate
linkage with a negative standard free energy
of hydrolysis, i.e. one that hydrolyses
spontaneously, releasing energy. The last
two phosphate bonds of ATP are examples.

Phosphate-bond energy is the major
method of transfer of chemical energy from
one enzymic reaction to another. For
example, the high-energy phosphate bond
of phosphoenolpyruvate can be broken to
supply energy for the formation of ATP
from ADP and phosphate. The ATP bond
energy may then be used to drive an energy-
requiring reaction, such as the formation of
glutamine from glutamic acid.

higher fungi fungi belonging to the taxa
Ascomycota, Basidiomycota, and Fungi
Anamorphici.

higher plants 1 vascular plants as opposed
to BRYOPHYTES.
2 vascular plants and bryophytes as opposed
to algae, which in some classifications are
still considered to be plants (they are now
usually classified as protoctists).

Hiller peat-borer see PEAT-BORER.

Hill reaction the light-induced transport of
electrons from water to nonphysiological
electron acceptors, e.g. potassium

ferricyanide, with the concomitant evolution of oxygen. The result is the reduction of the electron acceptor (*Hill reagent*) against the chemical gradient. Four moles of potassium ferricyanide are reduced for every mole of oxygen evolved. The reaction is named after Robin Hill, who discovered it in isolated chloroplasts in 1937.

hilum the scar on the seed coat (testa), comprising a corky abscission layer where the seed or ovule was attached to the FUNICULUS. In some species it may be large and highly coloured and may be used as a diagnostic character.

hip (hep) a type of PSEUDOCARP typical of members of the genus *Rosa* (roses). It consists of a large red cup-shaped hollow receptacle similar, except in colour and texture, to a POME. The receptacle contains a collection of flask-shaped achenes each covered with small hooklike hairs.

hispid covered with rough or stiff hairs as, for example, the leaves of the bristly hawk's-beard (*Crepis setosa*).

histidine a basic amino acid with the formula $C_3H_3N_2CH_2CH(NH_2)COOH$ (*see illustration at* amino acid). The occurrence of histidine in proteins is rare. Synthesis is via a complex pathway, starting from the condensation of ATP and 5-phosphoribosyl pyrophosphate, while breakdown is via glutamic acid. In bacteria, the synthesis of histidine has been found to be under genetic control, with histidine acting as its own REPRESSOR.

histochemistry the study of the distribution of certain chemical substances in cells and tissues by their colour reaction with specific chemicals. For example, cellulose will become bright blue when Schultze's solution is added.

histogen theory a concept of the organization and development of the APICAL MERISTEM, in which the meristematic region is differentiated into three main zones, the dermatogen, periblem, and plerome. The *dermatogen* is supposed to give rise to the epidermis, the *periblem* to the cortex, and the *plerome* to all the primary tissues internal to the cortex. This theory has largely been replaced by the TUNIC–CORPUS THEORY.

histogram a type of graph with vertical rectangular columns, the height of each column indicating the number of times that each class of result occurs in a particular sample or experiment. For example, the height of individual organisms in a given sample of a population can be plotted in this way.

histology the microscopical study of tissue structure.

histone any of the basic proteins confined to the nuclei of eukaryotic cells. Their basic properties are due to a high proportion of the basic amino acids lysine and arginine. These properties enable them to interact very strongly with DNA. There are five types of histones: H1, H2b, H2a, H3, and H4. Two molecules of each type (except H1) aggregate together to form 7–10 nm nodules around which DNA winds. H1 appears to clamp DNA into position on the nodules. Each entire DNA-histone complex is called a NUCLEOSOME. In forming nucleosomes, histones contribute to the packaging of relatively long DNA molecules into relatively minute chromosomes. Additionally, messenger RNA transcription only appears to be possible when H1 is released from the nucleosome. Histones may thus also play a role in the regulation of gene action. Histones are not associated with the genetic material of prokaryotes.

hnRNA see MESSENGER RNA.

Holarctic (Holarctica) a circumpolar BIOGEOGRAPHICAL REGION comprising Europe, Asia, and North America, components of the former supercontinent Laurasia, which fragmented into the present continents over a period spanning 246 to 66.4 million years ago. There are many similarities between the flora and fauna of these continents, reflecting their common origin in Laurasia.

holdfast (hapteron) a structure found at the base of many algae in flowing or tidal water that serves to attach the plant to a support. It is often dissected into many finger-like processes or rhizoids.

holistic (holological) describing an ecological approach that treats humans and the environment as a single system, studying ecosystems as a whole, rather than their component parts. Compare MEROLOGICAL.

hollow phase see HUMMOCK AND HOLLOW CYCLE.

Holobasidiomycetes a class of the BASIDIOMYCOTA (basidiomycetes) in which the hyphae form mycelia and usually have DOLIPORE SEPTA, often with mono-, di-, or polykaryotic compartments. The basidia are often large and fleshy, sometimes tough and woody or gelatinous. The unicellular basidia are borne in a hymenium and usually terminate in four sterigmata. The basidiospores are usually explosively discharged. The class contains some 27 orders, including the AGARICALES (mushrooms and toadstools), BOLETALES (boletes), CANTHARELLALES (chanterelles), LYCOPERDALES (puff balls and earth stars), NIDULARIALES (bird's-nest fungi), and PHALLALES (stinkhorns).

holocarpic describing fungi in which the entire fungal thallus is incorporated into the fruiting body.

Holocene see QUATERNARY.

holocentric describing CHROMOSOMES that have diffuse centromeres, in which the properties of the centromere are spread over the whole chromosome.

holoenzyme the catalytically active complex, consisting of enzyme plus COFACTOR, that is formed by enzymes requiring a cofactor for activity. In some cases the holoenzyme may be extremely complex. For example, pyruvate dehydrogenase contains three different cofactors organized in a multienzyme complex. See also COENZYME, APOENZYME.

holological see HOLISTIC.

holomorph a whole fungus in all its forms (morphs), i.e. with both anamorphic and teliomorphic forms, and phases. See STATES OF FUNGI.

holophyletic describing a TAXON all of whose members are descended from a common ancestor. See also MONOPHYLETIC.

holophytic designating organisms that obtain nourishment in the same manner as plants, i.e. that utilize the process of photosynthesis. The term is roughly equivalent to phototrophic. Compare HOLOZOIC.

holotype the sole specimen or other

element either used by or designated by an author as the nomenclatural TYPE of a species when he/she first published the description. Whenever a new taxon is described it is now imperative both to designate a holotype and state where it is deposited. Duplicates are known as ISOTYPES.

holozoic designating organisms that obtain nourishment in the same manner as animals, i.e. by the ingestion either of other organisms or their products, or other complex organic matter. The organic substances are digested, absorbed, and then assimilated. Holozoic organisms include, as well as animals, fungi and the insectivorous and parasitic green plants. The term holozoic is roughly equivalent to heterotrophic. Compare HOLOPHYTIC.

homeobox a nucleotide sequence in a regulator gene, about 180 base pairs long, that codes for an amino acid sequence called a *homeodomain*, which is thought to be involved in the binding of regulatory proteins to the DNA molecule. Homeoboxes are often almost identical in a wide range of eukaryotic organisms.

homeogene a gene that contains a HOMEOBOX. Such genes are important in regulating the embryonic development of eukaryotic organisms.

homeogenetic induction the influence of a differentiated cell on an adjacent undifferentiated cell such that it brings about similar differentiation in the adjacent cell. The process is seen when vascular tissues are cut. The severed ends of the vascular bundles induce differentiation in adjacent pith cells and thus re-establish a connection.

homeosis the phenomenon in which there is a continuous gradation of a feature between one organ type and another. For example, in certain parts of the range of features from flattened leaves to cladodes, it is almost impossible to define a feature as either a leaf or a stem.

homeostasis see HOMEOSTASIS.

homo- prefix denoting similar, the same.

homoallelic describing alleles of a gene that have different mutations at the same site within the a gene. Recombination between such *homoalleles* cannot produce a functional CISTRON. Compare HETEROALLELIC.

Homobasidiomycetidae see BASIDIOMYCOTA.

homoeology similarity between chromosomes derived from different but related species. Depending on the degree of homoeology, pairing between homoeologous chromosomes may be seen at meiosis in segmental allopolyploids. Such polyploids display meiotic behaviour intermediate between autopolyploids and amphidiploids (see AUTOPOLYPLOIDY, ALLOPOLYPLOIDY, POLYPLOIDY).

homoeostasis the maintenance of a relatively stable state of equilibrium in a biological system or in the metabolism of an organism.

homogametic sex the sex that has sex chromosomes of similar morphology, e.g. XX (as in human females). The gametes of this sex have only one kind of sex chromosome. Compare HETEROGAMETIC SEX.

homogamy the maturation of anthers and stigmas at the same time in the same flower so that the time of pollen presentation and reception coincides. Self pollination may be facilitated by this mechanism. Compare DICHOGAMY.

homogenate see HOMOGENIZATION.

homogenization the processing of a suspension of large solid particles to produce a suspension of uniformly distributed smaller particles (*homogenate*). The process is used, for example, to separate out cellular components. The cell consitutuents are ground up coarsely, then placed in a type of blender. The resulting suspension is then subjected to DIFFERENTIAL CENTRIFUGATION.

homoiomerous describing lichens in which the algal and fungal cells are evenly distributed throughout the thallus, as in *Collema*. Compare HETEROMEROUS.

homokaryon an organism that has genetically identical nuclei. See also HETEROKARYOSIS.

homologous 1 describing structures that have a common evolutionary or developmental origin. They may carry out similar or different functions. The stamens of angiosperms are generally said to be homologous with the MICROSPOROPHYLLS of lower vascular plants. Compare ANALOGOUS. See also DIVERGENT EVOLUTION.

2 see ISOMORPHIC.

homologous chromosomes (homologues) a pair of similar chromosomes, one of maternal and one of paternal origin. In the diploid cells of an organism, two sets of chromosomes are present, one set originating from the male gamete and one from the female gamete. When contraction and condensation of the CHROMATIN of the chromosomes occurs in PROPHASE, their distinctive morphological features become apparent. The positions of CENTROMERES and NUCLEOLAR ORGANIZERS, and the CHROMOMERE patterns are identical in homologous chromosomes and enable homologues to be identified. The genes in corresponding chromomeres influence the same characteristics and if one gene is dominant to the other, it will be expressed in the phenotype. Homologous chromosomes pair during prophase of the first meiotic division and subsequently separate to different daughter cells. Compare HOMOEOLOGY.

homologues see HOMOLOGOUS CHROMOSOMES.

homology (*adj.* homologous) 1 the similarity of a particular structure in different organisms due to their descent from a common ancestor. 2 the similarity of particular nucleotide or amino acid sequences in different nucleic acids, peptides or proteins.

homomeric see DIMER.

homonym a name identical in spelling to another, yet based on a different TYPE. The INTERNATIONAL CODE OF BOTANICAL NOMENCLATURE rules that all homonyms except the earliest are illegitimate. For example, *Viburnum fragrans* Bunge, described in 1831, is a later homonym of *Viburnum fragrans* Loisel., which was described in 1824. Since this was the first published name for the taxon, it must be adopted.

homoplastic describing PHENETIC similarity resulting from CONVERGENT or PARALLEL EVOLUTION rather than descent from a

recent common ancestor. Compare
PATRISTIC.

homoplasy the evolution of similar
structures in different organisms that do
not share a common ancestor. See also
CONVERGENT EVOLUTION, PARALLEL
EVOLUTION.

homopolysaccharide see POLYSACCHARIDE.

homospory the production of only one type
of spore by a species. The term is
commonly used of certain nonseed-bearing
vascular plants, such as the psilophytes (See
PSILOPHYTA) and most of the ferns, to
distinguish them from other vascular
plants that produce megaspores and
microspores (see HETEROSPORY). Compare
MONOPLANETIC.

homostyly the condition found in most
plants in which all the flowers have styles of
approximately the same length. Compare
HETEROSTYLY.

homothallic describing species in which the
thalli are morphologically and
physiologically identical and fusion can
occur between gametes produced on the
same thallus, as in some algae and fungi.
Compare HETEROTHALLIC.

homozygous describing an individual that
has been formed either from gametes
possessing identical alleles of a given gene
or from gametes resembling each other in
the arrangement of genes (i.e. not
possessing any chromosome inversions or
translocations). Homozygosity may be
inferred from the ability of an organism to
breed true for the characteristic under
investigation. Compare HETEROZYGOUS.

honeydew an exudate from the bodies of
aphids that is derived from plant sap.

honey fungus an agaric fungus (see
AGARICALES) (*Armillaria mellea*) that is a
serious parasite of broad-leaved and
coniferous trees and shrubs in temperate
regions. It is named for its honey-coloured,
toadstool-like fruiting bodies. It spreads by
means of rhizomorphs.

honey guides (nectar guides) lines or dots
on the petals that direct a pollinating insect
to the nectaries. An example is the orange
spot on the lower lip of the flower of the
common toadflax (*Linaria vulgaris*).

hordein a storage protein found in the
grains of barley. See PROLAMINE.

horizon see SOIL PROFILE.

horizontal gene transfer the transfer of
genetic material from one organism to
another by a process other than inheritance,
e.g. by infection with a VECTOR organism.

hormocyst a resting stage formed by certain
filamentous blue-green bacteria (see
CYANOBACTERIA) from side branches of the
filament.

hormogonium (*pl.* hormogonia) a short
filament of more or less spherical cells that
may be formed on germination of an
AKINETE in certain filamentous blue-green
bacteria (see CYANOBACTERIA). Although it
does not possess flagella it is able to move,
possibly by the streaming of mucilage along
the surface. When it comes to rest on a
suitable surface it gives rise to filaments.

hormone (growth substance, phytohormone)
any substance that has a marked and
specific effect on plant growth and that
produces this effect when present in very
low concentrations. Hormones may
stimulate plant growth by promoting cell
division (see CYTOKININ) or cell elongation
(see GIBBERELLIN) while others, such as
ABSCISIC ACID and ETHENE (ethylene),
inhibit certain developmental processes.
The term includes substances produced
within the plant and also artificial, often
structurally related, chemicals that have
similar effects. Endogenous hormones are
often produced in a particular region, for
example AUXIN is synthesized in shoot
apices. They are then transported from
these regions and take their effect at sites
far removed from the point of production.
Normal growth is only achieved when there
is a correct balance of the various
hormones. Abnormal growth and even
death may result from unusually high or
low concentrations of a hormone. This
reaction is exploited in the use of certain
weedkillers that contain synthetic auxins.

hornworts (horned liverworts) see
ANTHOCEROPHYTA.

horotelic see CHRONISTICS.

horsetails see SPHENOPHYTA.

hortus siccus see HERBARIUM.

hose-in-hose describing a flower that has an
extra set of petals within the normal
corolla. Such flowers are seen in some

azalea varieties, for example the evergreens 'Kirin' and 'Rosebud'.

host see PARASITISM.

hot spot an area that contains a particularly large number of rare or ENDANGERED SPECIES. Such areas are often earmarked for protection to conserve biodiversity. Compare GAP ANALYSIS.

hot spot of mutation a particular section of a DNA molecule that tends to mutate more often than other parts of the molecule. See MUTATION.

housekeeping gene a gene that must be expressed in every cell of an organism because it codes for proteins required for the basic functioning of the cell.

humic acid a mixture of organic acids derived from HUMUS.

humidity a measure of the moisture content of the atmosphere. There are several ways of expressing humidity. The *relative humidity* is the amount of moisture in the atmosphere relative to the maximum amount of moisture the air can hold at that particular pressure and temperature, usually expressed as a percentage. Relative humidity influences transpiration, which is more rapid when relative humidity is low. It is also an important factor affecting the outbreak and spread of many plant diseases (see BEAUMONT PERIOD). The *absolute humidity* is the total mass of water in 1 cubic metre of air. The *specific humidity* is the mass of water vapour in a given mass of air. Alternatively the water vapour pressure may be used.

humification the formation of HUMUS from dead organic material by the actions of saprobes. This is an oxidation process that results in the release of organic acids, or even simple inorganic molecules that can readily be taken up by plants. This is a vital process for the recycling of nutrients. See BIOGEOCHEMICAL CYCLE.

hummock and hollow cycle a cyclic change in the vegetation that takes place on a developing raised bog (see BOG). A pool of water on the peat is invaded by *Sphagnum* mosses and a low hummock is built up (the *building phase*). Other plants, such as *Calluna vulgaris* (ling) and *Erica tetralix* (cross-leaved heath), then establish themselves. When the ling dies the hummock is easily eroded (the

degenerate phase) and reduced to a water-filled hollow surrounded by other hummocks in various stages. The cycle can then be repeated. Similar cycles occur in the tundra where frost hummocks of peat are formed by frost action and subsequently eroded. In grassland, soil accumulates around a young growing grass seedling, forming a hummock, which builds up until the plant becomes fragmented, and lichens start to colonize it, leading to the degenerate phase as the hummock starts to erode.

humus (*adj.* humic) the soft moist amorphous black or dark brown organic matter in soil derived from decaying plant and animal remains (especially leaf litter) and animal excrement. The complex chemical reactions involved in the decomposition by soil organisms result in the formation of end products that can be used by plants. Humus has a very complex and variable chemical composition. Its colloidal properties enable it to retain water, making it a useful addition to sandy soils. It also helps the formation of soil crumbs (see CRUMB STRUCTURE), which improve aeration and drainage in clay soil and can also provide insulation in soil by reducing temperature fluctuations. There are two main types of humus. *Mor* humus (raw humus) is formed in acid conditions (pH 3–4.5) where the plant litter is low in bases. It is thus acid and this, together with the extremes of moisture or temperature that are often found in such habitats, inhibits the activity of the soil fauna. The slow decay is thus brought about by saprobic fungi. Bogs, heaths, and pine-forest soils have thick layers of mor humus. *Mull* humus (mild humus) is formed in less acid or alkaline soils, which are richer in calcium. Animals, e.g. earthworms, can consume it and the bacteria that bring about decay flourish. It may be found in warm or mild humid climates in such habitats as deciduous forest or grassland. An intermediate form of humus is termed *moder*. See also SOIL.

husk in some grasses, such as maize (*Zea mays*), a group of large scale leaves that enclose the female inflorescence.

hyaline glass-like, transparent or translucent.

hyaline cell any leaf cell without chlorophyll that is normally used for storing water or solutes.

hyaloplasm see CYTOPLASM.

hybrid an individual produced from genetically different parents. The term is often reserved by plant breeders for cases where the parents differ in several important respects. On the other hand, it may be used by geneticists for the F_1 generation of a MONOHYBRID cross, where the only difference is in a pair of alleles at a single gene. Interspecific hybrids, resulting from a cross between different species, are often sterile (HYBRID STERILITY) if derived from haploid gametes, but this may be overcome by polyploidy (see ALLOPOLYPLOIDY). Hybrids produced from different varieties of the same species (F_1 hybrids) are often more vigorous than either parent, so although they cannot breed true, they may be favoured by agriculturalists. See also HYBRID VIGOUR, CHIMAERA.

hybridization 1 natural or artificial processes that lead to the formation of a hybrid. The hybridization of normally self-pollinating species involves the removal of the anthers (emasculation) and the artificial transfer of pollen from another plant. A paintbrush is often used to place the foreign pollen on the stigma. Often hybridization techniques must overcome INCOMPATIBILITY systems. The development of new horticultural and agricultural cultivars involves hybridization techniques. See also INTROGRESSION, SELF POLLINATION.

2 the formation of hybrid molecules when DNA or RNA molecules from two genetically different sources are recombined (annealed) together. The stability of these hybrid molecules is an indication of the degree of relatedness of the two sources of nucleic acids. The term is also used when DNA probes bind to target. See also DNA HYBRIDIZATION, GENE PROBE, FLUORESCENE *in situ* hybridization.

hybrid sterility the reduced ability of some hybrids to produce viable gametes. This is caused by the absence of homologous pairs of chromosomes so that bivalents cannot form during meiosis. The resulting gametes are thus aneuploid. Hybrid sterility may be overcome by POLYPLOIDY. Sterility does not necessarily correlate with poor growth or inviability. Indeed many hybrids show increased vigour (see HYBRID VIGOUR).

hybrid swarm a continuous and variable series of hybrids between two original parental forms. It occurs where a barrier to reproduction between two distinct populations has broken down and there is a certain degree of ALLELE exchange between them.

hybrid vigour (heterosis) the exhibition by a hybrid of a more vigorous growth, greater yield, or increased disease resistance than either parent. The effect is thought to be due to an accumulation of DOMINANT alleles, each having additive effects, and masking the effects of deleterious recessive alleles. Thus, if a characteristic is affected by two genes A and B, and the parents are AAbb and aaBB, the hybrid would be AaBb. Selfing or crossing between identical F_1 HYBRIDS would produce an extremely variable F_2 generation with only half the heterozygosity of the F_1. Thus to maintain optimum vigour and uniformity the seed of some crop plants (e.g. Brussels sprouts) is produced each year by crossing two different parental pure lines (F_1 hybrids).

hybrid zone a geographical area in which hybrids between two geographical races occur. This usually happens at the boundary between two separate but closely related populations. See HYBRID SWARM.

hydathode a secretory structure that removes water from the interior of a leaf and deposits it on the surface. This process is known as GUTTATION and often occurs during damp humid nights when water absorption is high but transpiration is minimal. In some plants, e.g. runner bean, the hydathodes are glandular hairs and water is exuded by active secretion, while in others they are water pores and exudation is passive, the water being forced out by hydrostatic pressure. Both types of hydathode are usually situated at vein endings. Water pores are generally incompletely differentiated stomata and in certain plants, e.g. sea lavenders (*Limonium*), function as salt glands. See also GLAND.

hydric describing a habitat that is extremely wet. Compare MESIC, XERIC.

hydro- prefix denoting a liquid, usually water.

hydrocarbons naturally occurring organic compounds containing only carbon and hydrogen, e.g. butane. They include many of the components of coal, natural gas, and petroleum.

hydrochory the dispersal of seeds or spores by water.

hydrogen symbol: H. A colourless gaseous element, atomic number 1, atomic weight 1. It is an essential element in living tissues, and the most abundant element in the Universe. Hydrogen occurs mainly in water and hydrocarbons, and there are traces of molecular hydrogen in the upper atmosphere and in some natural gases. It is taken up by plants in water. The readiness with which hydrogen loses electrons to other molecules makes it an important component of molecules such as NAD, NADP, and FAD, which are involved in many oxidation–reduction reactions, and facilitates the formation of bonds betweeen molecules due to electrostatic forces (see HYDROGEN BOND). Many hydrogen-containing compounds, especially acids, tend to ionize in a polar environment such as water, releasing hydrogen ions (H^+). See also CHEMIOSMOSIS, PH.

hydrogenase an enzyme that catalyses reactions in which hydrogen is added to a substrate. Compare DEHYDROGENASE.

hydrogen bond a type of bond between molecules in which the attraction is due to electrostatic forces between a hydrogen atom in one molecule and an electronegative atom in another molecule. Hydrogen bonding gives water its characteristic polar properties, and its particular boiling point (which keeps it liquid at room temperature). Hydrogen bonds are important in holding large biological molecules in particular conformations, for example the double helices of DNA and RNA.

hydroid an elongated nonlignified water-conducting cell in the stems and leaves of certain bryophytes, such as *Polytrichum*, analogous to a TRACHEARY ELEMENT in vascular plants. Compare LEPTOID.

hydrolase any ENZYME that catalyses HYDROLYSIS reactions. Examples are the phosphatases, glycosidases, and peptidases, which catalyse the hydrolysis of phosphate esters, glycosidic linkages, and peptide bonds respectively.

hydrolysis (*adj.* hydrolytic) the breaking down of an organic compound by reaction with water. An example is the breakdown of starch to its component glucose units. Compare CONDENSATION.

hydromorphic soil see SOIL.

Hydromyxomycetes in some older classifications, a class of fungi that originally included the orders Labyrinthulales and Hydromyxales. The Labyrinthulales are now placed in the phylum LABYRINTHULATA in the kingdom Protoctista; in some other classifications these organisms are classed as a division of the kingdom CHROMISTA. The Hydromyxales are now considered to be members of the Protoctista.

hydronasty a nastic movement in response to changes in atmospheric humidity.

hydrophilic describing a molecule or surface that has an affinity for water. Such an affinity is characteristic of polar compounds such as proteins.

hydrophily POLLINATION by pollen carried by water. The pollen may be transported on the water surface, as in tasselweeds (*Ruppia*) and water starworts (*Callitriche*), in which case it needs to be light enough to float and water repellent and the stigmas must be exposed at the water surface. Movement of such pollen grains may be enhanced by a natural outer coating of oil, which alters the surface tension of the water. In tape grass (*Vallisneria spiralis*) the whole male flower is released and attaches itself to the female flower at the surface of the water.

Pollen may also be transported through water. The mechanisms for this are very variable and are thought to be derived from entomophilic mechanisms. In naiads (*Najas*), pollen grains are heavy and sink due to gravity onto the female flowers underneath. In eel grasses (*Zostera*) pollen grains transform into structures that resemble a pollen tube, which wrap themselves around the stigma. Compare ENTOMOPHILY, ANEMOPHILY.

hydrophyte (aquatic) a plant that is adapted to living either in waterlogged soil or partly or wholly submerged in water. Many hydrophytes absorb water and gases over the whole surface and have no stomata, e.g. the spiked water milfoil (*Myriophyllum spicata*), which is completely submerged in water. The mechanical and vascular tissue of many hydrophytes is reduced as water is plentiful and supports them. They often have large intercellular air spaces in their stems, roots, and leaves to overcome the difficulty of obtaining gases from the water. Hydrophytes that are partially submerged have floating leaves with stomata through which gases can be exchanged as in land plants. However, to prevent the leaves being flooded with water, the petioles may be very long or shaped like a corkscrew to adjust easily to changes in water level. In the giant water lily (*Victoria regia*), the enormous leaves have a vertical rim to prevent them from being flooded. Some species, e.g. water crowfoot (*Ranunculus aquatilis*), have both finely divided submerged leaves and floating leaves with stomata.

Hydrophytes are also a class in the RAUNKIAER SYSTEM OF CLASSIFICATION, and are defined as having their perennating buds (see PERENNATING ORGAN) under water. The buds may become detached and sink to the bottom (see TURION) or may be borne on submerged rhizomes as in water lilies (*Nuphar* and *Nymphaea*).

Compare MESOPHYTE, XEROPHYTE. See also HYDROSERE.

hydroponics (water culture) the growth of plants in a sterile medium, such as sand or vermiculite, to which nutrients are added in a balanced liquid fertilizer.

hydrosere a pioneer plant community that develops in water when the depth is decreased by silting. The species present depend on the chemical nature of the groundwater, typical species being various water lilies and pondweeds. As the water becomes more shallow, reeds develop. The organic matter builds up to form peat and then the hydrosere gives way to swamp. Eventually MESOPHYTES, such as coarse grasses, develop followed by shrubs and trees. See also SERE.

hydrosphere all the water in the atmosphere, oceans, seas, rivers, lakes, and rocks.

hydrostatic pressure a pressure due to the incompressibility of a fluid. The fluid exerts a pressure on the walls that contain it. TURGOR is due to hydrostatic pressure generated by the osmotic uptake of water into cells whose further expansion is prevented by the inelasticity of the cell wall.

hydrotropism a CHEMOTROPISM in which water is the orientating factor. Thus most roots show positive hydrotropism, growing towards moister regions of the soil, while hypocotyls and the reproductive organs of certain fungi exhibit negative hydrotropism. If water is in short supply hydrotropism exerts a stronger influence on the direction of root growth than gravitropism (geotropism).

hydroxyacetic acid see GLYCOLIC ACID.

2-hydroxybutanedioic acid see MALIC ACID.

hydroxyethanoic acid see GLYCOLIC ACID.

hydroxylation the addition of a hydroxyl group (-OH) to a molecule.

hydroxyproline a polar imino acid (see diagram) formed by the hydroxylation of PROLINE. The enzyme that catalyses this reaction only acts on proline residues that have been incorporated into a polypeptide chain. Hydroxyproline is rare in most proteins but common in cell wall proteins. Degradation of hydroxyproline does not follow that of proline; instead it is broken down to alanine and glycine.

Hydroxyproline

2-hydroxypropane-1,2,3-tricarboxylic acid see CITRIC ACID.

hygrophilous (hygrophilic) growing in or preferring moist habitats.

hygroscopic able to absorb water or water vapour from the surroundings.

hygroscopic water water that clings to the surface of soil particles by molecular attractions so that it is unavailable to plants. Compare CAPILLARY WATER. See also FIELD CAPACITY.

hygrotaxis the directional movement of a cell or organism in response to the stimulus of moisture or humidity.

hymenium in the fungi, a fertile layer consisting of asci or basidia. In the ascomycetes the hymenium lines the ASCOMA while in basidiomycetes it lines the BASIDIOMA.

Hymenomycetes in older classifications, a class of the BASIDIOMYCOTA that contains basidiomycetes with a well-developed basidioma and basidiospores that are usually forcibly discharged. It contains over 5000 species in some 675 genera. It is divided into two subclasses: the Phragmobasidiomycetidae (Heterobasidiomycetidae), which have septate basidia; and the Holobasidiomycetidae (Homobasidiomycetidae), with aseptate basidia. In more recent classifications the term Hymenomycetes is not used and the subclasses are raised to the rank of classes (PHRAGMOBASIDIOMYCETES and HOLOBASIDIOMYCETES, respectively).

hymenophore a spore-bearing body, such as a basidioma, or sometimes just the part of it that bears the hymenium.

hypanthium (*pl.* hypanthia) the flat or cup-shaped receptacle found in perigynous flowers. It is joined to the ovary when the ovary is inferior.

hyper- prefix denoting above, excessive(ly), extremely.

hyperaccumulator see BIOACCUMULATION.

hyperosmotic describing a solution that contains a greater concentration of osmotically active solutes than the cell, so that water leaves the cell by osmosis.

hyperplasia abnormal overdevelopment as a result of an increase in the number of cells in response to a disease-producing agent. Common manifestations of hyperplasia are WITCHES' BROOMS, CANKERS, GALLS, LEAF CURL, and SCAB. Compare HYPERTROPHY, HYPOPLASIA.

hypersensitivity increased sensitivity of a plant to attack by a particular pathogen so that the host tissue dies at the point of infection and thus prevents spread of the disease. The only symptoms are minute necrotic flecks (see NECROSIS). Plants showing this type of reaction to a particular pathogen are very resistant.

hyperthermophile an EXTREMOPHILE that thrives in an environment with a very high temperaure. Some ARCHAEA are unable to multiply at temperatures below 90°C, while others actually require a temperature of at least 105°C for growth. A few archaea can withstand temperatures of 113°C, e.g. *Pyrolobus.*

hypertonic having a higher osmotic pressure than an adjacent solution or a solution under comparison. Compare HYPOTONIC.

hypertrophy abnormal overdevelopment due to an increase in cell size. It is seen, for example, in the roots of crucifers (Brassicaceae) infected with club root (*Plasmodiophora brassicae*). Compare HYPERPLASIA, HYPOPLASIA.

hypha (*adj.* hyphal) a branched filament, many of which together make up a fungal MYCELIUM. Hyphal walls are usually made up principally of chitin laid down in microfibrils. The hyphae of most higher fungi are divided by incomplete cross walls (septa) into many uni- or multinucleate segments. Those of lower fungi are commonly COENOCYTIC. Hyphae may aggregate and anastomose to form a tissue-like mass of PLECTENCHYMA.

In certain fruiting bodies hyphae may be differentiated into thin-walled *generative hyphae,* thick-walled unbranched *skeletal hyphae,* and thick-walled branched *binding hyphae. Flexuous* describes the thin-walled branching hyphae found, for example, in pycnidia.

Hyphochytriomycota (hyphochytrids) a phylum of the kingdom PROTOCTISTA (or in some classifications a phylum of the kingdom CHROMISTA or a class, Hyphochytriomycetes, of the division PARAMYCOTA) containing parasitic or saprobic aquatic organisms that produce threadlike filaments that penetrate the host. Hyphochytrids are found in fresh water (a few occur in soil), living on organic detritus or as parasites on algae and fungi. The thallus is very simple, with rhizoids or a few feeding hyphae, which may or may not be divided by septa. In other species, only part of the thallus forms the reproductive organ. This is a zoosporangium, which releases flagellate zoospores, usually with an anterior hairy undulipodium. The zoospores develop into new thalli. Sexual reproduction and resting spores are unknown in this group.

Hyphomycetales the largest order of the HYPHOMYCETES, containing those anamorphic fungi in which the conidiophores are not compacted into specialized structures. It contains over 6000 species and some 635 genera including: *Penicillium* (common MOULDS), important as the source of penicillin; *Aspergillus*, many species of which cause spoilage of stored food, in some cases (e.g. *A. flavus*) producing toxic substances (aflatoxins); *Botrytis* and *Cladosporium*, common causes of leaf spot diseases; and *Fusarium*, one of the damping-off fungi.

Hyphomycetes a class of the FUNGI ANAMORPHICI (Mitosporic Fungi) containing anamorphic fungi that, if fertile, bear the conidia on hyphae, not in conidiomata. It contains some 11 000 species in about 1700 genera and is divided into the orders AGONOMYCETALES, HYPHOMYCETALES, Stilbellales, and Tuberculariales, based on the presence or absence of conidia and the degree of complexity of the conidia-bearing structures when present. In the latter two orders the conidiophores are compacted into specialized structures. This is an artificial class whose members are not phylogenetically related.

hypnospore a thick-walled resting spore.

hypnozygote a thick-walled dormant zygote.

hypo- prefix denoting under, beneath, less than normal.

hypobasal cell the inner of the two cells formed by the first division of the zygote. In plants showing EXOSCOPIC EMBRYOGENY it gives rise to the foot and in plants showing ENDOSCOPIC EMBRYOGENY it gives rise to the embryo. Compare EPIBASAL CELL.

hypocone see DINOMASTIGOTA.

hypocotyl the region of the stem derived from that part of the embryo between the cotyledons and the radicle. The transition from the stelar arrangement of the stem to that of the root occurs in the hypocotyl. The 'root' of certain crop plants, e.g. turnip (*Brassica rapa*), is actually a swollen hypocotyl. Compare EPICOTYL.

Hypocreales an order of fungi of the class ASCOMYCETES that includes the former Clavicipitales. Some classification systems place it in the PYRENOMYCETES. It comprises some 862 species in 115 genera and 3 families. Members produce small perithecia contained in the rounded heads of stromata (see STROMA). Each ascus contains eight threadlike asci. Most are saprobes or parasites of plants. The order includes the ergot fungus, *Claviceps purpurea*.

hypodermis (hypoderm) one or more layers of cells lying immediately beneath the epidermis in the leaves and other organs of many plants, differing morphologically from the underlying tissues. A true hypodermis develops from the GROUND MERISTEM and therefore has an origin different from that of the epidermis as is evidenced by the noncoincidence of the anticlinal walls of the two tissues. However, the term is often used to refer to the layer or layers of cells immediately inside the outermost layer, regardless of their derivation. The hypodermis often contains large quantities of SCLERENCHYMA to enable it to act as a strengthening tissue, as in the leaves of many conifers. See also EXODERMIS.

hypogeal describing seed germination in which the cotyledons remain underground as there is no great lengthening of the hypocotyl. The cotyledons gradually give up their contents to feed the developing plumule and radicle. Hypogeal germination is seen in the broad bean (*Vicia faba*). The term is also used of fruits that develop underground, such as those of the peanut (*Arachis hypogea*) and of the Bambara groundnut (*Vigna subterranea*). Compare EPIGEAL.

hypogean growing or living underground.

hypogyny the most commonly seen arrangement of floral parts in which the stamens, sepals, and petals are inserted below the ovary, giving a *superior ovary*. Compare EPIGYNY, PERIGYNY.

hypolimnion in a stratified lake, the lower layer of cold water. See also STRATIFICATION.

hyponasty a NASTIC MOVEMENT in which the plant responds by more rapid growth on the lower side of an organ, resulting in a curving upwards of that part. Compare EPINASTY.

hypo-osmotic describing a solution that contains a lower concentration of osmotically active solutes than the cell, so that water enters the cell by osmosis.

hypophloeodal growing or living within or under tree bark.

hypophloic haplostele see STELE.

hypoplasia underdevelopment as a result of disease or nutrient deficiency, giving dwarfed or stunted plants. In some cases particular parts of the plant may develop abnormally or not at all. Compare HYPERPLASIA, HYPERTROPHY.

hypopodium the part of a plant stem axis between the first leaf of an axillary bud and the main axis.

hypostasis the situation in which the expression of one gene (the hypostatic gene) is dependent on a second gene (the epistatic gene). The genes need not be on the same chromosome. See also EPISTASIS.

hypothallus 1 in certain lichens, a layer of fungal tissues situated below the thallus and extending beyond its edges. 2 in certain Myxomycota, a layer of tissue on which sporangium or its stalk rests.

hypothecium in an APOTHECIUM, the lower layer of hyphae below the HYMENIUM.

hypothesis an idea or theory that can be tested experimentally. In statistical testing of experimental data, the concept of the *null hypothesis* is used. The null hypothesis (H_o) is a preliminary assumption about the relationship between one or more parameters or distributions of data, and a states that any observed differences between two sets of data or variables is due to chance alone and not due to a systematic cause or relationship between them. The degree to which the data agree with the null hypothesis gives a measure of how likely the theory/conecept behind it may be. If H_o is rejected, then the statistical conclusion is that the alternative hypothesis (H_a), that any observed differences between the two sets of data or variables is not due to chance alone, is true.

hypotonic having a lower osmotic pressure than an adjacent solution or a solution under comparison. Compare HYPERTONIC.

I i

IAA see INDOLE ACETIC ACID.

IAA oxidase see INDOLE ACETIC ACID.

IBP see INTERNATIONAL BIOLOGICAL PROGRAMME.

ICBN see INTERNATIONAL CODES OF BOTANICAL NOMENCLATURE.

Ice Age a period during the late PLEISTOCENE characterized by cycles of cooling (glaciations) and warming (interglacials) of the Earth. This Ice Age has included at least four major glaciations during which ice caps spread towards the Equator from the poles. At its greatest extent, large areas of Britain, Europe, and North America were covered in ice. It is generally considered that the present time is part of an interglacial period, and that the ice may return at some date in the future. Similar ice ages have occurred in more distant geological time (see GEOLOGICAL TIME SCALE), but their causes are not fully understood. See also DEVENSIAN, FLANDRIAN, ATLANTIC PERIOD, DRYAS, GLACIAL PERIOD, LATE GLACIAL, POST-GLACIAL.

ice algae algae that live in close association with sea ice.

idio- prefix denoting separate, distinct, personal, belonging to oneself.

idioblast any cell markedly differing from the cells of the surrounding tissue. The BRACHYSCLEREIDS of pear fruits are an example.

idiochromosome a chromosome involved in the determination of sex.

idiogamy fusion of male and female gametes from the same individual.

idiogram see KARYOGRAM.

idiotype the total hereditary determinants of a plant or cell, comprising the genotype and all the extrachromosomal hereditary factors (the *plasmotype*).

i-gene see REGULATOR GENE.

Illiciales an order of flowering plants that contains some 159 species in 7 genera, found mainly in southeast Asia, eastern Australia, southern North America, and South America. It includes the families Illiciaceae, which contains the single genus *Illicium* (star anises); Winteraceae, including the winter's bark; and Schisandraceae. The Illiciales have simple alternate leaves without stipules; in the flower the sepals grade into petals with no clear demarcation between the two; the gynoecium has superior free carpels; and each seed contains a tiny embryo with copious endosperm. Many of these features are considered relatively 'primitive', and the order is placed in the subclass MAGNOLIIDAE.

illumination see DARK-GROUND ILLUMINATION, BRIGHT-FIELD ILLUMINATION.

illuviation the deposition in a particular soil horizon of dissolved or suspended material that has been leached from another part of the soil profile. See LEACHING.

imbibition the process by which a substance absorbs a liquid and, as a consequence, swells in volume but does not dissolve. Imbibition, which is usually reversible, is exhibited by many biological compounds, particularly by cell-wall constituents such as pectin, celluloses, and lignin. The testas of dehydrated seeds imbibe water just prior to germination and the process is also important in water uptake by roots. See also MATRIC POTENTIAL.

imbricate overlapping. The term may be used to describe the form of AESTIVATION in which the sepals and petals overlap in the bud and leaves overlap in vegetative buds (see VERNATION). It is also used of mature structures, e.g. the overlapping spikelets of foxtails (*Alopecurus*). Compare VALVATE.

imine a compound containing the *imino group* (NH) or its substituted form (NR), in which the H is replaced by a hydrocarbon or other non-acid organic group.

imino acid a molecule containing a CARBOXYLIC ACID group and an imino (NH) group. PROLINE and HYDROXYPROLINE, often described as amino acids, are in fact imino acids. Other imino acids are probably formed during the degradation of amino acids.

imino group see IMINE.

immersed in mosses, describing capsules that are borne on very short stalks, so that they are partly or wholly enveloped by the leaves.

immigration the flow of genes into a population as as result of individuals moving in from other populations and interbeeding with existing members of the population. See GENE FLOW. Compare REPRODUCTIVE ISOLATION.

immobilization the conversion of inorganic compounds into organic compounds by biological activity, mainly by microoranisms. The process often involves several different microorganisms and complex metabolic pathways. This activity removes the original compounds from the reservoir of nutrients available to plant roots. However, the process is also important in soil detoxification. For example, organic compounds in humus can irreversibly form strong complexes with toxic metals, preventing their uptake. See BIOACCUMULATION.

immunity complete RESISTANCE to a particular disease. Plants do not have an immune system, as animals do, but instead rely on physical barriers to prevent entry of a pathogen, or on various physiological reactions, e.g. HYPERSENSITIVITY or production of PHYTOALEXINS.

immunoelectrophoresis a serological technique used in CHEMOTAXONOMY to compare protein extracts from different plant species. An extract from each species (the antigens) is dispersed in a gel by ELECTROPHORESIS. Antisera, produced by injecting a laboratory animal with an extract from one particular species, is then allowed to diffuse through the gel in a direction at right angles to the electrophoretic separation. Precipitation arcs form where the concentrations of complementary antigen and antisera are greatest (see illustration). The degree of similarity between the antigens in the gel and the antigens of the species used to produce the antisera may be estimated by observation of the number, shape, clarity, etc. of the precipitation arcs. See also SEROLOGY.

imparipinnate describing a PINNATE leaf having a centrally placed unpaired terminal leaflet. Such leaves are seen in most milk vetches (*Astragalus*). *See illustration at* leaf. Compare PARIPINNATE.

imperfect flower a flower that lacks functional carpels or stamens.

imperfect fungi see FUNGI ANAMORPHICI.

imperfect state the asexual state of a FUNGUS, in which no sexual reproduction occurs. The fungus may or may not produce asexual spores. See FUNGI ANAMORPHICI.

importance value see DENSITY-FREQUENCY-DOMINANCE.

-inae suffix denoting a subtribe of plants. See HIERARCHICAL CLASSIFICATION.

inbreeding the production of offspring by the fusion of genetically closely related gametes. SELF FERTILIZATION is the most intense form of inbreeding. It is widespread in the plant kingdom, being inevitable in cleistogamous flowers and during lateral conjugation (e.g. in *Spirogyra*). Inbreeding increases homozygosity and hence decreases genetic variability. Thus when a population heterozygous for a particular gene, e.g. Aa, is selfed, 50% of the next generation will be homozygous for that gene (i.e. either AA or aa). Since the frequency of homozygous recessive lethals and semilethals increases by inbreeding, this can lead to a decline in vigour in species that are normally outbreeding. This is called *inbreeding depression.* Some species have developed systems (outbreeding mechanisms) to inhibit excessive inbreeding. Nevertheless, some degree of inbreeding can be an advantage, for example in a rapidly speciating population where it prevents the population being 'swamped' by foreign genes. Compare OUTBREEDING.

inbreeding depression see INBREEDING.

incipient plasmolysis see PLASMOLYSIS.

incompatibility 1 in seed plants, a complex mechanism that prevents fertilization and/or development of an embryo following pollination by the same individual or a genetically identical individual. It is achieved chiefly by incompatibilities between genes in the pollen grain and the stigma, which affect whether or not the pollen germinates on the stigma, or may affect the rate of growth of the pollen tube. See ALLOGAMY, PREZYGOTIC INCOMPATIBILITY, POSTZYGOTIC INCOMPATIBILITY, S ALLELES, SELF INCOMPATIBILITY. **2** in some fungi, the failure of gametes from genetically similar material to fuse due to physiological or morphological mechanisms. For example, in *Mucor* sexual fusion cannot occur between similar races or strains. **3** the rejection of grafts or other transplants between plants of different genetic makeup.

incomplete dominance (partial dominance) the partial expression of both alleles in a heterozygote so that the phenotype is intermediate between those of the two homozygotes. Thus in *Antirrhinum*, the homozygotes ++ and ww are red and white flowered respectively, while the heterozygote +w is pink. Where one allele is expressed to a greater extent than another, the former may be described as *partially dominant.* There are all degrees of dominance relationships between complete dominance and incomplete dominance: the dividing lines between the various categories are to some extent arbitrary. See also CODOMINANCE.

incubation period the phase following initial penetration of a pathogen when there are no visible symptoms of the disease. It may last several months in which case it is termed a LATENT INFECTION.

When the first visible symptoms appear the disease is said to have entered the infection stage.

incubous describing the leaf arrangement in leafy liverworts where the front edges of the leaves lie above the back edges of the leaves in front. Compare SUCCUBOUS.

incumbent resting on a support, bent over to rest on or touch an underlying surface.

indefinite growth (indeterminate growth) the ability of certain plants or plant organs to show unlimited growth. For example, trees continue to grow each year and the leaves of monocotyledons grow throughout the life of the plant. The term is also used of racemose inflorescences, in which growth of the inflorescence is not terminated by the production of an apical flower bud but continues indefinitely. Meristems may also be said to exhibit indefinite growth. Compare DEFINITE GROWTH.

indehiscent describing a fruit or fruiting body that does not open to disperse its contents. Indehiscent fruiting bodies are thought to have developed from suppression of the naturally occurring opening mechanisms found in dehiscent structures. The seeds or spores are released either when the surrounding wall decays or when it is eaten by an animal.

independent assortment the separation of a pair of alleles of one gene at meiosis independently of the separation of alleles of other genes. Thus, in a heterozygote AaBb, independent assortment would produce four kinds of gametes equally frequently: AB, Ab, aB, and ab. The *Law of Independent Assortment* was formulated by Mendel (Mendel's Second Law). However, it only holds true provided that genes are on separate chromosomes. If genes are on the same chromosome, then they tend to segregate together (LINKAGE). The term independent assortment can be applied to chromosomes as well as genes.

indeterminate growth see INDEFINITE GROWTH.

indicator a chemical substance that, by changing its colour, gives an indication of the pH of the solution to which it is added. Examples are litmus, phenolphthalein, sodium bicarbonate indicator, and *universal* *indicator* (a mixture of indicators that produces a distinct colour change for each unit of pH).

indicator species (ecoligical indicator) a species that by its presence or absence shows the type of environmental conditions that prevail. For example, the absence of lichen growth on trees is taken to indicate atmospheric pollution by sulphur dioxide. As some lichen genera, e.g. *Usnea*, are more sensitive to pollution than others, e.g. *Lecanora*, the degree of pollution can be ascertained by the relative abundances of the different lichens. Other examples of indicator species are *Gypsophila* species, which indicate alkaline conditions, and *Sphagnum* mosses and wavy hair grass (*Deschampsia flexuosa*), which indicate very acid conditions. Indicator species can also be used to reveal information about past environments. For example, ancient woodland indicator species are frequently cited as growing only in long-established woodland (pre-1600).

indicator species analysis a system of classification in which sites are ranked according to their species composition by a reciprocal averaging ordination and divided into two groups at the mid-point of the weighted data values of the ordinations. Several indicator species are identified as associated with each group, being associated exclusively or almost exclusively on one side or the other of this divide. The process may be repeated to further subdivide the groups, and a key is formed by the indicator species, making it relatively easy to place new sites in the classification without the need to do a lot more statistics. See MONOTHETIC, POLYTHETIC.

indifferent species see FIDELITY.

indigenous (native) describing an organism that occurs naturally in an area, i.e. that has not been introduced from another area. The term may be used generally for a large area such as a country, or it may be applied strictly to a particular locality or site. Compare EXOTIC, NATURALIZED.

individualistic hypothesis the hypothesis, proposed by H A Gleason in 1917 that vegetation is continuously varying in response to a continuously varying environment, so that no two plant communities are identical. From this it follows that rather than being discrete entities, communities form a continuum, most appropriately classified by ordination techniques. Compare ORGANISMIC CONCEPT.

indole (benzylpyrrole) a compound (C_8H_7N) formed by the condensation of a pyrrole ring with a benzene nucleus. It forms the basis of the auxin INDOLE ACETIC ACID (IAA).

indole acetic acid (IAA; indolylethanoic acid) the principal AUXIN of most plants. The other INDOLE compounds that also occur in plants probably owe their auxin-like activity to conversion to IAA at, or near to, the site of action. Thus, indole pyruvic acid, indole acetaldehyde, and other compounds may act as IAA reserves. The route of IAA synthesis is from the amino acid TRYPTOPHAN. Studies in vivo have shown that IAA is transported in greater quantity and inactivated or destroyed more quickly than synthetic auxins. This suggests that an important aspect in the control of auxin activity is the relative rates of synthesis, transport, and inactivation in different tissues. IAA is decomposed by light and by the enzyme IAA oxidase. It may also combine with other compounds to form an inert complex. The rapid oxidation of IAA by the plant's enzymes has limited its commercial use as a weedkiller, and synthetic auxins, e.g. 2,4-D, 2,4,5-T, and MCPA, are normally used.

indole alkaloids a group of alkaloids, the members of which contain an INDOLE nucleus. They are derived either from TRYPTOPHAN or PHENYLALANINE. Examples are strychnine, from *Strychnos nux vomica*, and yohimbine from the bark of *Corynanthe johimbe*. Certain alkaloids containing quinoline are derived from indole compounds, an example being quinine, obtained from the bark of *Cinchona* species.

indolylethanoic acid see INDOLE ACETIC ACID.

induced fit hypothesis see ACTIVE SITE.

inducer a substance that promotes the activity of a structural gene or block of genes. Inducers of enzymes in prokaryotes fall into two groups. In negative control systems (e.g. the lac OPERON of *Escherichia* *coli*) they may be substrates of the enzymes they induce or analogues, derivatives, or precursors of the enzyme substrates. In any event, they bind with a REGULATOR PROTEIN and so prevent the latter complexing with the OPERATOR GENE to inhibit mRNA synthesis (see OPERON). In positive control systems (e.g. the ara operon in *E. coli*) the inducer is a complex formed from the regulator-protein substrate. This complex must bind with the operator in order for mRNA synthesis to occur. In eukaryotes the nature of inducers is less well understood but appears more diverse. Steroid and other membrane-soluble growth substances, such as gibberellic acid, can act as inducers when complexed to cytoplasmic receptors.

inducible enzyme (adaptive enzyme) an enzyme that is synthesized in response to high concentrations of its substrate (see INDUCTION). When the substrate is absent the genes governing the enzyme's synthesis are repressed. Compare CONSTITUTIVE ENZYME. See also REPRESSOR, OPERON.

induction a process in which the *de novo* synthesis of a particular enzyme or group of enzymes is stimulated by the presence of the substrate, especially when the substrate is the cell's only carbon source. The induction of enzymes has been studied for the most part in bacteria. The classical example of enzyme induction is that of β-galactosidase in *Escherichia coli*. This enzyme rapidly increases in activity when *E. coli* cells are given lactose as the sole carbon source. This induction effect is not seen, however, if glucose is present, even if lactose is present in high concentrations.

In eukaryotic cells induction is not as common as in bacteria, nor is it so rapid. In primitive eukaryotes, such as *Saccharomyces* and *Neurospora*, induction of β-galactosidase can be observed, although at a much slower rate than in bacteria. In higher eukaryotes there are fewer proven examples of enzyme induction. One that has been studied in detail in white mustard (*Sinapis alba*) involves the synthesis of phenylalanine ammonia-lyase (PAL), which catalyses the formation of cinnamic acid (a precursor of the flavonoids) from phenylalanine. Transcription of the gene governing PAL

synthesis is induced by the P_{fr} form of PHYTOCHROME. See also OPERON.

indumentum (*pl.* indumenta) a dense or sparse covering, usually of hairs.

indusium (*pl.* indusia) a flap of tissue that partially or completely covers each SORUS in certain ferns. It is an outgrowth of the placenta. Indusia are seen on the undersurface of the fronds of *Dryopteris* as lines of brown kidney-shaped outgrowths.

infection 1 the invasion of plant tissues by a pathogenic fungus or microorganism. See PATHOGEN. 2 a disease caused by a pathogenic fungus or microorganism.

infection hypha see APPRESSORIUM.

inferior ovary see OVARY.

inflorescence any flowering system consisting of more than one flower. It is usually separated from the vegetative parts by an extended internode, and normally comprises individual flowers, bracts and PEDUNCLES, and PEDICELS. The change from vegetative to reproductive growth may be triggered by photoperiod or temperature and is thought to be hormonally controlled (see FLORIGEN, PHYTOCHROME, VERNALIN). The arrangement of flowers into inflorescences is believed to have evolved from solitary flowers. The increase in flower number has been accompanied by a decrease in the size of the individual flowers, and specialization for different roles.

Development of inflorescences may terminate vegetative growth (see DEFINITE GROWTH) or allow it to continue (see INDEFINITE GROWTH). The type of growth habit has considerable economic implications in such crops as the grain legumes where it directly affects total crop yield, and sequence and time of fruit ripening.

The architecture of inflorescences is varied (see illustration overleaf) and attempts to classify them have resulted in a proliferation of different names, of which the following are most commonly used. An inflorescence in which the stem terminates in a flower and subsequent flowers differentiate from side branches is a *cyme*. In *monochasial cymes* (*monochasia*), there is one branch per node, as in forget-me-nots (*Myosotis*). If the axis of a monochasial cyme

243

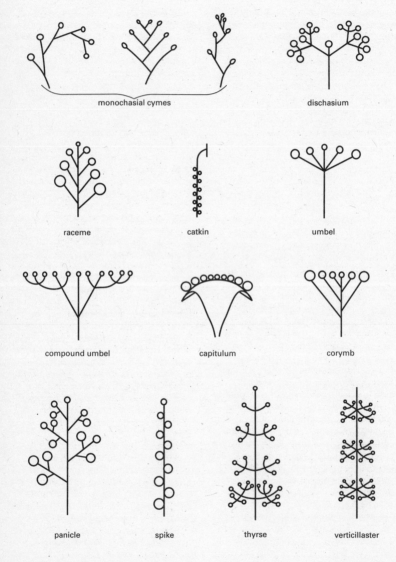

monochasial cymes

dischasium

raceme

catkin

umbel

compound umbel

capitulum

corymb

panicle

spike

thyrse

verticillaster

Types of **inflorescence**

is curved or coiled, it is called a *scorpioid cyme*. Dichasial cymes (*dichasia*) have two opposite branches, as in many of the pink family (Caryophyllaceae). In a *raceme* the apex continues growing and flowers develop sequentially up the stem, as in foxgloves (*Digitalis*). A *catkin* is a short dense raceme, characteristic of many trees (e.g. oak, *Quercus*; willow, *Salix*), with unisexual flowers in which the perianth is absent or vestigial. In an *umbel*, typical of the umbellifers (Apiaceae), many branches radiate out from the stem tip. A *capitulum* is a specialized dense inflorescence with sessile flowers, as in the daisy family (Asteraceae). Many species have compound inflorescences; for example, in the daisy family there may be racemes of capitula, and many umbellifers have umbels of smaller umbels. If the lower branches have elongated to give a flat-topped inflorescence, this is known as a *corymb*. Inflorescences with branched stalks are known as *panicles,/i>; those with sessile flowers are spikes*. These two terms are particularly associated with the arrangement of SPIKELETS in grasses: bents (*Agrostis*) have panicles; wheats (*Triticum*) have spikes. The terms *thyrse* and *verticillaster* are sometimes used for certain whorled inflorescences. A few families have specialized inflorescences, including the SPADIX of the arum family (Araceae), the SYCONIUM of figs (*Ficus*), and the CYATHIUM of the Euphorbiaceae.

Other classifications have been attempted but none satisfactorily cover the range and variety of inflorescences, and many earlier terms have fallen into disuse.

information analysis a method of classifying vegetation according to the extent to which members of different groups of plants differ from each other. It is based on the *information statistic*, which is zero when all the individuals in a group are identical. Individuals or groups showing the fewest differences from each other (least disorder or smallest change in information) are grouped together.

information index see SHANNON-WEAVER INDEX OF DIVERSITY.

information statistic see INFORMATION ANALYSIS.

infralittoral see SUBLITTORAL.

infralittoral fringe that part of the intertidal zone that is exposed only the lowest tides (which occur around the equinoxes). It forms a transitional region between the LITTORAL and SUBLITTORAL zones. On mud or sand it is characterized by *Zostera* (eelgrass) and *Laminaria* (kelps) species.

infrared gas analyser (IRGA) an instrument used for the quantitative measurement of the proportion of a particular gas in a mixture. Different gases absorb infrared at different characteristic frequencies. In the analyser a beam of infrared of the frequency known to be absorbed by the gas under investigation is passed through the gas sample. The instrument also contains a plate sensitized to detect infrared of this frequency. The amount of infrared detected by the plate will fall as the proportion of the gas in the sample increases. Infrared analysers are used for simple measurements, especially the continuous monitoring of air for a given pollutant. They are also used to measure rates of respiration or photosynthesis by recording changes in carbon dioxide level. More complicated analyses of gas mixtures can be made using a SPECTROPHOTOMETER.

infrared radiation electromagnetic radiation with a wavelength between about 700 nm and 1 mm, i.e. from the end of the visible spectrum at its red end to the start of the microwave range. It is invisible, but can be detected as warmth on the skin.

infraspecific occurring within a species or between individuals of the same species.

infraspecific variation differences in appearance exhibited by different members of the same species or breeding population. The variation may be a result of environmental influences, such as nutrition, crowding, seasonal effects, etc. It may also be due to genetic factors, principally the effects of genetic recombination and mutation. Whatever the ultimate cause, natural selection plays a major role in governing the extent of the variation and types of individuals present. The term may be applied to characteristics only apparent by advanced laboratory techniques, as well as to those readily visible to the naked eye.

infructescence an inflorescence that is bearing fruit.

inhibitor a substance that inhibits the activity of an enzyme. It may do this by binding at the ACTIVE SITE of the enzyme in place of the substrate (see COMPETITIVE INHIBITION). Such inhibitors, which are structurally similar to the normal substrate, are usually removable, so inhibition is not permanent. The degree of inhibition depends on the relative concentrations of substrate and inhibitor. Inhibitors that bind to sites other than the active site (*noncompetitive inhibitors*), altering the conformation (three-dimensional structure) of the enzyme and its ability to bind with the substrate (see ALLOSTERIC ENZYME, ALLOSTERIC SITE), are unaffected by substrate concentration and inhibition is irreversible. An example is cyanide, which combines with the metallic PROSTHETIC GROUPS of such enzymes as cytochrome oxidase and is a powerful inhibitor of aerobic respiration. Certain enzymes that contain an –SH group are particularly susceptible to this type of inhibition, usually by heavy metals. See also FEEDBACK INHIBITION.

initial (initiating cell, meristematic cell) any actively dividing cell in a MERISTEM. Each time an initial divides, one of the daughter cells retains its meristematic properties and remains in the meristem as an initial while the other differentiates to form a cell in the plant body. The latter daughter cell may undergo further divisions but in this event all daughter cells differentiate further. Initials commonly possess a dense cytoplasm. See also FUSIFORM INITIAL, RAY INITIAL, VASCULAR CAMBIUM.

initiation see TRANSLATION.

initiation codon see TRANSLATION.

initiation complex see TRANSLATION.

initiation factor see TRANSLATION.

inoculation see EXPLANTATION.

inoculum (*pl*. inocula) **1** the material that initiates disease in a previously uninfected plant. Under artificial conditions this may be a suspension of spores, which is sprayed onto the plant. In nature the inoculum is carried to a healthy plant from its source, e.g. an infected plant, by an agent, such as wind or an insect.

2 in microbiology or tissue-culture work, the cells that are introduced into a sterile medium to initiate a culture.

inositol a sugar alcohol formed by the complete hydroxylation of cyclohexane. The *myo* isomer is the only isomer with biological significance. *Myo*-inositol is synthesized in a two-stage reaction from glucose 6-phosphate. Its hexaphosphate ester, *phytic acid*, is an important storage product in seeds; in germinating seeds and young seedlings it is thought that the phytic acid is converted to *myo*-inositol. This then undergoes oxidation to form glucuronic acid; thus phytic acid acts as a source of URONIC ACIDS for synthesis of cell-wall components.

insecticide any chemical that kills insects. Loosely the term is also applied to chemicals that kill mites, nematodes, and other invertebrate pests. The substance often affects the nervous system and may be a stomach poison (see STOMACH INSECTICIDE). It may also act by CONTACT or as a FUMIGANT. See also BOTANICAL, CHEMOSTERILANT, CHLORINATED HYDROCARBON, DINITRO COMPOUNDS, ORGANOPHOSPHATE INSECTICIDE, SULPHUR DUST, SYSTEMIC.

insectivorous plant (carnivorous plant) a plant that is adapted to obtain food by digesting small animals (particularly insects) in addition to feeding by photosynthesis. Such plants are found in regions where the soil is deficient in certain nutrients, particularly nitrates. Examples are butterworts (*Pinguicula*) and sundews (*Drosera*) growing on heathland and moorland, and tropical plants, such as the Venus' flytrap (*Dionaea muscipula*) and the various pitcher plants (families Nepenthaceae and Sarraceniaceae). The plants are adapted in a number of ingenious ways to attract, trap, and kill the insects, which are then digested by PROTEOLYTIC ENZYMES secreted by the plant. Butterworts trap insects by having sticky infolded leaves. Venus' flytraps have folding leaves with spines along the edges that spring together and trap the insect. The leaves of pitcher plants form a long narrow container shaped like a pitcher, into which the insects fall and drown in the liquid at the bottom.

insect pollination see ENTOMOPHILY.

integument a protective structure that develops from the base of an ovule and encloses it almost entirely except for an opening, the micropyle, at the tip of the nucellus. In most angiosperms there are two integuments, which may or may not be fused. In gymnosperms and in most dicotyledons with fused petals there is only one integument. Occasionally a third integument is formed, which becomes conspicuous as an ARIL following fertilization. This is seen in the spindle tree (*Euonymous europaeus*). The integuments have a vascular supply continuous with the parent via the FUNICULUS.

intelligent design the idea that the diversity of living organisms, the occurrence of fossils, and the stucture of the universe result from an intelligent cause or agent, not an unguided process like evolution by NATURAL SELECTION. The evidence presented is mainly based on observations of patterns in nature and of gaps in evolutionary theory and in the fossil record. Most proponents claim the designer to be the Christian God. Intelligent design is not a scientific theory, since it cannot be tested by experiment. This idea is particularly popular in the United States, and has resulted in legal suits being filed in an attempt to compel schools to teach intelligent design alongside EVOLUTION in science classes.

intercalary meristem a MERISTEM positioned (i.e. intercalated) between more or less differentiated tissues, some distance from the APICAL MERISTEM. Intercalary meristems may contain not only INITIALS, but also relatively mature cells. This prevents it from being a structurally weak part of the organ. The primary function of an intercalary meristem is to facilitate longitudinal growth of a plant organ, independent of activity of the apical meristem. Examples are the meristems in the leaf sheaths and internodes of grasses and horsetails. The intercalary meristems of grasses account for their success when they are grazed by allowing the shoots to regrow after they are cut off.

intercellular located or taking place between cells.

interchange see TRANSLOCATION.

interfascicular cambium the part of the VASCULAR CAMBIUM that forms between the vascular bundles (fascicles) and joins up with the FASCICULAR CAMBIUM to form a continuous meristematic ring.

interference see CHIASMA INTERFERENCE.

interference microscope an advanced development of the PHASE CONTRAST MICROSCOPE in which the light from the condenser is split into two beams by a prism. The *object beam* passes through the specimen and the objective; the *reference beam* passes through a second matched objective without going through the specimen. The two beams are recombined before going through the eye piece. Interference between the beams produces a series of light and dark fringes in the field of view. These are the result of differences in refractive index of the specimen and allow detail of the specimen to be seen. The instrument is more sensitive than the phase contrast microscope and spurious effects (such as halos) are eliminated. With white light different parts of the specimen appear coloured as a result of interference.

intergenic suppression see REVERSE MUTATION.

interglacial see GLACIAL.

intermediate host a host essential to the life cycle of a PARASITE, but in which it does not reproduce sexually. Such hosts may act as reservoirs for future infection. For example, barberry is an intermediate host for wheat rust.

internal fertilization see FERTILIZATION.

International Biological Programme (IBP) an international research programme that took place from 1966–74 and aimed to understand the dynamics of the Earth's major terrestrial, aquatic and marine ecosystems (see BIOMES), to model ecosystem structure and productivity in order to assess the effects of past, present, and future environmental changes, and to develop a basis for managing ecosystems to improve productivity and environmental quality. The programme also encompassed human involvement in ecosystems and resource use and management. The IBP was instigated by the International Union of Biological Sciences, and was later taken over by the International Council of Scientific Unions. It was a forerunner of UNESCO's Man and Biosphere programme, and played a useful role in stimulating environmental awareness worldwide.

International Code of Botanical Nomenclature (ICBN) a periodically revised publication outlining the procedures for the scientific naming of plants, algae, blue-green bacteria (see CYANOBACTERIA), fungi, and slime moulds.

The rules have been drawn up to sort out errors and ambiguities arising from past misunderstandings and misidentifications and to ensure correct naming of new taxa. Fundamental to the code are the principles that naming of families and lower ranks is by reference to nomenclatural TYPES, that the first valid name published is maintained (for vascular plants this is the first name published on or after 1 May 1753, the date Linnaeus' *Species Plantarum* was published), and that once the circumscription, position, and rank of a group have been decided it can only have one correct name. However this does not preclude a group of plants being treated in different ways (e.g. being given a different position) by different authors, with each author being able to assign a different nomenclaturally correct name. For example, watercress is thought by some authors to belong to the genus *Nasturtium*, while others believe it is better placed in *Rorippa*. Following the rules of the Code, the names *Nasturtium officinale* and *Rorippa nasturtium-aquaticum* are equally acceptable. Names cannot be changed simply because they are inappropriate. However homonyms and superfluous names (names published for a taxon after that taxon already had a valid name) must be rejected. If two or more taxa are combined into one then the new taxon takes the name of the constituent that had the oldest valid name. If a taxon is split into two or more taxa, then one of the new taxa must be given the name of the old taxon.

The ICBN is only applicable to wild plants and because of the importance of many cultivated plants, a separate set of rules has been laid down in 57 Articles in the International Code of Nomenclature of Cultivated Plants or ICNCP.

International Union for the Conservation of Nature (IUCN) see WORLD CONSERVATION UNION.

internode a part of the stem lying between two adjacent NODES.

interphase the period in the CELL CYCLE following cytokinesis and preceding the next nuclear division.

interpositional growth see INTRUSIVE GROWTH.

interspecific describing something that occurs between or involves members of different species. Compare INTRASPECIFIC.

interstadial during a glacial period, a shorter period during which the climate becomes warmer. An interstadial is shorter and cooler than an interglacial period. See GLACIAL, ICE AGE.

intertidal the part of the littoral zone between the highest and lowest tide levels.

intine the innermost layer of the pollen grain wall, composed mainly of cellulose. It is less resistant than the EXINE and hence not preserved in geological deposits.

intracellular taking place or situated within the cell.

intrafascicular cambium see FASCICULAR CAMBIUM.

intragenic suppression see REVERSE MUTATION.

intraspecific describing something that occurs between or involves members of the same species. Compare INTERSPECIFIC.

intraspecific selection NATURAL SELECTION among individuals of the same species.

intrazonal soil see SOIL.

intrinsic rate of natural increase see BIOTIC POTENTIAL.

introgression (introgressive hybridization) the incorporation of genes from one species or subspecies into another related species or subspecies. It arises as a result of successful hybridization and subsequent backcrossing of the hybrids with one of the parental populations. Introgression is believed to have been a major factor in the evolution of many plants, especially crop plants. For example, the present wide range of the sunflower (*Helianthus annuus*) over North America is thought to have been achieved fairly recently by the introgression of genes from a number of *Helianthus* species, each providing characteristics enabling adaptation to different environments. Introgression is the basis of some plant breeding techniques. Compare GENE FLOW.

intron a noncoding polynucleotide (see NUCLEOTIDE) sequence between two coding regions (exons) of DNA. Introns are present in eukaryotic cells, but not in bacteria. The length and number of introns in a gene varies, but in some genes introns can contain more DNA than exons. In other genes they are absent altogether. mRNA initially synthesized on intron-containing genes is called *heterogeneous mRNA*, or hnRNA. While in the nucleus, probably in the nucleolus, introns are removed and the ends of the mRNA are annealed to produce mature mRNA. The latter is then translated into protein. How introns can be removed with such precision, and what happens to them after their removal, is obscure. Several functions have been ascribed to them. For example, they may speed evolution by enhancing recombination between exons. It has also been suggested that they play a role in differentiation. Introns should not be confused with REPEATED SEQUENCES, which occur between genes.

introrse describing anthers that release their pollen to the inside of the flower so promoting self pollination. Compare EXTRORSE.

intrusive growth (interpositional growth) the type of growth exhibited by certain cells when they intrude between other cells that are either not growing or growing at a slower rate. For example, certain members of the Liliaceae have vascular TRACHEIDS that elongate at both ends to achieve a length of between 15 and 40 times that of the original. These tracheids insinuate themselves amongst the surrounding cells, disrupting established PLASMODESMATA. Compare SLIDING GROWTH, SYMPLASTIC GROWTH.

intussusception the insertion of cellulose into spaces within the CELL WALL of an elongating cell. Water passes into the vacuole of the cell, creating an outward pressure on the wall, which consequently stretches. The new cellulose is then incorporated, the result being an increase in wall area. The next step is usually the thickening of the walls by APPOSITION.

inulin a storage polysaccharide composed of FRUCTOSE residues linked by $\beta(2-1)$ GLYCOSIDIC BONDS. The inulin chain is usually headed by a glucose residue and is 30–40 residues in length. Synthesis of inulin is from sucrose precursors, hence the glucose at the beginning of the chain. Inulin is an important reserve polysaccharide in the Asteraceae being found in high proportions in the genera *Helianthus*, *Dahlia*, *Inula*, and *Chicorium*. The tubers of Jerusalem artichoke (*Helianthus tuberosus*) may contain up to 58% inulin.

inverse analysis the analysis of vegetation in which species are grouped on the basis of presence/absence of certain attributes (e.g. presence or absence of particular species at different sample sites). It is sometimes referred to as a *species classification*. It is used especially in numerical classification, and the ordination methods used are described as either *plot* (individual) or *species* (attribute) *ordinations*. It is common practice to use both inverse analysis and normal analysis, then compare the degree to which the final groups produced by both methods coincide. A high degree of coincindence indicates an important 'type' community or *nodum*. See also Q TECHNIQUE. Compare NORMAL ANLAYSIS, R TECHNIQUE.

inversion 1 (chromosome inversion) a structural change in a chromosome in which a length of chromosome has detached and then rejoined the opposite way round. If the CENTROMERE is included in the inverted segment the inversion is termed *pericentric*. If the centromere is not in the centre of the inverted segment then the relative lengths of the chromatid arms either side of the centromere will be changed, altering the karyotype. *Paracentric* inversions, in which the centromere is not included, are the more common. An inversion is homozygous if the same segment is inverted in both HOMOLOGOUS CHROMOSOMES. Apart from the changed linkage arrangements of the genes and any POSITION EFFECTS, homozygous inversions behave much as normal homologous chromosomes. Often though the inversion only occurs in one chromosome giving an *inversion heterozygote*. These can be detected by the characteristic looped appearance during meiosis. Inversion heterozygotes often show reduced fertility because if a crossover occurs within the loop then a dicentric chromosome (a chromosome with two centromeres) and an acentric fragment will result (see diagram). Such pieces cannot move normally on the spindle and thus only half the gametes (those containing either of the two chromatids not involved in the crossover) will be viable. As the viable gametes will have the same linear sequence of genes as the parent, inversion heterozygotes have been termed *crossover suppressors*. However they do not suppress crossing over but the resultant recombinants, being inviable, are not apparent. Viable recombinants may be obtained if two crossovers occur in the inversion loop. Inversion heterozygotes may have an adaptive advantage in that any

useful combination of alleles in the inverted segment are likely to be held together.

2 in a GENE MUTATION, the transposition of two adjacent nucleotide bases in a CODON. Such a mutation will affect only a single codon, and hence a single amino acid in the resultant polypeptide if it is in a functional part of the gene.

invertase see SUCRASE.

invert sugar see SUCROSE.

in vitro describing experiments on biological processes that are carried out in laboratory apparatus. The literal meaning is 'in glass'. Compare IN VIVO.

in vivo describing experiments investigating biological processes that are carried out in living organisms. The literal meaning is 'in life'. Compare IN VITRO.

involucre **1** a whorl of BRACTS around or beneath a condensed inflorescence, such as a capitulum or umbel. It resembles and

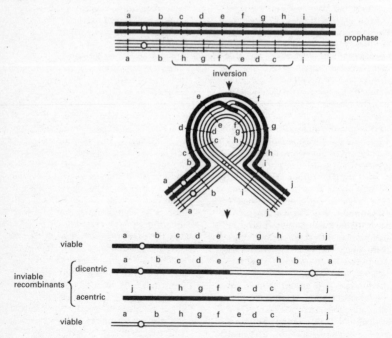

Inversion: consequences of crossing over in an inversion heterozygote

performs the function of the CALYX of a single simple flower.

2 the sheath surrounding a group of archegonia or antheridia in certain liverworts, such as *Sphaerocarpus* and its allies.

3 a tubular upgrowth of the thallus surrounding the base of the sporophyte in members of the Anthocerophyta (hornworts).

involute describing a kind of individual leaf folding (ptyxis) in which the leaf is curled towards the adaxial surface. *See illustration at* ptyxis.

iodine symbol: I. A nonmetallic element belonging to the halogen group. It forms an almost black, crystalline solid at room temperature, atomic number 53, atomic weight 126.90. It does not occur on its own in nature, but is always combined with other elements. It is found in ionic form in sea water and in various brines. Iodine is found in plants and animals, especially in seaweeds. It is an ESSENTIAL ELEMENT for many animals, including humans.

ion see ANION, CATION.

-ion see ZURICH–MONTPELIER SCHOOL OF PHYTOSOCIOLOGY.

ion-beam etching a method of specimen preparation for scanning electron microscopy (see ELECTRON MICROSCOPE) in which the specimen is placed under high vacuum and the surface is then eroded by a beam of ions to expose subsurface features. The specimen is then coated and examined. The technique is very successful with some materials but produces artefacts by uneven erosion of the surface of others.

ion-exchange chromatography a type of CHROMATOGRAPHY in which the components of mixtures are separated by differences in their acid–base behaviour. The chromatogram column is filled with a charged resin that is easily able to exchange ions for any charged molecules that are passed through the column. Amino acid separation is often achieved by this method and the resin commonly used is sulphonated polystyrene in which the sulphonic acid groups are charged with sodium ions. If an acidic mixture of amino acids is added to the column then, because the amino acids are mostly cations at acid

pH, they will tend to displace the sodium ions. The more basic amino acids, e.g. histidine, will be held more tightly to the resin than the more acid amino acids, e.g. glutamic acid. By gradually increasing the pH and sodium concentration of the mobile phase the amino acids are displaced from the resin. They emerge from the column in order of decreasing acidity.

IPA (isopentenyl adenosine) a natural CYTOKININ, similar in structure to ZEATIN. In addition to being a very active cytokinin, it has been found as a minor base in certain TRANSFER RNA molecules. In the tRNA specific for the amino acid serine, IPA is found following the anticodon and thus may serve to mark the end of the anticodon.

IRGA see INFRARED GAS ANALYSER.

iron symbol: Fe. A metal element, atomic number 26, atomic weight 55.85, required by plants as a macronutrient. It is important as a constituent of the CYTOCHROMES, FERREDOXIN, and of certain enzymes, e.g. succinate dehydrogenase. Deficiency of iron leads to CHLOROSIS since iron plays a part in chlorophyll formation. Iron is the second most abundant metal in the soil being present in such minerals as haematite, magnetite, and limonite. However in alkaline conditions iron may be precipitated as an insoluble ferric salt making it unavailable to plants. This can be overcome by adding sequestrols containing chelated iron. Ferrous compounds in the soil are also used by certain bacteria, which derive energy by oxidizing them to ferric compounds.

iron pan see HARDPAN.

irregular see BILATERAL SYMMETRY.

irritability (sensitivity) response by a living organism to external stimuli. In plants, the responses often take the form of various types of movements caused by the bending of the growing region. In addition, certain aquatic microorganisms and some motile gametes exhibit irritability by changing their physical position. See TROPISM, NASTIC MOVEMENTS, TAXIS.

isidium an outgrowth from a lichen thallus containing both algal and fungal cells that may break off and propagate the lichen vegetatively. Isidia show a variety of forms, e.g. the coral-like outgrowths of *Umbilicaria* or the budlike projections of *Collema*.

island biogeography the study of species distribution on islands, often extended to include islands of habitats surrounded by other, hostile environments. It aims to predict the rate of colonization of islands by species from elsewhere, and the survival and diversity of island species. Islands are typically rich in endemic taxa, but because it takes time for species to reach islands to recolonize, they are highly vulnerable to human interference and disturbance, especially by introduced species. The theory of island biogeography states that where immigration and extinction rates are equal, the number of species is directly proportional to the size of the island and inversely proportional to the distance of that island from the mainland. See ISLAND HOPPING. See also ADAPTIVE RADIATION, FOUNDER EFFECT, GEOGRAPHICAL ISOLATION.

island hopping the colonization of islands by species from adjacent islands, and especially the colonization of newly isolated islands in an island chain.

iso- prefix denoting the same or equal.

isochromosome a METACENTRIC chromosome in which the two arms are genetically identical. See also TRISOMIC.

isoelectric focusing a form of ELECTROPHORESIS that separates AMPHOTERIC substances (substances that display both acidic and basic properties) using their ISOELECTRIC POINTS. Proteins and amino acids may be separated in this way. The mixture of substances is passed down a polyacrylamide gel containing a mixture of buffers that form a pH gradient. Once a protein or amino acid reaches the pH of its isoelectric point, it will stop moving, so it will become concentrated at a particular level on the gel.

isoelectric point the pH value of a medium that results in a molecule having no net electric charge. At the isoelectric point the molecules will coalesce and precipitate, because there is no repulsion between them. See ISOELECTRIC FOCUSING.

isoenzyme (isozyme) any one of the multiple forms of a given enzyme, each with different kinetic characteristics. The different isoenzymes are usually formed by different combinations of the same two or more subunits. Isoenzymes are readily distinguishable using ELECTROPHORESIS. They are formed by different alleles and, within one species or a group of species, have become an enormously important tool in detecting genetic variation. They are used to determine relationships between populations and provide evidence for natural selection and genetic drift. One commercial application is in forestry, where local races of trees are especially well adapted to particular environmental conditions. Seed isoenzymes can be used to check that the seed supplied is really of the race it is claimed to be.

Isoetales an order of the LYCOPHYTA containing a single genus, *Isoetes* (quillworts), with about 10 species, predominantly aquatic perennial plants, found worldwide except for the Pacific islands. The stem is condensed into a fleshy rootstock, from the top of which arises a rosette of long narrow LIGULATE leaves. Seasonal cambial activity occurs in a zone around the primary vascular tissue but is highly unusual in that it differentiates cortex rather than STELE. Thus the oldest tissue occurs at the outer edge of the stem and is continually sloughed off as new material is generated from within. The megaspores are borne in the axils of leaves from early in the season, in the outer part of the rosette, while the later-formed leaves bear microsporangia. The antherozoids differ from those of the Selaginallales and Lycopodiales in having many undulipodia rather than just two. Some species show CRASSULACEAN ACID METABOLISM, and some of these (e.g. *I. andicola*) are unique in having no stomata and an internal structure similar to that of Cretaceous forms.

isoflavones see FLAVONES.

isogamy (*adj.* isogamous) the production or fusion of morphologically identical gametes (*isogametes*). It is found only in the more primitive algae and fungi and constitutes the simplest form of sexual reproduction. Although the gametes appear identical it has been found that two gametes produced by the same parent usually will not fuse (see HETEROTHALLIC).

Thus physiological differences exist and these are recognized by ascribing such gametes with plus and minus signs depending on their compatibility relationships (see MATING STRAIN). Compare ANISOGAMY, OOGAMY.

isogeneic (syngeneic) describing a graft involving genetically identical scion and stock. See GRAFTING.

isokont describing an organism having UNDULIPODIA similar in form and length, as have the motile algae in the Chlorophyta, hence the former name of the taxon, Isokontae. Compare HETEROKONT.

isolating mechanism SEE REPRODUCTIVE ISOLATION.

isoleucine a nonpolar branched chain amino acid with the formula $C_2H_5CH(CH_3)CH(NH_2)COOH$ (*see illustration at* amino acid). Isoleucine is synthesized from threonine. The pathway is similar to that for valine and leucine and many of the enzymes are common to all three. Degradation of isoleucine is also closely related to that of leucine and valine, leading to succinyl CoA. The final part of the degradation, from propionyl CoA, is common to valine, isoleucine, and methionine.

isomer (*adj.* isomeric) any one of two or more compounds that have the same molecular formula, but different structural formulae (*stereoisomers*) or different spatial arrangements of atoms, i.e. their atoms are arranged in different ways, for example glucose 6-phosphate and fructose 6-phosphate.

isomerase any ENZYME that catalyses the conversion of a molecule from one isomeric form to another. An example is glucose phosphate isomerase, which catalyses the conversion of glucose 6-phosphate to fructose 6-phosphate in glycolysis. See also EPIMERASE.

isomorphic 1 (homologous) describing a life cycle in which the alternating generations are morphologically identical as, for example, in the brown alga *Dictyota*. Individuals may be identified as sporophyte or gametophyte by observation of the reproductive process. Isomorphic life cycles are seldom seen in plants with ARCHEGONIA. An exception is *Psilotum*, in

which the subterranean gametophyte resembles, in anatomy, branching pattern, and in the production of multicellular gemmae, the rhizome of the sporophyte. See ALTERNATION OF GENERATIONS. **2** describing an organ or organism that exists in one form only. Compare DIMORPHISM, POLYMORPHISM.

isonome method a technique used to study the distribution of plants in a particular area where there are obvious variations in species distribution. Continuous QUADRATS are laid out to cover the chosen area and the ABUNDANCE of each species in each quadrat is recorded. The abundance values of each species are plotted on separate pieces of squared paper. Isonomes (which resemble contour lines) are then constructed by joining up roughly equal abundance values. Isonomes of certain environmental factors, such as the topography of the area, can also be constructed. By superimposing the isonomes of one species onto those of another it can be seen if there are any correlations between the distributions of the two species. Similarly, by superimposing the isonomes of a species onto the isonomes of an environmental variable it may be seen whether that particular environmental factor is correlated with the species distribution.

isopentenyl adenosine see IPA.

isophylly the state in which all the leaves on a plant have the same morphology. Compare heterophylly.

isoprene unit the fundamental unit from which many different hydrocarbons, particularly the terpenes (see TERPENOIDS), are constructed. It has the formula $CH_2C(CH_3)CHCH_2$.

Isoprene unit

isoquinoline alkaloids a group of ALKALOIDS, the structures of which are based on the isoquinoline nucleus. They are derived from the amino acids tyrosine and phenylalanine. Examples are morphine, from the opium poppy (*Papaver somniferum*), and curare, from the bark of *Strychnos* species.

isosmotic (iso-osmotic) describing a solution that contains the same concentration of osmotically active solutes as the cell, so that there is no net movement of water between the solution and the cell. See OSMOSIS.

isotonic having the same OSMOTIC PRESSURE as the solution under comparison. If two isotonic solutions are separated by a permeable membrane there will be no net migration of ions from one to the other. See also HYPERTONIC, HYPOTONIC.

isotope see ISOTOPIC TRACER.

isotopic dating see RADIOMETRIC DATING.

isotopic tracer a stable or radioactive *isotope* that can be used to label a metabolite and consequently follow its fate in an intact organism. Many elements found in living organisms have rare isotopes that are useful as tracers. For example, the most abundant form of carbon, carbon-12 (^{12}C), has six protons and six neutrons in its atomic nucleus. About 1.1% of carbon exists as the stable isotope ^{13}C, which has one extra neutron. ^{14}C, which has two extra neutrons, exists in minute amounts, and is radioactive, emitting beta rays. Both of the rarer isotopes are incorporated and function the same way as ^{12}C in an organism. The light-independent reactions of photosynthesis (see CALVIN CYCLE) were elucidated using ^{14}C-labelled carbon dioxide and finding the order in which substances incorporated the isotope. Tritium, a radioactive isotope of hydrogen having three neutrons instead of the usual one, has been used to label precursors of DNA, e.g. tritiated thymidine, to examine DNA synthesis.

isotype a specimen collected at the same time and from the same plant or localized population of plants as the HOLOTYPE. These duplicate specimens are often separated and deposited in several institutions. See also TYPE.

isozyme see ISOENZYME.

isthmus the narrow central portion connecting the two halves of a placoderm desmid.

iterative evolution the repeated evolution of similar structures in the same main lineage. This is often due to mutations of certain regulatory genes controlling morphogenesis.

iteroparous (polycarpic, polycarpous) describing a plant that reproduces more than once during its lifetime. Iteroparous plants include all perennials except SEMELPAROUS perennials.

IUCN see WORLD CONSERVATION UNION.

Jacob–Monod model (operon model) a theory to explain how gene activity is regulated, proposed by F. Jacob and J. Monod in 1960. The theory is well substantiated for prokaryotes but there is little evidence that it is directly applicable to eukaryotic cells. The operon model shows how messenger RNA synthesis, and thus the quantity of enzymes present, may be regulated. By controlling the latter, the system as a whole provides a method of coordinating and regulating cell metabolism at the level of transcription. A parallel model, taking into account the very different organization of chromosomal DNA, has since been proposed for eukaryotes (see REPEATED SEQUENCE). See also OPERON.

jelly fungi see TREMELLALES.

jorchette stem a horizontal stem with adventitious roots.

J-shaped curve see DENSITY INDEPENDENCE, EXPONENTIAL GROWTH.

jungle a colloquial term for a dense tropical rainforest or monsoon forest with many lianas and other climbers, palms, and bamboo scrub. It is a form of subclimax vegetation (see CLIMAX) typically found along river banks and fringing clearings where the extra light promotes vigorous growth.

junk DNA the part of DNA that does not code for proteins. Its function is not fully understood, but it is believed to be involved in controlling the timing and location of gene expression and protein synthesis. This appears to be achieved by the TRANSCRIPTION from this DNA of *microRNA* molecules, tiny nucleic acid molecules that bind to particular mRNA molecules either activating or inactivating/destroying them. during development, these microRNAs are transcribed at particular times and in specific locations in the plant. They may be potential tools for scientists to manipulate plant development.

Jurassic the middle period of the MESOZOIC era between about 195 and 136 million years ago. The climate was uniformly warm and humid and the flora was varied and abundant and dominated by the GYMNOSPERMS. See GEOLOGICAL TIME SCALE.

juvenility the condition expressed in immature organisms, which are generally smaller in size than the adults and lack reproductive capability. The seedling represents the juvenile stage of many plants. The term is particularly used when the young plant form is very different from that of the adult. For example in gorse (*Ulex*), the juvenile leaves are flat and trifoliate but the adult plant has spines. In many conifers the seedlings bear needle-like juvenile leaves, and specialized leaves, such as the scalelike leaves of cypress (*Cupressus*) and the dwarf shoots of pine (*Pinus*), are formed later. The change is not always from a simple to more complex form. For example, in ivy (*Hedera helix*) the juvenile leaves are lobed while the mature leaves are entire. The submerged leaves of many aquatic plants are similar to the juvenile leaves of terrestrial forms of the same species. In some varieties of ornamental plants the juvenile foliage persists in the adult plant (see also NEOTENY).

K k

Kampfzone see ELFIN FOREST.

karyogamy (caryogamy) the fusion of two nuclei during SEXUAL REPRODUCTION. Fusion normally occurs between haploid nuclei and follows immediately after PLASMOGAMY, resulting in the formation of a diploid zygote. In the higher fungi, especially the basidiomycetes, karyogamy may be delayed and occur separately from plasmogamy, resulting in a DIKARYON.

karyogram (idiogram) a drawing or photograph of the chromosomes present in an individual in which the chromosomes are arranged in homologous pairs in an agreed conventional sequence. See also KARYOTYPE.

karyokinesis the process of nuclear division that precedes cytoplasmic division or CYTOKINESIS. There are two main types of nuclear division: mitosis, in which daughter cells are exact replicas of the parent cell; and meiosis, in which daughter cells have half the number of chromosomes of their parent cell and in which the genetic material has been recombined.

karyoplasm see NUCLEUS.

karyotype the physical appearance of chromosomes from an individual or species as seen at METAPHASE of mitosis. Cells from the root meristem are commonly used, suitably stained, usually with acetic orcein or acetocarmine. The number, size, and shape of chromosomes in a set is usually highly characteristic of a species and often also constant within a genus. The karyotype is most conveniently studied by forming a KARYOGRAM.

keel (carina) 1 the two lower fused petals of a 'pea' flower, which form a boatlike structure around the stamens and styles. See also STANDARD, WING.
2 any ridgelike structure.

kelps see LAMINARIALES.

kennarten species (characteristic species) in PHYTOSOCIOLOGY, a collective term for species belonging to fidelity classes 3–5 (see FAITHFUL SPECIES), i.e. preferential, selective, and exclusive species.

keranga (heath forest) a kind of tropical FOREST that occurs on infertile and thin, peaty soils developed over siliceous rocks in Southeast Asia, the coastal regions of Gabon in central Africa, the Amazon basin, and Guyana. Kerangas are dominated by shrubs of the family Ericaceae (the heaths), and have a low, uniform, single-layered canopy with densely packed trees, and usually little ground flora. Many of the tree species have nitrogen-fixing bacteria housed in root nodules, e.g. *Gymnostoma nobile*, and there are many insectivorous plants and ant plants. Some kerangas have significant numbers of endemic species.

kerosene see PETROLEUM.

keto- prefix denoting that a molecule includes a ketone group – a carbon atom linked by a double bond to an oxygen atom and by single bonds to two other carbon atoms.

keto acid a CARBOXYLIC acid containing a ketone group, e.g. pyruvic acid.

α-ketoglutaric acid a five-carbon dicarboxylic keto acid with the formula $HOOCCO(CH_2)_2COOH$. α-ketoglutarate is an intermediate in the KREBS CYCLE. It is formed by the oxidation of isocitrate, with the concomitant reduction of NAD to NADH. In the next step of the Krebs cycle α-ketoglutarate undergoes oxidative decarboxylation to form succinyl CoA in a reaction similar to the formation of acetyl CoA from pyruvate.

α-ketoglutarate is also involved in amino acid metabolism (see TRANSAMINATION).

ketose any monosaccharide with the carbonyl (CO) group at a position other than on the terminal carbon atom, so forming a ketone group. The simplest ketose is the three-carbon dihydroxyacetone. Other ketoses include the five-carbon ribulose and xylulose and the six-carbon fructose and sorbose (see diagram). Compare ALDOSE.

dihydroxyacetone   ribulose

xylulose   fructose   sorbose

Ketose: some common ketoses

key 1 a list of characteristics drawn up to enable speedy identification of a specimen. Prominent contrasting features are given at each stage in the key so that by a process of elimination successively smaller groupings of organisms are split off until the specimen can be identified. Simple keys to flowering plants often rely solely on flower structure. For example, the first step in the key may be to form a number of groups on the basis of petal number. The second stage could be to subdivide these groups on the basis of flower colour and further subdivisions could then be made by petal shape, flower symmetry, etc. The groupings derived from such a key are artificial rather than natural and this method of identification has the disadvantage that if a mistake is made at any stage identification will be completely wrong.
2 see SAMARA.

kieselguhr see DIATOMACEOUS EARTH.

killing frost a frost that causes the freezing of aqueous solutions in plant cells, causing the cells to burst and leading to the death of the plant. Such a frost may not deposit ice crystals on external surfaces as does a hoar frost – it may be a 'black frost'. Different species of plants differ in their susceptibility to frost depending on the nature and concentration of solutes in the plant cells. See COLD TOLERANCE, GROWING SEASON.

kilobase a unit of length in nucleic acids. 1 kb = 1000 bases of base pairs.

kinase (phosphotransferase, phosphorylase) any TRANSFERASE enzyme that catalyses the transfer of phosphate to or from ATP or a related molecule. For example, hexokinase catalyses the phosphorylation of glucose and some other hexose sugars and is an important enzyme in the interconversion of hexose sugars, as most isomerases require a phosphorylated sugar as a substrate. Hexokinase is also one of three kinases in the glycolytic pathway (see GLYCOLYSIS).

kinesis a change in the rate of locomotion of a motile cell or organism in response to a particular stimulus. The direction of locomotion remains random and is not related to the direction of the stimulus. Compare TAXIS.

kinetin (6-furfurylaminopurine) a synthetic CYTOKININ. It was first isolated from DNA extracts of animal origin but has now also been synthesized commercially. When applied to certain leaves, it delays senescence and improves nutrient uptake.

kinetochore (spindle attachment) a point within a CENTROMERE consisting of paired brushlike filaments, 200–500 nm in length, lying parallel to each other. Kinetochores induce the formation of MICROTUBULES, approximately 50–60 from each, which join to the spindle fibres and so attach the chromosomes to the equator of the spindle during cell division. Formerly the term kinetochore was used synonymously with centromere.

kinetosome (basal body, blepharoplast) a cylindrical structure in the cytoplasm from which an undulipodium arises in motile eukaryotic cells. It is composed of a cylinder of nine microtubules linked by fine crossbridges. It transfers ATP and possibly other materials to the fibres in the shaft of the undulipodium. Striated fibrous extensions, termed 'rootlet fibres', may arise from the kinetosome and project backwards into the cytoplasm. These probably serve to anchor the undulipodium.

kingdom in most classification systems, the highest ranking category in the taxonomic hierarchy. The term was first coined in ancient times, when only the plant and animal kingdoms were recognized. Today, the most widely accepted classification system, the FIVE KINGDOMS CLASSIFICATION, recognizes five kingdoms, only four of which (the BACTERIA, FUNGI, PLANTAE, and Animalia) are true taxonomic groupings; the fifth, the PROTOCTISTA, represents unrelated organisms that do not fit into the other four kingdoms. Some taxonomists also recognize *superkingdoms*, the PROKARYA and EUKARYA, based on fundamentally differing cell structures. A few taxonomists suggest the higher grouping of *domain*, proposing three domains – the Eukarya and two bacterial (prokaryote) domains, the Archaea and Bacteria. Others refute this on the grounds that both the Archaea and the Bacteria share the feature of not having evolved as a result of symbiosis between different cell types.

kinin see CYTOKININ.

Kjeldahl method a method of determining the nitrogen content of organic material. It is used especially for calculating the protein content of human and animal food, and for analysing the content of nitrogen in water, effluent, fertilizers, and fossil fuels. There are three steps. First, the material is digested by boiling at 370–400°C with concentrated sulphuric acid, potassium sulphate, and a metal catalyst, such as copper, to break it down to simpler substances. Then sodium hydroxide is used to raise the pH. This converts the ammonium ions to ammonia, which can then be separated by distillation. The ammonia is absorbed into a solution of boric acid to form ammonium borate. Finally, the amount of nitrogen is determined by titration with hydrochloric acid in the presence of mixed indicator (bromocresol green and methyl red).

Klebsormidiales an order of green algae (CHLOROPHYTA) of the class CHAROPHYCEAE comprising solitary cells, clusters of cells, or unbranched filaments. Asexual reproduction may be by fragmentation or by transformation of the whole cell into a zoospore with two undulipodia.

klinostat (clinostat) an apparatus used to investigate plant growth in the absence of directional stimuli (TROPISMS). It consists of an electrically driven motor that slowly rotates a platform to which a seedling is attached in some manner. The rotation of the platform prevents any one side of the plant receiving an uninterrupted stimulus. Using a klinostat with a vertical rotating platform it can be demonstrated that a normally gravitropic (geotropic) root will grow horizontally when not subject to the influence of gravity.

Köhler illumination a form of BRIGHT-FIELD ILLUMINATION used with light microscopes, in which the light source is of irregular form or brightness and has an adjustable stop, similar to the aperture diaphragm on a camera. An image of the light source is focused on the condenser lens, which is then focused so that an image of the stop is seen in the plane of the specimen. The condenser is then the effective light source and the field is uniformly illuminated. Compare CRITICAL ILLUMINATION.

Kranz anatomy a type of organization of the photosynthetic tissues found in the leaves of c_4 PLANTS. In such plants the cells of the BUNDLE SHEATH are large and contain specialized elongated chloroplasts with no grana but capable of forming starch grains. The mesophyll chloroplasts are similar in structure to those of c_3 PLANTS but do not form starch grains. It is within the mesophyll chloroplasts that oxaloacetic acid is formed through the combination of carbon dioxide and phosphoenolpyruvate. This is then transported to the bundle sheath chloroplasts where the carbon dioxide is released and refixed by ribulose bisphosphate carboxylase.

Krebs cycle (tricarboxylic acid cycle, TCA cycle, citric acid cycle) a cyclic sequence of reactions, found almost universally in aerobic organisms, in which the acetyl portion of ACETYL COA is oxidized to carbon dioxide and hydrogen ions. The energy released by this oxidation is principally conserved by the concomitant reduction of NAD and FLAVOPROTEIN. The reduced cofactors eventually provide energy for ATP formation via the RESPIRATORY CHAIN and OXIDATIVE PHOSPHORYLATION.

Many different substrates can feed into the Krebs cycle. Pyruvate, from the breakdown of glucose, feeds into the cycle by an oxidative decarboxylation to form acetyl CoA; the enzyme is the pyruvate dehydrogenase complex.

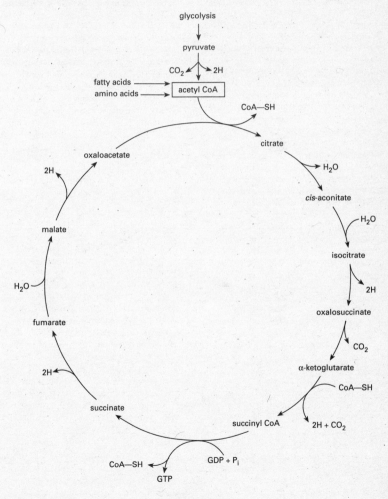

Krebs cycle

The Krebs cycle is not simply catabolic in function. It is in fact an amphibolic pathway, combining both anabolic and catabolic functions. The amino acids glutamate, aspartate, alanine, and glycine arise directly from the cycle, while many other amino acids, the hexose and pentose sugars, the purines and pyrimidines, and the porphyrin-based molecules are indirectly formed from Krebs cycle intermediates. Certain enzymatic reactions, termed *anaplerotic* reactions, produce Krebs cycle intermediates to replace those removed by synthetic pathways. An example is the formation of oxaloacetate by the carboxylation of pyruvate.

Krummholz see ELFIN FOREST.

k-selection see LIFE-HISTORY STRATEGY.

kymograph a slowly revolving drum, often covered with graph paper or smoked paper, on which a tracer records measurements of such activities as plant growth movements.

Ll

labelling the tagging of a compound with an ISOTOPIC TRACER so that its fate within an organism can be followed.

labellum (lip) 1 one of the three petals of an orchid flower. It differs in morphology and patterning from the two lateral petals and gives the flower its characteristic form. The laterals and the sepals may all be similar but the labellum is always distinct and often larger than the other perianth segments. It may serve as a platform for pollinating insects, which are attracted by its distinctive colours and markings. See also PSEUDOCOPULATION. *See illustration at* orchidaceae.
2 the platform formed by the lower petal or group of fused petals in various other lipped flowers, such as those of the Lamiaceae, Scrophulariaceae, and Fabaceae.

Labiatae see LAMIACEAE.

Laboulbeniomycetae see ASCOMYCOTA.

labriform lip-shaped.

Labyrinthulata (Labyrinthulomycota) a phylum of the PROTOCTISTA containing the slime nets and thraustochytrids. There are some 42 species in 13 genera and 2 orders. Some classification schemes place the Labyrinthulata in the kingdom CHROMISTA or regard them as a class (Labyrinthulomycetes) of the phylum MYXOMYCOTA. They are found in both salt and fresh water. These strange organisms form colonies of individual cells that live and move by gliding within a slime network that they secrete. Some colonies may be several centimetres long. The spindle-shaped cells can move only in the slime, which is produced by special organelles called BOTHROSOMES. Colonies may encyst

in unfavourable conditions. Some species undergo sexual reproduction in which motile (undulipodiate) isogametes are produced. The zygote germinates directly to produce a new slime net. The thraustochytrids were once considered to be fungi, but they produce cells with undulipodia; they are included in this phylum on account of their bothrosomes and extracellular matrix.

laciniate deeply divided into narrow lobes.

Lac operon the OPERON controlling the synthesis of the enzymes acetylase, β-galactosidase, and permease, which are involved in the metabolism of lactose in *Escherichia coli*. It was the subject of a detailed study by François Jacob and Jacques Monod in the late 1950s, which led them to propose the JACOB–MONOD MODEL of the control of gene activity. The lac operon consists of a cluster of adjacent structural genes coding for these three enzymes. These genes are transcribed as a unit, and their transcription is controlled by the action of a repressor molecule coded for by a separate regulator gene and an inducer (the substrate lactose or one of its derivatives).

lactate fermentation see GLYCOLYSIS.

lactic acid (2-hydroxypropanoic acid) a colourless, liquid 3-carbon carboxylic acid, $CH_3CH(OH)COOH$, formed from pyruvate during GLYCOLYSIS in animal cells, and also a major metabolic product of certain bacteria.

lactic acid bacteria a group of Gram-positive (see GRAM STAIN) anaerobic fermenting bacteria that are exploited for their ability to ferment sugars, especially lactose, to produce lactic acid (2-hydroxypropanoic acid), as well as acetate, formate, succinate, carbon dioxide, and ethanol. Such organisms (e.g. *Lactobacillus*, *Streptococcus lactis*) are used on a large scale in the food industry, especially as starter cultures in cheese making. Lactic acid bacteria are rod-shaped organisms belonging to the phylum Endospora that do not produce spores. See also LAC OPERON, JACOB–MONOD MODEL.

lactone an oganic molecule formed by the reaction of an –OH group on an alcohol with the –COOH group on a CARBOXYLIC ACID, forming a cyclic ester with the group –O.CO– as part of a ring in the molecule.

lacuna a space, gap, cavity, or depression, especially: **1** a cavity, usually air filled, between cells, i.e. an intercellular space. **2** a LEAF GAP.
3 a depression in the thallus of a lichen. Compare LUMEN.

lacunose having a surface with many indentations or cavities.

LAD see LAST APPERANCE DATUM.

laevigate smooth and shiny.

laevulose see FRUCTOSE.

laggard a chromosome that fails to get included in the daughter nuclei after ANAPHASE of nuclear division because its movement from the spindle equator to the poles is delayed or impeded.

lag phase the period, following the inoculation of a nutrient medium with microorganisms, in which there is much physiological activity but no increase in numbers. The rate of cell division accelerates when the organisms become adapted to the culture conditions and enters the LOGARITHMIC PHASE.

LAI see LEAF AREA INDEX.

lake forest a FOREST dominated by conifers, but in many ways transitional between boreal coniferous forest and temperate deciduous forest. Patches of lake forest survive in eastern North America.

Lamarckism the evolutionary theory put forward by Jean-Baptiste de Lamarck in 1809. He postulated that a characteristic that is acquired during the lifetime of an organism as a result of environmental pressures can be transmitted to the next generation. This is the theory of the inheritance of ACQUIRED CHARACTERISTICS

and Lamarck believed that new species could arise in this way. Erasmus Darwin also supported this view, but Lamarck was the first to give examples, one of which concerned the seeds of marsh plants. He suggested that if they reached high ground they became adapted to drier conditions by the development of new acquired characteristics, which could be inherited by the succeeding generation resulting in the development of a new species. The theory was later accepted by Darwin (see PANGENESIS) but challenged by Weismann. In the 20th century Lamarckism was almost entirely rejected in favour of Neo-Darwinism, but today it is enjoying something of a resurgence with the discovery of various poorly understood instances of inheritance of characteristics, especially of biochemistry and physiology, that appear to be related to the experience of the parent organism during its lifetime. See also WEISMANNISM.

lamella (*pl.* lamellae) a layer within the cytoplasm or within an organelle formed from a flattened membrane-bound vesicle or tubule. The vesicular cavity is therefore narrowed so that the lamella consists of two membranes lying close together. CHLOROPLASTS have a complex internal system of lamellae.

Lamiaceae (Labiatae) a large dicotyledonous family, commonly called the mint family, comprising some 6700 species in about 252 genera. Most species are shrubby or herbaceous. Labiates characteristically have stems that are square in cross section and simple leaves in opposite decussate (see LEAF) pairs. They are often covered in aromatic hairs and many species, such as mints (*Mentha*), sage (*Salvia officinalis*), and thymes (*Thymus*), are used as pot herbs. The flowers have five petals, which are usually fused into a tube that terminates in two distinct lips (exceptions being the genera *Mentha* and *Lycopus*). In many species, such as deadnettles (*Lamium*), the flowers appear as whorls (verticillasters) at each node. In others, such as woundworts (*Stachys*), the flowers are grouped in spikes. The fruit is a carcerulus or drupe. Many species are used in perfumes (e.g. lavender, *Lavandula*), and the species of the former Verbenaceae include important timber trees, such as teak (*Tectona*).

lamina (blade) **1** the usually flattened bladelike portion of a leaf, as distinct from the petiole and leaf base. The shape of the lamina and the nature of its margin are important taxonomic characters (*see illustrations at* leaf). Leaf laminas are usually the main photosynthetic organs and are structured accordingly (see PALISADE MESOPHYLL, SPONGY MESOPHYLL). They are placed so as to make the best use of incoming light with the minimum amount of either overlapping or wasted space. This is particularly evident in a tree canopy. The pattern so achieved is termed the *leaf mosaic* and is mainly a result of PHYLLOTAXIS. **2** any thin flat organ, such as a petal or the thallus of many macroscopic algae.

Laminariales (kelps) an order of brown algae (Phaeophyta) that contains the largest known algae, *Macrocystis* and *Nereocystis*. Most species occur below the low-tide mark. The sporophyte thallus is parenchymatous and shows complex differentiation into a holdfast, stipe, and lamina. Growth takes place in a special meristem region (*meristoderm*) on the surface of thallus, a feature characteristic of the Laminariales. The gametophyte generation is a much smaller, relatively simple, filament and shows well-developed OOGAMY. The gametophyte is capable of limited independent existence, supporting itself by photosynthesis. The sporophyte thallus produces zoospores in sporangia on the surface of the lamina. These germinate to produce the gametophyte thalli, which are either male or female, bearing antheridia or oogonia respectively. After the egg is extruded from the oogonium, it remains on the surface, releasing pheromones to attract sperm. A fertilized ovum develops into a young sporophyte, which remains on the gametophyte and eventually grows over it. Certain species of kelp produce ALGINIC ACID. See also ALTERMNATION OF GENERATIONS.

laminarin the main assimilation product of brown algae (Phaeophyta). It is composed chiefly of glucose units though mannitol may also be present.

laminate placentation a form of PLACENTATION in which the placentae arise from all over the inner surface of the ovary. It is seen in the flowering rush (*Butomus umbellatus*) and the white water lily (*Nymphaea alba*).

lanceolate narrow and tapering at both ends. *See illustration at* leaf.

land bridge a piece of land connecting two large land masses, which allows plants and animals to migrate between the land masses. Land bridges often come and go as sea level changes. For example, the land bridge where the Bering Sea now lies is thought to have been the route by which humans first colonized North America from Siberia. The land bridge at Panama, which was submerged for millions of years, so that North and South American species evolved separately in isolation, became land when the sea level fell in the Pliocene (see TERTIARY), and mixing of species began again.

landnam a characteristic stratigraphic horizon in which the amount of tree pollen compared to the amount of pollen of other plant species is much reduced. It occurs at different times in different locations, and is sometimes accompanied by a layer of charcoal. For example, in the Neolithic, the non-tree pollen shows an increase in species of arable weeds and cereal crops, suggesting that the landnam represents the start of widespread forest clearance by farmers. Across northern Europe, the landnam is dated to about BP 5000. See POLLEN ANALYSIS.

landrace an ancient or primitive cultivar of a crop plant. Landraces are often genetically very heterogeneous and contain numerous alleles that contributed to the survival of the organism under natural conditions. Since intensive plant breeding can result in the loss of these alleles, landraces are a source from which plant breeders can selectively reintroduce them into highly bred cultivars.

landscape architecture the design and construction of artificial outdoor landscapes such as gardens and parks to enhance the urban environment.

lanose woolly.

last appearance datum (LAD) the last

recorded occurrence of a particular taxon in the GEOLOGICAL TIME SCALE.

late blight see BLIGHT.

late glacial towards the end of the last ice age, the period between the initial rise in global temperature following the first minimum of the DEVENSIAN glaciation and the very rapid rise in temperature taken to mark the start of the post-glacial FLANDRIAN period.

latent infection an infection in which no symptoms are visible during the first phase of the infection. Growth of the pathogen stops soon after penetration but resumes at some later stage. For example, banana *anthracnose infection* (caused by the fungus *Colletotrichum musae*) originates in the field on unripe fruit but after penetration the fungus becomes latent as a subcuticular hypha for up to five months. The fungus resumes activity as the fruit nears maturity and typical black lesions develop on the ripe fruit. The period between infection and the production of spores by the pathogen is the *latent period*. See also INCUBATION PERIOD.

latent period see LATENT INFECTION.

lateral meristem a meristem arranged parallel to the sides of the organ in which it occurs and responsible for increase in girth usually by formation of secondary tissues (see SECONDARY GROWTH). The CAMBIA are lateral meristems. Occasionally the term is used of axillary meristems. Compare APICAL MERISTEM. See also PRIMARY THICKENING MERISTEM.

lateral root any ROOT that originates endogenously from the PERICYCLE of another root.

lateral shoot a shoot that develops from a vegetative bud in the AXIL of a leaf or from a NODE of a stem.

laterite (*adj.* lateritic) a hard crust that may develop on the soil surface in tropical regions with alternating wet and dry seasons. In the wet season the soluble mineral salts are washed down into the lower horizons (see SOIL PROFILE). In the dry season the soil solution moves back to the surface by capillarity and the aluminium and iron oxides accumulate and combine together to form the crust. Laterization results in soil called a *latosol*.

late wood (summer wood) the outer portion of a GROWTH RING, which is produced in summer when growth is relatively slow. It is denser wood than early wood, with vessels of smaller diameter. Compare EARLY WOOD.

latex a fluid produced by many higher plants and by certain agaric fungi (see AGARICALES), e.g. the milk caps (*Lactarius*). It is often white, but may be colourless, reddish, or yellowish. In the fungi the latex is produced in a latex duct consisting of anastomosing hyphae. In green plants latex is stored in laticifers. The fluid contains various substances either in solution or suspension, e.g. alkaloids, starch grains, sugars, mineral salts, etc. In some species there is a high concentration of rubber (caoutchouc) though rubber is limited in occurrence to the dicotyledons. Commercial rubber production utilizes the latex of Brazilian rubber trees (*Hevea brasiliensis*) and, on a smaller scale, Indian rubber trees (*Ficus elastica*). The latex of other species, e.g. *Palaquium gutta*, gives, on coagulation, gutta-percha. The formation of caoutchouc and gutta-percha appears to be mutually exclusive with no plant yet being found to produce both. Chicle, from the latex of *Achras zapota*, and balata, from members of the Sapotaceae, are other important latex products.

latex tube see LATICIFER.

laticifer (*adj.* laticiferous) a cell or a linked complex of cells containing a milky liquid called LATEX and penetrating various tissues in certain plants. A laticifer consisting of a single cell is referred to as *simple*, whereas a joined complex is termed *compound*. Laticifers may be further classified as *articulated* (branched) or *nonarticulated* (unbranched). Sometimes the end walls between the elements of a compound laticifer break down forming a continuous tube, sometimes referred to as a *laticiferous vessel*. A laticifer enclosed by a ray is termed a *latex tube*. Laticifers are present in many plants, such as the rubber tree, *Hevea brasiliensis*, rubber being the latex secreted by the laticifers and tapped by cutting through the bark.

latifoliate broad-leaved.

Latin square an experimental design in which the number of treatments is the same as the number of replications, and each treatment occurs once in every column and row. It is analogous to a RANDOMIZED BLOCK design in two directions. It is often used in fertilizer field trials and has the advantage of eliminating from the total variation environmental differences, such as soil fertility, that exist across and down the square experimental plot. The Latin square generally yields more useful information than would a RANDOMIZED BLOCK design of similar size. However it has the limitation that with a large number of treatments there must consequently be a large number of replicates and beyond a certain point the labour involved is not worth the information obtained. Also, with small squares, the method is insensitive because the number of degrees of freedom is low. This can be overcome by replicating the squares.

latosol see LATERITE.

Laurasia see CONTINENTAL DRIFT.

Law of Independent Assortment see INDEPENDENT ASSORTMENT, MENDEL'S LAWS.

Law of Segregation see MENDEL'S LAWS.

layering a method of plant propagation involving the pegging down of runners and stolons to the soil surface. Adventitious roots develop where a node touches the soil and a shoot develops from the lateral meristem. New daughter plants eventually establish. Carnations are commonly propagated in this way. See also AIR LAYERING.

LDP see LONG-DAY PLANT.

leachate see LEACHING.

leaching the washing out of soluble substances from the upper layers of the soil by water passing down the SOIL PROFILE. The substances (*leachate*) are either deposited lower down in the B horizon or removed completely. It takes place when the amount of rainfall exceeds the amount of water lost by surface evaporation. It may result in podsolization (see PODSOL) or the development of an impermeable layer of mineral salts at some point in the soil profile. Compare FLUSHING. See also ELUVIATION.

leader the tip of the main stem of a plant. The form of the leader (whether erect or drooping) can aid identification of certain conifers. For example, the western red cedar, *Thuja plicata*, has an upright leader, which distinguishes it from the similar Lawson cypress, *Chamaecyparis lawsonia*, which has a drooping leader. The term may also be used of the terminal segment of any main branch in distinction to any lateral branches arising from it. In those conifers in which the leaves are borne in clusters on short side shoots, e.g. cedars (*Cedrus*) and larches (*Larix*), the side shoots are termed spurs while the main shoot is the leader.

leading dominant in the WISCONSIN SCHOOL, the species with the highest importance value in a particular stand or quadrat.

leaf the main photosynthetic organ of most green plants, consisting of a lateral outgrowth from a stem and comprising LAMINA, PETIOLE, and leaf base. The shape of the lamina and nature of its margin are important taxonomic characters (see illustrations on next page). There are often a great number of leaves on any one plant, although these may be lost in the colder or drier months in deciduous plants. A leaf typically consists of conducting tissues and photosynthetic cells (the MESOPHYLL) often differentiated into PALISADE and SPONGY MESOPHYLL, surrounded by epidermis. The epidermis is perforated by stomata, usually more numerous on the abaxial (lower) side of the leaf (see STOMA). The epidermis is usually covered by a waxy CUTICLE. This prevents excessive water loss by transpiration. Undivided leaves are termed *simple leaves*, while leaves that are divided into series of *leaflets* are termed *compound leaves*. In many plants leaves may be reduced or even absent, as in many XEROPHYTES. Sepals, petals, and bracts are considered to be modified leaves and many believe the stamens and carpels are also derived from leaves. See also MICROPHYLL, MEGAPHYLL.

leaf area index (LAI) the ratio of the total surface area of a plant's leaves to the ground area available to that plant, i.e. LAI = leaf area/ground area. The LAI is of value when considering the number of plants that can be successfully cultivated on a given area of land. See also NET ASSIMILATION RATE.

leaf area ratio a value obtained by dividing the total leaf area of a plant by its dry weight. The ratio is useful in relating total photosynthetic to total respiratory material within the plant, thereby giving information concerning the plant's available energy balance.

leaf buttress a lateral prominence on the shoot apex, destined to differentiate into a leaf. It is the earliest stage in the development of a leaf PRIMORDIUM, and later forms the leaf base, the remainder of the leaf growing upwards and outwards from the buttress.

leaf culture a form of TISSUE CULTURE in which excised leaves, leaf material, or leaf primordia are grown on a sterile growth medium. Mature leaves can be kept healthy under culture conditions for considerable periods. Leaf PRIMORDIA have been used to study growth and DIFFERENTIATION processes. Experiments with the cinnamon fern (*Osmunda cinnamonea*) show that the smallest primordia usually develop into shoots when cultured but, as the size of the excised primordia increases, there is an increasing tendency for them to develop into leaves. This indicates that leaf primordia are not irrevocably committed to becoming leaves until a relatively late stage.

leaf curl a plant disease in which an increase in cells on either side of the midrib and extra growth of the palisade and spongy mesophyll cause curling and puckering of the leaves. In peach and almond this is

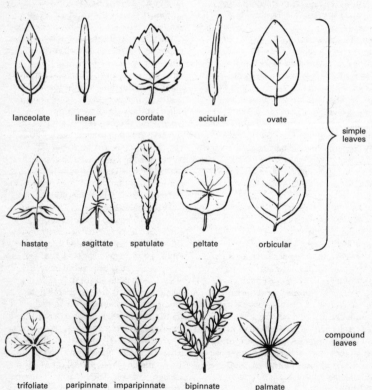

lanceolate linear cordate acicular ovate

simple leaves

hastate sagittate spatulate peltate orbicular

trifoliate paripinnate imparipinnate bipinnate palmate

compound leaves

Leaf: some common leaf shapes

caused by the fungus *Taphrina deformans* and in tobacco and cotton by tobacco and cotton leaf curl viruses respectively. These viruses are transmitted by whiteflies (*Bemisia tabaci*).

leaf fall the shedding of leaves as a result of the formation of a zone of ABSCISSION at the base of the petiole. Leaves on dead branches do not fall as no such zone is produced. In deciduous plants there is a continual shedding of older leaves throughout the growing season but with the onset of the winter or dry season there is a conspicuous shedding of all the remaining leaves. In evergreens leaves are shed and replaced continually (though often with a season of greatest activity) and there is no period when the plant is devoid of foliage. Leaf fall is associated with a drop in the AUXIN level of the lamina.

leaf gap (lacuna) a parenchymatous area in the STELE of many vascular plants, associated with and positioned immediately above a LEAF TRACE. Leaf gaps are characteristic of angiosperms but are also present in some gymnosperms and ferns.

leaflet see LEAF.

leaf mosaic see LAMINA.

leaf scar the scar left on a plant stem after ABSCISSION of a leaf. It takes the shape of a cross-section of the PETIOLE where present. The major veins often leave distinctive scars within the leaf scar.

leaf sheath the lower part of a leaf of some plants that encircles the stem, arising at the NODE and encircling the INTERNODE above, often overlapping where the two margins meet, e.g. the leaves of grasses (see POACEAE). In some species successive leaf sheaths overlap, conferring considerable strength on the stem. This is particularly important in grasses, which have meristems at the nodes that consititute structurally weak points.

leaf spot a plant disease in which the principal symptom is limited areas of NECROSIS on the leaves. There are numerous causes of leaf spots – mineral imbalance, insects, weather conditions, viruses, bacteria, and fungi. The leaf spots may be minor symptoms or a particular phase of a disease that develops other, more characteristic, symptoms. Bacteria and fungi are the usual causal agents of the diseases in which leaf spots are the main symptom. Angular leaf spot of cucumber is

entire

dentate

serrate

crenate

lobed

sinuate

undulate

spinose

ciliate

Leaf: examples of leaf margins

caused by the bacterium *Pseudomonas lacrymans*. Fungal leaf spots affect most crop plants and common pathogenic genera are *Septoria, Botryodiplodia, Colletotrichum, Gloeosporium, Cercospora, Alternaria*, and *Helminthosporium*.

leaf trace a vascular strand leading from the STELE to the leaf. See also LEAF GAP.

Lecanorales an order of the ASCOMYCOTA containing the majority of LICHENS (8000–10 000 species in some 347 genera and 40 families) and a few saprobic genera. The lichens bear apothecia on their upper surface from which ascospores are violently discharged. The asci have a single thick wall, usually with a thick caplike tip. The Lecanorales includes such genera as *Cladonia, Lecanora, Parmelia, Umbilicaria*, and *Usnea*. The fungal partners of these lichens are closely related to the LEOTIALES. In certain lichens, e.g. *Sphaerocarpus* species, the asci and paraphyses of the apothecium disintegrate at maturity to form a mass of spores and hymenial tissues. Such lichens are more usually placed in a separate order, the Caliciales. In other lichens the apothecium may be locular and elongate and contain pseudoparaphyses rather than true paraphyses. Such lichens are more usually distributed between a number of orders, including the unitunicate Ostropales and Rhytismatales and the bitunicate Arthoniales, Dothideales, and Patellariales.

lecanorine describing a lichen ASCOMA in which the margin cannot be distinguished from the rest of the thallus. The margin is said to be thalline and consists of both algal and fungal cells. Lecanorine ascomata are seen in *Lecanora*. Compare LECIDEINE.

lecideine describing a lichen ASCOMA having a thin margin, termed a proper margin, composed solely of fungal cells, as in *Lecidia*. Most lichen ascomata do have a proper margin but is is often hidden by the thalline margin (see LECANORINE) when this is present.

lecithin (phosphatidyl choline) a PHOSPHOLIPID widely occurring in plants and animals, especially in CELL MEMBRANES.

lectotype a specimen or other element (description, illustration, etc.) subsequently selected from the original material on which the name of a TAXON was based. Very often these original elements will be SYNTYPES. A lectotype is only necessary when the original author failed to designate a HOLOTYPE. See also TYPE.

leghaemoglobin a protein found in the centre of root nodules of leguminous plants infected with the nitrogen-fixing bacterium *Rhizobium* (see NITROGEN FIXATION). It has been found that the haemoglobin is coded for by a legume gene but that synthesis occurs only in the presence of the bacterium. The haemoglobin is believed to transport oxygen to the bacterium (which respires aerobically) in such a way that the activity of the nitrogen-fixing enzyme DINITROGENASE (which is destroyed on exposure to oxygen), is not affected. Haemoglobin is not found elsewhere in the plant kingdom.

legume (*adj.* leguminous) **1** (pod) a dry dehiscent fruit containing one or more seeds. It develops from a single carpel, which on ripening splits along the ventral and dorsal sutures to form two valves, each bearing seeds alternately on the ventral margin. Dehiscence is due to differential drying of the carpel wall, which in some species may result in explosive release of the seeds. The valves may also twist during dehydration, dislodging any remaining seeds. This type of fruit is typical of the Fabaceae (Leguminosae) but may also be found in other families. See also LOMENTUM. **2** any member of the Fabaceae.

Leguminosae see FABACEAE.

lemma (flowering glume) the lower of a pair of BRACTS beneath each floret (flower) in the inflorescence of a grass. The lemma is usually membranous to coriaceous, whereas the other bract, the PALEA (if present), is usually thinner and more delicate. Compare GLUME.

lenitic describing a freshwater ecosystem in which there is not a continuous flow of water, such as a lake or pond. Compare LOTIC.

lenticel a small elliptical pore containing loosely packed cells that is the means of gaseous exchange in the PERIDERM of plant axes. Lenticels are analogous to the stomata of primary tissues and vary in size from being almost microscopic to about 1 cm in length. Compare STOMA.

lenticular lens-shaped.

Leotiales (Helotiales, cup fungi) an order of the ASCOMYCOTA containing some 2036 species in about 392 genera and 13 families. Its members are similar to the PEZIZALES except that the asci lack an operculum. The ascomata are usually small and often brightly coloured. Most members of the Leotiales are saprobes, but some are plant parasites, e.g. *Sclerotina*, which causes brown rot of stored fruit. A few are lichenicolous.

Lepidodendrales an extinct order of the LYCOPHYTA containing arborescent forms, such as *Lepidodendron*, that flourished during the late Palaeozoic era. Species of *Lepidodendron* reached heights of up to 30 m. The trunks, some of which had a girth of over 3 m at their base, were unbranched for the main part but divided dichotomously at the top into numerous branches. The trunk was patterned with diamond-shaped leaf scars and from its base arose four dichotomously branching axes that gave rise to the root system. Strobili, similar (except in size) to those of *Selaginella*, were borne on the ends of the branches and it appears that the method of fertilization was basically similar to that of *Selaginella*.

lepidoid scaly.

lepidote describing a surface with tiny scales.

leprose 1 rough to the touch. 2 scaly. 3 describing a lichen thallus that is not organized into cortex, medulla, or PHYCOBIONT layers, and which produces powdery SOREDIA over its entire surface.

leptoid an elongated nutrient-conducting cell in the stems of *Polytrichum* (hair mosses) and related bryophytes, analogous to a sieve cell in vascular plants. Compare HYDROID.

leptokurtic describing a distribution of values that is similar in overall shape to a normal distribution, but which has a much higher central peak with a more flattened curve on either side. See BINOMIAL DISTRIBUTION.

leptosporangium a type of SPORANGIUM that is derived from one initial cell. It is typical of ferns of the order FILICALES (the

leptosporangiate ferns). The wall of a leptosporangium is usually only one cell thick and does not contain tissue derived from the archesporium. Its characteristic feature is that as the wall cells dry out, they flick the spores away explosively. Compare EUSPORANGIUM.

leptotene see PROPHASE.

lesion a visible area of diseased tissue on an infected plant.

lethal gene a mutant allele that results in the death of an organism. The mutation commonly takes the form of a small chromosomal deletion, so that the genetic code for a key protein is missing or disrupted. Such mutations do not necessarily take effect at fertilization but may manifest themselves at any stage of development. The chlorophyll-less mutant (chl$^-$) of barley, which is typically recessive, shows its lethal effects only after the food reserves in the germinating seed have been used up.

leucine a nonpolar branched chain amino acid with the formula

$$(CH_3)_2CHCH_2CH(NH_2)COOH$$

(*see illustration at* amino acid). It is synthesized from pyruvate and is degraded by a complex pathway leading eventually to acetoacetate. Many of the enzymes involved in the synthesis and degradation of leucine are the same as those involved in the metabolism of isoleucine and valine.

leucoplast (leukoplast) a colourless PLASTID found in the cells of roots and underground stems and storage organs. Leucoplasts have rudimentary LAMELLAE.

leucosin see CHRYSOLAMINARIN.

leukoplast see LEUCOPLAST.

liana (liane) a long-stemmed woody climbing plant that grows from ground level to the canopy of trees. Lianas abound in tropical forest and individual stems can be as long as 70 m. The plants climb up and over the tops of very tall trees and benefit from full sunlight at maturity. They may bind the trees together so if one dies it is held in position until it decays. They are rarer in most temperate regions, but a typical British example is old-man's beard (*Clematis vitalba*).

lichen a distinct type of organism in which the thallus is composed of both fungal and either algal or bacterial cells in symbiotic association (see SYMBIOSIS). The fungal partner (the *mycobiont*) is usually an ascomycete and is dominant to the photosynthetic partner (the *phycobiont*), which is usually a chlorophyte (green alga) or a blue-green bacterium (see CYANOBACTERIA). Occasionally the fungus is a basidiomycete or an anamorphic fungus. Formerly classified as plants (see LICHENES) and later assigned to an artificial phylum of fungi (see MYCOPHYCOPHYTA), they are now variously classified according to the nature of the fungal partner (most are placed in the ASCOMYCOTA).

The fungus forms the main part of the thallus, which is usually a stratified structure consisting of an upper and lower *cortex* of compact fungal tissue with a *medulla* in between of loosely woven hyphae. The phycobiont cells are in a layer between the medulla and upper cortex and are closely surrounded by hyphae. This stratified type of structure is termed *heteromerous*. In some lichens, e.g. gelatinous lichens, the thallus consists of loosely woven hyphae and algal cells scattered in a jelly-like matrix. This unstratified structure is termed *homoiomerous*. The fungus obtains carbohydrates from the photosynthetic partner, while the latter receives water and nutrients from the fungus. The phycobiont is also protected from desiccation by the surrounding fungal body. Lichen fungi are not found in the free-living state and can be cultured independently only with difficulty. Algae very similar to those in lichens are found growing independently though it is uncertain if these are identical to the lichen algae.

Three main types of lichen are recognized by their growth habit. *Crustose* lichens grow closely attached to the substrate and usually lack distinct lobes. *Foliose* lichens are generally attached loosely to the substrate by rhizinae and the thallus has lobed leaflike extensions. *Fruticose* lichens are either erect and bushy or hanging and tassel-like, and are only attached at one point. Some lichens are intermediate in form. For example, *Cladonia* species initially form a basal crust from

which arise erect branching structures (podetia).

Reproduction may be by dispersal of fragments of the thallus containing both fungal and algal cells, by special vegetative reproductive bodies (see SOREDIUM, ISIDIUM), or by fungal spores. The nature of the asci and ascospores produced by the fungal component are important in identification.

Some lichens are extremely tolerant of almost total desiccation and can colonize exposed bare areas where plants are unable to survive. They are important agents of rock disintegration. These are however very slow growing, some crustose lichens possibly only extending by 1 mm a year. Some are believed to live for up to 4000 years. They have few commercial uses though some yield dyes, while others contain antibiotics, e.g. usnic acid from *Usnea*. Some of the larger species of arctic zones, e.g. reindeer moss (*Cladonia rangiferina*), are important as food for deer. The distribution of certain lichen species i used as an indicator of atmospheric pollution (see INDICATOR SPECIES).

Lichenes an obsolete division of the Plantae containing the LICHENS. It was created before it was realized that lichens are symbiotic associations between fungi and algae. Lichens are now considered to be fungi especially adapted to obtain food from algae living within their tissues. They are consequently classified in the appropriate rank of the Eumycota, according to the nature of the fungus.

lichenicolous fungi fungi that grow on or in lichens. They may be parasites, commensals, or saprobes. Obligate lichenicolous fungi include some 1000 species in about 300 genera, ranging from Ascomycota and Basidiomycota to Fungi Anamorphici.

lichenin a storage glucan found in LICHENS, comprising 80–200 β-glucose units.

lichen woodland open woodland with sparsely distributed conifers and a ground layer dominated by lichens. It occurs in the boreal zone between the forest/tundra ecotone and the closed-canopy coniferous forest.

lichen zone one of a series of 10 zones

characterized by the presence/absence and differing proportions of particular lichen species, which provide an indication of the level and type of air pollution in an area. The standard zones in England and Wales range from zone 0 (complete absence of lichens) in highly polluted areas to zone 10, with many species, including bearded lichens such as *Usnea*, found in the absence of such air pollutants as sulphur dioxide.

Liebig's law of the minimum the concept, proposed by Justus von Liebig in 1840, that the health and growth rate of a plant and the size it attains depend on the availability of the scarcest of its essential nutrients. This has developed today into the concept of LIMITING factors. See ESSENTIAL ELEMENT.

life cycle the series of events from the production of gametes in one generation to the same stage in the subsequent generation (see diagram overleaf). In some plants the life cycle only involves the production of one type of individual (see HAPLOBIONTIC). In most however, the life cycle encompasses a haploid and a diploid generation, i.e. there is an ALTERNATION OF GENERATIONS.

life form the overall MORPHOLOGY of an organism, on the basis of which a species may be described as a tree, shrub, herb, succulent, etc. In describing an area of vegetation much information is conveyed by naming the type or types of life form that predominate. The RAUNKIAER SYSTEM OF CLASSIFICATION is often used for this purpose.

life-history strategy the strategy of growth, reproduction, and dispersal that is selected for in a particular competitive situation or environment. There are two main strategies. *k-Selection* occurs in stable environmental situations, and is selection for maximum competitive ability (see EQUILIBRIUM SPECIES), with fewer offspring but high survival rates and prolonged development to maturity. Populations of such species typically have an S-shaped growth curve. *R-selection* occurs in species in rapidly changing environments (see FUGITIVE SPECIES), newly exposed habitats in the early stages of succession, unstable environments such as deserts, or temporary

habitats such as ponds. It favours rapid rates of population increase under favourable conditions, efficient methods of long-distance dispersal, and short life cycles. See also BIOTIC POTENTIAL.

life zone a zone in which the flora and fauna are typical of a particular latitude or range of altitude. They are related to a range of interacting environmental gradients. On a world scale, life zones are equivalent to BIOMES. The term is more commonly used in relation to more local situations, such as the zones of communities on a mountainside.

ligase (synthetase) any enzyme that catalyses reactions involving bond formation using energy derived from the simultaneous hydrolysis of a nucleotide triphosphate. with concomitant cleavage of ATP. Examples include glutamine synthetase, which catalyses the formation of glutamine from glutamic acid, and DNA ligase, which closes single-strand breaks in DNA.

light (visible radiation) electromagnetic radiation with wavelengths ranging from roughly 400 nm (extreme violet) to 770 nm (extreme red). Light from the sun provides the energy to fuel the photosynthetic FIXATION of carbon dioxide by green plants and is thus the basis of life on earth. See also ACTION SPECTRUM, RED LIGHT, FAR-RED LIGHT.

light-and-dark-bottle technique see OXYGEN METHOD.

light-dependent reactions see LIGHT REACTIONS.

light green a light microscope stain that shows up cytoplasm and cellulose. It is often used as a counterstain with safranin to show cytoplasm.

light-independent reactions see DARK REACTIONS.

light microscope (optical microscope) an optical instrument that contains one or more lenses and is used in the laboratory to enlarge objects that are too small to be examined in detail by the naked eye. Its maximum resolution (the capacity to observe fine detail clearly), is about 0.3 μm (i.e. it can distinguish points only 0.3 μm apart) as compared to about 80 μm for the average human eye (see RESOLVING POWER). Light is passed through a condenser, which

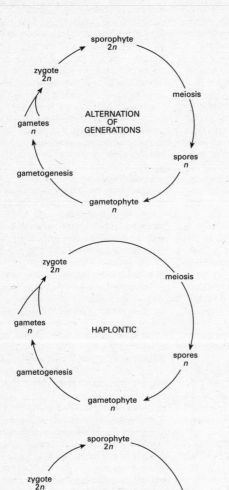

the type of life cycle
found in most plants,
in which there is a
definite development,
however reduced, of both
sporophyte and
gametophyte generations

a form of life cycle
found in certain algae,
e.g. the Ulotrichales,
in which there is no
sporophyte generation,
except as the zygote

a form of life cycle
found in certain algae,
e.g. the Fucales, in
which there is no
gametophyte generation.
This cycle is also
typical of animals

Types of **life cycle** found in plants and algae

converges the light rays onto the specimen, and then through one or more lenses. The earliest instruments were *simple microscopes* with a single lens (i.e. basically magnifying glasses). Such instruments are little used today as they have to be positioned close to the eye and have a limited field of vision. Also the illumination and mounting of the specimen is very tricky. The *compound microscope* has two lenses, one at either end of the body tube. The objective lens is at the base of the tube close to the specimen and the eyepiece or *ocular* at the other end close to the eye. The image is focused either by moving the body tube or by moving the microscope stage to which the specimen is secured. Most modern compound microscopes have several objective lenses of different magnifications, fitted to a revolving nosepiece. Some microscopes have a built-in light source while others are fitted with a mirror to reflect light from a lamp. The adjustment of an iris diaphragm reduces glare by limiting the part of the specimen that is illuminated. The modern compound microscope is used extensively for observing fine detail of microorganisms and thin sections through tissues and organs. Sections are often stained to increase contrast. The *stereoscopic binocular microscope* is used for work requiring lower magnifications, such as dissection, or for viewing detail of comparatively large specimens. It has two eyepieces that give a three-dimensional image, and the specimen is usually placed on a contrasting background. Compare ELECTRON MICROSCOPE. See also BRIGHT-FIELD ILLUMINATION, DARK-GROUND ILLUMINATION, PHASE CONTRAST MICROSCOPE, INTERFERENCE MICROSCOPE, CONFOCAL MICROSCOPY.

light reactions (light-dependent reactions) the sequence of photosynthetic reactions that, taking place on the THYLAKOID membranes of chloroplasts, produce the ATP and NADPH used in the carbon dioxide FIXATION of the subsequent DARK (light-dependent) reactions. Incident light is absorbed by pigments, distributed between two separate photosystems, I and II. The energy absorbed causes excitation of chlorophyll *a* molecules, which emit high-energy electrons that are passed down an ELECTRON TRANSPORT CHAIN, their energy being used to produce ATP and the reduced form of NADP+. Unlike respiration, the transfer of electrons in photosynthesis is from less electronegative molecules to more electronegative molecules. The energy needed to achieve this reversal of normal electron flow is provided by light. See PHOTOSYNTHESIS, PHOTOSYSTEMS I AND II.

light saturation point see COMPENSATION POINT.

ligneous woody.

lignicolous describing organisms that live on or in wood, such as bracket fungi.

lignification see LIGNIN.

lignin (*adj.* lignified) a complex carbohydrate polymer making up about 25% of the wood of trees and also found in the cell walls of sclerenchyma tissues and vessels, fibres, and tracheids at maturity. It increases the strength of such tissues making them more resistant to compression and tension. Lignin is formed by condensation of the phenolic compound coniferyl alcohol. Its distribution in tissues can be shown by staining with acidified phloroglucin, which turns lignin red. Th process of impregnation of cell walls with lignin is called *lignification*.

Lignosae in John Hutchinson's system of classification, a subgroup of the dicotyledons into which the predominantly woody families are placed. The predominantly herbaceous plants are placed in the Herbaceae. The division is considered unnatural by most taxonomists as it separates certain families generally thought of as closely related.

ligulate strap- or tongue-shaped as, for example, the outer ray florets of the inflorescences of plants in the family Asteraceae.

ligule (ligula) 1 an outgrowth from the top of the leaf sheath in grasses. It may be a scalelike flap of tissue or a ring of hairs and is often of great diagnostic value. Compare AURICLE. 2 the strap shaped elongation of the corolla tube in certain florets of the Asteraceae. The five teeth of the ligule represent the tips of five fused petals. Genera of the tribe Lactuceae, e.g. *Taraxacum*, *Hieracium*, have capitula consisting solely of ligulate florets. **3** An outgrowth from the upper side of the MICROPHYLL in species of *Selaginella*. The SPOROPHYLLS in the strobuli also possess ligules.

Liliaceae a large monocotyledonous family, commonly known as the lily family, containing about 4950 species in some 288 genera. Many of the Liliaceae possess swollen underground PERENNATING ORGANS, such as bulbs, corms, and rhizomes. The leaves are usually narrow and parallel-veined, and the flowers usually have two whorls of petals. The fruit is usually a capsule, and the seeds contain copious endosperm. Many species, including lilies (*Lilium* spp) and tulips (*Tulipa* spp), are important horticulturally for their showy flowers, which display a wide range of form and colour. Recent taxonomic preference is to separate species with the different forms of flower (and sometimes also of leaf) into different families, such as the Agavaceae (agaves), Alliaceae (onions), Amaryllidaceae, Convallariaceae (lily-of-the-valley), Hyacinthacaeae (hyacinths), and Trilliaceae (trilliums). These are now placed in the order Asparagales rather than Liliales (see LILIIDAE).

Liliidae a subclass of the monocotyledons containing mainly herbaceous plants, often with large flowers. It may contain, depending on the classification scheme referred to, up to six orders: the Haemodorales, with six families including the Pontederiaceae (free-floating hydrophytes, e.g. water hyacinth and pickerelweed), Haemodoraceae (kangaroo-paws), Taccaceae, and Pentastemonaceae; Asparagales, with 23 families, including the Agavaceae (agaves, hostas), Alliaceae (onions), Amaryllidaceae (daffodils, narcissi, snowdrops), Asparagaceae (asparagus, butcher's broom, lianas), Asphodelaceae (asphodels, aloes, red-hot pokers), Hyacinthaceae (hyacinths, grape hyacinths), Trilliaceae (trilliums, herb Paris), and Xanthorrhoeaceae (grass trees, black boys); Dioscoreales, with two families, the Smilacaceae (bamboo vines, sarsaparilla, lianas) and Dioscoreaceae (yams); Velloziales, with just the Velloziaceae; Liliales, with eight families, including the

Alstroemeriaceae (alstroemerias), Iridaceae (irises, gladioli, freesias), and LILIACEAE (lilies, tulips); and Orchidales, with the two families Corsiaceae (Burmanniaceae) and ORCHIDACEAE. In some classifications, the Iridaceae are placed in their own order, the Iridales. The Zingiberales and Triuridales are sometimes placed in the Liliidae.

Liliopsida see MONOCOTYLEDONAE.

Limes convergens a clearly defined boundary zone between two major, fairly uniform habitat types in which the underlying environmental change is relatively abrupt in space and/or time. It contains species typical of both the habitats on either side, e.g. a flood meadow, which has species typical of both wetlands and pastureland.

Limes divergens a poorly defined boundary zone between two rather variable habitat types, which gradually grade into one another. It is a more stable zone than the LIMES CONVERGENS zone, since environmental change is gradual.

limestone pavement a large, more or less level area of limestone that has been weathered into blocks (*clints*) by solution processes that have created a series of deep, smooth-sided clefts (*grikes*) in between the blocks. The grikes have a very distinctive flora that often includes many rare species.

lime sulphur see SULPHUR DUST.

limicolous growing in mud.

liming the addition of lime to a soil to combat acidity, to improve soil structure, or to remedy calcium deficiency. Quicklime (calcium oxide), slaked lime (calcium hydroxide), or ground chalk or limestone may be used. Liming makes heavy clay soils more workable as it encourages the formation of soil crumbs by a process termed *flocculation*.

limiting factor a factor in the environment (*ecological factor*) that by its presence or absence, or increase or decrease will govern the behaviour of an organism or a metabolic process within an organism. Most metabolic processes depend on more than one factor being present in order to proceed. When all other conditions are favourable, the factor nearest its minimum value is the limiting factor. For example, when a plant photosynthesizes on a warm sunny day in a moist environment, the amount of carbon dioxide available in the air will be the limiting factor. As evening approaches, light will become the limiting factor.

limiting layer see MERISTODERM.

limits of tolerance the upper and lower limits of the values of a range of environmental factors within which an organism or species is able to survive. The range of conditions so defined, or the geographical area in which they are found, constitute the *range* of the species.

limnetic zone in large, deep freshwater ecosystems, the part of the ecosystem beyond the LITTORAL ZONE that lies above the COMPENSATION DEPTH. This zone may be absent from very small, shallow lakes and ponds.

limnology the study of inland aquatic ecosystems.

Lindeman's efficiency a measure of ecological efficiency defined as the ratio of the energy assimilated at one TROPHIC LEVEL in a food chain to the energy assimilated at the preceding trophic level. Compare ECOLOGICAL EFFICIENCY.

linear describing leaves, such as those of grasses, that are elongated and parallel sided for much of their length. *See illustration at leaf.*

linkage the tendency for different genes to segregate together during meiosis because they occur on the same chromosome. For example, if the genes Aa and Bb occur on the same chromosome then a double heterozygote AaBb formed from Ab and aB gametes would yield a significantly larger number of Ab and aB gametes than AB or ab gametes. Similarly, if the parental gametes were AB and ab, the heterozygote would generate AB and ab gametes frequently, and Ab and aB gametes rarely. This is in complete contrast to INDEPENDENT ASSORTMENT where the genes are on different chromosomes and therefore produce all types of gamete equally frequently. When genes are linked, the two less frequent gametes are produced as a result of CROSSING OVER. A group of linked genes is called a *linkage group*, and in general one chromosome corresponds to one linkage group. However, two widely

separated genes on a long chromosome with frequent CHIASMATA may behave as though they are on different chromosomes. See also CHROMOSOME MAPPING.

linkage disequilibrium the nonrandom association between alleles at two or more loci such that certain combination of alleles are likely to occur more or less frequently in a population than would be expected from their distance from each other. Compare LINKAGE EQUILIBRIUM.

linkage equilibrium (gametic equilibrium) the situation that would theoretically exist in a population that is undisturbed by selection or migration, in which all possible combinations of groups of linked genes are present at equal frequency. Compare LINKAGE DISEQUILIBRIUM.

linkage group see LINKAGE.

linkage map a diagrammatic representation of the order of genes on a chromosome. Linkage maps can be constructed by crossing appropriate strains carrying different alleles on homologous chromosomes. Results show that the arrangement of genes is linear, although the distance between the genes is expressed not in units of length but in *crossover units* or *Morgan units* (named after T. H. Morgan who discovered linkage). The Morgan unit (*morgan*) is a unit of relative distance between genes on a chromosome, in which 1 morgan (1 M) represents a crossover value of 100%. A crossover value of 1% is a centimorgan (cM) and a value of 10% is a decimorgan (dM).

Infrequent crossing over is interpreted to mean that genes are near each other (closely linked), and is expressed quantitatively by a small crossover value (% crossing over). Conversely, genes are regarded as being further apart if the frequency of crossing over is higher, as indicated by a large crossover value. See also CHROMOSOME MAPPING, LINKAGE.

linkage value the proportion of parental types to RECOMBINANTS in the progeny of a cross.

linked describing alleles or genes showing less than 50% recombination. The more closely linked (the smaller the percentage recombination), the greater the likelihood of the linked genes or alleles being transmitted together to the daughter cells.

linoleic acid a diunsaturated eighteen-carbon FATTY ACID found in various vegetable oils, especially linseed oil. In plants it is formed from OLEIC ACID. However animals cannot synthesize it and, as a precursor of arachidonic acid and the prostaglandins, it is essential in their diets.

linolenic acid an eighteen-carbon triunsaturated FATTY ACID. It occurs in many plant lipids but is especially prevalent in the chloroplast, where it is the predominant fatty acid. Like LINOLEIC ACID, it is essential in animal diets.

lip see LABELLUM.

lipase any HYDROLASE enzyme involved in the breakdown of storage fats. Lipases catalyse the breakdown of TRIACYLGLYCEROLS into free fatty acids and glycerol. The fatty acids are then further degraded in the β-oxidation cycle. Lipase activity is very high in germinating oil-rich seeds, such as that of the castor bean.

lipid a water-soluble hydrocarbon that can be extracted from cells and tissues by nonpolar solvents. Two major classes of lipid can be recognized. *Complex* or *saponifiable lipids* comprise all those lipids that yield soaps on alkaline hydrolysis. These include ACYLGLYCEROLS, WAXES, PHOSPHOGLYCERIDES, and SPHINGOLIPIDS. *Simple lipids* are a heterogeneous group of compounds, including the STEROLS, CAROTENES, and XANTHOPHYLLS. Although they are called simple lipids they can be very complex structurally; natural rubber is a simple lipid but it is an extremely large and complex polymer.

The functions of lipids within the plant are varied. Storage triacylglycerols provide an energy store, especially in oil-rich seeds such as the castor bean. The phosphoglycerides are important structural components of membranes, while waxes and cutin, components of the cuticle, help control water loss from the plant.

lipopolysaccharide any molecule containing both lipid and carbohydrate elements. Lipopolysaccharides are the principal components of Gram-negative (see GRAM STAIN) bacterial cell walls. Such molecules have a repeating trisaccharide backbone, to which are attached side chains of oligosaccharides and the fatty acid 3-hydroxymyristic acid.

lipoprotein an association of lipids and proteins in a system that combines aspects of the properties of both types of molecule. Most membrane systems (see PLASMA MEMBRANE) are composed primarily of lipoprotein. The major lipids in plant membranes are PHOSPHOGLYCERIDES and GLYCOLIPIDS. (Glycolipids are especially abundant in the chloroplast membranes.) The protein component is not limited to any one type of protein but membrane proteins often have a slightly modified structure to improve their interaction with lipids. For example, there are relatively few DISULPHIDE BRIDGES in membrane proteins.

The forces binding lipid and protein together in a lipoprotein system are not conventional chemical bonds. The most important binding forces are hydrophobic interactions but polar and electrostatic forces are also important.

lirella a long APOTHECIUM.

lithocyst a cell that contains a CYSTOLITH.

lithophyte 1 a plant that grows on rock outcrops or on rocky or stony ground. An example is the fern *Asplenium ruta-muraria* (wall-rue). 2 an organism partly composed of siliceous or calcareous materials, e.g. the stoneworts (CHAROPHYCEAE).

lithosere see XEROSERE.

lithosphere the solid outer layer of the earth, including the upper part of the mantle and all the rocks of the crustal plates (see PLATE TECTONICS). It varies in thickness from about 1–2 km on the mid-oceanic ridges to 300 km under the oldest parts of the continental crust.

lithotroph an organism that derives its energy from the oxidation of inorganic compounds or elements. Compare ORGANOTROPH.

litter (L-layer) the remains of dead plants (e.g. leaves, flower, bark, twigs) on the soil surface.

littoral describing the seashore between low and high tide marks that is exposed alternately to air and water. It may be muddy, sandy, or rocky. The size of the zone varies with the slope of the coast and the height of the tides. Typical seaweeds are brown algae (phaeophyta) of the order FUCALES. The uppermost part of the littoral zone is called the *littoral fringe*. Below the littoral fringe is the *eulittoral* zone, which is characterized by shellfish such as mussels, oysters, and barnacles, and on sandy shores burrowing invertebrates such as polychaete worms and crabs, and by green algae such as *Ulva* and *Enteromorpha*.

littoral fringe see LITTORAL.

liverworts see HEPATOPHYTA.

living fossil a present-day species that has certain characteristics only found elsewhere in extinct groups of organisms. Such species have evolved very slowly. The maidenhair tree (*Ginkgo biloba*) is an example. It was discovered by Western botanists in Japan in the seventeenth century and subsequently in China but only in cultivation. The fossil members of the Ginkgophyta were widely distributed in the Mesozoic era. Similarly the dawn redwoods (*Metasequoia*) were only known from fossil remains until *M. glyptostroboides* was discovered in a remote part of China in the mid-1940s. See also BRADYTELIC.

llanos savanna GRASSLANDS found in the Orinoco basin of Venezuela and in northeast Colombia. The main species are swamp grasses and sedges, with a scattering of scrub oaks and dwarf palms, especially along rivers and at high altitude.

L-layer *See* litter.

loam a type of SOIL in which there is an even mixture of fine clay particles and coarser sand particles. It is the best type of soil for cultivation as the sand helps drainage and aeration but the clay prevents excessive water loss and binds the organic material. See also SOIL TEXTURE.

lobed describing a leaf that is divided into curved or rounded parts connected to each other by an undivided central area. *See illustration at* leaf.

loci see LOCUS.

lock-and-key theory a theory that explains the mechanism of enzyme action in terms of complementary sites on enzyme and substrate that fit together rather like a lock and key to form a temporary enzyme–substrate complex. The binding site on the enzyme is called an *active site*, and is generally complementary to the substrate in both shape and charge.

locule (loculus) a cavity within which specialized organs may develop, most usually the ovules or pollen grains. In an anther there are normally four locules. In a simple ovary there is one locule while in a compound ovary there may be one or many locules depending on how the carpels are fused.

loculicidal see CAPSULE.

Loculoascomycetae in some classification schemes, a class of the ASCOMYCOTA in which the asci are two-walled (bitunicate) and are contained in perithecia enclosed in a stroma (see PSEUDOTHECIUM). It includes some 530 genera and over 2000 species, some of which are important plant parasites, e.g. *Venturia inaequalis*, which causes apple scab.

loculus see LOCULE.

locus (*pl.* loci) the position occupied by a gene on a chromosome. ALLELES of a gene occupy the same locus on HOMOLOGOUS CHROMOSOMES.

lodicule either of the minute scales between the stamens and the GLUME of a grass flower. They are believed to represent the reduced perianth parts and may, through changes in turgor, be involved in the opening of the flower. Rarely there are three lodicules, while in a few species they are absent.

loess a fine-textured yellowish azonal SOIL that is widespread in central Europe, southern Russia, northern China, and Argentina. It consists of clay and silt particles that were deposited at the edge of the ice sheets during the last ice age. It is a fertile, often calcareous, soil and is the parent material of CHERNOZEM soils.

logarithmic phase the period during the growth of microorganisms in culture when the cells are dividing rapidly and there is a huge increase in cell number. If the logarithm of cell numbers is plotted against time then a straight line is obtained on the graph. This period usually ends quite abruptly when nutrient levels fall, the pH of the medium changes, or toxic wastes build up. The logarithmic phase may be maintained by transferring cells to a fresh culture medium or by adding more nutrients to the existing medium. Compare LAG PHASE. See also EXPONENTIAL GROWTH.

logistic curve see LOGISTIC EQUATION.

logistic equation (logistic model) a mathematical expression describing the S-shaped growth curve of a simple population in a confined space with limited resources.

$$dN/dt = rN(N-K/K)$$

where N is the number of individuals in the population, t is time, r is the biotic potential of the organism, and K is the carrying capacity for that particular species in that environment.

The exponential growth of the population is eventually curbed by resource limitations, and population growth slows almost to zero as the population reaches the carrying capacity. The resulting S-shaped curve is also known as the *logistic curve*. See also EXPONENTIAL GROWTH.

logistic model see LOGISTIC EQUATION.

lomasome a complex invagination of the plasma membrane into the cytoplasm. Lomasomes have been identified in some fungal hyphae and spore-producing structures, algal cells, and in some cells of higher plants. In the prokaryotic cells of cyanobacteria, lomasomes are thought to be the site of biochemical processes that in eukaryotic cells are associated with the membrane-bound organelles. See CHROMATOPHÓRE.

lomentum a dry dehiscent fruit, developed from a single carpel, that contains one or more seeds. It resembles a LEGUME but on ripening false septa divide the pod into one-seeded units or valves that fracture at maturity. Such fruits are seen in sainfoin (*Onobrychis viciifolia*).

long-day plant (LDP) a plant that appears to require long days (i.e. days with more than a certain minimum length of daylight) before it will flower. In actual fact, it requires a daily cycle with no long dark periods. For example, henbane (*Hyoscyamus niger*) does not flower when given cycles of 12 hours light followed by 12 hours of darkness, but will flower if given cycles of 6 hours light followed by 6 hours darkness. Compare DAY-NEUTRAL PLANT, SHORT-DAY PLANT. See PHOTOPERIODISM.

long shoot a shoot, especially a woody shoot, that has relatively long internodes and hence well-spaced leaves. Such shoots are often distinct from other shoots on the plant and often serve to colonize new ground, as in *Forsythia*.

loose smuts see SMUTS.

lophotrichous describing a bacterium that has a group of flagella at one end of the cell. Compare MONOTRICHOUS, PERITRICHOUS.

lorica an outer coating around a cell that differs in structure and composition from a cell wall and is often larger than the cell itself.

lotic describing a freshwater ecosystem in which there is a continuous flow of water, such as a river. Compare LENITIC.

Lotka–Volterra model a model of the dynamics of herbivore/plant or predator/prey interactions in species living in the same environment and the same space. Fluctuations in the primary producer/prey population are described by the equation

$$dN/dt - rN - aPN$$

where N is the number of producers/prey, t is time, and a the searching efficiency/attack rate.

Fluctuations in the herbivore/predator populations are described by

$$dP/dt = fa\,PN - qP$$

where f is the efficiency of the herbivore/predator at turning food into offspring and q is another constant. The model reinforces the competitive exclusion model (see COMPETITION), and also explains cyclic fluctuations of these populations, in which reduction in the numbers of herbivores/predators allows the producers/prey to increase in number, which in turn promotes an increase in numbers of the herbivores/predators.

lower fungi a general term for a group of phyla that includes the Chytridiomycota, Hyphochytriomycota, Myxomycota, Oomycota, Plasmodiophora, and Zygomycota.

lower plants the non-vascular plants, i.e. the BRYOPHYTES.

low-temperature scanning electron microscopy the preparation of specimens for the scanning ELECTRON MICROSCOPE, and their subsequent examination, at very low temperatures. The specimen is frozen with liquid nitrogen (the melting point of nitrogen is −210°C), coated at liquid nitrogen temperatures, and examined in a special specimen chamber similarly cooled. In this way frozen hydrated material can be examined. The results are similar to those obtained by CRITICAL-POINT DRYING but with fewer artefacts. The technique is particularly useful for delicate plant specimens.

luciferase see LUCIFERIN.

luciferin a pigment of some bioluminescent organisms that emits light when oxidised, the process being mediated by the enzyme *luciferase.*

lumen a cavity enclosed by a cell wall, such as the centre of a xylem vessel. Compare LACUNA.

luminescence see BIOLUMINESCENCE.

lunate crescent-shaped.

Lusitanian floral element a geographical component of the British flora that is believed to have migrated north early in the post-glacial period before the land bridge with continental Europe was inundated. It includes species with a disjunct distribution (see DISJUNCT SPECIES), occurring in western Ireland and the Iberian peninsula.

lutein the commonest XANTHOPHYLL pigment, an orange pigment found in the leaves of green plants and also in certain algae, such as the Rhodophyta.

lyase any enzyme that catalyses the nonhydrolytic cleavage of its substrate. An example is pyruvic decarboxylase, which catalyses the formation of acetaldehyde and carbon dioxide from pyruvic acid.

lycopene see CAROTENE.

Lycoperdales an order of basidiomycete fungi of the class HOLOBASIDIOMYCETES containing the puff balls and earth stars. It includes about 272 species in some 33 genera. The BASIDIOMA remains closed until maturity and its contents become dry and powdery. In the earth stars (e.g. *Geastrum triplex*) the outer wall of the basidioma splits into a number of rays, which peel back to surround the spores still enclosed by the thin inner wall. The whole fruiting body thus fancifully resembles a flower or star. The puff balls (e.g. *Lycoperdon pyriforme*) and earth stars discharge their spores in a series of puffs when disturbed.

The Lycoperdales are widespread, especially in soil and rotting wood. Most are saprobes, but a few may form mycorrhizas.

Lycophyta (Lycopodophyta) a phylum of vascular plants containing only six living genera but having a rich fossil record, being especially abundant in the Carboniferous. In older classifications, the group is called the Lycopsida and considered to be a subdivision or class of the Pteridophyta. There are about 1000 living species, mostly in tropical regions. Many are epiphytes. The dichotomously branching sporophyte is differentiated into a shoot and root, and the vascular tissue contains properly differentiated phloem. Shoots often arise from branching horizontal rhizomes. The shoot bears small, usually spirally arranged, leaves (microphylls) that leave no leaf gap in the stele. The sporangia are either borne singly in the axil of the sporophyll (a modified microphyll) or on the upper surface of the sporophyll near the base. In some species the sporophylls are photosynthetic, but in others they are scalelike and grouped together to form strobili. Both homosporous forms (e.g. *Lycopodium*) and heterosporous forms (e.g. *Selaginella*) exist. In heterosporous forms, separate male and female gametophyte thalli may be produced, or the microspores may produce antherozoids directly. In most species, archegonia and antheridia are formed and the sperm swim through a film of water to the eggs in the archegonia. Embryogeny is endoscopic.

The Lycophyta is divided into three extant orders: LYCOPODIALES, SELAGINELLALES, and ISOETALES. There are two extinct orders: the LEPIDODENDRALES and the Pleuromeiales, which is known from the Triassic and is intermediate in form between the Lepidodendrales and the Isoetales.

Lycopodiales an order of the LYCOPHYTA containing four genera, all homosporous: *Lycopodium* (club mosses), *Lycopodiella*, *Phylloglossum*, and *Huperzia*. There are some 170 species of *Lycopodium* distributed worldwide. They are all herbs and have long dichotomously branching stems bearing numerous small leaves with no ligules (compare SELAGINELLALES). The unilocular sporangia are borne singly in the axils of sporophylls, which in many species occur together in distinctive club-shaped strobili. The free-living gametophyte may, depending on the species, be either photosynthetic or saprophytic. It is always found in association with an ENDOMYCORRHIZA. The antheridia, which produce many antherozoids, each with two undulipodia, are located in the centre of the apex of the PROTHALLUS and are surrounded by a ring of archegonia. *Lycopodiella* (40 species) is very similar to *Lycopodium* and may form interspecific hybrids in the wild. *Phylloglossum* contains the single species *P. drummondii*, which is very different from *Lycopodium* in habit, consisting of a basal whorl of leaves and a single strobilus borne on a long stalk. *Huperzia* contains about 300 species and resembles *Lycopodium*, but the sporophylls are not apical; some species reproduce by GEMMAE.

Lycopsida see LYCOPHYTA.

lysigeny the formation of a space by the destruction of cells. This is often achieved by enzymatic dissolution. Compare SCHIZOGENY.

lysimeter a device for measuring actual evapotransiration. A lysimeter is a large container, up to several metres in diameter, for which measurements of water input and water loss are made over a period of time. The changing weight indicates the amount of water retained by the system; the evapotranspiration can be calculated by subtraction. A lysimeter may also comprise a block of vegetation-covered soil up to 1 m³, for which the amount of water added and the amount lost by runoff and percolation is measured. Again, evapotranspiration may be calculated, and also the rate of percolation of water through the soil and the soluble constituents removed in this water.

lysine a basic amino acid with the formula $NH_2(CH_2)_4CH(NH_2)COOH$ (*see illustration at* amino acid). Lysine is synthesized from aspartic and pyruvic acids via an aldol condensation reaction; control of synthesis is through feedback inhibition of this condensation reaction by lysine. Degradation of lysine occurs via acetyl CoA and the Krebs cycle.

Lysine is a precursor in the synthesis of the pyridine and piperidine alkaloids, e.g. the nicotine derivative anabasine.

lysis the death and subsequent breakdown of a cell. See also AUTOLYSIS.

lysogeny see BACTERIOPHAGE

lysosome an ORGANELLE, bounded by a single membrane, that contains hydrolytic enzymes (e.g. acid phosphatase, ribonuclease, β-galactosidase, protease) capable of degrading cellular components. Phago-lysosomes are formed in association with vacuoles containing ingested material (see ENDOCYTOSIS). Autolysis occurs when components of the cell, e.g. mitochondria or fragments of endoplasmic reticulum, are degraded by the activity of lysosomal enzymes. This process is important in the recycling of valuable nutrients from ageing plant tissue. Vesicles from the GOLGI APPARATUS are also involved in the intracellular translocation of lysosomal enzymes.

lytic cycle see BACTERIOPHAGE.

lytic response see BACTERIOPHAGE.

M m

macchia see MAQUIS.

maceration in microscopical preparation, the chemical dissolution, often by the application of strong acids, of the matrix binding parts of a specimen. The isolated pieces of the specimen may then be examined or subjected to further preparation.

machair herb-rich grassy plains occurring on calcareous shell sand that has accumulated behind coastal dunes over impermeable rock strata in north-west Scotland and outlying islands. The machair is highly fertile, but the harsh climate limits the crop that can be grown. It is also used for pasture. It is a product of thousands of years of human management, and contains many diverse MICROHABITATS that make it an important location for nationally rare invertebrates and bird species.

macrandry (*adj.* macrandrous) the formation of ANTHERIDIA or similar structures on normal filaments, i.e. the male sex organs are of normal size. Compare NANNANDRY.

macro- prefix denoting large or long.

macroalga an alga whose thallus is large enough to be seen with the naked eye.

macrobiota the larger organisms of a region. The term is applied especially to soil organisms such as moles, and to tree roots, and sometimes also to large invertebrates. The plant component of the macrobiota is termed the *macroflora*, the animal component the *macrofauna*. Compare MESOBIOTA, MICROBIOTA.

macrocarpous forming large fruits.

macroconsumer a large organism that feeds on whole living organisms or parts of organisms or particulate organic matter.

Macroconsumers are mostly animals, and are sometimes termed *phagotrophs*. See FOOD CHAIN. Compare microconsumer.

macrocyst 1 in the MYXOMYCOTA, a large number of myxamoebae enclosed in a cyst. 2 the resting form of a young PLASMODIUM. 3 a form of fruiting body found in some cellular slime moulds (Acrasiomycota).

macroelement see MACRONUTRIENT.

macroevolution evolution on a large scale, i.e. above the species level. The term encompasses the differentiation of new genera, families, orders, classes, and phyla. Compare MICROEVOLUTION.

macrofauna see MACROBIOTA.

macrofibrils fibrils within the cell wall that are large enough in some cases to be visible under the light microscope. They are formed from microfibrils lying parallel to each other.

macroflora see MACROBIOTA.

macrogamete the larger of a pair of conjugating gametes (see CONJUGATION), often assumed to play the role of the female gamete. Compare MICROGAMETE.

macrolichens the foliose, dimorphic, and fruticose LICHENS, which are relatively large and can be identified with the aid of a hand lens or dissection microscope.

macromolecule a very large molecule of high molecular weight (containing about 10 000 or more atoms). Examples include such polymers as starch and cellulose.

macromutation a MUTATION that causes a profound change in an organism. Such mutations are typically those of regulator genes controlling the expression of many structural genes.

macronutrient (essential element,

macroelement) any chemical element required by a plant in relatively large quantities for successful growth. Macronutrients include CARBON, hydrogen, and oxygen, which are obtained from carbon dioxide and water. The remaining macronutrients, NITROGEN, PHOSPHORUS, POTASSIUM, SULPHUR, MAGNESIUM, CALCIUM, and IRON, are obtained from the soil. In a culture medium these are provided by calcium and potassium nitrates, potassium and iron phosphates, and magnesium sulphates. Compare MICRONUTRIENT.

macrophyll see MEGAPHYLL.

macrophyte a plant or alga large enough to be seen with the naked eye. The term includes all the plants and the larger algae (macroalgae).

macrosclereid a short sclerenchyma cell (SCLEREID), somewhat columnar in shape. Macrosclereids form the outer layer of the seed coat in many plants. See also MALPIGHIAN CELL.

macrosporangium see MEGASPORANGIUM.

macrospore see MEGASPORE.

macrosporophyll see MEGASPOROPHYLL.

macrostylous bearing a long stamen.

macrothallus in the life cycle of an alga, a vegetative thallus large enough to be seen with the naked eye.

maculate spotted, blotched.

maculiform shaped like a spot.

made ground (made land) an area of dry land that has been created by humans, for example, by reclamation of marshes, lakes, or shores by infilling with soil, rubble, or refuse.

made land see MADE GROUND.

magnesium symbol: Mg. A metal element, atomic number 12, atomic weight 24.3, important as a macronutrient for plant growth. It forms part of the chlorophyll molecule and deficiency thus leads to CHLOROSIS. It is also a cofactor for kinase and phosphohydrolase enzymes.

Other symptoms of magnesium deficiency are NECROSIS, stunted growth, and in some plants a puckering and whitening of the leaves at the edges. Magnesium is added to soil as magnesium sulphate or magnesium oxide.

Magnoliales an order of primitive

dicotyledons of the subclass Magnoliidae, sometimes regarded as the basal group of the dicotyledons and monocotyledons. Members are characterized by simple alternately arranged leaves with petioles and large stipules that fall off as the leaf expands, leaving a characteristic leaf scar around the node. The flowers are usually bisexual, often large and showy. They are borne singly at the ends of branches or in the axils of leaves. There are at least two whorls of petaloid perianth segments and numerous spirally arranged stamens. The carpels are also spirally arranged. Each seed contains a tiny embryo and copious endosperm.

There are some 17 families in the Magnoliales. They include the Magnoliaceae (magnolias, tulip tree), Annonaceae (including the custard apples), Myristicaceae (including nutmeg and mace), and Lauraceae (including laurels, sassafras, camphor, cinnamon, and avocado). This last family is sometimes placed in an order of its own, Laurales. Many of the Magnoliales are sources of spices and medicines, and the order includes many ornamental shrubs and trees.

Magnoliidae a subclass of the dicotyledons containing plants in which the flower parts are numerous and often inserted spirally, and the gynoecium is APOCARPOUS. The number of orders included in this subclass varies considerably between classifications. Usually recognized are the MAGNOLIALES, including the Magnoliaceae, Annonaceae, Myristicaceae, and Lauraceae; the ILLICIALES, including the Winteraceae; the Aristolochiales, including the Aristolochiaceae, Hydnoraceae, and Rafflesiaceae; the Piperales, including the Piperaceae; the Nelumbonales; the Ranunculales, including the Berberidaceae, RANUNCULACEAE, Papaveraceae, and Fumariacae; the Nymphaeales, including the Nymphaeaceae and Cabombaceae; and the Ceratophyllales.

The order Sarraceniales, sometimes included in the Magnoliidae is more often placed in the DILLENIIDAE. The Rafflesiales, here included in the Aristolochiales, is sometimes placed in the Rosidae. The Nymphaeales may merit elevating to subclass rank. In some schemes the Ranunculales are placed in a subclass, Ranunculidae, which is separated from the Magnoliidae on the basis that its members are mostly herbaceous rather than woody plants and generally more advanced than the Magnoliidae. They lack the oil cells common to many members of the Magnoliidae and have TRICOLPATE pollen as compared to the MONOCOLPATE pollen of the Magnoliidae. Their stomata usually lack subsidiary cells whereas those of the Magnoliidae often have two subsidiary cells.

Magnoliophyta in certain classifications, a division containing the flowering plants (see ANTHOPHYTA). It is divided into the two classes Magnoliopsida (dicotyledons) and Liliopsida (monocotyledons).

Magnoliopsida see DICOTYLEDONAE.

major gene a gene whose effects are readily identifiable in the PHENOTYPE. The existence of different alleles of such genes results in instances of DISCONTINUOUS VARIATION in the phenotypes. All MENDELIAN GENES are examples of major genes. The definition excludes genes that modify the expression of other genes. Compare POLYGENES.

malate shuttle a method that has been proposed to explain the transfer of reducing power across the chloroplast membranes. NADPH, produced by the light reactions of photosynthesis, is not able to pass through the chloroplast membranes. However there is evidence that reducing power produced by photosynthesis is nevertheless used in the cytosol. It is suggested that oxaloacetate in the chloroplast stroma is reduced by NADPH to malate, which is readily transported across the chloroplast membranes. In the cytosol malate is oxidized to oxaloacetate with the concomitant production of NADH from NAD. The oxaloacetate then passes back into the chloroplast stroma to begin the cycle again.

male describing either the reproductive parts or a whole organism that bears the microspore-producing apparatus and does not nurture the developing embryo. More strictly, the term applies to the gametophyte that produces antheridia. Compare FEMALE.

See also ANTHER, ANTHERIDIUM, ANTHEROZOID.

maleic hydrazide a GROWTH INHIBITOR often used as a herbicide or to inhibit sprouting. In some species it promotes flowering by inhibiting vegetative growth.

male sterility a condition in which pollen production is prevented by mutation of one or more genes governing its formation. Male sterility has been employed by plant breeders as a method of ensuring cross pollination and hence F_1 HYBRID production. Male-sterile cultivars can be maintained by crossing ss (female parent) × SS (male parent) giving 50% ss and 50% Ss offspring (where s = male sterile).

malic acid (2-hydroxybutanedioic acid) a four-carbon dicarboxylic acid (see CARBOXYLIC ACID) with the formula HOOCCH(OH)CH$_2$COOH. Malate is a Krebs cycle intermediate. It is formed by hydration of fumarate and is oxidized to oxaloacetate with concomitant formation of NADH. Malate can also be formed from pyruvate by malic enzyme. Malate can be transported across certain membranes (see MALATE SHUTTLE). This is important in C_4 PLANTS, where malate is an intermediate in the HATCH–SLACK PATHWAY.

mallee see MAQUIS.

malonyl ACP the three-carbon dicarboxylic acid (see CARBOXYLIC ACID) malonic acid bound to ACYL CARRIER PROTEIN (ACP). Malonyl ACP, with MALONYL COA, is central to fatty acid synthesis. It is formed from malonyl CoA and ACP and then reacts with enzyme-bound acetate to form acetoacetyl ACP, with concomitant release of carbon dioxide. Acetoacetyl ACP is then reduced to butyryl ACP before the addition of another molecule of malonyl ACP to form a six-carbon chain. Malonyl ACP thus provides the two-carbon building blocks with which long-chain fatty acids are progressively synthesized.

malonyl CoA the coenzyme A derivative of the dicarboxylic acid (see CARBOXYLIC ACID, MALONIC ACID. It is synthesized from ACETYL COA and bicarbonate in a reaction catalysed by the enzyme acetyl CoA carboxylase. Malonyl CoA provides the carbons for the synthesis of fatty acids – the malonyl moiety reacts with acyl carrier protein (ACP) to form MALONYL ACP, which in turn combines with the growing fatty acid molecules, with elimination of carbon dioxide, to add two more carbons to the chain.

Malpighian cell an alternative term for a MACROSCLEREID when present in the testa of a leguminous seed.

Malpighian layer a protective layer or layers of cells in the testas of many seeds. It contains closely packed thick-walled columnar cells, often heavily impregnated with CUTIN or LIGNIN, with no intercellular spaces. The Malpighian layer is relatively impermeable to moisture and gases.

maltose (malt sugar) a disaccharide consisting of two glucose units linked by an $\alpha(1–4)$ GLYCOSIDIC BOND. Maltose is widely distributed in plants but does not seem to have a specific function; rather it is an intermediate product in the breakdown of starch to glucose. Concentrations of maltose are particularly high during seed germination when starch reserves are being rapidly broken down. Malt, essential to the brewing industry, is produced by allowing barley seeds to germinate then drying them slowly in a kiln. See also AMYLASE.

mammilla in the CACTACEAE, a tubercle on the swollen stem from which arise groups of spines called AREOLES. The areoles develop from buds in the axils of leaves whose tissues now form part of the mammilla.

manganese symbol: Mn. A metallic element, atomic number 25, atomic weight 54.94, needed in trace amounts by plants for successful growth. Manganese ions are required as COFACTORS by certain enzymes, e.g. KINASES and IAA oxidase. Certain deficiency diseases, e.g. blight of sugar cane and grey speck of oats, are attributed to manganese deficiency. More general symptoms include dwarfing and mottling of the upper leaves. Manganese is often added to the soil as manganous sulphate to treat such problems.

mangrove any of various halophytic trees and shrubs that form dense thickets, forests, and swamps along muddy coasts, estuaries, and salt marshes in tropical and subtropical climates, many of them inundated by the tide for at least part of the time. They are important for trapping sediment and building up the coast, and for protecting the hinterland against tidal surges and storms. They are also important pioneer species. Many species have PROP ROOTS that help to trap more mud and silt. Some species have PNEUMATOPHORES that rise up out of the mud to take in air. Both pneumatophores and prop roots have abundant LENTICELS for maximum diffusion of air to supply the waterlogged roots below. The nutrient-rich sediments and the shelter afforded by the mangroves make mangrove swamps important nursery grounds for fish, including many commercial species. The bark of many mangrove species is used for tanning, turning leather an intense red colour. Mangroves of the species *Rhizophora*, family Rhizophoraceae, especially *R. mangle*, are sometimes said to be 'true mangroves'. Other important mangrove genera include the genus *Bruguiera*, and certain members of the families Avicenniaceae, Sonneratiaceae, and Meliaceae. The coastal mangrove swamps and forests are generally of low height, up to about 10 m high, but taller trees may occur further inland. The stems and exposed roots are often covered in red algae (Rhodophyta). In some species the seeds germinate on the parent plant, and spear-shaped plantlets drop from the branches and either embed themselves in the mud below, or float away to colonize new sites.

maniciform shaped like a cuff.

man-induced turnover in a BIOGEOCHEMICAL CYCLE, the additional flow of a particular element that results from human activity, for example the carbon released into the environment by the burning of fossil fuels (see CARBON CYCLE).

mannan a polysaccharide in which the major monosaccharide subunit is mannose. Galactose and glucose are also often present, forming galactomannans, glucomannans, and galactoglucomannans. Mannans frequently occur in HEMICELLULOSE. They are also found as reserve polysaccharides in some higher plants, e.g. the ivory palm (*Phytelephas macrocarpa*), in which the extremely hard endosperm (known as vegetable ivory) is composed of mannans.

mannitol a common sugar alcohol formed by reduction of the carbonyl group of mannose or fructose to an alcohol. Mannitol is the principal soluble sugar in fungi and lichens and it accumulates in some algae much as sucrose accumulates in vascular plants. Mannitol is a major photosynthetic product in the brown algae (Phaeophyta), lichens, and some higher plant species.

mannose an aldohexose sugar (*see illustration at* aldose). In some plants, notably members of the Fabaceae (Leguminosae), mannose rather than glucose is the monosaccharide building block for reserve polysaccharides: the polysaccharides formed are known as MANNANS. Mannose is also found as a component of some HEMICELLULOSES. The reduction product of mannose, MANNITOL, is an important sugar in lichens and some algae.

mantle see MYCORRHIZA.

manubrium a tubular structure, a number of which arise on the inner walls of the antheridium of chlorophytes of the class CHAROPHYCEAE. A mass of filaments form from specialized cells at the tip of the manubrium. The cells of the filaments subsequently form antherozoids.

manure animal excreta, usually mixed with other material, especially straw, used to improve soil fertility. The term may be used in a wider sense to mean any sort of FERTILIZER.

map distance see CHROMOSOME MAPPING.

map unit see CHROMOSOME MAPPING.

maquis a stunted form of scrub vegetation found on poor soils in seasonally arid regions. It consists of a mass of tangled evergreen bushes and shrubs with leathery leaves, growing to a maximum height of about three metres (though usually shorter), interspersed with scattered twisted dwarfed trees such as olive (*Olea europaea*) and fig (*Ficus*). It is thought that some maquis is the result of agricultural clearance, but some may represent a plagioclimax (see CLIMAX), the result of natural fires or grazing pressure. The species include rockroses (*Cistus*), broom (*Cytisus scoparius*), gorse (*Ulex*), heaths and heathers (*Erica*), etc.,

together with many aromatic herbs, such as thymes (*Thymus*). Maquis is widespread in countries around the Mediterranean. In Italy it is called *macchia*, and in Spain, *mattoral*. Similar vegetation in California is called *chaparral*, in the Cape Province of South Africa *fynbos*, and in Australia *mallee*. See also GARIGUE.

Marattiales an order of the FILICINOPHYTA consisting entirely of large fleshy tropical eusporangiate ferns, commonly known as giant ferns. There are about 100 species in 6 genera. In most species the sporophyte consists of a broad short stem giving rise to large compound fronds. The sporangia are formed on the abaxial surface of fertile fronds that otherwise resemble the sterile fronds. The spores germinate to form a thallose gametophyte, somewhat similar to a liverwort, which may be quite long lived. The Marattiales have a rich fossil record going back to the Carboniferous.

Marchantiales an order of thallose liverworts (see HEPATOPHYTA) in which the thallus is usually flat, ribbon-like, and dichotomously branched and lies prostrate on the ground. Air chambers in the thallus open to the surface via small pores. The ventral surface often has small scales, and there are two types of rhizoids. The sporophyte capsule is only one cell thick, except at the apex. The order includes some of the most complex thalli found in the liverworts, in some species even showing rudimentary conducting tissues. Some species produce their reproductive organs on erect structures that resemble umbrellas.

marginal effect a phenomenon exhibited by the plants at the edge of a stand of vegetation whereby they grow more vigorously than those in the centre. It has been well documented in advancing areas of bracken (*Pteridium*), where the fronds at the edge are significantly taller than those behind them. This is believed to be due to increasing rhizome age at the centre rather than to depletion of soil nutrients. Marginal effects may play a part in the HUMMOCK AND HOLLOW CYCLE of vegetation.

marginal meristem a region of meristematic tissue located along the edges of a leaf PRIMORDIUM. Repeated divisions of

INITIALS in this area give rise to the mesophyll and epidermal tissues of the blade.

marginal placentation (ventral placentation) a form of PLACENTATION in which the placentae develop along the VENTRAL SUTURE of a simple ovary. It is seen in the pods of the Fabaceae.

margo 1 see BORDERED PIT. 2 in a pollen grain, the zone of thickened exine that occurs around a COLPUS.

marker gene 1 a gene of known function whose expression is readily identifiable, for example, the production of an enzyme or other protein, fluorescence, antibiotic resistance, or other visible or chemically testable effect. In GENETIC ENGINEERING, marker genes are inserted into an organism together with the gene being transferred. If the marker gene is expressed in the resulting cell/organism, this indicates that the desired gene has also been successfully transferred. See CHROMOSOME MAPPING, GENETIC ENGINEERING. 2 a gene bearing a particular mutation that aids in studying the inheritance of that gene.

marl a lime-rich clay, usually formed as an alluvial deposit (see ALLUVIAL SOIL).

marsh a region of vegetation where the water table is at or just beneath the soil surface. The soil is neither highly acid nor alkaline (compare BOG, FEN). Common marsh plants are reeds, sedges, and various grasses. A marsh is a stage in the development of a CLIMAX vegetation from a HYDROSERE.

Marsileales see FILICALES.

masked symptoms symptoms of a plant disease that are absent in certain environmental conditions (e.g. particular temperatures or light intensities) but appear when the plant is exposed to other conditions.

mass flow hypothesis the theory that translocation of sugars in the phloem is brought about by a continuous flow of water and dissolved sugars between sources and sinks. (A source is the site of production of sugars, usually leaves, and the sink is the site of their utilization, for example the root system.) At the source OSMOTIC PRESSURE is high due to the continuous formation of sugars and at the sink osmotic pressure is low as the sugars are used up. Thus water from the xylem enters the phloem at the source and leaves it at the sink, returning to the source via the xylem. This tends to drive the contents of the phloem towards the sink.

mass selection a form of ARTIFICIAL SELECTION in which individual plants from a population are selected for the desired phenotype, then bred together to produce the next generation.

mass spectrometry a method of identifying the chemical constituents of a substance by first vapourizing it, then separating the various gaseous ions according to their charge:mass ratio.

massulae 1 mucilaginous extensions of the TAPETUM that surround the microspores and megaspores of water ferns of the genus *Azolla*. Four massulae surround the megaspore and are believed to aid buoyancy. A variable number of massulae are developed in the microsporangium, each of which contains a number of microspores. See also GLOCHIDIUM. 2 the parts of the POLLINIUM in some orchid species. In many orchids, e.g. *Orchis* spp., the pollinium breaks into a number (around 20) of massulae, each of which contains numerous pollen grains and is the unit of pollination.

mast a fruit, usually of a forest tree, often formerly fed to pigs. The term is most commonly applied to fruits of beech and oak.

master chronology see DENDROCHRONOLOGY.

Mastigomycotina a former subdivision of fungi that produce motile cells (zoosporic fungi). In current classification schemes members of the Mastigomycotina are dispersed through the CHYTRIDIOMYCOTA (or CHYTRIDIOMYCOTINA), HYPHOCHYTRIOMYCOTA, OOMYCOTA, and PLASMODIOPHORA.

mastigoneme any of numerous fine hairlike rodlets attached to a tinsel flagellum (now more properly called a *mastigonemate undulipodium*). Mastigonemes are arranged in longitudinal rows and are orientated at an angle, giving the undulipodium a feathery appearance at high magnification. See also UNDULIPODIUM.

maternal effect any temporary effect or influence of the maternal genotype or phenotype on the first generation progeny.

maternal inheritance differences in phenotype that occur between individuals with identical genotypes as result of a non-nuclear gene inherited from the female parent. See CYTOPLASMIC INHERITANCE.

mating design the pattern in which individuals are cross-pollinated in an artificial pollination programme, for example how many crosses each individual is involved in.

mating strain a group of organisms within a species that are characterized by not being able to interbreed with each other. However they are able to breed with members of other physiologically different but morphologically identical groups. Different mating strains are usually denoted by the symbols + and − since male and female forms cannot be distinguished as such. The differences are genetically determined. Compare PHYSIOLOGICAL RACE. See HETEROTHALLIC.

matric potential (Ψ_m) That component of the WATER POTENTIAL of plants and soils that is due to capillary and imbibitional forces. Thus the water potential of cell walls and intercellular spaces is largely due to matric potential. Values of matric potentials are always negative and range from 0 bar in fully turgid tissues to −10 bar in slightly wilted plants. See also OSMOTIC POTENTIAL, PRESSURE POTENTIAL.

matrix 1 in a mitochondrion, the ground substance inside the organelle, which surrounds the internal membranes. It is bounded by the mitochondrial envelope. 2 any medium in which something is embedded; for example, the noncellulose substance in which the cellulose microfibrils are embedded in the plant cell wall.

mattoral see MAQUIS.

mazaedium in lichens, a fruiting body that disintegrates to form a powdery mass of spores.

m chromosome a very small chromosome, a number of which may be found in the nuclei of mosses. Their function is obscure.

MCPA (2-methyl-4-chlorophenoxyacetic acid) A synthetic AUXIN of the PHENOXYACETIC ACID group that is used as a selective weedkiller.

meadow a region of moist grassland maintained at the subclimax stage by mowing. Similar areas maintained at the subclimax stage by grazing are often termed *pastures*.

meadow steppe a form of STEPPE that occurs in wetter areas. It is characterized by a a lack of trees and large shrubs, and a very dense plant cover consisting of perennial grasses and a rich component of broad-leaved forbs (see BASIC GRASSLAND). It occurs near the margins of steppes and other temperate GRASSLANDS.

mean (arithmetic mean) the average of a series of quantities, obtained by adding up all the observed values and dividing by the total number of observations. If a normal curve were plotted from an infinitely large number of observations, then the mean (and the MEDIAN and MODE) would be the distance from the origin of the axes to the centre point of the curve, represented by the symbol μ. However when sample size is limited, as always occurs in practice, then the estimated mean, rather than the median or mode, gives the best estimate of the true mean.

mean deviation the average magnitude of deviations from the centre of a normal curve. Compare STANDARD DEVIATION.

mean square the estimated square of the STANDARD DEVIATION. See VARIANCE.

mechanical tissue (strengthening tissue, supporting tissue) any tissue consisting of cells with thickened cell walls, such as COLLENCHYMA and SCLERENCHYMA.

mediaeval woodland woodland known to have existed prior to 1600 or 1650, depending on the authority used. See also ANCIENT WOODLAND.

median in a series of observed values, the one quantity that has an equal number of observations on either side of it. Compare MEAN, MODE.

Mediterranean forest the climax vegetation of much of the Mediterranean region, a mixed evergreen forest, most of which has been destroyed centuries ago and has largely been replaced by MAQUIS and GARIGUE.

Mediterranean scrub see MAQUIS, GARIGUE.

medium the surrounding substance in which an organism exists, such as air or water, or (in an experimental situation) a mounting medium, staining medium, culture medium, etc.

medulla **1** in lichens, the layer of loosely arranged hyphae below the cortex and the algal layer.

2 in fungal fruiting bodies, the layer of longitudinal hyphae.

3 in algae, the inner layers of a multicellular thallus, made up of nonpigmented cells that often have a storage role.

4 see PITH.

medullary ray (primary ray) any of the radial extensions of the PITH (medulla), consisting of parenchyma and penetrating between the vascular bundles in the primary tissues of the stem. These primary rays run from the pith to the cortex. As well as providing a living link between the pith and the cortex, and a pathway for gaseous diffusion and transport of dissolved substances, they may also be used for food storage. Compare RAY (SECONDARY RAY).

medullated protostele see STELE.

megaphyll (macrophyll) a leaf typical of seed plants and ferns, often quite large and usually with LEAF GAPS associated with the LEAF TRACES. Megaphylls are thought to have evolved from early leafless plants, such as *Rhynia*, by the development of unequal dichotomies of the axis and 'overtopping' or dominance of the longer indeterminate shoot over the shorter determinate shoot. This is thought to have been succeeded by flattening (planation) of the branches of the determinate shoot and subsequent webbing between them by ground tissue to form typical megaphylls, although the precise order of these events remains somewhat unclear. Compare MICROPHYLL. See TELOME THEORY.

megasporangiophore a stalklike structure that bears MEGASPORANGIA. The ovule-bearing structures (MEGASPOROPHYLLS) making up the female cones of cycads (see CYCADOPHYTA) are termed megasporangiophores.

megasporangium (macrosporangium) a structure in which MEGASPORES are formed. In the seed plants it corresponds to the OVULE. Compare MICROSPORANGIUM.

megaspore (macrospore) the larger of the two types of haploid spores formed after meiosis in heterosporous species and usually designated female. It is immobile and contains food reserves for the gametophyte. In angiosperms, the cell that gives rise to the nuclei of the EMBRYO SAC is termed a megaspore, though it is usually smaller than a pollen grain (MICROSPORE).

megaspore mother cell (megasporocyte) a diploid cell that gives rise by meiosis to four megaspores. Often only one develops and the rest abort. Compare MICROSPORE MOTHER CELL.

megasporophyll (macrosporophyll) a leaflike structure that bears the MEGASPORANGIA. In angiosperms and gymnosperms it is represented by the CARPEL and OVULIFEROUS SCALE respectively. Compare MICROSPOROPHYLL. See also SPOROPHYLL.

megastrobilus a strobilus that bears only MEGASPORANGIA, for example the female cones of some conifers and cycads.

meiocyte a cell that divides by meiosis to produce HAPLOID spores. The MICROSPORE MOTHER CELLS (pollen mother cells) in anthers and the MEGASPORE MOTHER CELLS in the NUCELLUS tissue are the meiocytes of flowering plants.

meiosis (reduction division) the process by which a diploid cell divides to form four haploid cells. The process consists of two consecutive divisions, each with a sequence of stages similar to those of MITOSIS. During the first division, which is the actual reduction division, the pairing (SYNAPSIS) and subsequent separation of HOMOLOGOUS CHROMOSOMES into separate nuclei results in the reciprocal exchange of portions of maternal and paternal CHROMATIDS (see CROSSING OVER). It is in this pairing and separation of homologues that meiosis essentially differs from mitosis (see diagram). The two haploid nuclei resulting from the first division then divide for a second time, during which the chromatids are separated as in mitosis. Four haploid cells are therefore formed. Meiotic division is the process by which the haploid gamete-producing (gametophyte) phase in the life cycle of plants is established. See PROPHASE, METAPHASE, ANAPHASE, TELOPHASE,

CYTOKINESIS, ALTERNATION OF
GENERATIONS.

meiosporangium a SPORANGIUM in which
meiosis occurs. The resulting haploid
spores are termed *meiospores*.

meiospore see MEIOSPORANGIUM.

meiotic drive a process occurring during
meiosis that results in the two kinds of
gametes produced by a heterozygote not
being equally common. It is more often
seen in the formation of megaspores than
microspores. An example is seen in maize
plants that are heterozygous for a
chromosome mutation, in which one of the
homologues of chromosome 10 has an
abnormal terminal knob. The knob acts as a
centromere and causes the abnormal
chromosomes to move to the poles of the
spindle earlier in both the first and second
divisions of meiosis. The product of
meiosis is a linear tetrad in which the
abnormal chromosomes are more likely to
be in the outer two cells. Since all four cells
of a tetrad normally develop into pollen
grains, the abnormal chromosome is
recovered in equal numbers among the
microspores. However only the basal cell of
the tetrad develops into the megaspore.
Thus the abnormal chromosome is
recovered at higher frequencies (about 70%)
among the megaspores.

Melanconiales an order of the FUNGI
ANAMORPHICI (Mitosporic Fungi) with
acervular conidiomata. There are about
1000 species in some 120 genera. They
include many fungi parasitic on plant stems
and leaves, e.g. *Gloeosporium* and
Colletotrichum, both causal agents of
ANTHRACNOSE. They are placed in the class
COELOMYCETES. In some classification
schemes they are assigned to the order
Tuberculariales. The order is considered
obsolete by some taxonomists.

membrane see CELL MEMBRANE.

membrane potential the potential
difference across a membrane that results
from the action of proton pumps (see
CHEMIOSMOTIC THEORY), which use energy
from ATP to maintain an imbalance of
positive and negative ions on either side of
the membrane.

Mendelian genes genes that behave
according to MENDEL'S LAWS, i.e. they
determine clear-cut characteristics easily
recognized with the naked eye, they show
complete dominance, and (where more
than one gene is concerned) they are
located on different chromosomes, i.e. there
is no linkage.

Mendelian population an interbreeding
group of organisms that share a common
GENE POOL.

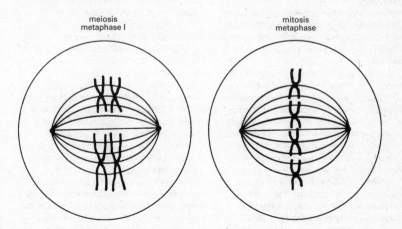

Meiosis: the basic differences in chromosome behaviour in meiosis and mitosis

Mendelian ratio the ratio of contrasting characteristics present in the offspring of a cross when the gene(s) determining the characteristic(s) behave according to MENDEL'S LAWS.

Mendelism the theory that inherited characteristics are governed by discrete factors or genes (PARTICULATE INHERITANCE), which are transmitted to the offspring in a regular and predictable way as described by MENDEL'S LAWS.

Mendel's laws two laws of inheritance attributed to Gregor Mendel in recognition of his experimental investigations with peas. Mendel's first law, the *Law of Segregation*, states that while an organism may contain a pair of contrasting ALLELES, e.g. Tt, these will segregate (separate) during the formation of gametes, so that only one will be present in a single gamete, i.e. T or t (but not both or neither). Mendel's second law, the *Law of Independent Assortment*, states that the segregation of alleles for one character is completely random with respect to the segregation of alleles for other characters. Thus a TtGg individual will produce equal numbers of all four possible kinds of gametes: TG, Tg, tG, tg. Mendel's first law is still applicable for all chromosomal genes. Mendel's second law is only true if the genes involved are on nonhomologous chromosomes. See also LINKAGE.

mericarp any of the one-seeded portions that result when a compound fruit divides at maturity. Mericarps may be dehiscent or indehiscent. See also CREMOCARP, REGMA.

meristele see STELE.

meristem (*adj.* meristematic) a region containing actively or potentially actively dividing cells, including INITIALS and their immediate undifferentiated derivatives. Some consider a meristem to consist only of initials but the difficulty in recognizing a boundary between initials and their immediate derivatives has led many to consider the term in the broader sense including undifferentiated nondividing cells. Some meristems, such as axillary buds, are inactive for much of the time, their cells being potentially capable of active division (see APICAL DOMINANCE). Meristems may be APICAL, axillary, LATERAL, MARGINAL, or INTERCALARY. Later meristems are *secondary meristems*, responsible for increase in girth of the plant during SECONDARY GROWTH.

meristem culture (shoot-tip culture) the culture of excised meristems on suitable nutrient media under aseptic conditions. The stem apex is usually used though axillary meristems may also be taken. Often GIBBERELLIC ACID must be added to the medium to promote normal growth into a plantlet. Once a suitable medium has been found for the regeneration of plantlets, the technique can be used as a means of rapid propagation since the plantlets can be divided at intervals into segments that can be grown on individually. Meristem culture may be used to propagate infertile plants, nursery stock, or F_1 HYBRIDS that do not breed true. As few virus diseases affect the plant apex, the technique can be used to produce virus-free stock.

meristoderm (limiting layer) the outer region of the stipe of members of the LAMINARIALES, an order of the Phaeophyta (brown algae), in which MERISTEM-like activity occurs. Continued division in the region below the lamina serves to replace tissues that are worn away by tide action.

merological describing a method of studying ecosystems in which the component parts of the ecosystem are studied in detail, the results then being used to produce a picture of the whole. Compare HOLISTIC.

meromictic lake a lake that never experiences complete vertical mixing of the water. Compare DIMICTIC LAKE, MONOMICTIC LAKE.

meroplankton organisms that form part of the PLANKTON for only a part of their lives

mesarch describing xylem maturation in which the older cells (protoxylem) are in the centre of the xylem strand because maturation has progressed both centrifugally and centripetally. Compare ENDARCH, EXARCH.

mesic describing an environment that is neither very wet (hydric) nor very dry (xeric).

meso- prefix denoting intermediate (as in mesic) or in the middle (as in mesocarp).

mesobiota a collective term for soil organisms of intermediate size, i.e.

invertebrates such as worms, arthropods, and mollusks and non-microscopic fungi. The plant, algal, and fungal components of the mesobiotia are the *mesoflora*, while the animals constitute the *mesofauna*).

mesocarp the middle layer of the PERICARP of an angiosperm fruit, positioned between the exocarp and endocarp. In many fruits, such as the peach (*Prunus persica*), the mesocarp is the fleshy part of the fruit. However, in some fruits no mesocarp is present, the pericarp consisting only of exocarp and endocarp.

mesocotyl in grasses and related plants, the part of the seedling axis between the shoot apex and the point of attachment of the SCUTELLUM, formed by the growing together of the HYPOCOTYL and the COTYLEDON.

mesofauna see MESOBIOTA.

mesoflora see MESOBIOTA.

mesogenous describing an angiosperm stomatal apparatus in which the SUBSIDIARY CELLS are derived from the same INITIAL as the guard cells. Compare PERIGENOUS. See also SYNDETOCHEILIC.

mesokaryotic see DINOMASTIGOTA.

mesopause see ATMOSPHERE.

mesophilic describing microorganisms that require moderate temperatures (30–40°C) for optimal growth. Compare THERMOPHILIC, PSYCHROPHILIC.

mesophyll the main tissue of a leaf, mostly differentiated as photosynthetic CHLORENCHYMA. In many plants, for example the mesophytes of temperate climates, the mesophyll is differentiated into PALISADE and SPONGY MESOPHYLL.

mesophyte a plant without adaptations to environmental extremes. Compare HYDROPHYTE, XEROPHYTE.

mesosome a structure found in bacterial cells formed by intrusions of the plasma membrane into the protoplasm. The respiratory enzymes are associated with the mesosome and it has the same function as the mitochondrion of eukaryotic cells. It is also attached to the DNA and is thought to control the separation of replicated DNA molecules during cell fission.

mesosphere see ATMOSPHERE.

mesotherm a plant that grows best in a warm temperate climate, where the mean temperature of the hottest month is at least 22°C and the mean temperature of the coldest month is not less than 6°C.

mesotony a pattern of development of shoot branches in which the branches towards the centre of the shoot elongate to a greater extent than those below and above them (see diagram). Compare ACROTONY, BASITONY.

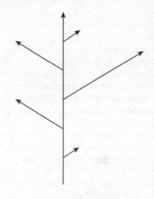

Mesotony

Mesozoic an era of geological time between about 225 and 65 million years ago. It is divided into the TRIASSIC, JURASSIC, and CRETACEOUS periods. See GEOLOGICAL TIME SCALE.

messenger RNA (mRNA) a linear single-stranded polynucleotide (see NUCLEOTIDES) of uridine, adenine, guanidine, and cytidine monophosphates. mRNA is synthesized on DNA by mRNA polymerase enzyme, and the sequence of bases in it is a complementary copy (transcript) of the bases in one strand of DNA, the *sense strand*. The molecule initially manufactured on DNA now undergoes modification prior to its involvement in PROTEIN SYNTHESIS. This modification, called *post-transcriptional processing*, occurs in the NUCLEOLUS and involves the addition of adenine polynucleotide (poly A) to one end and a guanine derivative to the other end. In eukaryotic cells, post-transcriptional processing may also include excision of INTRONS. Sometimes the term *heterogeneous RNA* (hnRNA) is given specifically to newly formed intron-containing mRNA, and the term mature mRNA is reserved for the molecule produced as a result of post-transcriptional processing. Mature mRNA moves to the ribosomes in the cytoplasm, where the triplets of bases in it act as a template and correspond to amino acids in proteins.

mestom sheath (mestome sheath) a thin-walled type of BUNDLE SHEATH present in some grasses.

metabolic pathway see BIOSYNTHETIC PATHWAY.

metabolism (*adj.* metabolic) the sum total of the enzymatic reactions occurring in a cell, organ, or organism. Metabolism serves four major functions: it obtains chemical energy from fuel molecules or from light; it converts exogenous nutrients into the precursors of macromolecular cell components; it assembles these precursors into macromolecules; and it synthesizes molecules to carry out specialized functions in a particular cell.

Metabolism can be broadly divided into catabolic and anabolic processes. CATABOLISM, the breakdown of complex biomolecules to release energy, occurs in three stages. Macromolecules are broken down in the first stage into their constituent subunits. These subunits are then converted into a few simple molecules ready for stage three, the complete oxidation of the simple molecules to carbon dioxide and water. Stage three, of which the Krebs cycle is an important component, is also the first stage of ANABOLISM (see BIOSYNTHETIC PATHWAY); hence stage three is both anabolic and catabolic and is termed *amphibolic*.

Particular metabolic processes are located in specific areas of the cell. Thus protein synthesis takes place on the ribosomes and the enzymes of the Krebs cycle are found in the mitochondria. There is a continuous turnover (see TURNOVER RATE) of the end products of cell metabolism. This metabolic turnover is a characteristic of all living systems.

metabolite a chemical that takes part in a metabolic reaction. See BIOSYNTHETIC PATHWAY.

metacentric describing a chromosome in

which the chromatid arms are of a similar length either side of the CENTROMERE so that the centromere appears to be near the centre of the chromosome. Compare ACROCENTRIC, TELOCENTRIC.

metachromatic stain a stain that dyes certain tissues a different colour to the dye solution. Examples are methyl violet and thionine. Thionin violet produces colours in sections ranging from blue to reddish violet and is particularly useful for distinguishing chromatin and mucin. See also STAINING.

metamer (phytomer) in plant architecture, a repeating unit made up of a node and its associated leaf and axillary buds plus part of the internodes on either side. Metamers in different positions on a plant may have different morphological characteristics.

metaphase the stage in nuclear division following PROPHASE that commences with the disintegration of the nuclear membrane and the formation of the spindle. In MITOSIS individual chromosomes, each divided into chromatids, gather around the equator of the spindle. At metaphase of the first division of MEIOSIS the BIVALENTS formed during prophase gather at the equator. The two centromeres of each bivalent are finally situated one on either side of the equator. This grouping of the chromosomes in a plane at the equator of the spindle is termed the *metaphase plate*. The placing of any one homologue on one or the other side of the equator is completely random (see INDEPENDENT ASSORTMENT), no factor operating to discriminate between 'maternal' and 'paternal' chromosomes. At metaphase of the second division of meiosis, spindles form in the haploid daughter cells resulting from the first division and the individual chromosomes become attached to the spindle equator, as in metaphase of mitosis.

The end of metaphase and the beginning of ANAPHASE, in mitosis, is marked by the division of the centromeres so that the chromatids are no longer held together. In meiosis, division of the centromeres does not occur until the end of metaphase of the second division.

metaphase plate see METAPHASE.

metaphloem late PRIMARY PHLOEM. The

metaphloem completes its elongation after the surrounding tissues have ceased elongating and therefore, unlike the PROTOPHLOEM, is not obliterated. However, in plants exhibiting SECONDARY GROWTH, the growth of the secondary phloem often results in obliteration of the metaphloem. Unlike those in the protophloem, these obliterated cells do not usually differentiate into fibres, although they may undergo sclerification (see SCLERENCHYMA). The SIEVE ELEMENTS of the metaphloem are usually wider and longer than those of the protophloem, and their SIEVE PLATES are usually more distinct. Compare METAXYLEM.

Metaphyta see PLANTAE.

metapopulation a population made up of a series of subpopulations linked by migration. This is typical of locations where there is a patchy distribution of suitable habitats, but where the distances between patches are small enough to allow relatively frequent migration.

metaxenia the effect of the pollen source on the maternal tissues of the fruit. Compare XENIA.

metaxylem late PRIMARY XYLEM. The metaxylem completes its elongation after the organ has ceased longitudinal growth and therefore its TRACHEARY ELEMENTS are not destroyed. Like the PROTOXYLEM, the metaxylem is composed mostly of tracheary elements and parenchyma cells, but SCLERENCHYMA fibres may also be present. The vessels or tracheids of the metaxylem, however, are generally wider and more numerous than those of the protoxylem and have pitted secondary cell walls. In plants not exhibiting secondary growth at maturity, the metaxylem is the only water-conducting vascular tissue. Compare METAPHLOEM.

methanogen an EXTREMOPHILE bacterium of the domain Archaea that produces methane as a byproduct of metabolism.

methanotroph a bacterium that uses methane as a nutrient. See also METHYLOTROPH.

methionine a nonpolar sulphur-containing amino acid with the formula $CH_3S(CH_2)_2CH(NH_2)COOH$ (*see illustration at* amino acid). Methionine is synthesized from aspartate via the nonprotein amino acid homoserine. Control of methionine synthesis is by feedback inhibition. Methionine is broken down to succinyl CoA, which can be further oxidized in the Krebs cycle. The methyl group of methionine is important in the formation of a large number of methylated biomolecules. The actual methyl transferring molecule is not methionine itself but its ATP-activated derivative, S-adenosyl methionine. Methionine has also been implicated in the formation of the hormone ETHENE (ethylene).

methylation the addition of a methyl group (-CH_3) to a compound. The methylation of certain nucleotides in DNA affects gene expression, as methylated DNA is not easily transcribed. Methylation is maintained during DNA replication by means of certain enzymes, thus methylation can play an important role during development by restricting the transcription of certain genes to particular cell lineages.

methylene blue a blue dye that is particularly important as a bacterial stain as it has an affinity for bacterial protoplasm.

methylotroph an organism that can use organic compounds containing only one carbon atom (e.g. methane, methanol) as its sole source of carbon and energy. See also METHANOTROPH.

mevalonic acid a precursor in the formation of TERPENOIDS and STEROIDS. It has the formula $CH_2OHCH_2C(OH)(CH_3)CH_2COOH$. Biosynthesis of GIBBERELLINS, CYTOKININS, and ABSCISIC ACID is from mevalonic acid. Certain GROWTH RETARDANTS, e.g. Amo-1618 and CCC (chlorcholine chloride), are believed to act by blocking the step from geranylgeranylpyrophosphate to kaurene in the mevalonic acid synthesis pathway, so preventing gibberellin synthesis.

micelle 1 an area within the MICROFIBRILS of a cell wall, about 50 nm in length, where the cellulose chains lie very closely parallel to each other. The resulting orderly organization of the cellobiose units (see CELLULOSE) results in a very regular arrangement of atoms in these areas so that the cellulose is essentially crystalline. **2** a clay-humus particle that is formed when finely divided mull HUMUS complexes with the clay particles in the soil. It is negatively charged and thus various positive ions are absorbed on its surface. These may be exchanged for other positive ions and the total amount of ions held in this way is termed the *exchange capacity* of the soil.

Michaelis constant (K_m) A measure of the kinetics of an enzyme reaction. K_m is the concentration of substrate (moles l^{-1}) at which the reaction proceeds at half its maximum rate. The value of K_m is different for each enzyme.

micro- prefix denoting **1** small, minute. **2** one millionth part of a particular unit, as in millimetre. **3** relating to a very small area, as in microclimate. **4** relating to something that is visible only with the aid of a microscope, as in microorganism.

Microascales an order of fungi of the ASCOMYCOTA whose members lack a stroma and produce solitary perithecial or cleistothecial ascomata. The ascomata are usually black and thin-walled, and the asci are formed in chains. The cleistothecia are characteristically beaked. There are about 79 species in some 15 genera. Most are saprobes living in soil; a few are pathogens. They include the species *Ceratocystis ulmi*, which causes Dutch elm disease. See also PLECTOMYCETES.

microbe see MICROORGANISM.

microbial genetics the study of the genetics of microorganisms. Because of the ease with which microorganisms can be cultured in the laboratory, and their rapid reproduction rate, they are ideal subjects for the study of heredity,

microbial mat a matlike community of microorganisms.

microbiology the study of the structure, biology, and ecology of microorganisms.

microbiota a collective term for soil organisms of microscopic size, such as bacteria, fungi and very small protoctists. Compare MACROBIOTA, MESOBIOTA.

microbody a small ORGANELLE in the cytoplasm, about 0.5 μm in diameter and surrounded by a single membrane. Microbodies contain oxidative enzymes. See GLYOXYSOME, PEROXISOME.

microclimate the climate of a very small

area, which may range in size from the ground under a log to a pond or a town.

microconsumer a microorganism that is a DECOMPOSER, e.g. many bacteria and fungi, which derive their nutrients from breaking down dead organic matter, releasing simpler organic and inorganic substances into the environment. See DECOMPOSITION, BIOGEOCHEMICAL CYCLE, SAPROBE. Compare MACROCONSUMER.

microcosm (microecosystem) a small-scale, simplified ecosystem created for experimental purposes, e.g. a sample of water cotaining living organisms in a jar or tank.

microcyst a resistant sporelike structure produced by a bacterial fruiting body, which may germinate to produce a new bacterium.

microecosystem see MICROCOSM.

microelement see MICRONUTRIENT.

microenvironment a relatively small, more or less isolated habitat, such as the underside of a stone.

microevolution evolution on a small scale, i.e. within a species. It is brought about by NATURAL SELECTION acting on the genetic variation between individuals of a population, which in turn is due to mutation and the RECOMBINATION of genes during sexual reproduction.

microfauna see MICROBIOTA.

microfibril a ribbon-like structure about 10–45 nm in width and visible only under the electron microscope. Microfibrils form the basic structural units of the cell wall. Each microfibril is formed from many long-chain cellulose molecules lying approximately parallel to each other. Hydrogen bonds between hydrogen and oxygen atoms in adjacent chains bind them together. The intermolecular spaces and the interfibrillar spaces are partly occupied by structural substances, e.g. pectic compounds and hemicellulose. As a result of imbibitional and capillary effects, these spaces are also filled with water so that the cell wall is normally heavily hydrated. In primary walls (see CELL WALL) water may account for up to 70% of the volume. In secondary walls, more efficient packing of the microfibrils leaves less room for water between them.

microfilament a microscopic filament, only

0.4–0.7 nm in diameter, consisting of two strands of subunits of the protein actin twisted together together in a helix. Microfilaments occur in eukaryotic cells, where they are involved in maintaining or changing the shape of the cell, cell movement (e.g. amoeboid movement), cyclosis and movement of organelles and vesicles (e.g. pinocytosis), and cytokinesis. They often form bundles or sheets just below the plasma membrane and at the interface between moving and stationary cytoplasm. Microfilaments can change length rapidly by polymerization or depolymerization of the actin subunits, and can also form complex networks. They are sometimes found associated with a myosin-like protein, as in muscle, so may also have a contractile role.

microflora small plantlike organisms, such as certain algae and fungi, found in a given area, e.g. on a leaf surface. The term is sometimes more loosely used for groups of microorganisms, including bacteria, for example in the gut of a mammal. See also MICROBIOTA.

microgamete the smaller of a pair of conjugating gametes (see CONJUGATION), often assumed to play the role of the male gamete. Compare MACROGAMETE.

micrograph a photograph taken through a microscope. *Photomicrographs* and *electron micrographs* are produced using light microscopes and electron microscopes respectively.

microhabitat see HABITAT.

microlichens the leprose, crustose, placodioid, and squamulose lichens, which are small and can be identified only with the aid of a microscope.

micrometer a device used in microscopy to measure the size of objects accurately. An eyepiece micrometer (*graticule*) is a round glass disc with an engraved scale, which is placed inside the eyepiece of a light microscope. Before using it to measure a specimen it is calibrated with a stage micrometer. This is normally a glass slide with a scale etched in, for example, one hundredths of a millimetre. The microscope is focused on the stage micrometer so the scale in the graticule can be seen just above the scale of the stage

micrometer. The size of the divisions of the eyepiece micrometer can then be calculated against the known scale of the stage micrometer.

micrometre symbol: μm. A unit of length (formerly called a *micron*, symbol μ) equal to one thousandth of a millimetre, i.e. 10^{-6} metre. It is used in the measurement of cells, cell inclusions, bacteria, etc.

micron see MICROMETRE.

micronucleus 1 during TELOPHASE of mitosis or meiosis, the smaller of the two daughter nuclei that is produced by lagging chromosomes or by chromosome fragments formed as as result of chromosome aberrations. The larger nucleus is termed the *macronucleus*. See CHROMOSOME MUTATION, NONDISJUNCTION. **2** in certain binucleate protoctists, the smaller of the two nuclei. The larger nucleus is termed the *macronucleus*.

micronutrient (microelement, trace element) any chemical element required by a plant in very small quantities for successful growth. Micronutrients thus do not fulfil major nutritional requirements but act as enzyme COFACTORS or as essential components of pigments, enzymes, etc. They include many heavy metals, such as COPPER, ZINC, MOLYBDENUM, MANGANESE, and cobalt. BORON is also required in trace amounts and it is possible that sodium, chlorine, vanadium, and SILICON may be necessary to some plants. Specific DEFICIENCY DISEASES may arise if a particular micronutrient is unavailable. Compare MACRONUTRIENT.

microorganism an organism so small that it can be seen only with the aid of a microscope.

micropalaeontology the study of microscopic fossils, including fossil microbes and microscopic parts of larger organisms.

microphyll a leaf typical of many lower plants, such as members of the Lycophyta and Sphenophyta (horsetails), that is usually small and is not associated with LEAF GAPS in the stele. Microphylls are thought to have evolved from early leafless plants by the development of *enations* (outgrowths from the axis), each with a single LEAF TRACE, although the leaf trace probably evolved after the ENATIONS. Compare MEGAPHYLL.

microplankton members of the plankton that are between 20 μm and 200 μm in diameter. Compare NANOPLANKTON.

micropropagation the production of large numbers of plants in the laboratory. This may involve seeds or, if genetically identical plants (clones) are required (e.g. when multiplying genetically engineered plants), tissue culture. Micropropagation is also important for producing disease-free plants, and for increasing the populations of rare plant populations for preservation in botanic gardens, for reintroduction to suitable habitats, or for study.

micropyle (*adj.* micropylar) the small channel that remains between the tips of the integuments at the apex of the ovule. The pollen tube usually enters the nucellus via the micropyle prior to fertilization. The micropyle may persist as a small pore in the seed through which water is absorbed prior to and during germination.

microRNA see JUNK DNA.

microsatellite DNA (microsatellite marker) a short rereated sequence of DNA, usually 1–6 nucleotides in length, in a DNA molecule, which occur next to a particular gene. The repeat may or may not be separated by other nucleotides. There may be different numbers of repeats in different members of the same species, leading to POLYMORPHISM. Such repeats also make it possible to have many different alleles of a gene at the same locus.

microsatellite marker see MICROSATELLITE DNA.

microscope an instrument for producing a magnified image. See LIGHT MICROSCOPE, ELECTRON MICROSCOPE.

microsere see SERE.

microsome (*adj.* microsomal) a small VESICLE that forms from ENDOPLASMIC RETICULUM. Microsomes separate out during centrifugation of cell homogenates. Fragments of membrane are seldom found after centrifugation, and, with their continuous membranes, microsomes thus demonstrate the ability of membranes to repair themselves.

microspecies one of a series of populations derived from the same parent, which are morphologically similar but rarely exchange genes.

Microsporales an order of unbranched filamentous green algae (CHLOROPHYTA) of the class CHLOROPHYCEAE that have uninucleate cells, each with a single large chloroplast without pyrenoids, and no intercellular connections. Some species are floating, while others are anchored by holdfasts. Reproduction is by FRAGMENTATION of an unusual kind: overlapping H-shaped pieces of cell wall are shared between neighbouring cells; as the cells divide, new H-pieces are interpolated between the existing ones.

microsporangiophore a stalklike structure bearing microsporangia. See also MICROSPOROPHYLL.

microsporangium a structure in which MICROSPORES are formed. In the seed plants it corresponds to the POLLEN SAC. Compare MEGASPORANGIUM.

microspore (androspore) the smaller of the two types of spore formed after meiosis in heterosporous species and usually designated male. In seed plants it is the POLLEN grain.

microspore culture see POLLEN CULTURE.

microspore mother cell (microsporocyte) a cell that gives rise by meiosis to four haploid cells that develop into MICROSPORES. Unlike the tetrad formed by meiosis of the MEGASPORE MOTHER CELL, all four cells usually develop into microspores. Pollen mother cells are examples of microspore mother cells.

microsporocyte see MICROSPORE MOTHER CELL.

microsporogenesis the development of MICROSPORES from a MICROSPORE MOTHER CELL.

microsporophyll a modified leaf that bears the MICROSPORANGIA. In angiosperms and gymnosperms it is represented by the stamens and male scales respectively. Compare MEGASPOROPHYLL.

microstrobilus a STROBILUS that bears only microsporangia, for example the male cones of some conifers, cycads, and ginkgos.

microthallus in algae, a vegetative stage in the life cycle that is visible only with a microscope.

microtome a device used for cutting thin sections (3–5 μm thick) of plant or animal material for examination under the microscope. The specimen is often supported, either by embedding it in a suitable medium, such as paraffin wax, or by freezing it. There are various types of microtome; all have a means of holding the specimen, a knife, and a mechanism for moving the specimen slowly towards the knife. *Ultramicrotomes* are used for cutting very thin sections (20–100 nm) needed for electron microscopy. Glass or diamond knives are used and the specimen advanced by minute increments by thermal expansion. The sections are floated in a water- or water-and-acetone-filled trough surrounding the knife and collected by transferring them onto a fine copper grid. When cutting frozen sections (*cryo-microtomy*), both the cutting knife and the specimen are enclosed in a cold chamber.

microtope the actively growing upper layer of a BOG, which consists of the living parts of sphagnum mosses (see SPHAGNALES) and the roots of vascular plants.

microtubule an unbranched tubule identified by electron microscopy in a wide variety of cells. Microtubule numbers vary considerably in different kinds of cell and through the life of a cell. They may either be arranged in an orderly manner or randomly distributed through the cytoplasm. Their outer diameter is about 25 nm and they have an apparently hollow core, 4 nm in diameter. The walls are composed of subunits of a protein, TUBULIN, which are arranged linearly to form protofilaments. Thirteen protofilaments arranged in a cylinder form the wall of each microtubule.

Microtubules form the axoneme fibres of eukaryotic cilia and undulipodia and the SPINDLE fibres in dividing cells. They are also thought to organize the microfibrils of cell walls and may be involved in the transport of materials within and between cells, e.g. through sieve tubes.

microtubule organizing centre (MTOC) a general term for KINETOSOMES, CENTRIOLES, and other structures involved in the formation and organization of microtubules or their derivatives.

micton a widely distributed species formed as a result of HYBRIDIZATION between two or more species, but in which APOMIXIS

does not occur, all individuals being cross-fertile.

middle lamella the first-formed layer of the primary wall (see CELL WALL) formed from the PHRAGMOPLAST. Compounds of pectic acid, e.g. calcium and magnesium pectates, are deposited into the cavities of the phragmoplast and these eventually coalesce to form a continuous opaque layer. Cellulose molecules, organized to form MICROFIBRILS, are deposited to build up the primary wall. In mature tissues the middle lamella cements the walls of contiguous cells.

midrib 1 the vein running down the middle of a leaf from the petiole or leaf base to the leaf tip, often dividing the leaf into similar halves or mirror images.
2 any thickened region of cells running down the centre of a thallus or lamina, such as the midrib of wracks (*Fucus*). See also NERVE.

migration 1 the movment of individual organisms or their propagules (e.g. plant parts such as buds) from one location to another. 2 in plant succession, the arrival of migrating propagules (termed *migrules*) at a newly exposed site.

migrule see MIGRATION.

mildew a plant disease in which growth of the causative fungus is seen on the surface of the host. Downy (or false) mildews penetrate deeply into their hosts, while powdery (or true) mildews live on the surface of their hosts in a similar way to MOULDS.

millimicron see NANOMETRE.

Millon's test a technique used to demonstrate the presence of the amino acid TYROSINE. Millon's A, a mixture of sulphuric acid and mercury(II) sulphate is added to the test solution. A yellow colour indicates the presence of protein. A drop of Millon's B, a sodium nitrite solution, is then added. A red colour indicates a positive reaction.

mineral cycle see BIOGEOCHEMICAL CYCLE.

mineralization the conversion of dead organic material to inorganic material as a result of decomposition by microorganisms. The substitution of inorganic material for organic consitituents is one of the main processes involved in fossilization, e.g. 'petrified forests'.

mineral soil SOIL which contains so little organic matter that its properties are determined mainly by its mineral content.

minerotrophic describing vegetation that receives dissolved nutrients from groundwater.

minimal area the smallest area that contains the species typical of a particular plant COMMUNITY. Its size will depend on the degree of heterogeneity of the community. If the species are sampled using increasing sizes of sample units, and the number of species is plotted against area (giving a *species-area curve*), the point at which the curve becomes horizontal is the minimal area.

Miocene see TERTIARY.

mire an area of wet, boggy ground, especially on peaty soil.

mirror yeasts (shadow yeasts) members of the order Sporobolomycetales, comprising ANAMORPHS of basidiomycetes that produce ballistospores. They were formerly grouped together with the Cryptococcales (which do not produce ballistospores) in the BLASTOMYCETES, but are now assigned to the polyphyletic phylum FUNGI ANAMORPHICI. Many (e.g. *Sporobolomyces* and *Tilletiopsis*) are found as epiphytes on leaves.

mismatch repair the enzyme-mediated repair of mispaired nucleotides produced during DNA replication. See DNA REPAIR.

missense mutatation a MUTATION in a gene, in which one of the codons of the genetic code has been altered, to that it codes for a different amino acid. The resulting protein is usually unable to function normally.

Mississippian see CARBONIFEROUS.

mitochondrial DNA (mtDNA) the DNA found in MITOCHONDRIA. It is a circular molecule and is not associated with histones or other proteins. mtDNA functions independently of nuclear DNA, coding for certain RNA components of ribosomes, and for certain respiratory enzymes that are synthesized on the mitochondrial ribosomes. Unlike animal mtDNA, plant mtDNA evolves very slowly, and is capable of breakage and recombination. Since mitochondria seldom pass into the male gametes, it is passed

from generation to generation in the cytoplasm of the female gametes.

mitochondrion an ORGANELLE, found in all eukaryotic cells, that provides an efficient apparatus for the production of ATP. The number present varies considerably, being greatest in metabolically active cells. Mitochondria vary in form being generally spherical or threadlike and there is consistency of form in any one cell type. They are approximately 1–3 μm across and are generally freely distributed in the cytoplasmic MATRIX, though tending to concentrate in regions of the cell where the demand for ATP is high. The mitochondrion is surrounded by two membranes. The outer one is smooth but the surface area of the inner one is very greatly increased by infoldings that extend to varying distances into the central compartment, forming shelflike structures, or *cristae*. The central compartment contains a colloidal matrix with a fine fibrillar structure. Ribosomes of the smaller 70S type and circular DNA molecules, both characteristic of prokaryotic cells (see ENDOSYMBIOTIC THEORY), are present, as are DNA polymerase enzymes and enzymes of the Krebs cycle. In cell cultures, mitochondria display twisting and wriggling movements and the cristae change their shape when respiratory activity is stimulated.

The inner surface of the crista has many knoblike structures uniformly distributed in the membrane. These are composed of a protein, F_1, capable of transferring phosphate to ADP (i.e. an ATP synthetase enzyme). Electron transfer compounds, required for the eventual formation of water from hydrogen ions and oxygen, are situated within the membrane of the crista. The energy made available by electron transfer is incorporated into ATP molecules formed by the activity of the F_1 protein particles. The outer membrane is freely permeable to water and soluble ions but the inner membrane has limited permeability and contains carrier compounds that transfer specific metabolites. One such specific carrier allows a molecule of ADP to enter only if a molecule of ATP passes in the reverse direction. Another concentrates calcium ions in the matrix.

Mitochondria can only arise by fission of preexisting ones and therefore must be transmitted to daughter cells during MITOSIS. The DNA and ribosomes they contain enable them to synthesize many of their constituent molecules.

mitogenic capable of inducing mitosis.

mitosis the process by which a cell divides to form two daughter cells, each having a nucleus containing the same number of chromosomes, with the same genetic composition, as that of the mother cell. The changes that occur in the structure of the chromosomes and in the cytoplasm are clearly visible with a light microscope and form a sequence of stages; PROPHASE, METAPHASE, ANAPHASE, TELOPHASE, and CYTOKINESIS. As a result of mitotic divisions, all the cells of the sporophyte of higher plants have diploid nuclei that are genetically identical to that of the fertilized egg cell. Similarly all the cells of a gametophytic plant body have haploid nuclei genetically identical to the gamete from which the plant arose.

Mitosporic Fungi see FUNGI ANAMORPHICI.

mitotic apparatus a collective name for the asters (CENTROSOMES), SPINDLE and the fibres that link the chromosomes to the centrosomes.

mitotic crossing over (somatic crossing over) a process that results in genetic recombination occurring in somatic cells. It was first discovered as a rare event in *Drosophila* (fruit flies) but has since been found to occur in other organisms, especially fungi. In the latter, it sometimes occurs at a frequency that suggests it is a major source of genetic variation. The mechanisms involved are uncertain, since chromosomal pairing (a supposed requirement of crossing over in meiosis) is normally believed not to occur during mitosis. Mitotic crossing over can result in genotypically and phenotypically different cell lines in the same organism.

mitotic index the proportion of cells, usually the fraction of cells in 1000 cells, that are undergoing MITOSIS in a given sample.

mitotic recombination the RECOMBINATION of genetic material during mitosis and

asexual reproduction. This is responsible for generating genetic VARIATION in heterokaryons.

mitotic spindle see SPINDLE.

mixed lactic fermentation (heterolactic fermentation) a form of ANAEROBIC RESPIRATION found in certain microorganisms, e.g. *Lactobacillus brevis*, in which the end products are a molecule each of lactic acid, ethanol, and carbon dioxide.

mixed layer in an ocean or large lake, the layer of water above the THERMOCLINE, which is subject to mixing by wind action.

mixed sorus see SORUS.

mixed woodland a woodland in which in minority of trees, but at least 20% of all the trees, are either broad-leaved or coniferous.

mixoploid describing a cell population in which different cells have different chromosome numbers. See MOSAICISM.

mixotrophy the possession of both autotrophic and heterotrophic nutrition.

mode the class of observations that occurs most frequently, for example the number 3 in the series 9, 3, 3, 1. Compare MEAN, MEDIAN.

moder see HUMUS.

modern synthesis see NEO-DARWINISM.

modifier a gene that modifies the expression of another gene.

modulator see ALLOSTERIC ENZYME.

moisture balance see MOISTURE BUDGET.

moisture budget (moisture balance) an expression of the water flux in an ecosystem when in a state of equilibrium:
precipitation = runoff + evapotranspiration + change in soil moisture storage

Over the course of a year, a given system will be more or less in equilibrium as winter rainfall and sometimes snowmelt replenish soil moisture lost by high evapotranspiration rates in summer.

moisture index 1 the proportion of total precipitation that is used by the vegetation. 2 a measure of the effectiveness of precipitation for plant growth that takes into account the weighted influence of water surplus and deficiency in relation to water and its variation with the seasons. For a given location, this is calculated by the formula

I_m = humidity index – 0.6 (aridity index) giving

$$I_m = (100s - 60d)/n$$

where I_m is the moisture index, s the water surplus, d the water deficiency, and n the water need. s and d are calculated as the difference between monthly precipitation and monthly potential evapotranspiration, and n is the annual potential evapotranspiration. In the above equation s is the total surplus from all the months in which a water surplus occurred, and likewise d is the total of all the monthly deficiencies.

mold see MOULD.

molecular clock a means of tracking the course of EVOLUTION using certain molecules that are assumed to mutate at known, regular rates. For tracking molecules over relatively short periods of evolutionary time, MITOCHONDRIAL DNA (mtDNA), which mutates about once every 2000–3000 years, is used. Chromosomal DNA cannot be used, as its mutations are often repaired by associated enzymes (see DNA REPAIR). mtDNA also has the advantage that the mitochondria are passed on in the cytoplasm of the egg, but not in the sperm/antheridia, so there is no mixing of genetic material at fertilization, making the evolutionary evidence easier to interpret. mtDNA is not so useful in plants, because plant mtDNA mutates much more slowly. For tracking changes over a longer period of time, the slower-ticking clocks of cytochrome c or ribosomal RNA are used. See also MUTATION.

molecular evolution the gradual changes in the composition of proteins that results from substitution of one amino acid for another as a result of random mutation. This is thought to result from GENETIC DRIFT rather than natural selection. See NEUTRAL THEORY, PROTEIN SYNTHESIS.

molecular-exclusion chromatography see GEL FILTRATION.

molecular genetics a branch of genetics dealing with molecular aspects of genetic mechanisms.

molecular marker a specific NUCLEOTIDE sequence that is linked to a particular gene locus or phenotypic characteristic, which can be detected and visualized by molecular techniques.

molecular systematics a form of TAXONOMY that seeks to derive evolutionary (phylogenetic) relationships from comparisons of functionally equivalent macromolecules from different organisms, for example amino acid sequences in proteins such as enzymes, or nucleotide sequences in nucleic acids (e.g. ribosomal RNA). The closer the composition of the macromolecules, the closer the relationship between two organisms.

Molisch's test see ALPHA-NAPHTHOL TEST.

molybdenum symbol: Mo. A metal element, atomic number 42, atomic weight 95.94, needed (under certain conditions) in trace amounts for plant growth. It is involved in nitrate reduction being a part of the FLAVOPROTEIN enzyme nitrate reductase. Plants absorbing nitrogen as ammonium ions do not need molybdenum. The DINITROGENASE system involved in NITROGEN FIXATION also requires molybdenum since one of the two proteins forming the dinitrogenase complex is a molybdenum–iron–sulphur protein.

monadelphous describing STAMEN filaments that are all fused for the greater part of their length, so forming a tube around the style.

Monera see BACTERIA.

Moniliales a group of anamorphic (asexual) fungi comprising the former BLASTOMYCETES and the HYPHOMYCETES.

moniliform resembling a string of beads.

monoallelic describing a polyploid in which all the alleles at a particular locus are identical.

monocarpellary see FRUIT, GYNOECIUM.

monocarpic see SEMELPAROUS.

monocaryon see MONOKARYON.

monocentric 1 describing a CHROMOSOME with a single centromere, as in most chromosomes. 2 describing a cultivated species with only one gene centre (centre of diversity).

monochasial cyme see INFLORESENCE.

monochasium see INFLORESCENCE.

monoclimax (see CLIMAX THEORY).

monoclinous see HERMAPHRODITE.

monocolpate describing a pollen grain having one COLPUS, as is commonly found in most petaloid monocotyledon species.

Monocotyledonae (Liliopsida, monocotyledons) a class of the

ANTHOPHYTA (angiosperms) containing all the flowering plants having embryos with one cotyledon. Most are herbaceous and do not possess a cambium, hence they lack secondary thickening although some families, e.g. Arecaceae (palms), have arborescent forms. Other features by which the Monocotyledonae can generally be distinguished from the DICOTYLEDONAE include: flower parts inserted in threes, or multiples thereof; a fibrous root system composed of ADVENTITIOUS roots; and numerous scattered vascular bundles (see ATACTOSTELE). There are exceptions to all of these features, however. The 60 or so families of the monocotyledons have been divided into a varying number of subclasses (or superorders) by different authorities. Most commonly five subclasses are recognized: ALISMATIDAE, ARECIDAE, COMMELINIDAE, ZINGIBERIDAE, and LILIIDAE.

monoculture the growing of a single species or variety of crop over a large area. This provides economies of scale and a uniform height of plant for mechanical harvesting. However, it makes the crop highly susceptible to pathogens and pests, as these can easily move from host to host.

monoecious (*n.* monoecy, monoecism) having the female and male reproductive organs separated in different floral structures on the same plant. Monoecy decreases the chances of SELF POLLINATION and is often associated with wind pollination, as in hazels (*Corylus*) and maize (*Zea mays*). The genetic control of sexual expression in monoecious plants is very complex and not clearly understood but it appears to involve combinations of genes that either promote or suppress the formation of male or female flowers. Monoecy is strongly associated with particular plant groups, e.g. Fagaceae and other temperate trees, sedges (*Carex* spp.), and spurges (*Euphorbia* spp.). Compare HERMAPHRODITE, DIOECIOUS.

monogenic describing a trait determined by the ALLELES at only one gene locus. Compare MULTIGENIC.

monogerm (*adj.* monogermous) a fruit that contains only one ovule/seed.

monohybrid an individual that is heterozygous in respect of a single gene. It is obtained by crossing parents that are homozygous for different alleles, e.g. a homozygous tall parent (TT) crossed with a homozygous short parent (tt) will produce monohybrid offspring (Tt). When selfed, a 3:1 ratio of dominant:recessive phenotypes, or some modification thereof, will be produced. This is called the *monohybrid ratio*. Compare DIHYBRID.

monohybrid heterosis see OVERDOMINANCE.

monohybrid ratio see MONOHYBRID.

monokaryon (monocaryon) a fungal hypha or mycelium that contains only one nucleus per cell.

monolete describing a spore that has a simple linear scar marking the point at which it was joined in the tetrad. The microspores of members of the ISOETALES are monolete, which sets them apart from other LYCOPHYTA. However their megaspores are trilete (see TRIRADIATE SCAR).

monomictic lake a lake in which vertical mixing of the water takes place only once a year. Compare DIMICTIC LAKE, MEROMICTIC LAKE.

mononucleotide see NUCLEOTIDE.

monophyletic describing taxa arising from the diversification of a single ancestor or ancestral population, i.e. natural groups in a classification. In cladistics only taxa that contain *all* the descendants of a common ancestor are considered monophyletic. An example of such a taxon would be the monocotyledons. Taxa that contain some but not all the descendants of an ancestor are termed PARAPHYLETIC. Compare POLYPHYLETIC.

monoplanetic describing fungi that produce only one type of zoospore (planospore). Compare DIPLANETIC.

monoploid see HAPLOID.

monoplontic 1 a haploid individual. 2 the haploid phase of a life cycle.

monopodial branching a type of growth exhibited by many plants in which secondary shoots or branches arise behind the growing point but remain subsidiary to the main stem, which continues to grow indefinitely. The largest secondary shoots are furthest from the apex of the main stem and the size of the shoots decreases regularly towards the top of the plant. This results in the pyramidal form of growth typical of many conifers, e.g. the spruce (*Picea*). The secondary shoots also tend to show the same pattern of branching along their length. Compare SYMPODIAL BRANCHING.

monosaccharide a carbohydrate with the empirical formula $(CH_2O)_n$; in organic compounds n is between three and seven. The carbons in a monosaccharide are usually arranged in an unbranched chain with all carbons but one being hydroxylated; the remaining carbon is either ketonic (KETOSE sugars) or aldehydic (ALDOSE sugars). The commonest monosaccharides are the *hexoses*, with six carbons (e.g. glucose and fructose), and the *pentoses*, with five carbons (e.g. ribose).

All monosaccharides except dihydroxyacetone are *chiral* molecules, i.e. they exhibit stereoisomerism. Most naturally occurring monosaccharides are in the D-form, although some L-isomers do occur. In aqueous solution most monosaccharides form a ring structure in which the aldehydic or ketonic carbon links with one of the hydroxylated carbons. Two forms of ring, the PYRANOSE and the FURANOSE, are commonly seen. The ring structure of a monosaccharide has two isomeric forms, designated α and β.

monosome a CHROMOSOME that lacks a homologue in a DIPLOID organism. See HOMOLOGOUS CHROMOSOMES.

monosomic an organism deficient in one chromosome from an otherwise diploid set, i.e. $2n - 1$. Monosomy typically arises by fertilization occurring between a normal gamete (n), and one deficient in the said chromosome $(n - 1)$. Monosomics are more common in polyploids where the deleterious consequences of missing a chromosome can be masked by chromosomes in other genomes. See ANEUPLOIDY.

monospermous bearing only one seed.

monospore a spore formed by certain red algae (see RHODOPHYTA/) of the BANGIOPHYCIDAE as a means of asexual reproduction.

monostele see STELE.

monostichous describing the arrangement of leaves on a stem in which there is only one leaf per node, arranged in a single row all on one side of the stem. Compare DISTICHOUS, TRISTICHOUS.

monotelic describing an inflorescence in which the main axis and all the lateral axes end in a flower. Compare POLYTELIC.

monothetic in a numerical classification scheme, the use of only one attribute or criterion as the basis for each subdivision, e.g. association analysis. Compare POLYTHETIC.

monotrichous describing a bacterium that has a single polar FLAGELLUM. Compare LOPHOTRICHOUS, PERITRICHOUS.

monotypic describing any taxon that includes only one subordinate taxon. Thus a family with just one genus or a genus with a single species are examples of monotypic taxa. An extreme case of monotypy occurs in the classification of the maidenhair tree, *Ginkgo biloba*. This species is the sole representative of the genus *Ginkgo*, the family Ginkgoaceae, the order Ginkgoales, and the phylum Ginkgophyta – each being monotypic taxa.

monovalent enzyme SEE ALLOSTERIC SITE.

monsoon forest see FOREST.

montane a mountain FOREST of temperate and tropical latitudes, whose species composition is more similar to that of lowland forests at higher latitudes than to forests at lower altitudes in the same latitude.

month degrees the number of degrees by which the average monthly temperature of a particular site differs from a standard temperature, such as the minimum temperature required for growth or flowering. A standard minimum temperature of 6°C is often used. Month degrees are useful when making comparisons between growing seasons, and for determining in which locations conditions may be suitable for growth.

Montpellier school of physociology see ZURICH–MONTPELLIER SCHOOL OF PHYTOSOCIOLOGY.

Montreal Protocol on Substances that Deplete the Ozone Layer an international agreement set up at a convention in Montreal, which came into force on 1 January 1989. Its aim was to agree international controls on emissions of substances that contribute to the depletion of the ozone layer (see OZONE HOLE). It was initially ratified by some 29 countries and the member states of the EC, which between them accounted for 82% of the world's consumption of ozone. By 1998 more than 265 countries had ratified the agreement.

moorland land that is found in wet exposed conditions where the soil water can seep laterally very slowly but is not stagnant as in BOGS. It has an acid peaty soil and in Britain is usually dominated by heather (*Calluna vulgaris*). Other common plant species found in Britain include mat grass (*Nardus stricta*) and purple moor-grass (*Molinia caerulea*). The top soil loses water in summer but the subsoil is permanently waterlogged. The peat layer is rarely more than 30 cm deep and most of the plants have their roots in the mineral soil horizon (see SOIL PROFILE). In many moorland areas this vegetation type is maintained by periodic fires.

mor see HUMUS.

mordant see STAINING.

morel the edible ascoma (fruiting body) of fungi of the genus *Morchella*, order PEZIZALES.

Morgan unit see CHROMOSOME MAPPING, LINKAGE MAP.

morph see STATES OF FUNGI.

morphactin any of a group of synthetic compounds, all derivatives of fluorene-carboxylic acid, that affect many aspects of plant growth and DIFFERENTIATION. Morphactins have little effect on mature tissues but in seedlings and growing shoots APICAL DOMINANCE is removed and internode elongation inhibited, giving a dwarfed bushy appearance. Expansion of the leaf lamina is also prevented and flowers, if produced, may be deformed. Seeds from morphactin-treated plants may contain abnormal embryos. Morphactins applied to seeds slow down germination and inhibit lateral root formation from the radicle. In legumes, ROOT NODULE formation is prevented. Other diverse effects of morphactins include the inhibition of PHOTOTROPISM and GRAVITROPISM, the induction of

parthenocarpy, and an alteration in sex expression in dioecious plants.

Morphactin is thought to induce dwarfism by affecting the orientation of the spindle in dividing cells. It also disrupts the formation of a middle lamella between daughter cells. Although morphactins superficially resemble gibberellins in structure their effects are not thought to be brought about by interaction wth gibberellin or any of the other natural hormones.

Morphactins may be used commercially to slow growth of, for example, mixed plant populations at roadsides.

morphogenesis the developmental changes that give rise to the adult form from the zygote.

morphology (*adj.* morphological) the study of form, particularly external structure. Compare ANATOMY.

morphosis modification of an individual organism's MORPHOGENESIS as a result of environmental changes.

morphospecies a group of organisms that differs in MORPHOLOGY from all other groups of organisms.

mosaic 1 an irregular pattern of small light and dark green areas on the foliage of a plant. Infection with a virus is the usual cause. Examples include tobacco mosaic and cucumber mosaic.

2 a leaf mosaic. See LAMINA.

3 an organism that is made up of cells that have different genotypes although they have developed from the same zygote. See also CHIMAERA.

mosaic evolution evolution in which different adaptive characteristics, e.g. leaves and flowers, evolve at different rates.

mosaicism the occurrence of different chromosome numbers or chromosome structures in different tissues of the same individual.

mosses see BRYOPHYTA.

mossy forest (moss forest) a tropical MONTANE forest, usually consisting of twisted low-growing or dwarf trees whose trunks, branches and twigs are draped with mosses, liverworts and lichens. These growths are particular luxuriant where mists and fogs are frequent.

mother cell a cell that divides to form other

differentiated cells and hence loses its identity, for example, the cell that gives rise to a sieve element and companion cell. See also POLLEN MOTHER CELL.

motile capable of independent locomotion.

mottle a viral disease of plants that causes leaves to develop patches or more diffuse areas of yellow.

mottling the production of patches of different colours, usually rusts and greys or blues, in soils due to repeated periods of anaerobic conditions, which affect the oxidation states of certain soil minerals, especially iron compounds.

mould a fungus that produces a distinct mycelium or spore mass, which often resembles a velvet-like pad, on the surface of its host. Moulds frequently occur on dead or decaying vegetable matter, such as food or stored fruits. Common examples are white bread mould (*Mucor*) and blue and green moulds of citrus fruit (*Penicillium italicum* and *P. digitatum*). See also MUCORALES, HYPHOMYCETALES, EUROTIALES.

The term 'mould' is also used loosely for some fungus-like protoctists.

mounting the final stage in the preparation of permanent slides for microscopical examination. The mount, which may be a section, wholemount, squash, or smear, is immersed in a mounting medium, usually a liquid that later solidifies. The medium permeates entirely through the specimen, leaves no air spaces, and does not support the growth of bacteria and fungi. Examples are glycerol, Canada balsam, and osmium tetroxide.

mtDNA see MITOCHONDRIAL DNA.

MTOC see MICROTUBULE ORGANIZING CENTRE.

mucilage (*adj.* mucilaginous) any substance that swells in water to form a slimy solution. Mucilages are usually concerned with water retention; for example, the pentosan mucilages produced in the interior of succulent xerophytes serve to increase the water-holding capacity of the cells and hence reduce the transpiration rate. Many seeds have a mucilaginous coating that aids water uptake during germination. Structurally, mucilages are very complex: linseed mucilage is a mixture of a complex polyuronide, proteinaceous matter, and cellulose. Mild hydrolysis of the polyuronide yields xylose and galactose residues and a more resistant fraction consisting of galacturonic acid (see URONIC ACIDS) and rhamnose residues.

mucocyst a vesicle containing mucilage.

mucopeptide see PEPTIDOGLYCAN.

mucopolysaccharide (glycosaminoglycan) any of a group of POLYSACCHARIDES that contain amino sugars (e.g. glucosamine). They are able to trap water to form an elastic gel. When mucopolysaccharide units are bound to protein they form a *mucoprotein*.

mucoprotein any complex consisting of proteins joined to MUCOPOLYSACCHARIDES (especially glycosaminoglycans). Mucoproteins are found in the cell walls of some bacteria. Some of them have been found to have anti-tumour activity.

Mucorales an order of the ZYGOMYCOTA containing about 300 species (56 genera) of mostly saprobic fungi, many of which cause spoilage of stored food. Its members generally form a dense mycelium from which arise sporangiophores or conidiophores. The tips of these may become pigmented, hence the common name 'pin mould' for *Mucor mucedo*. In the coprophilous fungus *Pilobolus* a specialized mechanism has developed to discharge the sporangia violently. ZYGOSPORES are formed by sexual reproduction and many species are HETEROTHALLIC.

mucronate having a small fine point (a *mucro*; see APICULUS) arising abruptly, usually at the tip.

mugeinic acid an organic acid exuded by the roots of some grasses (e.g rye), which acts as a chelating agent, and is involved in the uptake of heavy metal ions from the soil. See BIOACCUMULATION, CHELATION.

mulch organic material, such as peat, leaf mould, or shredded bark, that is spread on the ground to suppress annual weeds. It is also applied around the base of trees and shrubs to help absorb and retain water and add nutrients.

mule a plant hybrid that is sterile due to infertile pollen or undeveloped pistils.

mull see HUMUS.

Müllerian body see FOOD BODY.

multicellular describing an organism that is made up of more than one cell.

multifactorial inheritance see MULTIPLE-FACTOR INHERITANCE.

multigene family a collection of identical or almost identical genes in the same GENOME. The numbers of gene copies and their distribution between the different chromosomes can be highly variable even between quite closely related species, depending on the particular GENE FAMILY.

multigene variety a variety that carries several genes conferring resistance to a particular pathogen.

multigenic describing a trait controlled by many genes. Compare MONOGENIC.

multigerm an aggregate fruit that contains several ovules.

multilocular see OVARY.

multinucleate describing a cell that has more than one nucleus, or an organism whose cells have more than one nucleus each.

multiple alleles a series of alleles of a particular gene. In theory probably all genes can exist in more than two alternative forms since mutations can occur at any number of points along their length. In practice, natural selection may eliminate all but one or two of these in wild populations. Examples of naturally occurring multiple allelism are found in the INCOMPATIBILITY systems of plants. For example, in Brussels sprouts (*Brassica oleracea* var. *bullata*) there are some 35–40 different incompatibility alleles, termed *S* alleles.

multiple-factor inheritance (multifactorial inheritance) the determination of an inherited characteristic by POLYGENES, so that it shows approximately continuous variation from one extreme to another. The size of beans, ear-length in maize, and many other characteristics show multiple-factor inheritance, and do not, at first sight, behave like Mendelian characters. Compare SINGLE-FACTOR INHERITANCE. See CONTINUOUS VARIATION.

multiple fruit (composite fruit) a fleshy fruit that incorporates a complete inflorescence and is thus derived from the ovaries of many flowers. It may also incorporate other floral parts and the receptacles. The COENOCARPIUM, SOROSIS, and SYCONIUM are multiple fruits.

multiple land-use strategy a management strategy that allows a range of activities, such as nature conservation, camping, walking, flood control, forestry, to take place sustainably in a given area.

multiseriate filament a filament in which the cells are arranged in more than one row (series).

multivalent an association of three or more HOMOLOGOUS CHROMOSOMES observed during meiosis in polyploid or polysomic organisms. See TRIVALENT, QUADRIVALENT.

Munsell soil colour system a standard scheme for assigning names to colours in a soil, to help in distinguishing and identifying different soil horizons (see SOIL PROFILE).

muramic acid a substance found in the PEPTIDOGLYCAN walls of EUBACTERIA, but lacking from the ARCHAEA (archaebacteria).

murein a substance made up of polysaccharides cross-linked by amino acids to form a rigid framework, constituting a major component of bacterial cell walls. It is found only in prokaryotes, being the equivalent of the cellulose of plant cell walls.

muricate having a surface covered by sharp points or prickles or hard short projections.

muriform having a pattern like that of a brick wall.

Musci see BRYOPHYTA.

muscicolous growing on or among mosses (Bryophyta).

mushroom the usually umbrella-like fruiting body or sporophore of fungi in the AGARICALES. It is a pseudoparenchymatous structure (see PLECTENCHYMA) composed of numerous fused hyphae, and is differentiated into a stalk or *stipe* and a circular cap, the *pileus*. In immature mushrooms these structures are united by a membrane, the *universal veil*, but this breaks as the stalk elongates leaving a cuplike structure, the *volva*, around the base of the stalk. In some mushrooms the edge of the cap is united with the stalk by a second membrane, the *partial veil*. This ruptures to expose the underside of the cap leaving a ring, the *annulus*, towards the top of the stem. The lower side of the cap is composed either of GILLS or pores, whose surfaces are covered by the hymenium on which the basidiospores are produced.

Mushroom is often used in a wider sense to mean any macroscopic fungal fruiting body. The term *toadstool* is essentially synonymous with mushroom in both the narrow and broad senses, but is more often used of inedible species.

muskeg a peat bog characteristic of the northern coniferous forest of North America. Plants commonly found are pitcher plants, sundews, and black spruce (*Picea mariana*), which thrives in the moist conditions.

mutability the ability of a gene to undergo MUTATION. Different genes may differ in their susceptibility to mutation. See MUTABLE GENES.

mutable genes a class of genes that frequently undergo spontaneous mutation.

mutagen (*adj.* mutagenic) any agent that causes an increased frequency of MUTATION. Mutagens are typically short-wave electromagnetic radiations (e.g. ultraviolet irradiation, X-rays, and cosmic rays), ionizing radiations (e.g. α- and β-particles), and chemicals (e.g. nitrous acid and proflavin) that react with nucleotides. A fourth category of mutagens, BASE ANALOGUES, do not alter existing bases but are incorporated in place of normal bases during DNA replication. At subsequent replications base analogues may 'mispair', resulting in substitution of the 'wrong' base into DNA, and thus an altered codon.

The action of the various types of chemicals and radiation is more variable. Sometimes, as with nitrous acid, the base pairing specificity of the nucleotides is altered, subsequently resulting in incorporation of the 'wrong' base, as with base analogues. Sometimes, as with ultraviolet irradiation, adjacent nucleotides are caused to complex with each other instead of bonding to bases in the complementary helix. This may weaken the double helix and result in chromosome breaks.

mutagenicity the degree to which particular agents or situations are likely to induce mutations.

mutagenesis the process by which a mutation is produced.

mutant site the site on a CHROMOSOME at which a mutation has occurred.

mutation an inherited change in the genes as inferred from an inherited change in the appearance of the offspring. The term can be extended to include genetic changes occurring in somatic cells (*somatic mutations*). The latter can result in abnormalities of growth such as CHIMAERAS. They are not inherited unless the mutant tissue gives rise to a reproductive shoot.

There are two main categories of mutations: GENE MUTATIONS, which are alterations in a single gene; and CHROMOSOME MUTATIONS, which may involve large numbers of genes. Mutations are characteristically rare, harmful, and recessive. The frequency of spontaneous gene mutations differs with different gene loci. For example in maize, the mutation giving rise to shrunken endosperm occurs about once in every million replications of the gene locus. However the mutation giving a colourless aleurone layer occurs more frequently, about 110 times in every million replications of the locus. Reversions of a mutant gene to a normal gene occur less frequently. The low frequency of spontaneous gene mutations may be attributed to the great stability of DNA. One reason for the occurrence of spontaneous mutation is that all the bases can also exist in very rare isomeric forms. Thus while adenine normally pairs with thymine, in its rare isomeric form adenine will pair preferentially with cytosine. Over two replications this could result in a spontaneous switch from an AT pair to a GC pair in the DNA. The low spontaneous frequency of mutation can be increased by various agents called MUTAGENS.

The harmful effect of most mutations is due to the fact that they occur randomly. A purely random change to a normal allele (gene) is more likely to result in a defective protein than one that confers a selective advantage on its owner. For similar reasons most mutations are recessive. In a heterozygote the effect of a mutant allele that produces a nonfunctional protein is likely to be masked by the normal allele, which produces a functional protein. The former would therefore be described as recessive and the latter dominant.

Despite their infrequent occurrence and their frequently deleterious effects, mutations are of considerable biological significance because they are the ultimate source of all genetic variation. The occurrence of different alleles, e.g. T (tall peas) and t (short peas), of the same gene (pea height) may thus be attributed to mutations occurring sometime in the history of the species.

mutational hotspot a site within a gene or genome at which frequent MUTATIONS occur.

mutational load see GENETIC LOAD.

mutation map a diagram showing the frequency of mutations at different locations along the length of a chromosome.

mutation rate the number of mutation events per gene or species per unit of time. The mutation rate is often expressed in terms of mutations per gene per cell generation. This can be particularly important in unicellular organisms like bacteria, which divide very frequently, and is an important factor in the acquisition of immunity to antibiotics. In plants it affects the ability of a species or variety to develop resistance to particular diseases or herbicides. Mutations usually occur at a fairly constant very low rate, but certain parts of certain genomes may mutate unusually frequently. Mutation rates can also be greatly increased by high-energy radiation such as X-rays and gamma rays, and by treatment with carcinogenic chemicals. See also MOLECULAR CLOCK, MOLECULAR EVOLUTION.

mutation theory the theory, first proposed by Hugo de Vries in 1903, that suggests that new forms of organisms (mutations) arise suddenly in a population, and that evolution proceeds by natural selection operating on these. The theory was based on de Vries' observations of the frequent occurrence of markedly different forms of *Oenothera erythrosepala* (evening primrose). (It was later demonstrated that the different forms were triploids or tetraploids and

hence gave an exaggerated idea of the rate and effects of mutations.) The theory was first validated experimentally when it was found that X-ray induced heritable mutations occurred in *Drosophila*. It is now recognized that although new allelic forms can only arise by mutation, much of the variation on which natural selection acts is generated by crossing over and recombination during meiosis.

mutator gene a gene that is capable of increasing the spontaneous mutation rate.

muton see GENE.

mutualism (*adj.* mutualistic) an intimate relationship between two or more living organisms that is beneficial to all the participants. A lichen is an example of *obligatory mutualism* between an alga or a bacterium and a fungus since neither can survive without the other. The relationship between the bacterium *Rhizobium* and members of the family Fabaceae is an example of *facultative mutualism* (*protocooperation*), as the bacterium can survive independently of the plant. See also AMENSALISM, COMMENSALISM.

Mycelia Sterilia see AGONOMYCETALES.

mycelium a mass of branching HYPHAE that makes up the vegetative body of most true fungi. In the ASCOMYCOTA and BASIDIOMYCOTA the hyphae also anastomose to form a reticulate structure, and they may be more-or-less loosely interwoven.

mycetocyte in certain species of insect, a cell that contains symbiotic bacteria or fungi.

mycetozoans an old zoological term for slime moulds and related organisms (see MYXOMYCOTA).

mycobacteria a bacterium of the genus *Mycobacterium*.

mycobiont the fungal partner in a LICHEN, which is usually an ascomycete fungus. Compare PHYCOBIONT.

mycocecidium a plant gall formed as a result of of infection by a fungus.

mycology the study of fungi.

Mycophycophyta formerly, in the FIVE KINGDOMS CLASSIFICATION, an artificial phylum containing the lichens, which are partnerships (obligate mutualisms) between two organisms of different phyla, a fungus and either an alga or a blue-green bacterium (see CYANOBACTERIA), and are now classified according to the nature of the fungal partner, which is usually an ascomycete.

mycoplasmas (pleuropneumonia-like organisms, PPLO) extremely small parasitic microorganisms, some of which are less than 0.2 μm in diameter. In the FIVE KINGDOMS CLASSIFICATION they are placed in the phylum Aphragmabacteria of the EUBACTERIA. Mycoplasmas have no cell wall. They require sterols for growth; cholesterol makes up over 35% of the lipid content of their plasma membranes. They are thought to be responsible for certain YELLOWS diseases of plants and have been implicated in witches' broom of alfalfa and maize stunt. They cause death of tissue cultures in laboratories and various serious diseases in animals and humans, e.g. pleuropneumonia, bovine mastitis, and puerperal septicaemia. Because they have no cell wall, they are resistant to antibiotics, such as penicillin, that act by inhibiting cell wall synthesis.

mycorrhiza a symbiotic association between a fungus and the roots of a plant. It has been demonstrated that mycorrhizal roots take up nutrients better than uninfected roots and some of these nutrients, such as phosphate and water, are passed on to the host; the fungi obtain carbohydrates and possibly B-group vitamins from the tree. There are three main types – endomycorrhizae, ectomycorrhizae and vesicular–arbuscular mycorrhizae.

In an *endomycorrhiza (endophytic mycorrhiza, endotrophic mycorrhiza)*, the fungus penetrates and lives within the cells of the CORTEX, growth on the outside of the root being limited to finely branched HAUSTORIA (arbuscules). Within the cells of the outer cortex the HYPHAE form tightly coiled masses termed *pelotons*. Such associations are the oldest and most widely distributed form of mycorrhiza. They are found on many herbaceous species, especially orchids and heathers, and on certain woody plants, such as rhododendrons. The fungi involved are most commonly ZYGOMYCETES of the order GLOMALES, or species of *Rhizoctonia* on

orchids or *Phoma* on heathers. Plants that develop such associations grow less well in nutrient-poor soils without the appropriate fungus. Endomycorrhizas are sometimes classified according to the type of plant involved; for example, as arbutoid, ericoid, monotropoid, or orchid mycorrhizas.

In an *ectomycorrhiza (ectophytic mycorrhiza, ectotrophic mycorrhiza)* there is a well-developed MYCELIUM forming a *mantle* on the outside of the root. The hyphae penetrate the cortex and form an intercellular meshwork termed the *Hartig net*. The outer mantle, which replaces the PILIFEROUS LAYER of the root, subsequently arises from this net. Such associations are found on many deciduous north temperate trees and often a tree will not grow properly unless the appropriate fungus is present. It is thus a common practice when planting young trees to inoculate the soil with ectomycorrhizal fungi. These are frequently members of the Agaricales. Infected roots commonly show a characteristic *coralloid* form of growth in which the lateral roots fail to elongate and instead branch repeatedly, forming a swollen mass.

A *vesicular–arbuscular mycorrhiza (arbuscular–vesicular mycorrhiza)* is a form of endomycorrhiza in which the fungus lives between the cells of the cortex and forms temporary hyphal projections that penetrate the cortical cells. The projections may simply be swollen vesicles or may consist of finely branched hyphae called *arbuscules*. The fungi involved are zygomycetes of the order Glomales (e.g. *Glomus, Gigaspora*). They are probably the most ancient and certainly the most widespread fungi throughout the world in associations with plants.

mycosis a disease of humans or other animals that is caused by a fungus.

mycosterol see STEROL.

Mycota see FUNGI.

mycotoxin a toxin produced by a fungus.

mycotrophic describing a symbiotic association between a fungus and the whole of a plant. Such an association occurs when a mycorrhizal fungus extends into the aerial parts of a plant, as in certain heathers and orchids.

mycovirus a virus that infects a fungus and replicates within it.

myosin a contractile protein found in filaments involved in the motility of members of the PROTOCTISTA. (It is also involved in the contraction of muscles in animals.)

myrmecochory the distribution of seeds or other reproductive structures by ants. For example, the seeds of *Helleborus* species are distributed by ants, which are attracted to the plant by an oil-containing swelling on the raphe.

myrmecophily the condition when an organism (a myrmecophile) lives with a colony of ants. It is often seen in the animal kingdom but some flowering plants have specialized inflated organs in which to house ant colonies. For example, various tropical American *Acacia* species have specialized thorns, which provide shelter for the ants, and produce fatty FOOD BODIES (Beltian bodies) and extrafloral nectaries, which provide food. The ants in turn attack insect pests that would otherwise feed on the *Acacia* leaves and plants, such as climbers, that might overwhelm the *Acacia*. Similarly epiphytes belonging to the genera *Myrmecodia* and *Hydnophytum* in the Rubiaceae have large swellings on their roots, which house ants in a network of cavities. These ants are thought to provide extra nutrients for the plants.

myrmecophyte a plant that engages in myrmecophily.

myxamoeba a SWARM CELL that has lost its undulipodia. Myxamoebae may increase by division before acting as gametes.

myxobacteria (gliding bacteria, slime bacteria) flexible creeping bacteria that lack flagella and move by gliding across the substrate. Classified in the phylum Proteobacteria, they are Gram-negative (see GRAM STAIN) rods, ranging in diameter from 5 μm to less than 1.5 μm, and often form complex colonies. The cells produce a slime made from polysaccharides, in which they are embedded. The gliding movement seems to involve contractile filaments.

Myxobacteria are found in soil, dung, and rotting plant material; most are saprobes, some species producing CELLULASE enzymes. Asexual reproduction is normally by binary fission, but under conditions of nutrient or water shortage certain myxobacteria produce resting cells called *myxospores*, which are often borne in fruiting bodies and release large numbers of individuals when they germinate.

Myxobionta see MYXOMYCOTA.

Myxogastria see MYXOMYCOTA.

Myxomycetes (true slime moulds) a class of the MYXOMYCOTA in which the thallus consists of a naked mass of protoplasm termed a PLASMODIUM, which gives rise to various forms of sporangia. They number about 690 species divided among some 60 genera and are usually found growing on decaying vegetation. The various stages encountered in the life cycle (see SWARM CELL, MYXAMOEBA) show certain animal-like characteristics, notably the ingestion of solid particles of food.

In some classifications the Myxomycetes contains the orders Acrasiales and Plasmodiophorales but these are now usually elevated to class or phylum status as the ACRASIOMYCOTA and PLASMODIOPHORA.

Myxomycota (Gymnomycota, Myxobionta, Myxogastria, plasmodial slime moulds, acellular slime moulds) a phylum of the kingdom PROTOCTISTA (or, in some classification schemes, a division of the Fungi), containing those species that have amoeboid, plasmodial, or pseudoplasmodial thalli. The spores germinate to produce cells with one or two undulipodia. The phylum contains some 719 species in about 74 genera and is divided into two classes, the MYXOMYCETES and the PROTOSTELIOMYCETES. In some classification schemes, the ACRASIOMYCOTA and the PLASMODIOPHORA are also included in this phylum.

Myxophyceae, Myxophyta old names for the blue-green bacteria. See CYANOBACTERIA.

myxospore see MYXOBACTERIA

N*n*

n symbol for the haploid chromosome number.

NAA see NAPHTHALENE ACETIC ACID.

nacreous having a pearly lustre.

NAD (nicotinamide adenine dinucleotide, coenzyme I) a pyridine-based nucleotide that functions as a COENZYME in many oxidation–reduction reactions, according to the general reaction: NAD^+ + reduced substrate $\rightarrow$ NADH + H^+ + oxidized substrate. An example of such a reaction is the oxidation of citrate to α-ketoglutarate in the Krebs cycle. Reduced NAD can act as a reducing agent or it may be reoxidized in the RESPIRATORY CHAIN with coupled ATP formation. NAD is not usually tightly bound to the enzyme with which it is acting but instead acts as a second substrate, binding only during the reaction and being released into the medium with the reaction products.

NAD was formerly called *DPN*, an abbreviation for the trivial name *diphosphopyridine nucleotide*.

NADP (nicotinamide adenine dinucleotide phosphate, coenzyme II) a pyridine-based nucleotide that functions as a COENZYME to several different OXIDOREDUCTASES. NADP usually acts as an electron donor in enzymatic reduction reactions; in this it differs from its counterpart NAD, which acts principally as an electron acceptor in biological oxidations. NADP was formerly called *TPN*, an abbreviation for its trivial name *triphosphopyridine nucleotide*. See also PENTOSE PHOSPHATE PATHWAY.

naked cell a cell that lacks a cell wall or other covering, being surrounded only by the plasma membrane.

nannandry (*adj.* nannandrous) the existence in a species of male individuals that are considerably smaller than the females. For example, within the moss genus *Homaliadelphus* there is a species in which the male gametophyte is sufficiently small to grow as an epiphyte upon the leaves of the female gametophyte. The dwarf male filaments of certain members of the OEDOGONIALES are further examples. Compare MACRANDRY.

nano- prefix denoting one billionth part (10^{-9}) of something.

nanometre symbol: nm. A unit of length (formerly called a *millimicron*, symbol mμ) equal to one thousandth of a micrometre, i.e. 10^{-9} metre (1 Angström).

nanoplankton that portion of the plankton that cannot be trapped in a plankton net, i.e. organisms with a diameter of less than 25 μm.

naphthalene acetic acid (NAA) a synthetic AUXIN that has proved of great use in stimulating the rooting of cuttings and promoting flowering.

napiform shaped like a turnip.

NAR see NET ASSIMILATION RATE.

naringenin see FLAVANONES.

nastic movements (nasties) plant responses caused, but not directed, by external stimuli. Thus the opening and closing of many flowers are nastic movements, the trigger most often being certain conditions of light or temperature. The movements may be caused either by differential growth rates or by changes in water potential in specialized cells. Compare TROPISM, TAXIS. See also EPINASTY, HAPTONASTY, HYPONASTY, PHOTONASTY.

nasty see NASTIC MOVEMENTS.

natant floating or swimming in water.

National Biodiversity Network (NBN) a relatively new UK organization that aims to provide coordination and shared services for member organizations seeking to collect and retrieve information relating to BIODIVERSITY and the CONSERVATION of wildlife. Members can access its website to find sources of biodiversity information, dictionaries of species names, and to make enquiries.

national park an area of land set aside by a national government for the preservation of the natural environment for human enjoyment. The restrictions applied to national parks vary from country to country. In the UK they focus on preserving the landscape; in Africa some exist primarily for the CONSERVATION of game, human activity being strictly regulated, while in the USA both natural habitats and landscape are protected, but a certain amount of human activity is permitted. Where human settlements and wildlife coexist, the needs of habitats and wildlife have to be balanced against the demands of tourism, which brings in money that can be used to further conservation. This is often resolved by keeping some parts of the national park off-limits to people (see also BIOSPHERE RESERVE). Until the late 19th century, most NATURE RESERVES existed to preserve game for hunting. The first true national parks were created in Canada in the mid-1880s. Continental Europe began to follow suit after World War I, and the UK only after World War II. See also NATURAL HERITAGE SITE.

native see INDIGENOUS.

natural classification a CLASSIFICATION of organisms in which there is a high level of predictivity, as opposed to artificial classifications, which tend to be very unpredictable. Most taxonomists strive towards the production of natural classifications, although their approaches and interpretations of the data may vary greatly. Natural classifications can best be achieved by exploring the fields of cytogenetics and DNA structure, microanatomy, and phytochemistry in addition to morphology, anatomy, and plant geography, thus broadening the base on which classifications are founded.

Many phylogenists believe that only those classifications based on MONOPHYLETIC taxa are natural.

Natural Heritage Site a natural area that provides outstanding examples of the earth's evolutionary history or current geological processes, contains natural phenomena that are unique or of outstanding beauty, provides habitats for rare or endangered plants or animals, or is a site of exceptional BIODIVERSITY. Natural Heritage Sites are one of two main types of World Heritage Site (the other being Cultural Heritage Sites), designated as having outstanding universal value under the International Convention for the Protection of the World's Cultural and Natural Heritage, a convention adopted by UNESCO in 1972. All sites on the official list have strict legal protection by the relevant national government. Natural Heritage Sites include many NATIONAL PARKS.

naturalized describing a species that was originally introduced from another country, but is now so well established that it maintains itself independently of human assistance, and has invaded native (see INDIGENOUS) populations.

naturalized plant see EXOTIC.

natural order a historical category more or less equivalent to the modern FAMILY. Fifty eight natural orders were listed by Linnaeus in 1764, many corresponding closely to present-day families, e.g. Apiaceae (Umbelliferae) and Asteraceae (Compositae). However, the International Rules of Nomenclature state that natural orders cannot be used in place of family.

natural selection the action of the environment, as opposed to the actions of humans, on individual organisms such that certain genotypes will survive and reproduce more successfully than others, which will eventually die out. By this process the characteristics of a population will change according to the nature of the environmental pressures acting on them. Over a number of generations the population may diverge into a number of distinct groups each adapted to a particular microenvironment. This process will be hastened if there are barriers to gene flow between the groups (see REPRODUCTIVE ISOLATION). The concept of natural selection is the cornerstone in Darwin's theory of evolution (see DARWINISM). Compare CREATIONISM, LAMARCKISM, NATURAL DESIGN.

natural turnover rate the normal rate at which an element is transfered through a BIOGEOCHEMICAL CYCLE.

natural woodland a woodland that has not experienced any human intereference, and there has been no planting by humans.

nature and nurture the interaction between heredity and environment respectively on the phenotypic expression of a character or the development of an organism. See HERITABILITY.

nature conservation the active management of the Earth's natural resources and environments to ensure that they are not depleted or degraded, and to maintain the maxium genetic diversity (see BIODIVERSITY) of species and within species. Such management must also protect the abiotic resources – the BIOGEOCHEMICAL CYCLES, so that they can continue to sustain life. Of necessity conservation must at times accommodate sustained resource use and sustained yield from the biosphere. See BIOSPHERE RESERVE, NATURE RESERVE, SUSTAINABLE USE, WORLD CONSERVATION MONITORING CENTRE, WORLD CONSERVATION UNION, WORLDWIDE FUND FOR NATURE.

nature reserve an area of land set aside and legally protected by a government for the preservation of natural habitats, fauna, and flora. Some reserves exist to protect specific habitats, while others are managed in order to preserve particular species of plants or animals. Nature reserves are vitally important for saving endangered species. Nature reserves for the preservation of game animals for hunting have existed for many centuries, but their creation for the preservation of wild animals for their own sake began only in the 19th century. See also NATIONAL PARK.

N banding a method of staining CHROMOSOMES that reveals the pattern of the CHROMATIN, so can be used to identify chromosomes or particular segments of chromosomes.

Nearctic a biogeographical region that includes North America from north of tropical Mexico northwards to include Greenland and the Aleutian islands.

nearest-neighbour distance the average distance between individuals in a population. Assuming that individuals are randomly distributed, the expected nearest neighbour distance is calculated as $(\pi/n)/2$ m, where n is the density of individuals sper square metre. If this distance is less than 1 the individuals are considered to have a clumped distribution, if greater than one, they are fairly evenly spaced. 2.149 is the maximum possible value, indicating maximum possible spacing, which is attained by a hexagonal pattern.

near-natural community a COMMUNITY of plants consisting mainly of indigenous species that is still following its natural course of SUCCESSION despite some degree of human interference, for example coppiced woodland.

neck the slender portion of the ARCHEGONIUM through which the male gamete has to travel to reach the female gamete prior to fertilization.

necro- (necr-) prefix denoting relating to death or the dead.

necrology the study of processes that affect dead plant and animal material, for example decomposition, mineralization, and fossilization (see FOSSIL).

necron mud see GYTTJA.

necrosis the death of a plant cell or group of cells while the rest of the plant is still alive, particularly when the dead tissue becomes dark in colour. Necrosis is a common symptom of fungus infection and the shape of necrotic areas is often characteristic of the particular disease.

necrotroph (*adj.* necrotrophic) a parasitic organism that feeds on the dead tissues of its host.

nectar the sugary solution secreted by NECTARIES. It is most commonly produced in the flowers of entomophilous plants (see ENTOMOPHILY), but may be produced extraflorally. It acts as an attractant and reward for insects or other pollinators and normally contains 20–60% sugar (sucrose and glucose/fructose in variable quantities) depending, in part at least, on the main pollinators (e.g. bird-pollinated flowers have more dilute nectar than bee-pollinated flowers). Nectar often contains aromatic compounds, which may attract pollinators, and may also contain amino acids, important in the nutrition of some pollinators (e.g. butterflies), and other substances.

nectar guides see HONEY GUIDES.

nectary a GLAND secreting NECTAR. Nectaries are usually present at the base of a flower, sometimes in a spur, to attract pollinators. Nectaries may also be *extrafloral*, for example the gland spines of certain cacti. Here they may serve to attract ants, which may act as protectors of the plant (see MYRMECOPHILY). Others have no clear function.

negative feedback a kind of FEEDBACK INHIBITION in which one of the end products of the metabolic pathway inhibits an earlier stage of the pathway, often by acting as an allosteric inhibitor of an embryo.

negative staining a STAINING technique that relies on the property of certain components of the specimen not to take up the dye or *negative stain*. These then show up light against a dark background. For example, in light microscopy, bacteria can be made visible by mixing them with a dark dye such as nigrosin or Indian ink. The method is used widely in electron microscopy to study viruses and certain large molecules. The material is mixed with negative electron-dense stains, such as potassium molybdate or phosphotungstic acid (PTA). Any proteinaceous material present in the specimen does not take up the stain. In the resulting electron micrograph the electron-transparent proteinaceous material will appear dark against a light background.

neighbourhood a partly isolated subpopulation that shows some inbreeding.

nekton free-swimming animals of the pelagic zone of an ocean or lake, which are capable of maintaining their own positions in spite of water movements. They include shrimps, crabs, squid, fish and whales.

Many of them are prodigious feeders on phytoplankton (see PLANKTON).

nematode a microscopic soil-dwelling worm (phylum Nematoda) that may attack roots and other underground organs of plants, causing extensive damage.

nematophagous fungi fungi that are parasitic on or prey on nematodes. Most of them are also facultative SAPROBES, living in the soil and in rotting vegetatable matter.

nemorose living in woodland.

neocatastrophism see CATASTROPHISM.

Neo-Darwinism the expansion and modification of Charles Darwin's theory of evolution by natural selection (see DARWINISM) in the light of genetic studies and also relevant discoveries in molecular biology. The collection of such data into one coherent theory of evolution in the 1930s and 1940s has been termed the *modern synthesis*. Genetics explains how variations in a species arise by both chromosome and gene mutations, and how these variations are maintained by the RECOMBINATION and reassortment of different alleles.

neo-Lamarckism a modern version of Lamarck's theory of EVOLUTION by the inheritance of acquired characteristics. See LAMARCKISM.

neoteny the arresting of the normal development of all cells except for those in the germ line, resulting in a sexually mature organism with juvenile characteristics. Such a process, producing large morphological changes, could result from comparatively small genetic changes affecting growth rate. It is a method by which one species may arise virtually instantaneously from another. For example, it is thought most likely that *Lemna* (duckweeds) evolved from the floating water plant *Pistia* in this manner because the seedlings of *Pistia* are very similar to the adult form of *Lemna*. At a more general level it is believed herbaceous plants originated from trees by neoteny. The resemblance of an adult organism to the juvenile forms of its ancestors is termed *paedomorphosis*.

The arresting of development at a later stage may lead to large changes in the morphology of a particular organ. Thus the tubelike flower of *Delphinium nudicaule*, which resembles the buds of other

Delphinium species, may have evolved in this manner.

Neotropical floristic region see BIOGEOGRAPHY, FLORISTIC REGION, PLANT GEOGRAPHY.

neotype a plant specimen chosen to act as a standard taxonomic reference point following the irretrievable loss of all TYPE material. The neotype should match the original description as closely as possible and ideally should come from the same locality as the original material.

neritic see SUBLITTORAL.

nerve (costa) a narrow thickened strip of tissue running down the middle of a moss leaf. It is sometimes termed a MIDRIB.

net assimilation rate (NAR) a value that relates plant productivity to plant size. It is obtained by dividing the rate of increase in dry weight by leaf size (usually leaf area).

net community productivity see PRIMARY PRODUCTIVITY, OXYGEN METHOD.

net plankton that portion of the plankton that can be trapped in a plankton net, i.e. organisms with a diameter of more than 25 μm. Compare NANOPLANKTON.

net plasmodium a mass of non-living slimy material resembling a PLASMODIUM, which is laid down by a cellular slime mould (see ACRASIOMYCOTA).

neuston the collection of small and microscopic organisms that live in the surface layer of a body of water.

neutralism the situation where two populations of a species coexist without affecting one another.

neutral soil a soil with a pH of 6.6–7.3.

neutral stain see STAINING.

neutral theory of molecular evolution the theory that evolutionary change at the molecular level is caused by random drift of selectively equivalent mutant genes rather than by selection. See also GENETIC DRIFT.

neutrophilous describing an organism that grows best in a habitat that is neither acidic nor alkaline.

nexine the inner layer of the EXINE in a pollen grain.

niacin see NICOTINIC ACID.

niche see ECOLOGICAL NICHE.

nicotinamide adenine dinucleotide see NAD.

nicotinamide adenine dinucleotide phosphate see NADP.

nicotinic acid (niacin) a carboxylic acid, formula C_5H_4NCOOH, so named because it forms part of the ALKALOID nicotine. It is a member of the B group of VITAMINS and is formed in plants and most animals (but not humans) from various precursors, notably tryptophan. As the amide nicotinamide, $C_5H_4NCONH_2$, it is a constituent of the coenzymes NAD and NADP. Certain plant products, notably maize, are very poor in nicotinic acid. If these form the basis of a human diet, pellagra, caused by nicotinic acid deficiency, results.

Nidulariales an order of the BASIDIOMYCOTA containing some 60 species in 5 genera, known as the bird's-nest fungi. The BASIDIOMA is nestlike and contains a number of separate spore masses (the 'eggs'), which are exposed when the covering membrane dies back. The spores are dispersed by a splash-cup mechanism. An example is *Crucibulum laeve*. The order was formerly placed in the now obsolete class Gasteromycetes. Today it is usually placed in the class HOLOBASIDIOMYCETES.

night-break effect a phenomenon exhibited by many plants whereby a period of light, administered artificially during the night, even if only of short duration or low intensity, will interfere with the normal flowering rhythm of the plant. The precise effects vary with the time that has elapsed since the light phase but generally, under short-day conditions, a light treatment inhibits flowering in SHORT-DAY PLANTS and promotes it in LONG-DAY PLANTS. See PHOTOPERIODISM.

night temperature the temperature prevailing during the period of darkness, which can greatly affect the growth of many plants. For example, tomatoes have been shown to grow better if the night temperature is significantly lower than the day temperature. It is thought that low night temperatures inhibit respiration.

ninhydrin a colourless compound that is used in CHROMATOGRAPHY to locate amino acids and proteins on the chromatogram. Amino acids with a free amino group turn blue on heating with ninhydrin whereas those with a substituted amino group (proline and hydroxyproline) turn yellow.

nitrification see NITROGEN CYCLE.

nitrifying bacteria see NITROGEN CYCLE.

nitrogen symbol: N. A nonmetallic element, atomic number 7, atomic weight 14, important as a macronutrient for plant growth. It is essential for the formation of amino acids and the purine and pyrimidine bases, and consequently for protein and nucleic acid synthesis. It is also found in many other compounds, e.g. porphyrins and many coenzymes. Nitrogen is absorbed by roots as the nitrate (NO_3^-) ion or more rarely as ammonium (NH_4^+) or nitrite (NO_2^-) ions. Deficiency leads to spindly growth and yellowing of the leaves. Soil nitrogen is replenished by various natural processes (see NITROGEN CYCLE, NITROGEN FIXATION). It is also boosted by fertilizer applications of such compounds as urea, ammonium nitrate, ammonium sulphate, and nitrochalk. Many high-yielding crop varieties depend on high levels of nitrogen for optimum yields. However excessive nitrogen produces soft tissues with a high water content that are particularly prone to frost damage. Flowering and fruit set may also be reduced at the expense of vegetative growth, while potassium deficiency can be induced by high concentrations of nitrogen.

nitrogenase (dinitrogenase) the enzyme system isolated from certain bacteria and shown to be responsible for nitrogen fixation. In the presence of ATP and a suitable electron donor it reduces molecular nitrogen (N≡N) to ammonia. It can also reduce other compounds with triple bonds, such as acetylene (HC≡CH). The enzyme is destroyed by free oxygen and is inhibited by high concentrations of ammonia.

nitrogen cycle the circulation of nitrogen between living organisms and the environment. Atmospheric nitrogen is returned to the soil by nitrogen-fixing microorganisms (see NITROGEN FIXATION) and by electrical discharges in storms, which cause nitrogen and oxygen to combine. The oxides so formed dissolve in rain to form nitrous and nitric acids, which, in the soil, combine with mineral salts to form nitrites and nitrates. The nitrites and the ammonia in the soil derived from animal excretion and the decay of organic

matter, are converted to nitrates by the process of *nitrification*. Plants usually assimilate nitrogen as nitrates and the activities of *nitrifying bacteria* are thus essential to plant growth. The oxidation of ammonia to nitrite is carried out by *Nitrosomonas* species and the oxidation of nitrite to nitrate by *Nitrobacter*. The reverse process, DENITRIFICATION, is mediated by different bacteria. Nitrates may also be lost by LEACHING.

Human activities also affect the nitrogen cycle in various ways. Nitrogen is fixed industrially by combining it with hydrogen in the Haber process, and nitrogen is added to the soil in fertilizers. Nitrogen is removed by overcultivation, by sewage systems that prevent urine and faeces from reaching the soil, and by factory farming, which often results in animal wastes not being returned to the land in manageable form.

nitrogen fixation the fixation of atmospheric nitrogen either by lightning (see NITROGEN CYCLE) or by free-living or symbiotic microorganisms (*diazotrophs*). The association between many leguminous species (see FABACEAE) and the nitrogen-fixing soil bacterium *Rhizobium*, which lives in the ROOT NODULES of the legume, has a significant effect on soil fertility. Hence much use has been made of leguminous crops in agriculture. There are different strains of *Rhizobium*, each specific to one or a group of closely related species. The bacteria are attracted to the legume by a hormone and invade the root hairs. They then divide forming filaments that eventually reach and infect the plant cortex. The cortex is stimulated to enlarge and form a nodule the centre of which is red due to the presence of haemoglobin (see LEGHAEMOGLOBIN). The bacterium fixes nitrogen by means of the enzyme DINITROGENASE.

About 250 species of plants other than legumes also form symbiotic associations with nitrogen-fixing microorganisms. For example, bog myrtle (*Myrica gale*) and alder (*Alnus glutinosa*) have nitrogen-fixing root nodules that appear to contain plasmodiophorans. *Gunnera* and the water fern *Azolla* have blue-green bacteria (see

CYANOBACTERIA) in their roots. Free-living nitrogen-fixing microorganisms include the bacteria *Azotobacter*, *Klebsiella*, and *Clostridium* and most cyanobacteria and other photosynthetic bacteria. See also SYMBIOSIS.

nitrophilous preferring nitrate-rich habitats.

node a point on the stem from which one or more leaves arise. In the mature stem the nodes are usually well separated by internodes, which elongate through the action of INTERCALARY MERISTEMS. However in some plants, e.g. rosette plants and grasses, the nodes remain close together giving these species their characteristic growth form. The pattern of vascular connections between the stem and leaf at the node may be described as unilacunar, trilacunar, multilacunar, etc. depending on how many leaf gaps are left in the stele.

nodum any abstract unit of classification of vegetation, whatever its position in the hierarchy. The term is analogous to 'taxon' in plant taxonomy. Compare COMMUNITY. See also INVERSE ANALYSIS.

nomen in TAXONOMY, there are a series of specific Latin terms for names given to species or other taxa that do or do not conform to the relevant Code of nomenclature. These are (in alphabetical order) as follows: *nomen abortivum* a name that contravenes the current code; *nomen ambiguum* a name that is ambiguous, because different authors apply it to different taxa; *nomen conservandum* an old name that has been officially conserved, even though it does not conform to the rules; *nomen correctum* a name whose spelling is required or allowed to be altered under the rules; *nomen dubium* a name that cannot reliably be assigned to a particular taxon; *nomen illegitimum* a name that is illegitimate, and must be rejected; *nomen imperfectum* a name that, as it was originally published, conforms to the rules, but requires some correction, e.g. in spelling; *nomen invalidum* a name that has not been properly published, is unavailable, or in some other way invalid; *nomen inviolatum* a name that is acceptable and conforms to all the rules; *nomen neglectum* a name that was published in the past, but has since been overlooked; *nomen novem* a

name proposed as a replacement for an existing name; *nomen nudum* a name that, although published, does not conform to all the rules and must therefore be rejected, even if corrected; *nomen oblitum* a forgotten name, e.g. one that has not been used for at least 50 years (such names may be used with official permission); *nomen perfectum* a name that, as originally published, meets all the requirements, and needs no correction, but worse ending may be changed with a valid reason; *nomen substitutum* a name that is substituted for an earlier invalid one; *nomen translatum* a name derived by alteration of a previously published name in order to transer it to a different taxonomic level; *nomen triviale* a species name.

nomenclatural type see TYPE.

nominal scale a scale that provides a set of predefined values for scoring quantitative data, e.g. soil colour.

nomogenesis a theory proposed by L. S. Berg in 1922 that regards evolutionary change as due to orderly processes inherent in living organisms and independent of environmental influences and denies any role to chance. See INTELLIGENT DESIGN.

noncoding DNA a segment of DNA that does not code for a gene product.

noncompetitive inhibition the irreversible inhibition of the activity of an enzyme due to an INHIBITOR that is usually structurally unrelated to the normal substrate. The inhibitor may bind permanently to the active site of the enzyme, or it may cause an irreversible change in conformation of the enzyme so that it can no longer fit and interact with its substrate. Compare COMPETITIVE INHIBITION.

non-cyclic photophosphorylation see PHOTOPHOSPHORYLATION.

nondisjunction the failure of HOMOLOGOUS CHROMOSOMES to separate during anaphase 1 of meiosis. As a consequence, one daughter cell, and hence the two gametes formed from this, will receive both homologues. By corollary, the other daughter cell, and the gametes formed from this, will be deficient for the chromosome in question. Cells containing abnormal numbers of chromosomes are called aneuploids (see ANEUPLOIDY). See also MONOSOMIC, TRISOMIC.

non-hierarchical classification see HIERARCHICAL CLASSIFICATION.

non-Mendelian inheritance a form of inheritance in which the progeny show an unusual ratio of phenotypes that does not conform to Mendel's laws. This may be due to extrachromosomal inheritance.

nonpersistent describing pesticides that break down and become inactive fairly quickly. Such pesticides should be used if spraying soon before harvest. Examples are the natural insecticides derris and pyrethrum.

non-renewable resource a natural resource that has built up over a long period of geological time and cannot be renewed within a human lifespan. Examples include fossil fuels and many mineral deposits, which accumulated under conditions that no longer exist today, and water in aquifers that are being replenished too slowly or not at all at the present time. See also SUSTAINABLE DEVELOPMENT. Compare RENEWABLE RESOURCE.

nonsense mutation a mutation that causes a nonsense CODON to occur in a gene.

nonsense triplet a name given originally to triplets of bases (CODONS) that do not specify any amino acid and were thus once considered as having no function. There are three such triplets, UAG, UGA, and UGG, all of which have subsequently been found to act as 'stop signals', i.e. they define the end of a polypeptide chain. Polypeptide initiation also requires a 'start signal', the signals being AUG or GUG, which also code for valine or methionine (or a methionine derivative) respectively. See GENETIC CODE.

normal analysis the grouping of sites or classification of vegetation types based on analysis of the attributes of individuals, for example the grouping of sites according to species composition, rather than to the presence or absence of certain species. This is sometimes called a *plot* classification. See also R TECHNIQUE. Compare INVERSE ANALYSIS, Q TECHNIQUE.

normal curve (normal curve of errors) the bell-shaped curve obtained when a series of observations that shows a normal (Gaussian) distribution is plotted on a graph. The mean, median, and mode all occupy the same high middle point of the

curve, which is perfectly symmetrical around this point. See BINOMIAL DISTRIBUTION.

normal deviate symbol *c*. The ratio of the deviation, *d*, to the standard deviation, σ. It is used to assess experiments where the results do not fall into a limited number of classes but show a continuous range, for example the recording of crop yield in response to various fertilizer levels. To find whether any deviation is significant its probability is looked up in a table of normal deviates. See also T DISTRIBUTION.

normal distribution see BINOMIAL DISTRIBUTION.

normalizing selection see STABILIZING SELECTION.

Northern blotting a procedure similar to Southern Blotting, which transfers RNA from a gel to a carrier.

northern coniferous forest see FOREST.

nu body see NUCLEOSOME.

nucellar embryogeny a form of embryogeny in which the embryo arises directly from the NUCELLUS.

nucellus a rounded or oval mass of parenchymatous tissue in an ovule, containing the EMBRYO SAC. Its size and shape may be used as a diagnostic character for particular taxa. It is almost totally surrounded by the integuments except for a small channel, the MICROPYLE, through which the pollen tube may grow prior to fertilization. At fertilization the nucellus may be reabsorbed as the embryo develops or it may persist to form a nutritive PERISPERM in some seeds.

nuciferous bearing nuts.

nuclear cap in some of the CHYTRIDIOMYCOTA, a saclike structure enclosing part of the zoospore nucleus, which contains most of the cell's ribosomes.

nuclear division see KARYOKINESIS.

nuclear gene a gene that is located on a chromosome in the nucleus.

nuclear membrane (nuclear envelope) a double membrane surrounding the nucleus. Each layer has a typical membrane structure (see PLASMA MEMBRANE) and is 4 nm to 6 nm in width. The outer membrane is studded with ribosomes on the cytoplasm side while the surface of the inner membrane next to the nucleus is smooth. Electron micrographs of many cells reveal connections between the outer membrane and the endoplasmic reticulum. Between the membranes there is a clear space, the *perinuclear cisterna*. At various points the two membranes fuse and *nuclear pores* are formed. The number of pores varies with the degree of nuclear activity. Extensions of the outer granular or fibrous region of the nucleus form cylindrical *nuclear complexes* within the pores. Their precise function is obscure but they appear to be involved in the transfer of information from nucleus to cytoplasm. There is evidence that the nuclear pore complexes allow free diffusion across the membranes, facilitates the receptor-mediated transport of proteins, RNA, and ribonucleoprotein particles. The complexes may also be involved in the regulation of gene expression, chromosome positioning during mitosis, and also secretion.

nuclear pore see NUCLEAR MEMBRANE.

nuclear pore complex see NUCLEAR MEMBRANE.

nuclear sap a non-staining or only faintly staining liquid or semiliquid substance found in a NUCLEUS during interphase.

nuclear stain a BASIC STAIN such as methylene, methyl green, crystal violet, or azure B, which binds preferentially to the nucleus.

nuclease an enzyme that breaks down nucleic acids. See ENDONUCLEASE, EXONUCLEASE.

nucleic acid a complex organic acid comprising polymers of NUCLEOTIDES formed by CONDENSATION reactions that establish phosphodiester bonds between the component nucleotides. There are two types, DNA and RNA. The acidic properties of these compounds are due to phosphoric acid, a component of nucleotides. They are called nucleic acids because they were first associated only with the nucleus. However they have subsequently been found in chloroplasts and mitochondria.

nucleic acid hybridization see DNA HYBRIDIZATION.

nucleocytoplasm the part of a eukaryotic cell that consists of the nucleus and cytoplasm but excludes the organelles, such as mitochondria and plastids.

nucleohistone a complex formed between the polynucleotides of DNA and basic proteins called HISTONES, which only occur in the nuclei of eukaryotic organisms. Nucleohistone complexes are visible as NUCLEOSOMES.

nucleoid (nuclear region) the part of a PROKARYOTE cell that contains the genetic material (DNA). Unlike the nuclei of eukaryotic cells, it is not bounded by a membrane.

nucleolar chromosome the chromosome that bears the nucleolar organiser.

nucleolar organizer the region of chromosomal DNA that codes for RIBOSOMAL RNA. Such regions can be identified as secondary constrictions (the centromere being the primary constriction) often located towards the end of the chromosome. Like centromeres, nucleolar organizers are uncoiled regions of the chromosome and stain poorly. NUCLEOLI in the interphase nucleus have been shown to be associated with these regions.

nucleolus a structure within the nucleus that stains densely with basic dyes and consists of proteins associated with RNA. The number and distribution of nucleoli varies but is usually characteristic for any one cell type. Electron micrographs show a central area of short fibres surrounded by a matrix of protein material with granules embedded in the peripheral region. Nucleoli are closely associated with the regions of chromosomal DNA that code for ribosomal RNA (see NUCLEOLAR ORGANIZER). The TRANSCRIPTION of the code is dependent on a specific RNA polymerase found only in the nucleolus. A long precursor molecule is formed initially and is processed to produce two shorter molecules. The longer of these associates with proteins in the nucleolus to form the larger ribosomal subunits. The smaller molecules similarly associate with proteins to form the smaller subunits, possibly in the nucleolus or surrounding nucleoplasm. The formation of complete RIBOSOMES from the subunits only occurs when the latter reach the cytoplasm.

nucleomorph in the CRYPTOMONADA, a DNA-containing structure that lies in the space between the chloroplast and the associated endoplasmic reticulum.

nucleoplasm see NUCLEUS.

nucleoprotein a complex of DNA or RNA and associated protein. In eukaryotes the CHROMOSOMES are composed of nucleoprotein – DNA and HISTONES, with small amounts of other proteins. See also NUCLEOHISTONE.

nucleoside a general term for the category of substances formed when PURINE or PYRIMIDINE base combines with carbon–1 of a PENTOSE sugar. Such base–sugar complexes are much less common than base–sugar–phosphate complexes, which are known as NUCLEOTIDES.

nucleoside diphosphate sugars (NDP-sugars) compounds formed by the reaction of a phosphorylated hexose sugar with a nucleoside triphosphate (NTP) in the general reaction:

NTP + sugar 1-phosphate → NDP-sugar + pyrophosphate

NDP sugars are important as high-energy glycosyl donors in the synthesis of most polysaccharides, for example the enzyme starch synthetase utilizes uridine diphosphate glucose (UDP-glucose) and adenosine diphosphate glucose (ADP-glucose) as substrates.

UDP-glucose is the most important of the NDP-sugars. It can be synthesized both from glucose 1-phosphate using the reaction shown above, or from sucrose using the enzyme sucrose synthetase (see SUCROSE). Besides their importance as glycosyl donors NDP-sugars are important intermediates in the interconversion of monosaccharides and their derivatives. For many enzymes catalysing sugar interconversions, notably the EPIMERASES, NDP-sugars are substrates rather than the sugars themselves or their phosphate esters.

nucleosome (nu body) a nodule, some 7–10 nm in diameter, consisting of HISTONES, around which is wrapped a strand of DNA about 150 base pairs long. One type of histone, H1, is positioned between adjacent nucleosomes and may serve to clamp the DNA in position. By wrapping around histone beads, the total length of DNA is condensed by a factor of some seven times. Supercoiling of the

string of nucleosome beads results in a further condensation. Nucleosomes are thus of considerable importance in packaging enormous lengths of DNA into minute chromosomes.

nucleotide a compound consisting of three essential parts – a PURINE or PYRIMIDINE base linked to carbon–1 of a PENTOSE sugar (either ribose or deoxyribose), which in turn is esterified (see ESTER) on carbon–5 to phosphoric acid. *Polynucleotides* are polymers of nucleotides. There are two types: RNA, which contains ribose and the bases uracil, adenine, guanine, and cytosine; and DNA, which contains deoxyribose and thymine, adenine, guanine, and cytosine. Both play an informational role in the cell. In contrast, *dinucleotides*, such as nicotinamide adenine dinucleotide (NAD) and flavin adenine dinucleotide (FAD), function as hydrogen carriers, while *mononucleotides*, such as adenosine triphosphate (ATP), serve as energy carriers.

nucleus the part of a eukaryotic cell that contains the genetic material. It is enclosed in a NUCLEAR MEMBRANE. Prokaryotic cells have no nucleus as such. The interphase nucleus of eukaryotic cells contains a network of chromatin fibres, which become organized into CHROMOSOMES prior to cell division. Apart from chromatin the substance of the nucleus (the *nucleoplasm* or *karyoplasm*) contains complexes and enzymes necessary for the replication of DNA and the synthesis of RNA molecules. One or more NUCLEOLI are present associated with specific regions of the chromosomal DNA, the NUCLEOLAR ORGANIZERS.

nucule see CHARALES.

null hypothesis see HYPOTHESIS.

nullisomic see ANEUPLOIDY.

numerical method see NUMERICAL TAXONOMY.

numerical taxonomy (taxometrics) a classification method based on the numerical analysis of the variation of a large number of characters in a group of organisms. It is assumed that a classification will be more predictive the more characters on which it is based. It is also assumed that, to begin with, each character is of equal weight (see

WEIGHTING) although some characters may later be weighted. Initially, a matrix of data is compiled of OPERATIONAL TAXONOMIC UNITS (OTUs) against characters so that for every OTU the state of each of perhaps 50 or more characters is recorded. This matrix can be subjected to a variety of mathematical analyses, which provide a measure of the similarity or dissimilarity between all the OTUs. The end product is usually one or more DENDROGRAMS.

nurse tissue metabolically active tissue that is used in tissue culture work to stimulate the growth of single cells that cannot be grown using defined media. The nurse tissue, which is often CALLUS tissue, may be separated by a piece of filter paper through which hormones and nutrients can diffuse. Pollen grains have been induced to grow into haploid plants by culturing them on filter paper in contact with intact anthers acting as nurse tissue. It may not be necessary to separate the nurse tissue if it is first treated, for example with X-rays, to prevent cell division.

Growth media may be enriched with hormones by adding nurse tissue enclosed in a selectively permeable membrane. Using such *conditioned media* it is possible to achieve satisfactory growth using a much lower inoculum density than would normally be necessary. This is important in cloning work because cells can be plated out at much lower densities so increasing the chances that any resulting group of cells is derived from a single cell.

nut a dry indehiscent FRUIT that is usually shed as a one-seeded unit. It forms from more than one carpel but only one seed develops, the rest aborting. The pericarp is usually lignified and is often partially or completely surrounded by a CUPULE. True nuts include the acorn, hazelnut, and beechnut. The term is often loosely applied to any woody fruit or seed, such as the walnut (which is a drupe), the Brazil nut (which is a seed), or the peanut (or groundnut), which is a pod. *Nutlets* are small nuts and are typical of the Lamiaceae (Labiatae).

nutation see CIRCUMNUTATION.

nutlet see NUT.

nutrient cycle see BIOGEOCHEMICAL CYCLE.

nutrient medium see CULTURE MEDIUM.

nyctanthous describing a plant that flowers at night.

nyctigamous describing flowers that close by day and open at night. Many moth- and bat-pollinated flowers behave in this way.

nyctinasty a NASTIC MOVEMENT in which plant parts, especially leaves and flowers, assume a characteristic position at night. These *sleep movements* most often result in a folding together of leaflets, e.g. in the prayer plant (*Maranta leuconura*). In certain *Acacia* species the process seems to be a form of PHOTONASTY, as the assumption of the night position coincides with the onset of darkness. In other species the cycle of folding continues even if the plant is kept in the dark; the process is then clearly autonomic (see AUTONOMIC MOVEMENT) rather than nastic.

Oo

ob- a prefix meaning inverted.

obcordate describing a leaf shaped like an inverted heart.

obdeltate shaped like an inverted delta.

obligate having a specific requirement for a particular environmental factor or mechanism and being unable to function if it is not available. For example, obligate parasites cannot grow in the absence of their host; obligate outbreeders (see OUTBREEDING) require pollen transfer from another individual for fertilization. Compare FACULTATIVE.

obovate see OVATE.

obtuse having a blunt or rounded leaf apex.

obvolute describing an arrangement of leaves or scale leaves in a bud in which CONDUPLICATE leaves interleave. *See illustration at* vernation.

ochrea (ocrea) a tube around the stem formed by sheathing STIPULES at the leaf base. Ochreas are characteristic of the Polygonaceae giving the familiar swollen joints on the stems of such plants.

octadecanoic acid see STEARIC ACID.

Oedogoniales an order of green algae (CHLOROPHYTA) of the class CHLOROPHYCEAE comprising three genera of filamentous algae. These algae are characterized by an unusual kind of cell division, which results in parallel rings of closely spaced 'caps' of old cell wall material being formed at the anterior end of dividing cells. Prior to cell division a ring of wall material forms at the anterior end of

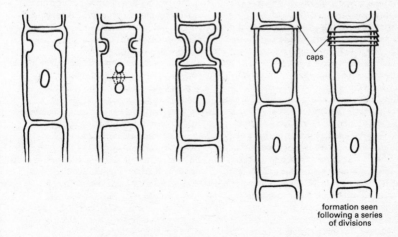

caps

formation seen
following a series
of divisions

Oedogoniales: stages of cell division in *Oedogonium*

the cell. Nuclear division then occurs and the two resulting nuclei become separated by a septum, which, however, is not contiguous with the outer wall at this stage. Meanwhile the ring of wall material at the top of the cell enlarges and the old cell wall to the outside of the ring ruptures. The new wall material is then free to extend and forms the outer walls of the new cell. Part of the old wall remains as a 'cap' at the upper end of this cell. The septum between the daughter nuclei moves up to fuse with the base of the new wall material and cell division is completed (see diagram).

The zoospores and sperm of the Oedogoniales are also peculiar in having numerous undulipodia forming an apical ring. Sexual reproduction involves an advanced form of OOGAMY. The reproductive structures, oogonia and ANDROSPORANGIA (rather than antheridia), form on normal filaments (macrandry). The androsporangia produce motile cells called ANDROSPORES, which are chemically attracted to the oogonia. They attach to the oogonia and form dwarf filaments, which release sperm.

offset a short shoot that arises from an axillary bud near the base of the stem and gives rise to a daughter plant at its apex. Examples of offsets are those produced by the houseleek (*Sempervivum tectorum*). The offset is a type of short RUNNER and, like runners, is a means of vegetative reproduction. The term is also applied to bulbils and cormlets that form at the side of the parent bulb or corm.

oidium (arthrospore) a spore that forms by the organized fragmentation of a HYPHA as seen, for example, in *Endomyces*.

oil a TRIACYLGLYCEROL that is liquid at room temperature. The major fatty acids in oils are the unsaturated oleic and linolenic acids. Most oils function as energy storage compounds and are especially important in some oil-bearing seeds such as the castor bean. A few plants, e.g. some orchids and species of the vine *Dalechampia*, secrete oil to attract insect pollinators. See also ESSENTIAL OIL.

oil body see ELAIOSOME.

oil gland a GLAND that secretes an ESSENTIAL OIL.

oil immersion a technique used in light microscopy in which a special oil-immersion objective lens is employed to increase the resolving power of the microscope. A specimen mounted on a glass slide is covered by a drop of clear oil placed on the coverslip. The oil has the same refractive index as the glass lens. When the lens is immersed in the oil, the amount of light entering the objective increases (i.e. the effective aperture is increased). Such a technique is particularly useful at high magnifications. Oil is traditionally used but other liquids (e.g. sugar solution) may also be suitable.

okadaic acid a toxin produced by some dinoflagellates (see DINOMASTIGOTA). It causes a type of shellfish poisoning.

Okazaki fragment see DNA LIGASE, DNA REPLICATION.

old-growth forest an American term for CLIMAX or near-climax FOREST that has existed since before European settlement. Such forests are considered to be primary forest.

oleic acid (*cis*-octadec-9-enoic acid) An eighteen-carbon monounsaturated FATTY ACID. It is the predominant unsaturated fatty acid in plant cells, except in some specialized organelles, such as the chloroplast, where LINOLENIC ACID predominates. Oleic acid is formed from STEARIC ACID in a desaturation reaction involving molecular oxygen, a reducing agent, and FERREDOXIN.

oligo- prefix denoting few.

Oligocene see TERTIARY.

oligomer a protein made up of only two or a few identical polypeptide subunits.

oligopeptide an unbranched peptide containing 2–10 amino acids.

oligosaccharide a CARBOHYDRATE composed of between two and ten polymerized MONOSACCHARIDE units. Most oligosaccharides contain two (disaccharides), three (trisaccharides) or four (tetrasaccharides) sugar units. The most important plant sugar, SUCROSE, is a disaccharide. CELLOBIOSE is another common disaccharide, being the basic repeating unit of cellulose. Other disaccharides include trehalose and gentiobiose. The commonest trisaccharide is RAFFINOSE and the commonest tetrasaccharide STACHYOSE. Oligosaccharides may be intermediates in polysaccharide metabolism or may serve as storage compounds.

oligospermous bearing only a few seeds.

oligotrophic poor in minerals and bases. The term is usually used of ponds and lakes with low nutrient levels but it can be applied in other contexts, for example to bog peats, which are usually low in nutrients. In an oligotrophic stretch of freshwater there is limited growth of aquatic plants. There is consequently little decaying vegetable matter and the absence of large numbers of DECOMPOSERS means there is sufficient oxygen to support fish and other aquatic animals. Compare DYSTROPHIC, EUTROPHIC.

oligotrophication the process by which nutrient cycling slows and/or nutrients become depleted in aquatic ecosystems as a result of pollution, especially acid rain. See also BIOGEOCHEMICAL CYCLES.

ombrogenous bog see BOG.

one gene:one enzyme hypothesis the theory that the principal function of a gene is to determine the structure of an enzyme molecule. This was based on evidence showing that the presence or absence of an enzyme in a METABOLIC PATHWAY was a genetically determined characteristic, inherited as a single gene. The theory was extended to include proteins other than enzymes. Subsequent analysis showed that where a protein consists of several polypeptide chains, it is more correct to say that one gene codes for one polypeptide chain. The genes coding for the component polypeptides of an enzyme need not be closely linked.

one gene:one polypeptide hypothesis see ONE GENE: ONE ENZYME HYPOTHESIS.

onomatologia the conventions used in naming plants.

ontogeny all the changes that occur during the life history of an organism. In a plant this would be the development from the zygote through to the production of gametes and, by the fusion of two gametes, a new zygote that will give rise to the subsequent generation. Compare PHYLOGENY.

oogamy (*adj.* oogamous) an extreme expression of ANISOGAMY, in which fertilization occurs with the fusion of a large nonmotile female gamete or egg and a small usually motile male gamete. Compare ISOGAMY.

oogenesis the formation of ova (eggs) and, in angiosperms, the embryo sac.

oogonium the female reproductive organ of certain algae and fungi. It is initially unicellular and usually thin walled, and produces one or more nonmotile OOSPHERES. These may be released prior to fertilization or remain in the oogonium, as in the fungus *Pythium*. Compare ARCHEGONIUM.

Oomycetes see OOMYCOTA.

Oomycota (oomycetes) a phylum of the kingdom PROTOCTISTA or, in some classification schemes, of the kingdom CHROMISTA; in other systems these organisms are considered to be a class (Oomycetes) of fungi of the division PARAMYCOTA. The phylum contains some 694 species in about 95 genera, including the water moulds (see SAPROLEGNIALES) and downy mildews. Members typically produce HETEROKONT zoospores with an anterior hairy (mastigonemate) undulipodium and a posterior smooth (whiplash) undulipodium. Oomycotes are coenocytic, producing aseptate hyphae. The normal vegetative state is diploid, or occasionally polyploid. The cell walls contain very little chitin, the main structural element being a form of cellulose. Sexual reproduction is oogamous; the gametes are nonmotile. Reproductive structures in which meiosis occurs develop at the tips of hyphae; an antheridium makes contact with an oogonium and produces a fertilization tube that penetrates it and down which the male nuclei migrate. Fertilization results in thick-walled dormant oospores. These may germinate directly into a new thallus, or they may produce diploid zoospores that then develop into thalli.

Nine orders are recognized: the Leptomitales, Myzocytiopsidales, Olpidiopsidales, PERONOSPORALES, Pythiales, Rhipidiales, Salilgenidiales, SAPROLEGNIALES, and Sclerosporales. Oomycotes are found in fresh or salt water and in soil and include saprobes and parasites, some economically important. Notorious examples are *Phytophthora infestans*, cause of the great potato blight famine in Ireland; *Plasmopara viticola*, which causes mildew of grapes; *Pythium*, the DAMPING-OFF fungus; and *Saprolegnia parasitica*, a parasite of fish often found in aquaria.

oosphere a large nonmotile gamete that is rich in nutrients and is normally designated as female. One or several oospheres may be formed within an OOGONIUM. They may be released unfertilized, as in the brown alga *Fucus*, or they may be retained within the oogonium and there form a zygote or OOSPORE following fertilization. The oosphere is equivalent to the egg cells formed in the ARCHEGONIA or EMBRYO SACS of land plants.

oospore a thick-walled zygote that is formed following the fertilization of an OOSPHERE. Compare ZYGOSPORE.

open canopy SEE CANOPY.

open population a POPULATION in which there are no barriers to gene flow. Compare CLOSED POPULATION.

open vernation an arrangement of leaves or scale leaves in a bud in which the edges of adjacent leaves do not touch. *See illustration at* vernation.

operational taxonomic units (OTUs) the entities whose affinities are studied by NUMERICAL TAXONOMY. Depending on the level and the type of investigation an OTU may be of any taxonomic rank or an individual organism.

operator (operator gene) a site on the DNA adjacent to a STRUCTURAL GENE that acts as part of a molecular 'switch' and determines whether or not transcription of the structural gene will occur. The switch may be turned 'off' by a REPRESSOR molecule binding to the operator, so preventing mRNA polymerase reaching the DNA. Conversely, absence of a repressor allows mRNA polymerase to bind to the DNA, i.e. the switch is 'on'. Other methods by which the operator genes function are also known. See OPERON.

operculum 1 the membranous cap covering the PERISTOME in the undehisced capsule of many moss sporophytes. It is displaced either by the pressure that builds up within the capsule or by the swelling of the cells of the annulus.
2 a lid covering the aperture in many pollen grains, which is pushed aside by the emerging pollen tube.
3 a cap covering the OSTIOLE in certain ascomycete fungi.

operon a group of adjacent genes coding for a set of enzymes in a particular BIOSYNTHETIC PATHWAY. They act as a single unit in that they are either all transcribed together or none are transcribed at all. In order for transcription to occur, mRNA polymerase, which catalyses mRNA synthesis, must first bind with the DNA at a site called the *promoter*. mRNA polymerase activity is prevented (negative control) or promoted (positive control) by the binding of a regulator protein between the promoter and the structural genes (the genes that actually code for the enzymes) at a site called the operator. The regulator protein is itself the product of a gene, the regulator, which may be some distance from the operon. A classic case of negative control is the lac (lactose) operon in the bacterium *Escherichia coli*. Here the regulator protein prevents transcription in the absence of an inducer (a substrate molecule, i.e. lactose, or some derivative). However, if an inducer is present, it complexes with the regulator, the protein formed by the latter no longer binds to the operator, mRNA polymerase activity is permitted, and the enzymes are synthesized. See also JACOB–MONOD MODEL.

Ophioglossales an order of the FILICINOPHYTA containing a small group of morphologically distinct eusporangiate ferns. The spores are not borne on the undersurface of the frond but in a separate stalked spike, and unlike all other ferns the fronds do not show circinate ptyxis. The order contains three genera: *Ophioglossum* (adder's tongues) containing about 30 species; *Botrychium* (moonworts) containing about 25 species; and *Helminthostachys*. They are highly polyploid; *O. reticulatum* has the highest chromosome number of any living organism, with $2n = 1260$.

opportunist species SEE FUGITIVE SPECIES.

opposite describing a form of leaf arrangement in which the leaves arise in pairs at each node. If each pair is at right angles to the pairs above and below it, as is usually the case, the arrangement is termed *decussate*. If the leaf pairs all arise in the same plane a *distichous* arrangement results. Compare ALTERNATE, WHORLED. *See illustrations at* phyllotaxis, vernation.

optical microscope see LIGHT MICROSCOPE.

orbicular almost circular and flattened, as are the leaves of the fringed water lily (*Nymphoides peltata*). *See illustration at* leaf.

Orchidaceae a large cosmopolitan family of monocotyledons, commonly known as the orchid family, containing about 18 500 species in some 788 genera. Most are epiphytes, but the family also includes many terrestrial herbs and a few lianas. There are also some subterranean forms and some parasites that lack chlorophyll. Many of the terrestrial forms have tubers or rhizomes, and the epiphytes often have fleshy *pseudobulbs*, comprising one or more thickened stem internodes. Many form mycorrhizal associations (see MYCORRHIZA). The leaves are simple, spirally arranged, occasionally reduced to whorls of scales. The flowers are usually bisexual and irregular in shape, epigynous, and with flower parts in threes (see illustration). They range from solitary flowers to racemose inflorescences. The perianth typically comprises an outer whorl of three petalloid sepals and an inner whorl of three petals. The adaxial (uppermost) petal, the labellum (or lip), usually appears to be abaxial (lowermost) in position as a result of twisting of the pedicel (resupination) or, in some single-flowered orchids, twisting of the ovary. This position enables the labellum to serve as a landing platform for pollinating insects. The stamen (there are only rarely two or more) usually unites with

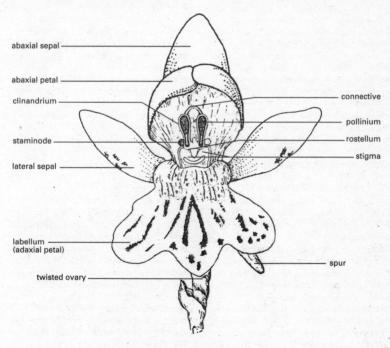

Orchidaceae: front view of an orchid flower

the style to form a *column*, the upper part of which bears the elaborate anther (*clinandrium*) and ROSTELLUM. When mature, the pollen grains become fused together to form structures called *pollinia*. These may actually break off and stick to insect pollinators, to be rubbed off on the stigma of the next orchid visited. The fruit is a capsule containing a large number of very small light seeds that lack endosperm. In many species the seeds are dispersed by wind, but in some species they are dispersed by ants, beetles, birds, bats, or even frogs. Orchid flowers are often highly complex in form and use many devices to attract pollinators, including elaborate traps and even intoxication. Some, such as the bucket orchid (*Coryanthes* sp.), secrete wax that attracts certain wasps, which then become trapped in the orchid and in finding the only way out brush against the pollinia. Other species emit scents resembling the pheromones (sexual attractant hormones) of female insects and may even have an insect-like appearance (e.g. the bee orchids, *Ophrys* spp., and the wasp-mimicking *Cryptostylis*). In the hammer orchid (*Drakaea fitzgeraldii*) part of the inflorescence pivots so that when a wasp alights on the flower it is thrown against the rostellum and pollinia.

Apart from the climbing *Vanilla*, especially *V. planifolia*, whose pods are the source of vanilla flavouring, and the dried tubers of *Orchis*, *Dactylorhiza*, and *Eulophia*, which form salep, used for culinary and medicinal purposes (and in Turkish delight) in parts of Europe and Asia, the orchids are of no great economic importance. Their main uses are as ornamental plants, many being grown in greenhouses. It is easy to produce hybrids and increase the variety of cultivated orchids.

order a major category in the taxonomic hierarchy, usually comprising groups of families thought to possess a degree of PHYLOGENETIC unity. Groups of similar orders are placed in classes. The Latin names of orders usually terminate in -ales, e.g. Rosales and Geraniales. See also INTERNATIONAL CODE OF BOTANICAL NOMENCLATURE.

ordered tetrads asci containing four ascospores in a linear order, which relates directly to the sequence of meiotic events. Analysis of these tetrads can provide valuable genetic information about crossing over. Ordered tetrads are produced by certain species of the ASCOMYCOTA, including various yeasts and species of *Neurospora*.

ordination a technique used in ecological work to relate the composition of different stands of vegetation to each other. It is used by those who consider that no two stands of natural vegetation are ever the same and thus they cannot be classified into distinct associations (compare CLASSIFICATION). Certain properties of each stand, namely data on the species present, are plotted against one or more axes, each axis representing some environmental gradient. The end result should be an arrangement in which the different vegetation stands are ordered in a manner that best reflects their similarities and differences.

Ordovician the second period of geological time in the Palaeozoic era, about 500–440 million years ago. There are no known terrestrial organisms in this period. See GEOLOGICAL TIME SCALE.

organ a specific part of an organism that is specialized to perform a particular function or group of functions. An organ contains several different kinds of TISSUES. Examples of organs are leaves, flowers, and roots.

organ culture the culture of excised organs in a suitable aseptic medium. Roots, leaves, embryos, meristems, and many other plant structures have been successfully maintained in culture. By observing growth in the presence and absence of various nutrients, valuable information about the physiology of the organ can be obtained. For example, chemical factors affecting sex expression have been studied using cultures of isolated flowers, and root nodule formation has been investigated in cultures of legume roots. See also MERISTEM CULTURE, OVULE CULTURE, EMBRYO CULTURE.

organelle a membrane-enclosed structure in the cytoplasm organized to carry out a specific process essential to the life of the cell. Examples are the MITOCHONDRIA, CHLOROPLASTS, and GOLGI APPARATUS.

organ genus see FOSSILS.

organic acid an organic compound that can release hydrogen ions (H^+) to a base, e.g. a carboxylic acid.

organic soil a soil composed mainly of organic matter, e.g. peat.

organismic concept the idea that communities or other groups of individuals behave in some way like a single living organism. They are sometimes called *supra-organisms*. The term is used particularly in relation to plant communities. For example, a CLIMAX community develops to maturity, and if it subsequently dies, another identical community will replace it. The high level of interdependence of species in a complex community is considered to resemble the interdependence of cells and tissues in a single organism. This approach favours a classification approach to vegetation analysis. Compare INDIVIDUALISTIC HYPOTHESIS.

organogenesis the developmental changes that occur during the formation of a particular organ.

organomercurial fungicide an organic FUNGICIDE containing mercury. Most of this group are phenylmercury derivatives, e.g. phenylmercuric acetate, and are used as SEED DRESSINGS. Mercury-containing fungicides are highly toxic to mammals and treated grain must be clearly labelled or even dyed to distinguish it from edible grain.

organophosphate insecticide an organic INSECTICIDE, such as malathion and parathion, often used to control aphids, mealybugs, mites, etc.

origin of life the way in which living organisms have developed from inorganic matter. It is generally believed that the earth is about 4600 million years old and that the earliest evidence of life dates from at least 3500 million years ago. Fossils of this age are interpreted as bacteria-like organisms. It is thought that in the early stages the earth's atmosphere was chemically reducing, composed of such gases as hydrogen, nitrogen, methane, ammonia, and water vapour. Such an atmosphere would be more favourable for the formation of complex organic molecules than the present oxidizing atmosphere (largely nitrogen and oxygen). Laboratory experiments in which mixtures of such gases are subjected to electric sparks or ultraviolet radiation show that simple organic molecules (e.g. amino acids and nitrogenous bases) can be formed. Most theories on the origin of life involve starting with a 'primaeval soup' containing such compounds. Possibly long-chain molecules (e.g. proteins and carbohydrates) were formed by catalytic reactions in which simple molecules adsorbed on regular mineral surfaces (e.g. mica clays). The way living organisms evolved from such molecules is still in question. Possibly it first involved the formation of primitive cells (COACERVATES) leading to simple heterotrophic prokaryotic organisms. From these came photosynthetic organisms resembling the cyanobacteria, which gradually formed an oxidizing atmosphere enabling aerobic organisms to develop. Aerobic eukaryotic organisms then appeared, possibly arising as symbiotic associations of prokaryotes (see ENDOSYMBIOTIC THEORY), and from these came multicellular organisms.

ornithine a nonprotein amino acid, formula $NH_2(CH_2)_3CH(NH_2)COOH$. It is an intermediate in the synthesis of ARGININE, and many ALKALOIDS, e.g. the tropane alkaloids, are derived from ornithine. It also occurs in the cell walls of some bacteria.

ornithocoprophilous preferring habitats rich in bird excrement.

ornithophily pollination by birds. Plants pollinated primarily by birds occur throughout the world, except temperate Eurasia. Some birds, such as hummingbirds, confined to the New World, are highly specialized for visiting flowers. Other important flower visitors are the Old World sunbirds and the mainly Australian honeyeaters. Members of many other bird groups are effective pollinators. Ornithophily is believed to be derived from pollination by insects (ENTOMOPHILY).

orophilous describing plants that grow in subalpine habitats.

orthogenesis evolution that proceeds in a constant direction for many generations, and was once believed to be due to some directing force within the organisms themselves, independently of environmental factors, and with no involvement of chance. Such trends are currently explained by directional selection. See SELECTION. Compare INTELLIGENT DESIGN.

orthophosphate a salt or ester of orthophosphoric acid (H_3PO_4).

orthotropism a TROPISM in which the growth response is directly towards or away from the source of the stimulus. Thus the vertical growth of tree trunks may be described as negative orthogravitropism and the growth of the gills of agaric fungi positive orthogravitropism.

orthotropous (atropous) describing a form of ovule orientation in the ovary in which the ovule develops in an upright position from the placenta (*see illustration at* ovule). This arrangement is rare but may be found in some members of the Polygonaceae. Compare ANATROPOUS, CAMPYLOTROPOUS.

osmium tetroxide a fixative (preservative) for proteins and phosphoglycerides that is used in the preparation of biological material for light microscopy and electron microscopy (see FIXATION). It is fat soluble and is reduced to black osmium dioxide by unsaturated fats. Saturated fats in a specimen may be dissolved out in dehydration or embedding, the blackened unsaturated fats remaining. It is therefore used to stabilize and stain cell constituents. Often a preparation is pre-fixed with glutaraldehyde and post-fixed with osmium tetroxide to increase the extent of preservation.

osmolarity the total molar concentration of solutes in a nutrient solution or medium that affects its osmotic potential.

osmometer an apparatus designed to demonstrate OSMOSIS and to measure osmotic pressure (WATER POTENTIAL). One form of osmometer consists of a porous pot, rendered selectively permeable by impregnation of its pores with copper ferrocyanide. The pot is filled with the solution under investigation and a piston placed on the solution. The pot is then immersed in a container of distilled water. Water passes by osmosis into the porous pot and lifts the piston until the force of the water entering the pot equals the

HYDROSTATIC PRESSURE created by the weight of the piston.

osmophilic describing organisms that grow best in habitats that have a relatively high osmotic pressure due to having a high concentration of sugars or salts.

osmoregulation collective term for the internal processes by which an organism controls the osmotic pressure of its cells, tissues, and body fluids.

osmosis the process by which solvent molecules migrate across a SELECTIVELY PERMEABLE MEMBRANE from a region of low solute concentration to a region of higher concentration. In biological systems, the solvent is invariably water and the migration tends to equalize solute concentrations on the two sides of a cell membrane. Osmosis is of great importance in the maintenance of cell TURGOR and also in vascular transport. See also OSMOTIC PRESSURE, WATER POTENTIAL.

osmotic potential (solute potential) the component of WATER POTENTIAL that takes into account the concentration of solutes in the cell. It is represented by the symbol Ψ_o or Ψ_s and is always negative. Osmotic potential includes a component due to the MATRIC POTENTIAL of colloidal substances and organelles in the cytoplasm.

osmotic pressure (Π) The force that has to be applied to a solution to prevent OSMOSIS occurring. It is equal to the DIFFUSION PRESSURE DEFICIT plus TURGOR PRESSURE. The term is going out of use because it does not take capillary and imbibitional forces into account, which can predominate in certain circumstances. Thus the term WATER POTENTIAL is now preferred.

osmotrophy a type of heterotrophic nutrition shown by SAPROBES, which involves the absorption of organic nutrients in solution. Compare PHAGOTROPHY.

ostiole a small pore found in the reproductive bodies of certain algae and fungi (e.g. the conceptacles of *Fucus* or the perithecia of certain ascomycete fungi) through which the spores are released.

OTUs see OPERATIONAL TAXONOMIC UNITS.

outbreeding (outcrossing) the production of offspring by the fusion of distantly related gametes. The consequences of outbreeding are opposite to those of INBREEDING and mechanisms have been developed by the majority of plants to promote it. The main outbreeding mechanisms are INCOMPATIBILITY factors, heteromorphism (e.g. HETEROSTYLY), PROTOGYNY, PROTANDRY, and dioecy (see DIOECIOUS).

outcrossing see OUTBREEDING.

outwelling the nutrient enrichment of seawater by the inflow of nutrient-rich estuarine water. This enhances the growth of phytoplankton (see PLANKTON), which form the basis of the local food chain.

ovary the swollen basal part of the carpel in angiosperms, which contains the ovules. The ovary is hollow and may contain one or many ovules in its locule, each attached by a FUNICULUS. When the carpels are partially or totally united a compound ovary is formed, which may be either unilocular or multilocular depending on whether or not the fused carpel walls break down. Depending on its position in the flower an ovary may be described as *inferior*, when the other floral organs are inserted above it, or *superior* when the other floral organs are inserted below (see EPIGYNY, PERIGYNY, HYPOGYNY). The ovary wall is usually thick and serves to protect the developing ovules. After fertilization the ovary wall forms the PERICARP of the fruit.

ovate describing an organ, such as a leaf, that is up to about four times as long as it is broad and tapers at both ends, with the broadest part below the middle (*see illustration at leaf*). *Obovate* indicates that the broadest part is above the middle.

overdispersion see DISPERSION.

overdominance (superdominance) in genetics, the situation where a character is expressed in a more pronounced way in the heterozygote than in either of the homozygotes. This can be a contributory factor in heterosis.

overgrazing excessive grazing, by wild or domestic animals, which results in the degradation of the vegetation, eventually exposing the soil to erosion. In extreme cases it may cause desertification.

overlapping DNA a single DNA sequence that can code for different genes according to the location of the READING FRAME. See also TRANSCRIPTION, TRANSLATION.

over-representation in pollen analysis, the occurrence of locally derived pollen in such quantities that it obscures the general pattern of regional change. For example, when studying peat cores, pollen from the peat bog plants themselves may dominate the sample.

overspecialization 1 the idea that an evolutionary trend might proceed so far that it was no longer adaptive, possibly leading to extinction. In fact, NATURAL SELECTION would operate before this point was reached, preventing further more extreme change. 2 the evolution of highly specialized organisms that are unable to adapt to environmental change.

ovule the female gamete and its protective and nutritive tissue, which develops into the dispersal unit or SEED after fertilization in seed plants. In angiosperms the ovule comprises a central EMBRYO SAC containing the gamete and other haploid nuclei, the surrounding NUCELLUS, and one or two protective INTEGUMENTS interrupted by a small opening, the MICROPYLE. The ovule is attached to the placental tissue (see

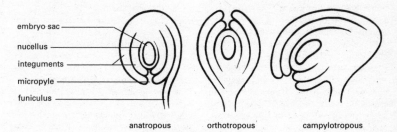

embryo sac
nucellus
integuments
micropyle
funiculus

anatropous   orthotropous   campylotropous

Ovule: types of ovule orientation

PLACENTATION) by means of the FUNICULUS. In most plants the ovule is completely bent around so that the micropyle faces towards the placenta. Such ovules are described as anatropous (see diagram). In some plants, e.g. certain members of the Caryophyllaceae, the ovule is at an angle to the funiculus, which appears to join the ovule midway between the micropyle and chalaza. This is the campylotropous form of ovule orientation. In a few plants, e.g. *Polygonum*, the ovule is erect or orthotropous. There may be one or many ovules in the locule of an ovary. In gymnosperms the ovule tends to be larger and is borne naked on an OVULIFEROUS SCALE rather than within an ovary. The formation of a cuticle around the ovule of seed plants is a significant development in the evolution of the terrestrial habit. It is thought to be derived from the megasporangium of nonseed plants.

ovule culture the culture of excised ovules on suitable media *in vitro*. Ovules have been cultured with pollen grains of the same species in order to observe the processes of fertilization and embryo development. Culture with pollen of different species may enable certain crosses to be made that are normally impossible due to incompatibility factors in the stigma. Some ovules in culture give rise to callus tissue that subsequently forms numerous embryo-like structures (embryoids). These can be separated and grown on to mature plants, so providing a means of plant propagation. See also EMBRYOGENY.

ovuliferous scale a scale that bears ovules and then seeds in the female STROBILUS or cone of the conifers. It is borne in the axils of the spirally arranged woody bracts that constitute the cone. The ovuliferous scale is a highly specialized MEGASPORQPHYLL and is homologous with a carpel.

ovum an unfertilized egg cell.

oxalic acid (ethanedioic acid) an organic acid, formula $(COOH)_2$, found in many plants e.g. rhubarb (*Rheum raponiticum*) and dumbcanes (*Dieffenbachia*). It helps remove excessive amounts of various cations, e.g. calcium, sodium, and potassium, by forming the respective oxalates, and is highly toxic.

oxaloacetic acid a four-carbon dicarboxylic keto acid with the formula $HOOCCOCH_2COOH$. Oxaloacetate is a KREBS CYCLE intermediate formed by oxidation of malic acid. It reacts with acetyl CoA to form the six-carbon citric acid. Oxaloacetate is also an intermediate in glucose synthesis, and in c_4 PLANTS of the HATCH–SLACK PATHWAY. The carboxylation of pyruvate to form oxaloacetate is a major reaction for replacing Krebs cycle intermediates that have been withdrawn to take part in synthetic reactions. The reverse of this reaction is an important route of glucose synthesis.

oxidase any of the flavin-linked DEHYDROGENASES in which the reduced coenzyme can be reoxidized by molecular oxygen, which forms hydrogen peroxide with water. An example is the enzyme L-amino acid oxidase, which catalyses the oxidative deamination of L-amino acids.

β-oxidation the principal route for the degradation of fatty acids to ACETYL COA. In β-oxidation, carbons are removed from the fatty acid chain two at a time; thus one molecule of acetyl CoA is formed on every turn of the spiral of β-oxidation (see diagram). The acetyl CoA resulting from this process can be further oxidized in the KREBS CYCLE to yield more energy or it can be converted to sugar via the GLYOXYLATE CYCLE. In germinating oil-rich seeds the enzymes of β-oxidation are located in the GLYOXYSOME along with the enzymes of the glyoxylate cycle.

oxidation–reduction reaction (oxidoreduction, redox reaction) a process in which electrons are transferred from an 'electron donor' or reducing agent to an 'electron acceptor' or oxidizing agent. Oxidation–reduction reactions are major energy-supplying reactions in both autotrophic and heterotrophic organisms. In aerobic organisms the final electron acceptor for oxidation–reduction reactions is oxygen. Anaerobic organisms however use an organic compound as a final electron acceptor, e.g. in yeasts acetaldehyde accepts electrons to form ethanol. In all organisms there are intermediate oxidation–reduction enzymes and COENZYMES between the fuel molecules and the final electron acceptor. The most important of these are the pyridine nucleotides (see NAD, NADP).

oxidative decarboxylation see KREBS CYCLE.

oxidative phosphorylation the formation of ATP from ADP and inorganic phosphate coupled to the movement of electrons down the RESPIRATORY CHAIN. ATP formation occurs at three sites on the respiratory chain: during the transfer of electrons from NAD to COENZYME Q; during the transfer of electrons from cytochrome *b* to cytochrome *c*; and during the transfer of electrons from cytochrome *aa*₃ to oxygen. The nature of the link between oxidative phosphorylation and the respiratory chain is not understood although there are several theories to explain it. The most widely supported theory suggests that the respiratory chain creates a pH gradient across the internal mitochondrial membrane and that the potential energy of this gradient is then utilized to convert ADP to ATP. See CHEMIOSMOTIC THEORY.

oxidoreductase any ENZYME that catalyses oxidation–reduction reactions. The two main groups are the pyridine-linked and the flavin-linked DEHYDROGENASES. Electron-transferring proteins, notably the iron–sulphur proteins and the CYTOCHROMES, may also be classed as oxidoreductases. See also OXIDASE.

2-oxopropanoic acid see PYRUVIC ACID.

oxygen symbol: O. A nonmetallic, colourless, odourless, tasteless gas, atomic number 8, atomic weight 16, the most plentiful element in the Earth's crust (about 40% by weight, found in carbonates (limestone), silicates (sandstones) and other minerals. Oxygen makes up 20.95% of the atmosphere. It is an essential component of living organisms, found in carbohydrates, fats, and proteins, and necessary for aerobic respiration. Normal molecular oxygen is O_2,

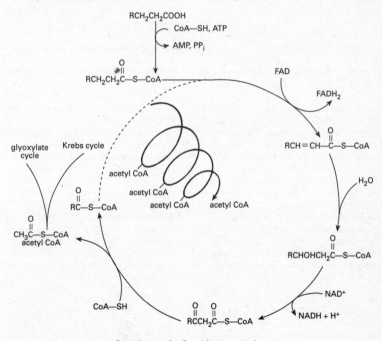

β-oxidation: the β-oxidation spiral

Oxygen also exists as O_3 (*ozone*), formed by the action of ultraviolet in the upper atmosphere. See also OZONE LAYER.

oxygenase an enzyme that catalyzes reactions in which oxygen is introduced to a molecule, e.g. ribulose bisphosphate oxygenase.

oxygen demand see BIOCHEMICAL OXYGEN DEMAND, BIOLOGICAL OXYGEN DEMAND, CHEMICAL OXYGEN DEMAND.

oxygen method a method for estimating gross PRIMARY PRODUCTIVITY by monitoring the rate of production of oxygen. It is used especially in aquatic ecosystems. Pairs of light and dark bottles are filled with water at a particular depth and suspended at this depth for several hours. The oxygen content of the dark bottle, which does not let light in, is taken to indicate the uptake loss of oxygen by respiration of the phytoplankton (see PLANKTON) and other microorganisms during this period. It is assumed that the respiration rate in the light bottle is the same. By substracting the loss of oxygen in the dark bottle from the gain in oxygen of the light bottle (due to photosynthesis of the phytoplankton), the gross primary productivity is obtained. This method is suitable only for communities in which the proportion of photosynthesizing organisms is relatively large, so that the rate of respiration from other microorganisms is insignificant. It is not suitable for use with larger plants, since some of the oxygen released by photosynthesis may be respired without leaving the plant.

oxygen quotient (QO_2) the volume (microlitres) of oxygen taken up in 1 hour by 1 milligram of plant tissue at normal temperature and pressure.

oxygen sag the fall in the percentage saturation of dissolved oxygen in the water of a stream or river downsteam from a source of pollution, and its subsequent recovery further downstream.

ozone see OXYGEN.

ozone hole a markedly thin region of the OZONE LAYER; ozone holes were first detected in the ozone layer above the polar regions in the late 1970s, since when they have increased in extent. This depletion of the ozone layer is thought to be due to complex photochemical reactions involving chlorofluorocarbons (CFCs; highly stable compounds used in aerosols, etc.), halons (gases used in industry and in some kinds of fire extinguishers), and nitrogen oxides from aircraft. The reaction of CFCs with ultraviolet light in the stratosphere releases chlorine and fluorine atoms, which are highly reactive and destroy ozone. One atom of chlorine or fluorine can destroy some 10^5 molecules of ozone. Continued depletion of the ozone layer due to man-made pollutants is likely to lead to increased cancer rates and to as yet unquantified effects on the marine plankton that forms the bases of ocean food webs, as well as destabilization of the world's weather patterns. See MONTREAL PROTOCOL ON SUBSTANCES THAT DEPLETE THE OZONE LAYER.

ozone layer (ozonosphere) an ozone-rich layer of the earth's atmosphere about 15–50 km above the earth's surface around the region of the stratosphere. The ozone is produced by the action of solar radiation on oxygen molecules. The ozone molecules absorb most of the sun's ultraviolet radiation, thus protecting the earth's living organisms from its harmful mutagenic effects. The absorption of ultraviolet radiation also warms the stratosphere, helping to stabilize it and prevent mixing with other layers, reducing winds in the lower atmosphere. It thus acts like a thermal blanket around the earth. Where the ozone layer is thin, there is an increased incidence of human skin cancers (see OZONE HOLE). See ATMOSPHERE.

ozonosphere see OZONE LAYER.

P p

P680 a special form of CHLOROPHYLL *a* with
a light absorption peak of 684 nm. It is one
of the constituents of the photosystem II
light-gathering centre of photosynthesis.
The absorption of light energy by the P680
molecules causes, among other things, the
formation of a strong oxidant that can then
oxidize water. This results in the evolution
of molecular oxygen and initiates electron
flow, which continues, via photosystem I,
through to the production of reduced NADP.
See also P700.

P700 a special form of CHLOROPHYLL *a* with
an absorption peak for light of 700 nm
wavelength, found within the photosystem I
light-gathering centre of photosynthesis.
Other ordinary chlorophyll *a* molecules are
also present and these absorb light energy
and pass it to P700, creating a charged
form, P700⁺. Electron transfer occurs from
photosystem II, electrons being passed on
ultimately to produce NADPH, the reduced
form of NADP. See also P680.

pachytene see PROPHASE.

Pacific coast forest the most dense
coniferous FOREST in the world, extending
along the coast and mountain ranges just
inland from California to southern British
Columbia. It is reknowned for its giant
conifers, including the coastal redwood
(*Sequoia sempervirens*), big tree
(*Sequoiadendron giganteum*), and Douglas fir
(*Pseudotsuga taxifolia*), which are among the
tallest trees in the world, some reaching
heights of 110 m. and girths in excess of 20
m. It includes some of North America's
most important OLD-GROWTH FOREST.

padang a type of KERANGA found on
PODSOLS in southeast Asia.

paedomorphosis see NEOTENY.

pairing see SYNAPSIS.

Palaearctic a BIOGEOGRAPHICAL REGION that
comprises Europe, Russia and the former
Soviet states, northern Arabia and the
Mediterranean coastal strip of north Africa.

palae- prefix denoting ancient, archaic, early,
or primitive.

palaeobotany see PALAEONTOLOGY.

Palaeocene see TERTIARY.

palaeoecology the study of extant and fossil
organisms with a view to establishing the
nature of the life of past ages, the
environmental conditions that prevailed,
and the interactions between them. Details
of the environment in which given fossil
organisms lived can be assumed from a
knowledge of the prevailing climatic
conditions in which the rocks that contain
them were formed. However, difficulties
arise in distinguishing between the
environment in which the organism lived
and the environment in which it was
buried. See also ASSEMBLAGE, PALYNOLOGY,
REFERENCE SPECIES.

palaeolimnology the study of the history
and development of freshwater ecosystems.

palaeontology the study of FOSSILS. The
branch dealing with the study of plant
fossils is termed *palaeobotany*, which
includes *palaeoethnobotany*, the study of fruit
and seed remains found in archaeological
sites. Palaeontology requires a knowledge of
the way in which fossilization takes place
and of the type of rock in which the fossil is
found, a knowledge of the environmental
conditions that prevailed when the rock was
laid down, and an appreciation of the

geological time scale so that the fossil can be dated.

palaeosol a soil that was not formed during the present period of PEDOGENESIS. It may simply have been formed earlier, or it may have been buried, or buried and later exposed.

palaeospecies a group of organisms known only from their fossils, which differs from all other known groups.

palaeotropical floristic region see BIOGEOGRAPHY, FLORISTIC REGION, PLANT GEOGRAPHY.

Palaeozoic the era of geological time between about 570 and 225 million years ago that succeeds the Precambrian and precedes the Mesozoic. It is subdivided into six periods, the CAMBRIAN, ORDOVICIAN, and SILURIAN (lower Palaeozoic), and the DEVONIAN, CARBONIFEROUS, and PERMIAN (upper Palaeozoic). See also GEOLOGICAL TIME SCALE.

pale- see PALAE-.

palea (pale, valvule) the upper of the pair of BRACTS beneath each floret (flower) in the inflorescence of a grass, the other being the LEMMA. The palea is usually thin, narrow, and parallel sided. It usually has two ribs, which may project as prominent keels. Compare GLUME. *See illustration at* poaceae.

palindrome see REPEATED SEQUENCE.

palisade mesophyll photosynthetic CHLORENCHYMA tissue composed of more or less elongate cells arranged in radial columns. Palisade mesophyll occurs with SPONGY MESOPHYLL as a ground tissue in the leaves of many MESOPHYTES. Here the palisade mesophyll is usually on the adaxial (upper) side of the leaf, whereas in many XEROPHYTES it forms the bulk of the mesophyll and is present on both sides of the leaf. When the stem has taken over much or all of the photosynthetic work of the plant, the cortex may consist largely of palisade parenchyma.

Palmae see ARECACEAE.

palmate (digitate, palmately lobed)
1 describing a compound leaf having four or more leaflets arising from a single point. The leaves of the horse chestnut (*Aesculus hippocastanum*) and of lupins (*Lupinus*) are examples. *See illustration at* leaf.
2 describing a form of VENATION in which

several equally prominent veins branch out from the base of a leaf blade. See ACTINODROMOUS.

palmelloid describing a type of colonial growth form in which an indefinite number of cells are held together after division in a mucilaginous matrix. Palmelloid forms are characteristic of certain cyanobacteria, e.g. *Merismopedia*, and various algae and other protoctists.

palmitic acid (hexadecanoic acid) a sixteen-carbon saturated fatty acid. It is the commonest saturated FATTY ACID in plants, forming a high proportion of plant fats. It is also on the synthetic route to other fatty acids, such as STEARIC and OLEIC ACIDS.

palms see ARECACEAE.

Palouse prairie a distinctive type of prairie on fertile loess soils in southeast Washington and neighbouring areas of Idaho USA, which supports a rich growth of native tussock grasses. Most of this prairie is now given over to wheat, and only 1% of the original vegetation remains.

palsa boggy tundra (see BOG), that contains *palsas*, i.e. mounds or ridges up to 35 m long, 15 m high and 7 m high, found in periglacial areas and composed mainly of peat and containing a permanent ice lens. Palsas occur in damp boggy areas. They are believed to form as a result of frost-heaving of soil, together with the growth of ice inside.

palustrine growing in boggy ground.

palynology the study of living and fossil pollen grains, spores, and similar structures. The diversity of such structures has led to an increasing use of palynological data in systematics, while their resistance to decay has made them an important feature of palaeobotanical and palaeoecological work. See also EXINE. include aearopalynology

pamirs ALPINE vegetation of the high arid plateaux of Tibet. It is composed mostly of sparse tufts of grass and large, cushion plants. Compare PUNAS.

pampas temperate South American GRASSLAND, especially extensive in Argentina. The original grassland was dominated by tussock grsses, with shorter grasses and xerophytic shrubs in the drier areas. Much of the wetter northeastern pampas is now cultivated.

pan a heavily compacted SOIL HORIZON that is either high in clay or cemented with minerals that have been leached out of the upper soil horizons. See HARDPAN, LEACHING, SOIL PROFILE.

panbiogeography an approach to biogeography first developed by Leon Croizat (1894–1982), that is a cartographic method of analysing the spatial structure of the distribution of species and other taxa in an attempt to better understand their origin in time and space.

pandurate fiddle-shaped.

Pangaea see CONTINENTAL DRIFT.

pangenesis a now disregarded theory formulated by Charles Darwin that suggested a mechanism by which acquired characteristics may be inherited. He postulated that there were particles (pangenes) carried in the body fluids from all the organs of the body to the reproductive cells. These particles then influenced the gametes that in turn influenced the characteristics of the succeeding generation.

panicle an inflorescence, associated particularly with grasses, in which the flowers are formed on stalks (peduncles) arising from the main axis. Each stalk is a raceme. The peduncles are long and spreading, as in oats (*Avena*). The term is also used to describe any type of branching inflorescence. *See illustration at* inflorescence.

panmixis unrestricted random crossing. It is probably fairly rare but the term may be applied to crossing that appears to be random with respect to a particular character or gene. Compare ASSORTATIVE MATING.

pantocolpate describing a pollen grain having many colpi (see COLPUS).

pantonematic flagellum see UNDULIPODIUM.

pantothenic acid a compound, formula $CH_2OHC(CH_3)_2CHOHCONHCH_2CH_2COOH$, that is the precursor of COENZYME A. It is synthesized by plants and bacteria but not by vertebrates, which consequently require it as a VITAMIN in the diet.

pantropical distributed throughout the tropics.

papain see PROTEASE.

paper chromatography a widely used chromatographic technique for separating the components of mixtures using absorbent paper as the stationary phase. A sheet or strip of paper, with a concentrated spot of the mixture on a pencil-drawn base line, is vertically suspended in a suitable solvent (the mobile phase), that seeps slowly upwards through the paper by capillary action. The compounds in the mixture travel upwards with the solvent and separate out at different levels. When the solvent reaches a level near the top of the paper, this point (the solvent front) is marked. The paper is removed and dried. Colourless compounds may then be developed either by spraying the paper (the chromatogram) with a suitable chemical, such as NINHYDRIN, or by viewing the paper in ultraviolet light. The compounds can then be identified either by comparing them with chromatograms of known standard solutions run at the same time, or by calculating the RF VALUES. See CHROMATOGRAPHY, PARTITION CHROMATOGRAPHY, TWO-DIMENSIONAL ANALYSIS.

papilla a projection from a cell, usually of the epidermis, often regarded as a kind of TRICHOME. Papillae are often swollen and covered with wax and in XEROPHYTES may serve to protect from sunlight and excessive water loss.

pappus a modified CALYX made up of a ring of fine hairs, scales, or teeth that persist after fertilization and aid the wind dispersal of the fruit, often by forming a parachute-like structure. It is seen in members of the Asteraceae (Compositae), e.g. dandelions (*Taraxacum*) and thistles (*Carduus*). The pappuses of an inflorescence may form a 'clock'.

paraflagellar body see PHOTORECEPTOR.

parallel evolution (parallelism) the development of similar adaptations in related organisms as a result of being subject to similar selection pressures. Thus the different species of a genus with a wide distribution may show the same adaptations to similar environments in widely separated regions. The development of divided submerged leaves in various species of crowfoot, e.g. water crowfoot (*Ranunculus aquatilis*) and river crowfoot (*R. fluitans*), is an example. However the phenomenon is not of widespread occurrence and some authorities deny its existence.

parallelodromous describing a form of leaf venation in which two or more primary veins originate at the base of the lamina and run to the apex in an essentially parallel manner. The leaves of many monocotyledons have this type of venation. In traditional terminology such venation is simply termed *parallel. See illustration at* venation.

parameter a quantity that characterizes and describes a population.

páramo a unique kind of ALPINE meadow and scrub found in parts of the Andes above the tree line that experiences frequent cloud and mist. Grasses and lichens are abundant, with many showy herbaceous plants and scattered taller plants, particularly *Espeletia* species, which grow up to 20 ft tall with large rosettes of pale, densely hairy leaves.

Paramycota in some modern classification schemes, a division of fungi characterized by walls containing β(1–4) glucan rather than chitin. They have a hyphal growth habit or a thallus with rhizoids, the main storage polysaccharide is mycolaminarin, sterols are cholesterols or alkylsterols, and spores are usually motile. Two classes are recognized, the Oomycetes (see OOMYCOTA) and the Hyphochytriomycetes (see HYPHOCHYTRIOMYCOTA). Compare EUMYCOTA. See also FUNGI.

paramylon see PARAMYLUM.

paramylum (paramylon) a polysaccharide that is the assimilation product of the EUGLENIDA, made up of β(1–3)-linked units of glucose.

parapatric describing species that live in adjacent but separate habitats.

parapatric speciation SPECIATION in which the new species forms a population within the ancestral species' geographical range, for example through reproductive isolation or behavioural mechanisms such as different flowering times or pollinators.

paraphyletic describing a group of organisms that does not contain all the descendants of a common ancestor. In plant classification, the dicotyledons are a paraphyletic group because their ancestors also gave rise to the more advanced monocotyledons.

paraphyllum in mosses, a tiny irregular green 'leaflet' among true leaves, e.g. *Thuidium* (tamarisk-leaved feather moss).

paraphysis a sterile unbranched usually multicellular hair borne amongst the reproductive structures of many bryophytes, algae, and fungi. Paraphyses are protective and in some cases are thought to aid in dehiscence.

parasexual recombination a rare form of RECOMBINATION seen in the progeny of certain heterokaryotic fungi (see HETEROKARYOSIS), which does not involve meiotic segregation. The sequence of events involved is termed the *parasexual cycle* and is as follows: two genotypically distinct nuclei in a heterokaryon fuse to give a diploid heterozygous nucleus; the diploid nucleus multiplies alongside the haploid parent nuclei in the heterokaryon; a homokaryotic diploid mycelium is established from a diploid CONIDIUM; recombination occurs during crossing-over in mitosis; a haploid nucleus is formed by progressive shedding of chromosomes from the diploid nucleus.

Such recombination has been observed experimentally in various ascomycetes, basidiomycetes, and anamorphic (imperfect) fungi (Fungi Anamorphici) and has also made feasible the study of the genetics and 'breeding' of anamorphic fungi. However, it is uncertain how widespread parasexuality is in natural populations of fungi.

parasitism the temporary or permanent relationship between two different species, in which one, the *parasite*, benefits by obtaining food and/or shelter at the expense of the other, the *host*. Some parasites have little effect on the host, some cause serious diseases, and some kill the host. *Ectoparasites* live on the surface of the host, while *endoparasites* live either inside or between the host cells. An *obligate parasite*, e.g. *Phytophthora infestans*, which causes late blight of potato, can only grow on its host.

However, a *facultative parasite*, such as the oomycote *Pythium*, which causes damping-off, can exist by feeding in a different way after the death of the host. A species of parasite may live on one particular host or a number of similar hosts, or may alternate between two or more different species. Some flowering plants are *partial parasites*, being able to live independently but becoming parasitic in certain circumstances. For example, some members of the Scrophulariaceae, such as eyebrights (*Euphrasia*), will photosynthesize and live independently but when their roots come into contact with those of grasses they become attached and absorb food. See also HEMIPARASITE.

parastichy an imaginary spiral around a stem apex joining adjacent leaf PRIMORDIA. Two parastichies can be visualized running in opposite directions (see diagram). If a plant is described as having, for example, (3+5) or 3/5 PHYLLOTAXIS, this means that the shallower of the two parastichies joins every third primordium formed at the apex (1, 4, 7, 10, etc.) and the steeper parastichy joins every fifth primordium (1, 6, 11, 16, etc.).

Parastichy: opposite parastichies at the stem apex

paratonic movement a movement exhibited by a plant or plant part in response to an external stimulus. TAXES, TROPISMS, and NASTIC MOVEMENTS and mechanical hygroscopic and turgor movements are all examples of paratonic movements. Compare AUTONOMIC MOVEMENT.

paratype a specimen other than the ISOTYPE or HOLOTYPE cited in the original publication of the name of a TAXON. A paratype can also be one of a collection of SYNTYPES, following the selection of a LECTOTYPE. Although of little significance nomenclaturally, paratypes are of considerable importance to the taxonomist, who may not be able to see other type material of a particular taxon. See also TYPE.

parenchyma unspecialized tissue usually composed of more or less isodiametric polyhedral cells with thin nonlignified cellulose cell walls and living protoplasts. Parenchyma cells are often present in great numbers, forming a *ground tissue* (ground parenchyma) in which the other tissues are embedded. See also COLLENCHYMA.

parichnos an area of tissue made up of disintegrated cells seen at the base of the sporophylls in the strobili of *Lycopodium* and in the leaves of various fossil relatives of *Lycopodium*, such as *Lycopodites* and *Lepidodendron*.

parietal placentation a form of PLACENTATION in which the placentae develop along the fused margins of a unilocular compound ovary, as in violets (*Viola*).

paripinnate describing a pinnate leaf in which all the leaflets are paired. Such leaves are seen in vetches (*Vicia*), the terminal leaflet having been replaced by a tendril. Compare IMPARIPINNATE. *See illustration at leaf.*

park 1 land set aside for public recreation and enjoyment, and artificially landscaped. 2 an enclosed piece of land on which deer, wild or semi-wild animals are kept. 3 In ancient farming systems, enclosed fields situated between the fields immediately next to the farm buildings and outer fields used only occasionally for pasture.

park woodland woodland with an open canopy, used for pasture.

paroecious describing a plant species in which each plant is bisexual, but the male and female reproductive organs occur separately on the same shoot.

parthenocarpy the production of a fruit without the setting of seeds. Parthenocarpy is most often seen in plants with numerous ovules in the fruit, e.g. fig and melon. In some species, e.g. cucumber, pollination is not required to stimulate parthenocarpy. Where pollination is necessary parthenocarpy may result either because the pollen tube does not reach the ovule or because the embryos abort following fertilization, as in seedless grapes. Triploidy, which results in sterile plants, can also lead to parthenocarpy, as in the banana.

parthenogenesis the development of an egg cell into an embryo without fertilization. This is usually due to defective meiosis, which results in an egg nucleus with an unreduced number of chromosomes, as is found in dandelions (*Taraxacum*). In some species, e.g. hawkweeds (*Hieracium*), the megaspore is replaced by a cell of the nucellus. A cell or individual produced by parthenogenesis is termed a *parthenote*.

parthenospore (azygospore) a thick-walled resting spore that develops by PARTHENOGENESIS from an unfertilized gamete. Parthenospores are formed by certain algae, e.g. *Protosiphon*, and fungi, e.g. *Entomophthora muscae*. They may give rise to more gametes or to a haploid individual. Compare ZYGOSPORE.

parthenote see PARTHENOGENESIS.

partial dominance see INCOMPLETE DOMINANCE.

partial parasite 1 see HEMIPARASITE.
2 see PARASITISM.

partial random sample see RANDOM SAMPLE.

partial veil see MUSHROOM.

particulate inheritance the determination of inherited characteristics by stable discrete factors that remain unchanged from one generation to another. Differences observed between parents and offspring are thus due to the RECOMBINATION rather than modification of such factors. The particulate nature of inheritance was established with Mendel's work. Earlier theories formulated without experimental evidence tended to the idea that the characteristics of parents blended together in the offspring (see BLENDING INHERITANCE).

partition chromatography a kind of CHROMATOGRAPHY in which the components of mixtures are separated according to their different partition coefficients. Any substance dissolved, at a given temperature, in two immiscible solvents has, at equilibrium, a certain characteristic proportion dissolved in each

solvent. The ratio of the amount of solute in one solvent to that in the other is the *partition coefficient*. In column partition chromatography the column contains some hydrophilic substance, e.g. silica or starch granules. Water is tightly bound to the surface of the granules and acts as the stationary phase. A solvent immiscible with water, e.g. butanol or phenol, and carrying the mixture to be separated is then passed through the column. This is the mobile phase. Those components of the mixture that are more soluble in water will pass through the column more slowly than those that are less soluble in water but more soluble in the mobile phase. The components are separated by collecting and analysing small amounts of the eluate (the liquid passing from the bottom of the column) as it emerges from the column. Mixtures of amino acids are often separated in this way.

PAPER CHROMATOGRAPHY is a particular kind of partition chromatography in which the hydrophilic cellulose of the paper holds the water and acts as the stationary phase.

partition coefficient see PARTITION CHROMATOGRAPHY.

passage cell any of the thin-walled cells found at intervals in the endodermis, generally opposite the PROTOXYLEM poles. It is believed that passage cells provide an easier route for water and dissolved solutes to pass through from the cortex to the protoxylem.

passive absorption see ABSORPTION.

passive chamaeophyte see CHAMAEOPHYTE.

passive dispersal see DISPERSAL.

pasture see MEADOW.

pathogen an agent able to cause disease. The term is often restricted to agents that are themselves living organisms.

pathogenicity see VIRULENCE.

patristic describing similarity between two or more taxa that is due to common ancestry. The concept of patristic relationship is purely PHYLOGENETIC and thus not used in the construction of PHENETIC classifications. See also CLADISTIC.

pattern analysis a method of detecting a non-random distribution of organisms. It is usually applied to the analysis of vegetation, where it involves comparing the observed number of individuals per quadrat with the number expected from a POISSON DISTRIBUTION (i.e. from a randomly distributed population). It can be tested for fit using the CHI-SQUARED TEST, or by comparing the variance:mean ratio for the collected data with that of a Poisson series. A value significantly greater or less than 1 indicates that the population is overdispersed (see DISPERSION) or underdispersed respectively.

PCR see POLYMERASE CHAIN REACTION.

PCS see PLANT CONSERVATION SUBCOMMITTEE.

PE see EVAPOTRANSPIRATION.

peat partially decomposed plant material. It builds up in areas with poor drainage, namely bogs and fens. Acid bog peat (or peat moss) is composed primarily of the remains of bog plants, such as *Sphagnum* mosses and sedges. It is used as a mulch and to improve soil and is also a constituent of potting composts. Concern about the removal of peat from peat bogs has led to alternatives for horticulture. Fen peat is more alkaline and contains considerably more nutrients than bog peat and if drained provides fertile agricultural land. It is composed mainly of sedges and grasses.

peat-borer a device for extracting peat cores without significantly disturbing the surrounding peat. One of the commonest is the *Hiller peat-borer*, which uses a short AUGER head to penetrate the peat, behind which is a sharp-edged chamber that can be opened and closed at the required sample depth.

peat moss see SPHAGNUM.

pectic acid an $\alpha(1-4)$-poly-D-galacturonic acid (see URONIC ACIDS) that forms the middle lamella of plant cell walls and is the basis of PECTIN, in which the carboxyl groups are masked by methyl esters.

pectic substances pectins and related polysaccharides, in which galacturonic acid (see URONIC ACIDS), GALACTOSE, or a derivative of these is a major component. Besides the pectins found in cell walls and middle lamellae, large quantities of pectic substances are found in some fruits, e.g. apples, pears, and citrus fruits. Pectins are

used in the food industry as gelling agents in jams, etc.

pectin a structural POLYSACCHARIDE found in plant cell walls. It is rich in $\alpha(1-4)$-linked galacturonic acid residues (see URONIC ACIDS), though other sugars, e.g. rhamnose, are present in small quantities. The carboxyl groups of the galacturonic acid residues in pectin may be methylated. Methyl esterification takes place after polymerization, the methyl donor for the reaction being S-adenosyl methionine (see METHIONINE).

pectinate resembling a comb.

ped see SOIL STRUCTURE.

pedalfer a SOIL in which the soluble lime has been leached from the surface layers by heavy persistent rain. It is thus acid and the rate of decomposition of soil litter is consequently slow. Iron and aluminium hydroxides are abundant. Pedalfers are one of the two main types of zonal soils (compare PEDOCAL) and include BROWN EARTHS, PODSOLS, and TUNDRA soils.

pedicel the stalk attaching individual flowers to the main axis (peduncle) of the inflorescence. It often develops in the axil of a bract and its internal structure is typically stemlike. The pedicel may act as a temporary storage organ for sugars prior to seed development.

pedigree method a plant breeding technique that is used to create entirely new varieties of plants that combine the best qualities of selected existing varieties. The method is limited to self-pollinating species (see SELF POLLINATION). Parents are selected and artificially crossed. A few (say six to ten) of the resulting F_1 hybrid seeds are sown widely spaced to encourage prolific seed set. Assuming the parents were HOMOZYGOUS, the F_1 PLANTS should all be identical. The seed from each plant is harvested separately and grown, again widely spaced, in separate 'family' plots. The genotypic variability becomes apparent in the F_2 and the characteristics of each plant are noted. The seed from a small number (perhaps only 1%) of promising plants is harvested to grow on to the F_3. The seeds from each plant are again sown in separate plots and in this way the 'pedigree' of each plant can be recorded. By noting the variability within each of the F_3 plots it may be determined which plants of the previous generation depended excessively on heterozygosity for vigour. At this stage whole families of plants may be discarded. As heterozygosity is approximately halved each generation, increasing emphasis is placed in succeeding generations on selection between rather than within families. If breeding a crop plant, then at about the F_6 generation seeds from selected plants are grown closely spaced as would occur in normal field conditions. Yield may then be determined. From this point selection is entirely between families. By the F_8 generation the remaining selected lines may be assumed to be over 99% homozygous and they are bulked up for variety trials. Compare BACKCROSSING.

pedocal a SOIL in which lime accumulates in the surface layers, which consequently become alkaline. Pedocals occur in regions with light rainfall and are usually fertile as they may be rich in potash and phosphates as well as lime. In persistently dry conditions calcium salts may be deposited on the soil surface. Pedocals are one of the two main types of zonal soils (compare PEDALFER), and include the CHESTNUT-BROWN SOILS, CHERNOZEMS, and RED DESERT SOILS.

pedogenesis the natural process of forming of SOIL from consolidated parent material. It involves processes such as weathering, humification, and leaching.

pedology the study of the natural formation, composition and distribution of SOILS.

pedon a sampling block that encompasses the whole of a soil profile, i.e. it extends from the surface to the parent material, and spreads laterally far enough to include all horizons and horizon boundaries.

peduncle the main stem of an inflorescence. See PEDICEL.

peg see ROOT COLLAR.

pelagic describing the organisms that live in the surface waters of a sea or ocean, such as PLANKTON.

pellicle (periplast) 1 a membrane surrounding the protoplast in some unicellular protoctists, in which it is highly structured and may be rigid or flexible, usually due to proteinaceous materials, e.g.

Euglena. **2** the scales, spines, and spicules that cover some unicellular algae, e.g. the Chrysomonada.

pellucid transparent.

peloton see ENDOMYCORRHIZA.

peltate describing a structure that is circular, with the stalk inserted in the middle, such as the leaf of the garden nasturtium (*Tropaeolum majus*), the hairs on the undersurface of the leaves of sea buckthorn (*Hippophae rhamnoides*), or the SPORANGIOPHORES of certain cycads. *See illustration at leaf.*

pendulous placentation see APICAL PLACENTATION.

penetrance the degree to which a gene is expressed in the phenotype. Most dominant alleles show complete penetrance. However a gene may be affected by environmental factors to the extent that not all the individuals with a particular genotype show the phenotype characteristic of that genotype. In some maize plants, for example, anthocyanin pigment only develops in those parts of the plant exposed to high light intensities. In shaded parts of the plant the gene is not expressed. A gene whose expression is affected in this way is said to show *incomplete penetrance*.

penetration tube see APPRESSORIUM.

Penicillium a widespread genus of FUNGI with a septate white MYCELIUM and aerial CONDIOPHORES bearing greyish- or bluish-green conidia with a powdery appearance. Most species are saprobes. Some strains produce antibiotics such as penicillin, others are used in the manufacture of certain cheeses.

Pennales (pennate diatoms) an order of DIATOMS that have markings arranged bilaterally along the long axis of each valve. Many also have a raphe, a slit running the length of the valve, which secretes mucilage to aid in motility. Alternatively, some pennate diatoms have a clear area, a *pseudoraphe*, in place of a raphe. The Pennales have fewer chloroplasts than the CENTRALES.

pennate diatoms see PENNALES, DIATOMS.

penni-parallel see CRASPEDROMOUS.

Pennsylvanian see CARBONIFEROUS.

pentamerous (5-merous) Describing flowers in which the parts of each whorl are inserted in fives, or multiples of five. This is the most commonly found arrangement in dicotyledons.

pentosan a polysaccharide in which the major monosaccharide subunits are pentoses. The most common monosaccharide components of pentosans are xylose and L-arabinose. Many HEMICELLULOSES are pentosans, although they are usually heteropolymers containing other sugars besides the pentoses.

pentose any five-carbon sugar. Common pentoses are RIBOSE, DEOXYRIBOSE, XYLOSE, and ARABINOSE. Pentoses are synthesized from hexoses in the PENTOSE PHOSPHATE PATHWAY, from carbon dioxide in the CALVIN CYCLE, or by decarboxylation of nucleoside diphosphate sugars. Polymers of pentoses are collectively termed PENTOSANS.

pentose phosphate pathway (pentase phosphate shunt, phosphogluconate pathway, hexose monophosphate shunt) a multifunctional metabolic pathway involving oxidation of glucose and formation of three-, four-, five-, six-, and seven-carbon sugars. In the pentose phosphate pathway glucose 6-phosphate is first oxidized and decarboxylated to a five-carbon sugar with the concomitant formation of NADPH. The five-carbon sugar then undergoes various combinations and transformations, similar to those of the Calvin cycle, leading eventually to the reformation of glucose 6-phosphate. Although the pathway can be used for the oxidation of glucose, it is more important as a source of five-carbon sugars, as a pathway for the oxidation of five-carbon sugars, and as a source of NADPH for such purposes as fatty acid synthesis.

Pentoxylales an order of extinct gymnosperms from the JURASSIC period, the affinities of which are uncertain.

PEP see PHOSPHOENOLPYRUVATE.

pepo a type of BERRY with a hard exterior derived either from the epicarp or noncarpellary tissue of the plant. In members of the Cucurbitaceae the hard exterior is formed from the receptacle of the flower.

peptidase any HYDROLASE enzyme that catalyses the hydrolysis of peptides. The

distinction between PROTEASES and peptidases is not a clear one – peptidases often have protease activity and vice versa. Several types of peptidase can be distinguished. *Exopeptidases* can only attack terminal peptide bonds while *endopeptidases* attack bonds within the peptide chain and are often specific to certain types of peptide linkage. *Aminopeptidases* attack the amino terminal of peptides and *carboxypeptidases* attack the carboxy terminal. *Dipeptidases* attack only dipeptides.

peptide a molecule consisting of a number of amino acids covalently joined by PEPTIDE BONDS. Small peptides containing only a few amino acids are known as oligopeptides while larger peptides are called POLYPEPTIDES. Most oligopeptides arise as breakdown products of proteins but a few have specific functions. For example, the tripeptide glutathione activates some enzymes and protects lipids from autooxidation.

peptide bond the bond that forms when two amino acids are joined together with the elimination of water in a CONDENSATION reaction. The peptide bond is a covalent carbon–nitrogen bond (see diagram overleaf). The amino acids that make up proteins are linked by peptide bonds. These bonds can be broken by HYDROLYSIS reactions.

peptidoglycan (mucopeptide) a macromolecule made up of chains of amino sugars (acetylglucosamine and acetylmuramic acid) linked to a tripeptide of alanine, glutamic acid, and lysine or diaminopimelic acid. It is found in the cell walls of most bacteria, where it confers shape and strength. One of the defining characteristics of the ARCHAEA (archaebacteria) is the lack of peptidoglycan in their cell walls.

perennating organ in biennial or perennial plants, a storage organ that enables the plant to survive unfavourable seasons, such as winter or a dry season. Examples include swollen taproots, tubers, rhizomes, bulbs, and corms. The plants usually die down and survive as underground perennating organs with resting buds. Perennating organs often also serve as a means of VEGETATIVE REPRODUCTION. For example, new bulbs

Peptide bond: formation of a peptide bond

and corms arise in the axils of scale leaves on existing bulbs and corms, and rhizomes may fragment and the adventitious roots put down at each node serve to support the new shoots as independent plants.

perennial a plant that lives for two or more years. There are two main types. *Herbaceous perennials* survive each winter as underground storage or perennating organs, such as BULBS, CORMS, RHIZOMES, and stem and root TUBERS. The foliage leaves and flowers die back in winter. *Woody perennials* are trees and shrubs whose aerial stems have woody tissues and persist above ground. They may be deciduous or evergreen. Normally SEMELPAROUS plants may become perennial if grazed or cut or if the flowers fail to set fruit; examples are *Medicago* and *Cirsium*. Compare ANNUAL, BIENNIAL, EPHEMERAL.

perfoliate describing a SESSILE leaf in which the base of the lamina extends either side of the node and joins together on the far side so the stem is completely encircled. This formation is also seen when pairs of sessile leaves unite at their bases, as in the perfoliate honeysuckle (*Lonicera perfoliatum*).

perforation plate the remains of the end walls between two adjacent vessel elements

in a VESSEL of the xylem, forming an opening between the cells, thus facilitating the free movement of water through the vessel. Perforation plates are present in some ferns, some gymnosperms (Gnetophyta), and most angiosperms, and are believed to have evolved independently in the three groups. Compare SIEVE PLATE.

perialgal vesicle in a symbiotic relationship involving an alga, a vesicle in a host cell that contains the alga.

perianth the structure that protects the developing reproductive parts of the flower. In dicotyledons it normally consists of two distinct whorls, the CALYX and the COROLLA. In many monocotyledons (such as the tulip) these whorls are not differentiated and the individual perianth units are then termed TEPALS. The perianth units tend to be simple, separate, and inserted spirally in the more primitive families, such as the Magnoliaceae and Papaveraceae. In the more advanced families, such as the Lamiaceae and Asteraceae, they are often fused and inserted in whorls. The opening and closing of some flowers is controlled by temperature and light, these stimuli being perceived by the perianth segments.

periblem see HISTOGEN THEORY.

pericarp the wall of a fruit, derived from the maturing ovary wall. In fleshy fruits the pericarp usually has three distinct layers, the outer toughened EXOCARP, a fleshy MESOCARP, and an inner ENDOCARP, which may be variously thickened or membranous. In dry fruits the pericarp tends to become papery or leathery. If the pericarp opens at maturity the fruit is described as dehiscent but if it remains closed the fruit is termed indehiscent.

pericentric inversion see INVERSION.

perichaetium any of the leaves or bracts surrounding the sex organs of bryophytes or the structure formed by such a whorl. Those enveloping the archegonia are sometimes termed *perigynia* while those around the antheridia are termed *perigonia*. See also PSEUDOPERIANTH.

periclinal parallel to the surface. The *periclinal wall* of a cell is thus one that is parallel to the surface of the plant body. A *periclinal division* is one that results in the formation of periclinal walls between daughter cells. Such a division results in an increase in girth of the organ. In cylindrical organs, such as stems and roots, the term *tangential* may be used in place of periclinal, usually in descriptions of cell walls. Compare ANTICLINAL.

periclinal chimaera a CHIMAERA in which tissues of one genetic type completely surround tissues of another genetic type. The ornamental shrub *Laburnocytisus adami* is a periclinal chimaera that originated from the grafting of *Cytisus purpureus* onto a *Laburnum anagyroides* stock. The epidermis is composed of *Cytisus* cells and the internal tissues of *Laburnum* cells. Externally it is not obvious that this *graft hybrid* is a chimaera but occasionally shoots arise consisting solely of tissues from one or other parent. The seeds, which derive from the internal tissues, are always *Laburnum*. Investigations into the stability of this chimaera led to the TUNICA-CORPUS THEORY of apical organization.

The variegated leaves of certain plants are examples of periclinal chimaeras that have originated from a plastid mutation preventing chlorophyll synthesis. The mutation may occur in either the second layer of the tunica (L2) or the corpus (L3).

Differences in colour appear between the leaf margins and the centre of the leaf because the margins are derived from the L2 while the centre is composed of both L2 and L3. Thus if the mutation occurs in the L2 then the margins are white and the centres green (the underlying green L3 tissues are not masked by the white L2 tissues). This type of variegation is seen in *Pelargonium*.

pericycle (pericyclic region) the outermost layer of the STELE, lying immediately within the endodermis. In most dicotyledons the pericycle, which is composed mostly of parenchyma cells, also includes modified PROTOPHLOEM often with the addition of sclerenchyma fibres. The pericycle is a more obvious structure in roots, lateral roots arising from this region.

periderm a protective tissue of secondary origin, comprising the phellogen, phellem, and phelloderm and replacing the epidermis as the outer cellular layer of the stem and root in plant axes exhibiting secondary growth. In such tissue the functions of stomata are performed by LENTICELS. Periderm also often develops at the site of a wound or other recently exposed surface, thus helping to prevent the entry of pathogens. Compare BARK.

peridinin a XANTHOPHYLL pigment characteristic of dinoflagellates (Dinomastigota).

peridium the two-layered outer wall of certain fungal fruiting bodies. It is most obvious in the LYCOPERDALES, in which it consists of an outer exoperidium and an inner endoperidium. In puffballs (*Lycoperdon*), the exoperidium sloughs off or breaks up into a number of scales, while in the earthstars (*Geastrum*) it peels back over the endoperidium to give the characteristic star-shaped fruiting body. A peridium is also seen in the aecium surrounding the aeciospores of rust fungi and in the ascoma of certain Erysiphales (powdery mildews).

perigenous describing an angiosperm stomatal complex in which the subsidiary cells are not derived from the same INITIAL as the guard cells. Compare MESOGENOUS. See also HAPLOCHEILIC.

perigonium see PERICHAETIUM.

perigynium see PERICHAETIUM.

perigyny (*adj.* perigynous) the arrangement of floral parts, intermediate between HYPOGYNY and EPIGYNY, in which the sepals, petals, and stamens are inserted on the receptacle at about the same level as the ovary. Perigynous flowers, e.g. cinquefoils (*Potentilla*), are seen when the receptacle becomes flattened or concave, in contrast to *hypogynous* flowers in which the receptacle is convex.

perinuclear cisterna see NUCLEAR MEMBRANE.

period see GEOLOGICAL TIME SCALE.

periplasmodium see TAPETUM.

periplast see PELLICLE.

pericentric inversion see INVERSION.

perisperm a nutritive tissue derived from the NUCELLUS (and therefore diploid) that is found in the seeds of certain plants in which the endosperm does not completely replace the nucellus. Many of the Caryophyllaceae contain such tissue in their seeds.

perispore an extra layer that may surround the spore in some species, particularly certain ferns.

peristome the ring of teeth surrounding the opening of a moss capsule, visible when the OPERCULUM is removed. It may consist of a double or single ring. The peristome is closed over the opening of the capsule in wet weather but as the teeth dry out they fold back allowing the spores to escape.

perithecium (pyrenocarp) the type of ASCOMA characteristic of certain fungi of the Ascomycota. It is a pear-shaped structure with a pore or ostiole opening to the exterior. See also PSEUDOTHECIUM.

peritrichous describing bacteria that have flagella all over the surface of the cell. Bacteria with flagella at one or both ends of the cell are termed *polar*. See also MONOTRICHOUS, LOPHOTRICHOUS.

permafrost permanently frozen ground that occurs in TUNDRA regions. In parts of Alaska and Siberia the permafrost extends to depths of 600 m and 1400 m respectively, a relic from the last glaciation. Permafrost is often overlain by ground that thaws out in summer, the 'active layer'.

permanent quadrat see QUADRAT.

permanent stain a stain that lasts for a long time without fading and does not damage the specimen. A single stain may be used, such as FAST GREEN for macerated tissues, or a tissue section may be double-stained, using, for example SAFRANIN and the counterstain light or fast green. Aqueous dyes, such as safranin, should be used before the section is dehydrated, and light green on completion of dehydration. CLEARING and MOUNTING follow. Compare TEMPORARY STAIN. See also STAINING.

permanent wilting percentage see PERMANENT WILTING POINT.

permanent wilting point the point at which the amount of water in the soil has dropped to such a level that the plants begin to wilt and will not recover, even if moved to a cool and dark place, unless more water is added to the soil. It occurs when the WATER POTENTIAL of the soil is the same as, or lower (more negative) than, the water potential of the plant. The *permanent wilting percentage (wilting coefficieint)* is the percentage of water remaining in the soil at the permanent wilting point. The amount of water retained in the soil when the permanent wilting point is reached varies depending on the plant species and the type of soil. See also FIELD CAPACITY, WILTING.

permeability 1 the ease with which a liquid or gas can pass through a substance, e.g. through the soil. 2 the degree to which a substance can diffuse across a barrier such as a membrane, or pass across it by other means, such as active transport. See SELECTIVELY PERMEABLE MEMBRANE.

permeability coefficient a measure of the rate at which a solute can pass across a membrane. It is defined as the net number of solute molecules crossing 1 cm^3 of membrane per unit time when the concentration of the solute is 10 moles l^{-1}.

permease a protein that acts as a carrier molecule aiding the transport of substances across membranes by active or passive means. It differs from an enzyme in that it can alter the equilibrium point of a reaction.

Permian the final period of the Upper Palaeozoic era about 280–225 million years ago. The climate supported luxurious vegetation but became increasingly variable with widespread periods of glaciation.

During this period, plants similar to *Ginkgo* and the cycads first appeared. See also GEOLOGICAL TIME SCALE.

Peronosporales an order of the OOMYCOTA containing both saprobic and parasitic species, many of which are serious plant pathogens. Its members, which number about 148 species in some 18 genera, are parasites found in damp terrestrial habitats and include: *Pythium* (many species of which cause DAMPING-OFF of seedlings); *Phytophthora infestans* (late blight of potato); *Plasmopara*, *Peronospora*, and *Bremia* (various DOWNY MILDEWS); and *Albugo* (white blister rusts). The Peronosporales are characterized by having a mycelium with haustoria, asexual reproduction by zoospores (or occasionally conidia), and sexual reproduction by zoospores produced from oospores.

peroxisome a MICROBODY that contains amino acid OXIDASES, urate oxidases, and CATALASE. The latter catalyses the decomposition of hydrogen peroxide, produced by the activity of the other enzymes, to oxygen and water. See PHOTORESPIRATION.

persistent describing pesticides that take a long time to break down and become inactive. Such pesticides give longer protection than nonpersistent pesticides but are not suitable for use soon before harvest. Some are harmful to the environment as they accumulate in food chains and poison wildlife. Examples are the weedkiller MCPA and the insecticide DDT.

pest any organism that damages crops or reduces their yield or that irritates or injures livestock. The term is more often used of animals that cause physical damage, for example locusts or rodents. Organisms that cause disease are more commonly termed pathogens or parasites.

pesticide any chemical used to kill pests, such as an INSECTICIDE, FUNGICIDE, or HERBICIDE.

petal an individual unit of the COROLLA. Petals are thought to be modified leaves with a simplified internal structure, having only one vascular bundle compared with the several normally found in leaves and sepals. Insect-pollinated plants tend to have large, often yellow or white, scented petals often with a nectary at the base and honey guides patterning the surface. The petals of wind-pollinated plants, when present, tend to be small and dull coloured. Compare SEPAL.

petalody see DOUBLE FLOWER.

petaloid resembling a petal.

petalostemonous see ANDROECIUM.

petiolate describing a leaf that has a PETIOLE (stalk).

petiole the stalk that attaches the leaf lamina to the stem. The point of attachment is often strengthened by a widening of the base of the petiole. Some leaves (sessile leaves) lack a petiole and are joined to the stem at the base of the lamina. Sessile leaves are characteristic of most monocotyledons. The petiole is generally similar in structure to a stem except that the vascular and strengthening tissues are asymmetrically arranged so as to bear the weight of the lamina. The different patterns of veins evident in petioles are useful taxonomically. Various modifications of the petiole are seen. Some are flattened and bladelike (see PHYLLODE). Others are inflated, as in *Eichhornia crassipes* (water hyacinth), where they aid buoyancy. In many species the base of the petiole sheaths the stem, as in *Heracleum mantegazzianum* (giant hogweed). The base of the petiole may be modified as a PULVINUS. The petioles of some climbing species, e.g. *Clematis*, are haptotropic. The base of a fern RACHIS is sometimes termed a petiole.

petiolule the stalk of a LEAFLET in a compound leaf.

petrified wood fossil wood formed by the infiltration of minerals, usually silica or calcite, into cavities within and between cells in the original wood. In some petrified wood the substituted minerals preserve details of cell structures as well as the external shape.

petrified forest 1 a fossilized forest. See PETRIFIED WOOD. **2** along a shoreline at low tide, an area of peat containing the remains of tree stumps from a former forest cover, indicating the sea level has risen or the land has sunk. Such forests are common in the estuaries of southwest England, following changes in land and sea level since the last glaciation.

petrophilous growing on rock. See also LITHOPHYTE.

Pezizales (cup fungi) an order of the ASCOMYCOTA, in older classification schemes placed in the class DISCOMYCETES, in which the asci are operculate (see OPERCULUM) and are contained within a hymenium. The hymenium lines the surface of the cuplike apothecium and may be brightly coloured, as in *Sarcoscypha coccinea*, the scarlet elf cup. The group also contains the edible morels (*Morchella*) and certain truffles. The order includes about 1029 species in some 177 genera, mostly saprobes found in soil, rotting wood, and dung; a few form mycorrhizas.

P$_{fr}$ the active form of the plant pigment PHYTOCHROME that has a peak of light absorption in the far red of the spectrum, i.e. at about 725–730 nm. It is interchangeable with the P$_R$ form.

PGA (phosphoglyceric acid) see GLYCERATE 3-PHOSPHATE.

pH a measure of the hydrogen ion concentration, [H$^+$], and hence of the acidity or alkalinity of an aqueous solution. It is the logarithm of the reciprocal of [H$^+$], i.e. $\log_{10}(1/[H^+])$. The concentration of H$^+$ in a litre of pure water (i.e. a neutral solution) is 10^{-7} mol dm^{-3}, giving a pH value of 7. One integer on the pH scale indicates a tenfold difference in H$^+$ concentration. The lower the pH value, the higher the concentration of H$^+$ and vice versa.

Phacidiales see RHYTISMATALES.

Phaeophyta (brown algae) a phylum of the PROTOCTISTA (or, in some classification schemes, a division of the Plantae, or a class, Phaeophyceae, of the CHROMOPHYTA) consisting predominantly of marine ALGAE and including many of the large seaweeds, such as the wracks (FUCALES) and kelps (LAMINARIALES). The organization of the thallus is more advanced than in other algae and growth is truly parenchymatous in many species; the thallus is often differentiated into outer photosynthetic layers (cortex) and inner nonpigmented layers (medulla), which have a storage function. The simpler members of the phylum are branching filaments. Growth

may be diffuse, or it may be restricted to a meristem region, which may be at the ends of branches (terminal or apical) or within branches (intercalary; see TRICHOTHALLIC GROWTH). Some species have a unique trichothallic meristem at the base of a terminal hair or trichome. The thallus may be floating or anchored by a holdfast. Like the CHRYSOMONADA, the Phaeophyta have chlorophylls *a* and *c*, but also possess CAROTENE and XANTHOPHYLL pigments, including FUCOXANTHIN. The main storage products are LAMINARIN, MANNITOL, and fucosan. The cell walls contain cellulose and the mucopolysaccharide ALGINIC ACID, which has many commercial uses. Alternation of generations is ISOMORPHIC in some species and HETEROMORPHIC in others, with the gametophyte usually filamentous and the sporophyte a complex parenchymatous or pseudoparenchymatous structure. The kelps are the most advanced, with no free-living haploid phase. Sexual reproduction is usually oogamous. Isomorphic forms (sometimes assigned to the class Isogeneratae) include *Ectocarpus* (order ECTOCARPALES) and *Dictyota* (order DICTYOTALES); heteromorphic forms (Heterogeneratae) include the kelps; and forms with no alternation of generations (Cyclosporae) include the wracks.

phaeoplast a brown plastid containing the photosynthetic pigments of a brown alga (see PHAEOPHYTA). The brown colour is mainly due to the carotenoid FUCOXANTHIN.

phage see BACTERIOPHAGE.

phagocytosis see ENDOCYTOSIS.

phagotroph see MACROCONSUMER.

phagotrophy a type of heterotrophic nutrition that involves the ingestion of particulate food. Compare OSMOTROPHY. See also CONSUMER.

Phallales an order of the HOLOBASIDIOMYCETES containing the stinkhorns. There are about 137 species in some 32 genera. The basidioma is bounded by a PERIDIUM covering a gelatinous layer. When mature, the basidioma wall of the stinkhorn is broken open by an expanding stalk that rapidly grows, carrying the foul-smelling spores at its tip. The stench attracts flies and other insects, which help to disperse the spores. A common species is *Phallus impudicus*.

phanerogam in early classifications, a plant whose reproductive organs are easily visible either as flowers or cones. Compare CRYPTOGAM.

phanerophyte a plant with perennating buds (see PERENNATING ORGAN) situated on upright stems well above soil level. The buds are therefore potentially exposed to wind, extremes of temperature, and drought. Phanerophytes are thus usually found in temperate moist regions. They include trees, shrubs, herbaceous and succulent plants, and vines. See also RAUNKIAER SYSTEM OF CLASSIFICATION.

Phanerozoic an EON that starts at the beginning of the CAMBRIAN period, 570 million years ago and continues to the present day.

pharming (biopharming) the genetic engineering of plants to produce pharmaceuticals, and their subsequent cultivation.

phase contrast microscope a microscope that enables living, and hence normally transparent, material to be viewed by converting (invisible) differences of phase in the transmitted light to (visible) differences of contrast. Light passing through a specimen changes its phase by different amounts due to differences in thickness or refractive index. The phase contrast microscope converts these differences to differences in amplitude, which can be seen as shades of grey (because the greater the amplitude of a wave the more bright or intense is the light). The effect is produced by using a special condenser together with a glass plate (the retardation plate) in the objective. The condenser produces a certain pattern of light, often a hollow cone. This is matched in the retardation plate by a similar pattern produced by a groove in the plate. Any light that is not diffracted by the specimen will pass through the groove in the retardation plate and remain unchanged in phase. Any light that has been diffracted by the specimen will pass through the thicker part of the retardation plate and hence be slowed even further. Contrast results from combination of the direct and refracted

light. If they are out of phase by half a wavelength the difference produces destructive interference of the light so producing contrast. When this technique is used in conjunction with a time-lapse ciné camera it enables such events as cell division to be filmed and later speeded up, a technique known as *cinemicrography*. See also INTERFERENCE MICROSCOPE.

phase diagram a graphical method for studying the stability of a biological system. It is used especially in the study of host–parasite interactions. The numbers of host and parasite are plotted over time. The form of the resulting curve indicates the stability of the system. By substituting different values for host and parasite populations, it is possible to get an indication of the limits beyond which the system becomes unstable.

phellem (cork) the compact protective tissue that replaces the epidermis as the outer cellular layer in plants with SECONDARY GROWTH. It is the outer layer of the PERIDERM and consists of numerous more or less isodiametric cells arranged in radial rows, with thick waxy suberized cell walls. Commercial cork is obtained from the cork oak *Quercus suber*, in which the same PHELLOGEN is active each year (see BARK). It consists of thin-walled unsuberized cork cells (*phelloids*) and is stripped from the tree approximately every nine years.

phelloderm (secondary cortex) PARENCHYMA tissue of secondary origin, derived centripetally from the PHELLOGEN. Cells of the phelloderm may be distinguished from those of the cortex by their arrangement in radial columns, reflecting their origin from the phellogen. See also PERIDERM.

phellogen (cork cambium) the CAMBIUM that arises external to the VASCULAR CAMBIUM in many plant axes undergoing SECONDARY GROWTH. It gives rise to PHELLEM and PHELLODERM. See also PERIDERM.

phelloid see PHELLEM.

phenetic (phenetic classification) describing relationships between organisms or groups of organisms that are assumed on the basis of overall similarities and differences. A phenetic classification takes into account as many characteristics as possible and, in contrast to a PHYLOGENETIC classification, does not attach more weight (see WEIGHTING) to characters that are assumed to be the result of evolutionary relationships. (Other characters may be weighted, especially those thought to be 'good' characters.) A phenetic classification does not aim to reflect evolutionary histories but may nevertheless well do so. It is most likely to diverge from a phylogenetic classification in instances of CONVERGENT and PARALLEL EVOLUTION.

Classifications published before or soon after the appearance of Charles Darwin's *Origin of Species* (1859) were obviously phenetic in nature since taxonomists were not thinking then in evolutionary terms. Most classifications drawn up after about 1880 tend to be phylogenetic. One phenetic system used as a basis of herbarium arrangement is that of George Bentham and Joseph Hooker, published as *Genera Plantarum* (1862–83). A main difference between this and more modern systems is the placement of the gymnosperms between the dicotyledons and monocotyledons. See NUMERICAL TAXONOMY.

phenetic classification see PHENETIC.

phenocopy a phenotype of one genotype that resembles the phenotype of a different genotype. The change in phenotype is normally brought about by unusual environmental conditions, such as abnormally high temperatures early in development.

phenogram see DENDROGRAM.

phenols see PHENOLICS.

phenolics (phenols) compounds containing a benzene ring directly substituted with one or more hydroxyl groups. The simplest member is phenol (carbolic acid), formula C_6H_5OH. Plant phenolics are widespread and extremely diverse and include the FLAVONOIDS. Simple phenolic compounds containing one or two rings are not as common as the flavonoids, though hydroquinone, COUMARIN, and the cinnamic acids are widely distributed. The amino acid TYROSINE also contains a phenolic ring. Less common are such phenolics as catechol, phloroglucinol, and pyrogallol. LIGNIN is a phenolic, probably

being formed from coniferyl alcohol, a derivative of cinnamic acid. TANNINS are also derived from phenolics.

phenology the study of the effects of climate on living organisms. It includes seasonal events such as flowering, migration, and growing seasons, and longer-term effects on, for example, the rate of growth of a species and the time taken to reach maturity.

phenolphthalein a hydrogen-ion indicator that is colourless in acids and red in alkalis.

phenon see DENDROGRAM.

phenon line see DENDROGRAM.

phenotype the expressed characteristics of an organism. The phenotype is determined by an interaction between the environment and the GENOTYPE and between dominance (see DOMINANT) and epistatic relationships (see EPISTASIS) within the genotype.

phenoxyacetic acids a group of compounds some of which show AUXIN-like activity. Various chlorine substituted forms cause a disruption in metabolism particularly at the meristems and have been widely used as selective weedkillers. An important factor in the use of phenoxyacetic acids as weedkillers is their immunity from the plant's endogenous IAA oxidizing system, which can normally inactivate superfluous auxins. Examples of such compounds are ZYMASES,4-D, ZYMASES,4,5-T, and MCPA.

phenylalanine an aromatic nonpolar amino acid with the formula $C_6H_6CH_2CH(NH_2)COOH$ (*see illustration at* amino acid). The biosynthetic pathway (see SHIKIMIC ACID) to phenylalanine from erythrose 4-phosphate and phosphoenolpyruvate is similar for all the aromatic amino acids. Phenylalanine and the other aromatic amino acids are precursors of a large number of aromatic compounds, mostly found only in plants. Deamination of phenylalanine yields cinnamic acid, important in the synthesis of lignin. Phenylalanine is also a precursor of such alkaloids as morphine and curare (the ISOQUINOLINE ALKALOIDS).

phialide see PHIALOSPORE.

phialospore a type of CONIDIUM found, for example, in many of the Eurotiales and Hypocreales. Phialospores develop in basipetal succession at the open-ended tips of specialized finger-like cells termed *phialides*.

phloem (bast) a vascular tissue whose principal function is the translocation of sugars and other nutrients. The phloem is composed mainly of SIEVE TUBES, SCLERENCHYMA cells, and PARENCHYMA cells, including COMPANION CELLS. It occurs in association with, and usually external to, the XYLEM. See also PRIMARY PHLOEM, SECONDARY PHLOEM.

phloroglucin a temporary stain that is used to detect the distribution of LIGNIN (which it turns magenta red) in thin sections of plant tissue. It is usually acidified with hydrochloric acid before use.

phosphatase an enzyme that catalyses the hydrolysis of ESTERS of phosphoric acid.

phosphatide see PHOSPHOGLYCERIDE.

phosphatidyl choline see LECITHIN.

phosphodiester bond a covalent bond that links a sugar and a phosphate group by means of an oxygen bridge. Such bonds hold together the sugar-phosphate backbone of nucleic acids. They are formed by CONDENSATION reactions and can be broken by HYDROLYSIS. *See illustration at* DNA.

phosphoenolpyruvate (PEP) an important phosphorylated intermediate of glucose metabolism. In GLYCOLYSIS PEP is the immediate precursor of pyruvic acid, while in the synthesis of glucose from acetyl CoA, PEP is formed from oxaloacetate generated in the GLYOXYLATE CYCLE. In C_4 plants PEP is an intermediate in the HATCH–SLACK PATHWAY. The aromatic amino acids, and hence a large number of other aromatic nitrogenous compounds, are also synthesized from PEP.

phosphoenolpyruvate carboxylase (PEP carboxylase) the enzyme that is responsible for carbon dioxide fixation in C_4 PLANTS and in plants exhibiting CRASSULACEAN ACID METABOLISM. It catalyses the carboxylation of phosphoenolpyruvate to oxaloacetic acid.

phosphofructokinase a TRANSFERASE enzyme that catalyses the transfer of a phosphate from ATP to fructose 1,6-phosphate, forming fructose 1,6-bisphosphate and ADP. Its activity is sensitive to the ATP/ADP ratio in the cell.

Phosphofructokinase is a key enzyme in GLYCOLYSIS and other reactions of carbohydrate metabolism.

phosphogluconate pathway see PENTOSE PHOSPHATE PATHWAY.

phosphoglyceric acid (PGA) see GLYCERATE 3-PHOSPHATE.

phosphoglyceride (phospholipid, phosphatide, glycerophosphatide) any of a group of complex lipids similar to acylglycerides (see ACYLGLYCEROL) except that the third hydroxyl group is phosphorylated, i.e. the backbone is glycerol 3-phosphate rather than glycerol. Common phosphoglycerides include phosphatidylcholine (see LECITHIN), phosphatidylethanolamine, and phosphatidylinositol (all found primarily in seed tissues), and phosphatidylglycerol (found in leaf tissue). Phosphoglycerides are the major lipids in most plant membranes, although GLYCOLIPIDS are important components of chloroplast membranes.

phospholipid see PHOSPHOGLYCERIDE.

phosphorescence the emission of light without significant accompanying emission of heat. In biological systems, this is often the re-emission of light that has previously been absorbed, although perhaps at a different wavelength. So-called phosphorescent seas are sometimes due to the luminescence of certain dinoflagellates (see DINOMASTIGOTA), especially *Noctiluca* species. They emit flashes of blue light in response to mechanical disturbance of the water. The light is produced in special vesicles (*scintillons*) by a enzyme reaction.

phosphorus symbol: P. A nonmetallic element, atomic number 15, atomic weight 30.9, important as a MACRONUTRIENT for plant growth. It is essential for the formation of nucleic acids, lipoproteins, and energy-carrying molecules and is involved in many aspects of metabolism, especially carbohydrate metabolism. Phosphorus occurs in soil minerals as the apatites and as calcium phosphate rock. It is absorbed by roots as the PO_4^{3-} ion, this ion acting as an important buffer in the cell sap. Deficiency results in poor root growth, bluish, bronzed, or purple leaves, and poor germination, ripening, and seed set. Phosphorus can be applied to soils as inorganic basic slag or superphosphate or in various organic materials, guano and bonemeal containing a particularly high percentage. It is often applied in combination with nitrogen and potash in compound fertilizers.

phosphorus cycle the BIOGEOCHEMICAL CYCLE by which phosphorus circulates between living organisms and the environment. Phosphorus is released as orthophosphates (PO_4^{3-}) from rocks and sedimentary deposits by weathering, leaching, and mining, and as organic phosphates during the decomposition of dead organisms. It also occurs in natural bodies of water in the form of soluble phosphates derived from runoff and from water that has percolated through the soil into rivers. Plant roots and soil organisms take up phosphates from the soil, and plankton and bacteria assimilate dissolved phosphates from water. The uptake of phosphorus from the soil is impeded by high concentrations of aluminium and iron, which may cause phosphates to precipitate out, rendering phosphorus unavailable to living organisms. In lakes the cycling of phosphorus is affected by stratification and seasonal turnover, while in the oceans it is affected by the presence or absence of upwelling currents, and by outwelling. See also EUTROPHICATION.

phosphorylase see KINASE.

phosphorylation the addition of a phosphate group to an organic molecule, which is under the control of a phosphorylase enzyme (see KINASE). Phosphorylation is one of the first steps in many metabolic pathways (see GLYCOLYSIS, ATP), in which a high-energy phosphate bond is formed that will later release energy at a strategic point in the pathway. Phosphorylation is also needed for the activation of some enzymes and the deactivation of others. See also OXIDATIVE PHOSPHORYLATION, PHOTOPHOSPHORYLATION.

phosphotransferase see KINASE.

photic zone (euphotic zone) the surface waters of seas, oceans, and lakes penetrated by light and inhabited by PLANKTON. In clear water light at the red end of the spectrum is absorbed and only blue light penetrates to lower levels. However in turbid water blue light is preferentially absorbed. The depth at which the rate of production of carbohydrates by photosynthesis is equalled by the rate of utilization by respiration is termed the *compensation depth*. Most of the plankton is found in the top 10 m.

photoautotroph an AUTOTROPH that obtains the energy for synthesizing organic food substances from inorganic components directly from sunlight. Photoautotrophs include most green plants and algae and some photosynthetic bacteria. Compare PHOTOHETEROTROPH.

photochemical reaction a reaction in which a chemical change is induced by light energy (see PHOTOSYNTHESIS). Such reactions are more or less independent of temperature reactions.

photoheterotroph a photosynthetic organism that requires a supply of organic compounds as a source of hydrogen. The purple nonsulphur bacteria are obligate photoheterotrophs and also require certain organic growth substances. Some of the purple sulphur bacteria are facultative photoheterotrophs in that under certain conditions they use organic acids rather than reduced sulphur compounds as a hydrogen source. See also HETEROTROPH.

photoinhibition 1 the stopping or slowing of a plant process by light, for example, the germination of some seeds is inhibited by light. **2** the reversible inhibition of photosynthesis in the leaves of certain species by light of a particular quality.

photokinesis a change in the speed of locomotion of a motile organism or cell in response to a change in light intensity.

photolithotroph a photosynthetic organism that uses an inorganic compound or element as an electron donor during photosynthesis. For example, green plants oxidize water during photosynthesis, while some bacteria oxidize hydrogen sulphide.

photolysis of water the photosynthetic splitting of water into gaseous oxygen and reducing equivalents. Two molecules of water are split to produce one molecule of oxygen, four electrons, which go through the electron transport chain, and four protons. The electrons and protons eventually reduce NADP to NADPH. This in turn is utilized in the reduction of carbon dioxide to carbohydrate.

photomicrograph see MICROGRAPH.

photomorphogenesis the effect of light on the regulation of plant morphology and growth. See also CRYPTOCHROME, ETIOLATION.

photonasty a NASTIC MOVEMENT induced by external light stimuli. Some flowers, e.g. those of the garden marigold (*Calendula officinalis*), exhibit these movements and open on exposure to light.

photo-organotroph a photosynthetic organism that oxidizes organic compounds during PHOTOSYNTHESIS. See also HETEROTROPH.

photooxidation the oxidation of a substance as a result of reactions triggered by the absorption of light energy. Many photosynthetic pigments, especially the CHLOROPHYLLS, can be oxidized ('bleached') and rendered nonfunctional by too high a light intensity. The CAROTENOID pigments, in particular, are thought to screen chlorophylls from light, preventing photooxidation.

photoperiod see PHOTOPERIODISM.

photoperiodism the alternation of light and dark periods that affects the physiological activities of many plants. As the day length changes through the year, various responses are shown by plants, depending on whether they are LONG-DAY or SHORT-DAY PLANTS. An example is the onset of flowering, triggered in a particular plant by a specific photoperiod. The length of the light period (*photoperiod*) is more important than the intensity of the light. It has also been shown that the effect of a long dark period can be nullified by a brief period of light. See also PHYTOCHROME.

photophile describing a phase during which light promotes flowering. In one theory advanced to explain PHOTOPERIODISM it is suggested that short-day plants are in the photophile phase during the day but at night they are in the *photophobe* (or *skotophile*) phase when light inhibits and darkness promotes flowering. The alternation between photophile and photophobe phases is seen as a type of

CIRCADIAN RHYTHM. If the period of daylight extends beyond the photophile phase into the photophobe phase, as would occur with a short-day plant on a long day, then flowering is inhibited.

photophobe see PHOTOPHILE.

photophosphorylation (photosynthetic phosphorylation) the production of ATP from ADP and inorganic phosphate utilizing light-induced electron transport during PHOTOSYNTHESIS as a source of energy. Two types of photophosphorylation occur, noncyclic and cyclic. In *noncyclic photophosphorylation* electrons are passed from water to $NADP^+$ via PHOTOSYSTEMS I AND II. During the transfer of electrons from the primary electron acceptor of photosystem II (termed C550) to P700 through plastoquinone, plastocyanin, and various CYTOCHROMES, one molecule of ATP is formed. In *cyclic photophosphorylation* only photosystem I is involved. Energy absorbed by P700 releases electrons, which are absorbed by a primary acceptor termed P430. Electrons then pass, via cytochrome b_6, into the chain connecting the primary electron acceptor of photosystem II to photosystem I. A molecule of ATP is formed as the electron passes down to P700. Cyclic phosphorylation does not produce reduced $NADP^+$. *See illustration at* photosynthesis.

photoreceptor any light-sensitive region of a plant. The *paraflagellar body*, visible as a swelling on the undulipodium of some euglenids, is an example.

photorespiration respiration that occurs in plants in the light. It differs from dark respiration in that the oxidation of carbohydrates occurs in the PEROXISOMES (rather than the mitochondria) and is not coupled to oxidative phosphorylation (see diagram). The rate of CO_2 release by photorespiration in C_3 PLANTS can be three to five times greater than that released by dark respiration. Since the process does not generate ATP it appears to be extremely wasteful. It has been estimated that photosynthetic efficiency could be improved by 50% if photorespiration were inhibited. In C_4 plants photorespiration is hardly detectable, possibly because synthesis of GLYCOLIC ACID, the substrate for photorespiration, is much lower in C_4 plants (about 10% of that of C_3 plants). This could be because the concentration of CO_2 in the BUNDLE SHEATH cells is so high that oxidation (instead of carboxylation) of ribulose bisphosphate is prevented.

Some C_3 plants, e.g. the grass *Panicum miliodes*, have a leaf structure comparable in some respects to the KRANZ STRUCTURE of C_4 plants. Such plants also tend to have reduced rates of photorespiration. These observations have stimulated research into the possibility of breeding C_3 plants with slower photorespiration rates.

photosynthate the chemical products of photosynthesis.

photosynthesis the sequence of reactions, performed by green plants, algae, and photosynthetic bacteria, in which light energy from the sun is converted into chemical energy and used to produce carbohydrates and ultimately all the materials of the plant. The photosynthetic reaction can be summarized as:

$$6CO_2 + 6H_2O + \text{light energy} \rightarrow$$
$$C_6H_{12}O_6 + 6O_2$$

There are two distinct phases in photosynthesis, the LIGHT (or light-

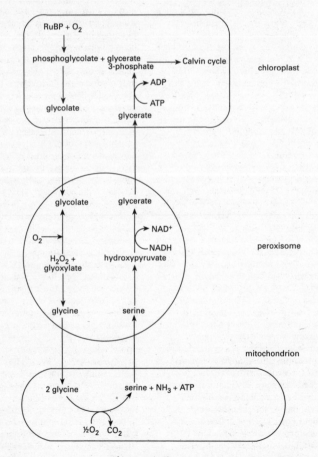

Photorespiration

dependent) reactions and the DARK (or light-independent) reactions. In green plants, algae, and cyanobacteria the light reactions involve the photolysis of water, producing hydrogen atoms and molecular oxygen. This oxygen, given off during photosynthesis, is the main source of atmospheric oxygen, essential for aerobic organisms. The hydrogen atoms produced are used to reduce $NADP^+$ to NADPH and the energy released also forms ATP from ADP and inorganic phosphate (photophosphorylation). This ATP and NADPH is used up during the dark (light-independent) reactions in which carbon dioxide is fixed (see FIXATION) into carbohydrates (see CALVIN CYCLE). See PHOTOSYSTEMS I AND II.

photosynthetic bacteria (phototrophic bacteria) bacteria that contain chlorophyll pigments, which enable them to photosynthesize. They fix carbon using the CALVIN CYCLE of reactions and (in different groups) derive their hydrogen from water or from hydrogen sulphide or other reduced sulphur compounds. Several groups of bacteria are photosynthetic. The green SULPHUR BACTERIA (Chlorobia), green nonsulphur bacteria (Chloroflexa), purple sulphur bacteria (e.g. *Thiocapsa*, *Thiospirillum*), and purple nonsulphur bacteria (e.g. *Rhodobacter*, *Rhodospirillum*) can photosynthesize only in anaerobic conditions. They contain BACTERIOCHLOROPHYLL; the purple bacteria (phylum Proteobacteria) also have additional red and brown pigments. The sulphur bacteria use hydrogen sulphide or hydrogen gas as hydrogen donor, while the nonsulphur bacteria use organic sources, such as lactate, pyruvate, or ethanol. The blue-green bacteria use chlorophyll *a* and water, releasing oxygen as a by-product of photosynthesis; the chloroxybacteria use both chlorophyll *a* and chlorophyll *b* and water, and also release oxygen during photosynthesis. Both these groups are placed in the phylum CYANOBACTERIA.

photosynthetic phosphorylation see PHOTOPHOSPHORYLATION.

photosynthetic quotient the volume of oxygen released during photosynthesis expressed as a proportion of the volume of carbon dioxide used in photosynthesis.

photosynthetic unit see CHLOROPLAST.

photosystems I and II (pigment systems I and II, PSI and PSII) two photochemical systems containing photosynthetic and accessory pigments and electron carriers that operate in sequence to perform the two light reactions of photosynthesis and so bring about PHOTOPHOSPHORYLATION and reduction of $NADP^+$ (see diagram). PSII contains chlorophyll *a* (P680) that absorbs light of 684 nm wavelength most efficiently. When activated by light it produces a strong oxidant that oxidizes water to oxygen, hydrogen ions, and electrons. The electrons are transferred via a primary electron acceptor (possibly C_{550}) and PLASTOQUINONE and the CYTOCHROME chain to PSI. The energy released when each electron is transferred is used to transform a molecule of ADP to ATP. PSI contains chlorophyll *a* (P700) that absorbs light of 700 nm wavelength most efficiently. When P700 is activated a strong reductant is produced that reduces $NADP^+$ to NADPH.

phototaxis a TAXIS in response to external light stimulation. Cells of *Chlamydomonas* swim freely towards a light source to enhance photosynthetic efficiency but will swim away if the source is too intense (*positive* and *negative phototaxis* respectively). Some chloroplasts also show phototaxis within cells, aligning themselves with respect to the incident light.

phototroph an organism that derives the necessary energy for the synthesis of organic compounds directly from sunlight. The term encompasses both PHOTOAUTOTROPHS and PHOTOHETEROTROPHS. Compare CHEMOTROPH.

phototropism (*adj.* phototropic) a TROPISM in response to light. Thus, shoots show *positive phototropism* by growing towards a light source. Work with oat (*Avena*) coleoptiles has shown that the stimulus is received in the tip of the seedling and that the response is effected by increased growth (elongation) on the relatively shaded side, triggered by AUXINS, which pass from the tip more readily on the shaded side.

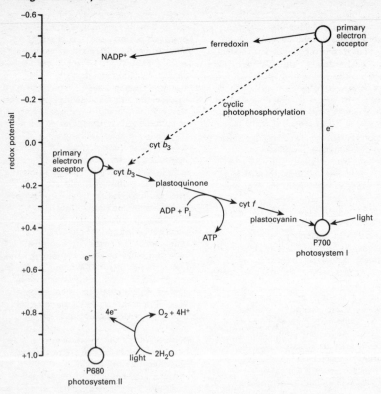

Photosystems I and II: the Z scheme depicting electron transport

Removal of the tip destroys the phototropic response. Most roots are *aphototropic*, showing no response to light.

Phragmobasidiomycetes a class of the BASIDIOMYCOTA or, in some classification schemes, a subclass (Phragmobasidiomycetidae) of the Holobasidiomycetes, in which the hyphae are septate, the large basidioma is often gelatinous, the basidia are usually in a hymenium and regularly divided by one or more septa, and the basidiospores are discharged violently. It includes the orders Agaricostilbales, Atractiellales, Auriculariales, Heterogastridiales, and TREMELLALES.

phragmoplast a system of fibrils that persists at the outer edge of the equator of the spindle at early telophase. As the CELL

PLATE forms in the equatorial region the fibrils become dispersed. The term is used by some authorities to mean the developing cell plate.

phreatic zone (groundwater zone, vadose zone) the zone below the water table in which the rocks and/or soil are saturated with water.

phyco- prefix denoting relating to algae.

phycobilin SEE PHYCOBILIPROTEINS.

phycobiliprotein (biliprotein, phycobilin) any of the blue or red pigments found in the blue-green bacteria (see CYANOBACTERIA) and the red algae (see RHODOPHYTA). Each consists of a protein bound to a porphyrin prosthetic group. Like the CAROTENOIDS they are accessory pigments in photosynthesis, but unlike the chlorophylls and carotenoids they are water soluble. Structurally they are very similar to the porphyrin part of the chlorophyll molecule, except that they contain no magnesium. The three classes of phycobiliproteins are the PHYCOERYTHRINS, PHYCOCYANINS, and allophycocyanins. See also PHYCOBILISOME.

phycobilisome a small particle, about 30 nm across, many of which are found on the membranes of the lamellar system of blue-green bacteria (Cyanobacteria). Phycobilisomes are thought to be aggregations of PHYCOBILIPROTEIN. They are also found in the plastids of red algae (Rhodophyta) between the photosynthetic lamellae.

phycobiont the photosynthetic partner in a LICHEN, which is either a unicellular green alga (often a member of the genus *Trebouxia),* or a blue-green bacterium *(see* CYANOBACTERIA). If a bacterium (e.g. *Nostoc, Rivularia, Gloeocapsa*) is present the lichen is usually gelatinous. The number of phycobiont genera involved in lichens (24 have so far been discovered) is relatively few compared to the large numbers of lichenized fungi. Compare MYCOBIONT.

phycocolloid a mucilage derived from the walls of certain brown and red algae (Phaeophyta and Rhodophyta), which is used in industry in gels and thickeners.

phycocyanin any of the blue pigments that form one of the two major classes of the PHYCOBILIPROTEINS. They are found in all blue-green bacteria (see CYANOBACTERIA) and in some red algae (see RHODOPHYTA) and act as ACCESSORY PIGMENTS in photosynthesis.

phycoerythrin any of the red pigments that form one of the two major classes of the PHYCOBILIPROTEINS. They are found in all red algae (Rhodophyta), and act as ACCESSORY PIGMENTS in photosynthesis. They absorb the dim blue-green light that reaches the lower levels of the ocean where the algae grow, and pass it on to chlorophyll so that it may be used in photosynthesis. The deeper in the sea an alga lives, the more phycoerythrin it contains in relation to chlorophyll.

phycology the study of algae.

Phycomycetes an obsolete class containing all the relatively unspecialized true fungi

(see FUNGI, EUMYCOTA), i.e. all true fungi except the ascomycetes, basidiomycetes, and FUNGI ANAMORPHICI. The members of this class are currently placed in several groups, including the CHYTRIDIOMYCOTA, OOMYCOTA, and ZYGOMYCOTA.

phycoplast an aggregation of spindle microtubules that are aligned parallel to the plane of cell division. The phycoplast may be involved in directing the laying down of a cell plate or ingrowth of a furrow.

phyletic see PHYLOGENETIC.

phyletic gradualism (gradualism) the concept that evolution is a gradual process of change over time. It considers that macroevolution is simply microevolution over a longer time scale. Compare PUNCTUATED EQUILIBRIUM.

phyllid a moss or liverwort 'leaf'.

phylloclade see CLADODE.

phyllocladium on a lichen, a squamose or granular outgrowth containing algae, e.g. *Stereocaulon.*

phyllode a more or less flattened PETIOLE resembling and performing the functions of a leaf. Vertically expanded petioles are seen in a number of *Acacia* species. The leaves of many monocotyledons are believed to be of phyllode origin. Compare CLADODE.

phyllody the transformation of parts of a flower into leaflike structures. This may occur in response to some infections. For example, maize infected with head smut (*Sphacelotheca reiliana*) develops phyllody of both male and female flowers.

phyllosphere the microenvironment on the surface of a living leaf.

phyllotaxis (phyllotaxy) the arrangement of leaves on a stem. In most plants the leaves are inserted singly and spirally on the stem. Sometimes two or more leaves are formed at each node giving an opposite or whorled arrangement (see illustration). The phyllotaxis of a given plant may be described by giving the angle of divergence between successive nodes. This varies between species but if many PRIMORDIA are being formed it is normally not less than 137.5°, the so-called *Fibonacci angle.* Phyllotaxy may also be described by the GENETIC SPIRAL or by PARASTICHIES. Many theories have been advanced to account for

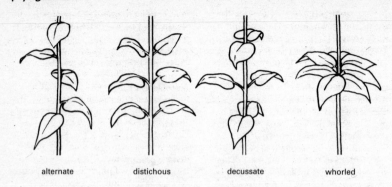

alternate   distichous   decussate   whorled

Phyllotaxis: types of phyllotaxis

the regular nature of phyllotaxis (see AVAILABLE SPACE THEORY, REPULSION THEORY).

phylogenetic (phyletic) describing the study of the stages in the evolutionary history of groups of organisms. It is based on a study of the fossil record, comparative anatomy, etc., and provides a basis for classification. Compare PHENETIC. See also CLADISTICS.

phylogeny the evolutionary history of an organism. A PHYLOGENETIC classification aims to reflect this.

phylum in the taxonomic hierarchy, the category of highest magnitude within a KINGDOM. In many modern classification systems phylum has replaced DIVISION. The names of plant phyla terminate in -phyta, e.g. Hepatophyta, Anthophyta.

physical factor a factor in the physical environment that affects plant and animal life.

physiognomy the form and structure of natural communities.

physiographic factor any of the factors, apart from CLIMATIC, BIOTIC, and EDAPHIC FACTORS, that affect the prevailing conditions within a habitat and the distribution of the plants and animals. Such factors include the topography of the area, altitude, drainage conditions, degree of erosion, slope of the land, etc.

physiological drought a drought condition suffered by plants that occurs despite there being sufficient water in the soil. It may occur when the concentration of solutes in the soil water is equal to or higher than that in the root cells so water cannot enter the plant by osmosis. This situation may be found in salt marshes and coastal mud flats. It also occurs during cold spells when there is an increase in the resistance of the root to water movement into the plant. Root resistance increases in response to low temperatures since the permeability of endodermal cells decreases rapidly below 5°C.

physiological race (forma specialis) a population that is physiologically distinct but morphologically indistinguishable from other members of the species. Many fungi have large numbers of physiological races, reflecting considerable genetic diversity, even in nonsexual species. Examples are the various races of certain pathogenic fungi, each of which attack different host species or varieties.

physiology (*adj.* physiological) the study of the processes and functions associated with life.

physode in brown algae (see PHAEOPHYTA), a vesicle containing polyphenols (see PHENOLICS).

phyt- prefix denoting relating to plants.

phytic acid (hexaphosphoinositol) the hexaphosphoric ester of INOSITOL. It is common as a phosphate storage compound in many seeds in which it exists as the calcium–magnesium salt, *phytin*.

phytoalexin a chemical produced by a host plant that inhibits the growth of a pathogenic fungus. For example, the phenolic compound pisatin has been shown to accumulate in pea tissues in the presence of various fungi. Similarly the degree of resistance shown by sweet potato (*Ipomoea batatas*) to attack by the fungus *Ceratostomella* has been correlated with the concentration of the terpenoid ipomeamarone in the plant tissue.

phytobenthos see BENTHOS.

phytochemistry the chemistry of plants, their metabolic processes, and products.

phytochrome a protein pigment that mediates in photoperiodic responses (see PHOTOPERIODISM) and certain other photoreactions, e.g. light-stimulated germination and the removal of the symptoms of etiolation. It exists in two interchangeable forms, P_r, which absorbs in the red part of the spectrum (660 nm), and P_{fr}, which absorbs in the far red (730 nm). Following exposure of a plant to red light P_r changes to P_{fr}, while after exposure to far-red irradiation P_r is reformed from P_{fr}. This reversion of P_{fr} to P_r may also occur in the dark in some plants, a process that is inhibited by low temperatures. P_{fr} may also be lost by decay.

In practice an illuminated plant receives a mixture of red and far-red light and the proportion of P_r to P_{fr}, and hence the response of the plant, will depend on the relative proportions of far-red and red light. Sunlight and fluorescent light have a high proportion of red light as compared to far-red and so promote formation of P_{fr} except in deep shade, where far-red will predominate, and most of the phytochrome will be in the P_r form.

It is thought that the change from P_r to P_{fr} and vice versa is brought about by a shift of two hydrogen atoms in the nonprotein part of the chromatophore, resulting in mutually antagonistic effects, the response of the plant often depending on the last type of radiation received. If the last exposure is to far-red these responses will not occur. The reversal effect of far-red light is only seen if the far-red is given soon after the red light. This is especially true in the case of rapid light-mediated responses, e.g. leaf movements in *Mimosa pudica*. Exposure of short-day plants to a short period of red light in the dark period inhibits flowering, but far-red irradiation reverses this. In long-day plants a short period of red irradiation in the dark period promotes flowering.

Phytochrome is found in very low concentrations but is extremely sensitive even to very short flashes of weak red light. Thus exposure of an etiolated plant to moonlight can induce normal growth. It is uncertain how these responses are brought about though it has been noted that gibberellin and cytokinin levels increase following exposure to red light as do the levels of certain enzymes. The transformation of phytochrome may bring about conformational changes in the molecule that expose an active site in the P_{fr} form that is able to bind to membranes and change their function. See also PHOTOPERIODISM.

phytoestrogen a compound synthesized by plants that has properties similar to those of oestrogen.

phytoflagellate an organism with undulipodia that contains chloroplasts.

phytogenetics plant genetics.

phytogeography see PLANT GEOGRAPHY.

phytohormone see HORMONE.

phytol an oxygenated diterpene (see TERPENOIDS) that comprises the major side-chain of chlorophyll. It can be extracted by hydrolysis from chlorophyll, and is used in the manufacture of synthetic vitamins E and K_1.

phytolith a calcareous or opaline stone secreted by certain plants or algae, often found fossilized.

Phytomastigophora in certain outdated classification systems, a class of undulipodiate photosynthetic microscopic organisms that included unicellular green algae (Chlorophyta), Chrysophyta, Cryptomonada, Diatoms, and Euglenida.

phytomer see METAMER.

phytopathology the study of plant diseases.

phytophagous describing an organism that feeds on plants.

phytoplankton see PLANKTON.

phytosociology (*adj.* phytosociological) the classification of plant communities according to their species composition and distribution. It also takes into account geographical distribution and the organization, development, and interdependence of communities. Modern phytosociology uses quantitative methods and computer analysis. See also ASSOCIATION, FIDELITY, KENNARTEN SPECIES, UPPSALA SCHOOL OF PHYTOSOCIOLOGY, ZURICH–MONTPELLIER SCHOOL OF PHYTOSOCIOLOGY.

phytosterol see STEROL.

phytotoxic describing a substance that is poisonous to plants.

phytotoxin a poison produced by a plant.

phytotron a large plant growth chamber or group of rooms in which many plants can be raised under strictly controlled conditions of light, photoperiod, temperature, humidity, etc. Often many phytotrons are housed in a specially designed building. Entrance to each unit is by a hermetically sealed door.

pico- prefix denoting a trillionth (10^{-12}) part of something, e.g. picosecond, or just an extremely small thing (e.g. picornavirus).

picoplankton plankton organisms that are less than 2 μm in diameter.

pigment any coloured compound. Pigments are coloured because they absorb certain types of light. Thus in solution they will have a characteristic ABSORPTION SPECTRUM. Because light is a form of energy, pigments that absorb light also absorb energy. The first reaction in PHOTOSYNTHESIS is the absorption of energy by accessory and photosynthetic pigments. Pigments also produce the characteristic colours of flowers and fruits.

pigment systems I and II see PHOTOSYSTEMS I AND II.

pileorhiza see CALYPTRA.

pileus (*adj.* pileate) the cap of a mushroom or toadstool. The upper surface is often covered in flaky scales, which are the remnants of the universal veil. The undersurface is composed of gills or pores.

pili see PILUS.

piliferous layer (root-hair zone) the absorbing region of the root epidermis, covered with ROOT hairs. It is situated about 4–10 mm from the root tip (beyond the zone of elongation). In young plants this region is responsible for most of the plant's water uptake. However in larger older plants and in conditions of water stress a greater proportion of water moves in across

the suberized (see SUBERIN) regions of the root further back from the root tip.

pilose having soft long hairs. Compare PUBESCENT.

pilus (fimbria) a projection from the surface of a bacterium. It is finer than a flagellum and also differs in having a hollow core. Pili are often found in large numbers and are thought to help attach bacteria to other cells rather than serve a locomotive function. Pilus formation may be initiated by a PLASMID, in which case it serves as a conjugation tube for the transmission of a copy of the plasmid to a suitable recipient cell. Plasmids with this ability are termed sex factors or F factors.

pin see HETEROSTYLY.

Pinales see CONIFERALES.

pine barren an area of pine FOREST dominated by almost pure stands of small or medium-sized trees, mainly pines with various oaks in drier areas, and white cedar in boggy places. The term is applied mainly to pine forests on poor, sandy, often marshy soil in the eastern USA, and represents a plagioclimax (see CLIMAX) due to centuries of burning.

pin flower see HETEROSTYLY.

pin-frame a device for getting a quantitative measure of vegetation cover. It consists of a frame that supports cross bars with a grid of pinholes and is mounted on adjustable legs. Pins of standard diameter are pushed gently through the pin-holes, and the plants touched by the tips are recorded. If parts of more than one plant lie below the pin, both top and bottom cover may be recorded. A quadrat sampled by the pin-frame method is called a *point quadrat*.

Pinicae see PINOPHYTA.

pin mould the colloquial name for moulds that bear sporangia on straight sporangiophores that resemble pins, e.g. *Mucor* and other members of the ZYGOMYCOTA.

pinna one of a number of first order leaflets in a compound leaf, such as is typical of many ferns. Compare PINNULE.

pinnate describing a compound leaf in which the leaflets (*pinnae*) are arranged in two rows, one on each side of the midrib. Such leaves are common in the Fabaceae. See also IMPARIPINNATE, PARIPINNATE.

pinnatifid (pinnatisect) deeply cut into lobes but not so far as the midrib. Examples are the pinnatifid leaves of the fen sowthistle (*Sonchus palustris*).

pinnatisect see PINNATIFID.

pinnule one of a number of second order leaflets in a compound leaf, such as is typical of many ferns, where the PINNAE are themselves divided into leaflets.

pinocytosis see ENDOCYTOSIS.

pinocytotic vesicle see ENDOCYTOSIS.

Pinophyta in certain classifications, a division containing the GYMNOSPERMS. It is divided into the following subdivisions: Cycadicae, containing the orders Cycadofilicales, Caytoniales, Bennettitales, and Cycadales; Pinicae, containing the orders Cordaitales, Coniferales, Taxales, and Ginkgoales; and Gneticae, containing the orders Ephedrales, Welwitschiales, and Gnetales. The Cycadicae are differentiated from the Pinicae on the basis that they have palmlike compound leaves, motile antherozoids, radially symmetrical seeds, and loose soft wood. The Pinicae in contrast have simple leaves, nonmotile sperm nuclei (except in *Ginkgo biloba*), bilaterally symmetrical seeds, and denser wood. The Gneticae resemble the Pinicae in having simple leaves, nonmotile sperm, and flattened seeds. However they differ from both the Pinicae and Cycadicae in possessing ovules with two integuments, rather than one, the inner of which is elongated into a tubular micropyle, and in having compound microstrobili.

pioneer community see SUCCESSION, ECESIS, SERE.

pioneer phase see HUMMOCK AND HOLLOW CYCLE.

pioneer species a species that colonizes a newly exposed environment, such as ground exposed by retreating glaciers and ice sheets, bare mud along coasts, and areas where forests have recently been felled. They constitute the first stage in a plant SUCCESSION. Most are fast-growing and relatively short-lived, and produce large numbers of small seeds, spores or other propagules that are readily dispersed over long distances, e.g. by wind. Many pioneer species modify the environment so that it becomes suitable for species in the next

stage of the succession, that will eventually replace the pioneer species. For example, lichens colonizing bare rock produce organic acids that begin to break down the rocks and produce the beginnings of mineral soil, which is further improved by decomposition of the remains of dead lichens. Mangroves and other pioneers of coastal muds trap more sediment, raising the level of the land so that it becomes drier and less subject to tidal inundation.

piperidine alkaloids a group of ALKALOIDS, the structures of which are based on the piperidine ring (a six-membered ring with no double bonds and containing one nitrogen atom). Most are derived from the amino acid lysine. An example is nicotine, found in many plants, especially *Nicotiana tabacum* (tobacco) and other *Nicotiana* species.

pisatin see FLAVONOIDS.

pistil a term used ambiguously to describe either a single CARPEL (simple pistil) or a group of fused carpels (compound pistil).

pistillate flower a flower possessing female parts (a pistil) but no male parts. Compare STAMINATE FLOWER. See also MONOECIOUS, DIOECIOUS.

pistillode in a male-only flower, a nonfunctional gynoecium-like structure.

pit a cavity in the secondary CELL WALL, allowing exchange of substances between adjacent cells. A pit consists of a *pit cavity* (the aperture in the secondary wall) and a *pit membrane* (the primary wall material adjacent to the cavity). Pits usually occur in pairs (*pit pairs*), the members of the pair being situated in adjacent cell walls. There are two main types of pit, SIMPLE and BORDERED. Though a pit in a secondary cell wall is distinct from a cavity in a primary cell wall (PRIMARY PIT FIELD), it is often difficult to distinguish between the two and then the term *pitted* is used for either structure. Pits that develop between cells that were not originally connected, or between daughter cells following cell division, are termed *secondary pits*. See *also* plasmodesmata.

pitcher plant see INSECTIVOROUS PLANT.

pit connection in plants and algae, a protoplasmic link between adjacent cells.

pith (medulla) a region of parenchymatous

tissue found in the centre of many plant stems to the inside of the STELE. A pith is also often seen in the roots of herbaceous plants but rarely in the roots of woody plants.

pit pairs see PIT.

pitted thickening a type of secondary wall patterning in TRACHEARY ELEMENTS in which the secondary cell wall is more or less continuous, the continuity only broken by PITS and PERFORATION PLATES (if present). The pits may be arranged in an opposite or alternate pattern (*see illustration at* tracheary element). If the pits are greatly elongated transversely and parallel to each other, they are said to be *scalariformly pitted*. Like RETICULATE and SCALARIFORM THICKENINGS, pitted thickening permits little further extension and is therefore found in tissues that have finished elongating at maturity, such as metaxylem and secondary xylem. Compare ANNULAR THICKENING, SPIRAL THICKENING.

placenta the tissue by which spores, sporangia, or ovules are attached to the maternal tissue. It is usually mostly undifferentiated but contains vascular tissue. See also PLACENTATION.

placentation the pattern of attachment of an ovule to the ovary wall by the PLACENTA. If there is only one ovule in the ovary then it is usually attached either at the base (BASAL PLACENTATION) or the apex (apical placentation). In a simple ovary the ovules may be attached either along the ventral suture (MARGINAL PLACENTATION) or, rarely, all over the inner surface of the ovary wall (LAMINATE PLACENTATION). In a compound ovary the ovules may be attached either on a central axis or on the wall along the junctions of the carpels (PARIETAL PLACENTATION). Placentation on the central axis is termed AXILE in a multilocular ovary and FREE-CENTRAL in a unilocular ovary. It is believed that free-central and parietal placentation are derived from axile placentation. The type of placentation is an important diagnostic character in many taxa.

placoderm desmids see GAMOPHYTA.

placoid describing a crustose lichen that is circular and platelike, with a lobed margin.

plagioclimax see CLIMAX.

plagiogeotropic see GRAVITROPISM.

plagiogravitropic see GRAVITROPISM.

plagiotropism a TROPISM in which the plant part is aligned at some angle to the direction of the stimulus. A plagiotropic response to gravity (*plagiogravitropism*) is exhibited by lateral shoots and roots. Compare ORTHOTROPISM. See also DIATROPISM.

plakea in green algae (see CHLOROPHYTA) belonging to the VOLVOCALES, a multicellular stage during asexual reproduction that inverts to bring the cells into the right orientation with their undulipodia facing outwards. A plakea may be a platelike structure that simply becomes slightly convex, or it may be a spherical structure that literally turns inside out.

plane ptyxis describing a kind of individual leaf folding (PTYXIS) in which the leaf remains flat while in the bud.

plankton (*adj.* planktonic) the large community of microorganisms that floats freely in the surface waters of oceans, seas, rivers, and lakes. They are moved passively by wind, water currents, or waves, having little or no powers of locomotion themselves. The *phytoplankton* includes many microscopic algae, particularly diatoms, and also cyanobacteria. They form the base of the food chain in water, being eaten by the *zooplankton* (animal plankton), which in turn provides food for fish. It has been estimated that 90% of the world's photosynthetic activity is carried out by the phytoplankton. A feature of the plankton is the huge variation in composition at different times of year (see BLOOM).

planktonic geochronology the use of the remains of planktonic organisms in dating sediments laid down in marine waters.

planogamete a gamete with UNDULIPODIA.

planospore a motile spore.

planozygote a zygote with UNDULIPODIA.

Plantae (Metaphyta) the plant kingdom. In the oldest classifications it was taken to include all organisms except animals and was divided into two divisions or subkingdoms, THALLOPHYTA and EMBRYOPHYTA. The kingdom was later restricted by excluding the bacteria and fungi, and many classifications now also exclude the algae. The FIVE KINGDOMS

CLASSIFICATION excludes the algae and blue-green bacteria (see CYANOBACTERIA), restricting the kingdom to the following phyla: BRYOPHYTA (mosses), HEPATOPHYTA (liverworts), ANTHOCEROPHYTA (hornworts), PSILOPHYTA (whisk ferns), LYCOPHYTA (club mosses), SPHENOPHYTA (horsetails), FILICINOPHYTA (ferns), CYCADOPHYTA (cycads), GINKGOPHYTA (*Ginkgo*), CONIFEROPHYTA (conifers), GNETOPHYTA, and ANTHOPHYTA (Angiospermophyta, flowering plants). The first three of these phyla are grouped together as the Bryata (nonvascular plants, or BRYOPHYTES); the remaining phyla form the Tracheata (vascular plants).

Plants are distinguished from animals (Animalia) by a number of factors. Most are autotrophic, making their food from organic starting materials by photosynthesis. Animals by contrast are heterotrophic. Certain parasitic higher plants, however, are heterotrophic. Plants are usually attached to a surface and not able to move around freely like animals (but neither can sponges and corals). Plants can generally respond to external stimuli only by growth movements. Most plant cells are surrounded by cellulose cell walls and starch is a common storage polysaccharide. Animals do not have cell walls and carbohydrates are commonly stored as glycogen. They do not have plastids or chlorophyll, and they form blastulas during development. Perennial plants tend to grow indefinitely, while in animals increase in size usually ceases at maturity. Plants are distinguished from ALGAE in producing embryos, and from the FUNGI in having a regular alternation of diploid and haploid generations and by their plastids and cellulose cell walls.

plant association see ASSOCIATION.

plantation a dense stand of one or two species of trees planted and managed by humans and harvested as a crop. The term is not applied to fruit orchards.

plant breeding the improvement of plants for agricultural, horticultural, or medical purposes. Probably all cultivated plants are the result of selection practices that originated some 10 000 years ago. Modern breeding techniques include artificial control of pollination, the generation of variability by artificial hybridization and MUTATION, and selection procedures such as the PEDIGREE METHOD and BACKCROSSING.

Plant Conservation Subcommittee (PCS) a subcommittee of the WORLD CONSERVATION UNION (IUCN) consisting of about 20 experts from many different countries. It meets annually and aims to help the IUCN provide advice on plant conservation to international conventions, governments, and other organizations involved in protection of the environment. The Committee collects information, develops policies, and attempts to set priorities and stimulate action on plant conservation. It works closely with the CONVENTION on International Trade in Endangered Species (CITES) and also helped to produce the *IUCN Red List of Threatened Plants* (see RED DATA BOOK). A recent extension of the Committee has been the establishment of a Medicinal Plant Specialist Group.

plant geography (phytogeography) the study of the geographical distribution of plants and their interrelationships with one another and with the environment. Many aspects overlap with the science of ecology but plant geography places more emphasis on the influence of the environment. See also VICARIANCE.

plantibodies antibodies produced in TRANSGENIC plants that are expressing the antibody-producing genes of an animal.

plantlet 1 during *in vitro* culture, the stage immediately before a whole plant is produced. **2** a colloquial name given to plants produced by adventitious buds on the margins of leaves of the parent plant, e.g. *Bryophyllum*.

plant sociability in the description of plant communities, a subjective measure of the distribution pattern and degree of organization of a species. In the ZURICH–MONTPELLIER PHYTOSOCIOLOGICAL SCHEME a 5-point scale is used: 1 occurring singly/only once in a place; 2 grouped/growing in tufts; 3 in small patches or cushions; 4 in extensive patches or carpets; 5 in very large patches or as a pure population. See also PHYTOSOCIOLOGY. Compare FIDELITY.

plaque a clear area in a plate of bacteria caused by the lysis of bacterial cells following infection by a bacteriophage.

plasmagel see CYTOPLASM.

plasmagene a GENE present in any structure other than a chromosome in the nucleus. The term commonly refers to genes present in organelles such as mitochondria and chloroplasts. See also CYTOPLASMIC INHERITANCE.

plasmalemma see PLASMA MEMBRANE.

plasma membrane (plasmalemma) the outer layer of the protoplasm below the cell wall. Like most membranes it is formed by the orderly orientation of protein and PHOSPHOGLYCERIDE molecules. Plasma membranes range from 7.5 nm to 10 nm in thickness and are composed of approximately 60% protein and 40% phosphoglyceride. Membranes of different species contain characteristic types of polar lipids in proportions that are probably genetically determined.

Davson and Danielli, in 1935, proposed that membranes are made up of a central region consisting of phosphoglycerides and an outer denser region composed of proteins. The phosphoglyceride molecules were believed to be arranged in two rows with their hydrophilic polar heads towards the outer edges and their hydrophobic hydrocarbon tails in the centre. Although it is still accepted that the two rows of phosphoglyceride molecules form the backbone of the membrane it is believed that globular proteins, rather than forming a distinct outer layer, actually penetrate the whole width of the membrane in places (see diagram). Both the phosphoglyceride

molecules and the proteins are thought to be able to move laterally giving the membrane fluid-like properties. This hypothesis of membrane structure is termed the *fluid–mosaic model*.

Plasma membranes are selectively permeable, controlling the passage of materials into and out of the cell. The proteins of the membrane include enzymes and compounds of the ACTIVE TRANSPORT system. Some of these proteins are thought to act as hydrophilic pores through which ions and polar substances can pass. Some of these channels may have a particular charge distribution that controls which substance can pass through. Other membrane proteins are carrier proteins, bonding to certain molecules and transporting them through the membrane, often using energy from ATP. Water and nonpolar molecules pass readily through the lipid part of the membrane. Time-lapse photography of living membranes reveals almost constant movement and it is probable that there is continual replenishment of membrane constituents. See CELL MEMBRANE.

plasmasol see CYTOPLASM.

plasmid any small autonomously replicating piece of DNA found in the cytoplasm of bacteria. Examples of plasmids are the R *factors* (R = resistance), which can carry genes conferring antibiotic resistance. Some plasmids may reversibly insert themselves into the bacterial DNA ('bacterial chromosome'), in which case they are called *episomes*. Some episomes are best regarded as bacterial viruses. Another example of an episome is the sex factor, or F *factor*. Possession of the F⁺ factor confers

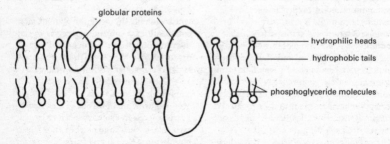

globular proteins

hydrophilic heads
hydrophobic tails
phosphoglyceride molecules

Plasma membrane: the proposed arrangement of phosphoglyceride and protein molecules

maleness upon a bacterium, as evidenced by the ability to transmit DNA through a sex pilus (conjugation tube) to a female (F⁻) bacterium.

plasmodesmata strands of cytoplasm that pass through CELL WALLS and connect the PROTOPLASTS of adjoining cells. They may be concentrated in pit pairs or may be distributed throughout the walls. They allow the passage of materials between cells.

plasmodial slime moulds see MYXOMYCOTA.

Plasmodiophora (Plasmodiophoromycota) a phylum of obligate microbial parasites or symbionts in the cells of plants, in which they often induce HYPERTROPHY. They feed by absorbing food directly across the cell membrane. The feeding stage is a multinucleate PLASMODIUM without cell walls. However, the zoospores may sometimes feed by engulfing food particles. During mitosis the nuclear membrane does not disintegrate, and later invaginates to separate the two daughter nuclei. Plasmodiophorans produce uninucleate resting spores that germinate in the soil to form biflagellate zoospores, which infect new hosts. Inside the host a zoospore divides to form a small primary PLASMODIUM that forms a wall, within which it produces further zoospores, which leave the plasmodium and escape into the soil. Here they are thought to act as gametes and fuse in pairs, eventually reinfecting the plant. Then a secondary plasmodium is formed, which eventually produces a uninucleate cyst by meiosis.

There are some 46 species in 16 genera. The plasmodiophorans are sometimes classified as fungi (the class *Plasmodiophoromycetes* or the order Plasmodiophorales of the MYXOMYCOTA). In the FIVE KINGDOMS CLASSIFICATION, they belong to the PROTOCTISTA. Some are serious parasites of crop plants, including *Plasmodiophora brassicae*, which causes CLUB ROOT in brassicas, and *Spongospora subterranea*, which causes powdery or corky scab in potatoes.

Plasmodiophorales see PLASMODIOPHORA.

Plasmodiophoromycetes see PLASMODIOPHORA.

Plasmodiophoromycota see PLASMODIOPHORA.

plasmodium the motile multinucleate amoeba-like mass of protoplasm that makes up the thallus of the plasmodial slime moulds (MYXOMYCETES). It develops from the zygote and in favourable conditions forms sporangia. In adverse conditions it may change into a dry resting body, the *sclerotium*.

plasmogamy the fusion of the protoplasts of two haploid cells during SEXUAL REPRODUCTION. Plasmogamy is usually followed immediately by karyogamy (nuclear fusion), but in some ascomycetes and basidiomycetes the two processes are separated so that binucleate mycelia are formed (see DIKARYON). See also SOMATIC HYBRIDIZATION.

plasmolysis the withdrawal of the cytoplasm from the cell wall because of the outward movement of water from the cell vacuole due to OSMOSIS. The cytoplasm eventually forms just a small central mass still enclosing a vacuole. Plasmolysis is seen when tissues are placed in solutions of lower water potential than that of the cell. By varying the concentration of external solutions and observing when a solution is just strong enough to induce plasmolysis in 50% of the cells the concentration of solutes in the cell vacuole can be determined. When working with single cells, the concentration that brings about *incipient plasmolysis*, i.e. the first signs of plasmolysis, is taken as being equal to the concentration of the cell vacuole.

Plasmolysis normally only occurs in experimental systems where the external solution can pass through the cell to fill the space between the cell wall and the cytoplasm. Thus the cells of a wilted plant do not undergo plasmolysis.

plasmon all extrachromosomal hereditary determinants, such as extrachromosomal DNA and plasmids.

plasmotype see IDIOTYPE.

plasticity the capacity of an organism to change its form in response to varying environmental conditions. For example, a plant transferred outside from a warm poorly lit greenhouse may produce smaller paler leaves and shorter internodes. Such changes can only occur in new tissues and the existing mature parts of the plant remain unchanged. In taxonomic work it is important to establish how plastic a particular character can be before placing too much reliance on it. It is also necessary in breeding work to establish what proportion of variation is caused by phenotypic plasticity.

plastid an organelle found in the cytoplasm of the majority of plant cells. Plastids are surrounded by a double membrane and show a wide variety of structure, ranging from minute PROPLASTIDS, less than 1.0 μm in diameter, to pigmented CHROMOPLASTS, 10 μm long with a complex internal arrangement of lamellae. These give colour to plant tissues. Plastids are generally interconvertible, one type being derived from another and capable of differentiating into at least one other form. Many plastids, e.g. amyloplasts and ELAIOPLASTS, are storage organelles. CHLOROPLASTS, the most important plastids, contain chlorophylls, carotenoids, electron transfer compounds, and enzymes, providing a highly organized and efficient photosynthetic organelle.

plastid inheritance SEE PLASTOGENE.

plastocyanin one of the electron transfer components of chloroplasts, mediating transfer of electrons from photosystem II to photosystem I. It is a blue copper-containing protein, and it is the reduction and oxidation of the copper atom that allows plastocyanin to accept, then donate, electrons. The position of plastocyanin in the electron transport chain is thought to be very close to photosystem I, probably between cytochrome f and photosystem I. See PHOTOSYSTEMS I AND II.

plastogene a PLASMAGENE present in a chloroplast or other plastid. The transmission of characteristics by such genes is termed *plastid inheritance*. See CYTOPLASMIC INHERITANCE.

plastoglobuli droplets of lipid, 10–500 nm in diameter, found singly or in groups in the stroma of chloroplasts. They do not have an enclosing membrane but nevertheless retain their form when isolated from the stroma. Plastoglobuli from chloroplasts of actively photosynthesizing leaves contain various lipid and lipophilic compounds, e.g. galactolipids, quinones, and polyisoprenols, but no chlorophylls or carotenoids. As

leaves become senescent, chloroplasts change to CHROMOPLASTS and changes in the plastoglobuli occur. They accumulate carotenoid pigments and enlarge considerably until they become the dominant features in the chromoplasts.

plastome (plastidome) a general term for all the PLASTIDS of a cell, or for the genetic information contained in the plastid DNA. See CYTOPLASMIC INHERITANCE, EXTRACHROMOSOMAL DNA.

plastoquinone one of the electron transfer components of chloroplasts, mediating transfer of electrons from photosystem II to photosystem I. It is situated between cytochrome b_3 and cytochrome f and the transfer of electrons to cytochrome f is accompanied by the formation of a molecule of ATP. See PHOTOSYSTEMS I AND II.

platyspermic describing the flat bilaterally symmetrical seeds characteristic of the CORDAITALES. Compare RADIOSPERMIC.

pleated sheet (β-pleated sheet) a protein conformation in which POLYPEPTIDE chains lie antiparallel, held together by (hydrogen bonds between nitrogen-bound H atoms and electrons from the oxygen atoms in the carbonyl (-C=O) groups of adjacent chains.

Plectascales SEE EUROTIALES.

plectenchyma a form of 'tissue', commonly found among the higher fungi, composed of a mass of interwoven anastomosing hyphae. It is termed *prosenchyma* when formed from long fused hyphae and *pseudoparenchyma* when it has a cellular appearance due to regular divisions in the hyphae. Pseudoparenchymatous tissue is also seen in the thalli of certain red algae (see RHODOPHYTA).

Plectomycetes in some classification schemes, a class of the ASCOMYCOTA containing those ascomycetes that produce a CLEISTOTHECIUM. It numbers about 2300 species in some 160 genera. It is not a natural class as it contains both fungi with single-walled asci and others with double-walled asci. In addition, some are primitive while others are regressive (see REGRESSION). It contains the orders EUROTIALES, MICROASCALES, ERYSIPHALES, and MELIOLALES.

plectostele see STELE.

Pleiocene see TERTIARY.

pleiomorphic (pleomorphic, pliomorphic) describing organisms that exhibit two or more different forms in their life cycle. For example, many ascomycetes have two conidial states.

pleiotropism (*adj.* pleiotropic) the control of several apparently unrelated characteristics by a single gene. Thus in tobacco a single gene is responsible for long anthers, calyces, capsules, and petioles. Compare POLYGENES.

Pleistocene see QUATERNARY.

pleomorphic see PLEIOMORPHIC.

plerome see HISTOGEN THEORY.

plesiomorphy a primitive character state whose origin can be traced back to a remote ancestor. Plesiomorphies shared by different taxa (*symplesiomorphies*) are not used in the construction of cladograms or phylogenies. Compare APOMORPHY. See CLADISTICS.

plethysmothallus a juvenile filamentous stage in the sporophytes of some brown algae (PHAEOPHYTA).

pleurocarpous describing mosses in which the reproductive organs are produced laterally and the main axis is usually creeping. Compare ACROCARPOUS.

Pleuromeiales see LYCOPHYTA.

pleuropneumonia-like organisms see MYCOPLASMAS.

plicate 1 arranged in or having folds as, for example, the mesophyll cells of many gymnosperms and certain angiosperms, which have inward foldings of the cell wall. 2 describing a kind of leaf folding in which the leaf is folded like a concertina. *See illustration at* ptyxis.

Pliocene see TERTIARY.

pliomorphic see PLEIOMORPHIC.

ploidy the number of compete chromosome sets in the nucleus of a eukaryotic cell.

plot ordination see INVERSE ANALYSIS.

plumule the embryonic shoot, derived from the EPICOTYL. In dicotyledons the plumule is situated between the cotyledons. If germination is EPIGEAL the plumule is protected during its passage to the soil surface by the cotyledons. If germination is HYPOGEAL the elongating plumule has a hooked tip.

plurilocular sporangium a sporangium that is divided by septa into many compartments. Plurilocular sporangia are seen, for example, on lateral branches of the diploid generation in the brown alga *Ectocarpus*. Compare UNILOCULAR SPORANGIUM.

pneum- prefix denoting relating to air or gas, e.g. pneumatophores, which allow air to reach submerged roots.

pneumatocyst in some brown algae (see PHAEOPHYTA), the hollow part of the stipe, which acts as a float.

pneumatophore (aerophore, breathing root, respiratory root) an erect root that protrudes some distance above soil level. Pneumatophores are formed in large numbers by certain plants, e.g. *Sonneratia* and some mangrove species, growing in areas with waterlogged badly aerated soils. The surface of the pneumatophore is perforated by numerous LENTICELS that promote gaseous exchange.

Poaceae (Gramineae) a monocotyledonous family containing the grasses, mostly annual or perennial herbs but including a few woody genera (e.g. bamboos). It contains about 9500 species in some 668 genera distributed worldwide. Grasses generally have long narrow parallel-veined leaves inserted distichously on a round hollow stem, often with basal and INTERCALARY MERISTEMS that allow the plant to resume growth after being grazed. Each leaf usually has a base that sheaths the stem, a lamina, and often a flap (LIGULE) where the sheath joins the lamina. The inconspicuous flowers are usually borne in a terminal panicle or spike consisting of a number of SPIKELETS. Each inflorescence is surrounded by two bracts called GLUMES. Each spikelet is made up of one or more florets comprising two scales (the LEMMA and PALEA) partly enclosing three stamens, with long filaments and large flexible anthers, and an ovary with a single ovule and two long feathery stigmas (see diagram). The pendulous anthers and feathery stigmas are adaptations to wind pollination (see ANEMOPHILY), which is predominant in the family. The fruit is a CARYOPSIS. Many grasses are polyploid, and APOMIXIS, CLEISTOGAMY, and VIVIPARY are sometimes used to overcome reproductive difficulties and inclement climates.

Grasses are the dominant vegetation in savannas, prairies, and steppes (see GRASSLAND). Economically they are the most important family of plants as they contain all the cereals, which are the staple diet of people in most parts of the world. Wheat (*Triticum*), maize (*Zea mays*), rice (*Oryza sativa*), barley (*Hordeum vulgare*), oats (*Avena sativa*), rye (*Secale cereale*), sugar cane (*Saccharum officinarum*), and sorghums (*Sorghum*) are all grasses. They are also widely planted for pasture and fodder. The bamboos have many economic uses as foods (bamboo shoots) and building materials and in furniture-making.

poculiform cup-shaped.

pod see LEGUME.

podetium a specialized spore-bearing structure that develops on the thalli of certain lichens, e.g. the erect cylindrical podetia of *Cladonia*.

podsol (podzol) a type of infertile ACIDIC SOIL (a PEDALFER) found in regions of heavy rainfall and long cool or cold winters. The soils of the northern coniferous forest zone of North America and Eurasia are typical examples. The soil is heavily leached by the sudden snow-melt in the spring that washes out lime and iron compounds. Thus beneath the humus layer in the A horizon (see SOIL PROFILE) there is a bleached horizon composed of quartz sand. Iron

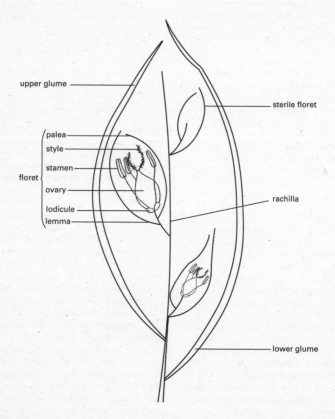

Poaceae: structure of a single spikelet of a grass flower, with detail of a single floret

compounds accumulate in the B horizon, which stains brown and has a clayey texture. These compounds may form an impermeable *hard pan* that, if developed sufficiently, can stop water percolating from above, resulting in the A horizon becoming waterlogged. Such soils may be termed GLEY podsols. The leaf litter in such regions also tends to be highly acid and maintains the acidity of the soil. The processes by which such soils are formed are collectively termed *podsolization*.

podsolization see PODSOL.

point mutation see GENE MUTATION.

point quadrat see PIN-FRAME.

Poisson distribution the type of distribution of data characterized by having a variance equal to its mean. It occurs when the probability of an event occurring on any particular occasion is extremely small but the very high number of occasions recorded makes it likely that the event will occur reasonably often overall. An example is the measurement of the density of a particular plant species using quadrat counts. The likelihood of that species occurring in any particular square of the quadrat is small but if sufficient quadrats are thrown the species will be recorded eventually.

polarilocular describing an ASCOSPORE that consists of two cells separated by a thick septum through which a narrow canal passes.

polarity the condition that results from the establishment of a definite orientation during the differentiation of a cell, tissue, or organ. Polarity is evident in the early growth of plants, as in the bipolar development of an embryo from the zygote. In later growth it is evidenced in the separate development of roots and shoots, as well as various phenomena at the cellular level.

polarizing microscope a microscope that uses polarized light to illuminate the specimen. Certain crystalline molecules transmit polarized light in a way that depends on the orientation of the molecules in the crystal. The technique has been used, for example, to investigate the arrangement of cellulose molecules in plant cell walls.

polar molecule a molecule in which the electrons are not equally shared between the nuclei, with the result that one end of the molecule bears a slight negative charge, and the other end a slight positive charge. For example, in water the oxygen atoms are slightly negatively charged and the hydrogen atoms slightly positively charged. Such molecules attract each other and other polar molecules.

polar nuclei the two haploid nuclei found in the centre of the EMBRYO SAC after division of the MEGASPORE, forming a binucleate *central cell*. The nuclei may fuse to form a diploid definitive nucleus before fusing with the male gamete to form the triploid primary endosperm nucleus. See also DOUBLE FERTILIZATION.

pollarding a severe form of pruning in which all the younger branches of a tree are cut back virtually to the trunk, which encourages new bushy growth. It is performed particularly on riverside willows, and sometimes on urban avenues of trees to provide summer shade. The tips of the branches of pollarded trees are characteristically club shaped due to the growth of wound wood. Compare STOOLING.

pollen the MICROSPORES of seed plants, which are produced in vast numbers, usually in a POLLEN SAC. They are formed as a result of meiosis of somatic POLLEN MOTHER CELLS. In primitive forms they are formed in unspecialized structures with little protection. In intermediate forms they are produced in specialized stamens, which are numerous and develop symmetrically in the flower. In the most advanced forms, pollen is formed in fewer stamens, which are more strategically placed with regard to the pollination mechanism. Pollen structure may be related to methods of pollination. In insect-pollinated species the pollen is often sticky or barbed, whereas wind-pollinated species usually produce light smooth pollen.

Pollen grains contain concentrated mitochondria, endoplasmic reticulum, and Golgi apparatus. The number of nuclei present in the cell at any given time may be diagnostic. On germination of the pollen grain a pollen tube pushes its way through an aperture in the pollen grain wall and the various nuclei migrate into the pollen tube (see also GENERATIVE NUCLEI, VEGETATIVE NUCLEUS).

The pollen grain wall consists of a resistant outer EXINE, which may be very highly sculpted and differs markedly between families, genera, and even species, and an inner INTINE, which may protrude through pores in the outer layer. These pores, apart from facilitating germination, may also be instrumental in water regulation and in controlling compatibility systems. Although most pollen is short lived, some pollen may be stored at low temperatures for long periods. See also PALYNOLOGY.

pollen analysis see EXINE, PALYNOLOGY.

pollen chamber a cavity at the micropylar end of the nucellus in some gymnosperms in which pollen grains lodge after pollination. The pollen may be immature at this stage but it ripens in the pollen chamber prior to germination and fertilization of the egg. See also POLLINATION DROP.

pollen diagram in PALYNOLOGY, a graphical representation of the pollen record for a particular location. Time is plotted on the vertical axis, and absolute amounts or relative proportions of the different pollen types are plotted as a histogram, each type or group being plotted as a vertical bar that varies in width with the quantity of pollen found at a particular depth/time. Alternatively, the quantities of pollen may be plotted on a continuous curve.

pollen mother cell (PMC) a SOMATIC CELL that, after meiosis, forms a TETRAD of pollen grains. Many are found closely packed within the pollen sacs of angiosperms and gymnosperms. In angiosperms the pollen mother cell and the resulting pollen grains have been shown to obtain proteins and their precursors from the disintegrating TAPETUM that surrounds them. Compare MEGASPORE MOTHER CELL. See also SPORE MOTHER CELL.

pollen sac a chamber in which POLLEN grains (microspores) are formed in the angiosperms and gymnosperms. It is homologous with the MICROSPORANGIA of the pteridophytes. In angiosperms there are usually four pollen sacs in an anther,

arranged in two lobes, either side of the connective tissue. In the conifers there may be numerous pollen sacs formed on microphylls in the axils of the male strobilus. Compare OVULE. See also STAMEN.

pollen tube an outgrowth of the INTINE of the pollen grain that, on germination, emerges through an aperture in the exine and grows towards the egg, carrying the male gametes with it. It represents the reduced male gametophyte in seed plants. In angiosperms pollen tube growth in compatible stigmatic tissue is usually rapid and may reach 1–3 mm/h. In gymnosperms growth is arrested at the nucellus and may not recommence until the next growing season. In *Cycas* and *Gingko* the gametes are motile. In other gymnosperms and in angiosperms the gametes are nonmotile and the mechanism of their movements down the pollen tube is not fully understood.

In angiosperms the pollen tube may reach the ovule from the stigma either by growing down the stylar canal or by enzymatically digesting its way between or through the individual cells of the style. It then grows through the ovular cavity and reaches the egg apparatus, usually via the micropyle (see also CERTATION) but sometimes through the chalaza (see CHALAZOGAMY). There, the contents of the pollen tube are discharged. The VEGETATIVE NUCLEUS disintegrates and one of the gametes fuses with the egg cell to form the zygote and the other with the polar nuclei or definitive nucleus to form the endosperm. See also SELF INCOMPATIBILITY.

pollen zone in PALYNOLOGY, a characteristic assemblage of species or types of pollen and spores typical of a particular climate or geographical region. In post-glacial Britain and Europe eight major pollen zones are recognized, the first three (Dryas, Allerød, and younger Dryas) representing the late glacial period. By defining zones for specific sites, local changes, including those due to human activity, can be analysed.

pollinarium see POLLINIUM.

pollination the transfer of pollen from the male reproductive organs to the female in seed plants. This involves transfer from the anthers to the stigma in angiosperms and from the microsporangiophores to the micropyle in gymnosperms. The process is usually effected by intermediary agents such as insects (see ENTOMOPHILY), wind (see ANEMOPHILY), or water (see HYDROPHILY). In flowering plants direct pollination may occur by gravity or contact where the stamens and stigmas are juxtaposed and mature simultaneously (see HOMOGAMY, CLEISTOGAMY).

pollination drop a drop of sugary fluid that is secreted through the micropyle in gymnosperms. Wind blown or insect transported pollen falls into the pollination drop and is drawn through the micropyle to the ovule as the drop is reabsorbed.

pollinium the structure formed when individual pollen grains remain massed together and are transported as a unit during pollination. They may be held together only by sticky secretions or they may be retained within the pollen sac wall, as in many orchids (*see illustration at* orchidaceae). When several pollen sacs remain together and are transported as a pollination unit they are called a *pollinarium*.

pollution a damaging or unwanted change in the environment resulting from human activity. Pollution affects the atmosphere, rivers, lakes, and seas, and the soil.

Air pollution is caused by the domestic and industrial burning of fossil fuels, by vehicle exhaust fumes, and by emissions from industrial processes. Carbon dioxide produced by the burning of fossil fuels and by vehicle exhausts accumulating in the atmosphere contribute to a global increase in atmospheric temperatures (see GREENHOUSE EFFECT). Emissions of sulphur dioxide cause ACID RAIN. The release of chlorofluorocarbons (CFCs) into the atmosphere from aerosols, refrigeration plants, and other sources causes the destruction of ozone in the stratosphere (see OZONE HOLE). CFCs are also potent greenhouse gases. Vehicle exhausts also release carbon monoxide, nitrogen oxides, other harmful chemicals, and fine particulate material that can affect breathing and coat plant leaves, reducing photosynthesis.

Waterways are polluted by sewage effluent and by nitrates and other nutrients derived from fertilizers that leach from cultivated land. A build-up of these leads to EUTROPHICATION. However, these pollutants are eventually biodegradable. Nonbiodegradable pollutants such as chlorinated hydrocarbon pesticides (e.g. DDT) and heavy metals from industrial effluent, persist and accumulate in the environment. Even low concentrations of heavy metals are toxic to plants, inhibiting enzymes, damaging cell membranes, and interfering with the uptake of nutrients and/or water.

polyadelphous describing stamen filaments that are fused into three or more groups.

polyandrous having separate stamens freely inserted on the receptacle.

polyarch describing primary xylem that is made up of several strands.

polycarpellary see FRUIT.

polycarpic see ITEROPAROUS.

polycentric describing a THALLUS that possesses several reproductive centres.

polychore distribution see WIDE DISTRIBUTION.

polyclimax see CLIMAX THEORY.

polycyclic stele see STELE.

polyembryony the formation of more than one embryo in an ovule. These may develop by division of the fertilized zygote (see CLEAVAGE POLYEMBRYONY) and thus be a product of sexual reproduction. They may also arise, alongside a zygote, from somatic tissue (see ADVENTIVE EMBRYONY). Embryos derived in this manner will have the same genetic constitution as the maternal parent. See also APOMIXIS, PARTHENOGENESIS.

polygenes several nonallelic genes all affecting the same character and approximately additive in their effects. Polygenic systems, also called *polymeric systems*, were first noted in red-kernelled maize. The intensity of the colour is governed by three unlinked genes and depends on the number of alleles for redness that are present. Compare PLEIOTROPISM.

polygenic character a character (phenotype) that is determined by several interacting genes, and therefore shows CONTINUOUS

VARIATION rather than occurring in a small number of discrete forms.

polymer a compound built up of a series of similar units joined together. Examples are starch, which is composed of repeating units of glucose, and nucleic acids, which are made up of nucleotides. The combining of units to form a polymer is called *polymerization*.

polymerase an enzyme that catalyses the addition of units to a polymer. Polymerases are usually named according to their substrate, e.g. DNA polymerase. Important polymerases include DNA polymerases and RNA polymerases. DNA polymerases catalyse the addition of complementary nucleotides to an existing DNA strand (which acts as a template) during DNA REPLICATION. They require a primer from which to begin polymerization. RNA polymerases (types I to III) catalyse RNA synthesis using an existing RNA strand or a DNA strand (DNA-dependent RNA polymerase) as a template; no primer is needed. RNA synthesis involves type I RNA polymerase to produce ribosomal RNA, type II for messenger RNA, and type III for transfer RNA. RNA-directed DNA polymerase (REVERSE TRANSCRIPTASE) catalyses the synthesis of DNA using RNA as a template.

polymerase chain reaction (PCR) a technique used to replicate a fragment of DNA to produce a large number of copies. It can be used to clone genes and to amplify genetic material for analysis, including GENETIC FINGERPRINTING. The DNA is 'unzipped' into two strands by heating. These strands are then used as primers, being added to a solution containing free nucleotides and DNA POLYMERASE (a heat-resistant form derived from a heat-resistant bacterium). With repeated cycles of heating and cooling, the DNA doubles with each cycle.

polymeric system see POLYGENES.

polymerization see POLYMER.

polymictic describing a lake whose water is continuously circulating, and which therefore does not experience STRATIFICATION. Such lakes are found mainly at high altitudes in the tropics.

polymorphism 1 the existence of a number of different forms within a species, whether caused by genetic or environmental factors. 2 in population genetics, the existence of many different forms in a population at the same place and time, such that the frequency of the rarest form cannot be explained simply on the basis of recurrent mutation. Though mutation is the ultimate source of all genetic variation, in *stable* or *balanced polymorphism* the frequency of the different phenotypes is maintained either because different forms are most successful in different microhabitats or in different years or because there is some advantage to the heterozygotes not possessed by either of the homozygotes. In *transient* or *unstable polymorphism*, one or more of the morphs (phenotypes) is eliminated or otherwise lost, so that the population tends towards *monomorphism* (one form).

polynucleotide see NUCLEOTIDE.

polypeptide a PEPTIDE chain containing a large number of amino acid residues (most polypeptides have a molecular weight greater than 5000). PROTEINS are made up of one or more polypeptides.

The amino acid sequence of a polypeptide chain determines the three-dimensional CONFORMATION of that molecule under physiological conditions. This conformation can be disrupted by heating or extremes of pH (see DENATURATION). See also CONJUGATED PROTEIN.

polypetalous having separate petals freely inserted on the receptacle. Compare GAMOPETALOUS.

polyphyletic describing taxa derived from two or more ancestral lines. Polyphyletic groups are thought by some to contain taxa whose resemblances are based on shared advanced (derived) characters that have arisen by CONVERGENT EVOLUTION (see CLADISTICS). Compare MONOPHYLETIC, PARAPHYLETIC.

polyploidy the condition in which an organism has three or more complete sets of chromosomes (see GENOME) in its nuclei. Polyploids originate when gametes containing more than one chromosome set fuse. Such gametes are formed when chromosomes fail to separate during anaphase 1 of meiosis. Consequently gametes are diploid instead of haploid, and fertilization results in triploid or tetraploid individuals. Polyploid individuals may be produced as a result of multiplication of chromosome sets from one species (AUTOPOLYPLOIDY) or by combining sets of chromosomes from different species (ALLOPOLYPLOIDY). In either case the polyploid offspring may be incapable of reproducing with their parents and so constitute new species. Polyploidy is common in flowering plants (40% of dicotyledons and 60% of monocotyledons are polyploid) and has probably contributed significantly to their evolution. In polyploid organisms harmful recessive alleles are more likely to be masked by normal dominant alleles. Polyploids also have a greater store of genetic variation and thus evolutionary potential is higher in polyploid populations.

In contrast to plants, polyploidy is extremely rare in animals because the sex-determining mechanism often depends on chromosome numbers and ratios. These are upset by polyploidy, and consequently polyploid animals are inviable or sterile. See also HYBRID, TRIPLOID, TETRAPLOID.

Polypodiales see FILICALES.

Polypodiopsida see FILICINOPHYTA.

Polyporales see APHYLLOPHORALES.

polypore see APHYLLOPHORALES.

polyribosome see POLYSOME.

polysaccharide (glycan) a high-molecular-weight polymer of monosaccharides or monosaccharide derivatives. The major functions of polysaccharides are as energy storage molecules (reserve polysaccharides) or as structural elements in cell walls and intercellular spaces (structural polysaccharides). Starch and cellulose are the most abundant plant polysaccharides. Glucose is the most commonly occurring monosaccharide residue in polysaccharides; both starch and cellulose are made up exclusively of glucose subunits. Other sugars important in polysaccharides include galactose, mannose, fructose, xylose, and glucuronic and galacturonic acids (see URONIC ACIDS).

Polysaccharides differ in the nature of their monosaccharide units, in the types of bonding between units, in chain length, and in degree of chain branching.

Polysaccharides containing only one type of sugar residue are known as *homopolysaccharides*; these include glucans (glucose polymers), MANNANS (mannose polymers), GALACTANS, and fructans. *Heteropolysaccharides* contain two or more different monosaccharides; examples are the HEMICELLULOSES and some PECTIC SUBSTANCES.

polysepalous having separate sepals freely inserted on the receptacle. Compare GAMOSEPALOUS.

polysiphonous describing a thallus made up of filaments arranged in regular tiers.

polysome (polyribosome) a string of RIBOSOMES attached together by a single molecule of messenger RNA. The strings may be folded or coiled into complex configurations, often spirals. The ribosomes move along the mRNA and as they pass each base triplet a transfer RNA molecule adds the corresponding amino acid to the base of a growing polypeptide chain. When a ribosome moves off the end of the mRNA it releases its completed polypeptide chain and is ready to start at the beginning again. This enables the code to be used highly efficiently.

polysomy the presence of several additional copies of one particular chromosome in a cell or organism. See ANEUPLOIDY.

polystely see STELE.

polytelic describing an inflorescence in which none of the major axes ends in a flower. Compare MONOTELIC.

polythetic in a numerical classification scheme, the use of several or all possible attributes or criteria as the basis for each subdivision. Compare MONOTHETIC.

polytypism (*adj.* polytypic) the occurrence of phenotypic variations between geographically separated populations of the same species. Such populations may eventually form CLINES or subspecies. See ALSO GEOGRAPHICAL ISOLATION. Compare POLYMORPHISM.

polyvalent enzyme see ALLOSTERIC SITE.

pome a type of fleshy PSEUDOCARP in which the succulent tissues are developed from a greatly enlarged urn-shaped receptacle, which encloses the real fruit at its core. The pome is typical of the Rosaceae, the apple and pear being examples.

population a local community of potentially interbreeding organisms. In asexual organisms, the term normally refers to a local community of physiologically or morphologically similar individuals of the same species. In plants it is poorly defined because of the variety of growth forms and breeding systems.

population dynamics the study of the changes in the numbers of individuals in a population and the attempted correlation of these with physical or chemical changes in the environment, biotic factors, or the genetic make up of the population. Such studies are carried out over a number of years as there may be seasonal and/or long-term changes. Most studies of plant population dynamics have been carried out on annuals rather than perennials. Factors that have been investigated include seed size, dormancy, thickness of seed coat, method of seed dispersal, leaf size, and stem height.

population ecology see AUTECOLOGY.

population genetics the study of the number, variety, and distribution of genes in a population or species, and of the factors that influence these. Population genetics has considerable implications for research interests as diverse as evolution, ecology, and plant breeding. See also HARDY–WEINBERG LAW.

porate see APERTURATE.

pore (porus) 1 a circular or slightly elliptic germinal aperture in a pollen grain. Compare COLPUS. 2 a small interstice in soil or rock, which is filled with air and/or water. The total volume of continuously interconnected pores is called the *pore space*. The percentage of the total bulk volume of a body of rock or soil that is taken up by the pore space is its *porosity*.

pore space see PORE, def. 2.

poricidal see CAPSULE.

porogamy see CHALAZOGAMY.

porometer an apparatus designed to measure the resistance to the flow of air through a leaf in different external conditions. An example is Meidner's porometer, which consists of a Perspex clamp attached to a bulb pipette. The bulb is squeezed flat and the leaf is inserted into, and covered by, the clamp. The time taken for the air to pass through the leaf and inflate the bulb is proportional to the resistance of the leaf and gives an estimate of the degree of opening of the stomata.

porosity see PORE, def. 2.

porphyrin any compound containing four PYRROLE groups joined into a ring by methene (–CH=) groups between their α carbons. Porphyrins form an important group of pigments that includes the chlorophylls, cytochromes, and haemochromes.

position effect 1 a modification in the normal expression of a gene due to a change in its position on the chromosome following an inversion.
2 the determination of the appearance of an organism according to whether two nonallelic mutations of a single gene are in the *cis* arrangement or the *trans* arrangement. The position effect is the basis of the CIS-TRANS test.

post-climax in the monoclimax model (see CLIMAX THEORY) of the development of climax vegetation, a community that differs from the typical climatic climax because local conditions are cooler and/or wetter than the average regional climate. For example, communities growing at high altitudes, especially on windward slopes. Compare PRE-CLIMAX.

post-glacial the period since the last glaciation 10 000 years ago, also termed the Holocene epoch.

postical relating to the lower surface of a DORSIVENTRAL shoot. Compare ANTICAL.

post-transcriptional processing see MESSENGER RNA.

postzygotic incompatibility incompatibility in which the zygote fails to develop.

potash see POTASSIUM.

potassium symbol: K. A soft alkali metal, atomic number 19, atomic weight 39.09, required as a MACRONUTRIENT by plants. It is the most abundant cation in plant tissues and is believed to have a role in chlorophyll and protein synthesis and in carbon dioxide fixation. Potassium deficiency is most likely to occur in siliceous and peaty soils, symptoms being poor root growth and a characteristic red or purple coloration of the foliage. Growing points are especially affected and flower and fruit formation is poor. The potassium content of compound fertilizers is called *potash* and is usually measured as the proportion of potassium oxide, K_2O, present.

potassium–argon dating see RADIOMETRIC DATING.

potato blight a fungal disease of potatoes and related plants. There are two types. The most serious is late blight, caused by *Phytophthora infestans* (see OOMYCOTA), forms brown patches on the leaves, and often a white mould on the lower surfaces. In wet weather the foliage may collapse. The tubers also develope brown lesions, which spread to form a brown rot. This disease caused the great Irish potato famine in the 1840s. The other blight, early blight. caused by *Alternaria solani* (see FUNGI ANAMORPHICI), is less serious, but can lead to wilting and necrosis.

potential evapotranspiration see EVAPOTRANSPIRATION.

potometer an apparatus designed to measure the rate of water uptake by a cut leafy shoot or whole plant and hence, indirectly, the rate of transpiration. One version consists of a glass tube that bends upwards at one end and supports the plant. The other end of the tube is connected to a capillary tube with an attached scale. The whole apparatus is filled with water from a reservoir joined via a tap to the tube and then made airtight. The plant takes up water and the flow of water is measured by observing the progress of an air bubble along the capillary tube. If the diameter of the capillary tube is known the volume of water taken up by the plant can also be measured. The apparatus is usually used to compare the rate of water uptake when the plant is subjected to certain changes in the external conditions, such as moving air, different light intensities, differences in humidity, etc. It may also be used to compare the rate of water uptake by different plants in the same conditions. An *atmometer* is a similar apparatus and is used to measure the rate of evaporation from a nonliving wet surface, such as a porous pot. By comparing water loss from a potometer with that from an atmometer under similar conditions, the rate of water evaporation from a leaf, which is controlled by the leaf to some extent, can be compared to the rate of uncontrolled evaporation.

powdery mildew a plant disease, caused by fungi of the order ERYSIPHALES, in which the pathogen grows as a white powdery coating (of mycelium and conidiophores) on leaves and stems. Examples include powdery mildew of cereals (*Erysiphe graminis*), gooseberry and blackcurrant mildew (*Sphaerotheca mors-uvae*), and apple mildew (*Podosphaera leucotricha*). Compare DOWNY MILDEW.

PPLO see MYCOPLASMAS.

P_r the inactive form of the plant pigment PHYTOCHROME that has a peak of light absorption in the red part of the spectrum, i.e. at about 655–665 nm. It is interchangeable with the P_{FR} form.

prairie a large expanse of temperate GRASSLAND in North America, much of it now given over to wheat. The natural vegetation consists of a variety of dominant, mostly XEROMORPHIC grass species and various perennial herbs. The topography is generally flat or consists of gently rolling hills, and the species composition of communities varies with topography, soils, and drainage.

Prasinophyceae a class of the CHLOROPHYTA (green algae) whose members all have undulipodia and are usually unicells covered in organic scales, made mainly of carbohydrates, which are formed in the Golgi vesicles then deposited on the outer surface of the cell. The undulipodia are also covered in scales. Prasinophyceans live among marine and freshwater plankton. Each cell contains a single chloroplast, which in *Micromonas* is known to contain chlorophyll *c*. In *Pyramimonas* and *Tetaselmis* (formerly *Platymonas*) the chloroplast contains a pyrenoid and an eyespot. Sexual reproduction is rare. Reproduction is usually asexual, by mitosis, the cells separating by FURROWING. In some species cell division resembles that of the CHAROPHYCEAE, while in others it is more similar to that of the CHLOROPHYCEAE.

preadaptation an adaptation that evolved in one habitat but is also advantageous for survival in an adjacent habitat, allowing the organism to colonize it.

Precambrian the earliest and longest era of geological time between about 4600 and 570 million years ago (mya). It precedes the Palaeozoic era. The Precambrian is subdivided into three *eons*: the *Hadean* (from the origin of the Earth to about 3900 mya), *Archaean* (3900-2390 mya) and the *Proterozoic* (2390-570 mya). Precambrian rocks are found in North America, Australia, and South Africa and what fossil material there is has mostly been collected from these regions. Fossils resembling blue-green bacteria (see CYANOBACTERIA), fungal spores, and fungal hyphae have been preserved. The earliest fossils date to about 3500 mya, and appear to be the remains of organisms that resemble bacteria. The earliest remains of EUKARYOTES appear much later, around 1400 mya, and the first multicellular animal fossils occur in the last 100 million years of the Precambrian.Calcareous algae have been found in certain late Precambrian rocks in Labrador and Montana. Some cherts of the gunflint formation of Ontario (about 2000 million years old) and older cherts of Australia and Africa have been shown to contain STROMATOLITES. There is much controversy about the existence of certain of these fossil remains (especially the CHEMICAL FOSSILS) and about the reason for the apparent gap between the few questionable life forms in the Precambrian and the comparative wealth of life forms in the Cambrian period. See GEOLOGICAL TIME SCALE.

preferential species see FIDELITY.

prefloration see AESTIVATION.

prefoliation see VERNATION.

preformation theory see EPIGENESIS.

pressure potential (turgor potential) a component of WATER POTENTIAL, represented by the symbol Ψ_p. Pressure potentials may be negative, in which case they represent tensions. Tensions arise as a result of transpiration and are caused by resistance of the tissues to water flow. In extreme conditions very low (−150 bar) pressure potentials can develop. Pressure potential gradients are responsible for the upward movement of water in the xylem

(see COHESION THEORY). Positive pressure potentials are termed HYDROSTATIC PRESSURES.

prevailing climax the CLIMAX COMMUNITY that occupies the greatest area in a particular region, and thus represents the vegetation best adapted to the current (prevailing) climate. This is a useful concept, since it applies only to one particular time, and therefore does not need to take into account changing environmental variables such as climate change or weathering and erosion.

prezygotic incompatibility compatibility in which pollen germination or pollen tube growth are prevented.

prickle a short pointed outgrowth from the epidermis of a plant. It may be simply protective, as in *Gunnera*, or it may also help the plant become hooked to a support, as do the recurved prickles of roses and brambles. A prickle is a modified multicellular TRICHOME. Compare SPINE, THORN.

primary consumer see CONSUMER, FOOD CHAIN.

primary endosperm nucleus the triploid nucleus that results from fusion of the POLAR NUCLEI or definitive nucleus of the embryo sac with one of the male gametes released from the pollen tube. It develops into the ENDOSPERM. See also DOUBLE FERTILIZATION.

primary forest see FOREST.

primary growth size increase due to cell division at the apical meristems and subsequent cell expansion. The tissues so produced are termed the *primary plant body* and comprise all the tissues of a young plant. The gymnosperms, most dicotyledons, and some monocotyledons exhibit increased thickening of stem and root later in life by a process of SECONDARY GROWTH.

primary phloem PHLOEM derived from the PROCAMBIUM in the primary plant body. In nonwoody plants the primary phloem, consisting of the PROTOPHLOEM and METAPHLOEM, is the only food-conducting tissue, whereas in mature plants exhibiting secondary growth this function is usually performed by the SECONDARY PHLOEM.

primary pigment see REACTION CENTRE.

primary pit field an area of greatly reduced thickness in the primary CELL WALL of a plant cell, often penetrated by PLASMODESMATA. Primary pit fields enable relatively easy transfer of materials between cells, thus having a similar function to PITS. Cells lacking secondary walls can possess only primary pit fields, whereas those possessing secondary walls may also possess pits, which are often, but not always, positioned directly over the primary pit fields.

primary plant body see PRIMARY GROWTH.

primary productivity the total biomass (organic matter) synthesized by green plants in an ecosystem in relation to the area covered. The usual units of primary productivity are grams per square metre. Alternatively, primary productivity may be expressed in terms of the heat of combustion of the biomass (equivalent to energy assimilated by the plants), as kilojoules per square metre. The rate of primary productivity can be calculated as $g\ m^{-2}\ year^{-1}$. *Gross primary productivity* is the rate at which plants assimilate light energy, while *net primary productivity* is the rate at which energy is incorporated into plant tissue (the rate of photosynthesis minus the rate of respiration). Methods used for measuring primary productivity include the: aerodynamic method (def. 2), carbon dioxide method, diurnal curve method, gas exchange method, harvest method and oxygen method.

primary ray see MEDULLARY RAY.

primary structure the first level of structure in a protein molecule: the sequence of amino acids. See also CONFORMATION, SECONDARY STRUCTURE, TERTIARY STRUCTURE, QUATERNARY STRUCTURE.

primary thickening meristem a MERISTEM that is found in certain monocotyledons, such as palms, below the leaf primordia at the apex. It serves to increase the width of the stem behind the apex by cutting off rows of cells by periclinal divisions. Prolonged primary growth, due to the activity of this meristem, allows some monocotyledons to attain considerable stature.

primary tissue tissue formed from a primary meristem, such as the procambium, protoderm, or ground meristem. In woody plants primary tissue is formed before secondary tissue (see SECONDARY GROWTH). In herbaceous plants that possess no secondary meristems, all tissues are primary. The primary xylem and primary phloem are examples of primary vascular tissues.

primary wall see CELL WALL.

primary xylem (primary wood) XYLEM derived from procambium in the primary plant body. In nonwoody plants the primary xylem, consisting of the PROTOXYLEM and METAXYLEM, is the only water-conducting vascular tissue, whereas in mature plants exhibiting secondary growth this function is performed largely by the SECONDARY XYLEM. As well as differing from secondary xylem in origin, the primary xylem usually also has longer TRACHEARY ELEMENTS, which are often arranged in a random fashion, although these differences are not always reliable. Primary xylem also consists of an axial system only and therefore does not contain rays, although rays may be present between the vascular bundles.

primer a short segment of a nucleic acid that provides a free 3'-OH end that can be linked to another nucleotide in the 5' to 3' direction.

primordium any immature part of a plant destined to differentiate into a certain cell, tissue, or organ. The term is usually used of a part of the APICAL MERISTEM that later differentiates further. Thus a leaf primordium later differentiates into a leaf. In early stages of development the leaf primordium appears as a microscopic projection (LEAF BUTTRESS) from the shoot apex.

prisere see SERE.

probability the expectation that over a series of observations a certain kind of observation will occur regularly and form a given proportion of the total number of observations. For example, if a plant heterozygous for height, Tt, is selfed, the probability of finding the double recessive, tt, in the progeny is ¼. The greater the number of progeny, the more likely it is that the actual number of double recessives will approach 25%. See also CHI-SQUARED TEST.

procambium (provascular tissue) the part of an APICAL MERISTEM, the derivatives of which give rise to the primary vascular tissues. Compare PROTODERM, GROUND MERISTEM.

procaryote see PROKARYOTE.

prochlorophytes see CYANOBACTERIA.

procumbent describing a plant or parts of a plant that trail loosely along the ground. An example of a procumbent plant is the heath bedstraw (*Galium saxatile*). Compare DECUMBENT, PROSTRATE.

producer an organism that is the first stage in a FOOD CHAIN. Producers include green plants and those bacteria that synthesize organic molecules from inorganic materials by photosynthesis or chemosynthesis. They are eaten by primary CONSUMERS. Compare DECOMPOSER.

production ecology the study of energy flow and nutrient cycling in ecosystems.

production efficiency the percentage of the energy assimilated by an organism that becomes incorporated into new BIOMASS.

production/respiration ratio (P/R ratio) the ratio of gross production to total respiration in a community. If the P/R = 1, the community is in a steady state, and organic matter is neither accumulating nor being depleted; if P/R is persistently greater than one, organic matter is gradually accumulating, and if persistently less than one, it is being depleted. In tropical rainforest the community may be in a steady state on a daily basis, but in more seasonal climates the average P/R over the course of a year may be a better indicator of stability.

productivity see PRIMARY PRODUCTIVITY.

proembryo the young plant individual after fertilization but before tissue differentiation into embryo and suspensor tissue.

profile diagram see TRANSECT.

profundal describing the region of a lake or pond below a depth of 10 m, where there is little light, oxygen, or warmth. Heterotrophic organisms, such as bacteria, fungi, molluscs, and insect larvae, live in this region but few green plants. Compare PHOTIC ZONE.

progressive succession the normal sequence of succession of communities, in which pioneer communities with few species and low productivity (see PRIMARY PRODUCTIVITY) give way to progressively more complex communities with higher productivity. Compare RETROGRESSION.

progymnosperm any of certain plants of the Devonian period that show various characteristics apparently intermediate between nonseed-bearing and seed-bearing vascular plants. They have heterosporous reproduction, and certain anatomical features such as woody fibres in the cortex and secondary phloem, that are also found in the early gymnosperms.

Prokarya in the FIVE KINGDOMS CLASSIFICATION, a superkingdom that contains only one kingdom, the BACTERIA. Members of the Prokarya are characterized by being PROKARYOTES. See also KINGDOM.

Prokaryotae see BACTERIA.

prokaryote (procaryote) an organism in which the nuclear material is not separated from the rest of the protoplasm by a nuclear membrane. BACTERIA are prokaryotes; the term is not used of viruses. As well as lacking a defined nucleus, prokaryotes also lack nucleoli, plastids, mitochondria, vacuoles, Golgi apparatus, and endoplasmic reticulum. Ribosomes are present but are smaller (70S) than those of eukaryotes (80S) though similar in size to those of chloroplasts and mitochondria. This observation has led to speculation that eukaryotes may have evolved as symbiotic associations of prokaryotic organisms (see ENDOSYMBIOTIC THEORY). The cells themselves are also much smaller (about 1 μm in diameter) than eukaryotic cells (about 20 μm in diameter) and cytoplasmic streaming is not apparent. The genetic material is a circular strand of DNA, which, unlike that of eukaryotes, is not complexed with histone proteins. Cell division is amitotic.

The biochemistry of prokaryotes is essentially similar to that of eukaryotes. However, sterols are conspicuous in their absence from prokaryotes and prokaryotic cell walls characteristically contain MURAMIC ACID, a sugar acid not found among eukaryotes. Some unusual amino acids, e.g. diaminopimelic acid, are also associated with the cell wall structure and certain familiar amino acids, e.g. alanine and aspartic acid, occur as their D-isomers. Peptides containing D-amino acids are resistant to hydrolysis by peptidase enzymes though lysis of Gram-positive bacteria (see GRAM STAIN) can be brought about by lysozyme enzymes.

The basic differences between prokaryotes and eukaryotes are regarded as the most fundamental between living organisms and have led to the bacteria being placed in a separate superkingdom, Prokarya.

prolamellar body a three-dimensional regular lattice found in ETIOPLASTS. It is composed of a continuous system of tubules but when exposed to light the symmetrical arrangement is rapidly lost as tubules become pinched off into two-dimensional sections of lattice. These form perforated sheets of membrane that move apart, extend and increase, finally establishing the typical granal and intergranal lamellae of the mature CHLOROPLAST. PLASTOGLOBULI in the etioplasts are dispersed as their lipid contents are utilized in membrane formation.

prolamine any of a group of simple plant proteins that are soluble in 70–90% alcohol but insoluble in water and absolute alcohol. Prolamines contain a high proportion of proline and glutamic acid but only small amounts of basic amino acids. Examples are gliadin, hordein, and zein, found as storage proteins in wheat, barley, and maize respectively. Compare GLUTELIN.

prolification see VIVIPARY.

proline a nonpolar IMINO ACID, formula C_4H_8NCOOH (*see illustration at* amino acid). Synthesis of proline is from glutamic acid, while breakdown occurs by reversal of the synthetic pathway. Cell walls are rich in 4-hydroxyproline, a derivative of proline.

promeristem in an APICAL MERISTEM, a collective name for the cells thatt initiate growth and the cells immediately derived from them.

promoter see OPERON.

promycelium the BASIDIUM of the UREDINALES and USTILAGINALES. It is formed on germination of the resting

spores of these fungi and is usually divided by cross walls into a number of cells.

propagation the increase of plant numbers by bulbs, corms, seeds, cuttings, grafting, etc. See also PROPAGULE, VEGETATIVE REPRODUCTION.

propagule any structure that functions as a unit of dispersal and propagation, e.g. spores, seeds, cuttings, tissue-cultured plantlets.

propane-1,2,3-triol see GLYCEROL.

prophage a BACTERIOPHAGE that has become integrated in the bacterial DNA and is replicated along with it when the bacterium divides. In this state it is quiescent but it may excise itself from the DNA at any time and, following replication, cause lysis of the host bacterial cell.

prophase the initial stage of nuclear division (see KARYOKINESIS). In both MITOSIS and MEIOSIS, the chromosomes become coiled and recoiled, and, in mitosis, the chromatids can be identified. As they shorten and thicken the distinctive features of the individual chromosomes can be identified with the light microscope. In meiosis, prophase of the first division can be divided into five substages, but there is no clear demarcation between them, the whole process being continuous. The first substage, *leptotene*, is the period during which shortening and thickening occurs, but, although it is known that DNA replication has occurred, the chromosomes do not appear to be divided into chromatids. During the next two substages, *pachytene* and *zygotene*, homologous chromosomes are attracted to each other and SYNAPSIS takes place. This is in contrast to the situation in mitotic prophase where homologues remain entirely separate from each other. The paired chromosomes continue to contract and coil around each other to form a composite structure called a *bivalent. Diplotene*, the fourth substage, begins as the mutual attraction between the chromosomes of the bivalents lapses and is replaced by mutual repulsion, commencing at the centromeres. The chromatids at this stage are clearly visible, and, as the chromosomes separate, it becomes apparent that they are held together at various points, where chromatids from opposite chromosomes have crossed over and are linked together (see CHIASMA). By *diakinesis*, the chromosomes are fully contracted. As the centromeres of the homologues continue to move apart, each pulls its attached chromatid pair with it and the regions of cross-over move towards the ends of the chromosomes (*terminalization*). By the end of prophase in both mitosis and meiosis, the nucleoli have dispersed and the nuclear membrane has broken down. In prophase of the second meiotic division, there is only one set of chromosomes i.e. one member only of each homologous pair. See METAPHASE, ANAPHASE, TELOPHASE.

proplastid a small PLASTID, less than 1.0 μm in diameter, with rudimentary internal structure. The inner of the two surrounding membranes is often extended into finger-like projections. Small membrane-bound vesicles and a few starch grains and PLASTOGLOBULI may be present in the matrix. Proplastids are present in the cells of meristematic tissue and are thought to differentiate into mature plastids.

prop root (columnar root) any of the ADVENTITIOUS roots that arise from the lower nodes of the stem in certain plants and serve to provide additional support. Such roots are seen in maize (*Zea mays*). The woody prop roots formed by certain trees are sometimes termed *stilt roots*. See also BUTTRESS ROOT, MANGROVE.

prosenchyma 1 any tissue composed of more or less elongated cells with tapering ends. The component cells are called *stereids*. Prosenchyma is an obsolete term, sometimes used in contrast with the more or less isodiametric PARENCHYMA cells.
2 see PLECTENCHYMA.

prosthetic group a COENZYME that is tightly bound to the enzyme with which it acts. An example is the biocytin coenzyme of acetyl CoA carboxylase, an enzyme of fatty acid synthesis. The biocytin has a terminal $CH(NH_2)COOH$ group, which covalently binds to the protein as part of the peptide.

prostrate describing a plant that grows closely along the ground, such as the white stonecrop (*Sedum album*).

prot- see PROTO-.

protandry the maturation of the male reproductive organs before those of the

female. For example, in many members of the Asteraceae and Fabaceae the pollen is released from the anthers before the stigma in the same flower is receptive. Protandry is a consequence of the normal centripetal development of the floral parts and is the most frequently encountered form of DICHOGAMY. Compare PROTOGYNY.

protease (proteinase) any HYDROLASE enzyme that catalyses the hydrolysis of polypeptide chains. Specific proteases act only on certain peptide (CO–NH) linkages. For example, the bacterial protein thermolysin hydrolyses only those peptide bonds in which the amino group is donated by leucine, isoleucine, or valine. The best known plant protease is *papain*, from the latex of the papaw or papaya tree, *Carica papaya*, which is used commercially as a meat tenderizer. See also PEPTIDASE.

protein (*adj.* proteinaceous) a complex biological MACROMOLECULE consisting of one or more POLYPEPTIDE chains. Two major classes of protein are recognized: *globular proteins*, most of which are enzymes; and *fibrous proteins*, which are usually structural or contractile in function. The structure of a protein can be divided into primary, secondary, tertiary, and quaternary components. Primary structure, the amino acid sequence of the polypeptide chain, determines the three-dimensional shape of the protein. The SECONDARY STRUCTURE is the coiling or pleating of the polypeptide chain. In globular proteins the coiled polypeptide chain is further folded into a three-dimensional shape, which is maintained by weak hydrophobic and polar interactions between amino acid residues (see TERTIARY STRUCTURE). QUATERNARY STRUCTURE is the combination of more than one polypeptide chain to form an *oligomeric* protein molecule. See also CONJUGATED PROTEIN, CONFORMATION.

proteinaceous endosperm layer see ALEURONE LAYER.

proteinase see PROTEASE.

protein sequencing the determination of the primary structure of PROTEINS – the type and sequence of amino acids in the polypeptide chain. The commonest method is *Edman degradation*, which uses a series of three chemical reactions, each requiring a specific pH, that remove successive amino acids from the amino terminus of the protein. The amino acids can then be identified by ION-EXCHANGE chromatography. The process is automated in a amino acid analyser, and is capable of analysing samples of only 5–10 picomoles of protein.

protein synthesis the multistage process by which information contained in the cell's genetic material is expressed as the amino acid sequence of a protein. All proteins are constructed from about twenty amino acids, the number, type, and sequence of which is unique to a particular protein. The arrangement of amino acids is predetermined by the sequence of bases in the genetic material, DNA (or RNA in some viruses). DNA is mostly confined to the nucleus in eukaryotic cells and contains four types of bases. Sequences of three bases (triplets) correspond to particular amino acids. Hence sequences of triplets (the genetic code) correspond to sequences of amino acids in proteins. The genetic code is first copied (transcription) by the formation of a complementary molecule called MESSENGER RNA (mRNA). After various modifications in the nucleus, mRNA moves to the RIBOSOMES in the cytoplasm where protein synthesis occurs. Amino acids in the cytoplasm bind to specific transfer RNA (tRNA) molecules at one end of which are three bases (the anticodon). Which amino acid is incorporated into the protein is determined by whether the three bases on tRNA are complementary to, i.e. can bind with, the triplets on mRNA (the codon). The matching of codons with anticodons and the addition of an amino acid to an existing polypeptide chain is called TRANSLATION. See also RIBOSOMAL RNA.

proteolysis the breakdown of proteins to their constituent amino acids by proteolytic enzymes. Much information about the action of specific PROTEASES and PEPTIDASES has been gained from studies on animal digestive tracts, but little is known of the mechanisms by which proteins are broken down within the cell. Radioactive tracer studies have shown that turnover of proteins within the cell is extremely rapid, indicating that intracellular proteolysis is a very efficient process.

proteolytic enzyme an enzyme that hydrolyses proteins. See PROTEASE, PEPTIDASE.

proteome see PROTEOMICS.

proteomics the systematic characterization of all the proteins in an organism – its *proteome* – and their structure, post-translation modifications, interactions, expression, function, and regulation.

proteoplast a colourless protein-storing plastid. See LEUCOPLAST.

Proterozoic eon see PRECAMBRIAN.

prothallial cell the smaller sterile cell formed along with the antheridial cell by the first division of the microspore in certain lower vascular plants, e.g. *Selaginella*, and in gymnosperms. The prothallial cell does not usually undergo any further division and represents the only vegetative tissue of the gametophyte generation.

prothallus the free-living gametophyte of certain lower vascular plants, e.g. ferns. It is usually poorly differentiated and in many species superficially resembles a thallose liverwort. The female gametophyte of gymnosperms is sometimes termed a prothallus.

Protista (protists) a kingdom containing all the unicellular organisms. It includes the bacteria, unicellular algae and fungi, and protozoans. In some classification schemes, the bacteria are placed in a separate kingdom, the Prokaryota or Monera. In the FIVE KINGDOMS CLASSIFICATION, the bacteria are placed in the kingdom BACTERIA, and the other members of the Protista are included in a new kingdom, PROTOCTISTA, together with the multicellular algae and various plasmodial forms. The justification for the new name is that Protista traditionally excludes multicellular organisms that are close relatives of certain unicellular organisms; for example, the stoneworts, which are closely related to unicellular green algae (such as *Chlamydomonas*), would be excluded from the Protista.

proto- (prot-) prefix denoting first, giving rise to, or parent/ancestor.

protoalkaloids see ALKALOIDS.

protobiont see COACERVATE.

protocooperation see MUTUALISM.

Protoctista (protoctists) in the FIVE KINGDOMS CLASSIFICATION, a kingdom that contains all those organisms that cannot be classified as bacteria, animals, plants, or fungi. Members are eukaryotic, most have mitochondria and aerobic respiration, and most possess undulipodia at some stage of the life cycle. The Protoctista differs from the traditional kingdom PROTISTA in that it includes some multicellular forms, particularly the large filamentous and parenchymatous algae. As well as the algae, it includes the amoebae and other protozoans, the slime moulds, and the slime nets. While protoctists show a wide range of nutrition, some being photoautotrophs, some heterotrophs, phagotrophs, or mixotrophs, they are less diverse biochemically than the bacteria.

The number of phyla recognized ranges from 27 to around 50. The taxa described in detail in this dictionary are those containing organisms that at some time or another have been classed with the algae or fungi: DINOMASTIGOTA (dinoflagellates), CHRYSOMONADA (golden algae), HAPTOMONADA, EUGLENIDA, CRYPTOMONADA, XANTHOPHYTA (yellow-green algae), Eustigmatophyta (eustigs), DIATOMS, PHAEOPHYTA (brown algae), RHODOPHYTA (red algae), GAMOPHYTA (conjugating green algae, including desmids), CHLOROPHYTA (green algae), Labyrinthulata (slime nets, thraustochytrids), ACRASIOMYCOTA (cellular slime moulds), MYXOMYCOTA (plasmodial slime moulds), PLASMODIOPHORA, HYPHOCHYTRIOMYCOTA, CHYTRIDIOMYCOTA (fungus-like organisms), and OOMYCOTA (oomycetes).

protoderm the outermost layer of the APICAL MERISTEM, the derivatives of which give rise to the EPIDERMIS and sometimes also to associated subepidermal tissues. Compare PROCAMBIUM, GROUND MERISTEM.

protogyny the maturation of the female reproductive organs before those of the male. For example, in some members of the Rosaceae and Brassicaceae the stigma becomes receptive before the anthers in the same flower release their pollen. Protogyny is contrary to the normal centripetal development of the floral parts and is not so frequently encountered as PROTANDRY. See DICHOGAMY.

protohemicryptophyte see HEMICRYPTOPHYTE.

protonema 1 the juvenile form of a moss or liverwort that develops on germination of a spore. In mosses and foliose liverworts it is usually a branched green filament, resembling a filamentous green alga, but in *Sphagnum* it becomes thallose and in thallose liverworts (see HEPATOPHYTA) it is not clearly differentiated. The familiar adult form develops from buds on the protonema. **2** the erect green filament that develops on germination of the zygote in algae of the CHARALES (stoneworts).

proton pump see CHEMIOSMOTIC THEORY.

protophloem early PRIMARY PHLOEM. The SIEVE ELEMENTS of the protophloem, which often lack companion cells, are functional for a brief period only and are usually later obliterated, being unable to keep pace with the elongation of the surrounding cells. These obliterated cells often differentiate into fibres.

protoplasm the substance of the PROTOPLAST of cells. It is composed of about 90% water in highly active cells although the figure may be as low as 10% in dormant cells. Proteins and amino acids account for approximately 65% of the dry weight, lipid material about 16%, and simple sugars about 12%. There are also small quantities of a wide variety of other organic compounds and mineral salts. Physically protoplasm is a colloidal system of at least three phases; an aqueous solution of organic and inorganic compounds, a disperse phase consisting of oil droplets forming an emulsion, and a framework of protein molecules forming fine fibrils and tubules. Proteins are capable of changing form and so altering viscosity. In addition protein and phosphoglyceride molecules are organized into CELL MEMBRANES, which are the basis of the structure of all ORGANELLES. The substance of the nucleus is termed *nucleoplasm*, while the remainder of the protoplasm is called CYTOPLASM.

protoplast the living part of a cell. In a plant cell this includes the CYTOPLASM, NUCLEUS,

CELL MEMBRANES, and ORGANELLES in their highly organized condition, but it does not include the cell wall or vacuole.

protoplast fusion the induction of the fusion of naked (wall-less) plant cells under culture conditions. Such cells or protoplasts can be produced by enzymatic digestion of cell walls, osmotic shock, etc. The technique has potential use for hybridizing unrelated or incompatible species.

protostele see STELE.

Protosteliomycetes (Protostelida, Protostelia, protostelid slime moulds) a class of plasmodial slime moulds (see MYXOMYCOTA) in which the trophic phase comprises amoeboid cells with slender unbranched pseudopodia. The sporocarps are on slender stalks, and the spores germinate to produce eight haploid motile spores. The group comprises some 29 species in 14 genera, found mainly on dead and decaying plants, rotting wood, and dung.

protoxylem early PRIMARY XYLEM. The protoxylem matures before the organ completes its longitudinal growth and is thus often distorted or destroyed as the surrounding tissues elongate. In higher plants, the protoxylem of the stem occurs at the innermost edge of the vascular bundles whereas in the root it occurs external to the METAXYLEM as protoxylem poles. The protoxylem is composed mostly of parenchyma cells with a relatively small number of tracheary cells. The TRACHEARY ELEMENTS of the protoxylem usually possess spiral, annular, or sometimes reticulate thickening and nonpitted secondary cell walls. Compare METAXYLEM, PROTOPHLOEM.

protoxylem lacuna see CARINAL CANAL.

provascular tissue see PROCAMBIUM.

provenance the geographical source and/or place of origin of a plant, variety, batch of pollen, seeds or other propagules.

provirus a viral genome that has been incorporated into a host genome.

proximal denoting the region of an organ that is nearest to its point of attachment.

P/R ratio see PRODUCTION/RESPIRATION RATIO.

pruinose having a whitish bloom, e.g. the fruits of the Oregon grape (*Mahonia aquifolium*).

pruning the cutting back of some or all of the branches of a woody plant. Pruning may be necessary for a number of reasons, e.g. the removal of dead or diseased wood or to train the plant into a special shape. Usually however pruning is performed to promote the vigour of a plant and, in the case of fruit trees, to maintain a balance between vegetative growth and fruit production. The time of pruning depends on whether the plant flowers early in the season on the previous season's wood, in which case pruning is carried out after flowering, or in the summer on the current season's wood, in which case pruning should be done in early spring to encourage new growth. Examples of the first group of plants are *Prunus* species and winter jasmine (*Jasminum nudiflorum*) and of the second group butterfly bush (*Buddleia davidii*) and lemon-scented verbena (*Lippia citriodora*). See also COPPICING, POLLARDING, STOOLING.

psammo- prefix denoting relating to sand.

psammon the organisms that live between the sand particles on a lake or sea shore. Diatoms and other algae provide food for the bacteria, protozoans, and other heterotrophic organisms of the psammon.

psammophile an organism that lives in sand.

psammosere see SAND DUNES.

pseudo- prefix denoting false.

pseudoalkaloids see ALKALOIDS.

pseudoalleles two or more mutations having similar effects on the phenotype but occurring in different parts of the same gene. Being in the same cistron they will be closely linked and recovery of recombinants from heterozygotes will be rare. See also CIS-TRANS TEST.

pseudocarp 1 (false fruit) a fruit that incorporates tissues other than those derived from the gynoecium. It may be derived from a single flower and include the receptacle or bracts, or may be derived from a complete inflorescence (see MULTIPLE FRUIT). See also POME, HIP.

2 a particular type of pseudocarp consisting of a number of achenes embedded in the outer surface of a fleshy

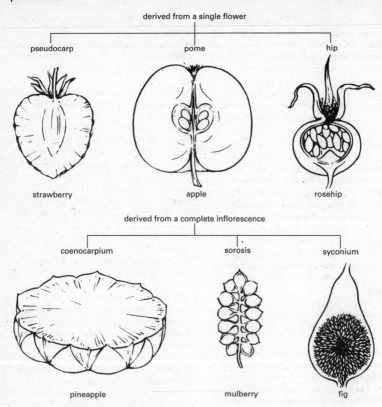

derived from a single flower

pseudocarp   pome   hip

strawberry   apple   rosehip

derived from a complete inflorescence

coenocarpium   sorosis   syconium

pineapple   mulberry   fig

Pseudocarp: various forms of pseudocarp

receptacle, e.g. strawberry (*Fragaria*) fruits. See diagram.

pseudocilium see PSEUDOFLAGELLUM.

pseudocopulation the attempt by a male insect to mate with a flower, or parts of a flower, which it mistakes, because of the colouring, shape, or scent of the flower, for a female of the same species. During pseudocopulation pollination of the flower is achieved. This pollination mechanism can be observed frequently in the orchid family, e.g. in *Ophrys*. See also ENTOMOPHILY.

pseudocyphella an irregularly shaped breathing pore on a bryophyte thallus.

pseudoendosperm 1 the development of an endosperm without fusion of the POLAR

NUCLEI or DEFINITIVE NUCLEUS with a male gamete. It may arise spontaneously from the definitive nucleus or from somatic tissue and is therefore usually diploid.

2 in gymnosperms, the tissues of the female gametophyte, which nourish the embryo. It is not homologous with the endosperm of angiosperms, being haploid rather than triploid.

pseudoflagellum (pseudocilium) in some green algae (Chlorophyta) of the order TETRASPORALES, a rigid protrusion from the cell that resembles an undulipodium but does not appear to be involved in motility.

pseudogamy a form of APOMIXIS in which a diploid embryo forms without fertilization though a stimulus from the male gamete is required and thus pollination is necessary. See also PARTHENOGENESIS.

pseudogene a gene that has become 'switched off' during the course of evolution. As it has no active function, it is not under any selection pressure and therefore evolves at a constant rate. Pseudogenes in related organisms can be compared to provide a standard against which the rate of evolution of other genes can be measured, to give an indication of the selection pressure on those genes.

pseudomonads a group of EUBACTERIA placed in the phylum Proteobacteria in the Five Kingdoms classification. They are Gram-negative (see GRAM STAIN) rod-shaped bacteria with polar flagella. They do not form spores. They metabolize a wide range of organic substrates. Many are facultative autotrophs. Pseudomonads carry out aerobic respiration, although some are capable of anaerobic respiration using nitrate as the terminal electron acceptor (see DENITRIFICATION). Many pseudomonads are pathogenic to plants, e.g. *Pseudomonas tabaci* and *P. angulata*, which cause wildfire and angular leaf spot of tobacco, and the xanthomonads (genus *Xanthomonas*), which are responsible for various blights and cankers.

pseudoparaphyses sterile, thread-like filaments found in the perithecia and pseudothecia of certain groups of fungi, They arise above the level of the asci, and grow down between them, eventually becoming attached to the floor of the cavity.

pseudoparenchyma see PLECTENCHYMA.

pseudoperianth a relatively late-developing membranous sheath surrounding the young sporophyte in certain liverworts, e.g. *Fossombronia*. It is sometimes termed a perigynium (see PERICHAETIUM).

pseudoplasmodium a small sluglike aggregation of myxamoebae that acts as a unit but within which each MYXAMOEBA remains distinct, with its own intact cell membrane. The fruiting structure is a *sorocarp*, which is sometimes borne on a slender cellulose stalk, the *sorophore*. This type of organization is found in the ACRASIOMYCOTA.

pseudopodetium the stalked part of a lichen

thallus, bearing APOTHECIA, SOREDIA, or PHYLLOCLADIA.

pseudopodium a leafless stalk that serves to raise the capsule in those mosses (e.g. *Sphagnum*) that lack a seta.

pseudoraphe see PENNALES.

pseudospore see ACRASIOMYCOTA.

pseudostem a false stem formed from swollen leaf bases, e.g. banana (*Musa* spp.).

pseudosteppe steppe-like GRASSLAND that occurs outside Europe and Asia, for example the temperate grassland of south-western Australia and grasslands along the southern edge of the Sahara desert.

pseudothecium the type of ASCOMA characteristic of the LOCULOASCOMYCETAE. Pseudothecia are superficially similar to PERITHECIA but contain two-walled asci and also differ in the way they develop. They are often loosely called perithecia.

pseudowhorl an arrangement of leaves that appear to arise at the same level, but are actually in an extremely tight spiral.

Psilophyta (whisk ferns) a phylum of VASCULAR PLANTS containing two living genera, *Psilotum* and *Tmesipteris*, that together constitute the order Psilotales. In older classifications these plants are classified as a subdivision or class, Psilopsida, of the PTERIDOPHYTA. The fossil forms, which are known chiefly from rocks of the Devonian period, are placed in the order Psilophytales. These fossils include *Psilophyton* and, from the RHYNIE CHERT, *Rhynia*. The living psilopsids are considered the most primitive of the vascular plants, but their relationships are uncertain and they may have more affinity with ferns. The sporophyte is dichotomously branching and bears RHIZOIDS rather than true roots. Small leaflike appendages are often present and the vascular tissue consists of tracheids and poorly defined phloem. The spores, which are homosporous, give rise to a subterranean gametophyte. Fertilization of the gametes relies on the presence of a film of water and embryogeny is EXOSCOPIC. In some classifications the living and fossil forms are separated into two subdivisions or classes, the Psilotopsida and Psilophytopsida respectively.

Psilophytales see PSILOPHYTA.

Psilopsida see PSILOPHYTA.

psychrophilic describing microorganisms that require temperatures below 20°C for optimal growth, e.g. the fungus *Cladosporium herbarum*. Compare MESOPHILIC, THERMOPHILIC.

psychrotroph describing microorganisms that are able to grow at low temperatures (such as below −15°C), but which grows better at higher temperatures. Compare PSYCHROPHILIC.

pter- prefix denoting winged.

Pteridophyta in older classifications that considered the possession of vascular tissue of equivalent significance to the seed-bearing habit, a division containing all the nonseed-bearing VASCULAR PLANTS (compare TRACHEOPHYTA). Its members show a heteromorphic alternation of generations and the gametophyte is often nutritionally independent of the sporophyte. The sporophyte differs from that of seed-bearing plants (see SPERMATOPHYTA) in lacking vessels in the xylem. Some species exhibit homospory in contrast to spermatophytes, which are always heterosporous. The Pteridophyta contained the subdivisions or classes Psilopsida, Lycopsida, Sphenopsida, and Pteropsida. In the Five Kingdoms classification, members of the Pteridophyta are assigned to the phyla PSILOPHYTA, LYCOPHYTA, SPHENOPHYTA, and FILICINOPHYTA.

Pteridospermales see CYCADOFILICALES.

pterochory the dispersal of winged seeds by the wind.

Pterophyta see FILICINOPHYTA.

Pteropsida a subdivision or class of VASCULAR PLANTS. In some classifications it includes only the ferns (see FILICINOPHYTA), while in others it also includes the gymnosperms and angiosperms. The Pteropsida differ from the Psilopsida, Sphenopsida, and Lycopsida in having large leaves often with a highly branched vascular system and leaf traces that leave a gap where they depart from the stele.

pteroylglutamic acid see FOLIC ACID.

ptyxis the pattern of folding of individual leaves (see illustration), as contrasted with VERNATION, which is the pattern of folding of leaves in a bud. The descriptive terms used for these different patterns may be combined. For example, a leaf may be conduplicate/plicate, in which case it is both folded towards the adaxial surface (conduplicate) and pleated like a concertina (plicate).

pubescent having fine short hairs. Compare PILOSE.

puff balls see LYCOPERDALES.

pulse the edible seeds of a leguminous plant (see FABACEAE).

pulse labelling the use of radioisotopes to measure rates of assimilation and biosynthesis in living cells, in which a suspension of cells or organelles is exposed to a small quantity of a radioisotope for a brief 'pulse' of time (a few seconds or minutes), after which the suspension is flooded with large amounts of the stable nonradioactive isostope, which competes with and thus blocks further uptake of the radioisotope. Over time, measurements are made of the level of radioisotope in samples from the suspension to follow its rate of uptake and the rate and/or pattern of its metabolism (i.e. of its incorporation into metabolites). For example, a suspension of photosynthesizing algal cells may be exposed to a pulse of $^{14}CO_2$ to follow the process of photosynthesis. See also ISOTOPIC TRACER.

pulverulent being or having the appearance of being covered in fine powder.

pulvinate cushion-like, convexly curved, or swollen.

pulvinus 1 a prominent swelling at the base of a petiole or pinna. It consists of cells (motor cells) that can rapidly move water into or out of their vacuoles. Such changes in turgor alter the position of the leaf or leaflet and are responsible for certain sleep movements (see NYCTINASTY) and haptotropic movements (see HAPTOTROPISM). Pulvini are found in many members of the Fabaceae.
2 a thickened region in the stem or leaf sheath of grasses, often containing an INTERCALARY MERISTEM.

punas ALPINE vegetation of the high, dry plateaux on the western side of the Andes. The vegetation is mainly tough, tufted grasses, especially bunchgrass, dwarf shrubs, and some hardy cushion plants. Compare PAMIRS.

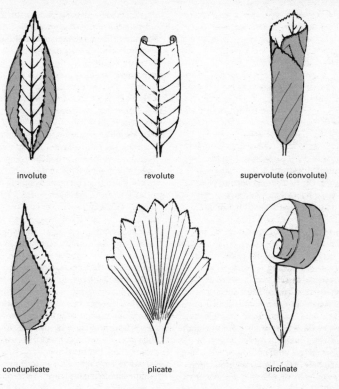

involute   revolute   supervolute (convolute)

conduplicate   plicate   circinate

Ptyxis: types of ptyxis

punctate describing a surface that has many pores or small, point-shaped depressions.
punctiform dot-like.
punctuated equilibrium an evolutionary theory proposing that species often arise rapidly, in terms of geological time, rather than by gradual change. It envisages that, following a short period (less than 100 000 years) of rapid speciation, there is a considerable period of stasis lasting several million years.
Punnet square a chequerboard diagram, attributed to the geneticist R. C. Punnet, used to illustrate how the gametes involved in a particular cross result in various genotypes in various frequencies (see diagram overleaf).
pure breeding see BREEDING TRUE.
pure culture see AXENIC CULTURE.

pure line (pure strain) a succession of generations recognized by their ability to produce genotypically identical offspring when selfed or crossed between themselves. The members of a pure line are said to breed true. By inference, such individuals are deemed to be HOMOZYGOUS.
purine a nitrogen-containing organic base with a double-ring structure (see diagram overleaf) synthesized mainly from amino acids. Two purines, ADENINE and GUANINE, are common constituents of nucleotides.
purple membrane see BACTERIORHODOPSIN.
purple non-sulphur bacteria see PHOTOSYNTHETIC BACTERIA.
purple sulphur bacteria see SULPHUR BACTERIA, PHOTOSYNTHETIC BACTERIA.
pustule a blister-like mass of spores developing below the epidermis of an

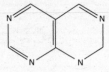

Purine

infected plant. It usually breaks through the surface at maturity.

pusule see DINOMASTIGOTA.

puszta a type of arid STEPPE grassland found in Hungary. The vegetation resembles the North American prairie.

putrefaction the anaerobic decomposition of protein-rich material by bacteria, with the emission of foul-smelling gases.

pycnidiospore see PYCNIDIUM.

pycnidium 1 (pycnium, spermogonium) a flasklike structure CONIDIOMA with a round or slit-like OSTIOLE, produced by certain fungi (e.g. *Puccinia* and *Leptosphaeria*) in which asexual *pycniospores* (pycnospores, spermatia) are formed. The inner surface is lined partially or entirely with cells that give rise to conidia. **2** a spore-producing body formed by the fungal component of certain lichens, e.g. *Parmelia physodes*. It is a flask-shaped structure sunken in the thallus and opens to the surface by a pore. Inside the pycnidium are a number of conidiospores,

which produce *pycnoconidia*. The functions of the pycnoconidia are uncertain, though it appears that in some species they may be, by behaving as male gametes, and initiate ascoma formation.

pycniospore see PYCNIDIUM.

pycnocline a zone of rapid change in density with depth in a body of water.

pycnoconidium see PYCNIDIUM.

pycnospore see PYCNIDIUM.

pyramid of biomass a diagram shaped like a pyramid that shows the amount of living material, measured by total dry weight, at each feeding stage (trophic level) of a FOOD CHAIN. The BIOMASS depends on the amount of carbon that can be fixed by green plants. See also FIXATION, PYRAMID OF NUMBERS.

pyramid of energy a diagram showing the energy contained in the organisms at each feeding stage (trophic level) of a food chain. It shows that the energy decreases from the lower to the higher trophic levels, energy being lost during transfer from one level to the other through incomplete digestion and assimilation, defecation and urination, respiration, heat loss, energy of movement, and so on. See also PYRAMID OF BIOMASS, PYRAMID OF NUMBERS.

pyramid of numbers a diagram showing the numbers of living organisms present at

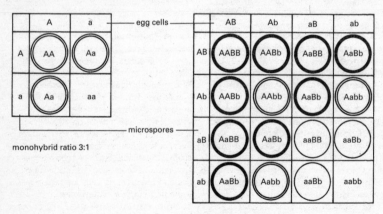

monohybrid ratio 3:1

dihybrid ratio 9:3:3:1

Punnet square: showing the gametes produced by the F_1 and the resulting zygotes of the F_2 in a monohybrid and dihybrid cross

each feeding stage (trophic level) of a FOOD CHAIN. The green plants (producers) are usually more numerous and form the base of the pyramid. The successive consumer levels usually decrease in numbers at each stage and form the tiers of the pyramid. There is typically one or a few predators at the top. Normally there are three or four tiers in the pyramid and rarely more than six. If the producers are large (e.g. trees) the pyramid will be partially inverted, as the base is smaller than one or more of the successive tiers. See also PYRAMID OF BIOMASS, PYRAMID OF ENERGY.

pyranose ring a six-membered ring containing five carbon atoms and an oxygen atom. Pyranose rings are formed by any ALDOSE sugars with five or more carbon atoms. In solution the aldehyde group at the first carbon atom reacts with the hydroxyl group on the fifth carbon atom. This reaction renders the first carbon atom asymmetric. The pyranose forms of aldoses can thus exist in two isomeric forms. For example, D-glucose in solution forms α-D-glucopyranose and β-D-glucopyranose, which differ in a number of properties, e.g. optical activity, solubility in water, and melting point. Compare FURANOSE RING.

pyrenocarp see PERITHECIUM.

pyrenoid a darkly staining proteinaceous body in the chloroplasts of many algae and of the ANTHOCEROPHYTA. In most cases it has a dense granular matrix and is surrounded by tightly packed starch plates. In euglenid algae (see EUGLENIDA) the core is traversed by lamellae similar to those in the chloroplast and the pyrenoid is surrounded by PARAMYLUM granules.

pyrenolichens see PYRENULALES.

Pyrenomycetes a class of the ASCOMYCOTA containing those ascomycetes that produce a PERITHECIUM. With over 6000 species and some 640 genera it is the largest group of ascomycetes and includes the orders HYPOCREALES, XYLARIALES, Sordariales, Coryneliales, and PYRENULALES. The class is now not widely recognized.

Pyrenulales (pyrenolichens) an order of the Ascomycota containing LICHENS in which the fungal partner produces PERITHECIA (pyrenocarps). These are buried in the thallus and only visible as dots on the surface. There are over 1666 species in the Pyrenulales, in about 47 genera including *Dermatocarpon*, *Pyrenula*, and *Verrucaria*. Lichens with fungi that form bitunicate rather than unitunicate perithecia may be placed in a separate order, the Dothideales.

pyridine alkaloids a group of ALKALOIDS, the structures of which are based on the *pyridine* nucleus (a six-membered ring containing five carbon atoms and a nitrogen atom, and with three double bonds). They include the hemlock alkaloids, e.g. coniine, which is present in *Conium maculatum* (hemlock).

pyridoxine (vitamin B_6) a pyridine derivative, essential in its activated forms, pyridoxal phosphate and pyridoxanine, as a coenzyme in various TRANSAMINATION reactions. See VITAMIN.

pyriform shaped like a pear or a tear drop.

pyrimidine a nitrogen-containing organic base composed of a single heterocyclic ring (see diagram), synthesized mainly from amino acids. Two pyrimidines, CYTOSINE and THYMINE, are constituents of DNA, while URACIL replaces thymine in RNA. Thiamine (vitamin B_1) is also partly constructed from a pyrimidine.

pyroclimax see CLIMAX.

Pyrimidine

pyrophosphate a salt or ester of pyrophosphoric acid ($H_4P_2O_7$).

pyrophyte a plant that is capable of tolerating fire, that needs fire to stimulate flowering, recycling of nutrients, and/or removal of dead or senescent vegetation, or that gains a competitive advantage from fire. An example is laurel sumac (*Rhus laurina*), which has a root crown from which branches erupt after fire; it seeds germinate only after being exposed to the heat from a fire – they benefit from extra nutrients, space, and light after neighbouring vegetation has been burned down. Many woody pyrophytes have exceptionally thick bark.

pyrrole an organic nitrogen-containing ring compound that forms the basis of the structure of PORPHYRINS. The ring contains four carbon atoms and one nitrogen atom (see diagram). Pyrrole rings are part of chlorophyll and haemoglobin molecules.

Pyrrophyta see DINOMASTIGOTA.

pyruvic acid (2-oxopropanoic acid) A three-carbon carboxylic keto acid with the formula $CH_3COCOOH$. In glucose metabolism pyruvate is an intermediate in GLYCOLYSIS. Under anaerobic conditions it is reduced to form ethanol or some other fermentation product, while in aerobic conditions it is decarboxylated to form acetyl CoA. In C_4 plants pyruvate is an intermediate of the HATCH–SLACK PATHWAY. In amino acid metabolism pyruvate is the degradation product of several amino acids, including alanine and cysteine, while other amino acids are

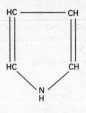

Pyrrole

synthesized from pyruvate. Pyruvate can also be carboxylated to form oxaloacetate or malate, both intermediates in the KREBS CYCLE. These reactions, known as *anaplerotic* reactions, replenish Krebs cycle intermediates.

pyxidium a CAPSULE that dehisces by means of a circular split, so that the top of the capsule comes off like a lid, as in henbane (*Hyoscyanus niger*).

Qq

Q_{10} effect see Q_{10} RATIO.

Q enzyme see STARCH.

QO_2 see OXYGEN QUOTIENT.

Q_{10} ratio (Q_{10} effect) a coefficient showing the multiple by which the rate of a chemical reaction increases with every 10°C rise in temperature. Thus if the reaction rate doubles for every 10°C rise in temperature, the Q_{10} = 2. For most enzyme-mediated reactions the Q_{10} is between 2 and 3.

Q technique a method of organizing data for analysis in which the observations (N) form the vertical columns and the variables (attributes, n) form the rows in a table. Compare R TECHNIQUE. See also INVERSE ANALYSIS, NORMAL ANALYSIS.

quadrat a small area, often 0.25 or 1.0 m² in open areas, though larger in woodland, marked out with a frame that can be used in an ecological survey to study in detail the distribution and abundance of different species. It can be used for random sampling within an area or for systematic sampling at regular intervals. A *permanent quadrat* can be used to record changes in the distribution and numbers of species over a number of years. A quadrat may be used in conjunction with a metal 'comb' with teeth at regular intervals that allows detailed sampling at equally spaced points. A hundred such points would estimate the percentage of a particular species present. See also TRANSECT.

quadrivalent the structure formed by the pairing of four HOMOLOGOUS CHROMOSOMES in a TETRAPLOID during prophase and metaphase of the first division of meiosis. Compare BIVALENT.

qualitative variation see DISCONTINUOUS VARIATION.

quantasome see CHLOROPLAST.

quantitative genetics a branch of genetics dealing with the inheritance of quantitative traits. See CONTINUOUS VARIATION.

quantitative variation see CONTINUOUS VARIATION.

quantum evolution (quantum speciation) rapid speciation occurring in small populations isolated from their ancestral population, in which the FOUNDER EFFECT and GENETIC DRIFT have a significant effect. It occurs if the population is in a new niche, especially if this niche is unoccupied. See also ADAPTIVE RADIATION, PUNCTUATED EQUILIBRIUM.

quantum speciation see QUANTUM EVOLUTION.

Quaternary the second period of the CENOZOIC era from two million years ago to the present day. It is divided into the Pleistocene and Holocene (Recent) epochs. At its commencement, the climate was cool and temperate and there were several glacial periods, the evidence for which is borne out by the distribution of fossils. Many plants and animals became extinct. In the Holocene epoch, beginning about 10 000 years ago, the rise of civilization saw the origin of crop plants and their associated weeds. See GEOLOGICAL TIME SCALE.

quaternary structure the fourth level of structure in a PROTEIN molecule: the linking together of two or more polypeptide chains to form a protein. This involves hydrogen bonds, ionic bonds, and hydrophobic interactions. See also

CONFORMATION, PRIMARY STRUCTURE, SECONDARY STRUCTURE, TERTIARY STRUCTURE.

quercetin a yellow crystalline flavonoid pigment ($C_{15}H_{10}O_7$) that usually occurs as a glycoside in the bark and 'skin' of many plants, and it is believed to underlie the medicinal properties of certain herbs. It is best known for anti-inflammatory and anti-allergy properties, and is also an antioxidant and PHYTOESTROGEN. Many other flavonoids are derived from quercetin. The red colour of the skin of apples and red onions is due to quercetin. It is usually produced commercially from cyanobacteria. Qeurcetin derived from the bark of the black oak (*Quercus velutina*) is used as the basis of many yellow, orange, and red dyes.

quiescence see DORMANCY.

quiescent centre the region in the APICAL MERISTEM of a root where little or no cell division occurs. The cells in the quiescent centre are capable of assuming meristematic activity if the INITIALS in the meristem are damaged, and thus act as a reservoir of potential initials, protected from damage by their relative inactivity. The quiescent centre may also be a site of AUXIN synthesis.

quillworts see ISOETALES.

quincuncial aestivation see AESTIVATION.

quinone (cyclohexadiene-1,4-dione, benzoquinone) an organic compound ($C_6H_4O_2$) containing two carbonyl groups in a 6-carbon unsaturaed ring structure. Quinones are found in bacteria, higher plants, and some fungi. Some are biological pigments, others act as electron carriers in mitochondria and chloroplasts.

quinone fungicide any fungicide developed from chlorinated quinones, such as chloranil, which is used as a seed dressing, and dichlone, which is used as a foliar fungicide.

Rr

race in taxonomic classification, a category below the rank of species, of uncertain position but occasionally used in floras in place of, or subordinate to, form. The term is also used in lieu of ecotype, implying a category between subspecies and variety covering geographical groupings of plants. Races are often uniform in respect of ecological preference, physiological requirements, and topographical distribution. See also PHYSIOLOGICAL RACE.

raceme see INFLORESCENCE.

racemose see INFLORESCENCE.

rachilla (rhachilla) 1 the secondary axis of a pinnately compound leaf, such as that of a fern.

2 the main axis of a sedge or grass SPIKELET.

Compare RACHIS.

rachis (rhachis) 1 the main axis of a compound leaf possessing pinnae, such as a fern frond.

2 the main axis of an inflorescence.

Compare RACHILLA.

rad (radiation absorbed dose) a measure of the amount of ionizing radiation absorbed by a tissue: 1 rad = 100 ergs of energy absorbed per gram of tissue.

radial 1 describing a longitudinal section that passes through the centre of a cylindrical organ (i.e. along a radius or diameter). In a radial section down a woody stem the growth rings appear parallel and flecks of rays may be seen. Compare TANGENTIAL.

2 see ANTICLINAL.

radial symmetry the arrangement of parts in an organ or organism such that any cut taken through the centre divides the structure into similar halves. Most stems and roots exhibit such symmetry. Radial symmetry in flowers is usually termed *actinomorphy*, and such flowers are described as *regular*. The flowers of relatively primitive angiosperm families, e.g. Ranunculaceae, are usually actinomorphic. In a floral formula, actinomorphy is represented by the symbol ⊕.

radiation absorbed dose see RAD.

radiation ecology the study of the pathways by which radioactive substances are dispersed through an ecosytem and their effects on the living organisms. It is concerned mainly with substances released as a result of human activity.

radical describing leaves arising close together at the base of the stem, as in a rosette plant. Compare CAULINE.

radicle the embryonic root, which in the seed is directed towards the MICROPYLE. It is normally the first organ to emerge from the testa on germination. The radicle may persist to form a TAPROOT. Alternatively it may be replaced either by lateral or ADVENTITIOUS roots.

radioactive dating see RADIOCARBON DATING, RADIOMETRIC DATING.

radioactive isotope see ISOTOPE.

radioactive tracer see ISOTOPIC TRACER.

radiocarbon the radioactive isotope of carbon, carbon-14, which is used in radiocarbon dating and as an ISOTOPIC TRACER.

radiocarbon dating a method used to determine the age of organic materials up to about 40 000–70 000 years old. It relies on the fact that the ^{14}C isotope of carbon is unstable and decays, emitting beta rays, to

^{14}N, with a half-life of about 5700 years. Plants incorporate ^{14}CO$_2$ into their tissues during photosynthesis but when they die the concentration of ^{14}C starts to fall at a rate related to the half-life. By comparing a specimen of unknown age with a sample of zero age, the age of the specimen may be calculated by measuring the amount of ^{14}C using a mass spectrometer.

The method assumes that the ^{14}C:^{12}C ratio in the atmosphere has always remained constant. Certain discrepancies between age determinations based on radiocarbon dating and DENDROCHRONOLOGY show there have been systematic variations in this ratio.

radioisotope see RADIOMETRIC DATING.

radiometric dating the determination of the age of rocks and minerals, and hence of the fossils they contain, by measurement of the levels of certain radioactive elements (*radioisotopes*). Two common methods are *potassium–argon* (K–Ar) dating and *rubidium–strontium* (Rb–Sr) dating. The technique employs the fact that radioactive elements decay to other stable elements at a constant rate. Hence by measuring the ratio of the stable daughter element to the radioactive parent element the age of the rock may be determined. The decay of potassium-40 to argon-40 has a half-life of 11.8×10^9 years, while that of rubidium-87 to strontium-87 is 48.8×10^9 years. Both techniques could theoretically be used on even the oldest of the earth's rocks since these are only some 4.6×10^9 years old. Such techniques have enabled geologists to construct an absolute geological time scale.

radiospermic describing the radially symmetrical seeds of the Cycadofilicales, Cycadales, Bennettitales, Caytoniales, and Anthophyta. Compare PLATYSPERMIC.

raffia (bast) a fibrous tissue derived from the young leaves of the palm *Raphia vinifera*, which is used for tying up plants, and as 'straw' for hats, mats, and other items.

raffinose a common plant trisaccharide found particularly in cotton seed and sugar beet. It is a tasteless nonreducing sugar composed of galactose, glucose, and fructose units. Raffinose is thought to be involved in resistance to environmental stress, by protecting membrane-bound proteins, for example during the drying out of seeds at maturation.

rainforest see FOREST.

rain-shadow area an area that lies in the lee of mountains or other highland, and which therefore receive much less rainfall than the leeward side of the high land. As clouds rise over mountains, they cool and shed much of their moisture as rain. The rain-shadow effect may occur on a regional scale, e.g. the eastern side of the Rocky Mountains in North America, or on a very local scale.

raised bog see BOG.

ramentum (*pl.* ramenta) Any of the small brown scales that cover the young fronds of a fern. The ramenta are shed as the frond unfurls but some persist on the RACHIS.

ramet any individual belonging to a known CLONE. Studies of phenotypic PLASTICITY in relation to environmental factors are more quantifiable if ramets are used rather than genetically mixed individuals.

ramiflorous born on branches.

Ramsar Convention see CONVENTION ON WETLANDS OF INTERNATIONAL IMPORTANCE.

random amplified polymorphic DNA (RAPD) the production of mulitple copies of segments of DNA by means of the polymerase chain reaction, in which the segments multiplied lie beween two identical binding sites for a particular primer, usually about 10 base pairs long. There may be several different segments of DNA so defined; after amplification these are separated and identified by electrophoresis. The numbers and locations of these sites vary from individual to individuals, providing a kind of genetic fingerprint. See GENETIC FINGERPRINTING, MOLECULAR SYSTEMATICS.

random assortment see INDEPENDENT ASSORTMENT.

randomization the allocation of experimental units to completely random positions in an experiment, normally by using tables of random numbers. This form of experimental design is used when the units are very similar and they are being observed under particularly controlled conditions as, for example, a pure line of plants in a growth chamber. It has the advantage over RANDOMIZED BLOCK and LATIN SQUARE designs in that it increases the number of degrees of freedom, which consequently reduces the error variance. If, however, the experimental units or the environment are variable then a completely random design is less likely to pick up significant differences between treatments.

randomized block a type of experimental design in which each block contains one representative of each treatment, the treatments being allocated to random positions in the block. Each block is thus a complete replication but allocation of treatments to columns is random. The columns are therefore not complete replications since they may contain two or more representatives of one treatment and none of another. Randomized blocks may be used, for example, when dealing with identifiable inherent variation. Compare LATIN SQUARE.

random sample a sample in which the selection and location of each individual organism or quadrat sampled is independent of all other individuals and also independent of any physical features of the environment. Thus, each individual of the population has an equal chance of being included in the sample. For vegetation sampling the usual practice is to place a grid over the sample area, and use random number tables to select coordinates for points at which to place quadrats. Where the community is markedly heterogeneous, a *stratified random sample (partial random sample)* method may be used, in which the grid is further subdivided before taking random samples within each segment or 'stratum' of the grid. The strata are selected so as to minimize variability within a stratum and maximize variability between strata.

rank the level or position occupied by a category in the taxonomic hierarchy, e.g. class, family, genus, etc. (see HIERARCHICAL CLASSIFICATION). Plant taxa belonging to ranks between and including the levels subkingdom and subtribe are usually indicated by characteristic suffixes. Hence, taxa of the rank subkingdom end in -bionta or -ata, phylum (or division) in -phyta, subdivision in -phytina, class in -opsida or -ae, subclass in -idae, order in -ales, suborder

in -ineae, family in -aceae, subfamily in -oideae, tribe in -eae, and subtribe in -inae.

Ranunculaceae a large family of dicotyledonous plants commonly known as the buttercup family. It is cosmopolitan in distribution though best represented in temperate and cold latitudes. There are some 62 genera and over 2450 species, most of which are herbaceous. The leaves are often divided or lobed and usually arise either from the base of the stem or in a spiral arrangement. The flowers may be solitary, e.g. *Anemone*, or borne in a raceme or cyme. The flowers are usually actinomorphic with the parts arranged spirally or, less commonly, in whorls. There are numerous stamens and normally five prominent petals. Some genera, e.g. *Consolida* (larkspurs), *Aconitum* (monkshoods, wolfbanes), and *Aquilegia* (columbines), have zygomorphic flowers. The fruits are usually follicles, achenes, or berries, rarely capsules (e.g. *Nigella*). The seeds have a tiny embryo and copious endosperm. The family contains many ornamentals, e.g. *Clematis*, *Caltha*, and *Nigella*, but there are no important food crops. See also MAGNOLIIDAE.

Ranunculidae see MAGNOLIIDAE.

RAPD see RANDOM AMPLIFIED POLYMORPHIC DNA.

raphe 1 a longitudinal ridge on the outer integument or seed coat in ANATROPOUS ovules where the funiculus becomes fused with the integument.
2 the longitudinal fissure in the cell wall of motile bilaterally symmetrical diatoms. It is believed to play a part in the movement of such organisms by allowing contact between the cytoplasm and external medium. See PENNALES.

raphide a needle-shaped crystal, often of calcium oxalate, found in clusters in plant cells. The presence and form of raphide clusters are sometimes useful as diagnostic characters. For example, the subfamily Rubioideae of the Rubiaceae is distinguished from the other three subfamilies, Cinchonoideae, Ixoroideae, and Guettardoideae, by the presence of raphides in the leaves.

Raphidophyceae (raphidomonads, chloromonads) a class of the

CHRYSOMONADA (or, in some classification schemes, the CHROMOPHYTA) containing unicellular algae with two undulipodia at the anterior end, one of which is trailing. Their disc-shaped plastids contain chlorophylls *a* and *c*, but have no fucoxanthin and no pyrenoids; the main storage compounds are fats. Instead of a cell wall or scales, they have a periplast of organic plates associated with the cell membrane. The Raphidophyceae are found in both freshwater and marine habitats, and some species, such as *Heterosigma*, sometimes form blooms that are toxic to fish. Some systematists believe the class should be elevated to phylum status.

Raunkiaer system of classification a method of plant classification devised by the Danish botanist, C. Raunkiaer. Plants are divided into groups depending on the position of perennating buds and the degree of protection that they give in cold conditions or drought (see diagram). Generally the closer the buds are to the ground the more protected they are. It is particularly useful for dividing given areas of vegetation into groups and comparing the occurrence of these groups in different climatic regions, especially in temperate plant communities. See CHAMAEPHYTE, CRYPTOPHYTE, HEMICRYPTOPHYTE, PHANEROPHYTE, THEROPHYTE, PERENNATING ORGAN.

ray (secondary ray) a radial line of parenchyma cells in the SECONDARY XYLEM and SECONDARY PHLOEM, derived from the RAY INITIALS of the vascular cambium. The ray system of gymnosperms often also contains TRACHEIDS, distinguished from the ray parenchyma by their BORDERED PITS and lack of protoplast. Rays may be *uniseriate* (one cell thick in tangential longitudinal section) or *multiseriate* (several cells thick in tangential longitudinal section). They provide a pathway for gaseous diffusion and transport of dissolved substances and may also be used for food storage. Compare MEDULLARY RAY.

ray floret see CAPITULUM.

ray fungi see ACTINOBACTERIA.

ray initials more or less cubical INITIALS in the vascular cambium that give rise to the components of the RAY (radial) system of the secondary xylem and secondary phloem (e.g. ray parenchyma cells). Compare FUSIFORM INITIALS.

R/B ratio see RESPIRATION—IOMASS RATIO.

RDP see RIBULOSE 1,5-BISPHOSPHATE.

reaction centre in photosynthesis, a hypothetical site at which energy from absorbed light is used to transport electrons for PHOTOPHOSPHORYLATION. There are two types of reaction centres in thylakoids, PHOTOSYSTEMS I AND II. They contain chlorophyll *a* molecules in a specific association with certain proteins and other cell components, as well as ACCESSORY PIGMENTS and a specific molecule of chlorophyll *a*, the *primary pigment*, that is the actual electron-transferring pigment and is P700 in photosystem I and P680 in photosystem II. The chlorophyll molecules and accessory pigments transfer the absorbed energy to P700 or P680, which ejects an electron to the ELECTRON TRANSPORT CHAIN.

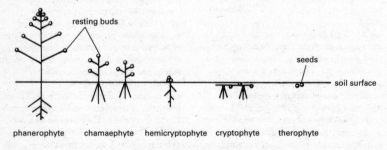

phanerophyte   chamaephyte   hemicryptophyte   cryptophyte   therophyte

Raunkiaer system of classification

reaction wood structurally abnormal wood formed in response to various stresses. In gymnosperms it is termed *compression wood* and tends to form on the lower side of branches. Compression wood is dense in structure due to the heavy lignification of TRACHEID walls. *Tension wood* forms in angiosperms on the upper side of the branches and the fibres tend to be gelatinous rather than lignified.

reading-frame see GENE MUTATION.

recalcitrant seed a seed that loses its viability if stored for any length of time, even under conditions that are normally conducive to seed longevity, i.e. low moisture content and low temperatures. Many tropical plants, e.g. coconut, rubber, and tea, have recalcitrant seeds. Such seeds can often be kept for only a year or less.

recapitulation theory a theory formulated by Ernst Haeckel that proposed that ONTOGENY is a short recapitulation of PHYLOGENY, i.e. an organism in its development goes through a series of stages that resemble the adult forms of its ancestors in evolutionary sequence. The theory has been largely discarded.

Recent see QUATERNARY.

receptacle 1 (thalamus, torus) the expanded region at the end of a PEDUNCLE to which the floral parts are attached. It is usually convex but may become flattened or concave. Such modifications (see HYPOGYNY, PERIGYNY, EPIGYNY) alter the position of the gynoecium in relation to other floral parts.
2 the point on a leaf or thallus where reproductive organs are borne. Examples are the swollen tips of the thallus of certain algae in the FUCALES, which bear the conceptacles.

receptor site in the cell wall of a bacterium, a set of reactive chemical groups that are complementary to a set on the tailpiece of a BACTERIOPHAGE, allowing the phage to recognize the bacterium and attach to it.

recessive describing an ALLELE that is expressed only when HOMOZYGOUS. Recessive alleles are in most cases considered to be the mutant form of DOMINANT alleles. They are often rarer in the population and usually lower the fitness of the individual. Their effect is often due to their failure to produce a normal functional protein. They are masked (i.e. recessive) in a heterozygote because the normal (dominant) allele will produce a normal protein and hence a normal phenotype.

reciprocal cross a cross in which the source of the male and female gametes is reversed. Thus in the cross yellow peas × green peas, the pollen would originate from the yellow-seeded parent in one cross, then from the green-seeded parent in the reciprocal cross. A reciprocal cross should determine whether there are any maternal (or paternal) factors influencing the inheritance of the characteristic.

reciprocal genes see COMPLEMENTARY GENES.

reciprocal translocation see TRANSLOCATION.

recombinant an organism that contains a combination of alleles different from that of either of its parents. It is the product of a genetic cross in which linked genes have recombined as a result of CROSSING OVER between them. The further apart the genes are on the chromosomes, the greater is the chance of crossing over occurring between them, and the greater the number of recombinants likely to be produced. See also CHROMOSOME MAPPING.

recombinant DNA 1 genetic material in which crossing over or chromosome reassortment has occurred by any natural means. See also GENETIC ENGINEERING.
2 genetic material that contains novel gene sequences produced using techniqes of GENETIC ENGINEERING (recombinant DNA technology).

recombinant DNA technology see GENETIC ENGINEERING.

recombination the formation of new combinations of genes during meiosis by CROSSING OVER and by reassortment of whole chromosomes into new sets. As a result of recombination, the gametes that an organism produces differ from the gametes from which it arose. Consequently the offspring, called RECOMBINANTS, will vary genetically and hence phenotypically from the parents. Recombination is a major source of VARIATION.

recombination frequency see CHROMOSOME MAPPING.

recon see GENE.

recurrence surface in a profile of a peat deposit, a sudden transition from highly humified peat (see HUMIFICATION) to fresh unhumified peat, which represents the recurrence of wetter conditions and resumption of peat growth following a drier period. Such profiles are typical of the sequence of hummocks and pools of a growing BOG surface. See HUMMOCK AND HOLLOW CYCLE.

recurrent parent see BACKCROSSING.

recurved describing leaves that are curved or rolled backwards, as in the moss *Tortula ruralis* (hairy screw moss).

red algae see RHODOPHYTA.

red-and-yellow forest soil a type of zonal acid soil (a PEDALFER) formed in wet subtropical regions. LEACHING results in the accumulation of salts in the B horizon. The red colour is due to iron oxides in the A horizon though if the soil is sandy, it tends to be yellow. Such soils contain little HUMUS and are soon exhausted of nutrients when cultivated. They can be easily eroded.

Red Data Book a loose-leaf updatable binder containing information on rare and endangered species worldwide, published at regular intervals by the Species Survival Commission of the WORLD CONSERVATION UNION (IUCN). The 1997 IUCN *Red List of Threatened Plants* is a recent addition to the Red Data Book series. Compiled by the WORLD CONSERVATION MONITORING CENTRE, it lists 33 798 species of vascular plants as being in danger of extinction. This represents 12.5% of the estimated total of 270 000 species of vascular plants and includes a number of species of known medicinal value (e.g. 75% of species in the Taxaceae).

red desert soil a type of coarse zonal soil rich in salts and lime (a PEDOCAL) but poor in humus. They are formed in hot deserts and can be cultivated under irrigation.

red/far-red effects those physiological responses exhibited by many plants as a result of being illuminated by either RED LIGHT or FAR-RED LIGHT. Usually a period of red light will reverse the effect of a period of far-red light and vice versa. Thus the response illicited in the plant depends on which treatment was given most recently. See PHYTOCHROME.

red light electromagnetic radiation of approximately 630 nm wavelength. Light in this part of the spectrum is the most effective in initiating various light-dependent reactions in plants. For example, short-day plants are prevented from flowering if given a short burst of red light during the dark period. Similarly red light is best for promoting germination in light-sensitive seeds and for reversing ETIOLATION. Compare FAR-RED LIGHT. See PHOTOPERIODISM, PHYTOCHROME.

Red List of Threatened Plants see RED DATA BOOK.

redox potential the tendency of a reducing agent to lose electrons (or an oxidizing agent to gain electrons) as measured against a known standard, usually hydrogen. By convention the reaction is written in the direction

$$\text{oxidant} + ne^- \rightleftharpoons \text{reductant}$$

where n is the number of electrons transferred. Compounds with a redox potential more positive than that of hydrogen tend to donate electrons to H^+; compounds with a more negative redox potential tend to accept electrons from H_2.

redox reaction see OXIDATION–REDUCTION REACTION.

red snow a prolific growth of the green unicellular alga (Chlorophyta) *Chlamydomonas nivalis* on snow, which colours it pink. The pink colour is due to carotenoid pigments that help protect the cell from photooxidation by the intense ultraviolet radiation at high altitude. See PHOTOOXIDATION.

red tide water that has been coloured red by the presence of large numbers of dinoflagellates (DINOMASTIGOTA) or other microorganisms. Such BLOOMS often produce toxins that are fatal to fish and shellfish, which in turn may become poisonous or even fatal to humans.

red tropical soil a type of zonal SOIL formed in rainforests, and in savanna with alternating dry and wet seasons. It is heavily leached but a balance is maintained between the nutrients produced by decomposition and those taken up by the plants. The red colour is due to the presence of iron, magnesium, and aluminium oxides. Some red tropical soils may form a hard crust of chemicals on the surface.

reducing sugar a sugar capable of reducing an oxidizing agent. To do this it must have a potentially active aldehyde or ketone group. All monosaccharides have such a group and consequently they are all reducing sugars. For a disaccharide to be reducing one of the reducing groups of either of the two component monosaccharides must be left intact. Maltose, which consists of two glucose units linked by an $\alpha(1-4)$ glycosidic bond, is reducing because the second glucose residue can undergo oxidation, having an aldehyde at carbon five. Sucrose however is nonreducing because the component glucose and fructose units are linked by their aldehyde (carbon one of glucose) and ketone (carbon two of fructose) groups. Various solutions, e.g. Benedict's and Fehling's, are used to detect reducing sugars. See also BARFOED'S REAGENT, BENEDICT'S REAGENT, FEHLING'S SOLUTION.

reduction division see MEIOSIS.

reduction oxidation a reaction in which an atom or molecule gains one or more electrons or hydrogen atoms or loses an oxygen atom. See OXIDATION–REDUCTION REACTION.

reduction potential see ELECTRODE POTENTIAL.

reference species a species whose fossils can be used to date stratigraphic deposits or to determine the age of a species lineage. Such species are usually extinct.

reflexed describing a structure that is bent sharply backwards, such as the petals of *Cyclamen*.

reforestation the planting of trees on land from which trees have previously been removed.

refugium (*pl.* refugia) see DISJUNCT DISTRIBUTION.

regeneration 1 the regrowth of tissues and organs from differentiated tissues of the plant. It is seen following damage or when an organ is removed from the parent plant. The technique of GRAFTING relies on the regeneration of vascular tissue, while vegetative propagation techniques of taking shoot, root, or leaf cuttings depends on the ability of such segments to regenerate complete plants. See also TISSUE CULTURE, VEGETATIVE REPRODUCTION, WOUND HORMONE. 2 recruitment into a population, usually from seedlings, allowing that population to persist. It can also apply to new vegetative growth from living roots or coppiced tree.

regeneration complex see HUMMOCK AND HOLLOW CYCLE.

regma a type of schizocarpic CAPSULE formed from several fused carpels, that breaks explosively into one-seeded units or *mericarps* at maturity, as in members of the Geraniaceae.

regolith a general term for the unconsolidated weathered material that overlies solid bedrock. It includes SOIL and rock fragments. Regolith reaches its greatest development (to depths of several hundred metres) in the tropics, where weathering is rapid.

regular flower see RADIAL SYMMETRY.

regular sample (systematic sample) one of a series of samples taken at regular intervals, as opposed to a random sample. Regular sampling usually involves a line or grid of regularly spaced quadrats, and is particularly suited to sampling across environmental gradients. The main source of error with this method is that the spacing interval used may coincide with a regular, unexpected variation in the environment.

regulator gene (i-gene) a gene whose product can prevent or promote the TRANSCRIPTION of other genes. The other genes may or may not be adjacent to the regulator gene. See also INDUCER, OPERON.

regulator protein a protein produced by a regulator gene to control the transcription of DNA by DNA polymerase. Such proteins usually act as REPRESSORS.

reiteration the process in which a dormant bud develops to form a shoot with the same architectural pattern as its parent shoot. Reiteration may occur in response to injury or to a change in the environment, for example if there is an increase in available light on that side of the tree.

relative dating see DATING.

relative density see DENSITY–FREQUENCY–DOMINANCE.

relative dominance see DENSITY–FREQUENCY–DOMINANCE.

relative frequency see DENSITY-FREQUENCY-DOMINANCE.

relative humidity see HUMIDITY.

relative importance value (RIV) the mean of the relative frequency, relative density, and relative cover percentage for each species in a community/treatment. This gives an idea of the importance of a particular species relative to the others present, while minimizing problems arising from patchiness of distribution. See DENSITY–FREQUENCY–DOMINANCE.

relative pollen frequency (RPF) in pollen analysis, a measure of the relative importance of each species, genus, or other taxon of pollen in a sample/assemblage. It expresses the pollen count of each species/taxon as a percentage of the total pollen count or the total tree pollen counted.

release factor see TRANSCRIPTION.

relevé in PHYTOSOCIOLOGY, the basic unit of vegetation recorded in the field that is considered representative of the community type being investigated. It should have a uniform floristic composition, with uniform relief and soil type. The actual size of the unit will vary with the type of community and should be at least as large as the largest minimal area for any of the species present.

relict describing a population or community that has a very restricted range today, but which is all that is left of a formerly much larger and more widely distributed population or community. The term is also used to describe a species that survives when related species have become extinct. For example, in temperate regions species and communities that were once more widespread in immediate post-glacial times now survive only at high altitudes where they do not face competition from species better adapted to warmer climates.

relict soil see PALAEOSOIL.

renaturation see DENATURATION.

rendzina (calcisol) a type of shallow intrazonal soil rich in lime formed from the underlying limestone or chalk rocks. The A horizon is brown or black and rich in humus. The B horizon, if one is distinguishable, is grey or yellowish with limestone fragments. BROWN EARTHS develop in similar climatic conditions from other parent rocks. See SOIL PROFILE. Compare TERRA ROSSA.

renewable resource a resource whose availability/supply rate is likely to equal or exceed any expected rate of consumption, for example an aquifer in an area of high rainfall. Compare NONRENEWABLE RESOURCE.

reniform shaped like a kidney.

repair synthesis see DNA REPAIR.

repeated sequence any long base-pair sequence found at many different places throughout the chromosomes. This excludes DNA coding for STRUCTURAL GENES, which with few exceptions are present as single copies. Repeated sequences account for more than 50% of all the DNA present and fall into three categories.

Repetitive DNA usually consists of widely scattered sequences about 300 base pairs long, which are often adjacent to structural genes. It has been proposed that these sequences are functionally equivalent to the OPERATOR GENES of bacteria. Suppose that sequence A is adjacent to three unlinked genes X,Y, and Z, and that A′ is an INDUCER of A. The presence of A′ in the cell would 'switch on' all genes to which A was adjacent, in this case X,Y, and Z. An appropriate mix of repeated sequences, say A,B,C, and D next to structural gene X would make possible an elaborate transcriptional control system as demanded in a complex organism.

Highly repetitive or *satellite DNA* is often about 20 base pairs long and is usually found near the centromere. *Palindromes*, which are often a few hundred base pairs long, are scattered throughout the chromosomes. Such sequences read the same in both directions, i.e. are palindromic. The functions of satellite DNA and palindromic DNA are uncertain.

repetitive DNA see REPEATED SEQUENCE.

replica plating 1 a technique used in electron microscopy to examine the surfaces of specimens that are too electron dense, too delicate, or too robust for sectioning. The specimen surface is coated with a thin film of material, such as carbon, in order to make a cast or replica. This is removed by soaking the specimen in 10% sodium hydroxide solution. It may be subjected to SHADOWING procedures before examination. The development of the scanning electron microscope has tended to make replica plating redundant, although it is still used in freeze fracturing.

2 a technique used in recombinant DNA technology (see GENETIC ENGINEERING) to determine which bacteria colonies contain recombinant DNA. The desired DNA is transferred to a bacterium by recombining it with the DNA of a bacterial plasmid into which a gene for antibiotic resistance has been inserted. It is inserted into the plasmid DNA in such a way that it interrupts the antibiotic resistance gene. Bacteria that contain a plasmid with the recombinant DNA will not be resistant to that antibiotic. A copy (replica plate) of the plate containing bacterial colonies is made by placing a piece of velvet on the surface of the plate, then pressing it into the agar of another (the replica). The antibiotic is added to this replica plate, and colonies that are killed by the antibiotic correspond to recombinant colonies on the original plate.

replicase an enzyme involved in catalysing the replication of DNA or RNA. See DNA REPLICATION.

replicate end wall in certain conjugating green algae (GAMOPHYTA), a fold in the end wall that allows for extra expansion after cell division, so pushing the daughter cells apart and aiding fragmentation.

replication the formation of exact copies (replicas). The term has been applied both to DNA and asexual organisms, both of which produce essentially identical copies/offspring. See also SEMICONSERVATIVE REPLICATION.

replication fork during DNA REPLICATION, the Y-shaped region that forms where the double-stranded DNA molecule is being 'unzipped'. At this site complementary strands of DNA are being synthesized.

replum in an ovary or fruit, a false septum formed by ingrowth from the placenta rather than from the carpel walls.

repressor a protein product of a REGULATOR GENE that prevents transcription of other genes by binding to the DNA and blocking attachment of messenger RNA polymerase.

reproduction the process by which new individuals of a species are formed and by which the species is perpetuated. See ASEXUAL REPRODUCTION, SEXUAL REPRODUCTION, VEGETATIVE REPRODUCTION.

reproductive isolation the prevention of gene flow between all the members of a population due to the development of reproductive barriers (*isolating mechanisms*), which result in the formation of distinct noninterbreeding groups that may in time develop into separate species. Isolating mechanisms can arise in many ways. Populations may be separated geographically (see ALLOPATRIC). Populations occupying the same area (see SYMPATRIC) may not interbreed because they have different flowering times or different pollinators, or because they occupy different niches. Such populations may also be isolated by the inviability or sterility of any hybrids that do form, often following POLYPLOIDY.

reptant describing a plant stem that creeps along the ground and takes root.

repulsion see TRANS arrangement.

repulsion theory (field theory) the theory that the pattern of origin of leaf PRIMORDIA at the shoot apex is regulated physiologically by inhibitory substances synthesized by the apex and the older primordia. A new primordium arises in a position where the concentration of these substances has fallen below a certain threshold level. Although no such inhibitors have yet been isolated various surgical and modelling experiments support the theory. Compare AVAILABLE SPACE THEORY. See also PHYLLOTAXIS.

rescue effect in island biogeography, the reduction of the risk of a species becoming extinct that results from immigration from the mainland of individuals of the same species.

residual a variation in the value of a variable that is not accounted for by a particular statistical test. Such variability may suggest the effect of an environmental factor that has not been taken into consideration.

resin (*adj.* resinous) a mixture of high-molecular-weight compounds, mainly polymerized acids, esters, and terpencids, exuded by certain plants particularly when wounded. Resins are insoluble in water but soluble in ethanol. On exposure to air the volatile components evaporate leaving a solid or semisolid residue protecting the damaged area. Resins are particularly prevalent in conifers, which contain specialized RESIN CANALS. Pine resin yields the essential oil turpentine ($C_{10}H_{16}$) on distillation and the solid residue rosin, used in lacquers. Other commercially important resins are dammar (from trees of the genera *Shorea* and *Agathis*), kauri (from the New Zealand conifer *Agathis australis*), jalap (from the Mexican convolvulaceous plant *Exogonium purga*), and mastic (from the evergreen anacardiaceous tree *Pistacia lentiscus*). Semisolid mixtures of resins and essential oils are often termed *balsams* and include Canada balsam from *Abies balsamea* and frankincense from species of *Boswellia*.

resin canal (resin duct) a longitudinal resin-containing channel in the SECONDARY XYLEM and leaves of many gymnosperms. Resin canals usually form by separation of parenchyma cells, which later form a lining to the canal termed an epithelium.

resistance the ability of a plant to restrict the activities of a pathogen so that its growth is not significantly affected. Complete resistance is termed IMMUNITY. Resistance may be specific, i.e. effective against a particular strain of the pathogen, or nonspecific, i.e. effective against all strains. Specific resistance is usually controlled by one or a few major genes, e.g. resistance of barley to mildew and of flax to rust are inherited as single genes. Nonspecific resistance is conferred by the combined action of a number of minor genes. Such multigenic or *field resistance* is usually more stable and less likely to be overcome by mutants or recombinants of the pathogen. Environmental factors may influence the expression and degree of resistance. Compare SUSCEPTIBILITY. See also HYPERSENSITIVITY, PHYTOALEXIN.

resolving power the capacity of a microscope to enable clear observation of the fine details of a specimen. This differs from the *theoretical resolution*, i.e. the capacity to distinguish between two individual points despite their close proximity, as the resolving power of a microscope depends on the perfection of the lens, the light source, etc. The theoretical resolution is limited by the wavelength of light and is greater with decreasing wavelength (see ULTRAVIOLET MICROSCOPE).

respiration the oxidative breakdown of food substances within the cells of living organisms, resulting in the liberation of energy for subsequent use in growth, etc. The process usually involves the absorption of molecular oxygen, and water and carbon dioxide are typically end products. The reactions involved occur in two stages, GLYCOLYSIS and the KREBS CYCLE. See also AEROBIC RESPIRATION, ANAEROBIC RESPIRATION.

respiratory chain a series of membrane-linked OXIDATION–REDUCTION REACTIONS in which electrons are transferred from reduced cofactors, formed in KREBS CYCLE and other DEHYDROGENASE reactions, to oxygen, which then combines with hydrogen ions to form water. At present seven different electron-transferring enzymes or cofactors are known intermediates between NAD, the starting point of the chain, and oxygen. The REDOX POTENTIAL is most negative at the start of the chain, and rises as the electrons flow towards oxygen. For every two electrons travelling down the respiratory chain, three molecules of ATP are generated by the process of OXIDATIVE PHOSPHORYLATION, which is linked to respiratory electron transfer. See ELECTRON TRANSPORT CHAIN.

respiratory enzyme see CYTOCHROME.

respiratory quotient (RQ) the ratio of the volume of carbon dioxide evolved during respiration to the volume of oxygen absorbed, i.e. RQ = volume CO_2/volume O_2. The complete respiratory breakdown of sugars by AEROBES often gives an RQ value of one, but it rises very considerably under anaerobiosis. When fats or proteins are used in respiration, the RQ value is less than one as less carbon dioxide is produced.

respiratory root see PNEUMATOPHORE.

restitution nucleus see ANEUSPORY.

restriction endonuclease (restriction enzyme) an enzyme that recognizes and binds to specific base sequences of double-stranded DNA and cuts the DNA at or near this point. Such enzymes are produced by certain bacteria in response to invasion by a bacteriophage – they destroy the virus by cutting up its DNA. Restriction enzymes are a useful tool in recombinant DNA technology (see GENETIC ENGINEERING). There are two different types: some make a 'clean cut', cleaving both strands of the molecule at the same base pair; others leave a protruding single strand of unpaired bases – a so-called 'sticky end'. Pieces of DNA that have been cut with the same restriction enzyme to produce complementary sticky ends readily bind together with the assistance of a sticky end DNA LIGASE. Restriction enzymes are used to cut open plasmids and segments of desired DNA so that they can be joined (annealed) by DNA ligase. For some genetic engineering purposes, in which heating is involved, restriction enzymes from heat-resistant bacteria are used.

restriction enzyme see RESTRICTION ENDONUCLEASE.

restriction fragment length polymorphism (rflp) the diversity in the sites at which particular RESTRICTION ENDONUCLEASE enzymes cleave the dna of different individuals of the same species due to random base changes. the fragments produced by the restriction enzymes differ, and this can be used to identify individuals (see GENETIC FINGERPRINTING). Mutations that produce defective genes or genes with new properties can also be detected by analysis of RFLPs.

restriction mapping a map of a gene or genome deduced from the sequences of restriction fragments produced by restriction endonuclease enzymes that cut at specific base sequences. See RESTRICTION FRAGMENT LENGTH POLYMORPHISM.

resupination (*adj.* resupinate) in orchid flowers (see ORCHIDACEAE), the twisting of the ovary, which turns the flower through

180° and causes the upper petal (the lip) to appear lowermost. A few orchids (e.g. *Hammarbya*) have a twist through 360° to return the lip to its original (upper) position.

reticulate describing something, such as the VENATION of a leaf, that has a netlike pattern.

reticulate chloroplast a complex plastid that forms a cylindrical network in the peripheral region of the cells of some green algae, e.g. *Oedogonium*.

reticulate evolution the pattern of EVOLUTION resulting from hybridization, recombination and/or allopolyploidy, so that ancestral lineages intersect repeatedly in the evolutionary tree. Horizontal gene transfer complicates the tracing of phylogenies.

reticulate thickening a type of secondary cell wall patterning in TRACHEARY ELEMENTS in which the secondary cell wall is laid down to form a network. Unlike ANNULAR THICKENING and SPIRAL THICKENING, reticulate thickening permits little further elongation and therefore occurs in tissues such as metaxylem, which have already completed most or all of their elongation. Compare PITTED THICKENING, SCALARIFORM THICKENING.

reticulodromous describing net or reticulate leaf venation as, for example, seen in *Rhododendron* leaves. *See illustration at* venation.

retrogression (retrogressive succession) the reversal of plant SUCCESSION, when a disturbance causes a reversion to an earlier stage of succession, with fewer species and less complex communities.

retrogressive succession see RETROGRESSION.

retrovirus a virus whose nucleic acid is RNA and which employs the enzyme REVERSE TRANSCRIPTASE to make a DNA copy of its genome, using its RNA as a template.

reverse mutation (reversion) a mutation that restores the original condition of the gene, so that it can now produce a functional protein again. The mutation must occur at the same locus as the original mutation. Functionality may also be restored by a *suppressor mutation* at another part of the gene (*intragenic suppression*) or in a different gene (*intergenic suppression*), which suppresses the effect of the first mutation or masks its effect in the phenotype.

reverse transcriptase an enzyme present in some microorganisms that reverses the normal transcription sequence and synthesizes DNA from messenger RNA. It is widely used in research for producing INTRON-free DNA. See also GENETIC ENGINEERING.

reversion SEE REVERSE MUTATION.

revolute describing a form of leaf folding in which the leaf is rolled towards the abaxial surface. *See illustration at* ptyxis.

R factor see PLASMID.

RFLP SEE RESTRICTION FRAGMENT LENGTH POLYMORPHISM.

Rf value (relative front) in CHROMATOGRAPHY, a value that is calculated by dividing the distance moved by the solute spot on the chromatogram by the distance moved by the solvent front. The value is constant for a particular molecule.

rhachilla see RACHILLA.

rhachis see RACHIS.

rhamnose a methylated PENTOSE sugar, rarely occurring free, but common as a constituent of many GLYCOSIDES, e.g. the flavonol glycoside quercitin, isolated from the bark of the oak *Quercus tinctoria*. Rhamnose is also found in certain gums (e.g. gum arabic and the gum of flax seeds), mucilages, and in bacterial polysaccharides. It has been identified in the free state in poison ivy (*Rhus toxicodendron*).

rheotaxis a change in direction of locomotion at of a motile organism or cell in response to a current in the medium in which it is moving.

rheotropism a tropic response to a water current. See TROPISM.

rhizina in lichens, a tuft of hyphae projecting from the underside of a thallus, used mainly for attachment to the substrate.

rhizodermis see EPIDERMIS.

rhizoid a threadlike outgrowth from a thallus, as seen, for example, in the gametophyte generation of mosses, liverworts, and ferns. Usually rhizoids serve to anchor the plant and absorb water and nutrients. In some algae, e.g. *Ulva*, they are incorporated into and strengthen the lamina. Mosses can be distinguished from liverworts by their multicellular, usually brownish rhizoids. In liverworts rhizoids are unicellular and more or less colourless, resembling the root hairs of higher plants. Compare ROOT.

rhizome an underground stem that grows horizontally and, through branching, acts as an agent of vegetative reproduction when they root at intervals, when the intervening piece of rhizome rots away. In some plants the rhizomes are cordlike, as in nettle (*Urtica dioica*), while in others, e.g. Solomon's seal (*Polygonatum multiflorum*), they are fleshy and also serve as PERENNATING ORGANS. Compare CORM, ROOTSTOCK, TUBER.

rhizomorph a tough cordlike mass of parallel-growing fused HYPHAE that distributes a fungus from one favourable location to another, across unsuitable substrates. Rhizomorphs are seen, for example, in the tree parasite *Armillaria mellea* (the boot-lace or honey fungus) and the dry-rot fungus *Serpula lacrymans*.

rhizomycelium see CHYTRIDIALES.

rhizophore any of the leafless branches, seen in many species of *Selaginella*, that arise from the stem at points of forking. Usually two rhizophores form at each fork but only one continues to develop. The rhizophore grows towards the soil, often branching repeatedly before reaching the soil surface. On contact with the ground, roots are produced from the swollen tips of the rhizophore.

rhizoplane the root surface.

rhizoplast a part of the CYTOSKELETON that forms a fibrous root-like proteinaceous structure linking a KINETOSOME with the nucleus.

rhizopodium a thin outgrowth of cytoplasm formed by certain algae that lack rigid cell walls. Rhizopodial algal cells are seen in the CHRYSOMONADA, XANTHOPHYTA, and DINOMASTIGOTA.

rhizotaxis the arrangement of lateral roots arising from the primary root system. It includes the number of longitudinal rows of lateral root primordia and the degree of regularity or the spacing between them. Compare PHYLLOTAXIS.

Rhodophyta (red algae) a phylum of the PROTOCTISTA (in some classification schemes a division of the Plantae) containing the most complex organisms in the kingdom. The red algae have no motile stages. The main photosynthetic pigments are chlorophyll *a* and the PHYCOBILIPROTEIN pigments PHYCOCYANIN and PHYCOERYTHRIN, which confer the characteristic red colour, although freshwater species (which contain relatively less phycoerythrin) are often grey-green. Red algae have a starchlike carbohydrate reserve (floridean starch) resembling AMYLOPECTIN. The cell wall may be of cellulose, MANNANS, or XYLANS and may be associated with mucopolysaccharides, such as AGAR and CARRAGEENAN. The cells may have more than one nucleus. Most red algae are multicellular. The thallus ranges from single cells to branching filaments and pseudoparenchymatous and parenchymatous structures.

The red algae are found worldwide, mainly in marine habitats ranging from beaches and rocky shores to quite deep water. There are some 4000 species in a single class, Rhodophyceae, which is divided into two subclasses, the BANGIOPHYCIDAE (Bangioideae) and the FLORIDEOPHYCIDAE (Florideae). Sexual reproduction is less complex in the Bangiophycidae, the reproductive structures are less differentiated, and this group lacks PIT CONNECTIONS between the cells, which are a prominent feature of the Florideophycidae. Certain members of the Florideophycidae secrete a calcareous covering and play a part in the formation of coral reefs. Mucilages from red algae are of commercial importance in the manufacture of ice cream and as stabilizers and gels, such as agar. Some species, such as *Chondrus crispus*, are eaten.

Rhynie chert a siliceous rock of the DEVONIAN period in Scotland, in which well preserved psilophyte fossils are found. In some, anatomical details can be clearly seen. The plants are similar to *Psilotum*, a living psilophyte, in having aerial branches, tracheids, an epidermis with stomata, and sporangia at the ends of the branches. Examples are *Rhynia* and *Horneophyton*.

These extinct psilopsids are placed in the order Psilophytales of the PSILOPHYTA, named after the Devonian fossil *Psilophyton*, discovered in eastern Canada in 1859. Another important fossil in the Rhynie chert, *Asteroxylon*, resembles a lycophyte.

rhythmic growth a pattern of growth in which the plant has clearly defined periods of dormancy or resting phases. Growth does not always cease during the resting periods, but it may be reduced, with shorter internodes, as in the sand sedge (*Carex arenaria*). Rhythmic growth is common in environments with alternating seasons, but may also occur under relatively uniform climatic conditions, with periods of cell division and differentiation alternating with periods of elongation, or leaf production alternating with flower production, as in tea shoots (*Camellia sinensis*). Compare SYLLEPTIC GROWTH.

rhytidome see BARK.

Rhytismatales (Phacidiales) an order of fungi of the ASCOMYCOTA, in older classifications placed in the class DISCOMYCETES. Its members, comprising about 411 species in some 71 genera, have inoperculate asci. Most are saprobes and parasites found on bark, wood, and leaves. They include the plant pathogen *Rhytisma acerinum*, which causes tar spot of sycamore leaves.

ribitol a SUGAR ALCOHOL that forms part of the structure of flavins, flavine adenine dinucleotide (FAD), flavine mononucleotide (FMN), and riboflavin.

riboflavin (vitamin B$_2$) a FLAVIN pigment that has been implicated both in the perception of the phototropic stimulus and in PHOTOOXIDATION of endogenous auxins. It consists of an isoalloxazine ring substituted with the sugar alcohol RIBITOL. In its active phosphorylated form riboflavin phosphate, otherwise known as FMN, it acts as a coenzyme for various oxidizing enzymes. Riboflavin is not synthesized by animals. See also FAD, VITAMIN.

ribonuclease (RNase) an enzyme that catalyses the hydrolysis of the phosphodiester bonds between adjacent NUCLEOTIDES in RNA, producing mono- and oligonucleotides. Endoribonucleases cleave bonds within the molecule, and exoribonucleases cleave nucleotides from one or both ends of the RNA molecule. RNase is implicated in the incompatibility system that prevents pollen with certain genotypes from germinating on the style of a flower.

ribonucleic acid see RNA.

ribonucleotide a NUCLEOTIDE composed of ribose sugar bound to a purine or pyrimidine base and bound to a phosphate group through one of the hydroxyl groups of the ribose. It is the basic unit of ribonucleic acid.

ribose an aldopentose sugar (*see illustration at* aldose). It is a component of nucleotides (e.g. AMP, ATP), dinucleotides (e.g. NAD, FAD), and ribonucleic acid (RNA). Unlike DEOXYRIBOSE ($C_5H_{10}O_4$), the empirical formula of ribose ($C_5H_{10}O_5$) is that of a typical five-carbon monosaccharide.

ribosomal RNA (rRNA) a form of RNA that is confined entirely to the RIBOSOMES. Synthesis of rRNA takes place on distinctive parts of the chromosomes associated with the nucleolus, called NUCLEOLAR ORGANIZERS. The base composition of rRNA is very similar in all species. After synthesis it combines with protein to form nucleoprotein. The partly constructed ribosomes then leave the nucleus, via nuclear pores, and enter the cytoplasm where synthesis is completed. The absence of fully constructed ribosomes in the nucleus may account for the absence of protein synthesis there. Compare MESSENGER RNA, TRANSFER RNA.

ribosomes protoplasmic particles that are sites for the assembly of amino acids into the polypeptide chains of protein molecules in the order dictated by the genetic code of messenger RNA. During the process mRNA, formed in the nucleus, and TRANSFER RNA, which brings the required amino acids to their correct positions, become attached to the ribosomes, together with various factors involved in amino acid chain initiation, elongation, and termination. Ribosomes are about 20 nm in diameter. There are about 10 000 in most bacterial cells, and tens of thousands in eukaryotic cells. In prokaryotic cells, e.g. *Escherichia coli*, they occur freely scattered throughout the protoplasm, but in eukaryotic cells most are associated with the membranes of the endoplasmic reticulum. They have also been identified in mitochondria and chloroplasts.

The ribosomes of prokaryotic and eukaryotic cells differ. When sedimented in a centrifuge (see SEDIMENTATION COEFFICIENT), two types of ribosomes settle out – 70S and 80S ribosomes. 80S ribosomes are found in the cytoplasm of eukaryotic cells, and 70S ribosomes in prokaryotic cells and in chloroplasts and mitochondria (see ENDOSYMBIOTIC THEORY). The ribosomes of prokaryotic cells contain 60–65% ribosomal RNA and 35–40% protein, while those of eukaryotic cells contain approximately equal quantities of each. Ribosomes consist of two parts or subunits, unequal in size and each containing RNA and a number of proteins. During protein synthesis both the developing polypeptide chain and the ribosome are translocated along the mRNA molecule as the genetic code is translated. Usually several ribosomes move simultaneously along the mRNA. Such a chain of ribosomes is called a POLYSOME or polyribosome.

ribozyme an RNA molecule that catalyses changes to its own structure: it is capable of self-splicing. See GENE SPLICING.

ribulose a five-carbon ketose sugar (*see illustration at* ketose). In the phosphorylated form, ribulose 5-phosphate, it is an intermediate in the PENTOSE PHOSPHATE PATHWAY and the CALVIN CYCLE. As RIBULOSE BISPHOSPHATE it is important as the carbon dioxide acceptor in photosynthesis.

ribulose bisphosphate (RuBP) a phosphorylated form of RIBULOSE that acts as the initial carbon dioxide acceptor in C_3 photosynthesis and as the eventual acceptor in C_4 photosynthesis and crassulacean acid metabolism. Addition of carbon dioxide occurs between the second and third carbon atoms and results in the formation of two molecules of glycerate 3-phosphate. These feed into the Calvin cycle, which allows the fixed carbon dioxide to be converted into carbohydrate and regenerates the ribulose bisphosphate. See also C$_3$ PLANT, C$_4$ PLANT, CARBOXYLATION, FIXATION.

ribulose bisphosphate carboxylase (rubisco, RuBP carboxylase, carboxydismutase) an enzyme that catalyses the carboxylation of ribulose bisphosphate to two molecules of glycerate 3-phosphate. This is the first step in the CALVIN CYCLE. Rubisco is the *fraction 1 protein* of chloroplasts. It is very abundant in chloroplasts and may account for up to 15% of the total protein. This enzyme can also act as an oxygenase, forming a molecule of glycerate 3-phosphate and a molecule of phosphoglycolic acid from ribulose bisphosphate and oxygen. This reaction forms the basis of PHOTORESPIRATION.

ribulose diphosphate (RuDP) an outdated term for the initial carbon dioxide acceptor in photosynthesis that has now been replaced by RIBULOSE BISPHOSPHATE. The terms di- and bis- both refer to the presence of two groups in a molecule. In diphosphates the second phosphate group is added directly to the first (e.g. adenosine diphosphate). In bisphosphates the second phosphate is on a different carbon atom to the first (e.g. ribulose bisphosphate or fructose bisphosphate).

rimiform shaped like a fissure or cleft.

ring-diffuse wood see ANNUAL RING.

ringing experiments a means of investigating the routes of water and carbohydrate translocation in plants by removing a ring of the outer tissues of the stem containing the phloem. This results in sufficient water still reaching the upper shoots to keep them turgid (see TURGOR) but photosynthetic products from the leaves are not translocated to regions below the ring and tend to accumulate above it. Such observations confirm that water is transported within XYLEM and photosynthates within PHLOEM vessels, the former being undisturbed and the latter disrupted by the ringing process.

ring-porous wood see ANNUAL RING.

Rio Summit see CONVENTION ON BIOLOGICAL DIVERSITY.

riparian relating to a river bank or lake shore. Such areas often have lusher vegetation than neighbouring areas, with greater plant density, biodiversity, and primary productivity.

ripening a stage in the development of a fruit in which the tissues become softer, sweeter, and less acid, and flesh and skin may become a different, often more intense colour. This is important in fruits that are dispersed by animals. Ripening is often accompanied by an increase in respiration rate, cell expansion, loss of chlorophyll, and the production of ETHENE.

RIV see RELATIVE IMPORTANCE VALUE.

riverine forest see FOREST.

RNA (ribonucleic acid) a single-stranded nucleotide polymer, each nucleotide being constructed from phosphoric acid, ribose, and an organic base. The base may be adenine or guanine (the purine bases), or cytosine or uracil (the pyrimidine bases). The polynucleotides are held together by phosphodiester bonds between the phosphate group of one nucleotide and the sugar of an adjacent nucleotide. The RNA molecule may be linear, as in messenger RNA, or fold back on itself to form a three-dimensional clover-leaf-shaped molecule, as in transfer RNA. RNA is the principal agent for the TRANSCRIPTION (copying) and TRANSLATION (conversion) of the genetic code during protein synthesis. Various types of RNA are associated with these processes, namely messenger RNA, transfer RNA, and RIBOSOMAL RNA. Viruses are exceptional and may not conform to the above generalizations. In many plant viruses, such as tobacco mosaic virus, the genetic material is RNA not DNA. In some viruses, such as φ6 bacteriophage, the RNA is double stranded, like DNA. See also INITIATOR, MESSENGER RNA, MICRORNA, RIBOSOMAL RNA, RNA POLYMERASE, TRANSFER RNA, TRANSCRIPTION, TRANSLATION.

RNA polymerase an enzyme that catalyses the synthesis of RNA, using a DNA strand as a template. It is used to catalyse the formation of messenger RNA during TRANSCRIPTION and to synthesize RNA in vitro. See POLYMERASE.

RNase see RIBONUCLEASE.

RNA splicing during TRANSCRIPTION in eukaryote cells, a process whereby noncoding sections (INTRONS) of the primary RNA transcript are removed and the remaining coding sections (EXONS) are spliced together to form the functional mRNA molecule. This is usually catalysed by SPLICEOSOMES consisting of small nuclear ribonucleoprotein particles (*snurps*), but is sometimes done by the RNA molecule itself (see RIBOZYME). The same mRNA transcript may be spliced in different ways in different cell types, at different stages of development, or in response to various biological signals. Mutations affecting RNA splicing are responsible for about 15% of all genetic diseases. Because of splicing one gene may code for more than one species of mRNA, and hence more than one protein, allowing the genome to consist of fewer genes than might be expected from the number of proteins present in the organism. See EXON, INTRON, TRANSCRIPTION.

rod cell see MACROSCLEREID.

rogue 1 any plant that varies from the rest of the crop and is consequently not wanted. Examples are wild oats (*Avena fatua*) growing in fields of cultivated oats (*A. sativa*), plants growing from self-pollinated seed in a field of F, hybrids, and diseased plants, such as wheat plants affected with covered smut.

2 to remove and destroy such plants.

root the usually underground part of the plant axis, specialized for anchorage, absorption, and sometimes food storage. It may usually be distinguished from a stem by the absence of chlorophyll and buds. (Exceptions are the aerial roots of certain epiphytes, e.g. *Taeniophyllum*, that can develop chlorophyll when illuminated. Adventitious buds may form on roots and give rise to suckers.) The root system may include an obvious main root derived from the RADICLE (see TAPROOT) or it may be a FIBROUS ROOT SYSTEM due to repeated branching of the radicle. Branches of the main root are called *secondary roots*. Roots that arise other than by branching of the primary root are called *adventitious roots*. Some root systems may include or consist solely of adventitious roots arising from the base of the stem.

The vascular tissues of the root normally form a solid central STELE, which is better able to resist the tensions and pressures exerted on a root than would be a hollow stelar cylinder, typical of most stems. The absorptive properties of the root are enhanced by the formation of ROOT HAIRS behind the tip. Beyond these the root branches to form lateral roots. Unlike stem branches, these do not arise superficially but develop from the outer tissues of the stele and grow through the root cortex. Lateral root formation is thus said to be endogenous.

Roots are usually positively hydrotropic and gravitropic. A root cap (see CALYPTRA) at the root tip protects the root as it grows down through the soil. Numerous modifications of roots exist (see AERIAL ROOT, CLIMBING ROOT, CONTRACTILE ROOT, PNEUMATOPHORE, PROP ROOT). Symbiotic associations between plant roots and various fungi and bacteria are common (see MYCORRHIZA, ROOT NODULE).

root bud a bud found on a root, which is capable of developing into a shoot. Such buds form from wound callus if the root is injured. Root buds may remain dormant for long periods of time and may give rise to clones of trees, as in poplar (*Populus*).

root cap see CALYPTRA.

root collar (peg) a bulging ring of tissue at the junction between root and shoot.

root frequency in measurements of the frequency of vegetation, records that include only the species rooted in a quadrat, rather than those whose foliage occurs inside the quadrat but which are rooted outside it.

root hair a TRICHOME originating from the PILIFEROUS LAYER of the root. Root hairs are projections from single epidermal cells in direct contact with the soil and serve to increase the surface area for absorption. They are responsible for the first stage in absorption of water and solutes from the soil. They are also thought to help retain contact between the root tip and the soil when dry conditions tend to cause the soil to contract away from the roots. Some plants with MYCORRHIZAS have no root hairs, or very few, and the mycorrhizas absorb the water and solutes.

root-hair zone see PILIFEROUS LAYER.

root nodule a tumorous growth that

develops on the roots of leguminous and certain other plants in response to infection by symbiotic microorganisms. In legumes the symbiont is always a bacterium of the genus *Rhizobium*. In nonleguminous plants with root nodules the symbiont appears to be either a member of the Plasmodiophora, e.g. in the roots of bog myrtle (*Myrica gale*) and alder (*Alnus glutinosa*), or a blue-green bacterium (see CYANOBACTERIA), e.g. in the roots of some *Gunnera* species. In legumes, following invasion of the root tissues, the bacteria induce a localized proliferation of the host tissues. Like the induction of crown gall (see TUMOUR-INDUCING PRINCIPLE), the ability of *Rhizobium* to induce nodules appears to be controlled by a PLASMID. The plasmid also controls the host specificity of different *Rhizobium* strains. NITROGEN fixation is carried out by the microorganisms, which assume a characteristic shape and size within the host cells (see BACTEROID).

root pressure the pressure that can build up in the root systems of plants so as to force water upwards through the xylem vessels. The force is a function of the OSMOTIC POTENTIAL of the root cell contents and is evidenced by the continuing flow of water from the cut surface of a recently severed stem. Positive root pressures tend to build up at night, when the rate of transpiration is very low, and lead to the process of GUTTATION. Root pressure is not of great importance in water uptake and varies seasonally, being lowest in magnitude during the summer.

root rot a plant disease in which there is disintegration of the root tissues. Above-ground symptoms are often initially similar to those of nutritional disorders or bad drainage. Annual plants appear sickly and eventually wilt and die as a result of shortage of water and nutrients. Affected trees usually show symptoms of gradual decline and die-back of the crown. Numerous different species of fungi and protoctists cause root rots. *Armillaria mellea* (the honey fungus) is a ubiquitous root pathogen affecting many forest and garden trees. Other common root-infecting genera include the basidiomycete *Rhizoctonia*

(*Corticium*), the oomycete *Phytophthora*, and the ascomycete *Fusarium*. See also TAKE-ALL.

root–shoot ratio the ratio of shoot tissues to root tissues in a plant. This represents the balance between the capacity for photosynthesis and the ability to exploit a large volume of soil for water and nutrients. Low root–shoot ratios are characteristic of plants in the early stages of succession, while high root–shoot ratios are more typical of plants in the later stages. In climax vegetation, with large mature plants such as forest trees, a higher proportion of roots is also needed for anchorage. Shoot and root volume are to some extent interdependent, as shoot growth depends on an adequate supply of minerals and water, while root growth depends on photosynthate from the shoots.

rootstock 1 a short erect underground stem or RHIZOME, as seen in various angiosperms, e.g. plantains (*Plantago*), in quillworts (*Isoetes*), and in certain ferns (*Osmunda*).

2 any underground part of a plant. See STOCK.

root tuber a TUBER derived from a swollen root. Examples include tubers derived from ADVENTITIOUS roots in certain orchids. Unlike stem tubers, root tubers may have a root cap (see CALYPTRA) and lateral roots. In the celandine (*Ranunculus ficaria*) the root tuber is formed from an adventitious root at the base of an adventitious bud at the lower end of the stem, and both root tuber and bud together form a detachable tubercle that serves as a means of vegetative reproduction.

Rosaceae a large cosmopolitan dicotyledonous family containing about 2825 species in some 95 genera. It includes both woody and herbaceous plants. The leaves are usually spirally arranged and, except in *Spiraea* and some other members of the subfamily Spiraeoideae, have two stipules at the base. The flowers are usually actinomorphic and often large and showy. There are numerous stamens arranged in whorls and the carpels are also usually numerous and free. An EPICALYX is often present. Many different types of fruit are formed, the pome being characteristic of the subfamily Maloideae, e.g. apples (*Malus*), pears (*Pyrus*), and quinces (*Cydonia*), and a drupe characteristic of the subfamily Prunoideae, e.g. plums, cherries, and apricots (*Prunus* spp). Achenes, e.g. burnet (*Sanguisorba*), follicles, e.g. strawberry (*Fragaria*), or drupelets, e.g. blackberry (*Rubus*), are found in the subfamily Rosoideae, and follicles in the subfamily Spiraeoideae, e.g. *Spiraea*.

The family contains most of the important orchard fruits, e.g. apricots, cherries, peaches, plums, pears, and loquats; many bush fruits, e.g. blackberries and raspberries; and other soft fruits, e.g. strawberries. It also includes many important ornamental genera, notably *Rosa*, of which there are estimated to be over 5000 cultivars. Other ornamentals include *Cotoneaster*, *Kerria*, *Potentilla*, and *Spiraea*.

rosette plant any plant with its leaves radiating outwards from a short stem at soil level. The rosette habit enables such plants to survive grazing and trampling and to be more successful in competing with other species for space. Examples are dandelions, daisies, and plantains.

Rosidae a subclass of the dicotyledons containing both woody and herbaceous species. It members usually have bisexual flowers that often contain numerous stamens, which develop centripetally. The Rosidae may contain some 14–22 orders, depending on the classification scheme used. The inclusion of the following orders is generally agreed: Rosales, including the ROSACEAE, Crassulaceae, and Saxifragaceae; Fabales, including the FABACEAE (Leguminosae); Proteales, including the Elaeagnaceae (e.g. oleaster) and Proteaceae (proteas); Podostemales; Haloragidales, including the Gunneraceae; Myrtales, including the Myrtaceae (e.g. eucalypts, guava), Onagraceae (e.g. willowherbs, fuchsias), and Melastomataceae; Rhizophorales (mangroves); Cornales (e.g. dogwoods); Santalales (e.g. sandalwoods); Celastrales (e.g. spindles); Rhamnales, including the Rhamnaceae (e.g. buckthorns) and Vitaceae (e.g. grapes, currants); Linales (e.g. linseed, flax); Polygalales (e.g. milkwort); Sapindales, including the Hippocastanaceae (e.g. horse chestnuts), Aceraceae (e.g. maples, sycamores), Anacardiaceae (e.g. cashew,

pistachio, mango, sumach), Rutaceae (e.g. orange, lemon), and Balsaminaceae (e.g. *Impatiens*); and Apiales, including the Araliaceae (e.g. ivy, ginseng) and APIACEAE (Umbelliferae; e.g. carrot, parsley).

Other orders that are sometimes included in the Rosidae are the Geraniales (here placed in the Dilleniidae), Connarales (here included in the Rosales), Rafflesiales (here included in the Aristolochiales in the Magnoliidae), Juglandales (here placed in the Hamamelidae), Oleales (here included in the Scrophulariales in the Asteridae), Euphorbiales (here placed in the Dilleniidae), and Nepenthales (also placed in the Dilleniidae).

rostellum a flap of sterile tissue that separates the stigmatic surface from the anthers in the column (the structure formed by the fusion of the sex organs) of an orchid flower. It is a modified stigma.

rot a disease in which there is disintegration of plant tissues. Rots are particularly important as postharvest diseases. See also DRY ROT, WET ROT, FOOT ROT, SOFT ROT, ROOT ROT.

rotenone see FLAVONOIDS.

rough endoplasmic reticulum see ENDOPLASMIC RETICULUM.

RPF see RELATIVE POLLEN FREQUENCY.

RQ see RESPIRATORY QUOTIENT.

rRNA see RIBOSOMAL RNA.

R technique the commonest method of statistical analysis in which data are plotted on a table or matrix with variables or attributes (n) on one axis and observations from different sample sites or individuals (N) on the other axis. Compare Q TECHNIQUE. See also INVERSE ANALYSIS, NORMAL ANALYSIS.

Rubiaceae a large family of dicotyledonous plants containing about 10 200 species in some 630 genera. It is mainly tropical in distribution, all the tropical species being woody. Temperate representatives are always herbaceous and include such species as *Galium saxatile* (heath bedstraw) and *Asperula arvensis* (field madder). The temperate species characteristically have square stems and leaves borne in whorls. All members, both temperate and tropical, bear stipules and the flowers are usually actinomorphic and hermaphrodite. The ovary is normally inferior and in most members contains two carpels. The family contains numerous heterostylous species, and many tropical species have long corolla tubes, pollinated by butterflies, moths, or (more rarely) birds.

Commercially important products of the Rubiaceae include coffee, from *Coffea arabica* and *C. canephora*, and quinine, from species of *Cinchona. Gardenia* species are much planted as ornamentals.

rubidium–strontium dating see RADIOMETRIC DATING.

rubisco see RIBULOSE BISPHOSPHATE CARBOXYLASE.

RuBP see RIBULOSE BISPHOSPHATE.

RuBP carboxylase see RIBULOSE BISPHOSPHATE CARBOXYLASE.

ruderal a plant that grows on or around human dwellings, agricultural land, or wasteland. Examples include many weeds that require relatively high concentrations of nutrients but cannot tolerate competition.

RuDP see RIBULOSE DIPHOSPHATE.

rufous reddish.

rugose wrinkled. Rugose leaves resemble a quilted mattress.

ruminate describing ENDOSPERM that is irregularly grooved or ridged and so appears chewed as, for example, seen in many members of the Myristicaceae.

Rumpomycetes see CHYTRIDIOMYCOTA.

rumposome see CHYTRIDIOMYCOTA.

runcinate saw-toothed.

runner a creeping stem that arises from an axillary bud and runs along the ground, giving rise to plantlets at the nodes, as in the creeping buttercup (*Ranunculus repens*), or apex, as in the wild strawberry (*Fragaria vesca*). Runners are formed by many rosette plants. They often differ greatly from the normal stem of the plant and usually possess greatly lengthened internodes. See also STOLON, VEGETATIVE REPRODUCTION.

rupestrine growing among rocks. Compare PETROPHILOUS.

russetting the development of brown corky patches on certain fruits, especially apple varieties, in response to various chemicals. Russetting of leaves, tubers, and other plant organs may also occur in response to damage or infection.

rust fungi see UREDINALES.

rusts plant diseases caused by fungi of the order UREDINALES. Rust diseases are easily recognized by streaks of dark pustules of spores (UREDINIOSPORES) on the leaves or stems. Autoecious rusts complete their life cycle on one host whereas heteroecious rusts have an alternate host for some of their spore stages. Many rust fungi cause economically important diseases. Rust of currant is caused by the heteroecious rust *Cronartium ribicola* whose alternate host is pine. Black stem rust of cereals (*Puccinia graminis*) has barberry (*Berberis*) as an alternate host.

ruthenium red a temporary stain that turns mucilage and certain gums pink. It is used as a test for PECTIN in the middle lamella of plant cells.

Ss

S *Symbol for* **1** sedimentation coefficient. **2** selection coefficient.

saccate resembling a sac or bag.

saccharide see CARBOHYDRATE.

Saccharomycetales (Endomycetales) an order of the ASCOMYCOTA in which the mycelium, which usually has septa with a series of minute pores, is poorly developed or even absent. The walls usually have chitin only around bud scars. Asexual reproduction is by budding or fission. There are no ascomata; the asci arise singly or in chains. The order contains some 273 species in about 75 genera. It includes the family Saccharomycetaceae, the yeasts, which contains the commercially important species *Saccharomyces cerevisiae* (brewer's yeast), used in bread and beer making, and *S. ellipsoideus*, used in wine making.

saccharose see SUCROSE.

saccoderm desmids see GAMOPHYTA.

sac fungi see ASCOMYCOTA.

safranin a red permanent stain that is used especially to stain nuclei in plant cells. It also stains lignified and cutinized tissues red and chloroplasts pink. It is usually used with a green (e.g. FAST GREEN) or blue (e.g. HAEMATOXYLIN) counterstain.

sagenogen (sagenogenetosome) see BOTHROSOME.

sagittate describing a structure shaped like an arrowhead. Sagittate leaves have two barbs extending back behind the point where the petiole is joined to the leaf. They are characteristic of the genus *Sagittaria*. The marsh arrow-grass (*Triglochin palustris*) is so named because of its sagittate dehisced (see DEHISCENCE) fruits. *See illustration at leaf.*

Sahel zone the semi-arid edge of the Sahara in western Africa, which is currently suffering from the advance of the desert as a result of overgrazing and removal of scrub vegetation. It supports a xerophytic type of savanna vegetation with scattered thorn trees.

salicylic acid ($C_7H_6O_3$) a carboxylic acid (see PHENOL) used to treat various skin conditions. The acetate ester, acetylsalicylic acid forms the basis of aspirin (initially derived from boiled willow (*Salix*) bark but now synthesized in the laboratory), while the methyl ester, methyl salicylate forms oil of wintergreen, used in fragrances and flavourings. The phenyl ester, phenyl salicylate, is used as an antiseptic and antipyretic. It occurs naturally in many plants and forms part of the defensive response, apparently being required for the production of proteins needed to resist pathogens.

salination (salinization) the accumulation of salts of magnesium, potassium, and sodium in water or soil. Salinization of soils occurs in arid and semiarid regions where water rises to the surface as result of capillarity powered by high rates of evaporation. As the water evaporates, its dissolved salts remain in the upper layers of the soil. This greatly reduces the number of plants that can grow there (see PHYSIOLOGICAL DROUGHT), thus contributing to desertification. Another cause is saline groundwater infiltrating the soil near coasts, often where too much water has been extracted for irrigation. Salinization of bodies of water occurs where rates of evaporation exceed precipitation.

saline relating to salt.

salinity a measure of the concentration of dissolved salts or ions in a given volume of water, measured after all the organic matter has been completely oxidized, all bromide and iodide has been converted to chloride, and all carbonate converted to oxide. Salinity is usually expressed in parts per thousand by weight. 1 part per thousand (ppt) = 1 *practical salinity unit* (psu). Fresh water has a salinity of zero, the average salinity of estuarine water ranges from 0 to 35 psu, and ocean water averages 35 psu. Much higher salinities may be recorded in inland lakes and seas, e.g. the Red Sea. An increase in salinity raises the relative density, lowers the freezing point and has a profound effect on on the plant density and species composition of ecosystems.

salinization see SALINATION.

S **alleles** a multiple allelic series governing INCOMPATIBILITY reactions in certain plant species. Plants possessing the same two *S* alleles are incompatible. Plants sharing one *S* allele are semicompatible in that about half the pollen grains will be able to germinate on the style and subsequently achieve pollination. Such plants are usually indistinguishable from fully compatible plants (plants possessing two different pairs of *S* alleles) since pollen is usually produced in sufficient quantity to mask the fact that 50% is inviable.

   S alleles have been found in many fruit crops (e.g. *Prunus avium*), in several grasses, in clover, and in brassicas. There may be more than 40 such alleles in a species and complex dominance relationships are found.

salt absorption the uptake of inorganic ions across the SELECTIVELY PERMEABLE MEMBRANES of plants. If the absorption is against a diffusion gradient, as is often the case across the TONOPLAST, then ACTIVE TRANSPORT is involved. In addition, the uptake is often selective. For example, in sea water the concentration of sodium ions is around forty times higher than that of potassium ions. However within the cells of marine algae the potassium concentration is often about six times greater than that of sodium.

salt gland a glandular structure on a plant leaf that excretes excess salt. It is often a modified HYDATHODE, e.g. *Limonium* spp.

salt marsh a region of vegetation, consisting of HALOPHYTES, found in sheltered river estuaries subject to frequent covering by the tides. Salt marshes are often highly productive and may be used as grazing land. Dominant vegetation may be grasses (such as *Puccinellia* spp.) or dwarf shrubs (e.g. *Halimione*), with pioneer species, such as *Salicornia* and *Spartina*, in the parts more frequently inundated. See also HALOSERE.

salt stress osmotic stress experienced by plants growing in saline conditions such as salt marshes. See OSMOSIS, PHYSIOLOGICAL DROUGHT.

Salviniales see FILICALES.

samara a type of ACHENE with a pericarp extended into a membranous wing, which aids wind dispersal of the seed. The winged fruits or *keys* of the ash (*Fraxinus excelsior*) are an example. The double samara, typical of sycamores and maples (*Acer*), is a kind of SCHIZOCARP.

sampling see MINIMAL AREA, PIN-FRAME, PLOTLESS SAMPLING, QUADRAT, RANDOMIZATION, RANDOMIZED BLOCK, RANDOM SAMPLE, REGULAR SAMPLE, SYSTEMATIC SAMPLE.

sand mineral particles, usually consisting mainly of quartz, with a diameter of 0.05–2.0 mm. A sandy soil is defined as one containing at least 85% sand and not more than 10% clay.

sand dune an accumulation of blown sand found on the coast and in inland desert areas. Coastal dunes are formed by inshore winds that carry sand particles and deposit them on various obstacles in their path, such as pieces of seaweed. The bare dunes facing the shore are known as *foredunes*. On the shore itself, small objects such as piles of seaweed washed up by the tide also trap sand and start the formation of *embryo dunes*. The same thing happens in deserts. In regions that are relatively undisturbed by the tides, seeds of XEROPHYTES germinate on the foredunes and establish a pioneer community termed a *psammosere*. The plants tolerate abrasion by blown sand, high winds, high temperatures in the day, and some salt from sea spray. Sand twitch (*Elytrigia juncea*) helps to stabilize the dune with its extensive network of rhizomes. It is also tolerant to limited immersion in sea water. Marram grass (*Ammophila arenaria*) then becomes established, although in some cases it may itself be the pioneer species. Like sand twitch, marram grass has a network of rhizomes that stabilizes the dune but it is intolerant to immersion in sea water. In this early stage of SUCCESSION, the plants help trap more sand, and also, by their death and decay, help to begin soil formation, preparing the environment for species of the next seral stage. The landward slopes of the dunes are more sheltered and less steep than the seaward slopes and other plants grow including sea spurge (*Euphorbia paralias*), sea holly (*Eryngium maritimum*), etc. A dune system is established, the innermost series being stable and termed *fixed dunes*. Behind these, dune pasture occurs with typical dune plants together with species commonly found in pasture. In some locations, large mature dune systems develop a *dune heath* vegetation dominated by drought-resistant ericaceous shrubs (see ERICACEAE). In between the dunes are damp hollows called *dune slacks*, often containing pools or lakes. These may support a rich flora. See *also* xerophyte, xerosere.

Sanger method see GENE SEQUENCING.

Sanger's reagent a reagent used to detect and quantify amino acids, peptides, and proteins in CHROMATOGRAPHY. It consists of 1-fluoro-2-4-dinitrobenzene, and reacts with certain amino groups to form yellow dinitrophenyl derivatives.

sap the liquid, consisting of mineral salts and sugar dissolved in water, that is found in xylem and phloem vessels. The term is also used of the fluid in the cell vacuole (*vacuolar sap*).

sapling a young tree not yet large enough to be of commercial use.

sapogenin see SAPONIN.

saponin any of a class of bitter-tasting GLYCOSIDES in which the aglycone portion (the *sapogenin*) is a steroid alcohol. Saponins are soluble in water, characteristically producing a foam, hence their name. Many are toxic to animals and have been used as fish poisons. Some saponins are important in the commercial production of steroid hormones, notably diosgen, obtained from various yam (*Dioscorea*) species. The function of saponins in the plant is obscure though they may serve to deter predators.

saprobe (saprophage, saprophyte, saprotroph, saprovore) (*adj.* saprobic) An organism that feeds by the external digestion of dead organic material, thus bringing about decay. Many fungi and bacteria are saprobes and play an important part in the recycling of matter, as the inorganic by-products of their digestion can be rebuilt into organic compounds by green plants. See also DECOMPOSITION.

Saprolegniales (water moulds) an order of the OOMYCOTA (or, in some classifications, of the fungal class Oomycetes) containing some 140 species in about 20 genera. They are aquatic, mainly saprobic or parasitic, fungus-like microorganisms (e.g. *Saprolegnia*), often with DIPLANETIC zoospores. Some species show another form of asexual reproduction and produce CHLAMYDOSPORES (gemmae), pieces of protoplast that break off from the mycelium and germinate into hyphae bearing sporangia. Some water moulds are parasitic on fish and their eggs, e.g. *S. parasitica*, which often attacks salmon and goldfish.

saprophage see SAPROBE.

saprophyte see SAPROBE.

saprotroph see SAPROBE.

saprovore see SAPROBE.

sap stain a stain, often in bluish or blackish wedge-shaped patches, in the sapwood of freshly cut timber, usually appearing within hours or days of cutting. It is caused by the action of various fungi.

sapwood (alburnum) the outer functional part of the SECONDARY XYLEM cylinder, as compared to the central nonfunctional HEARTWOOD.

SAR see SODIUM ABSORPTION RATIO.

sarcotesta a fleshy outer layer of the testa, e.g. in cycads.

sarment (*adj.* sarmentose) a slender, prostrate runner, e.g. strawberry (*Fragaria*).

satellite (trabant) a spherical body seen attached to one end of a chromosome arm by a narrow filament at mitotic metaphase. Often a pair of satellites are apparent. Usually satellites are only found on one or two chromosomes but occasionally more are found. The number and position of the satellites are one of the features of the KARYOTYPE. The satellite stalk functions as a NUCLEOLAR ORGANIZER.

satellite DNA see REPEATED SEQUENCE.

saurochory dispersal of seeds or spores by lizards or snakes.

savanna see GRASSLAND.

savanna woodland a savanna (see GRASSLAND) in which the trees and shrubs are sufficiently dense to form a light canopy. Most of the trees and shrubs are stunted and deciduous, but there are some evergreens. Savanna woodland must withstand frequent fires, so many of the woody plants have thick, fire-resistant bark.

saxicolous growing on rocks.

saxitoxin a toxin produced by some dinoflagellates (DINOMASTIGOTA), which causes paralytic shellfish poisoning.

scab any plant disease having conspicuous raised scablike lesions that develop as a result of the formation of cork layers. The host response to the disease is similar to that found in CANKER but occurs to a lesser degree. Scab diseases are very varied. Examples include common scab of potatoes caused by the bacterium *Streptomyces scabies* and apple scab caused by the fungus *Venturia inaequalis*.

scabrous having a surface that is rough to the touch.

scalariform resembling a ladder as, for example, certain secondary xylem elements possessing parallel bands of thickening or some perforation plates that have several pores separated by parallel bars of tissue.

scalariform thickening (scalariform–reticulate thickening) a type of RETICULATE secondary cell wall patterning in TRACHEARY ELEMENTS in which the network is broken by elongate unthickened areas arranged in a more or less parallel fashion. Whereas ANNULAR THICKENING and SPIRAL THICKENING permit further elongation, reticulate and scalariform thickenings allow little further extension, and are therefore to be found in tissues, such as METAXYLEM and SECONDARY XYLEM, that do not elongate after maturation. Compare PITTED THICKENING.

scale 1 in some algae and other protoctists, a platelike structure made of organic material on the surface of a cell.

2 see BUD, SCALE LEAF.

scale leaf (cataphyll) a small scalelike structure, often without chlorophyll, that arises in the same position as a leaf, and is therefore classified as a modified leaf. Scale leaves often protect vegetative or floral apical meristems. They are also common on rhizomes, where they subtend axillary buds. Scale leaves associated with inflorescences are called BRACTS, bracteoles, or hypsophylls. A few scale leaves are very large: the bracts of certain palms are woody and may reach lengths of over a metre.

scandent climbing.

scanning electron microscope see ELECTRON MICROSCOPE.

scape the leafless stem of a solitary flower or inflorescence, such as that of the dandelion (*Taraxacum*) inflorescence.

scarification the abrasion or chemical treatment of the surface of a HARD SEED to make it permeable to water and so hasten germination.

scarious having a dry membranous appearance, but fairly stiff.

scavenger see DETRITIVORE.

Schiff's reagent a colourless solution that is produced by the reduction of basic fuchsin (a magenta dye) with sulphurous acid. It is used in histochemical tests to detect aldehyde and ketone groups in certain compounds, which oxidize the reagent and restore its magenta colour. See also FEULGEN'S TEST.

schizocarp a dry fruit that is derived from two or more one-seeded carpels that divide into one-seeded units at maturity. The one-seeded units may be achenes, berries, follicles, mericarps, nutlets, or samaras. This form of fruit is intermediate between the dehiscent and indehiscent types. See also CREMOCARP. *See illustration at* fruit.

schizogeny the formation of a space by the separation of cells. Compare LYSIGENY.

Schizomycetes (fission fungi) formerly, a class of the fungi containing the bacteria. With the realization that bacteria have no affinities with fungi, the name is now obsolete.

Schultze's solution (chlor-zinc iodide, CZI) a temporary stain containing a mixture of zinc dissolved in hydrochloric acid added to iodine dissolved in potassium iodide. It is used to detect the presence of cellulose in plant tissue, which it turns blue. Lignified walls stain blue-green, lignin and suberin yellow, and starch blue-black.

sciaphilic (sciaphilous) describing an organism adapted to live in shade.

scion a shoot or bud taken from one plant and joined by GRAFTING or BUDDING onto another plant with roots, the STOCK.

sclereid a relatively short SCLERENCHYMA cell, usually formed when the wall of a parenchyma cell undergoes secondary thickening and, often, lignification (see LIGNIN). The simple pits of sclereids are often more conspicuous than those of fibres. Sclereids are often present as IDIOBLASTS in other tissues. They may however be present in large numbers, as, for example, in those plants where they form the testa of the seed. See also ASTROSCLEREID, BRACHYSCLEREID, MACROSCLEREID, SECONDARY GROWTH.

sclerenchyma strengthening tissue composed of relatively short cells (SCLEREIDS) and/or relatively long ones (FIBRES) with thick, often lignified (see LIGNIN), cell walls and usually lacking a living protoplast at maturity. Sclerenchyma cells usually possess simple unbordered PITS, although FIBRE-TRACHEIDS may have pits with a slightly raised border. Sclerenchyma may form by thickening (*sclerification*) of the secondary cell walls of parenchyma cells, often involving lignification, or it may develop directly from meristematic tissue.

sclero- prefix denoting hard.

sclerophyll (*adj.* sclerophyllous) a stiff, leathery leaf usually containing a lot of SCLERENCHYMA.

sclerophyllous vegetation a type of scrub and woodland characterized by hard, leathery evergreen foliage adapted to minimize water loss, e.g. pines (*Pinus*), olive (*Olea europea*), holly (*Ilex*). It it typical of areas with a Mediterranean climate of hot dry summers and mild, wet winters.

sclerotium 1 the pseudoparenchymatous, often rounded, resting body of certain fungi, which lacks spores and can survive long periods of adverse conditions to produce either a mycelium or fruiting bodies. Sclerotia are important sources of inoculum for various root-rot diseases (e.g. *Sclerotinia* spp.) The hard dark club-shaped structures seen in the ears of cereals and grasses affected with ergot (*Claviceps purpurea*) are sclerotia.

2 see PLASMODIUM.

scorpioid cyme see INFLORESCENCE.

SCP see SINGLE-CELL PROTEIN.

Scrophulariaceae a large family of dicotyledonous plants, numbering about 5100 species in some 269 genera. The family, commonly called the foxglove or figwort family, is distributed worldwide though the greatest concentration of species is in northern temperate regions. Most of its members are herbaceous, but some are climbers, shrubs, or trees. The leaves occur in a variety of forms but always lack stipules; they are usually spirally or oppositely arranged. There is also a range of flower and inflorescence types. The flowers are usually irregular and often two-lipped, as in the snapdragons (*Antirrhinum, Misopates*) and toadflaxes (*Linaria, Cymbalaria*). However in the mulleins (*Verbascum*) and speedwells (*Veronica*) the flower is almost actinomorphic. In many flowers there is a reduction in the number (assumed originally to have been five) of floral parts. For example, figworts (*Scrophularia*) only have four stamens, and speedwells have four sepals and four petals and only two stamens. The fruit is usually a dehiscent capsule.

Few members of the family are of economic importance though many genera, e.g. *Calceolaria, Hebe, Mimulus, Antirrhinum*, include ornamental varieties. Some species are parasitic or, more commonly, hemiparasitic on the roots of other angiosperms, especially grasses, and may be serious weeds, e.g. *Striga* (witchweeds). The foxglove (*Digitalis purpurea*) is the source of the drug digitalin (see CARDIAC GLYCOSIDE).

scrub a general term for vegetation dominated by shrubs. Scrub is often the natural vegetation in the transition zone between forest and grassland or heath and may also be part of the succession to woodland.

scutellum an intermediate absorbing organ

between the embryo and the nutritive endosperm in the grass fruit (caryopsis). During germination it secretes enzymes involved in the digestion of the endosperm. It is thought to be a modified cotyledon.

Scytosiphonales an order of brown algae (PHAEOPHYTA) containing algae with tubular parenchymatous gametophyte thalli, in some species constricted at intervals, and diffuse meristems. The lamina has a colourless central MEDULLA surrounded by layers of smaller photosynthetic cells. According to the environmental conditions, in some species, e.g. *Scytosiphon*, the thallus may be upright or encrusting.

SDP see SHORT DAY PLANT.

seaweed any of various macroscopic marine ALGAE found on rocky coasts or free floating in the sea. Species occurring in the intertidal zone are periodically exposed to the air and have developed leathery mucilaginous thalli to prevent desiccation. Most seaweeds belong to the PHAEOPHYTA, RHODOPHYTA, or CHLOROPHYTA.

secondary cell wall see APPOSITION, CELL WALL.

secondary consumer see CONSUMER, FOOD CHAIN.

secondary cortex see PHELLODERM.

secondary dormancy see DORMANCY.

secondary growth (secondary thickening) the increase in diameter of a plant organ resulting from cell division in a CAMBIUM, more specifically from secondary or lateral meristems – the vascular cambium and the cork cambium. Secondary growth results in the formation of *secondary tissues*, and generally produces an increase in girth rather than in height. For example the vascular cambium gives rise to secondary xylem, secondary phloem, and the cork cambium (PHELLOGEN) gives rise to the phellem (cork) and phelloderm (cork cortex). The phelloderm arises from the phellogen as radial columns of cells, readily distinguishable from the cells of the primary cortex. The phellogen, phellem, and phelloderm are collectively called the *periderm*. Periderm may also develop at the site of a wound, forming a barrier to the entry of pathogens. PRIMARY TISSUES, on the other hand, are derived from primary

meristems. The parts of the plant formed by secondary growth are collectively called the *secondary plant body*. Secondary growth occurs in most dicotyledons and gymnosperms and a few monocotyledons. Secondary tissues have an important supporting role; wood often consists almost entirely of secondary xylem. There are relatively few living representatives of lower vascular plants exhibiting secondary growth but it was common in many extinct species. See also GROWTH RING, MERISTEM.

secondary phloem PHLOEM derived from the vascular cambium in plants exhibiting SECONDARY GROWTH. As in the SECONDARY XYLEM, the secondary phloem consists of two systems: the *axial* (vertical) *system*, derived from the FUSIFORM INITIALS of the vascular cambium and consisting mainly of SIEVE ELEMENTS, their associated COMPANION CELLS, and some phloem fibres; and the *ray* or *radial* (horizontal) *system*, derived from the RAY INITIALS of the vascular cambium and consisting mostly of ray parenchyma.

secondary ray see RAY.

secondary root see ROOT.

secondary forest see FOREST.

secondary meristem see MERISTEM.

secondary pit see PIT.

secondary plant body see SECONDARY GROWTH.

secondary productivity the rate at which consumers convert the chemical energy of their food into their own biomass.

secondary structure the second level of structure in a PROTEIN: the basic shape of the amino acid chain. There are two main forms. The α-helix is an extended spiral held in shape by hydrogen bonds between adjacent C=O and NH groups. There is one turn of the helix for every 3.6 amino acids. Further cross-linking by DISULPHIDE BRIDGES confers varying degrees of hardness and elasticity on the protein. The other type of secondary structure is the β-pleated sheet, in which the amino acid chains are more extended. Adjacent chains are arranged antiparallel to each other and linked by hydrogen bonds formed between the C=O and NH groups of one chain and the NH and C=O groups of the next. Such a structure has high tensile strength but is also flexible. Often a protein is made up of regions of both α-helices and β-pleated sheets. See also CONFORMATION, PRIMARY STRUCTURE, TERTIARY STRUCTURE, QUATERNARY STRUCTURE.

secondary succession see SUCCESSION.

secondary thickening see SECONDARY GROWTH.

secondary tissue see SECONDARY GROWTH.

secondary wall see CELL WALL.

secondary xylem XYLEM derived from the vascular cambium in plants exhibiting SECONDARY GROWTH. Secondary xylem is regarded as consisting of two systems: the *axial* (vertical) *system*, derived from the FUSIFORM INITIALS of the vascular cambium and consisting mainly of TRACHEARY ELEMENTS and FIBRES; and the *radial* or *ray* (horizontal) *system*, derived from the ray initials of the vascular cambium and consisting mostly of ray parenchyma, forming the RAYS. As well as differing from PRIMARY XYLEM in origin, secondary xylem usually also has shorter tracheary elements, often arranged in regular rows, although these differences are not always reliable. Commercial wood is secondary xylem. There are two main types of commercial wood, *softwood* from gymnosperms and *hardwood* from angiosperms. However, these terms are somewhat misleading as there are hard gymnosperm woods and soft angiosperm woods. Softwoods contain only TRACHEIDS as the water-conducting elements, whereas hardwoods usually contain vessels.

section a rank in the taxonomic hierarchy above series but subordinate to genus. Genera containing a large number of species are often subdivided into sections. The section name is printed in italic with a capital first letter and either has the same form as the generic name or, if a plural adjective, agrees in gender with the generic name. For example, the genus *Chenopodium*, which contains about 110 species, is subdivided into the four sections *Agathophytum, Chenopodium, Pseudoblitum,* and *Morocarpus*. The section name is preceded by the abbreviation sect. to show its rank, thus *Chenopodium* sect. *Pseudoblitum*.

sectorial chimaera a CHIMAERA that results from a cell mutation in a meristem that subsequently gives rise to a group or sector of mutant cells. Such chimaeras are usually unstable.

secund describing a moss in which the leaves are all turned in the same direction, e.g. *Dicranum* (fork moss).

sedentary soil see SOIL.

sedimentary soil see SOIL.

sedimentation coefficient symbol: s. The rate of sedimentation in centimetres per second that occurs when a solution is subjected to a centrifugal force of one dyne (10^{-5} newton). These rates are extremely low and are usually multiplied by 10^{13} to give a reasonable figure. In this modified form the unit is termed the *Svedberg unit*, symbol S. Different molecules and cellular inclusions tend to have characteristic sedimentation rates depending on the weight and shape of the molecule. For example, the ribosomes of prokaryotes have a value of 70S and those of eukaryotes 80S.

seed the structure that develops from the fertilized ovule in seed plants. It usually contains one EMBRYO together with a food supply, which may be contained in a specially developed ENDOSPERM or in the COTYLEDONS of the embryo. The whole is surrounded by a protective coat, the TESTA. In gymnosperms the seed remains naked and unprotected but in angiosperms it is enclosed within the ovary wall. A seed may germinate immediately on ripening or may have special DORMANCY mechanisms to prevent germination under unfavourable conditions.

Although seed production is observed today only in angiosperms and gymnosperms, there are extinct forms possibly intermediate between these and the nonseed-bearing vascular plants (see CYCADOFILICALES). The development of the seed-bearing habit has released seed plants from dependence on the availability of water for their reproductive phase, thus opening up new habitats for colonization. In addition the seed allows for wide dispersal of the plant and, in annual and ephemeral plants, serves as a PERENNATING ORGAN.

Some plants produce seed without fertilization taking place (see APOMIXIS).

seed bank a collection of seeds made for the purpose of research or to conserve rare, threatened, or endangered species.

seed dressing a chemical applied to seeds to protect them against fungal diseases or insect attack. The chemicals are applied in dusts, slurries, or concentrated solutions. Treated seed is toxic and must be distinguishable from untreated seed to prevent it being used for feed. See also ORGANOMERCURIAL FUNGICIDE, QUINONE FUNGICIDE.

seed ferns see CYCADOFILICALES.

seed leaf see COTYLEDON.

seedless fruit a fruit that develops without fertilization occurring so no seed is formed. See PARTHENOCARPY.

seedling a young plant. The seedling often differs significantly from the mature plant in morphology and habit. See JUVENILITY.

seedling blight a disease of germinating seedlings. The term is sometimes applied to post-emergence DAMPING-OFF but generally it is used when well established seedlings are affected. Growth slows and foliage may become yellow or wilt. Roots of affected plants are found to be rotten and there may be lesions at soil level. The causal organisms include the oomycotes *Pythium* and *Phytophthora* and the fungi *Rhizoctonia* (*Corticium*), *Fusarium*, and *Helminthosporium*. Correct planting conditions, seed dressings, and soil sterilization are the usual preventative measures.

seed plants flowering plants (see ANTHOPHYTA) and GYMNOSPERMS. See also SPERMATOPHYTA.

segmental allopolyploid see ALLOPOLYPLOIDY.

segregation the separation of homologous chromosomes at ANAPHASE 1 of meiosis. As a consequence of segregation, alleles that were paired in the somatic cells of the organism become separated, so that only one allele of each kind is present in a single gamete.

segregational load see GENETIC LOAD.

segregation ratio the proportion of one type of offspring to another that occurs as a result of separation of alleles at meiosis (see DIHYBRID RATIO).

seismonasty a NASTIC MOVEMENT in response to shock. The result, as in the collapse of *Mimosa pudica* leaflets after subjection to shaking or singeing, is often the assumption of the night position as normally found after the completion of NYCTINASTY. See also HAPTONASTY.

Selaginellales an order of the LYCOPHYTA containing one extant genus, *Selaginella*, the 700 or so species of which are mainly tropical in distribution. In gross morphology *Selaginella* resembles *Lycopodium* (see LYCOPODIALES) except that it has ligulate microphyll leaves and the sporophylls are always grouped in STROBILI. The leaves are usually inserted spirally but in some species they are arranged in four ranks, the upper two consisting of small leaves adpressed to the stem while the lower two ranks contain expanded leaves. The roots are borne at the end of leafless branches (see RHIZOPHORE). Internally, selaginellas are distinguished by their trabeculate endodermis, which consists of filamentous cells that traverse a continuous cavity between the cortex and the pericycle, so suspending the stele in the centre of the stem.

The strobili bear both mega- and microsporangia and produce the largest megaspores of any spore-producing plant. The megaspore contains a large food supply, which enables the female gametophyte to develop independently of external food sources. The microspore develops into an extremely reduced male gametophyte consisting of a prothallial cell and an antheridium, which produces antherozoids with two undulipodia. Fertilization relies on a megaspore and microspore being in close proximity and a film of water is needed for the antherozoids to swim to the egg cell of the archegonium. Fertilization occurs before the megaspore is shed from the parent.

selection the differential reproduction of one phenotype as compared to others in a population in response to abiotic and/or biotic factors in the environment or to conditions deliberately imposed by humans (*artificial selection*) either by manipulating the environment or by choosing which individuals are allowed to interbreed. This affects the genotypes that persist from one generation to the next, and hence the allele frequency of the population. The probability that a particular phenotype will survive to be reproduced from one generation to the next is termed its *fitness* (*Darwinian fitness*). See also DIRECTION SELECTION, DISRUPTIVE SELECTION, NATURAL SELECTION, SELECTION PRESSURE, STABILIZING SELECTION.

selection coefficient (*s*) symbol: *s*. A measure of the relative lack of fitness of a genotype compared to another genotype in the population. If 1 in 100 individuals of a particular genotype fails to reproduce, *s* = 100.

selection differential the difference between one generation of a population and those individuals that have been selected to form the next generation in the average value of a quantitative character.

selection pressure the intensity with which the environment tends to alter the frequency of a particular allele in a given population. It is not possible to give an absolute value for selection pressure. However, by comparing the survival rates of individuals with different alleles, a measure of the FITNESS of one relative to another may be derived. See NATURAL SELECTION, SELECTION.

selective describing herbicides that kill some plants and not others. Usually broad-leaved plants are killed while narrow-leaved plants are unaffected. Selective weedkillers are therefore useful for cereal crops and lawns. Some selective weedkillers, e.g. Simazine, kill germinating plants but not well established shrubs. See also SYSTEMIC.

selective absorption see SALT ABSORPTION.

selective advantage the situation where individuals possessing a particular GENOTYPE in a population are more likely to produce viable progeny than individuals with other genotypes.

selective herbicide a herbicide that acts only on certain kinds of plants, for example on weeds but not crop plants.

selectively permeable membrane (**semipermeable membrane**) a membrane that acts as a barrier to certain substances but will allow others to pass through. In many cases a selectively permeable membrane will allow the passage of solvents, almost invariably water in

biological situations, but not certain solutes. Such a membrane usually lets through small molecules but not larger ones. See also OSMOSIS.

selective species see FIDELITY.

selective value see ADAPTIVE VALUE.

selenotropism a tropic response to the Moon.

self-compatibility the capacity of an organism for self-fertilization. Many plants can self-fertilize if pollination fails, e.g. dandelion (*Taraxacum*) and wood sorrel (*oxalis acetosella*), and some do it regularly, including many cereal crops.

self fertilization see AUTOGAMY.

self incompatibility (**self sterility**) the failure of gametes from the same plant to form a viable embryo. In angiosperms this is usually due to complex interactions between the pollen and stigmatic tissues. There are three main classes of self incompatibility, depending on the site of interaction. In gametophytic systems, probably the commonest and most primitive type, the pollen tube is blocked by callose as it grows down the stigma; in sporophytic systems the pollen grain fails to germinate or is blocked before it penetrates the stigma surface; in late-acting systems the pollen tube reaches the ovule but then aborts. These differences are not clear cut, and there are different forms of each system; some plants, such as those exhibiting heterostyly, operate some combination of these systems. Although self incompatibility is known to be genetically controlled (see *S* ALLELES) the exact method by which the genetic information manifests itself is not clearly understood. Enzymes either on the pollen or stigma, antibodies and antigens, hormones, or variable cell morphology may account for differential treatment of pollen grains on the stigma. Incompatibility is rarely total and usually develops as the stigma matures. Artificial crosses of normally incompatible lines can therefore be made by in-bud pollination, chemical treatment, or by stigma excision.

selfing 1 the process whereby members of a particular generation resulting from a specific cross are allowed to breed among themselves, but not with individuals from other crosses. For example, the F_1

GENERATION of a cross is selfed to produce the F₂ GENERATION, which may in turn be selfed to produce the F₃ generation.

2 see SELF POLLINATION.

selfish DNA DNA that is capable of moving around an organism's GENOME or inserting copies of itself at various locations in the genome without apparently providing any advantage to the organism. All it appears to achieve is to increase replication of DNA – hence the term 'selfish' DNA. The commonest form of selfish DNA is the TRANSPOSON.

selfish genes a concept advocated by Richard Dawkins which proposes that the gene and not the organism, is the unit of selection in evolution, i.e. organisms serve as agents for the replication of genes, rather than vice versa. In other words, they claim that NATURAL SELECTION acts on the genes rather tham on the individual organism. See also SELFISH DNA, TRANSPOSON.

self pollination (selfing) the transfer of pollen from the anthers to the stigma of the same flower, or to a flower on the same plant. This is achieved by HOMOGAMY and appropriate positioning of the reproductive parts so that pollen can be transferred, usually by insects, gravity, contact, or rain splash. Self fertilization and the consequent formation of seeds may or may not follow. Self pollination does not necessarily preclude CROSS POLLINATION except in cleistogamous plants.

self sterility see SELF INCOMPATIBILITY.

self-thinning the decrease in numbers of plants as density increases, as a result of competition for resources and other interactions, e.g. allelopathy.

Seliwanoff's test a test for ketose sugars, such as fructose, in a solution. The reagent is made by dissolving a few crystals of resorcinol in equal amounts of concentrated hydrochloric acid and water. In the presence of ketoses it forms a red colour on heating.

selva a term used in relation to Central and South America to denote tropical RAINFOREST and similar vegetation.

SEM see ELECTRON MICROSCOPE.

semantide an information-carrying molecule. DNA, as the carrier of the genetic code, is termed a *primary semantide* and is regarded by taxonomists and evolutionists as a primary source of information in drawing up classifications and evolutionary trees. RNA, as the transcription of the genetic code, is termed a *secondary semantide*, and proteins as the end product of this information transfer, *tertiary semantides*. See also DNA HYBRIDIZATION.

semelparous (monocarpic, hapaxanthic) describing a plant that reproduces only once during its lifetime. Semelparous plants include annuals, biennials, and monocarpic perennials, such as the century plant (*Agave Americana*) and some bamboos. Compare ITEROPAROUS.

semi- prefix denoting half, part, incomplete.

semicell see GAMOPHYTA.

semiconservative replication the process by which DNA makes exact copies of itself. During this process, double-stranded DNA uncoils and the separated polynucleotides act as templates for the formation of new complementary polynucleotide chains. Hence any given DNA molecule will consist of one old strand and one new one and the original molecule is semiconserved during the manufacture of an identical copy (replica). See also DNA REPLICATION.

semi-desert scrub a type of vegetation found around the edges of deserts, in areas that are still too dry to support the denser growth of thorn forest or savanna. Most of the plants are shallow-rooted shrubs and succulents, which can make the most of occasional rainfall.

semi-evergreen seasonal tropical forest a kind of forest found adjacent to rainforests in areas with a pronounced dry season. The dominant species are a mixture of evergreen and deciduous broad-leaved trees.

seminal root a root developed from a root primordium that was present in the embryo before germination.

semi-natural community a community that has been altered by human activity in the past, but which has undergone a long period of succession and now appears natural, e.g. chalk GRASSLAND, which is adapted to withstand grazing. The term also applies to woodlands that have undergone natural regeneration after being managed or cropped.

semipermeable membrane see SELECTIVELY PERMEABLE MEMBRANE.

semi-species a group of organisms that is intermediate between a race and a species, and which is probably in a late stage of SPECIATION. Outbreeding and gene flow between the group and other groups is restricted, but reproductive isolation is incomplete.

senescence the period between maturity and death of a plant or plant part. A gradual deterioration occurs and, in the case of senescent leaves and fruits, the end point is usually ABSCISSION from the plant. Senescence is characterized by an accumulation of waste metabolic products, a decrease in dry weight as reusable substances are withdrawn from the affected part, and a rise in the respiration rate (see CLIMACTERIC).

sense strand see MESSENGER RNA.

sensitive plant (*Mimosa pudica*) a plant of the family Fabacae (pea family) whose leaves are divided into a series of small leaflets (pinnae). When the leaf is touched, the pinnae progressively collapse and droop. This is achieved by changes in the turgor of special cells in swellings in the petiole at the base of each pinnae (see PULVINUS). See also NASTIC MOVEMENTS.

sensitivity see IRRITABILITY.

sensitivity analysis in mathematical modelling of an ecosystem, an analysis of the stability of the system being modelled and its sensitivity to changes in its parts. This will allow predictions to be made about its future behaviour and response to change.

sepal an individual unit of the CALYX. It is usually green and often hairy but in some species, e.g. marsh marigold (*Caltha palustris*), the sepals are brightly coloured and assume the function of PETALS. In such plants the petals may be absent, as they are in marsh marigold, or reduced, as in the Christmas rose (*Helleborus niger*), where they form small tubular nectaries. Sepals are supplied by several vascular bundles and are thought to be modified leaves.

sepaloid resembling sepals.

separation layer see ABSCISSION.

septicidal see CAPSULE.

septifragal see CAPSULE.

septum (*pl.* septa, *adj.* septate) any partition, whether within a cell, as in a septate fibre, or in an organ, such as a fruit. In fruits, such a separating wall is often termed a *dissepiment*.

sequestration the formation of a complex with an ion in solution, which prevents the ion having its usual activity. Ions are commonly sequestered by chelating agents. See CHELATION.

sequestrol a preparation of MICRONUTRIENTS used to treat DEFICIENCY DISEASES, particularly those caused by lack of iron or magnesium. The elements are chelated to promote their uptake by plant roots and prevent them becoming bound to the soil particles. A preparation containing a particular element is named accordingly, e.g. sequestering iron. See also CHELATION, SEQUESTRATION.

seral stage see SERE.

sere (seral community, seral stage) any plant community in a succession leading to a CLIMAX. The nature of a sere is influenced by that of the preceding sere and itself influences the development of the succeeding sere. The initial community (pioneer community) in a succession is known as a *prisere*. *Subseres* are the stages in the development of a secondary succession, while *microseres* are the stages in microhabitats, such as small puddles or animal droppings. See also CLISERE, HALOSERE, HYDROSERE, XEROSERE.

serial bud see ACCESSORY BUDS.

series a rank in the taxonomic hierarchy above species but subordinate to section.

serine a polar amino acid with the formula $HOCH_2CH(NH_2)COOH$ (*see illustration at* amino acid). Serine is synthesized from GLYCERATE 3-PHOSPHATE. It is also formed during photorespiration and probably during the catabolism of CYTOKININS. Breakdown of serine occurs by oxidative DEAMINATION to pyruvic acid. It is a component of several PHOSPHOGLYCERIDES.

serology the study of blood serum. Following invasion of an animal body by foreign materials (*antigens*) the serum contains *antibodies* produced to combat the antigens. Serum containing antibodies produced against a particular kind of antigen is termed *antiserum*. Each antigen stimulates the production of a specific antibody active against that antigen alone or particles very like it. The specificity of the immune response has been exploited in the identification of viruses and in CHEMOTAXONOMY, where it is used to compare protein extracts from different plant species. This field of plant taxonomy is termed *serotaxonomy*. A laboratory animal, usually a rabbit, is injected with a plant protein extract. A few weeks later when the animal has produced sufficient antiserum, a blood sample is taken. The antiserum is then mixed *in vitro* with antigens from the same species used to raise the antiserum (the homologous species extract) and subsequently against antigens from a range of other species. The degree of similarity between antiserum and antigen is measured by the turbidity produced in a liquid by the antibody–antigen precipitation reaction. The turbidity should be greatest when the antiserum is mixed with the homologous species extract. Antigens from different species will produce varying amounts of precipitation, the more similar species producing greater turbidity than the less similar species.

serotaxonomy see SEROLOGY.

serotinal in late summer.

serotiny (*adj.* serotinous) the retention of seeds in cones or pods on a tree for long periods, often many years, until some event, such as a fire, causes them to be released. Examples include the jack pine (*Pinus banksiana*) and many *Eucalyptus* species, whose seeds benefit from germinating in soil enriched by ash from the fire in a environment from which competitors have been eliminated.

serpentine barrens scrub or heath vegetation on areas of serpentine rocks, which on weathering release large amounts of magnesium into the soil, preventing the development of a natural climax community and reducing plant biodiversity. In places, local magnesium-tolerant races of certain species have evolved, and there may also be uncommon species that are tolerant enough to survive, and thus benefit from the lack of competition.

serrate describing a leaf margin that is toothed, with forward-pointing notches (*see*

illustration at leaf). Leaf margins finely toothed in this manner are termed *serrulate*.

serrulate see SERRATE.

sesquioxides a general term for hydrated oxides of aluminium and iron. The prefix 'sesqui-' denotes a 2/3 ratio, referring to the molecular formula, e.g. Al_2O_3.

sessile 1 Unstalked, as a leaf with no petiole, a flower with no pedicel, a stigma with no style, or a seaweed with no stipe. 2 Attached to a substrate, non-motile.

seta 1 the stalk of a bryophyte sporophyte that supports the capsule. At maturity, the seta usually elongates rapidly, bearing the capsule to a height suitable for spore dispersal. A seta often contains vascular tissue, enabling the sporophyte capsule to obtain nutrients from the parent gametophyte (see ALTERNATION OF GENERATIONS). See also APOPHYSIS.
2 a nervelike extension to a leaf as seen in certain species of *Selaginella*.
3 in certain of the BASIDIOMYCOTA, a stiff hair projecting from the hymenium in the basidioma.

setose bristly.

Sewall Wright effect see GENETIC DRIFT.

sex chromosome a chromosome that carries sex-determining genes. All other chromosomes are called AUTOSOMES. Sex chromosomes are widespread in the animal kingdom, associated with the fact that most animals have separate sexes. However most plants are hermaphrodite and sex chromosomes are correspondingly rare in the plant kingdom, although they are found in dioecious plants. In some such plant species sex is determined by one pair of alleles. However in certain angiosperm genera (e.g. *Salix*, *Silene*, *Cannabis*) cytologically identifiable sex chromosomes are seen.

sexine the outer layer of the EXINE in a pollen grain.

sex-limited inheritance the restriction of a trait to one sex as a result of incomplete PENETRANCE rather than allelic differences.

sex linkage (sex-linked inheritance) the tendency for a certain character to appear more often in one sex than the other. It occurs because the gene controlling that character occurs on the same chromosome as the gene or genes involved in sex determination.

sexual apex the tip of a thallus that is differentiated into structures associated with sexual reproduction. Such thalli are common in brown algae (Phaeophyta) and red algae (Rhodophyta). For example, the tips of the thalli of many wracks (*Fucus*) are differentiated into gametangia. Compare VEGETATIVE APEX.

sexual dimorphism see DIMORPHISM.

sexual reproduction the formation of new individuals of a species by the fusion of two normally haploid gametes to form a diploid ZYGOTE. The gametes may be derived from the same parent (see AUTOGAMY) or from two different parents (see ALLOGAMY). In certain unicellular organisms the whole individual may participate in the process, as seen when two haploid yeast cells fuse to form a diploid cell. In multicellular organisms the gametes are formed and often fuse in specialized organs (see GAMETANGIUM). Compare ASEXUAL REPRODUCTION.

shade plant a plant that is able to flourish in conditions of low light intensity. Some shade species are sensitive to intense light and cannot live in the open. Shade plants usually have thinner epidermal and palisade layers and fewer stomata than normal. They also tend to have a short compensation period, i.e. the food reserves used in respiration at night are quickly replaced by photosynthesis in the day. Shade plants grow in woodlands where the trees, when in leaf, form a canopy, cutting out much of the light. Green light, the part of the spectrum least absorbed by plants, is predominant beneath a closed canopy. Examples of shade plants are mosses, enchanter's nightshade (*Circaea lutetiana*), certain violets (*Viola*), and dog's mercury (*Mercurialis perennis*). Some trees are also tolerant of shade. For example, yew trees can tolerate the deep shade in beech woods but ash trees cannot. However, ash trees can survive in oak woods where the shade is not as dense. See also COMPENSATION POINT. Compare SUN PLANT.

shadowing (heavy-metal shadowing) a technique used in electron microscopy to determine the scale of surface features on a specimen. The specimen is partially coated with a thin layer of a heavy metal, e.g. platinum, palladium, or gold, which is evaporated in a vacuum chamber. The source of the metal ions is situated at a known angle to the specimen. Only the 'windward' side will be coated since metal atoms travel in straight lines. On placing the specimen in a beam of electrons, the uncoated 'leeward' side will allow freer passage of electrons and hence appear darker (a 'shadow') on the photographic plate. By measuring the length of the shadow cast it is possible, knowing the angle of the metal source, to calculate the height of the object casting the shadow. The technique has been used to gain more accurate measurements of very small specimens such as viruses.

shadow yeasts see MIRROR YEASTS.

Shannon-Weaver index of diversity (Symbol: H, information index) a logarithmic measure of species diversity that is weighted by the abundance of each species. See also DIVERSITY INDEX.

sheath 1 a slimy layer surrounding a bacterium, usually made up of MUCOPOLYSACCHARIDES.
2 a gelatinous layer of MUCILAGE surrounding the cells of certain algae. The sheath may help to retain moisture, hold the cells together, attach them to the substrate, or facilitate gliding movement.

shelf fungus a colloquial name for a fungus with a horizontal shelf-like fruiting body that protrudes from its substrate, e.g. a bracket fungus (see APHYLLOPHORALES).

Shelford's law of tolerance a law proposed by V. E. Shelford, which states that the presence and success of an organism depends on the extent to which a complex set of conditions lie within its limits of tolerance.

shifting cultivation see SLASH-AND-BURN.

shikimic acid an aromatic acid, abundant in certain higher plants, that is an intermediate in the synthesis of the aromatic amino acids phenylalanine, tyrosine, and tryptophan. It is also an intermediate in the synthesis of lignin and of the electron carriers COENZYME Q (ubiquinone) and plastoquinone. The shikimic acid pathway is the normal route for the synthesis of PHENOLICS and is one of the few pathways in which aromatic compounds are formed from aliphatic precursors.

shikimic acid pathway a metabolic pathway by which PHOSPHOENOLPYRUVATE (from GLYCOLYSIS) and erythrose 4-phosphate (from the PENTOSE PHOSPHATE PATHWAY) are combined to form a 7-carbon compound that is converted in a series of steps into SHIKIMIC ACID.

shingled describing an overlapping arrangement of liverwort (Hepatophyta) leaves that resembles roofing tiles (shingles).

shoot see STEM.

shoot-tip culture SEE MERISTEM CULTURE.

short-day plant (SDP) a plant that appears to require short days (i.e. days with less than a certain maximum length of daylight) before it will flower. In actual fact it requires a daily cycle with a long period of darkness. Examples of short-day plants are the autumn-flowering (e.g. *Chrysanthemum*) and spring-flowering (e.g. *Fragaria*) species of temperate latitudes. Compare LONG-DAY PLANT, DAY-NEUTRAL PLANT. See PHOTOPERIODISM.

short shoot a shoot, especially a woody shoot, that has relatively short internodes and hence closely spaced leaves. Such shoots often produce spines or flowers. EPICORMIC BRANCHING can also produce short shoots.

shrub a woody perennial that is smaller than a tree (i.e. less than 10 ft tall) and branches into several main stems close to the ground so that it has no obvious main trunk. Its above-ground parts persist all year round.

sibling species see SIBS.

sibs plants derived either by SELFING or by crossing between genetically similar parents. In the production of F_1 HYBRID varieties occasional crossing may occur within rather than between the parent lines giving a proportion of nonhybrid seed. The plants that grow from such seed are called sibs.

The term *sibling species* is sometimes used to describe species that are almost indistinguishable in the field because of recent common ancestry.

sieve area a specialized part of the primary wall (see CELL WALL) of a SIEVE ELEMENT, perforated by a number of pores, derived from a PRIMARY PIT FIELD. These pores are lined with CALLOSE and penetrated by cytoplasmic strands. The sieve areas of gymnosperm sieve cells are relatively unspecialized, whereas those of angiosperm SIEVE TUBES are usually differentiated as SIEVE PLATES.

sieve cell 1 a vascular cell in the PHLOEM of the lower vascular plants and gymnosperms, whose main function is the TRANSLOCATION of sugars and other nutrients. Sieve cells have relatively unspecialized sieve areas compared to those of SIEVE TUBE ELEMENTS in angiosperms and are longer and narrower in shape. **2** in certain brown algae (PHAEOPHYTA), a specialized cell used to transport organic material.

sieve element a vascular cell in the PHLOEM whose main function is the transport of sugars and other nutrients from the site of production to the site of storage or utilization. This is a general term covering both SIEVE CELLS and SIEVE TUBE ELEMENTS.

sieve plate a highly specialized SIEVE AREA, or collection of sieve areas, perforated by large pores and usually located on the end walls of SIEVE TUBE ELEMENTS. A sieve plate composed of a single sieve area is called a *simple sieve plate*. A number of sieve areas may be arranged collectively to form a *compound sieve plate*, as in *Nicotiana*. The pores of a sieve plate, so named for its resemblance to a sieve, are lined with CALLOSE and penetrated by cytoplasmic strands. Compare PERFORATION PLATE.

sieve tube a continuous longitudinal tube composed of numerous SIEVE TUBE ELEMENTS.

sieve tube element (sieve tube member) a relatively advanced vascular cell in the phloem of angiosperms, characterized by the presence of a SIEVE PLATE, a nonlignified (see LIGNIN) secondary cell wall, and a living enucleate protoplast. Sieve tube elements are joined together at their ends to form SIEVE TUBES, their end walls having become modified to form the sieve plates. Sieve tube elements usually have specialized parenchyma cells associated with them. These are known as COMPANION CELLS, the protoplast of the companion cell being connected to that of the sieve tube member by plasmodesmata. Compare SIEVE ELEMENT, SIEVE CELL.

sigmoid growth curve see S-SHAPED GROWTH CURVE.

silage cattle feed produced by the anaerobic decomposition by microbes of green plant matter, such as freshly cut grass, mixed with diluted molasses and sometimes chemical additives in a pit, silage tower, or sealed plastic sack.

siliceous containing silicon.

silicle see SILICULA.

silicon symbol: Si. A grey metalloid element, atomic number 14, atomic weight 28. It is not considered essential to plant growth but nevertheless accumulates as silica (SiO_2) in high proportions in the cell walls of some plants. The cell walls of the DIATOMS are largely composed of silica and certain members of the XANTHOPHYTA and CHRYSOMONADA also accumulate silica. Among higher plants, the stems of *Equisetum* are reinforced with silica and the leaves of many grasses contain substantial amounts. In grasses the details of silica deposition differ between species; this can be used as a taxonomic character, particularly useful as it is often well preserved in archaeological sites. The presence of silica bodies can be demonstrated by carbolic acid solution, which turns such bodies pink. Other crystals remain colourless.

silicula (silicle) a dry dehiscent fruit derived from two carpels fused together to form a flattened pod with two locules separated by a false septum. The seeds are exposed on the septum as the two valves separate from it, from the base up. Siliculae are similar to SILIQUAE, except that they are as broad or broader than they are long. The fruits of shepherd's purse (*Capsella bursa-pastoris*) are an example.

siliqua (silique) a dry dehiscent fruit similar to a SILICULA except that it is long and narrow. It is typical of members of the genus *Brassica*.

silique see SILIQUA.

silt mineral particles having a diameter of 0.002–0.05 mm. A silt soil is defined as

containing more than 80% silt and less than 12% clay and has a characteristic smooth soapy feel. See SOIL TEXTURE.

Silurian the third period of geological time in the PALAEOZOIC era from about 440 to 395 million years ago, during which there is the first evidence of the invasion of the land by plants and animals. Thick-walled spores with triradiate markings have been found. The fossil *Cooksonia* from the Upper Silurian with its slender branched stem, lack of lateral appendages, terminal sporangia, and cutinized spores, resembles *Rhynia*, a psilophyte from the RHYNIE CHERT. The fossil *Nematophyton* resembles the stipe of a seaweed. Marine life was similar to that of the ORDOVICIAN, and the climate is thought to have been uniformly warm, becoming drier towards the end of the period. See GEOLOGICAL TIME SCALE.

silviculture the management of woodland or forest for the production of timber and/or other wood products.

simple describing a leaf that is not divided,

simple pit a PIT lacking a border. Simple pits are found, for example, in certain parenchyma cells, extraxylary fibres, and sclereids. Compare BORDERED PIT.

simple sorus see SORUS.

single-cell protein (SCP) protein produced by unicellular organisms for animal and human consumption. Microorganisms are considerably more efficient in producing protein than crops and farm animals. Thus yeast and bacteria can double their mass in a minimum of 20 minutes while cereals would take 7–14 days and young cattle 4–8 weeks to achieve this. Considerable research has consequently been directed towards developing suitable industrial plants for the large-scale cultivation of certain bacteria, fungi, and algae. Many products are already available, e.g. Pruteen from ICI. The substrates used to grow the microorganisms can be low-cost agricultural wastes, e.g. straw, molasses, and animal sewage. The technology thus has potential applications in pollution control as well as food production.

single-factor inheritance the determination of a character by one major gene, although the gene may exist in various allelic forms. MENDELIAN GENES are examples of single-factor inheritance. Compare MULTIPLE-FACTOR INHERITANCE.

sink a site within a plant or cell where a demand exists for particular substrates or catalysts. Thus the mitochondria are sinks for oxygen and respiratory substrates while the chloroplasts require carbon dioxide. At a higher level of organization fruits, roots, and shoot apices are sinks for photosynthates. See MASS FLOW HYPOTHESIS.

sinker 1 see DROPPER.
2 a structure developed from the stem or hypocotyl of a parasitic plant, which embeds itself in the tissues of the host plant. See also HAUSTORIUM.

sinuate describing a leaf margin divided by irregularly spaced narrow notches into wide lobes. *See illustration at* leaf.

sinus a cavity or hollow.

siotropism a tropic response to shaking (see TROPISM).

siphonaceous (siphoneous) describing the tubular growth habit seen in certain algae, e.g. *Vaucheria.*

siphonales see CAULERPALES.

siphoneous see SIPHONACEOUS.

siphonogamy the use of a pollen tube to convey nonmotile male gametes to the egg. Compare ZOIDOGAMY.

siphonostele see STELE.

siphonous in algae, describing a thallus that is tubular or that is not divided into discrete cells by septa (i.e. it is coenocytic).

siphonous green algae (siphonaceous green algae) green algae (see CHLOROPHYTA) of the orders CLADOPHORALES, CAULERPALES, and DASYCLADALES. These have tubular, club-shaped, or filamentous thalli that are coenocytic. They are thought to have evolved from the ULOTRICHALES.

sister chromatids CHROMATIDS derived from the same parent chromosome. Sister chromatids are held together at their centromeres at metaphase of meiosis, but separate in anaphase to pass into different daughter cells.

SI units (Système International d'Unités) an international system of units now used for all scientific purposes. Derived from the m.k.s. system (based on the metre, kilogram, and second), SI units have now replaced imperial units. There are seven base units, two dimensionless units (formerly called supplementary units), and 18 other units derived from these nine units. Each unit has an agreed symbol; there are no plural forms of these symbols (e.g. 3 kg; not 3 kgs). decimal multiples of the units are indicated by a set of prefixes. See appendix.

skeletal hyphae see HYPHA.

skotophile see PHOTOPHILE.

slash-and-burn (shifting cultivation) a small-scale system of agriculture in which a small area of natural vegetation is cleared by cutting and burning, cultivated for several years, then abandoned when soil fertility declines and crop yields diminish, the cultivators perhaps returning at intervals for a few years to harvest fruit or to hunt the game that comes to feed on the remains of the crops and fruit trees. The area is soon colonized by pioneer species and eventually succession returns the climax vegetation. Once widespread across the world, slash-and-burn is now practised mainly in remote parts of tropical rainforests where the local population is still free to roam over large areas. Once population density increases (usual by immigration) or land ownership spreads, the area available is reduced and the sites are revisited too often, resulting in permanent degradation of the soil.

sleep movements see NYCTINASTY.

sliding growth a type of cell growth in which an area of the cell wall expands and slides over that part of the wall of the adjacent cell with which it was originally in contact. This results in the severance of plasmodesmata between adjacent cells but does not lead to cell disruption. Compare INTRUSIVE GROWTH, SYMPLASTIC GROWTH.

slime bacteria see MYXOBACTERIA.

slime moulds see ACRASIOMYCOTA, MYXOMYCOTA.

slime nets see LABYRINTHULATA.

smooth endoplasmic reticulum see ENDOPLASMIC RETICULUM.

smuts plant diseases caused by fungi of the order USTILAGINALES, so named because of the black spore masses usually produced on the host. In the *covered smuts,* such as covered smut of barley (*Ustilago hordei*), the mature spore mass in the ovary remains within the sorus for a while, often until the sorus is dispersed from the host. However in the *loose smuts*, such as loose smut of wheat (*U. nuda*), the spores form an uncovered mass of black powder. Seed treatment with fungicides is the usual control method. See also BUNT.

snag the standing part of a tree trunk that has broken off. Snags are important microenvironments, rich in species of DECOMPOSERS and other invertebrates, and of lower plants. They provide shelter and nest sites for a wide variety of animals, some of them now rare, such as the spotted owl of western North America. The presence of many snags and rotting logs in OLD-GROWTH FORESTS contributes to their species diversity.

snow algae see RED SNOW.

snow-patch vegetation a characteristic type of vegetation found under late-melting shallow snow patches and around the edges of deeper ones. It is often more luxurious than the vegetation outside the snow patch, or the vegetation under the deeper snow patches, striking a good balance between being insulated by the snow from the vicissitudes of winter weather, yet having a reasonably long growing season.

snurp see RNA SPLICING, SPLICEOSOME.

sobole a RHIZOME comprising a single horizontal underground stem that turns upwards at the tip.

soboliferous clump-forming.

sociability scale a subjective five-point scale used to represent the degree to which individual plant species are clumped. Species are ranked from 1, indicating an isolated shoot, to 5, indicating shoots growing in large carpets or in pure populations.

society a minor community within a CONSOCIATION, with a particular dominant species. It arises because of some variation in conditions within a habitat. For example, in a consociation such as an oak wood there may be a society of bluebells or primroses. See also ASSOCIATION.

sodic soil 1 a SOIL containing more than 15% exchangeable sodium (see cation exchange capacity). 2 a soil with a sodium content high enough to interfere with the growth of most plants.

sodium symbol: Na. A metal element, atomic number 11, atomic weight 23, found in all terrestrial plants, though not apparently an essential nutrient most, excepting plants with CRASSULACEAN ACID METABOLISM, and some halophytic C_4 PLANTS. As sodium chloride, it determines much of the salinity of the oceans and estuarine waters, playing an important role the habitats of plants, seaweeds, and planktonic algae. See also HALOPHYTE, PHYSIOLOGICAL DROUGHT.

sodium bicarbonate indicator a mixture of sodium bicarbonate solution and the dyes cresol red and thymol blue. It is used as an indicator to detect small changes in pH. It changes from red to orange and yellow with a slight increase in acidity.

soft rot a disease caused by fungi or bacteria, in which the tissues of the affected parts become soft. Soft rots are most notable as postharvest diseases. Bacteria of the genus *Erwinia* and fungi of the genus *Rhizopus* cause soft rots of stored vegetables and fruits – particularly in humid badly ventilated conditions. The tissues become soft as the enzymes from the pathogen break down the cell walls.

softwood see SECONDARY XYLEM.

soil the superficial layer that covers large areas of the earth's crust. It consists of mineral particles, decaying and decayed organic material, living organisms, air, and water. It is the medium in which most plants grow, supporting them and supplying them with nutrients, and is also a habitat for numerous animals and microorganisms.

The mineral content of the soil is derived from the mechanical or chemical weathering of exposed rocks. A soil may be formed from the underlying bedrock (*sedentary soil*) or it may result from the deposition of mineral particles transported by such agents as water or wind, resulting in a mixture of particles of different origins (*sedimentary soil*). The proportions of the various sizes of mineral particles (silt, clay, sand, gravel, etc.) have a profound effect on the SOIL TEXTURE and SOIL STRUCTURE. This is further influenced by the amount of organic matter present, most of which is concentrated in the top 230–300 mm of soil. This is gradually decomposed and converted into a colloidal material called *humus* (see HUMIFICATION, HUMUS).

The air in the spaces between the soil particles (see PORE SPACE) contains the same gases as the atmosphere above ground, but in different proportions, being generally richer in carbon dioxide and poorer in oxygen.

Living organisms affect the soil in a number of ways. The larger animals, notably earthworms, mix and aerate the soil and billions of microorganisms feed on and decompose the dead organic matter, thus making nutrients available to the plants. The water in the soil contains dissolved mineral substances and gases that can be taken up by plant roots. The level at which the soil becomes saturated with water (the WATER TABLE) rises and falls depending on the amount of precipitation. Its level affects soil aeration. The soil particle size also affects drainage, which in turn affects aeration. Soils with relatively large particle sizes, such as sandy soils, are more free-draining than those with smaller particles, such as clay soils.

There are a number of different methods used to classify soil. Generally soil can be divided into three main groups: *zonal soils*, where the soil type reflects the prevailing climatic conditions; *intrazonal soils*, where some other factor, such as the nature of the parent rock, has more influence on soil development; and *azonal soils*, which are immature newly formed soils, e.g. alluvial soils. Zonal soils may be further subdivided into PEDALFERS and PEDOCALS. Intrazonal soils can be subdivided into *hydromorphic soils* (with excessive moisture), *halomorphic soils* (with a high salt content), and *calcimorphic soils* (rich in lime from the parent rock).

A section through soil from the surface to the bedrock (a soil profile) usually shows a series of more or less distinct layers (*horizons*).

The soil colour may vary in different horizons and in different types of soil. Soil colour may be used as an indication of soil composition, amount of aeration, drainage, etc. For example, badly aerated soils are often bluish grey whereas well aerated soils are reddish because of the high

concentration of iron oxides. See also MUNSELL SOIL COLOUR SYSTEM, SOIL STRUCTURE.

soil-borne diseases any plant disease that originates from inoculum in the soil. DAMPING-OFF and CLUB ROOT are typical soil-borne diseases. Control can be difficult and methods include crop rotation, fumigation, chemical treatment, soil sterilization, the use of seed dressings, and improved cultural methods (e.g. better drainage). See also FUMIGANT.

soil factors see EDAPHIC FACTORS.

soil horizon see SOIL PROFILE.

soil-moisture index see MOISTURE INDEX.

soil profile the arrangement of the layers that can be seen in a section of SOIL extending from the surface to the bedrock. The layers are *soil horizons* and there are differences in colour, texture (see SOIL TEXTURE), and composition between them. The upper layer, horizon A or the zone of eluviation, contains the most humus, is most exposed to weathering, and is the zone in which most decomposition by microorganisms takes place. The underlying horizon B or zone of illuviation accumulates salts and colloids washed down from horizon A. Horizon C (the *subsoil*) consists of the mantle rock produced mainly by the mechanical weathering of the parent rock. In horticulture, the term subsoil is taken to mean the soil beneath the humus-rich horizon A (*topsoil*). Some soils do not have such horizons and may be more or less uniform throughout. They are generally found in mature soil in moist cool temperate climates. In arid and semiarid climates where the rate of evaporation exceeds the amount of precipitation there is a tendency for the upward movement of water by capillarity, and mineral salts, particularly calcium salts, are deposited in the upper layers resulting in calcification. See also LEACHING.

soil structure the arrangement of the mineral particles in soil. Particles may exist individually, e.g. sand particles, or they may be grouped together in larger units (*peds*). The products of decomposition of organic matter are often sticky and glutinous thus binding the particles together. The aggregates can be grouped in various ways and may be described as crumby, blocky, platy, etc. The structure provided by these aggregates makes the soil more fertile as the channels between the peds aid aeration and drainage. Soil structure can be broken down by overcultivation and removal of too much organic matter. Often a clay soil forms large sticky *clods* making it difficult to work. Clay soils can be improved by adding lime, which causes the individual clay particles to stick together (*flocculate*) in smaller aggregates.

soil texture the relative sizes of the different mineral particles in a SOIL that affects its aeration, capillarity, porosity, water absorption, and ease of cultivation. Gravel and stones are defined as being larger than 2 mm in diameter. On the same scale sand particles are 0.05–2.0 mm, silt 0.002–0.05 mm, and anything smaller than 0.002 mm is termed clay. Thus a sandy soil with larger particles has larger spaces and is well aerated and drained but dry. A clay soil is badly drained and becomes waterlogged. However clay particles tend to group together and also chemically attract other substances such as HUMUS. The *clay–humus complex* is chemically active and thus capable of holding plant nutrients in the soil, maintaining fertility. Cultivation affects soil texture, e.g. ploughing, harrowing, and rolling will break down large aggregates into smaller particles. Over cultivation, however, can lead to soil erosion.

Solanaceae a large family of dicotyledonous plants commonly called the potato or nightshade family. It contains about 94 genera and some 3000 species have been described. However the actual number of species may be closer to 2000 since the great INFRASPECIFIC VARIATION seen in many species has led to taxonomic confusion and some species being given a variety of different names.

The Solanaceae are cosmopolitan in distribution though there are concentrations of genera in Central and South America and Australia. The majority are herbaceous. The leaves lack stipules and the flowers are usually ACTINOMORPHIC (regular) and borne in a cyme. Most commonly there are five sepals and petals, more or less fused, and five anthers. In the largest genus, *Solanum* (about 1500 species), the flowers characteristically possess a column of touching, but not fused, stamens, which are particularly prominent because of the downward-turned petals. The fruit is usually a berry, though sometimes a capsule, as in henbane (*Hyoscyamus niger*) and thorn apple (*Datura stramonium*).

Many solanaceous plants are poisonous due to their possession of alkaloids, e.g. atropine found in deadly nightshade (*Atropa belladonna*) and nicotine found in tobacco (*Nicotiana tabacum*). Others are important as food plants, especially the potato (*Solanum tuberosum*) and tomato (*Lycopersicon esculentum*). The aubergines and the various capsicums also belong to this family. The genus *Nicotiana* includes the important cash crop tobacco and various ornamental species. Other ornamentals include species of *Petunia*, *Schizanthus* (butterfly flowers), and *Salpiglossis* (velvet flowers).

solarization (heliosis) the inhibition of photosynthesis due to the PHOTOOXIDATION of some of the chloroplast components at very high light intensities.

solenostele (amphiphloic siphonostele) see stele.

soligenous mire a mire that is supplied with water by rainfall and runoff. See also OMBROGENOUS BOG.

solod see SOLONCHAK.

solonchak (white alkali soil) a type of intrazonal infertile SOIL with a high salt content (halomorphic). It is found in arid and semiarid inland regions and in areas where sea spray has blown inland resulting in the deposition of salt. The evaporation of the surface water results in salts from the underlying layers being drawn upwards and deposited on the surface. Solonchaks are usually associated with lime-rich soils (PEDOCALS) and are commonly found in dry continental interiors. If leaching occurs either through heavier rainfall or by irrigation, the salts are washed down into the B horizon thus giving rise to a *solonetz* soil. If there is continued leaching and improved drainage a *soloth* or *solod* is formed, which is weakly acid.

solonetz see SOLONCHAK.

soloth see SOLONCHAK.

soluble RNA (s-RNA) an alternative name for TRANSFER RNA, so-called because it is more soluble in acids than other types of RNA.

solute potential see OSMOTIC POTENTIAL.

soma (*adj.* somatic) the part of an organism's body whose genetic elements are not passed on from generation to generation, i.e. the body excluding the gametes or asexual reproductive structures, such as spores, gemmae, or bulbils. See also WEISMANNISM.

somatic cell any cell of the body, i.e. any cell other than the spores, gametes, or their precursors.

somatic crossing over see MITOTIC CROSSING OVER.

somatic hybridization the production of cells, tissues, or organisms by fusion of nongametic nuclei. The phenomenon may be induced under laboratory conditions in cells that never normally fuse together and used as a plant breeding or genetic tool. It may also occur naturally, especially in fungi. See also PARASEXUAL RECOMBINATION, PROTOPLAST FUSION.

somatic meiosis meiosis that occurs during vegetative growth and is not directly related to reproduction. It occurs in certain red algae (RHODOPHYTA), e.g. *Batrachospermum*, producing thalli that are part diploid, part haploid. The haploid parts of the thalli later develop reproductive structures.

somatic mutation see MUTATION.

sooty mould a dark, soot-like fungal growth on honeydew, the sugary exudate produced by sap-sucking insects such as aphids and scale insects. The fungi belong to the order Dothideales.

soralia see SOREDIUM.

sorbitol a common sugar alcohol found especially in certain algae and in the fruits of many higher plants, e.g. mountain ash (*Sorbus aucuparia*). It is formed by the reduction of D-glucose or L-sorbose. In some plants, e.g. crab apple (*Malus sylvestris*), sorbitol, rather than sucrose, is the main form in which sugar is translocated in the phloem.

soredium a small segment of a lichen THALLUS consisting of a number of algal cells loosely surrounded by a few fungal filaments. Soredia serve as agents of vegetative reproduction and are seen as a powdery dust on the lichen surface. They are often formed in specialized structures termed *soralia*.

sorocarp see PSEUDOPLASMODIUM.

sorogenesis the production of sori (see SORUS), as in the Filicinophyta (ferns), or sorocarps (see PSEUDOPLASMODIUM), and in the ACRASIOMYCOTA (cellular slime moulds).

sorophore see PSEUDOPLASMODIUM.

sorosis a MULTIPLE FRUIT derived from the ovaries of several flowers, as in mulberries (*Morus*). Compare COENOCARPIUM.

sorus (*pl.* sori) a reproductive structure made up of a collection of SPORANGIA. Sori are usually prominent on the undersurface of fertile fern fronds, where they are often borne in a regular pattern, each sorus being covered by an INDUSIUM. Three different kinds of sori are recognized, according to the order in which the sporangia ripen. In a *simple sorus* all the sporangia ripen at the same time, e.g members of the Osmundaceae. This is considered to be a primitive state. In a *gradate sorus* the sori ripen in sequence from the apex to the base of the sorus, e.g. members of the Hymenophyllaceae. In a *mixed sorus*, late-ripening sori are mixed up with early-ripening ones, e.g. members of the Polypodiaceae. This is considered an advanced condition.

The reproductive areas, consisting of a mass of unilocular sporangia, found on the blades of the brown alga *Laminaria* (Phaeophyta) are also termed sori as are the various spore-forming bodies of the rust fungi (e.g. the aecium, telium, and uredosorus).

Southern blotting a method named after E. M. Southern (1938–) (who invented it) for extracting selected DNA fragments from an electrophoresis gel for analysis using DNA PROBES. The DNA is digested with restriction endonuclease enzymes, and the resulting mixture of fragments is separated according to their lengths by ELECTROPHORESIS on an agarose gel. The double-stranded DNA is then denatured by sodium hydroxide to form single-stranded DNA. A nitrocellulose filter is pressed against the gel, and the single-stranded

DNA fragments stick to it and are fixed permanently in position by heating. The DNA probe is then applied to locate the desired DNA fragment. Meanwhile, the electrophoretic separation pattern of the DNA is preserved. Compare WESTERN BLOTTING.

Southern Oscillation see EL NIÑO.

SP see DIFFUSION PRESSURE DEFICIT.

spadix a specialized INFLORESCENCE in which the flowers are sessile and borne on an enlarged fleshy axis, often with a sterile part extending beyond the inflorescence. The spadix is typical of the Araceae (arum family) and is usually surrounded by a large bract (SPATHE).

spathe a large BRACT enclosing a SPADIX, which may be highly coloured and petaloid, as in *Anthurium*. In such cases it functions in attracting insects to pollinate the plant.

spathulate see SPATULATE.

spatulate (spathulate) describing structures that have a broad apex and a long narrow base, such as the leaves of the daisy (*Bellis perennis*). *See illustration at* leaf.

spawn a piece of mycelium that establishes a fungus culture in a new location. The term is used particularly by mushroom growers.

special creation see CREATIONISM.

specialization the degree to which an organism is adapted to its environment. Species that are highly specialized have a narrow habitat range or NICHE, usually as a result of interspecific competition. See also ADAPTATION.

speciation the formation of new species. This typically involves the establishment of barriers (isolating mechanisms) that prevent two populations from interbreeding (see ALLOPATRIC, SYMPATRIC). NATURAL SELECTION may then occur, taking the two populations along different evolutionary paths so that they become progressively different from each other. Speciation is often regarded as a relatively slow process, especially in the animal kingdom (but see PUNCTUATED EQUILIBRIUM). In the plant kingdom, however, a major contribution to speciation is POLYPLOIDY. This may establish a new population (species) in a single generation that is incapable of reproducing with either parent. Polyploidy has thus been cited as an example of 'instant evolution'. See ALLOPATRIC, GEOGRAPHICAL ISOLATION, REPRODUCTIVE ISOLATION, SYMPATRIC, VICARIANCE. See also ADAPTIVE RADIATION, NEOTENY, STASIPATRIC SPECIATION.

species the fundamental unit of study in taxonomy, comprising all the POPULATIONS of one breeding group that normally are permanently separated from other such groups by marked discontinuities. This discontinuity is normally characterized by morphological differences. There have been many attempts to define a species, with the *biological species concept* defining a species as a population or group of populations whose members have similar morphological, physiological, and biochemical features and can interbreed to produce fertile offspring. This implies a breeding barrier of some kind with other species. This is broadly accepted, but does not work fully in many groups of plants. In some species there is a reproductive barrier such that any hybrids are sterile, but in others such hybrids retain some fertility; this is characteristic of certain plant groups, such as orchids, in which there is often complete cross fertility within a genus and sometimes across genera. HYBRID SWARMS may occur, involving two or more species, frequently in disturbed areas, leading to *aggregate species*, in which it is almost impossible to distinguish between the different species, e.g. bramble (*Rubus fruticosus* agg., where both hybridization and APOMIXIS occur). Polyploidy can cause a further complication, rendering hybrids fertile. Some so-called species that are normally separated by geographical barriers (see DISPERSAL BARRIER) may interbreed with each other if brought together. Species are normally regarded as distinct if they remain so for the bulk of their range and retain a recognizable suite of distinct morphological characters. In some plant groups, such as most mosses, species are defined purely on morphological discontinuities. See also SPECIATION.

In botanical nomenclature a species has two names, the genus name (generic name), which starts with a capital letter), and the species name (the *specific epithet*), which comes second and is always written in lower case. (See Appendix for the meanings of some common specific epithets.) The ending of the specific epithet always agrees with the gender of the generic name. Unlike the situation in zoology, it is not legitimate for the specific epithet to repeat the generic name. There are several INFRASPECIFIC categories (taxons below the level of species) in the taxonomic hierarchy, but SUBSPECIES is probably the most widely used. Groups of similar species are placed in genera (see GENUS). See also BINOMIAL NOMENCLATURE, HIERARCHICAL CLASSIFICATION.

species-area curve see MINIMAL AREA.
species classification see INVERSE ANALYSIS.
species diversity see DIVERSITY.
species group see SUPERSPECIES.
species ordination see INVERSE ANALYSIS.
specific epithet see SPECIES.
specific humidity see HUMIDITY.
spectrophotometer an instrument for investigating quantitatively the way in which electromagnetic radiation of different wavelengths is absorbed by a specimen. Typically, it consists of a source of the radiation (infrared, visible, or ultraviolet) from which a particular wavelength can be selected by means of a monochromater. A beam of this radiation is passed through the specimen (or reflected from it) and the intensity measured by a detector (e.g. a photocell). As the wavelength of incident radiation is changed, an ABSORPTION SPECTRUM is produced showing how the intensity of transmitted (or reflected) radiation varies with wavelength. Typically, substances absorb the radiation over particular bands of wavelength, corresponding to energy-level changes in their molecules. Spectrophotometry is used extensively as a means of analysis and in the study of how radiation interacts with matter (e.g. the absorption of light in photosynthesis). The speed of a chemical reaction can also be followed in cases where the products of the reaction absorb radiation of a particular wavelength that the reactants do not.
sperm see SPERM CELL, ANTHEROZOID.
spermatangium in the red algae (see RHODOPHYTA), the organ that produces nonmotile male gametes (spermatia).

spermatiophore a hypha that bears a SPERMATIUM.
spermatium (*pl.* spermatia) 1 a small nonmotile male cell developed on the mycelia of certain ascomycete or rust fungi (and in lichens containing certain species of fungi) and capable of fertilizing the TRICHOGYNE of an ascogonium or the receptive hyphae of rusts. 2 see PYCNIDIUM.
spermatocyte see ANTHEROCYTE.
Spermatophyta (seed plants, spermatophytes) in older classifications that considered the seed habit of equal or greater significance than the possession of vascular tissue, a division of the Plantae containing all the seed-bearing vascular plants. It contained the two classes Gymnospermae (gymnosperms) and Angiospermae (flowering plants). See also FIVE KINGDOMS CLASSIFICATION, TRACHEOPHYTA.
spermatozoid see ANTHEROZOID.
sperm cell one of the two male gametes formed in the pollen tube of gymnosperms after mitotic division of the BODY CELL. In some species, e.g. *Cycas* and *Ginkgo*, they have undulipodia and swim from the pollen tube to the archegonia. Such antherozoids, in contrast to the nonmotile GENERATIVE NUCLEI of angiosperms, are associated with a large amount of cytoplasm and in *Cycas* reach a diameter of 300 μm. In other gymnosperms, e.g. the Gnetophyta and Coniferophyta, the sperm cells are not motile. The term sperm cell is sometimes applied to GENERATIVE NUCLEI.
spermogonium see PYCNIDIUM.
Sphacelariales an order of brown algae (PHAEOPHYTA) that have filamentous or parenchymatous thalli, which grow by division of a single apical cell. Vegetative reproduction involves the production of propagules, special branches that detach from the parent and germinate when they reach a suitable substrate. These algae also reproduce by means of spores produced in unilocular sporangia. Sexual reproduction involves the seasonal production of plurilocular gametangia.
Sphaeriales see XYLARIALES.
Sphaeropsidales an obsolete term used for

some FUNGI ANAMORPHICI whose
conidiophores are formed within pycnidia.
sphaeroraphide see DRUSE.
Sphagnales see BRYOPHYTA.
sphagnum (bog moss, peat moss) any moss
of the genus *Sphagnum*, commonly found in
wet PEAT, and whose remains often form a
major part of the peat. Sphagnum moss is
rather like a sponge: there are air spaces in
the cells that can fill with water so the moss
can hold over twenty times its own dry
weight of water. Sphagnum peat is
harvested for fuel and horticultural use.
Where sphagnum grows over impermeable
rocks in areas of high rainfall it can form
extensive BOGS, such as those of northern
Scotland and the Great Dismal Swamp in
North Carolina and Virginia in the USA,
which once extended over 5700 sq km, but
today occupies just 1940 sq km.
S phase see CELL CYCLE.
Sphenophyta (Equisetophyta, Arthrophyta) a
phylum of spore-bearing VASCULAR PLANTS
containing a single living genus, *Equisetum*
(horsetails), with about 15 species, but
having a rich fossil record. In older systems
the group is regarded as a subdivision or
class, Sphenopsida or Equisetopsida, of the
PTERIDOPHYTA. The most conspicuous
feature of the sphenophytes is the aerial
jointed stem, which bears whorls of scale
leaves and green branches inserted at each
joint. The stems arise at intervals from a
perennial creeping rhizome and may be
either all photosynthetic or differentiated
into two types, green sterile stems and
colourless fertile stems. The stems are
grooved, with siliceous ribs, and consist of a
central cavity surrounded by a ring of
smaller cavities in the cortex (see
VALLECULAR CANAL). These alternate with
still smaller cavities associated with the
protoxylem (see CARINAL CANAL). The thick-
walled sporangia are borne on
sporangiophores in whorls in terminal
STROBILI. The spores give rise to separate
male and female gametophytes (prothalli),
although sex determination is partly
controlled by environmental factors, and
female prothalli may produce antheridia as
they age. The antherozoids have multiple
undulipodia.

The Sphenophyta is divided into four

orders: the Equisetales and the extinct
Calamitales, Sphenophyllales, and
Pseudoborniales. The arborescent
Calamitales, which flourished in the
CARBONIFEROUS, most closely resemble the
Equisetales.
Sphenopsida see SPHENOPHYTA.
spherosome an ORGANELLE 0.5–1.0 μm in
diameter with a single membrane and a
fairly granular matrix containing
triacylglycerols (triglycerides). Spherosomes
are abundant in cells in which lipids are
stored and they contain the hydrolytic
enzyme, LIPASE. They probably have a role
in the mobilization of stored lipids when
they are required for cell metabolism.
sphingolipid a type of complex lipid
containing a long-chain amino alcohol,
such as sphingosine, joined by an amide
linkage to a long-chain fatty acid.
Sphingolipids are important as membrane
components in both animal and plant cells.
In yeasts and in higher plants the amino
alcohol is commonly phytosphingosine.
spicate arranged in a spike (see
INFLORESCENCE).
spicule a small, hard spine or needle.
spike an inflorescence in which the flowers
are sessile and borne on an elongated axis,
as in wheat. *See illustration at* inflorescence.
spikelet the basic unit of a grass
inflorescence. It consists of a short axis or
RACHILLA, two bracts or GLUMES (rarely one
or none), and one or more florets and their
bracts (the LEMMA and PALEA). Spikelet
characteristics, such as size, number of
florets, or how it fractures at maturity, may
be used as diagnostic traits for the genus.
The whole spikelet is the unit of dispersal
for such grasses as *Holcus* and *Setaria*. *See
illustration at* poaceae.
spindle a fibrillar structure formed in the
cytoplasm from MICROTUBULES at the
commencement of METAPHASE. When
present, CENTRIOLES move so they are
diametrically opposite each other on the
disintegrating nuclear MEMBRANE. In the
majority of plant cells, where there are no
centrioles, poles still become established at
opposite points. The fibrils radiate from
each pole towards the centre or equator.
The attached microtubules direct the
chromatids (or chromosomes in the first

meiotic division) towards opposite poles. Evidence tends to support the view that this is achieved by the microtubules becoming progressively shortened as a result of depolymerization in the regions nearest to the poles.

spindle attachment see KINETOCHORE.

spindle polar body a CENTRIOLE or any other microtubule-containing organelle associated with the nucleus.

spine a modified leaf or part of a leaf forming a sharp pointed structure, often with a vascular trace leading to its base. If some or all of the leaves are modified into spines then leaf area, and hence transpiration, is reduced. Spines are thus common in XEROPHYTES, such as cacti, their points sometimes acting as nuclei on which water droplets can condense, run down to the soil, and provide moisture for the plant. In such plants the functions of the leaf may be taken over by the stem (see CLADODE). Spines also protect against herbivores and excessive sunlight. Compare THORN, PRICKLE.

spinney a small woodland dominated by thorn trees, e.g. hawthorn (*Crataegus*).

spinose bearing spines. *See illustration at* leaf.

spiral thickening (helical thickening) a type of secondary cell wall patterning in TRACHEARY ELEMENTS in which the secondary cell wall is laid down in a helical pattern. Spiral thickening, like ANNULAR THICKENING, permits further extension of the cell. It is therefore found in tissues such as PROTOXYLEM, which develop in young organs that are still elongating. Compare PITTED THICKENING, RETICULATE THICKENING, SCALARIFORM THICKENING.

spirillum any corkscrew-shaped bacterium. Compare BACILLUS, COCCUS, VIBRIO.

Spirochaetae (spirochaetes, spirochetes) a phylum of long, spirally twisted bacteria with a flexible cell wall. They swim in a corkscrew motion by means of two to more than 200 internal flagella (*axial filaments*) in the cell wall. Spirochaetes live in mud and water. Many can tolerate anaerobic conditions. The phylum includes pathogens causing such diseases as syphilis and relapsing fever.

spirodistichous describing a form of PHYLLOTAXIS in which the leaves are arranged in two rows (distichous), but the plane of these rows forms a spiral.

spiromonostichous describing a form of PHYLLOTAXIS in which the leaves are all on one side of the stem (monostichous), but successive leaves are arranged in a shallow helix.

spliceosome a complex of small ribonucleoproteins called *snurps*, made up of RNA combined with proteins. Its function is to remove the noncoding RNA (introns) from the initial RNA transcript during TRANSCRIPTION and splice the RNA molecule back together again. The RNA molecules in the spliceosome act like RIBOZYMES, catalysing the assembly of the spliceosome as well as the splicing process itself. See RNA SPLICING, TRANSCRIPTION.

splicing see RNA SPLICING.

spongy mesophyll the part of the leaf MESOPHYLL that is composed of variously shaped, often conspicuously lobed, cells with large intercellular spaces. It is sometimes regarded as a type of AERENCHYMA. A humid atmosphere is maintained in the intercellular spaces, thus facilitating gaseous exchange between the atmosphere and the mesophyll for photosynthesis and respiration.

spontaneous generation the theory that living organisms form directly and spontaneously from nonliving material. It was conclusively disproved by Louis Pasteur in 1862. A distinction should be drawn between this idea and the concept of abiogenesis and the origin of life.

sporangiolum a sporangium that contains only one or a very small number of spores.

sporangiophore a structure bearing one or more SPORANGIA. It may be a simple stalk, such as that formed on germination of the ZYGOSPORE of such fungi as *Rhizopus* and *Mucor*. Alternatively it may be multicellular or branched as in certain lower vascular plants. See also SPOROPHYLL.

sporangium a structure in which spores are formed. The sporangium may be simple and unicellular, as in the algae and fungi, or multicellular, as in the bryophytes and vascular plants. In higher plants the sporangium may be protected by a thickened outer wall. The entire protoplasmic contents may be converted into motile or nonmotile spores, which are usually liberated by rupture of the sporangium wall. See also MEGASPORANGIUM, MICROSPORANGIUM.

spore a simple asexual (initially at least) unicellular reproductive unit. Spores are produced by the SPOROPHYTE generation following meiosis and are thus usually haploid. On germination they develop into the haploid gametophyte. Sometimes the suffix '-spore' is used in terms, e.g. CARPOSPORE, ZYGOSPORE, that describe a diploid cell. See also ZOOSPORE, MEGASPORE, MICROSPORE.

sporeling a juvenile plant that has developed from a spore, just as a seedling is a juvenile plant that has developed from a seed.

spore mother cell (sporocyte) a cell that undergoes meiosis to form haploid spores. The term is usually used of homosporous species in which megaspore and microspore mother cells are not distinguished.

spore print a pattern of spores obtained by placing the cap of a sporulating fungus with its gills or pores face down on a sheet of paper allowing it to shed its spores. It reveals not only the pattern of the gills or pores, but also the colour of the spores.

spore wall the outer layer surrounding a spore. It is usually composed of SPOROPOLLENIN in pollen grains (see EXINE) or of PEPTIDOGLYCAN impregnated with dipicolinic acid in bacteria.

Sporobolomycetales see MIRROR YEASTS.

sporocarp a hard usually globose multicellular structure that contains the spores in water ferns, e.g. *Pilularia*. It is formed by the fusion of the margins of a fertile frond. The spores are not released until the structure decays.

sporocyte SEE SPORE MOTHER CELL.

sporodochium a form of CONIDIOMA in which a mass of CONIDIOPHORES is borne on a cushion of pseudoparenchyma and short conidiophores.

sporogenesis the formation of SPORES. See also SPORE MOTHER CELL.

sporogenous producing, containing or supporting spores.

sporogonium the SPOROPHYTE generation in mosses and liverworts, which develops after sexual reproduction and produces haploid spores. See ALTERNATION OF GENERATIONS, BRYOPHYTA, HEPATOPHYTA.

sporophore the fruiting body of ascomycete and basidiomycete fungi.

sporophyll a modified leaf that bears the SPORANGIA. In the lower plants, e.g. *Lycopodium*, it may be a leaf that has retained its normal structure and function and shows no great specialization. In the seed plants the leaf has become highly modified and lost its photosynthetic function, as in the carpels and stamens of the angiosperms. See also MEGASPOROPHYLL, MICROSPOROPHYLL.

sporophyte an individual of the diploid generation in the life cycle of a plant. The sporophyte arises from the fusion of two haploid gametes from the GAMETOPHYTE or haploid generation. In bryophytes, the sporophyte constitutes the minor part of the life cycle, and is either partially or completely dependent on the gametophyte for anchorage and nutrition. In the vascular plants the sporophyte is the dominant generation and is independent of the gametophyte. The sporophyte is more specialized than the gametophyte and by the development of such structures as a cuticle, conducting tissue, and stomata is adapted to a wider range of environments. See also ALTERNATION OF GENERATIONS.

sporopollenin the highly resistant material making up the EXINE of spores and pollen grains. It is a polymerized CAROTENOID, capable of withstanding temperatures approaching 300°C and strong acids.

sport an atypical form of an individual or part of an individual, due to MUTATION or SEGREGATION. The term is normally restricted to new forms and is most commonly used in horticulture.

sporulation the formation of spores.

spring wood see EARLY WOOD.

spur 1 a short thin side shoot from a branch, especially one that bears fruit or, in conifers, the shoots that bear the leaves. 2 a tubular projection from a flower, usually from the base of a perianth segment. Well developed spurs are seen in columbine (*Aquilegia vulgaris*) and larkspur (*Consolida ambigua*).

squamulose describing a THALLUS that is made up of small leafy lobes. In LICHENS it is a form intermediate between crustose and foliose.

squarrose describing a rough, hairy, or scaly surface.

squash a microscopical preparation in which the material is flattened before examination. It is commonly used in the study of chromosomes. Some tissues, e.g. anthers, may be squashed directly while others, e.g. root tips, must first be softened. This is commonly achieved by adding a CELLULASE enzyme.

s-RNA see SOLUBLE RNA.

S-shaped growth curve see DENSITY DEPENDENCE, EXPONENTIAL GROWTH.

stabilizing selection (normalizing selection) NATURAL SELECTION that tends to maintain the same genetic makeup of a species over many generations. Extremes of phenotypes are selected against, resulting in a restricted range of phenotypes. Stabilizing selection tends to operate when the population is already well adapted to the prevailing environmental conditions and competition is not too severe.

stachyose a nonreducing tetrasaccharide made up of two galactose units and one unit each of glucose and fructose. It is found in many of the Lamiaceae, especially the tubers of *Stachys tuberifera*, and in some of the Fabaceae.

staining the treatment of biological specimens with dyes (stains) to colour part of the structure so as to make details more clearly visible through a microscope. Most laboratory stains are synthetic organic dyestuffs, which are absorbed by or bind to particular types of tissue. They are applied by immersing the specimen in a solution of the dye. *Nonvital staining* is the colouring of dead tissue; *vital staining* is the staining of living tissue without harming or killing the cells.

Most stains are organic salts consisting of a positive and negative ion. In *acid stains* the colour comes from an organic anion (negative ion). Such stains (e.g. eosin) tend to colour the cytoplasm of cells. In *basic stains* the colour comes from an organic cation (positive ion). Such stains, e.g. HAEMATOXYLIN, tend to colour the nuclei. *Neutral stains* are mixtures of acid and basic stains and are used to stain both nucleus

and cytoplasm. An example is Leishman's stain, which is made by mixing the acid stain EOSIN with the basic stain METHYLENE BLUE in alcohol. Materials in cells can be described as *acidophilic* if they are receptive to acid dyes; *basophilic* if receptive to basic dyes; and *neutrophilic* if receptive to neutral dyes. Certain dyes need an additional substance (a *mordant*) to bind them to the tissue. The combination of the dyestuff and the mordant is called a *lake*. Iron alum, for instance, is used as a mordant for haematoxylin. In its absence, the tissue will not take up the stain.

Various complex techniques are used for staining different types of tissue for light microscopy. *Counterstaining* (or *double staining*) is the application of two stains in sequence so as to colour different parts of the specimen. For example, safranin is often counterstained with light green. Overstaining occurs when too much stain is taken up by the specimen. In such cases, part of the stain is removed by a solvent – a process known as *differentiation*.

By analogy with stains for the light microscope, *electron stains* are substances used with specimens for electron microscopy. They act by hindering the transmission of electrons (i.e. they are 'electron dense'). Examples are lead citrate, uranyl acetate (UA), and phosphotungstic acid (PTA). See also ACID-FAST STAIN, HISTOCHEMISTRY, METACHROMATIC STAIN, NEGATIVE STAINING, PERMANENT STAIN, TEMPORARY STAIN, VITAL STAIN.

stalk cell one of two cells, the other being the BODY CELL, formed by division of the GENERATIVE CELL in the male gametophyte of certain gymnosperms.

stamen the male reproductive organ of the flowering plant, many of which together make up the ANDROECIUM. It is a highly modified MICROSPOROPHYLL. A typical stamen is differentiated into ANTHER, FILAMENT, and CONNECTIVE. However in the very primitive flowering plants, e.g. *Magnolia* and *Degeneria*, distinct anthers and filaments are absent and the stamen is relatively broad and slightly flattened. In such cases the POLLEN SACS are borne on the surface of the microsporophyll. The more primitive flowering plants usually possess numerous separate stamens, as seen in the Ranunculaceae. The more advanced families have fewer stamens, which are often fused in some way. For example, there are usually four stamens in flowers of the Scrophulariaceae. Where the number of stamens is reduced they are strategically placed to maximize the chances of successful POLLINATION. Compare CARPEL.

staminate flower a flower possessing male parts (stamens) but no female parts, as in the male flowers of holly (*Ilex aquifolium*). Compare PISTILLATE FLOWER. See also MONOECIOUS, DIOECIOUS.

staminode a sterile stamen. It may remain rudimentary or partially developed, as in figworts (*Scrophularia*) and orchids, or become highly specialized and modified, as in *Iris*, where it forms a conspicuous part of the flower.

stand 1 a group of plants growing in a continuous area, e.g. a stand of beech trees. 2 in the classification of vegetation types, a recognizable plant association with a distinctive floristic composition. A stand is sometimes given the suffix -etum preceded by the generic (genus) name of the dominant species, e.g. pinetum.

standard 1 the broad often erect petal at the top or posterior side of the flower of plants belonging to the subfamily Papilionoideae of the Fabaceae (i.e. 'pea' flowers). See also KEEL, WING.

2 any of the three erect twisted inner petals of an *Iris* flower. See also FALL.

3 (specimen tree) a tree that is allowed to reach its full height.

4 a single-trunk tree large enough to be cut for timber.

5 a cultivated plant that has been grafted onto a strong, upright stem, e.g. a standard rose.

standard deviation the average magnitude of deviations from the centre of a NORMAL CURVE, obtained by squaring all the deviations, calculating their mean, and then finding the square root of the mean. It differs from the MEAN DEVIATION in that it removes difficulties introduced by sign, i.e. it does not matter which side of the centre of the curve any particular deviation is situated. The value so estimated, σ, is the point of maximum slope either side of the

central line of the normal curve. If it were possible to make an infinitely large number of observations then the mean deviation would be as good a way of finding the true value of σ as the standard deviation. However with a limited number of samples, as always occurs in practice, the standard deviation provides a better estimate. Often the estimated standard deviation is represented by the Roman letter *s* while the true standard deviation is represented by its Greek equivalent, σ.

standing crop see BIOMASS.

starch (amylum) the most abundant and important reserve polysaccharide in plants. Starch is an early end product of the photosynthetic reduction of carbon dioxide in chloroplasts: in the dark the starch thus formed is rapidly broken down to sucrose and transported to other organs. The bulk of the starch in plants is found in storage organs. Commercial extraction of starch is from such organs. Common sources are the roots of cassava (*Manihot esculenta*), which yield tapioca, the rhizomes of arrowroot (*Maranta arundinacea*), the stem pith of the sago palms (species of *Metroxylon*, *Arenga*, and certain other genera), the tubers of potato, and the grains of various cereals, especially maize, wheat, rice, and sorghum.

Starch consists of two structurally different fractions, AMYLOSE and AMYLOPECTIN. The relative amounts of amylose and amylopectin in a given starch sample depend on the species of plant from which the starch was obtained, amylose usually making up 20–30% of the starch granules. The shape of the starch grains formed by different species also differ (see HILUM) and this variation has been used in taxonomic work. Starch is synthesized from ADP-glucose by the enzyme starch synthetase. A branching enzyme, known as Q *enzyme*, is responsible for formation of the $\alpha(1–6)$ glycosidic bonds in amylopectin. Starch breakdown is catalysed by a class of enzymes known as amylases. These enzymes break down starch to the disaccharide MALTOSE, which is then further degraded by the disaccharidase maltase. Amylase activity is high during periods of rapid growth such as seed germination or sprouting of tubers.

starch sheath see ENDODERMIS.

starch–statolith hypothesis a possible explanation for the way plants perceive gravity. Most gravity-sensitive organs contain specialized cells called STATOLITHS in their cytoplasm. The statoliths gradually fall to the lower side of the cell under the influence of gravity. The settling out of these grains in reoriented cells is said to provide an internal stimulus, which is transmitted to the growing region of the organ so that appropriate amendments to the growth pattern can be made. These growth movements restore the original position of the cells and hence of the whole plant part. See also GRAVITROPISM.

star sclereid see ASTROSCLEREID.

stasigenesis the persistence of a lineage over a long period of time without notable change.

stasipatric speciation instantaneous speciation caused by polyploidy.

stasis a period of little or no evolutionary change, for example, the 'equilibrium' periods that alternate with periods of rapid speciation in the theory of PUNCTUATED EQUILIBIRIUM.

states of fungi the different forms (*morphs*) that a particular fungus may take. Many fungi have several states, producing both asexual spores and several different types of sexual spores. The state characterized by sexual spores, such as an ascoma or basidioma, has been termed the 'perfect' state, while that characterized by the presence of asexual spores or the absence of spores is the *anamorphic* or 'imperfect' state (the use of the terms 'perfect' and 'imperfect' in this context is becoming obsolete). See also ANAMORPH, HOLOMORPH, TELIOMORPH, FUNGI ANAMORPHICI.

statistics the mathematical analysis of experimental results by various methods that are designed to extract the maximum amount of useful and relevant information while at the same time taking into account the effects and potential errors produced by unknown or uncontrollable variables. The end result of any particular statistical treatment should be to reduce a mass of often incomprehensible raw data to a few figures that give an accurate summary of the results and usually an indication of how often such results could be expected to occur by pure chance. The application of statistical considerations to experimental design aims to ensure that results obtained from a necessarily limited sample can be seen to have general application.

statocyst a resistant silicified stage formed by freshwater chrysomonads. See CHRYSOMONADA.

statocyte any gravity-sensitive cell that contains STATOLITHS.

statolith any inclusion in the cytoplasm, such as a starch grain, that moves in response to gravity and is believed to be involved in the gravitropic response. See STARCH-STATOLITH HYPOTHESIS.

stearic acid (octadecanoic acid) an eighteen-carbon saturated FATTY ACID having the formula $CH_3(CH_2)_{16}COOH$. Stearic acid is formed by elongation of the smaller PALMITIC ACID. It occurs in many lipids and is also an intermediate in the synthesis of OLEIC, LINOLEIC, and LINOLENIC ACIDS.

stele (vascular cylinder) (*adj.* stelar) the central core of the stems and roots of vascular plants, comprising the vascular tissue, ground tissue (see PARENCHYMA), such as pith and medullary rays, and the PERICYCLE. The pericycle is considered to represent the outermost layer of the stele at the boundary with the CORTEX. The many different types of stele are thought to reflect the direction of the evolution of the vascular system in plants (see illustration). Different types of stele may be found in different regions of the same plant. In dicotyledons the root characteristically has a solid central core of vascular tissue (see PROTOSTELE), making it better able to resist pulling stresses, while the stem has a ring of discrete vascular bundles (see EUSTELE) and is better able to resist bending stresses. Steles are classified according to the arrangement of their tissues.

A stele consisting of a cylinder of phloem and pericycle surrounding a central core of xylem and lacking a central pith is termed a *protostele*. This type of stele is regarded as being the most primitive both ontogenetically and phylogenetically. A *haplostele* is a protostele in which the central core of XYLEM is circular when seen in transverse section. This is believed to be the most primitive type of protostele and was present in the earliest vascular plants, e.g. *Rhynia*. In a *hypophloic haplostele* (*collateral protostele*) the xylem in cross-section appears as a single strand with the phloem beneath it. This is a type of *monostele* – a stele that consists of a single vascular vessel. An *actinostele* is a protostele in which the central core of xylem is star-shaped or somewhat lobed as viewed in transverse section. This type of stele is exhibited in various species of *Psilotum* and *Lycopodium* and in the roots of higher plants. Depending on the number of xylem lobes, usually between two and eight, the stele is described as *diarch*, *triarch*, etc., and *polyarch* if the number of lobes exceeds eight. A *plectostele* is a protostele in which the xylem consists (in transverse section) of several plates of tissue surrounded by phloem. This type of stele is exhibited in some species of *Lycopodium* and is thought to have evolved from an actinostele. A *medullated protostele* is a protostele in which the central core of xylem consists mainly of tracheids interspersed with numerous parenchyma cells. It thus resembles the medulla (PITH) of a siphonostele (hence the name). The medullated protostele shows how the siphonstele may have evolved from a protostele, with the pith originating in the xylem itself. Medullated protosteles are exhibited in some primitive ferns, e.g. *Gleichenia*.

In a *siphonostele* there is a central core of pith internal to the xylem (compare PROTOSTELE, above). The pith may be extrastelar in origin, i.e. originating from invagination of cortical parenchyma, or it may originate in the xylem itself. In a simple siphonostele the LEAF TRACES do not leave any gaps in the vascular cylinder. A siphonostele may be *ectophloic*, i.e. possessing phloem external to the xylem, or *amphiphloic*, i.e. possessing phloem both external and internal to the xylem as in a solenostele. Siphonosteles are found in the stems of many ferns. A *solenostele* (*amphiphloic siphonostele*) is a siphonostele in which there is also a cylinder of phloem internal to the xylem. When an endodermis is present, there is also an endodermal layer internal to the internal phloem. Leaf gaps

are present, but are sufficiently spaced longitudinally so that only one is visible in any one transverse section. This type of stele is found in certain ferns, e.g. *Marsilea* and *Adiantum*, and is often quoted as evidence for the hypothesis that siphonosteles originated through invasion of the xylem by other extrastelar tissues. A *dictyostele* is essentially similar to a solenostele, except that the AMPHIPHLOIC cylinder of the solenostele is reduced to a meshwork by the occurrence of numerous overlapping leaf gaps. It forms a dissected siphonostele in which the vascular tissue (as viewed in transverse section) is divided into a number of AMPHICRIBAL vascular bundles called *meristeles*. These are separated by parenchymatous areas, which are usually associated with large closely placed leaf gaps. Dictyosteles are typical of many ferns, e.g. *Dryopteris*. Another type of dissected siphonostele is the *eustele*, typical of dicotyledons, in which the vascular cylinder (stele) appears (in transverse section) as a ring of COLLATERAL or BICOLLATERAL vascular bundles separated by MEDULLARY RAYS.

Monocotyledon stems typically have an *atactostele*, in which the vascular bundles are arranged more or less irregularly in the ground tissue (compare EUSTELE, above). A more regular arrangement of vscular bundles is seen in a *polycyclic stele*, in which the meristeles or vascular bundles are arranged in concentric rings. Polycyclic steles are present in some ferns.

In some species there may be more than one stele. *Polystely* is the condition of having a number of independent steles. For example, the aerial axes of many *Selaginella* species have several protosteles ascending the stem, separated by the cortex. In *distely* the vascular system is divided into two separate steles. This condition is exhibited in certain species of *Selaginella*, e.g. *S. kraussiana*.

stellate star-shaped, having a radiating pattern.

stem the part of the plant axis that is usually above ground and bears the leaves, reproductive parts, and buds. The terms *shoot* and stem are virtually synonymous when referring to parts above ground; an

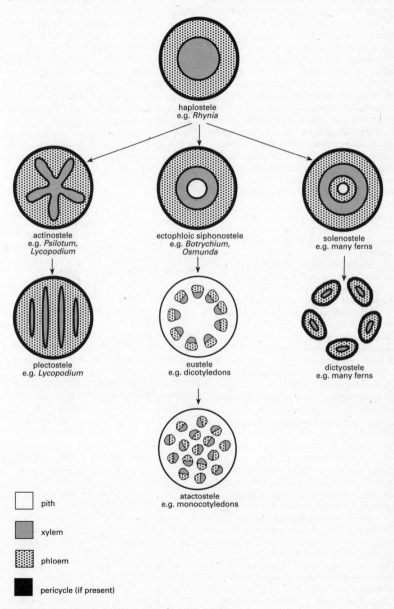

Proposed scheme of evolution of the various stelar types

underground stem, such as a CORM or RHIZOME, would not be described as a shoot. The vascular system of the stem conducts water and nutrients from the roots to the aerial parts and photosynthates from the leaves to any regions where they are needed. The vascular tissues of higher plant stems are usually arranged as a cylinder (as in most dicotyledons) or dispersed irregularly through the cortex (as in monocotyledons).

The stem is usually elongated and branched so the leaves are separated and displayed to best advantage and the flowers are in suitable positions for pollination. There may be one main stem or trunk with side branches, as in many trees, or a number of equally prominent stems. Compare ROOT.

steno- prefix denoting narrow, little.

stenoecious describing an organism that can survive in only a narrow range of habitats.

stenohaline (stenohalic) describing an organism that is unable to tolerate a wide range of salinity or osmotic pressure.

stenospermocarpy a process by which seedless fruits are produced, in which the embryo that develops after fertilization aborts.

stenothermal SEE STENOTHERMOUS.

stenothermous describing an organism that is unable to tolerate a wide range of temperature.

stephanokontous describing a cell that has a ring of undulipodia.

steppe temperate GRASSLAND dominated by perennial species of drought-tolerant grasses and scattered herbs, many with underground PERENNATING ORGANS. It is found in regions of loess soil, typically on CHERNOZEM soils, in a zone some 8000 km wide across Eurasia from Hungary and southern Russia to Central Asia and China. Across this region the climate is continental, with very hot summers and very cold winters, temperatures extremes increasing and rainfall decreasing eastward. The species composition varies with the local climate. Typical steppe grasses include feathergrasses (*Stipa*), sheep's fescue (*Festuca ovina*), bunchgrass (*Schizachyrium*) and bluegrass (*Poa*). In the western steppe there are many ephemeral species that germinate after rain, before the grasses grow back from their underground perennating organs. Herbs with bulbs, corms, and other storage organs also occur, e.g. *Tulipa*, *Allium*. In the colder central region *Artemisia* is common. Much of the former steppe is now under cultivation, especially for wheat and other grains. See also MEADOW STEPPE.

stereid SEE PROSENCHYMA.

stereoisomer SEE ISOMER.

stereotaxis SEE THIGMOTAXIS.

sterigma (*pl.* sterigmata) One of usually four projections at the apex of a BASIDIUM, each one of which bears a basidiospore. The cultivated mushroom is unusual in having only two sterigmata on each basidium, hence the name *Agaricus bisporus*.

sterile 1 describing an organism that is unable to produce offspring.
2 describing a culture of microorganisms or a tissue culture that is not contaminated by other organisms.

steroid any of a group of compounds characterized by possession of the perhydrocyclopenteneophenanthrene skeleton, which is made up of a series of four carbon rings (see diagram). Steroids may be distinguished from certain terpenoids with a similar skeleton by their possession of methyl groups at carbons 10 and 13. All natural steroids also have an oxygen group at carbon 3. The majority of steroids are alcohols (see STEROL). Other types of steroid include certain of the pseudoalkaloids (see ALKALOIDS) and the aglycone portions of the SAPONINS and CARDIAC GLYCOSIDES.

sterol a STEROID alcohol. The sterols found

Steroid: the steroid skeleton

in yeasts and fungi are called *mycosterols*, while those found in higher plants are known as *phytosterols*. Ergosterol is an example of a mycosterol; stigmasterol and sitosterol are common phytosterols, as are SAPONINS, CARDIAC GLYCOSIDES, and steroid alkaloids. Their side chains contain extra carbon atoms that are important in reactions involving adenosylmethionine, which is a donor of methyl groups (see METHYLATION) in many metabolic processes.

stigma (*adj.* stigmatic) **1**The receptive tip of the carpel, which receives pollen at pollination and on which the pollen grain germinates. The stigma is adapted to catch and trap pollen, either by combing pollen off visiting insects or by various hairs, flaps, or sculpturings. The stigmas of certain plants show HAPTOTROPISM. For example, the monkey flower (*Mimulus guttatus*) has a two-lobed stigma, which closes together when touched, so removing pollen from a visiting insect. The stigmas of wind-pollinated plants tend to be feathery or branched to increase the chances of pollination. The stigma may secrete sticky substances, which may act as a pollen trap but may also be involved in the complex pollen/stigma compatibility interactions. The stigma and style act as physiological filters in controlling cross fertilization (see SELF INCOMPATIBILITY). In members of the primitive angiosperm family Winteraceae the stigma is found along the margins of the ventral suture. In the closely related *Degeneria vitiensis* the margins of the carpel are not fused but are held together by interwoven PAPILLAE, which form the stigmatic surface. Studies of such plants have thrown light on the evolution of angiosperms.

2 See EYESPOT.

stilt root see PROP ROOT.

stimulus any influence that, however received and whether acting directly or indirectly on an organism, gives rise to some form of response within that organism. A stimulus may arise externally, as do light, gravity, and chemical stimuli, or it may be generated within the plant itself as in the case of AUTONOMIC MOVEMENTS.

stinkhorns see PHALLALES.

stinking smut see BUNT.

stipe a stalk, especially: **1** the part of the thallus of certain brown algae (PHAEOPHYTA), e.g. *Laminaria*, joining the lamina to the holdfast.

2 the stalk of a mushroom or toadstool.

stipitate having a stipe or stalk.

stipulate having a stipule.

stipule (*adj.* stipular) one of a pair of leaflike structures, spines, glands, or scales at the leaf base or along a petiole. Stipules are believed to protect the developing leaf in the embryonic state. They may occasionally be modified to enclose a bud, as in beeches (*Fagus*). Large green stipules are seen in the garden pea (*Pisum sativum*) and replace photosynthetic tissues lost by the modification of the leaflets into tendrils.

stock **1** (rootstock) a plant onto which shoots or buds (scions) of another plant are grafted. The stock affects the size, vigour, and time of flowering and fruiting, of the scion, and specially bred rootstocks are selected accordingly. Where necessary rootstocks may be chosen that confer frost hardiness or resistance to certain soil-borne pathogens. See GRAFTING.

2 a maintained culture of a particular organism, the origins and characteristics of which are known.

stolon a long branch that is unable to support its own weight and consequently bends down to the ground. Where nodes on the stolon touch the soil a new plant may develop from the axillary bud. Examples of stolons are the long shoots of currants and gooseberries (*Ribes*). Often ordinary shoots will behave like stolons if pegged to the ground, which is the basis of the LAYERING method of vegetative propagation. See also RUNNER, VEGETATIVE REPRODUCTION.

stoma (*pl.* stomata) A pore in the EPIDERMIS of aerial parts of a vascular plant, providing a means for gaseous exchange between the internal tissues and the atmosphere. The stoma is surrounded by GUARD CELLS, which control the size of its aperture. Guard cells are often the only epidermal cells to have significant numbers of chloroplasts. The term *stomate* refers to the opening itself, but is often misused to mean the entire *stomatal apparatus*, which includes the guard cells. The modified epidermal cells (usually one or two) adjacent to each guard cell are called SUBSIDIARY CELLS (or accessory cells).

Stomata are often more numerous on the lower surface of leaves than on the upper surface, where evaporation rates would be highest. Grasses, whose leaves are often almost vertical, have roughly equal numbers of stomata on each surface, while aquatic plants have no stomata at all. In many species, especially those of arid places, the guard cells may be sunk in pits in the epidermis, often protected from passing air movements by a barrier of hairs. This reduces the evaporation (TRANSPIRATION) that accompanies the stomatal opening needed to allow gaseous exchange for photosynthesis and respiration.

In dicotyledons the guard cells are typically kidney-shaped, while in grasses they are usually more dumbbell-shaped. Stomatal opening is achieved by the guard cells taking up water and swelling. Because of the arrangement of cellulose microfibrils in the guard cell walls, this causes the guard cell to increase in length rather than in width; since the guard cell is attached to adjacent guard cells at each end, it curves open.

stomach insecticide any insecticide that must be taken into the alimentary canal to be effective. Stomach poisons are often inorganic compounds containing such elements as arsenic, mercury, or fluorine.

stomatal apparatus (stomatal complex) see STOMA.

stomate see STOMA.

stomium the point at which rupture occurs in a SPORANGIUM or POLLEN SAC to release the spores or pollen. It is typically an area of thin-walled cells that rupture on the drying out of the surrounding tissue. The stomium is found under the ANNULUS in ferns and between the pairs of pollen sacs in seed plants.

stone cell see BRACHYSCLEREID.

stone plants succulent plants of the genus *Lithops*, whose leaves (in opposite decussate pairs) are swollen and almost globose and come just to the surface of the soil, so that they appear like stones. Their coloration usually blends with that of surrounding stones. The upper part of each leaf is

actually a transparent chamber of aqueous tissue, which screens the chloroplasts in cells below from the high radiation of the desert sun.

stoneworts see CHARALES.

stool 1 a stump or group of stumps of a tree capable of producing shoots or suckers, or a shoot growing from such a stump.
2 the persistent base of a coppiced tree. See COPPICING.

stooling the cutting back of a tree or shrub to ground level to encourage new growth from the base. See also COPPICING.

storied (storeyed) describing the axial cells and rays of wood that are arranged in horizontal series as viewed in tangential section.

strain any group of very similar or identical individuals, such as a PURE LINE, CLONE, PHYSIOLOGICAL RACE, or MATING STRAIN.

stramineous straw-coloured.

strangler a plant that uses another plant for physical support, and eventually suppresses its growth by smothering it with branches or aerial roots. Such plants are common in tropical rainforests, e.g. the strangler fig.

stratification 1 the arrangement of some factor in layers or strata. Thus vegetation in a woodland consists of an upper layer, the tree canopy, and a lower layer of herbs at ground level, often with one or more intermediate layers of shrubs and saplings. Other factors, such as temperature and light, show stratification, this being especially important in the study of aquatic ecosystems.
2 the practice of placing seeds between layers of moist sand or peat and exposing them to low temperatures, usually simply by leaving them outside during the winter. The treatment is necessary for those seeds that require a period of chilling before they will germinate. It is thought that low temperatures may block the action of a germination inhibitor. Compare VERNALIZATION.
3 the existence in a lake or other large body of water of layers of water of different density. For example, in a lake there is often a warm low-density layer (epilimnion) separated from a colder denser layer (hypolimnion) by a THERMOCLINE. In a few

situations the density differences are due to differences in salinity.

stratified random sample see RANDOM SAMPLE.

stratopause see ATMOSPHERE.

stratosphere see ATMOSPHERE.

strengthening tissue see MECHANICAL TISSUE.

striate marked with fine lines, grooves, or ridges.

strobilus (cone) 1 a well defined group of closely packed SPOROPHYLLS bearing SPORANGIA arranged around a central axis, as found in the gymnosperms and in certain lower vascular plants (e.g. Lycopodiales, Selaginellales, and Equisetales). In conifers it constitutes the commonly recognized 'cone' and comprises spirally arranged ovules and lignified OVULIFEROUS SCALES.

2 any of various conelike structures seen in angiosperms, e.g. the aggregate fruits of alders (*Alnus*) and hops (*Humulus*).

stroma 1 the hydrophilic proteinaceous matrix of CHLOROPLASTS. It contains the enzymes and reagents necessary for the dark (light-independent) reactions of photosynthesis.

2 (*pl.* stromata) a solid mass of PLECTENCHYMA that may bear PERITHECIA either on its surface, as in *Nectria cinnabarina* (coral-spot fungus), or embedded within the tissue with only the ostioles appearing at the surface, as in *Xylaria hypoxylon* (candle-snuff fungus).

stromatolites structures found preserved in many rocks, especially carbonate rocks, that were formed by the activities of blue-green bacteria (see CYANOBACTERIA). They are among the oldest organic structures to have been recognized. Stromatolites have been identified in the gunflint cherts of Ontario, which are about 2000 million years old, and in PRECAMBRIAN rocks in Australia and Africa some 3000 million years old. Often they consist of a number of white concentric rings, the outer of which may be up to a metre across. Microscopic examination has shown these rings to be composed of numerous blue-green bacteria (see CYANOBACTERIA) that have been preserved due to their ability to secrete calcium carbonate and form large stony

cushion-like masses. Stromatolite formation may be observed today in certain coastal zones, tropical streams, and by mineral springs where blue-green bacteria can flourish due to the absence of grazing invertebrate animals.

strophiole a fleshy outgrowth on the RAPHE of a seed, often found on bird-dispersed seeds.

structural gene a length of DNA that codes for an enzyme or other protein. It is equivalent to a CISTRON. See also REGULATOR GENE, OPERATOR GENE, OPERON.

structuring method a method of organizing data to reveal patterns therein, e.g. ORDINATION, CLASSIFICATION.

struma see ABSCISSION JOINT.

stub a tree that is intermediate between a STOOL and a POLLARD.

style the sterile portion of the carpel between the ovary and the stigma, which may be elongated or feathery, especially in wind-pollinated species, so that the stigma is presented in an effective place for pollination. In primitive forms the styles of individual carpels are separate. In more advanced forms they tend to be fused. In some plants, notably traveller's joy (*Clematis vitalba*) and the pasque flower (*Pulsatilla vulgaris*), the style elongates after pollination and remains attached to the fruit. This is an adaptation to promote wind dispersal of the seed.

stylodious see GYNOECIUM.

sub- prefix denoting below, beneath, under.

subclimax vegetation see CLIMAX.

suberin a fatty acid polyester (see ESTER) found in the cell walls of the endodermis (see CASPARIAN STRIP) and of bark. It renders the tissue resistant to decay and entry of water. Such a tissue is said to be *suberized*.

suberized tissue see SUBERIN.

sub-formation in the UPPSALA SCHOOL OF PHYTOSOCIOLOGY, a geographically distinct unit of a major FORMATION.

subhymenium a layer of tissue lying immediately beneath the HYMENIUM in certain basidiomycete and ascomycete fungi.

sublittoral (infralittoral, neritic, subtidal) describing a narrow zone in the sea or ocean that extends from low-tide mark (the edge of the littoral zone) to the edge of the continental shelf. The depth increases to about 200 m. The water is well oxygenated and light penetrates to the bottom. The temperature and salinity are fairly constant except near the shore line. In shallower regions sessile algae such as *Ulva* and *Laminaria* are found.

submerged forest see PETRIFIED FOREST.

submergence marsh the lower part of a salt marsh, between the mean high tide level and the mean high tide level of neap tides. It is usually submerged at least 360 times a year, usually for an hour of daylight each day, and never suffers prolonged periods of continued exposure.

subsere see SERE.

subshrub a shrub-like plant that has persistent woody parts only at its base.

subsidiary cell (accessory cell) one of a group of morphologically differentiated epidermal cells immediately surrounding the GUARD CELLS. Stomatal complexes lacking subsidiary cells are termed *anomocytic*. The number, arrangement, and development of subsidiary cells are often useful taxonomic characters. See also MESOGENOUS, PERIGENOUS, HAPLOCHEILIC, SYNDETOCHEILIC.

subsoil see SOIL PROFILE.

subspecies the rank subordinate to species in the taxonomic hierarchy. The category has been variously defined but usually refers to a regional or ecologically distinct form of a species that is distinguishable in morphology. The term subspecies is used when two or more populations are separated in some way (e.g. ecologically or morphologically) throughout their range. However they are not usually genetically isolated. There can therefore be a continuous intergrading of subspecies, thus making their delimitation more arbitrary than that of the species. Generally, if 90% or more of a group of infraspecific individuals are recognizably distinct from another similar group, then each may be ranked as subspecies. This is often referred to as the '90% rule'. Subspecies may share many similar attributes but retain essential differences, thus indicating that they are merely regional representatives of one species, sharing a common origin. The

abbreviation 'subsp.' or 'ssp.' is used to indicate a subspecies, e.g. *Daucus carota* subsp. *gummifer*. The name of the subspecies that includes the type is always the same as the specific epithet.

substitional load see GENETIC LOAD.

substomatal cavity an air space immediately below the stomatal pore, surrounded by SPONGY MESOPHYLL cells. The surfaces of the mesophyll cells are the main site of gas exchange with the air in the substomatal cavity and also the site of evaporation of water. The air in the substomatal cavity is in direct contact with the atmosphere when the stomatal pore is open, providing a continuous diffusion pathway for the mesophyll cells to the atmosphere. The stomata of some xerophytic plants are sunk into this cavity to minimize transpiration: saturated air builds up in the cavity, reducing the diffusion gradient to the atmosphere. See STOMA.

substrate 1 the molecule or molecules on which an enzyme exerts its catalytic action. An enzyme may be specific for only one substrate, e.g. aspartase, which catalyses the interconversion of aspartic and fumaric acids, and is strictly specific for the L-isomer of aspartic acid. Other enzymes can act on a range of substrates; alkaline phosphatase can hydrolyse many esters of phosphoric acid. The substrate reflects the structure of the enzyme ACTIVE SITE in that the substrate must have a binding group, by which it can bind to the active site, and a susceptible bond at which the enzyme can attack the substrate. These groups must have complementary structures in the active site. **2** (substratum) the surface on which an organism grows or to which it is attached.

substratum see SUBSTRATE, def. 2.

subulate shaped like an awl, i.e. narrow, pointed, and more or less flattened, e.g. the leaves of *Subularia aquatica* (awlwort).

succession a series of changes in the composition of the plant and animal life of an area over time. This may begin, in a *primary succession*, with the colonization of bare rock, sand, or soil by such organisms as algae, lichens, and mosses (a pioneer community) and gradually change towards a CLIMAX community, which is in some form of dynamic equilibrium. This may take many years and may be deflected at any stage, for example by a natural disaster, such as a fire or hurricane. There is normally a complex interaction between vegetation succession and associated animals, particularly grazers and browsers. After a natural disaster, subsequent succession may be known as *secondary succession*. See also SERE.

succinic acid (butanedioic acid) a four-carbon dicarboxylic acid (see CARBOXYLIC ACID) with the formula $HOOC(CH_2)_2COOH$. Succinate is an intermediate in the Krebs cycle, formed from succinyl CoA with concomitant formation of GTP. It is oxidized to fumarate by succinate dehydrogenase, a flavoprotein. In the reaction the FAD prosthetic group of succinate dehydrogenase is reduced to $FADH_2$. Succinate is also a product of the GLYOXYLATE CYCLE, and succinyl CoA is an important precursor in PORPHYRIN synthesis.

succinyl coenzyme A an intermediate in the decarboxylation and oxidation of α-ketoglutaric acid to succinic acid in the KREBS CYCLE, in a reaction coupled to the formation of energy-rich guanosine triphosphate (GTP). Succinyl CoA is also involved in acylation reactions and porphyrin synthesis.

succubous describing the leaf arrangement in leafy liverworts where the front edges of the leaves lie below the back edges of the leaves in front. Compare INCUBOUS.

succulent a plant that lives in places where water is either in short supply (physical drought), e.g. deserts or sand dunes, or where there is plenty of water but it is not easily obtainable (physiological drought), e.g. in salt marshes and mudflats. A succulent plant conserves water by storing it in large parenchyma cells in swollen stems and leaves. Many succulents reduce water loss by having rolled leaves, leaves reduced to spines, sunken stomata, etc. Some succulents conserve water by opening their stomata at night and closing them during the day (see CRASSULACEAN ACID METABOLISM). Examples of succulents are desert cacti, many saltmarsh plants, and

Sedum species. See also HALOPHYTE, XEROPHYTE.

sucker an adventitious shoot that develops from the root, often coming up some distance from the parent plant. Suckering shoots can be separated from the parent once they have developed their own root system. When suckers develop in grafted plants (see GRAFTING) they are of the same constitution as the STOCK. If this is different from the SCION, as is usually the case, the suckers should be removed before they affect the vigour of the scion.

sucrase (invertase) an enzyme that breaks down sucrose into glucose and fructose.

sucrose (cane sugar, beet sugar, saccharose) a nonreducing disaccharide of glucose and fructose linked through a high-energy bond between carbon two of fructose and carbon one of glucose. Sucrose is the major transport sugar in higher plants. It is formed in the chloroplasts and transported through the phloem to other organs and tissues, where it is either metabolized for energy or utilized in the synthesis of reserve and structural polysaccharides. Present evidence indicates that in vivo synthesis of sucrose is a two-step reaction involving fructose 6-phosphate and glucose. These react to form sucrose 6-phosphate, which is then dephosphorylated by the enzyme sucrose phosphatase. The enzyme sucrose synthetase can catalyse the direct manufacture of sucrose from UDP-glucose and fructose, but in the cell sucrose synthetase is thought to be the major route of sucrose breakdown, allowing direct formation of NUCLEOSIDE DIPHOSPHATE SUGARS from sucrose. The enzyme invertase (sucrase) also splits sucrose, to D-glucose and D-fructose, but its physiological significance is not clear. The mixture of D-glucose and D-fructose formed on hydrolysis by invertase is often termed *invert sugar* since the optical activity is inverted from dextrorotatory (sucrose solution) to laevorotatory (D-glucose and D-fructose mixture). Invert sugar is found in many fruits.

Table sugar is composed of sucrose crystals. Production is mostly from sugar cane (*Saccharum officinarum*), grown in the tropics, and sugar beet (*Beta vulgaris*), grown in temperate zones. Small amounts are obtained from other sources, e.g. sugar maple (*Acer saccharum*) and sweet sorghum (*Sorghum bicolor*).

suction pressure see DIFFUSION PRESSURE DEFICIT.

Sudan stains any of various temporary aniline stains used to colour fats and waxes. Examples are Sudan IV, Sudan blue, and Sudan black. They are used to stain cutinized and suberized tissues. Thus the Casparian strip readily turns orange on staining with Sudan III or Sudan IV.

suffrutescent habit a growth form in which the plant is woody at the base but has herbaceous branches. It is seen in alpine willows.

suffruticose chamaeophyte see CHAMEOPHYTE.

sugar any of the lower molecular weight carbohydrates, namely MONOSACCHARIDES, smaller OLIGOSACCHARIDES, and derivatives of these, most commonly SUCROSE (cane or beet sugar). The term is sometimes used synonymously with monosaccharides; thus SUGAR ALCOHOLS are reduction products of monosaccharides, while SUGAR ACIDS are monosaccharide oxidation products.

sugar acid an oxidized derivative of a monosaccharide. There are two biologically important types of sugar acid, the *aldonic* acids and the URONIC ACIDS. In aldonic acids the aldehyde group of a monosaccharide is oxidized to a carboxylic acid. The aldonic acid of glucose, gluconic acid, is an intermediate in the PENTOSE PHOSPHATE PATHWAY. Ascorbic acid (vitamin C) is also an aldonic acid.

sugar alcohol a monosaccharide in which the aldehyde group is reduced to an alcohol. Thus D-glucose yields the sugar alcohol SORBITOL while the alcohol from mannose is MANNITOL. Other important sugar alcohols are GLYCEROL, a central compound in lipid metabolism, and INOSITOL, an important intermediate in cell wall polysaccharide synthesis. Sugar alcohols, other than glycerol, are limited in nature to the plant kingdom.

sugar phosphate a phosphate derivative of a monosaccharide. Sugar phosphates are important intermediates in carbohydrate metabolism. The pentose phosphate

RIBULOSE BISPHOSPHATE reacts with carbon dioxide in the first reaction of the Calvin cycle. All the subsequent steps of the Calvin cycle also involve sugar phosphates as do the reactions of glycolysis.

sulcal see SULCATE.

sulcate (sulcal) marked with ridges, grooves, or furrows.

sulcus a groove or furrow, such as the longitudinal groove in the lower half of the test of a dinoflagellate, in which lies an undulipodium. See DINOMASTIGOTA.

sulphur symbol: S. A nonmetallic solid yellow element, atomic number 16, atomic weight 32.06. It is essential to growth, being found in the amino acids CYSTINE and METHIONINE, in the iron–sulphur proteins, e.g. FERREDOXIN, which are important in electron transport, and in COENZYME A. Sulphur is also found in various secondary metabolites, e.g. the mustard-oil glycosides. Sulphur is absorbed from soil as the sulphate (SO_4^{2-}) ion. Various nitrogen and potassium fertilizers are applied as sulphates and sulphur deficiency is not commonly a problem.

Sulphur is used to control various pathogens and pests (see SULPHUR DUST, DITHIOCARBAMATE FUNGICIDE). Sulphur dioxide, a by-product of many industrial processes, is an important atmospheric pollutant (see also ACID RAIN).

sulphur bacteria bacteria that live by oxidizing sulphides. They are found mainly in sulphur-rich muds and mineral-rich springs, including geysers and deep-sea hydrothermal vents. Such bacteria as the green sulphur bacteria (Chlorobia) and purple sulphur bacteria (certain members of the Proteobacteria) use hydrogen sulphide instead of water as a source of electrons in photosynthesis, releasing sulphur rather than oxygen. This sulphur may be stored in the cells as elemental sulphur globules, excreted, or further oxidized to sulphates or other sulphur-containing compounds before being excreted. Other bacteria oxidize different sulphides, thiosulphates, polythionates, and sulphites. The sulphur bacteria therefore play a role in the cycling of sulphur through the ecosystem (see BIOGEOCHEMICAL CYCLE). Examples of bacteria that oxidize inorganic sulphur compounds include *Thiobacterium*, which forms aggregates of rods in a gelatinous matrix, and the flagellate genera *Thiobacillus*, *Thiospira*, *Thiovulum*, and *Macromonas*. A few archaebacteria (see ARCHAEA), such as *Sulfolobus*, which live at high acidity in hot springs, oxidize elemental sulphur; most of them can also oxidize organic compounds.

There are also sulphur-respiring anaerobic bacteria that utilize organic carbon compounds, transferring electrons from the carbon compounds to the sulphur compounds. *Desulfovibrio* and *Desulfacinum* are examples. They release such gases as hydrogen sulphide into the sediments. The reaction of this hydrogen sulphide with iron compounds present in the muds is thought to have been the source of pyrite (iron sulphide) deposits.

sulphur dust powdered sulphur, used as an insecticide (especially against mites and scale insects) and as a fungicide, particularly in the control of mildew. It gives off sulphur dioxide when ignited and is thus often used as a FUMIGANT. It is also mixed with soil to kill soil-borne pathogens. Elemental sulphur, in combination with calcium hydroxide and water, forms *lime sulphur*, an orange-coloured liquid used to kill various mites and to control peach leaf curl, scab, and powdery mildew.

summer wood SEE LATE WOOD.

sun plant a plant that is able to flourish only in conditions of high light intensity. Sun plants usually have wide epidermal and palisade layers and numerous stomata on the lower surface of the leaf. Examples are the dominant trees of many woodland communities. Compare SHADE PLANT.

super- prefix denoting over, above, upon, higher, greater, more than, extra, superior.

superdominance SEE OVERDOMINANCE.

super gene (complex gene) a group of closely linked genes that determine different aspects of the same character and tend to act as a single functional unit. The S/s gene, which determines the 'pin' (s) and 'thrum' (S) style character of numerous *Primula* species, is an example (see HETEROSTYLY). Besides controlling style length, it also has loci for stigma

morphology, incompatibility reactions, pollen surface characteristics, and anther position. Very occasionally crossing over occurs between these loci resulting in a plant with both pin and thrum characteristics.

superior ovary see OVARY.

superkingdom see KINGDOM.

supernumerary buds see ACCESSORY BUDS.

supernumerary chromosome see B-CHROMOSOME.

superposed buds see ACCESSORY BUDS.

superspecies a complex of closely related species that exist in different geographical areas. Within the group only certain pairs of species are capable of hybridization together. See also ALLOPATRIC.

supervolute (convolute) describing a kind of individual leaf folding in which the leaf is rolled towards the adaxial surface to such an extent that one side of the lamina rolls over the other. *See illustration at* ptyxis.

supporting tissue see MECHANICAL TISSUE.

suppressor mutation see REVERSE MUTATION.

supra- prefix denoting above, over, transcending.

supralittoral zone the part of the seashore immediately above the uppermost limit of the LITTORAL ZONE. It is affected by sea spray, but under normal conditions is never submerged at high tide.

surface area/volume ratio the ratio of the surface area of a cell, organ, or organism to its internal volume. The higher the ratio the greater the possibility for gaseous exchange and absorption of dissolved minerals. Thus the size of unicellular organisms is limited by the rate at which oxygen and carbon dioxide can diffuse into and out of the cell. The form of higher plants, especially the possession of numerous flat leaves, serves to increase the surface area/volume ratio. In this way a ratio of 30 cm^2 of surface to 1 cm^3 of tissues can be achieved. This ratio is greatly increased if intercellular spaces are taken into account.

surface tension a phenomenon apparent at the boundary of any liquid whereby an elastic film appears to be stretched over the surface. It is caused by the attractive forces between molecules at the surface of the liquid. The surface tension of water is particularly strong due to the orientation of the hydrogen bonds in the water molecule. This property slows down the vaporization of water and helps prevent plant desiccation. More importantly surface tension is responsible, with the strong adhesive forces of water, for CAPILLARITY.

survival of the fittest see DARWINISM, NATURAL SELECTION.

susceptibility the condition of a plant such that it is likely to succumb to attack by a pathogen. Extreme susceptibility, in which the host cells die soon after invasion by a pathogen, is termed HYPERSENSITIVITY and this reaction in fact confers considerable resistance. Susceptibility is often found where a crop cultivar has relied on one major gene for resistance to a particular pathogen. If a mutant of the pathogen develops that can overcome this major-gene resistance, then the cultivar is often seen to have no other means of preventing the invasion and spread of the pathogen. Susceptibility, like RESISTANCE, varies according to environmental conditions. For example, excessive application of nitrogen fertilizer may lead to lavish growth that is more easily invaded by a pathogen.

suspended placentation see APICAL PLACENTATION.

suspension culture the system of growing single cells and small cell aggregates in a liquid growth medium that is kept agitated by means of bubbling, shaking, or stirring so the cells do not settle out. Microorganisms and cells derived from callus tissues may be grown in this way. Growth is maintained by providing continuous aeration and by either transferring portions of the suspension to fresh medium or replacing a part of the culture with fresh medium. Suspension cultures of plant cells derived from friable masses of CALLUS show a similar growth curve to cultures of microorganisms in that there is a lag phase and a logarithmic phase of growth. Cells of many plant species can be induced to form embryoids (see EMBRYOGENY) in suspension culture, which, if removed from the culture and given appropriate conditions, will develop into complete plants.

suspensor the line of cells that differentiates from the PROEMBRYO by mitosis and anchors the embryo in the parental tissue. It also conducts nutrients to the embryo.

sustainable development the management of economic growth while at the same time preserving the quality of the environment for future generations. It aims to promote the sustainable use of renewable resources so that they do not become depleted. It may sometimes have to balance exploitation with conservation, as for example in ecotourism, where tourism may have a negative impact on local ecosystems, but also raises awareness of conservation issues, may help to fund conservation, and may also provide an income for local communities so that they do not have to depend on less sustainable resources. See also CONVENTION ON BIOLOGICAL DIVERSITY, NONRENEWABLE RESOURCE, WORLD CONSERVATION MONITORING CENTRE, WORLD CONSERVATION UNION.

Svedberg unit see SEDIMENTATION COEFFICIENT.

swamp a region of vegetation that develops in stagnant or slow-flowing water, such as that around a lake margin. Initially in temperate regions there may be a layer of peat supporting numerous reeds. When this builds up above the water level many other species become established, including sedges, irises, and mosses (see SUCCESSION). In some places grasses or trees may dominate. Swamps may also develop on water-retentive clays in desert and semidesert areas.

Mangrove swamps are found along many tropical coasts. Mangroves are HALOPHYTES and grow in a tangled mass to a height of 10 metres or more. The stems and the exposed roots are often covered with red algae (see RHODOPHYTA). Some species have stilt roots (see PROP ROOT), which anchor the trees in the soft mud and help to retain silt, thus gradually enabling other species to grow. Other mangrove species have PNEUMATOPHORES.

swarm cell in the MYXOMYCOTA, an amoeba-like cell that is produced on germination of a spore. Swarm cells have two anterior smooth undulipodia and lack a cell wall.

Usually one, but occasionally up to four such cells emerge from one spore. They ingest food, reproduce by division, and then become gametes, in some instances becoming myxamoebae first. The gametes fuse in pairs to form a zygote that develops into the PLASMODIUM. In adverse conditions the swarm cells can encyst.

syconium (syconus, synconium) a type of pseudocarp or inflorescence in which achenes develop on the inside of a hollow receptacle. The fruit and inflorescence of figs (*Ficus*) is an example. *See illustration at* pseudocarp.

syllepsis see SYLLEPTIC GROWTH.

sylleptic growth (syllepsis) a pattern of growth in which there are no dormant or resting periods, and leaves and axillary shoots are produced continuously. Compare RHYTHMIC GROWTH.

symbiont an organism that is a partner in a symbiotic relationship (see SYMBIOSIS).

symbiosis an intimate relationship between two or more living organisms (*symbionts*). It may be used in the narrow sense to mean only relationships in which all the partners benefit. In this sense it is synonymous with MUTUALISM. In its wider sense it covers other relationships, such as PARASITISM and COMMENSALISM. Two forms of symbiosis are recognized: *conjunctive symbiosis*, in which the two partners form a single body, e.g. the PHYCOBIONT and MYCOBIONT in a lichen, and *disjunctive symbiosis*, in which there is no direct physical contact between the two, e.g ants and myrmecophytes (see MYRMECOPHILY).

sympatric (*n.* sympatry) describing two or more populations, living in the same place, that could interbreed but do not usually do so because of various differences, e.g. in time of flowering or type of pollinator. These populations may, through natural selection, become so distinct that they eventually become unable to interbreed and may be regarded as separate species (sympatric SPECIATION). This is *sympatric evolution*. Compare ALLOPATRIC.

Sympetalae see ASTERIDAE.

sympetalous see GAMOPETALOUS.

symplast the continuum of cell protoplasts throughout the plant, linked by PLASMODESMATA. Resistance to water flow

through the symplast is far greater than through the APOPLAST and the symplast is consequently only a secondary route for water movement.

symplastic growth a type of growth in which adjacent cells grow at the same rate so that the SYMPLAST is not disrupted. Compare INTRUSIVE GROWTH, SLIDING GROWTH.

symplesiomorphy see CLADISTICS.

sympodial branching a type of GROWTH seen in some plants, e.g. elm and lime, in which the apical bud withers at the end of the growing season and growth is continued the following season by the lateral bud immediately below. Compare MONOPODIAL BRANCHING.

synangium a compound fruiting unit developed from the lateral fusion of individual sporangia, as seen in some ferns, e.g. *Marattia*, and gymnosperms, e.g. *Welwitschia*.

synapomorphy see CLADISTICS.

synapsis the pairing of homologous chromosomes during PROPHASE of the first division of meiosis. The nature of the force that causes homologues to be drawn to each other is not known, but pairing is a very exact process, commencing with the pairing of the CENTROMERES and continuing with pairing of CHROMOMERES at various points along the length of the homologues. During the process, contraction of the chromosomes (condensation) continues and they become closely coiled around each other to form BIVALENTS.

synaptonemal complex a structure that forms during prophase I of meiosis and is involved in the intimate pairing between HOMOLOGOUS CHROMOSOMES.

syncarp a compound fruit comprising two or more carpels.

syncarpous describing a gynoecium in which the carpels are fused. The degree of fusion (*syncarpy*) varies from only being joined at the ovary base (*semicarpous*), as in the Lamiaceae, to fusion of the ovary and style (*synstylovarious*), as in the Caryophyllaceae, to fusion of the whole carpel (*syncarpous*), as in the Brassicaceae. This fusion may or may not persist in the fruit at dispersal. The degree of fusion of the carpels is usually taken as an indication of the degree of specialization in the flower. Compare APOCARPOUS.

syncarpy see SYNCARPOUS.

synchronous culture a culture of cells in which all the individuals are at approximately the same stage in the CELL CYCLE. This can be achieved by means of various kinds of shocks, or by the use of drugs. Such a culture is a very useful tool for biochemical and physiological research.

synconium see SYCONIUM.

syncytium a mass of protoplasm containing many nuclei. This term was formerly used to distinguish a structure resulting from the fusion of protoplasts from a coenocytium, which is derived from nuclear division without accompanying cytoplasmic cleavage; today the term coenocyte is preferred for both structures.

syndetocheilic describing a gymnosperm stomatal complex in which the SUBSIDIARY CELLS are derived from the same initial as the guard cells, as occurred in the Bennettitales. Compare HAPLOCHEILIC. See also MESOGENOUS.

syndiploidy a doubling of the chromosome number through a fault in the mitotic process. Commonly there is either failure of spindle formation at anaphase or failure to form a dividing wall between the two daughter cells. If such cells continue to divide normally an AUTOPOLYPLOID segment of tissue may arise in the plant. If this develops flowers, autopolyploid seeds may result.

synecology the study of the interactions between all the living organisms in a natural community, such as an oak wood or a pond, and the effect upon them of nonliving (ABIOTIC FACTORS) in the environment. Compare AUTECOLOGY.

synergidae (synergids) the two haploid nuclei at the micropylar end of the EMBRYO SAC that do not participate in the fertilization process. Together with the egg nucleus they constitute the egg apparatus. Their function is not known and they abort soon after fertilization.

synergism the phenomenon whereby the effect of two substances acting together is greater than the sum of their individual effects. Combinations of different

hormones often act synergistically, for example, auxin and gibberellin combined bring about considerably greater internode elongation than either could individually.

syngamy see FERTILIZATION.

syngeneic see ISOGENEIC.

syngenesious describing an ANDROECIUM in which the anthers are fused, as in composite plants (see ASTERACEAE). Compare ADELPHOUS.

synonym in TAXONOMY, a name other than the official name. Such names are usually older names that do not conform to the current rules of naming.

synthase see SYNTHETASE.

synthetase (synthase) an enzyme involved in the synthesis of organic molecules.

synthetic theory a theory of evolution that combines Darwin's theory of natural selection with Mendelian genetics (see MENDELISM).

syntype any one of two or more specimens or other material (descriptions, illustrations, etc.) designated by an author in the original publication of the name of a TAXON. As a taxon can only have one type, a LECTOTYPE can subsequently be chosen from a group of syntypes to serve as the nomenclatural TYPE.

Synurophyceae a class of the CHRYSOMONADA (or, in some classification schemes, of the CHROMOPHYTA) including mainly single undulipodiate cells or colonies of undulipodiate cells that have paired undulipodial swellings and no eyespots. The Synurophyceae are distinguished from the CHRYSOPHYCEAE by having siliceous scales and containing only chlorophylls a and c_1. Examples include *Synura* and *Mallomonas*.

systematics the scientific study and description of the variation in living organisms and the relationships that exist between them (the term is often used synonymously with TAXONOMY). It is frequently preceded by a prefix, such as chemo- or bio-, denoting specialized fields of study that are valid in their own right. Several distinct phases in the development of systematics can be recognized. Pre-Darwinian systematics was primarily based on morphological and anatomical data. An understanding of natural selection and the behaviour of populations led to new interpretations and the inclusion of PHYLOGENETIC considerations. Rapid developments in technology during recent decades have greatly aided the fields of chemosystematics and NUMERICAL TAXONOMY.

systematic sampling see SAMPLING.

systemic (translocated) describing a chemical that is absorbed by a plant and transported throughout the tissues. Systemic fungicides and insecticides are applied to render plant tissues toxic to fungi or insects. Their effects are longer lasting than those of contact pesticides. Systemic herbicides, such as ZYMASES,4-D, ZYMASES,4,5-T, and MCPA, are used as selective weedkillers, killing most broad-leaved plants but leaving grasses and cereals unaffected. Such herbicides are also used to kill deeply rooted weeds.

Tt

2,4,5-T (2,4,5-trichlorophenoxyacetic acid) a
synthetic AUXIN of the PHENOXYACETIC
ACID type with three substituted chlorine
atoms. It has mainly proved of use as a
selective weedkiller and as a defoliant.
However, contamination with the highly
toxic chemical dioxin has led to measures
to restrict its use.

tachytelic see CHRONISTICS.

tactic movement see TAXIS.

taiga see FOREST.

take-all a disease of grasses and cereals
caused by the fungus *Gaeumannomyces*
(*Ophiobolus*) *graminis*. The pathogen survives
in the soil on crop debris and can infect the
crop at any stage. The dense mycelium on
the roots and stem bases gives them a
typical blackened appearance. If the plants
survive until ear emergence the ears have
little grain and appear bleached and are
therefore called 'whiteheads'.

tandem DNA see REPETITIVE DNA.

tandem repeat a chromosome MUTATION in
which two or more identical chromosome
segments bearing the same gene sequence
lie next to each other.

tangential 1 describing a longitudinal
section cut down a cylindrical organ that
does not pass through the centre of the
organ. A tangential section through a woody
stem is at a tangent to the growth rings and
the growth-ring pattern appears paraboloid.
Compare RADIAL.

2 see PERICLINAL.

tannin (tannic acid) any substance capable of
precipitating the gelatin of animal hides as
an insoluble compound, so changing the
hide to leather, resistant to putrefaction. All
tannins are obtained from plants and most

are polyphenols. Common sources of
tannin include tea (*Camellia sinensis*), sumac
(*Rhus* spp.), and the bark and gallnuts of oak
(*Quercus* spp.).

tapetum the food-rich layer of cells that
surrounds the SPORE MOTHER CELLS in
vascular plants. In some plants it breaks
down to form a fluid termed the
periplasmodium, which is absorbed by the
developing microspores. In others the
tapetum remains intact until shortly before
anther dehiscence and secretes substances
into the locule.

Taphrinales an order of the ASCOMYCOTA
containing many plant parasites, with most
of the 95 species being placed in the genus
Taphrina. There are three other genera. The
Taphrinales form a limited dikaryotic
mycelium, with a two-layered wall, that
gives rise to asci. The ascospores reproduce
asexually by budding before giving rise
again to hyphae. These fungi commonly
induce hyperplasia, as in peach leaf curl,
caused by *T. deformans*, and witches' broom
of birch, caused by *T. betulina*.

taproot a persistent robust primary root,
often penetrating some depth below
ground level and sometimes specialized for
storage. Swollen taproots are produced by
many BIENNIAL plants, e.g. carrot (*Daucus
carota*).

tautonym a specific or infraspecific name
that exactly repeats the generic name (see
SPECIES), e.g. *Magnolia magnolia*. This is not
allowed in botanical nomenclature but is,
however, acceptable in zoology.

Taxales an order of the CONIFEROPHYTA
containing 4 extant genera (including *Taxus*,
yews) and about 16 species, all in the family

Taxaceae. They are evergreen or deciduous trees, with needle-like leaves that are spirally arranged but often have the appearance of being in two ranks. The male flowers are borne in globular cones, and the pollen grains are not winged. The Taxales differ from the CONIFERALES in having solitary ovules borne on short terminal shoots, but not in cones. The ovules bear ARILS, which often form brightly coloured fleshy cups around the seeds. The embryo has two cotyledons. The Taxales are economically important as timber and ornamental trees and as a source of the drug tamoxifen, used in the treatment of breast cancer. They are found in the northern hemisphere and New Caledonia. Their fossil record goes back to the TRIASSIC. See also TAXOL.

taxis (tactic movement; *pl.* **taxes)** a free directional locomotor movement exhibited by whole organisms that change their physical position in response to external stimuli. Taxis is shown by many small, generally unicellular, organisms, such as *Euglena*, which moves using an undulipodium. It is also shown by bacteria and the reproductive cells of some plants and algae. See also CHEMOTAXIS, PHOTOTAXIS. Compare TROPISM, NASTIC MOVEMENTS.

taxol a diterpene (see TERPENOIDS) originally extracted from the bark of the Pacific yew (*Taxus brevifolia*) but now made artificially, that provides a treatment for solid tumours, especially cancers of the breast, brain, ovary and lung, and some leukaemias. It contains alkaloids that destabilize microtubules, inhibiting mitosis.

taxometrics see NUMERICAL TAXONOMY.

taxon (*pl.* taxa) A named taxonomic group of any rank. Thus at the family level taxa may be represented by the Rosaceae and Lamiaceae, while *Rosa* and *Lamium* are examples of generic taxa. The term was coined to replace clumsy phrases such as taxonomic entity and taxonomic unit. Furthermore, the organisms contained within a rank (e.g. genus, order, or species) can also be referred to as taxa.

taxonomic characters features, such as form, physiology, structure, and behaviour, that are assessed in isolation from the rest of the plant by taxonomists, in order to make comparisons and interpretations. It is important to distinguish between characters and character states. For example, leaf width may be a character, while leaves 4 mm wide are an expression of that character, i.e. its character state. Characters are often referred to as 'good' or 'bad' but this is strictly relative. Thus a good diagnostic character, such as compound leaves in a group of plants that mainly have entire leaves, would be bad for separating taxa in a group in which leaf divisions were either variable or often compound. See also WEIGHTING.

taxonomy the study of the principles and practices of classification. The term taxonomy is strictly applied to the study and description of variation in the natural world and the subsequent compilation of classifications. However, it is often used more loosely and includes the part of biological science referred to as SYSTEMATICS. Taxonomy is a vast subject and various sections can be recognized within the discipline. In dealing with the flora of an area several phases can be recognized. The first phase is mainly concerned with identification and is sometimes referred to as exploratory or pioneer. Study of many tropical areas is still in this stage. Once material is better known and taxonomists have a good understanding of local and regional variation of the species it moves into the consolidation phase (the flora of Europe comes into this category). These two phases are jointly described by some as 'alpha' taxonomy. Once cytological or biosystematic data are available these can be added to existing data. Taxonomy in which all available evidence is considered is described as the encyclopaedic phase or 'omega' taxonomy. Some authors make further distinctions between classical (mainly intuitive) and experimental approaches, the latter including biosystematic, chemosystematic, and numerical procedures. See also BIOSYSTEMATICS.

TCA cycle see KREBS CYCLE.

t distribution the ratio of the deviation, *d*, to the estimated standard deviation, *s*. It is used in experiments where the number of observations is small (31 or less and hence 30 or less degrees of freedom) and consequently the estimated standard deviation may differ widely from the true standard deviation. To find whether any deviation is significant, its probability is looked up in 'Student's' t tables, which take the number of degrees of freedom (N) into account. The greater the value of N, the smaller the value of t for any given level of probability.

teleutosorus see TELIUM.

teleutospore see TELIOSPORE.

teliomorph the sexual ('perfect') form, or morph, of a fungus, possessing reproductive structures, such as ascomata or basidiomata. Compare ANAMORPH. See also STATES OF FUNGI.

Teliomycetes a class of the BASIDIOMYCOTA containing those fungi in which a sorus (telium) develops in place of a basidioma and produces sori containing thick-walled resting spores (TELIOSPORES) that act as probasidia. The mycelium is usually limited and intercellular, often with haustoria. The hyphae are septate and have simple pores. Several different types of spores may be produced during the life cycle. Most Teliomycetes are parasites of angiosperms and ferns. The class contains the orders UREDINALES (rusts) and Septobasidiales.

teliospore (teleutospore) a dark two-celled thick-walled binucleate spore formed towards the end of the growing season by rust and smut fungi (see RUSTS, SMUTS). Teliospores develop in a black sorus, the *telium* (or *teleutosorus*), and are the form in which the fungus overwinters. The two nuclei in each cell fuse and the zygote nuclei then undergo meiosis so that on germination a haploid promycelium grows from each cell. This subsequently forms BASIDIOSPORES. The teliospores of smut fungi are termed ustilospores (see USTOMYCETES). See HETEROECIOUS.

telium (teleutosorus) a sorus in which TELIOSPORES develop.

telocentric describing a chromosome that has the CENTROMERE at, or very close to, one end so that only one chromosome arm is visible. Compare ACROCENTRIC, METACENTRIC.

telomere the region of a chromosome at the tip of an arm. It consists of repeated sequences of DNA that appear to ensure that each cycle of DNA REPLICATION is completed. Each time the cell divides, some of the telomere sequence is lost, and after about 60–100 cell divisions the cell dies. Cell death, which increases in frequency during ageing, is thought to be related to loss of the telomeres.

telome theory the proposition that certain plant organs are derived from modified reduced branch systems. Thus MEGAPHYLLS are believed to be derived by the flattening of a branch system and the subsequent development of ground tissue (webbing) between the branches. Some believe stamens may also have originated in this manner, rather than from microsporophylls.

telophase the stage in nuclear division following ANAPHASE, in which the separated chromatids or the separated homologous chromosomes of bivalents collect at the poles of the spindle and the nuclei of the daughter cells are formed. In MITOSIS and the second division of MEIOSIS, the chromatids (now complete single-stranded chromosomes) become surrounded by vesicles as they reach the poles and these eventually merge to form the nuclear membrane. The chromosomes lengthen as they uncoil and the CHROMATIN assumes its interphase condition. Nucleoli reappear as they are formed at the nucleolar organizers.

At telophase of the first division of meiosis, the daughter haploid nuclei that form at the poles contain one member only of each homologous pair of chromosomes present in the original mother cell. Each chromosome is divided into two chromatids (genetic recombinants as a result of chiasma formation) held together by a CENTROMERE. The duration of the following interphase is variable but there is no further DNA replication and the chromosomes remain in a relatively contracted condition. The two daughter haploid nuclei then enter the second division of meiosis.

As daughter nuclei become organized at the poles, a CELL PLATE begins to form in the cytoplasm in the equatorial region. This marks the beginning of cell wall formation. In meiosis a cell wall may not form at the end of the first division, the daughter cells being separated by a cell plate only.

TEM see ELECTRON MICROSCOPE.

temperate forest see FOREST.

temperate grassland see GRASSLAND, PAMPAS, PRAIRIE, STEPPE.

temperate phage see BACTERIOPHAGE.

temporary stain a type of stain used for immediate observation through the light microscope. Such stains often damage the section, or the colour of the stain fades after a short time. For example, when plant sections are stained with SCHULTZE'S SOLUTION and mounted in 50% glycerol, the stain eventually causes swelling and dissolves the walls. Some specimens can be mounted directly in the stain, for example, when using RUTHENIUM RED. Compare PERMANENT STAIN. See also STAINING.

tendril tendrillate) A modified leaf, leaflet, branch, or inflorescence of a climbing plant that coils around suitable objects, such as other nearby plants, and helps support and elevate the plant. Examples are the branch tendrils of white bryony (*Bryonia dioica*), the leaf tendrils of yellow vetchling (*Lathyrus aphaca*), and the inflorescence tendrils of virginia creepers (*Parthenocissus*). The tendrils of some species terminate in disclike suckers while others, e.g. those of grapevines (*Vitis*), resemble climbing roots in being negatively phototropic and hence growing into dark cracks on the support.

Tenericutes a division of the EUBACTERIA containing bacteria that lack rigid cell walls.

tension wood see REACTION WOOD.

tent pole in the Cycadofilicales (seed ferns) and *Ginkgo biloba*, an extension of the female gametophyte towards the pollen chamber.

tepal an individual perianth part in flowers that have no distinct calyx and corolla, as occurs in many monocotyledons. In the tulip, lilies, etc. the tepals are all highly coloured.

teratology the study of malformations and abnormal growth. See also WITCHES' BROOM, GALL, LEAF CURL, CLUB ROOT.

terete smooth, cylindrical, and tapering as, for example, a grass stem.

terminal bud see BUD.

terminalization see PROPHASE.

termination in PROTEIN SYNTHESIS, the addition of the final amino acid to a polypeptide chain and the detachment of the finished chain from the ribosomes.

terminator sequence a sequence of bases that acts as a signal for the termination of TRANSCRIPTION.

ternate arranged in threes or subdivided into threes.

terpene see TERPENOID.

terpenoid any of a large class of compounds derived from multiples of the unsaturated hydrocarbon isoprene. They include the *terpenes*, which are all hydrocarbons. Terpenoids based on one, two, three, or four ISOPRENE SUBUNITS are called hemiterpenoids (e.g. apiose), monoterpenoids (e.g. camphor), sesquiterpenoids (e.g. zingiberine), and diterpenoids (e.g. phytol) respectively.

The terpenoids are a very diverse class of compounds. They may be cyclic, e.g. cannabidiol, or acyclic, e.g. phytol, a precursor of chlorophyll. Many give the characteristic odour or flavour to a plant oil, e.g. limonene (lemon oil), pinene (pine oil), and menthol (mint oil). The fat-soluble vitamins A, E, and K are terpenoids as are the ubiquinones and plastoquinones and the hormones gibberellic acid and abscisic acid. The carotenoids (tetraterpenoids) and sporopollenin, gutta-percha, and natural rubber (polyterpenoids) are further examples of this group.

Terpenoid distribution is considered potentially valuable in taxonomic work. For example, various changes have been suggested at the tribe level in the Asteraceae following work on the occurrence of the bitter-tasting sesquiterpene lactones. In the genus *Pinus*, the distribution of turpentines has been used to justify placement of *P. jeffreyi* in the Macrocarpae. Traditionally it was placed in the Australes on the basis of its resemblance to *P. ponderosa*, a typical member of the Australes.

terra rossa a type of intrazonal soil rich in lime (calcimorphic) from the underlying limestone. It is a clayey soil with a low humus content and its bright red colour results from large amounts of iron oxides. It may be found in regions of Spain, southern France, and southern Italy. Compare RENDZINA.

terricolous living on the ground or in the soil.

Tertiary the first period of the CENOZOIC era from about 65 to 2 million years ago. It began with the Palaeocene epoch, which was followed by the Eocene, Oligocene, Miocene, and Pliocene (Pleiocene) epochs. The climate was warm and wet to the end of the Oligocene, becoming cooler and drier towards the end of the period as the oceans contracted and the present mountain ranges were formed. Abundant plant fossils have been found, many showing similarity to modern-day plants. Changes in the distribution of the flora can be related to changes in climate throughout the period. Siliceous remains of diatoms (the diatomaceous earths, kieselguhr) are widespread, blue-green bacteria (see CYANOBACTERIA) contributed to the formation of oil shales, green and red algae (see CHLOROPHYTA, RHODOPHYTA) contributed to marine limestone deposits, and the fossil stoneworts (see CHAROPHYTA) contributed to the formation of freshwater limestone beds. The angiosperms were established by the beginning of the Tertiary but evolved and expanded rapidly during this period, with all the more specialized flowers appearing in an explosive adaptive radiation. This was accompanied by the evolution of specialist pollinating insects, particularly bees, butterflies, and moths. Grasses became more abundant in the Pliocene, possibly because of a change to a drier climate. Modern genera of conifers appeared, but other gymnosperm groups declined during the Tertiary. There was an adaptive radiation of leptosporangiate ferns (Filicales). See GEOLOGICAL TIME SCALE.

tertiary structure the third level of structure in a protein: the mode of folding of the polypeptide chain (i.e. of the α-helix or β-pleated sheet). This involves hydrogen bonds, DISULPHIDE BRIDGES, and hydrophobic interactions (the protein tends to fold so that hydrophobic side groups are protected from the aqueous surroundings of the cell fluids, while the hydrophilic side groups protrude into the cell solution). It is this folding that is often important in

conferring the shape of the ACTIVE SITE. See also CONFORMATION, PRIMARY STRUCTURE, SECONDARY STRUCTURE, QUATERNARY STRUCTURE.

test 1 see TESTA. **2** the outer covering of the unicells of certain PROTOCTISTA.

testa (test) the protective outer covering of a seed, derived from the INTEGUMENTS of the ovule after fertilization. In primitive forms the outer integument may remain fleshy and the inner become lignified to form a *sarcotesta* and *sclerotesta* respectively. In more advanced forms both the integuments fuse and become hard and dry except for a small unthickened area at the MICROPYLE, which facilitates radicle emergence at seed germination. In some species the outer surface of the testa becomes covered with mucilage, hairs, or fibres to aid seed dispersal. The fibrous nature of the cotton seed has long been exploited commercially.

test cross a cross between an individual of uncertain genetic constitution and a homozygous recessive. The latter is often a parent, in which case the test cross is a specific form of BACKCROSS. The purpose of a test cross is to ascertain the unknown genotype. For example, if T (tall peas) is dominant to t (short peas) then any tall peas could be either TT or Tt. If tall plants (TT) were test crossed to short plants (tt), all the offspring would be tall plants. If the tall plants were Tt, then short (tt) peas would be among the offspring.

test of significance a way of finding whether the results obtained from an experiment differ from the hypothesis the experiment was constructed to test because of sampling error or because the original hypothesis is invalid.

tetra- prefix denoting four, or having four parts or components.

tetrad a group of four cells formed by meiosis. All four cells may survive to form spores, as occurs in the formation of pollen grains. Alternatively some of the cells abort, as in the angiosperm ovule, where one cell develops into the embryo sac and the remainder abort. The four cells of the tetrad may remain joined together. This provides a useful system for the study of the nature and extent of recombination, a procedure termed TETRAD ANALYSIS. The shape of the tetrad, e.g. whether it is linear, tetrahedral, tetragonal, or rhomboidal can be a useful taxonomic character and is a major factor determining pollen morphology. See also TRIRADIATE SCAR.

tetrad analysis the genetic analysis of the four cells resulting from a meiotic division in order to gain information on the nature and extent of recombination. Tetrad analysis is only possible when the meiotic products remain together in groups (tetrads). This does occur in certain organisms, notably ascomycete fungi, e.g. *Neurospora*, and some nonvascular plants, e.g. the liverwort *Sphaerocarpus*.

tetramer a PROTEIN made up of four polypeptide chains or units.

tetramerous (4-merous) Describing flowers in which the parts of each whorl are inserted in fours, or multiples of four, as in the willowherbs (*Epilobium*) and members of the Brassicaceae.

tetraploid an organism or cell with four times the haploid number of chromosomes. The fusion of diploid gametes will result in the formation of a tetraploid. Such gametes initially arise because of the failure of homologous chromosomes to separate at meiosis. Alternatively a tetraploid segment of a plant may arise because of the failure of chromatids to separate at mitosis. If this segment develops reproductive structures then diploid gametes will be formed. See AUTOPOLYPLOIDY, ALLOPOLYPLOIDY.

tetrapyrrole a molecule consisting of four joined pyrrole rings, each made up of of four CH units and one NH unit. The pyrrole may be arranged in a line (linear) or in a ring (cyclic). Tetrapyrrole-based pigments are found in most organisms. In plants they include many light-sensitive pigments, including chlorophyll. See also PORPHYRIN.

tetrarch primary XYLEM made up of four strands.

tetra-allelic describing a POLYPLOID that has four different alleles at a particular gene locus.

tetrasomic see ANEUPLOIDY.

Tetrasporales an order of green algae (CHLOROPHYTA) of the class CHLOROPHYCEAE in which the cells form PALMELLOID clusters and colonies, some up to several centimetres long; the cells are nonmotile and have eyespots and contractile vacuoles. Although they have no undulipodia, they have basal bodies. *Tetraspora* cells have two PSEUDOFLAGELLA.

tetrasporangium a sporangium in which four spores (the *tetraspores*) are formed as a result of meiosis. It is seen in some of the red algae (see RHODOPHYTA) of the subclass FLORIDEOPHYCIDAE, e.g. *Polysiphonia*.

tetraspore see TETRASPORANGIUM.

tetrasporophyte see FLORIDEOPHYCIDAE.

thalamus see RECEPTACLE.

Thallophyta (Thallobionta) a former division or subkingdom of the PLANTAE containing all nonanimal organisms without differentiated stems, leaves, and roots, i.e. the algae, fungi, lichens, and bacteria. It was later restricted to the algae and lichens. Compare EMBRYOPHYTA.

thallose describing an organism that has a THALLUS. See HEPATOPHYTA.

thallus (*adj.* thalloid, thallose) **1** a plant or algal body that is not differentiated into true leaves, stems, and roots. It is often a flattened structure, e.g. the gametophyte generation of the thallose liverwort *Pellia*. **2** the vegetative body of a lichen.

theca the outer covering of the cells of certain green algae (CHLOROPHYTA) and dinoflagellates (DINOMASTIGOTA). In the Chlorophyta the theca is composed of fused scales. In the Dinomastigota it consists of cellulose plates.

thermoacidophile an archaebacterium (see ARCHAEA) that thrives in highly acidic environments at high temperatures.

thermocline in a body of water, a zone of rapid temperature change with depth. It usually lies between a turbulent upper layer and a stratified lower layer of water. See also STRATIFICATION.

thermoduric describing a microorganism that can survive high temperatures. The term is used especially in relation to bacteria that can survive pasteurization.

thermonastic movements see THERMONASTY.

thermonasty a NASTIC MOVEMENT in response to a change in temperature. For example, when subjected to a temperature increase of between 5 and 10°C, *Crocus* flowers open in a few minutes due to a relative increase in growth rate on the inner side of the petals.

thermoperiodic response the response of a plant to a diurnal fluctuation in temperature. For example, growth of tomato plants is best when a certain day/night temperature regime is experienced. Meristematic activity and extension growth are often affected by thermoperiodicity. Some species flower earlier and more profusely if they experience low night and high daytime temperatures.

thermophilic describing microorganisms (*thermophiles*) that require high temperatures (45–65°C) for growth. Such organisms are found in rotting vegetable matter and hot springs. Compare MESOPHILIC, PSYCHROPHILIC.

thermosphere see ATMOSPHERE.

therophyte an ANNUAL or EPHEMERAL plant that survives unfavourable conditions in the form of a seed. Therophytes are common on cultivated land and in desert regions where perennial plants cannot establish themselves. See RAUNKIAER SYSTEM OF CLASSIFICATION.

thiamin (thiamine, vitamin B₁) a substituted PYRIMIDINE joined through a methylene bridge to a substituted thiazole. In its active form, thiamin pyrophosphate, it acts as a coenzyme in reactions involving the transfer of aldehyde groups in carbohydrate metabolism. Thiamin is synthesized by plants but not by some microorganisms or most vertebrates, which thus require it in the diet.

Thiessen polygons (Voronoi networks, Delaunay triangulations) a method of analysing plant competition by plotting the position of each plant, then using straight lines to join the points to their neighbours. Each line segment is then bisected by a line a right angles, which continues until it hits another line. This creates a series of closed polygons termed Thiessen polygons. The area contained in each polygon is closer to the point on which the polygon is based than to any other point in the dataset. The size of the polygon indicates the density of the plants – small sizes indicating higher densities. The method is also used to

calculate various properties of watersheds using measurements from scattered weather stations (the points). It can be applied to calculate the area of influence of any series individual events or features that can be represented as points.

thigmotaxis a TAXIS in which the movement is parallel to lines of stress in the substrate. It is seen, for example, in myxobacteria.

thigmotropism see HAPTOTROPISM.

thin-layer chromatography a widely used, fairly fast chromatographic technique for separating the components of mixtures. A glass plate is covered with a thin layer of cellulose, silica gel, or alumina. This represents the solid stationary phase. A line is scratched on the base of the plate and a small spot of the mixture is applied to the base line using either a wire loop or a capillary tube. The plate is then suspended in a suitable solvent (the mobile phase), which rises up the plate by capillary action, carrying and separating the mixture into its constituents. The remainder of the procedure is similar to that used in PAPER CHROMATOGRAPHY.

thiocarbamide see THIOUREA.

thiol see SULPHYDRYL.

thiourea (thiocarbamide) a chemical, formula NH_2CSNH_2, widely used to stimulate germination in seeds that have a light requirement. It is believed to act by increasing GIBBERELLIN levels.

thorn a sharply pointed woody structure formed from a modified reduced branch and connected to the vascular system of the plant. Examples are those of the hawthorn (*Crataegus monogyna*) and sloe or blackthorn (*Prunus spinosa*). Compare SPINE, PRICKLE.

thorn forest see FOREST.

thorn scrub see FOREST.

thorn woodland see FOREST.

thraustochytrids see LABYRINTHULATA.

threonine a polar amino acid with the formula $CH_3CH(OH)CH(NH_2)COOH$ (*see illustration at* amino acid). The biosynthetic pathway of threonine follows that of METHIONINE from aspartic acid to the formation of homogentisic acid, at which point the two pathways diverge. Breakdown occurs via glycine and acetyl CoA. ISOLEUCINE is synthesized from threonine.

thrum see HETEROSTYLY.

thylakoid any one of the layers making up the grana in CHLOROPLASTS. It consists of a channel surrounded by a pigmented membrane.

thymidine see THYMINE.

thymine a PYRIMIDINE base, characteristically present in DNA, and absent from RNA. In DNA it pairs specifically with adenine in the complementary strand, so that the thymine:adenine base ratio is 1:1. Thymine is linked to other bases of the same strand through a sugar-phosphodiester backbone, the nucleotide of thymine being called *thymidine* (thymine + deoxyribose sugar + phosphate). Thymidine is an essential growth factor for many microorganisms. Thymine is more fully described as 5-methyl-2,4-dioxypyrimidine and is derived from sugars and amino acids. Compare URACIL.

thyrse see INFLORESCENCE.

tidal flat an expanse of intertidal mud, sand, and marsh formed in a lagoon, bay, or estuary in locations with a large tidal range, e.g. the Wash in eastern England. In tropical areas such habitats are usually colonized by mangroves, while in parts of Europe, North Africa, and North America they are invaded by species of the grass *Spartina*.

tiller a shoot that develops from axillary or adventitious buds at the base of a stem, often in response to injury of the main stem. Tillering is seen when the trunk of a tree is lopped, as in the practice of COPPICING. It is also characteristic of the growth of grasses, giving the tufted appearance of many species. Tillering enables grasses to withstand grazing. A reasonable amount of grazing can actually enhance the growth of grasses by increasing the numbers of shoots produced.

tilth the physical state of a soil, especially the degree of aggregation of the soil particles, in relation to its suitability for cultivation and crop growth.

timber line (Waldgrenze) the more or less clearly defined region at high altitudes or latitudes beyond which normal dense tree growth does not occur. The trees growing between the timber line and the TREE LINE tend to be dwarfed and deformed and are termed ELFIN FOREST.

tinsel flagellum see UNDULIPODIUM.

tissue an assemblage of similar cells in a multicellular organism that work together to perform one or more specific functions. Examples are palisade and spongy mesophyll tissues. Compare ORGAN.

tissue culture the growth of isolated plant or animal cells or small pieces of tissue under controlled conditions in a STERILE growth medium. The medium is designed to meet the requirements of the tissue involved. Experiments varying media composition have yielded information on the nutritional needs of particular cells, this not being readily ascertained in an entire organism. Whole plants can often be regenerated from small segments of tissue or even single cells. This feature has been used commercially in the propagation of various plants, notably orchids. The regeneration of plants from meristem explants has been used to free crops of virus infection since the virus particles do not penetrate meristematic tissues. Tissue culture also has various applications in plant breeding work. For example, crosses can be made between normally incompatible plants (see INCOMPATIBILITY) if the embryo is 'rescued' and grown in culture before abortion can take place. Tissue culture provides a means of propagating new varieties of crop plants as it can be used to produce CLONES, whereas using seeds involves the mixing of alleles that accompanies sexual reproduction. There is also no delay in waiting for suitable seasons and seed set. See also ANTHER CULTURE, PROTOPLAST FUSION, SUSPENSION CULTURE, CALLUS.

toadstool see MUSHROOM.

tocopherol (vitamin E) a terpenoid-like substance found in high concentrations in certain seeds, notably cereals. It is believed to prevent the oxidation of lipids and so prolong seed viability. In a few species applications of tocopherol have been shown to replace the VERNALIZATION requirement. Deficiency in certain animals appears to cause infertility. See VITAMIN.

tolerance the ability of a plant to survive an attack by a pathogen without serious loss of yield.

toluidine blue a general purpose alkaline

ANILINE STAIN commonly used to give an even coloration to the specimen and distinguish it from the embedding material. It is frequently used to stain the 1 µm sections cut on ultramicrotomes as an adjunct to transmission electron microscope studies.

tomentose describing a surface densely covered in short hairs, such as that of the leaves of hoary mullein (*Verbascum pulverulentum*).

tomentum a thick covering of short hairs.

tonoplast the membrane separating the cell VACUOLE from the protoplasm. It has the same structure as the PLASMA MEMBRANE and mediates between the protoplasm and the vacuolar sap.

topogenous mire (topogenous peat) a BOG that is restricted to a location where precipitation is concentrated, such as a valley bottom. Such bogs form in locations that experience relatively low rainfall and humidity and summer drought.

topogenous peat SEE TOPOGENOUS MIRE.

topophysis the situation in which a bud or a cutting from a particular branch or shoot will develop only the morphology of that branch or shoot: its fate is fixed. For example, in trees that have different types of branches with different leaf shapes and orientation, a cutting taken from one type will never develop into another type. Similarly, a cutting taken from a juvenile plant will retain juvenile characteristics. By comparison, some buds and shoots can develop in different ways according to their environmental history.

toposequence a sequence of SOILS whose distinctive characteristic are related to topography and geographical location.

topsoil SEE SOIL PROFILE.

torus 1 the thickened central part of the pit membrane in the BORDERED PITS of many gymnosperms, consisting of primary cell wall material. The torus is believed to act as a valve, blocking the pit when there is an inequality of pressure on either side of the pit, as is the case when an adjacent tracheary element is damaged.
2 SEE RECEPTACLE.

totipotency the capacity, exhibited by certain types of isolated differentiated plant cell, to regenerate whole plants. The phenomenon is seen as evidence for the theory that all nucleated plant cells possess all the genes necessary to direct the formation of a complete plant. To realize this potential the cell must be removed from the inhibiting influence of the rest of the plant body and given the appropriate stimuli, namely the correct balance of nutrients and hormones. See also COMPETENCE, REGENERATION.

toxin a poisonous substance, usually of microbial origin, that stimulates the production of antitoxins in an animal body. Many bacterial diseases are due to the release of bacterial toxins. These may be *endotoxins*, which are formed within the bacterium and released on the death and disintegration of the bacterial cell, or *exotoxins*, which are secreted through the bacterial cell wall. The exotoxins, many of which are highly active enzymes, generally have more severe effects. Several fungi, e.g. death cap (*Amanita phalloides*) and ergot (*Claviceps purpurea*), also produce toxins. *Aflatoxins*, produced by the mould fungus *Aspergillus flavus*, can cause severe liver damage. Higher plant toxins are less common, an example being ricin, an albumin in castor oil (*Ricinus communis*) seeds, that causes agglutination of red blood cells.

T4 phage a BACTERIOPHAGE commonly used as source of DNA POLYMERASE, LIGASE, and POLYNUCLEOTIDE KINASE.

TPN SEE NADP.

trabant SEE SATELLITE.

trabecula (*adj.* trabeculate) a barlike structure extending across a lumen or lacuna. For example, in some plants extensions of the secondary cell wall thickening form trabeculae across the lumina of xylem tracheids. In the Selaginellales (LYCOPHYTA) the STELE is suspended between the CORTEX and the PERICYCLE by trabeculae.

trace element SEE MICRONUTRIENT.

tracer SEE ISOTOPIC TRACER.

trachea a xylem vessel.

tracheary element a water-conducting cell in the xylem, i.e. a VESSEL element or TRACHEID. The secondary cell walls of the tracheary elements show various types of thickening (see illustration). Annular and spiral patterns of thickening allow further extension of the tracheary element and tend to be found in PROTOXYLEM. The other forms of thickening are seen in METAXYLEM. Although bryophytes do not posses xylem, some do possess water-conducting cells known as HYDROIDS. Some consider these are also tracheary elements.

Tracheata see VASCULAR PLANT.

tracheid a tracheary cell in the xylem of many plants that has a lignified secondary cell wall and usually lacks a living protoplast at maturity. Tracheids may be distinguished from VESSELS as they lack perforation plates and are generally smaller in diameter and greater in length. The only perforation in tracheids are the PITS in the secondary cell wall, which are most concentrated on the end walls. In the Cycadophyta, Ginkgophyta, and Coniferophyta and in some primitive angiosperms that lack vessels, the tracheids are the only cells specialized to transport water up the plant. Compare FIBRE, FIBRE-TRACHEID.

Tracheophyta in older classifications that consider the possession of vascular tissue of greater significance than the seed-bearing habit, a division of the Plantae containing all the VASCULAR PLANTS. It is divided into the subdivisions Psilopsida (see PSILOPHYTA), Lycopsida (see LYCOPHYTA), Sphenopsida (see SPHENOPHYTA), and PTEROPSIDA (including the ferns, gymnosperms, and angiosperms). The psilophytes are considered the most primitive as they lack true roots and a well defined phloem. These features are properly developed in the other subdivisions. Also, the psilophyte gametophyte resembles the sporophyte in being dichotomously branched and possessing vascular tissue, while in the other subdivisions the gametophyte is comparatively reduced.

trait any detectable phenotypic characteristic.

trama the central tissue in the GILL of a fungus, consisting of loosely packed hyphae.

tramp species a species that has been unintentionally spread to new locations by human activity. For example, the brown seaweed japweed or strangleweed (*Sargassum muticum*) which becomes entangled in outboard motors and blocks underwater pipes, was probably introduced from Japan in consignments of oysters.

transaminase see TRANSAMINATION.

transamination the transfer of the amino group of an amino acid to a keto acid, usually α-ketoglutaric acid, to form glutamate and the keto acid derived from

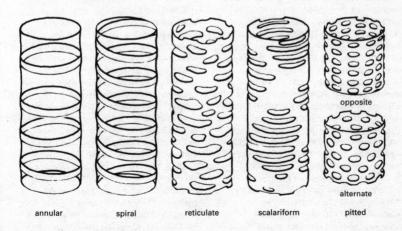

annular   spiral   reticulate   scalariform   pitted

opposite

alternate

Tracheary element: types of secondary wall thickening

the original amino acid. Transamination reactions are catalysed by enzymes called *transaminases*, such as aspartate transaminase, which catalyses the reversible formation of oxaloacetic acid and glutaric acid from aspartic acid and α-ketoglutaric acid. Transamination is central to amino-acid metabolism as it is one of the steps in the biosynthesis and the degradation of nearly all amino acids.

trans arrangement (repulsion) the situation in which an individual is HETEROZYGOUS for two linked GENES (either two units of function or two mutations in the same functional unit), and the recessive (or mutant) allele of one gene is on the same chromosome as the dominant (or normal) allele of the other gene. The homologous chromosome will thus possess the dominant allele of the first gene and the recessive allele of the second gene. If the recessive alleles are in the same functional unit then their effects will show in the phenotype because there is no unchanged functional unit to mask them. If they are in different functional units, then the phenotype will be normal. This is the basis of the CIS-TRANS TEST. Compare CIS ARRANGEMENT.

trans-butanedioic acid see FUMARIC ACID.

transcription the part of protein synthesis that involves the formation of a complementary copy of the genetic code by MESSENGER RNA synthesis. Transcription occurs on the genetic material itself: principally nuclear DNA in eukaryotic organisms. As in DNA REPLICATION, the DNA unwinds with the aid of an enzyme, and complementary bases are added to one strand – the coding stand – to form messenger RNA. However, instead of the nucleotide base thymine pairing with adenine, the base uracil pairs up instead. The nucleotides are joined together by the enzyme RNA polymerase. Once the RNA polymerase reaches the end of a gene, it releases the RNA from the template.

The newly synthesized RNA is then spliced to remove pieces of non-coding RNA (introns). This is done by units in the nucleus called SPLICEOSOMES. After splicing, the mature mRNA passes out of the nucleus through the nuclear pores into the cytoplasm, where TRANSLATION will occur. See also RNA SPLICING.

transcriptome all the DNA transcripts – mainly messenger RNA molecules – being produced by an organism. It represents the DNA coding for the expressed genome. See also EXON, INTRON, JUNK DNA, SPLICEOSOME, TRANSCRIPTION.

transduction the transfer of genes from one bacterium to another by the action of a TEMPERATE PHAGE acting as a vector (transmitting agent). It occurs when a phage excises from a bacterial genome in a faulty manner, taking some of the bacterial genes with it and leaving some of its own behind. The transferred genes (the transducing material) are incorporated into the bacterial genome of a new host cell and result in the transduced cell showing a permanent genetic change. *Abortive transduction* is sometimes seen, in which the transferred genes do not become incorporated into the genome but remain as a plasmid in the cell. The recipient bacterium then only shows the new properties until the plasmid is lost from the cell. Transduction occurs naturally at low frequency but this can be increased substantially under laboratory conditions. It is an important tool in genetic engineering.

transect a line across an area along which a study can be made of the distribution and abundance of plant species. The line is marked out with a tape and its position indicated on a map of the area. A *profile transect* is made along a slope and the species present at various levels are recorded. The plants actually touching the tape may be recorded or QUADRATS may be placed along the transect at regular intervals. A *belt transect* is a strip of ground, 1 m or 0.5 m wide, between two parallel transect lines. The species are recorded a section at a time. Scale diagrams, such as *profile diagrams* or *belt transect histograms*, can be constructed from the recorded results and the distribution of the species related to any habitat factors that have also been recorded.

transferase any ENZYME that catalyses reactions in which a functional group is transferred from a donor to an acceptor molecule. Six classes of transferase are distinguished, including enzymes that transfer phosphate groups (phosphotransferases or KINASES) and enzymes that transfer acyl groups (e.g. phosphate acyltransferase, which catalyses the transfer of an acyl group from phosphate to coenzyme A).

transfer cell a parenchyma cell usually found in close association with transport tissues such as xylem and phloem. Its cell wall has many protruberances that increase the surface area in contact with adjacent cells, and is interrupted by many PLASMODESMATA that allow movement of solutes.

transfer RNA (tRNA) a type of RNA molecule that binds to amino acids in the cytoplasm and assists in their incorporation into polypeptide chains at ribosomes. The tRNA molecule is shaped like a clover leaf. Each amino acid binds only to a specific type of tRNA molecule. At one end of a tRNA molecule is a characteristic sequence of three bases, the ANTICODON, which will temporarily pair with a complementary triplet, the CODON, in messenger RNA. At the same time the amino acid will detach from its tRNA and form a peptide bond with the amino acid that preceded it. By this means the polypeptide grows by the addition of amino acids one at a time. The amino acid that is incorporated will be determined by the specificity of an amino acid for a particular type of tRNA molecule and the specificity of pairing between the anticodon and codon. See TRANSLATION.

transformation see BACTERIAL TRANSFORMATION.

transfusion tissue the tissue immediately surrounding at least part of the vascular bundles in the leaves of the Coniferophyta, e.g. *Pinus*. Transfusion tissue is composed of TRACHEIDS (transfusion tracheids) with conspicuous BORDERED PITS, and parenchyma cells (transfusion parenchyma) containing tannin-like substances and sometimes starch. The main function of the transfusion tissue is believed to be the transport of materials between the vascular bundles and the mesophyll.

transgenic describing organisms that contain genetic material from one or more other species. For example, crop plants have been engineered to include herbicide resistance genes, and other plants to produce pharmaceuticals or vitamins. See also PHARMING, PLANTIBODIES.

transient polymorphism see POLYMORPHISM.

transition a mutation caused by substitution in the DNA of one purine base for the other, or one pyrimidine for the other. Hence adenine might replace guanine and vice versa, or thymine might replace cytosine and vice versa. It is thought that mutagens such as 5-BROMOURACIL induce transitions. Compare TRANSVERSION.

transition region 1 the part of the plant body between the stem and the root in which there are intermediate arrangements of the tissues. *Vascular transition* takes place in this region, ensuring continuity of vascular tissues throughout the plant. See HYPOCOTYL.
2 an INTERCALARY MERISTEM located between the bladelike portion and the narrow stemlike portion of the plant body in the LAMINARIALES.

translation the part of PROTEIN SYNTHESIS that involves the decoding of the base sequence of MESSENGER RNA, and formation of a corresponding polypeptide. In EUKARYOTES translation occurs in RIBOSOMES in the cytoplasm. Messenger RNA is synthesized in the nucleus, and moves to the cytoplasm through the pores in the nuclear membrane. The proteins are synthesized from the amino terminal. The process is highly endergonic requiring four ATP equivalents per amino acid residue. Five stages may be recognized: activation, *initiation*, elongation, and termination/release and modification.

The binding of one amino acid to another (i.e. the formation of a peptide bond) requires energy, so before they participate in translation, amino acids must be activated – provided with the energy they will need for the reaction. During activation the energy required to activate an amino acid is transferred to it from ATP with the aid of a specific aminoacyl-tRNA synthetase enzyme. The same enzyme then joins the activated amino acid to its appropriate TRANSFER RNA (tRNA) with a high-energy bond (the aminoacyl bond). Each tRNA is specific for one amino acid.

During the first stage, initiation, the two subunits of the ribosome assemble and become attached to the mRNA molecule near a specific codon (the *initiation codon*), which denotes the start of the message. This initiation codon on the mRNA attracts a particular aminoacyl tRNA (the *initiator*), which always carries the amino acid N-cormyl methionine. The initiator binds to the mRNA at the small subunit of the ribosome. Various proteins called *initiation factors* are also involved. The whole structure forms the *initiation complex*. The start codon AUG can now interact with the initiator. A large ribosomal subunit attaches, and the ribosome is read to begin protein synthesis.

The small subunit has two tRNA binding sites, the P site, to which the initiator is bound and the A site. As elongation commences, a tRNA bearing the amino acid specified by the next codon on the mRNA binds to the A site. A peptide bond forms between the amino acids on the P and A sites, and the initiator detaches from the ribosome, leaving its amino acid attached to the amino acid on the second tRNa. This amino acid now moves to the newly vacated P site, pulling the mRNA with it. The third specified amino acid now arrives at the A site on its tRNA, and the elongation process continues.

When the full polypeptide chain is formed, a termination codon on the mRNA causes the release of the completed protein from the final tRNA molecule. The last tRNA and the mRNA then dissociate from the ribosome. Usually several ribosomes are associated with the mRNA molecule at the same time, forming a POLYSOME. Compare TRANSCRIPTION.

translocated see SYSTEMIC.

translocation 1 the conduction of soluble materials from one part of the plant to another. The process includes the movements of food substances in the PHLOEM tubes, the transfer of hormones from their point of production, and the upward flow of dissolved salts in the TRANSPIRATION STREAM. Translocation of food materials is often preceded by enzyme action, which converts the substance concerned from an insoluble to a soluble

form, as from starch to sugar. See also MASS FLOW HYPOTHESIS, TRANSPIRATION. **2** a chromosome mutation in which a chromosome segment has become detached and reattached to a different (nonhomologous) chromosome. The most commonly occurring translocations are known as *reciprocal translocations* or *interchanges*, which involve the mutual exchange of segments from nonhomologous chromosomes. See also DELETION, DUPLICATION, INVERSION.

transmission electron microscope see ELECTRON MICROSCOPE.

transmitting tissue (conducting tissue) the specialized thin-walled tissue that constitutes the central part of the STYLE in some angiosperms, through which the pollen tube grows down to the funiculus. It may be involved in self-incompatibility systems.

transpiration the loss of water by evaporation from a plant surface. In a leaf it has been shown that although the combined area of stomatal pores (see STOMA) is on average only 1–2% of the total leaf area, the amount of transpiration they allow is 90% of the transpiration that occurs from a water surface the same area as the leaf. 5% is lost directly from the epidermal cells.

Transpiration rates are greatest when the leaf cells are fully turgid and when the external RELATIVE HUMIDITY is low. Water forms a film around the mesophyll cells and evaporates into the substomatal cavity from where it diffuses into the air. The degree of opening of the stomata (stomatal resistance) is of prime importance in governing the rate of water loss. The width of the BOUNDARY LAYER at the leaf surface is also important. In dry conditions, transpiration can cause WILTING and so the plant may develop features such as waxy cuticles to minimize the problem. See also ANTITRANSPIRANT, COHESION THEORY.

transpiration stream the flow of water from the roots to the leaves via the xylem vessels that is caused by TRANSPIRATION at the leaf surface. See also COHESION THEORY, ROOT PRESSURE.

transposable element see TRANSPOSON.
transposon a genetic unit that can move

from one part of a chromosome to another. Transposons can cause chromosome mutations of adjacent genes. They are common in bacteria, in which they usually contain only one or two genes. Some PLASMIDS act as transposons, carrying genes for such traits as antibiotic resistance between bacteria. Similar units in eukaryotes are called *transposable elements*. They are common in certain genera, such as *Zea* (maize).

transversion a mutation caused by the substitution of a PURINE base for a PYRIMIDINE base (or vice versa) in the DNA. Compare TRANSITION.

trap crop any crop planted in or around another crop to attract pests away from the more valuable crop. Trap crops are usually heavily treated with pesticides, burnt, or ploughed in.

traumatic acid a straight chain dicarboxylic acid (see CARBOXYLIC ACID), formula $COOHCH:CH(CH_2)_8COOH$. It has been isolated from green bean pods and is believed to act as a WOUND HORMONE.

tree a large woody perennial with a single main stem (trunk) that remains more or less unbranched near the ground and whose aerial parts persist above ground all year round. Certain trees may have more than one main trunk, e.g. oak (*Quercus*), but no branches leave these trunks near the ground. Compare HERB, SHRUB.

tree ferns see FILICALES.

tree line (Baumgrenze) the more or less clearly defined region at high latitudes (the taiga/tundra boundary) or high altitudes (the subalpine/alpine boundary) beyond which trees do not grow. Occasionally inverted tree lines are seen in which trees do not grow below a certain line as, for example, in frost hollows. The limits to tree growth are presumably set by some climatic factor though it is by no means certain which particular factor is operating. Numerous theories have been advanced based on different climatic variables such as heat, light, or carbon dioxide deficiency, excessive wind or snow depth, etc. A recent theory suggests that, beyond the tree line, leaves become desiccated in winter because the summer growing season is not long enough to allow the leaves to mature fully

and resist water stress. Compare TIMBER LINE.

tree ring see ANNUAL RING.

tree-ring analysis see DENDROCHRONOLOGY, DENDROCLIMATOLOGY.

tree-ring index a series of annual ring widths that has been standardized to represent particular dates/ages. Such indices can be used to infer palaeoclimate or to date and correlate samples from different sites in the same locality.

tree veld see VELD.

Tremellales an order of the PHRAGMOBASIDIOMYCETES containing fungi that produce gelatinous basidiomata, hence the common name gelatinous or jelly fungi. These number about 256 species in some 60 genera, usually found growing as saprobes on dead wood. They have characteristic rounded basidia divided by septa into four cells. Typical examples are witches' butter (*Exidia glandulosa*) and the yellow brain fungus (*Tremella mesenterica*).

tri- prefix denoting three, made up of three parts, or three times.

triacylglycerol (triglyceride) an ester of GLYCEROL and three long-chain carboxylic acids or fatty acids. Triacylglycerols may be triesters of (see ESTER) either the same fatty acid, or more commonly of two or three different fatty acids (mixed triacylglycerols). The most common fatty acids found in triacylglycerols are PALMITIC, STEARIC, OLEIC, and LINOLENIC ACIDS.

Triacylglycerols are synthesized from glycerol phosphate and fatty acyl CoAs. The glycerokinase that phosphorylates glycerol is a soluble enzyme, but the other enzymes of triacylglycerol synthesis are associated with the MICROSOMES. Breakdown of triacylglycerols to glycerol and free fatty acids is achieved by the action of lipases.

Triacylglycerols are important energy storage molecules especially in seeds, many of which (e.g. rape, linseed, castor bean, coconut) are important commercially as sources of fats and oils. Fats and oils are solid and liquid triacylglycerols respectively. See also B-OXIDATION.

triallelic describing a POLYPLOID that has three different alleles at a particular gene locus.

triarch primary XYLEM that comprises three strands.

Triassic (Trias) the first period of the MESOZOIC era between about 225 and 195 million years ago. In the early Triassic, arid conditions prevailed though later the climate became mostly temperate, but variable. The early conditions have resulted in fossils of the period being generally rare. The fossil lycophyte *Pleuromonia*, found in Triassic sandstone, is thought to be an intermediate form between the Lepidodendrales and the Isoetales. Members of the sphenophyte order Sphenophyllales became extinct at the beginning of the period. The now extinct BENNETTITALES arose in this period and common fossils are *Bennettites* and *Williamsoniella*. Members of the extinct order CAYTONIALES are found and the GINKGOPHYTA are well represented in the Upper Triassic (e.g. *Ginkgoites*). See GEOLOGICAL TIME SCALE.

tribe the rank subordinate to family but superior to genus in the taxonomic hierarchy. The term is applied to assemblages of similar genera within large families. The Latin names of tribes have the ending -eae. An example in the family Apiaceae is the tribe Saniculeae, which embraces the genera *Eryngium*, *Astrantia*, *Alepidea*, and *Sanicula*. Similar tribes may be grouped together in subfamilies. Tribes may be split into subtribes, the Latin names of which have the ending -inae.

Tribophyceae a class of the yellow-green algae (XANTHOPHYTA) that are pale green and whose cells typically have two chloroplasts, which contain chlorophyll *a* but not fucoxanthin. PYRENOIDS are present, and there are cytoplasmic vesicles containing chrysolaminarin. The cell wall is usually of cellulose or pectic substances. The cells are usually nonmotile, and the most common types of thallus are coccoid cells (e.g. *Botrydiopsis*), filaments (e.g. *Heterococcus*), and pseudoparenchymatous and coenocytic forms (e.g. *Vaucheria*). They are found mainly in fresh water. Asexual reproduction is usually by production of aplanospores; sexual reproduction is rare.

tricarboxylic acid see CARBOXYLIC ACID.

tricarboxylic acid cycle see KREBS CYCLE.

2,4,5-trichlorophenoxyacetic acid see 2,4,5-T.

trichoblast in the red algae (RHODOPHYTA), a very thin tapering colourless branch or hair.

trichocyst in dinoflagellates (DINOMASTIGOTA), a vesicle containing a crystalline rod that can be discharged. Trichocysts are thought to be used for defence.

trichogyne a receptive, often hairlike, uni- or multicellular structure that projects from the female sex organ in some algae, ascomycetes, and lichens. It serves to attract and receive the male gamete or nucleus prior to fertilization. See also CARPOGONIUM.

trichome 1 any outgrowth, such as a root hair, from a plant epidermal cell. Trichomes are very varied in form and function, their morphology often yielding important taxonomic characters. They may be elongate, scalelike or peltate, glandular or nonglandular, and unicellular or multicellular.

2 in bacteria, a string of connected cells, such as the cells that make up the filament of a blue-green bacterium.

Trichomycetes a class of the ZYGOMYCOTA containing parasitic fungi that live in the guts of arthropods. Trichomycetes characteristically have a simple thallus attached to the host by a basal holdfast. Some species have chitin-impregnated walls. Asexual reproduction involves the production of spores in conidia or sporangia. Orders include the Harpellales and Asellariales.

trichothallic growth a form of growth seen in certain brown algae (Phaeophyta), e.g. *Ectocarpus* and *Cutleria*, in which cell division is restricted to well defined intercalary regions (regions located within the thallus rather than at its tip).

tricolpate describing a pollen grain having three COLPI, as is commonly found amongst most dicotyledon species.

trifoliate describing a compound leaf having three leaflets arising from the same point, such as the clovers (*Trifolium*) and wood sorrels (*Oxalis*). *See illustration at leaf.*

trifurcate forming three forks or branches.

triglyceride the former name for a TRIACYLGLYCEROL.

trigonous having a triangular cross-section.

trihybrid an organism that is heterozygous for three genes, e.g. AaBbCc. Trihybrid crosses, especially of the type AaBbCc × aabbcc, are commonly used devices in chromosome mapping. By comparing the frequencies of the phenotypes, the chromosomal arrangements of the three genes A/a, B/b, and C/c can be deduced. This is technically easier than setting up three DIHYBRID crosses AaBb × aabb, AaCc × aacc, and BbCc × bbcc, which would only achieve the same result.

trilete see TRIRADIATE SCAR.

trilocular having three LOCULES.

trimer a PROTEIN made up of three polypeptide chains or sub-units. Compare DIMER.

trimerous (3-merous) Describing flowers in which the parts of each whorl are inserted in threes, or multiples of three. This arrangement is characteristic of many monocotyledons.

trioecious describing a species that has male, female, and hermaphrodite flowers in separate individuals.

triose phosphate a three-carbon phosphorylated sugar. Two commonly occurring triose phosphates are GLYCERALDEHYDE 3-phosphate and dihydroxyacetone phosphate (DHAP), which are both intermediates in the synthesis and breakdown of glucose. Interconversion of these two triose phosphates is catalysed by the enzyme triose phosphate isomerase. Triose phosphates are important in other metabolic processes. Glycerol is formed from DHAP, and glyceraldehyde 3-phosphate is an intermediate in the PENTOSE PHOSPHATE PATHWAY.

triphosphopyridine nucleotide see NADP.

tripinnate describing a BIPINNATE leaf in which the secondary leaflets are further subdivided, as occurs in certain ferns. *See illustration at* leaf.

triplet a sequence of three nucleotides on a DNA or messenger RNA molecule. Most of the 64 possible triplets (see GENETIC CODE) code for amino acids (see CODON). However a few act as 'start' signals while others act as 'stop' signals to begin and end a polypeptide chain (see NONSENSE CODON).

triplet code hypothesis see GENETIC CODE.

triploid an organism, tissue, or cell that possesses three complete sets of chromosomes per nucleus. A triploid may arise from the fusion of a haploid and a diploid nucleus, or from the fusion of three haploid nuclei. Triploid organisms are normally infertile, since meiosis results in gametes that have more or less than the normal haploid number of chromosomes. They can spread by vegetative means (see VEGETATIVE REPRODUCTION), and some crops, such as cultivated bananas, are normally triploid, being spread from offsets. The endosperm tissue of angiosperms is typically triploid (or pentaploid in some monocotyledons).

triquetrous having a triangular cross-section with acute angles at each corner.

triradiate branching in three directions, e.g. a Y-shape.

triradiate scar a scar on the surface of a spore that marks the point at which it was joined to the other three spores making up the TETRAD. Triradiate scars are seen when the tetrad shows tetrahedral symmetry. It is conspicuous because the spore wall is fairly thin at this point while the rest of the wall is thickened with cutin. In the MEGASPORES of *Selaginella kraussiana* the archegonia break through the megaspore wall at this point. Spores bearing a triradiate scar are termed *trilete*. See MONOLETE.

trisomic an organism with an extra chromosome in addition to the normal complement, i.e. 2n+1. Trisomy typically arises when a normal gamete (n) fuses with one containing an extra chromosome (n+1). The extra chromosome may be a simple copy of one of the chromosomes (primary trisomic), an ISOCHROMOSOME (secondary trisomic), or may include parts of two nonhomologous chromosomes (tertiary trisomic). Trisomics often have abnormal phenotypes and are frequently less fertile than diploids. See ANEUPLOIDY. See also HOMOLOGOUS CHROMOSOME.

tristichous describing a form of alternate leaf arrangement in which successive leaves arise 120° around the circumference of the stem giving three vertical rows of leaves. Compare DISTICHOUS, MONOSTICHOUS.

trivalent the temporary association of three homologous chromosomes in a triploid or trisomic organism. Such associations are observed between mid prophase and late metaphase of the first meiotic division.

tRNA see TRANSFER RNA.

tropane alkaloids a group of ALKALOIDS derived from the amino acid ornithine. They include hyoscyamine and its stereoisomer atropine, both solanaceous (see SOLANACEAE) alkaloids. These are prepared from *Datura stramonium* (thorn apple) and *Atropa belladonna* (deadly nightshade) respectively. The cocaine alkaloids, derived from the leaves of *Erythroxylum coca* and *E. novagranatense* (coca), are also examples.

trophic level see FOOD CHAIN.

tropical forest see FOREST.

tropical rainforest see FOREST.

tropic movement see TROPISM.

tropism (tropic movement) a directional response of a plant or plant part to an external stimulus from a specific direction. Thus there are *positive* and *negative tropisms* depending on whether growth or 'movement' is towards or away from the source of stimulation. The movement or bending is accomplished by unequal rates of growth on the two sides of the organ, usually in response to increased AUXIN levels in the tissues. Compare NASTIC MOVEMENTS, TAXIS. See AEROTROPISM, CHEMOTROPISM, DIATROPISM, GRAVITROPISM, HAPTOTROPISM, HELIOTROPISM, HYDROTROPISM, ORTHOTROPISM, PHOTOTROPISM, PLAGIOTROPISM, RHEOTROPISM, SELENOTROPISM, SIOTROPISM.

tropopause see ATMOSPHERE.

tropophyte a plant that is adapted to survive in a climate where there are alternating wet and dry seasons, by having a resting phase during the dry season. For example, many trees in monsoon forests shed their leaves in the dry season.

troposphere see ATMOSPHERE.

true alkaloids see ALKALOIDS.

true fungi see EUMYCOTA.

true mosses see BRYOPHYTA.

truffle the subterranean ASCOMA of a fungus belonging to the PEZIZALES or Elaphomycetales, or a BASIDIOMA of fungi of the order Hymenogastrales. There are about 180 species that form truffles.

truncate describing a leaf that is squared off at the apex.

tryptophan a nonpolar aromatic amino acid with the formula

$C_8H_6NCH_2CH(NH_2)COOH$ (*see illustration at* amino acid). Tryptophan is synthesized via the SHIKIMIC ACID PATHWAY and broken down to alanine and acetyl CoA. As with the other aromatic amino acids, tryptophan is the precursor of many aromatic compounds. Examples are the auxin indoleacetic acid and certain of the indole alkaloids, such as strychnine and yohimbine.

tube cell (tube nucleus) one of the cells contained within the microspore of gymnosperms, the others being the generative cell and the prothallial cell or cells. In *Cycas* it enters the pollen tube but its function is uncertain. In angiosperms the cell at the tip of the pollen tube that precedes the generative nuclei is sometimes termed the tube cell but it is more often called the VEGETATIVE NUCLEUS.

tube nucleus see TUBE CELL, VEGETATIVE NUCLEUS.

tuber a swollen part of a stem or root, usually modified for storage, and lasting for one year only, those of the succeeding year not arising from the old ones, nor bearing a position relative to them. Examples of such perennating organs are the stem tubers of potato (*Solanum tuberosum*) and the ROOT TUBERS of *Dahlia*. Root tubers develop from ADVENTITIOUS roots (compare TAPROOT). A stem tuber may be distinguished from a root tuber by the presence of buds or 'eyes'. See also VEGETATIVE REPRODUCTION.

Tuberales an obsolete order of ascomycetes that contained fungi whose ascomata are formed underground and resemble tubers. It included truffle-forming fungi, e.g. *Tuber*. Members of the former Tuberales are now placed in the PEZIZALES and Elaphomycetales.

tuberculate having a surface covered with small warty projections as, for example, the rind of ridge cucumbers. Pollen covered in wartlike structures (verrucae), e.g. that of ivy (*Hedera helix*), is normally described as *verrucose*.

tubulin a protein that forms the MICROTUBULES of the CYTOSKELETON and the SPINDLE. It is made up of subunits that can be assembled and disassembled with the aid of specific enzymes.

tumour-inducing principle a PLASMID carried by the bacterium *Agrobacterium tumefaciens*, which causes CROWN GALL disease in plants. The plasmid is necessary for the transformation of normal host tissue into tumour tissue. It is believed that this is brought about by the incorporation of the plasmid into the plant genome. Tissue removed from crown gall tumours can be grown in culture without the addition of auxin and cytokinin.

tundra a major regional community (BIOME) in which the vegetation is poor, the few species being mainly lichens, mosses, heaths, sedges, grasses, and some herbaceous plants, but no trees. It is a region of cold DESERT, the temperature rarely exceeding 10°C. The topsoil is frozen for about nine months of the year and the subsoil is subjected to permafrost. The plants are thus subjected both to extreme cold and physiological drought as the soil water is frozen. The topography, type of soil, degree of shelter, etc., give rise to differences in vegetation from one locality to another. Arctic tundra is found north of the tree line of North America and Eurasia in a band of varying width circling the Arctic Ocean. In the most northerly zone of tundra, the high Arctic tundra, there is typically an incomplete vegetation cover except in a few very sheltered habitats. The vegetation is usually marshy, with lichens on more exposed sites. The southern limits of the Arctic tundra, the low Arctic tundra, is characterized by a continuous vegetation cover. Here the plant communities form a vegetation mosaic determined by the local microhabitats. In the Antarctic there are only small scattered areas of vegetation, consisting of mosses and lichens on some of the islands. See also ALPINE.

tunica see TUNICA–CORPUS THEORY.

tunica–corpus theory a concept of the organization and development of the APICAL MERISTEM, in which the meristematic region is differentiated into an outer peripheral layer or layers, termed the *tunica*, and an inner mass of cells, termed the *corpus*. The tunica is

characterized by chiefly ANTICLINAL divisions and the corpus mainly by PERICLINAL divisions. The corpus gives rise to the interior part of the plant body and the tunica differentiates the outer layers including the epidermis. Compare HISTOGEN THEORY.

turbid (*n.* turbidity) describing a body of water that is opaque as a result of suspended sediment or pollutants.

turbinate 1 shaped in a close spiral, like a spinning-top, e.g. The fruits of some figs (*Ficus*). 2 describing swellings on the vegetative thallus (turbinate organs or turbinate cells) of certain chytrid species.

turf algae algae that appear as a fuzzy growth on underwater surfaces.

turgid see TURGOR.

turgidity see TURGOR.

turgor (turgidity) the state existing in a plant cell when, due to the intake of water by OSMOSIS, the protoplast exerts an outward pressure on the cell wall. When the cell is fully *turgid*, although the pressure is sufficient to make the wall bulge, the wall is strong enough to prevent more expansion and the further ingress of water. Turgidity is the main factor in maintaining rigidity and support in unlignified (see LIGNIN) parts of the plant. See also HYDROSTATIC PRESSURE.

turgor potential see PRESSURE POTENTIAL.

turgor pressure in calculations of OSMOTIC PRESSURE, the HYDROSTATIC PRESSURE exerted by the contents of the cell against the cell wall.

turion a type of perennating bud formed by certain aquatic plants, e.g. frogbit (*Hydrocharis morsus-ranae*), that is shed from the plant and lies dormant on the pond or river bed until the spring. See DORMANCY, PERENNATING ORGAN.

turnover 1 the proportion of a population that is lost by deaths and emigration or gained by births or immigration over a given period of time. 2 the changing species composition of a community or habitat as certain species go extinct and other species (either immigrants or newly evolved local species) move in to occupy their niches. 3 the proportion of its BIOMASS that is taken into a community through PRIMARY PRODUCTIVITY each year, i.e. the ratio of productivity to biomass. **4** (turnover rate) in a BIOGEOCHEMICAL CYCLE, the rate at which a nutrient flows into or out of a nutrient pool divided by the quantity of that nutrient in the pool. This indicates the importance of the nutrient flux in relation to the pool size. **5** the number of molecules of substrate turned into product by an enzyme in a given time (*turnover number*). **6** the breakdown of the STRATIFICATION of a lake due to wind action, which in temperate regions usually happens in autumn. This allows mixing of nutrients from bottom sediments (some derived from the decomposition of the remains of organisms that have sunk down from the upper layers during the period of stratification), so allowing for renewed growth of plankton when conditions are favourable.

turnover number see TURNOVER, def. 5.

turnover rate see TURNOVER, def. 4.

turnover time a measure of the movement of an element round a biogeochemical cycle, indicating the time taken to fill or empty a nutrient reservoir of a particular nutrient. It is the reciprocal of the turnover rate: the quantity of nutrient present in a particular nutrient pool divided by its flux (the rate at which it is moving into/out of the pool).

two-dimensional analysis a technique used to separate mixtures of closely related molecules. One method involves two-dimensional PAPER CHROMATOGRAPHY, in which a chromatogram produced by a solvent running in one direction is then subjected to a second separation by placing the chromatogram at right angles in a second solvent. Another method involves separation by ELECTROPHORESIS followed by paper chromatography.

two-way table see CONTINGENCY TABLE.

tyloses bladder-like ingrowths that protrude into the tracheary elements of older wood eventually causing blockage. They originate from adjacent parenchyma cells via paired PITS in the cell walls. Tyloses often become filled with tannins, resins, gums, or various pigments, so giving the heartwood its characteristic darker colour. These substances also help to preserve and strengthen the wood, while some of the pigments are important commercially as dyes (e.g. haematoxylin). Tyloses are sometimes found in the vessels of herbaceous plants, in which their function is unclear. They may act to seal off damaged vessels. See also DUTCH ELM DISEASE.

type (nomenclatural type) the single element (e.g. illustration, specimen, etc.) on which the description associated with the original publication of a name was based. The type of a taxon can be a HOLOTYPE, LECTOTYPE, or NEOTYPE, as appropriate. The nomenclatural type of a species or infraspecific taxon of a vascular plant is usually a HERBARIUM specimen, but some species are typified by an illustration or description, in the absence of herbarium material. The names of some lower plants are based on type cultures, where the type is a living specimen. The type of a genus or infrageneric taxon (i.e. subgenus, section, series, etc.) is a designated species, while that of a family, subfamily, or tribe is a genus. The type of an order is a family, for example, the family Rosaceae is the type of the order Rosales.

Because of its importance, most type material is kept in different coloured folders, thereby enabling its rapid recovery in large collections.

type specimen see HOLOTYPE.

tyrosine an aromatic polar amino acid with the formula $HOC_6H_4CH_2CH(NH_2)COOH$ (*see illustration at* amino acid). The biosynthesis of tyrosine is similar to that of PHENYLALANINE, diverging only in the last few reaction steps. Breakdown to acetyl CoA and fumaric acid occurs via phenylalanine. Like phenylalanine, tyrosine is a precursor of the ISOQUINOLINE ALKALOIDS. It is also a precursor of various phenolic (see PHENOL) inhibitors, e.g. coumaric acid.

U u

ubiquinone see COENZYME Q.

UDP See NUCLEOSIDE DIPHOSPHATE SUGARS.

Ulotrichales an order of green algae (CHLOROPHYTA) of the class ULVOPHYCEAE containing simple unbranched filamentous algae whose cells are uninucleate and haploid (e.g. *Ulothrix*). They are found in both marine and freshwater environments. Each cell of the filament contains one bandlike chloroplast with pyrenoids around its periphery and cell division is in one direction only. Vegetative propagation is by FRAGMENTATION. Asexual reproduction is by various kinds of zoospores or, more rarely, aplanospores. The zoospores, each of which has four undulipodia, germinate directly into new filaments. Sexual reproduction varies from isogamy to the well-developed oogamy seen in *Cylindrocapsa*. Motile gametes have two undulipodia. The zygote may pass through a CODIOLUM STAGE before producing haploid zoospores that develop into new filaments.

ULR see UNIT LEAF RATE.

ultracentrifuge see CENTRIFUGE.

ultramicrotome see MICROTOME.

ultrastructure (fine structure) structural details of cells below the limit of resolution of the light microscope and only revealed by the electron microscope.

ultraviolet microscope a microscope that uses ultraviolet radiation to illuminate the specimen and form the image. This increases the resolution because ultraviolet radiation has much shorter wavelengths (about 300 nm) than visible light, and resolution increases as wavelength decreases. Ultraviolet microscopy is

difficult and complex and is rarely used since the development of the electron microscope.

Ulvales an order of green algae (CHLOROPHYTA) of the class ULVOPHYCEAE containing parenchymatous thalloid algae with uninucleate cells (e.g. *Ulva*). They are most common in marine and brackish waters. The sheetlike thallus is formed by division in two planes of a filament that initially resembles those of the ULOTRICHALES. Some species have more tubelike thalli. These algae are unspecialized: any cell can photosynthesize and can also reproduce. There are no distinct reproductive structures. Some species have an ISOMORPHIC alternation of generations. A few go through a CODIOLUM STAGE.

Ulvophyceae a class of green algae (Chlorophyta) containing large algae that have a closed mitotic spindle (the nuclear envelope does not break down) and no microtubule organization during cell division. The basal bodies have microtubular roots in a cross-shaped pattern. Haploid and diploid phases alternate in the life cycle (see ALTERNATION OF GENERATIONS), which may be ISOMORPHIC or HETEROMORPHIC. The thalli range from multicellular filaments in the ULOTRICHALES (e.g. *Ulothrix*) to simple parenchymatous sheets (e.g. *Ulva*) and tubes (e.g. *Enteromorpha*) in the ULVALES or coenocytic tubes in the CLADOPHORALES, CAULERPALES, and DASYCLADALES. They are found mainly in shallow marine habitats, but a few live in fresh water. See also CODIOLUM STAGE.

umbel an inflorescence in which the flowers are borne on undivided pedicels originating from a common node on the main axis. The outermost flowers are borne on the longest pedicels so that the whole inflorescence is flat topped and gives the appearance of an umbrella. *Compound umbels* may be borne on branched pedicels so that numerous smaller umbels constitute the whole inflorescence. These inflorescences are typical of the Apiaceae (Umbelliferae) and provide landing platforms for pollinating insects. *See illustration at* inflorescence.

Umbelliferae SEE APIACEAE.

umbilicate 1 describing a structure (e.g. the PILEUS of a mushroom or toadstool) that has a central depression or pit.
2 describing a lichen thallus that is approximately circular and is attached to the substrate by a central point on its lower surface, e.g. *Umbilicaria*.

umbilicus the organ of attachment of some foliose lichens.

umbo (*adj.* umbonate) a convex swelling in the middle of the PILEUS of a mushroom or toadstool.

uncoupling agent any chemical that uncouples the process of respiration from that of phosphorylation. For example, in GLYCOLYSIS the formation of glycerate 3-phosphate from glyceraldehyde 3-phosphate normally yields two molecules of ATP. However arsenate prevents the formation of ATP at this point. Arsenate also acts as an uncoupling agent in OXIDATIVE PHOSPHORYLATION. Many other uncouplers of oxidative phosphorylation are known, most of which contain an aromatic ring and an acidic group. Examples are 2,4-*dinitrophenol* (DNP) and dicumarol. DNP has been used to investigate many aspects of plant physiology. For example, in research on phloem transport it has been shown that DNP inhibits translocation of assimilates. This implies there is some energy-requiring metabolic component in phloem transport and that the process is not purely due to osmotic potential and turgor pressure. DNP has also been used to investigate the theory that the respiratory CLIMACTERIC that occurs during fruit ripening is due to the accumulation of natural uncoupling agents (uncoupling agents promote respiration as they cause a build-up of ADP).

underdominance SEE POLYMORPHISM.

understorey (understory) in a woodland or FOREST ecosystem, the layer of vegetation below the canopy but above the ground layer or shrub layer. It contains shade-tolerant trees and the saplings of canopy species and emergents, as well as many creepers and climbers.

undulate describing a leaf margin that is wavy *See illustration at* leaf.

undulipodium (*pl.* undulipodia, *adj.* undulipodiate) a eukaryotic cilium or 'flagellum', used mainly for locomotion or feeding. Undulipodia have the same structure, which differs from that of prokaryotic flagella. Therefore, the term FLAGELLUM is now usually restricted to the equivalent structure in bacteria. The term cilia (*sing.* cilium) is sometimes applied to undulipodia that are not more than 10 mm long. Cilia are usually numerous; their movements may create currents that carry extracellular material over the cell surface or into a gullet, or they may beat in a coordinated fashion to produce locomotion. Longer undulipodia occur singly or in pairs and their activity moves the cell. Both arise from a KINETOSOME in the cytoplasm. Internally they consist of 11 fibres running lengthwise and constituting the *axoneme*, surrounded by a membrane continuous with the plasma membrane. Nine of the fibres form a peripheral outer cylinder and the remaining two are located in the centre, surrounded by a central

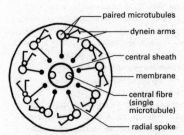

Undulipodium: cross section of an undulipodium

sheath (the so-called 9 + 2 arrangement, see diagram). Each central fibre is a single MICROTUBULE, but the outer ones are paired microtubules (A and B microtubules fused together) of the protein TUBULIN, one of each pair (the B microtubule) being slightly wider than the other. At regular intervals along the length of the narrower A microtubule, short paired projections (arms) arise, each positioned in a clockwise direction around the axoneme when viewed from the base. These consist of *dynein*, a protein with ATPASE activity. The A microtubule is linked to the central fibres by a radial spoke. The production of force in undulipodia depends on the sliding movements of one peripheral fibre against its neighbour, energy being derived from dynein activity. Localized sliding of this kind, occurring in sequence around and along the axoneme, produces linear forces that slightly distort the shape of the undulipodium and induce bending to bring about the typical wave from base to tip. It is thought that the first stage in bending of an undulipodium is the linking of the two dynein arms of the A microtubule with the B microtubule of an adjacent pair of microtubules. Hydrolysis of ATP provides the energy to break this link, and another forms further down the B microtubule. Adjacent pairs of microtubules thus slide past each other in a kind of ratchet process. Magnesium and calcium ions are also implicated in this mechanism.

Undulipodia may beat in a single plane, or with a corkscrew motion that causes the organism to spin round as it travels, following a helical path. Most undulipodia are at the posterior end of the organism and propel it through the water. Others may pull it from in front. In this case they are often *mastigonemate undulipodia* (tinsel, flimmer, or pantonematic flagella), being covered in tiny hairlike structures called MASTIGONEMES. Mastigonemate undulipodia are found in certain protoctists.

unequal crossing over a crossover that results in one chromatid with one copy of a chromosome segment and another with three copies. This is due to CROSSING OVER after improper pairing between HOMOLOGOUS CHROMOSOMES that are not exactly aligned.

uni- prefix denoting one, single.

uniaxial describing a plant that has a single central axis with no branches.

unicarpellous see GYNOECIUM.

unicell (*adj.* unicellular) an organism that consists of a single cell in its normal vegetative state.

unicentric distribution an extremely localized distribution, with only one or two isolated populations. Such a distribution is typical of RELICT species that have survived, for example from the last Ice Age, on isolated mountain tops or sheltered gorges. See also REFUGIA.

unifacial describing a leaf that is more-or-less cylindrical and lacks the two distinct sides of a DORSIVENTRAL leaf. Such a leaf is radially symmetrical.

unilateral a kind of panicle in which all the branches are turned to one side.

unilocular carpel a carpel that retains its own discrete cavity (loculus), which is not fused with those of other carpels.

unilocular sporangium a SPORANGIUM that is not divided and comprises only one locule or compartment. Unilocular sporangia are seen in the sori of the brown alga *Laminaria*. Compare PLURILOCULAR SPORANGIUM.

uninucleate describing a cell that contains only one nucleus or an organism whose cells contain only one nucleus.

uniramous unbranched.

uniseriate filament a filament whose cells are arranged in a single row.

unisexual describing a flower or plant in which the male organs (stamens) or female organs (ovules) are present, but not both. A unisexual plant is said to be DIOECIOUS, while unisexual flowers may occur on either dioecious or MONOECIOUS plants.

unit leaf rate (ULR) the rate of photosynthesis per unit leaf area.

unitunicate describing an ASCUS in which the layers of the ascus walls do not separate during the release of ascospores. Compare BITUNICATE.

univalent a single unpaired chromosome present at MEIOSIS when bivalents are also present, for example, the sex chromosome of an XO male.

universal indicator see INDICATOR.

universal veil a sheet of tissue that encloses the developing fruiting body of many basidiomycete fungi (see MUSHROOM). It ruptures as the stalk elongates, the remnants forming the volva at the base of the stalk and the flecks of tissue seen on the upper surface of the cap.

univoltine describing a species in which one generation reaches maturity each year.

Uppsala school of phytosociology a group of botanists led by G. E. Du Rietz and colleagues at Uppsala, Sweden, from 1921 on, who devised a series of methods for classifying vegetation based on its floristic composition. In contrast to the Zurich–Montpellier school, it emphasizes exclusive constant or preferential species rather than FIDELITY. In recent times the two schools have tended to converge. See PHYTOSOCIOLOGY.

upstream describing a sequence of DNA in a DNA molecule or protein which lies away from the direction of synthesis of the DNA or protein molecule.

upwelling a vertical movement of water from the deep ocean to the surface. Such water is often rich in nutrients and supports a rich growth of plankton, the basis of the ocean FOOD CHAIN.

uracil a PYRIMIDINE base, characteristically present in RNA and absent from DNA. The positions held by uracil in RNA are occupied by thymine in DNA; uracil is thus said to 'replace' thymine during RNA synthesis. Uracil, more properly called 2,4-dioxypyrimidine, is derived ultimately from sugars and amino acids.

uranyl acetate a widely used electron stain (see STAINING) suitable for high-resolution transmission electron microscopy. It stains nucleic acids and proteins thus enhancing the contrast obtained in observations of membranes and membrane-bound organelles.

urceolate flask-shaped.

Uredinales an order of the TELIOMYCETES containing the rust fungi (causing RUSTS), which number over 7000 species in some 160–170 genera. The TELIOSPORE is terminal, and the BASIDIOSPORES develop on sterigmata and are actively discharged from the promycelium. The mycelium is usually intercellular, often with haustoria, mainly in the aerial organs of its host. Several different hosts are often infected successively in the life cycle, with up to five types of spores being produced in each – spermatia (see PYCNIDIUM), AECIOSPORES, UREDINIOSPORES, teliospores, and basidiospores. Compare USTILAGINALES.

urediniospore (uredospore, urediospore) a dikaryotic spore formed in a spore cluster (a *uredosorus* or *uredium*) by certain rust fungi (UREDINALES). The urediniospores of *Puccinia graminis* are formed on the main host, wheat, and are responsible for the characteristic rusty streaks of infected plants. On release they infect more wheat plants. See HETEROECIOUS.

urediospore see UREDINIOSPORE.

uredium see UREDINIOSPORE.

uredosorus see UREDINIOSPORE.

uredospore see UREDINIOSPORE.

uridine the NUCLEOSIDE composed of uracil joined by a β-glycosidic bond to D-ribose.

uridine diphosphate see NUCLEOSIDE DIPHOSPHATE SUGARS.

uridylic acid the NUCLEOTIDE formed from uracil.

urn the part of a moss capsule that contains the spores.

uronic acids a class of sugar acids in which the carbon atom carrying the primary hydroxyl group (i.e. the CH_2OH end of the molecule) is oxidized to a carboxyl group (COOH). Uronic acids are biologically the most important group of the sugar acids and are common constituents of polysaccharides. For example, *glucuronic acid*, derived from glucose, is a common constituent of gums and mucilages. *Galacturonic acid* is a component of pectic substances while *mannuronic acid* is found in the seaweed gum ALGINIC ACID.

usnic acid an acid found in certain species of lichens, which has antibiotic activity against some Gram-positive bacteria (see GRAM STAIN), and has been used in ointments to treat skin infections.

Ustilaginales an order of the USTOMYCETES containing the smut fungi (causing SMUTS), which number some 950 species in some 50 genera. The TELIOSPORE (ustilospore, smut spore, or brand spore) is not terminal and BASIDIOSPORES develop directly on the promycelium rather than on sterigmata. The basidiospores are not actively discharged. The smuts are endoparasites, mainly of angiosperms, forming mycelia in the host tissues, often with haustoria and sometimes also with CLAMP CONNECTIONS. When grown in culture they produce yeastlike forms. Compare UREDINALES.

ustilospore see USTOMYCETES.

Ustomycetes a class of the BASIDIOMYCOTA containing facultative plant parasites with mainly intercellular mycelia composed of septate hyphae in which the rim of the central pore is somewhat flattened; CLAMP CONNECTIONS may be present. Diploid *ustilospores* are produced in sori. These germinate to produce a linear basidium, which may be septate; there is no basidioma. The class contains over 1000 species in some 63 genera and includes the orders Cryptobasidiales, Cryptomycocolacales, Exobasidiales, Graphiolales, Platyogloeales, Sporidiales, and USTILAGINALES (smut fungi).

utricle (bladder) 1 an ovoid compartment lined with sensitive hairs that serves to trap and digest small animals. It is seen in species of *Utricularia* (bladderworts) and is a modification of the leaf.

2 an indehiscent bladder-like fruit formed by certain plants in the Cyperaceae (sedges), Chenopodiaceae, and Amaranthaceae. It is a type of achene.

V v

vacuolation see CELL EXTENSION.

vacuole a fluid-filled cavity within the cytoplasm and separated from it by a membrane, the TONOPLAST. In newly formed cells, when division has ceased, vacuoles are formed from small detached parts of the endoplasmic reticulum. As fluid (cell sap) accumulates they enlarge and coalesce to form a single vacuole, pushing the protoplasm against the cell wall. The surface area of the wall is increased, by stretching and the formation of additional material, to accomodate the increased volume of the cell. The vacuolar sap is a solution of organic and inorganic compounds. These may include sugars, soluble polysaccharides, soluble proteins, amino acids, carboxylic acids, red, blue, and purple anthocyanins, and mineral salts. Starch grains, oil droplets, and crystals of various kinds may also be present. These constituents of the sap probably represent metabolic by-products and reserve food material.

vadose zone see PHREATIC ZONE.

vagile describing a plant or protoctist that is motile. Compare SESSILE.

vaginule in a moss sporophyte, a ring or sheath of tissue enveloping the seta.

valine a nonpolar amino acid with the formula $(CH_3)_2CHCH(NH_2)COOH$ (*see illustration at* amino acid). Valine is synthesized from pyruvate via α-ketoisovalerate. The degradation of valine is extremely complex, leading eventually to proprionyl CoA and thence to succinyl CoA. Methionine, leucine, and isoleucine also degrade to proprionyl CoA; the last

sequence of reactions to succinyl CoA is thus common to all four amino acids.

vallecular canal a longitudinal channel in the stem internode of *Equisetum* and some of its fossil relatives, positioned radially opposite a longitudinal furrow (valley) between the stem ridges. The vallecular canals are arranged roughly between the vascular bundles. Compare CARINAL CANAL.

valley bog see BOG.

valvate describing the arrangement of leaves or scale leaves in a bud in which the edges of adjacent leaves touch but do not overlap. *See illustration at* vernation.

valve **1** one of the segments into which a fruit or a spore capsule splits.
2 in a DIATOM, the frustule, or either of the two 'halves' of the silica cell test.

valvule see PALEA.

Van Valen's law a law proposed by L. Van Valen, which states that when the age classes of different taxa are plotted against the logarithm of the number of survivors at the start of each interval, the plot is a more or less straight line, i.e. there is a more or less constant rate of extinction, and the probability of a taxon becoming extinct should not vary with respect to its age. See also POPULATION DYNAMICS.

variance (mean square) symbol V or σ^2. The square of the STANDARD DEVIATION. The mean square is the estimated variance.

variation the occurrence of differences between individuals. Such differences may be due to inherited (genetic) and environmental factors. Genetic variation is commonly due to RECOMBINATION in sexually reproducing organisms, although the ultimate source of all genetic variation is MUTATION (see GENE). Environmental variation may be caused by various factors, such as population density, nutritional status, light intensity, etc. A characteristic that is largely influenced by environmental factors is said to show phenotypic PLASTICITY and exhibit low HERITABILITY. Variation may be due to differences in kind (see DISCONTINUOUS VARIATION) or differences in degree (see CONTINUOUS VARIATION). See also INFRASPECIFIC VARIATION.

variegation the occurrence of patches or streaks of different coloured tissues in a plant organ, usually a leaf or petal. It may be due to infection, particularly viral infection, mineral deficiency, or physiological or genetic differences between the cells. The variegated petals of certain tulip varieties (e.g. Rembrandt tulips) are due to a virus infection. The variegated leaves of *Coleus* are caused by groups of cells developing different pigment combinations. This is a form of somatic variegation and heritable differences between different coloured parts of the leaf are not evident. In contrast the variegated leaves of, for example, *Pelargonium* are due to genetic variation arising from mutation (see PERICLINAL CHIMAERA).

variety a rank subordinate to species but above the category form in the taxonomic hierarchy. Varieties are morphological variants, which may or may not have a clear geographical distribution. Sometimes they represent only a colour or habit phase. The variety of one author may be designated a subspecies or form by another. See also CULTIVAR.

varve see VARVE DATING.

varve dating a method used to determine the age of a particular sediment and of the fossils that it contains. A varve is a layer of sediment that has been deposited in a glacial lake by the melt waters in spring and summer. Because the particles brought down in the spring are much coarser than those deposited later in the year, annual layers can be distinguished. Varves were formed in the Pleistocene when the ice was retreating, and by counting them, the chronology for parts of northern Europe

has been established. See also DENDROCHRONOLOGY.

vascular describing plants that have a vascular system that includes XYLEM and PHLOEM vessels.

vascular bundle one of a number of strands of primary vascular tissue constituting the vascular system of the plant. Vascular bundles consist mainly of xylem and phloem, which may be separated by a FASCICULAR CAMBIUM. The relative position of xylem and phloem determines the type of bundle (COLLATERAL, BICOLLATERAL, CONCENTRIC, AMPHICRIBAL, or AMPHIVASAL), and is often important taxonomically. The vascular bundles in a dictyostele are called MERISTELES. See also STELE.

vascular cambium a LATERAL MERISTEM, found in those vascular plants exhibiting secondary growth, that gives rise to SECONDARY XYLEM and SECONDARY PHLOEM mostly by PERICLINAL cell divisions. The vascular cambium contains FUSIFORM INITIALS, which give rise to the axial system, and RAY INITIALS, which give rise to the radial (ray) system of the secondary tissues. Compare PHELLOGEN. See also CAMBIUM, FASCICULAR CAMBIUM, INTERFASCICULAR CAMBIUM.

vascular cryptogam a vascular plant that reproduces by spores rather than by seeds.

vascular cylinder see STELE.

vascular plant any plant containing conducting tissue, i.e. xylem and phloem. Vascular plants are usually terrestrial or epiphytic and the sporophyte, which is the dominant generation, is differentiated into stem, leaves, and roots. They also differ from nonvascular plants (see BRYOPHYTES) in possessing stomata. Fossil forms intermediate between the nonvascular and vascular plants have not been discovered and it seems probable that the two groups have evolved separately.

Depending on the relative importance attached to the possession of vascular tissue as compared to the production of seeds, the vascular plants were formerly classified either in one division, the TRACHEOPHYTA, or two, the PTERIDOPHYTA and the SPERMATOPHYTA. In most modern classifications, including the FIVE

KINGDOMS CLASSIFICATION, the vascular plants (Tracheata) are placed in nine phyla: the PSILOPHYTA, LYCOPHYTA, SPHENOPHYTA, FILICINOPHYTA (Pterophyta), CYCADOPHYTA, GINKGOPHYTA, CONIFEROPHYTA, GNETOPHYTA, and ANTHOPHYTA (Angiospermophyta).

vascular system the continuous network of *vascular tissue*, i.e. XYLEM and PHLOEM, throughout a plant body. See also VASCULAR BUNDLE.

vascular transition see TRANSITION REGION.

vector 1 an agent that carries a pathogen. Strictly this includes wind, rainsplash, infected tools, etc. but more usually the term is applied to animal vectors and in particular to insects. Man can also be a vector – for example, a scientist examining an outbreak of disease may carry fungal spores on his clothing when moving to a healthy field. Insects are particularly important in the transmission of virus and mycoplasma diseases, the most common vectors being aphids, whiteflies, and leafhoppers. The insects acquire the pathogen while feeding on infected plants and transmit it when they move on to a healthy plant. Mites and nematodes are also vectors of virus diseases.

2 (cloning vector) see GENE CLONING.

vegetation mosaic the distribution pattern of different plant communities or of different stages of the same community. For example, the cyclical changes in the hummock -and- hollow cycle of raised bogs lead to a pattern of bogs at different stages of the cycle

vegetative relating to cell lines that do not lead to reproductive cells.

vegetative apex the tip of a thallus or shoot that is not sexually differentiated. Compare SEXUAL APEX.

vegetative cell see VEGETATIVE NUCLEUS.

vegetative growth growth of the vegetative parts of a plant.

vegetative nucleus (vegetative cell) the large nucleus formed within the POLLEN grain of angiosperms along with one or two smaller GENERATIVE NUCLEI. After germination of the pollen grain it migrates to the tip of the pollen tube where it may be termed the *tube nucleus*. It is thought to control the growth and development of the pollen tube and

disintegrates when the pollen tube penetrates the nucellus.

vegetative propagation see VEGETATIVE REPRODUCTION.

vegetative reproduction (vegetative propagation) a form of asexual reproduction in which specialized multicellular organs formed by the parent become detached and generate new individuals. Such parts may include bulbs, corms, gemmae, rhizomes, stems, tubers, etc. The regenerative capacities of various plant organs have been exploited through a number of techniques used in agriculture and horticulture to multiply stocks. These methods include budding, layering, air layering, cutting, and pegging, and may be enhanced by the careful control of microclimate or by growth regulators.

vegetative state 1 a stage in a plant's life cycle when it reproduces asexually. See VEGETATIVE REPRODUCTION.
2 a stage in a bacteriophage life cycle when its genome multiplies in the host cell directing synthesis of the components of more phages, but during which it does not cause lysis of the host cell and thus become infective.

veil see MUSHROOM.

vein a VASCULAR BUNDLE, or a group of closely associated bundles, in a leaf. In a leaf vein, the xylem is almost invariably positioned adaxially and the phloem abaxially, although there is sometimes an additional layer of adaxial phloem. Veins are sometimes surrounded by bundle sheaths of collenchyma, sclerenchyma, or parenchyma, which may extend to the leaf epidermis. The pattern formed by the veins in a leaf is called the VENATION.

veination see VENATION.

velamen the multiple EPIDERMIS of the aerial roots of many orchids, aroids, and other monocotyledons. It consists of densely packed cells that lack living protoplasts and have thickened walls. In wet weather these cells become filled with water but it is not certain whether or not the velamen performs an absorptive function.

veld (veldt, bushveld) southern African savanna, often with scattered trees (tree veld) and shrubs (bush veld) at lower elevations and thorn bushes (thorn veld) in more arid areas. Much of the veld is used for pasture and farming. The grass veld is dominated by species of red grass (*Hyparrhenia*), and in drier regions, by drought-resistant grasses with a less complete ground cover. In more arid areas there is a scattering of acacias, aloes, and scrub. The *tree veld* of higher regions is a woodland savanna with open or light tree cover, dominated by *Brachystegia* trees (Fabaceae) that are fire-resistant, and by tall perennial grasses with herbs. At lower elevations the tree veld supports scattered acacia and marula trees (*Scelerocarya birrea*) with red grass; in some regions succulents such as euphorbias take over from red grass, and baobabs (*Adansonia digitata*), mopane trees (*Colophospermum mopanei*), and fan palms become common. Tree veld is probably maintained by fire, and in its absence may well revert to forest or scrub. See GRASSLAND.

velum 1 see ANNULUS.

2 the flap of tissue that protects the sporangia in certain lycophytes, e.g. *Isoetes*.

venation (veination) the pattern formed by the veins of a leaf, as viewed from above or below. The many different types of venation (see illustration) are useful diagnostically, especially in identifying fragmentary material. As a very general rule, the leaves of monocotyledons tend to have many parallel veins of more or less equal width while those of dicotyledons may be divided into one or a few primary veins, with secondary and tertiary veins branching off giving a net-veined or reticulate pattern, but there are exceptions in both groups.

venter the swollen flask-shaped base of an ARCHEGONIUM that contains the megaspore.

ventral 1 in THALLOSE organisms, the lower

craspedromous   parallelodromous

campylodromous

brochidodromous

actinodromous

eucamptodromous

reticulodromous

Venation: some common forms of leaf venation

surface next to the ground or other substrate.

2 in lateral organs, ADAXIAL.

ventral canal cell one of the products that is formed along with the egg cell when the primary ventral cell of an ARCHEGONIUM divides. It has no cell wall and lies at the base of the neck. When the archegonium is mature the ventral canal cell becomes mucilaginous and may produce chemicals to attract the male gametes.

ventral placentation see MARGINAL PLACENTATION.

ventral suture the line of fusion where the margins of the MEGASPOROPHYLL join to form the characteristic tubular shape of the ovary as found in angiosperms. This is often one of the first lines of weakness along which DEHISCENCE occurs at fruit ripening.

vernal relating to late spring.

vernalin a hypothetical plant hormone that, it has been suggested, is formed in meristematic regions of a plant subjected to cold. It is thought to be a GIBBERELLIN. It is probable that there is no one substance formed by VERNALIZATION and that the biochemical basis of vernalization is different in different cold-requiring species. Thus applying gibberellin to seeds replaces the vernalization requirement in some species but not others. Other substances that have partly or completely replaced the vernalization requirement in different species include auxin, kinetin, RNA, and vitamin E.

vernalization the promotion of flowering by exposure of young plants to a cold treatment. For example, the winter varieties of wheat, barley, oats, and rye will normally only flower in early summer if they were sown before the onset of winter. However in areas experiencing very harsh winters, as in the Soviet Union, this may not be possible. The plants are thus given an artificial cold treatment and planted in the spring. See VERNALIN.

vernation (prefoliation) the arrangement of leaves or scale leaves in a bud (see illustration). The terms given to each pattern of vernation also apply to the arrangement of perianth segments in a flower bud (AESTIVATION). Compare PTYXIS.

verrucose see TUBERCULATE.

versatile describing an anther that is joined to the filament about half way along its length and can move freely. Compare BASIFIXED, DORSIFIXED.

verticillaster see INFLORESCENCE.

verticillate arranged in one or more whorls.

vesicle a general term for a cavity within the cytoplasm that is surrounded by a membrane. Vesicles vary considerably in shape, being tubular, spherical, discoid, ovoid, etc. They may contain particles, e.g. endocytotic particles (see ENDOCYTOSIS), or

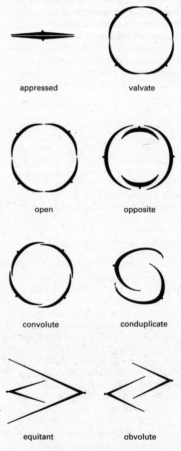

appressed   valvate

open   opposite

convolute   conduplicate

equitant   obvolute

Vernation: types of vernation

fluids, e.g. secretory products from the GOLGI APPARATUS. The membrane isolates the contents from the cytoplasm.

vesicular–arbuscular mycorrhiza see MYCORRHIZA.

vessel a continuous longitudinal tube composed of advanced tracheary cells (VESSEL ELEMENTS). Vessels are present in the xylem of some ferns, most angiosperms, and in the Gnetophyta and are the main water-transporting cells of these plants. Compare TRACHEID.

vessel element (vessel member) a tracheary cell in the XYLEM of some ferns, most angiosperms, and the Gnetophyta, characterized by the presence of a PERFORATION PLATE and a lignified secondary cell wall and lacking a living protoplast at maturity. Vessel elements are joined end to end to form VESSELS, their end walls having broken down to give the perforation plates. Vessel elements are generally shorter and broader than TRACHEIDS and also differ in having perforation plates and often different types of wall pitting.

vestigial organ an organ that has no function, and is often reduced in size, but which resembles certain fully functioning organs in related organisms. Vestigial organs are thought to be evolutionary relicts – present and functioning in some ancestral species. Examples include the reduced stamens in female flowers of DIOECIOUS species, and the scale leaves of parasitic flowering plants, thought to have evolved from fully functioning leaves.

vibrio any comma-shaped bacterium. Compare BACILLUS, COCCUS, SPIRILLUM.

vicariad see VICARIANCE.

vicariance (vicariance biogeography) the splitting up of an original biota into several isolated biotas (*vicariants*) by past geological or climatic events. The vicariants then develop independently and different species evolve. The process is presented as an alternative to the traditional theories of speciation by dispersal that have previously been used to explain biogeographical patterns. The existence of areas that contain many species different from but related to species in another distant area is taken by many as evidence for vicariance. Such

species are called *vicariads*. One major event that is postulated as a means by which vicariants could have arisen is the fragmentation of the ancient land masses into the present-day continents (see CONTINENTAL DRIFT). This could explain the disjunct distributions of many genera, e.g. *Liriodendron* (tulip trees) and southern beeches (*Nothofagus*). Compare DISPERSALIST BIOGEOGRAPHY. See ALLOPATRIC, GEOGRAPHICAL ISOLATION.

vicarious distribution the situation where the distributions of two closely related species derived from a common ancestor differ geographically but not ecologically. For example, pairs of closely related plant species one of which occurs in western Europe and the other in North America.

vicarious species see VICARIANCE.

villous having a shaggy appearance due to a covering of long soft curly TRICHOMES.

violaxanthin a XANTHOPHYLL pigment that absorbs light in the blue region of the spectrum. Violaxanthin may in certain circumstances serve as a precursor of ABSCISIC ACID. It may also, with riboflavin, be involved in the PHOTOOXIDATION of endogenous auxins.

virion the inert extracellular phase of a VIRUS, consisting of a strand of DNA or RNA surrounded by a protein coat.

viroid an extremely small infectious agent consisting solely of RNA with no enclosing coat or capsid. Viroids have been isolated from various plants in which they are able to replicate and cause characteristic disease symptoms. Examples are potato spindle tuber viroid, chrysanthemum chlorotic mottle viroid, hop stunt viroid, and avocado sunblotch viroid.

virulence (pathogenicity) the capacity of a pathogen to cause disease. A pathogen often exists in a number of PHYSIOLOGICAL RACES, which, with respect to one particular crop cultivar, may be either virulent or avirulent (non-virulent). The release of resistant crop cultivars imposes a selection pressure on the pathogen to develop new genes for virulence. It has been shown there is a gene-for-gene relationship between crop resistance and virulence of the pathogen.

virulent phage see BACTERIOPHAGE.

virus a small infectious agent that is only able to replicate by modifying the genetic machinery of living host cells. Outside the host cell a virus consists of DNA or RNA surrounded by a protein shell (see CAPSID). In this inert state a given virus has a characteristic size and shape (e.g. polyhedral, spherical, rod-shaped, etc.). Some of the simpler viruses, e.g. tobacco mosaic virus (TMV), can be crystallized.

Approximately 400 plant viruses are known, most of which are single-stranded RNA viruses. Some, e.g. TMV and cucumber mosaic virus, have a wide host range while others are limited to a few species. Viruses can produce a variety of symptoms, e.g. mosaics, leaf spots, and deformed growth of certain organs. The broken flower colours of certain ornamentals, e.g. Rembrandt tulips, are due to virus infection. Some viruses are symptomless though they may still markedly reduce yield. Viruses are transmitted by vectors and by infected seed and pollen. Control is by using virus-free seed or by breeding for hypersensitivity. Viruses are generally not found in meristematic shoot tips and virus-free stocks of certain species can be obtained by tissue culture of meristem explants. CROSS PROTECTION is successful in some crops.

A virus that infects bacteria is termed a BACTERIOPHAGE.

viscotaxis a change in the direction of locomotion of a motile cell or organism in response to a change in the viscosity of the medium in which it is moving.

visible mutation a MUTATION that reveals itself in the phenotype, as opposed to a LETHAL MUTATION, which can only be inferred from the absence of an expected class of individuals in the progeny of an experimental cross.

vital stain a stain used to dye living tissue without harming or killing the cells. Examples include trypan blue and vital red. *Intravital staining* involves the injection of a stain into an organism, some of the living cells taking up the dye. *Supravital staining* involves the removal of living tissue from a multicellular organism and its subsequent staining. See also STAINING.

vitamin any compound, essential in trace amounts for the normal functioning of an organism, but not synthesized by some heterotrophs, which thus need to obtain it from plants or microorganisms. All vitamins with the exceptions of vitamins C (ASCORBIC ACID), E (TOCOPHEROL), and K, function as COENZYMES. They can be divided into two groups, the water-soluble vitamins, which contain all the B-group vitamins (see THIAMIN, RIBOFLAVIN, NICOTINIC ACID, PANTOTHENIC ACID, PYRIDOXINE, BIOTIN, FOLIC ACID, CYANOCOBALAMIN) and vitamin C, and the fat-soluble vitamins, including vitamins A, D, E, and K.

vitamin A an isoprenoid compound (see ISOPRENES) important in many aspects of animal growth, an early symptom of deficiency being nightblindness. It is not present in plants but carotene pigments when ingested are cleaved into two molecules of vitamin A.

vitamin B$_1$ see THIAMIN.

vitamin B$_2$ see RIBOFLAVIN.

vitamin B$_6$ see PYRIDOXINE.

vitamin B$_{12}$ see CYANOCOBALAMIN.

vitamin C see ASCORBIC ACID.

vitamin E see TOCOPHEROL.

vitamin K a fat-soluble quinone found in most plants and many microorganisms. It is believed to play a part in the transfer of electrons from photosystem II to photosystem I in photosynthesis. In animals deficiency affects the normal blood-clotting mechanism.

vitta a resin canal or oil cavity. Vittae are often found in the fruits of plants in the Apiaceae. The number and position (whether between or in the primary ridges of the fruit) are important diagnostic characters in the family.

vivipary (*adj.* viviparous) **1** the premature germination of seeds or spores in situ on the maternal plant before they have been released. This occurs in the mangroves where seeds may develop into sizeable seedlings before they are shed from the parent tree. **2** (false vivipary, prolification) the differentiation of young plants or BULBILS at the floral axils, instead of flowers. This is seen, for instance, in viviparous fescue (*Festuca vivipara*) and in the spider plant (*Chlorophytum comosum*).

volunteer a plant that has grown from self-sown seed. Volunteers may be important inoculum sources for some diseases.

volva the cuplike structure that encircles the base of the fruiting bodies of many basidiomycete fungi. It is part of the remnants of the UNIVERSAL VEIL.

Volvocales an order of green algae of the class CHLOROPHYCEAE containing motile unicells (e.g. *Chlamydomonas*), colonial forms (e.g. *Gonium*), and coenobial forms (e.g. *Volvox*). They are planktonic, being common in freshwater lakes and ponds. Members of the Volvocales have one chloroplast in each cell, often containing an eyespot and a pyrenoid. Usually asexual reproduction is by zoospores, while sexual reproduction varies from ISOGAMY to OOGAMY. In some classifications, the unicellular forms are placed in a separate order, Chlamydomonadales. *Chlamydomonas* has glycoproteins rather than cellulose in its cell wall, and some unicellular species, such as *Phacotus*, produce a LORICA. Under certain conditions, such as lack of water, PALMELLOID stages forms may occur. The zygote produced during sexual reproduction has four undulipodia and is called a *planozygote*. The colonial forms show a range of forms, from platelike colonies (e.g. *Gonium*) through convex plates to spherical colonies (e.g. *Pandorina*). In more complex colonies, such as *Volvox*, only certain large cells (gonidia), which do not have undulipodia, are involved in asexual reproduction. In *Volvox*, sexual reproduction involves the production of small clusters of sperm that swim to a female colony and fertilize the eggs there. The zygotes form resting spores.

Voroni networks SEE THIESSEN POLYGONS.

W w

Waldgrenze see TIMBER LINE.

Wallace effect the idea, proposed by the naturalist A. R. Wallace, that reproductive barriers within a species may be developed and subsequently improved by selection. If there are a number of optimum phenotypes favoured by selection, intermediate organisms less fitted to the environment will be eliminated by NATURAL SELECTION. Thus hybrids between such phenotypes will be at a disadvantage and hybridization will be selected against.

wall pressure the force exerted upon cell contents by the cell wall. It is equal and opposite to the TURGOR PRESSURE.

Warburg effect the inhibition of carbon dioxide assimilation and photosynthesis by atmospheric oxygen, described by O. Warburg in 1920. This phenomenon was later discovered to be due to PHOTORESPIRATION.

water bloom see BLOOM.

water culture see HYDROPONICS.

water moulds see SAPROLEGNIALES.

water potential symbol Ψ. A measure of the energy available in an aqueous solution to cause the migration of water molecules across SELECTIVELY PERMEABLE MEMBRANES during osmosis. Values of water potential cannot be calculated absolutely but are highest in pure water, which for convenience is given the value zero, and fall with increasing solute concentration. Water always tends to move from areas of high (less negative) to areas of low (more negative) potential. This principle governs water conduction in plants. Water potential is the sum of OSMOTIC POTENTIAL, PRESSURE POTENTIAL, and MATRIC POTENTIAL. The term is replacing the concept of OSMOTIC PRESSURE and differs from osmotic pressure in taking capillary and imbibitional forces into account.

water table the surface below which pores and cracks in the rocks are saturated with water. It usually follows the ground surface topography, but with less exaggerated rises and falls. Where the water table rises above ground level a spring, river, or lake forms. See also PHREATIC ZONE.

Watson–Crick model a molecular model for the structure of DNA, formulated by James Watson and Francis Crick in 1953. It comprises a double helix of intertwined sugar-phosphate chains, with parallel rungs of flat base-pairs linking them together. An essential part of this model was the suggestion of specific base-pairing between the nucleotide bases adenine/thymine and cytosine/guanine. This model also provides a theory for how DNA (and hence genes) might replicate itself. See DNA, DNA REPLICATION.

wax a mixture of esters of higher FATTY ACIDS with higher monohydric alcohols or STEROLS. Waxes may also contain odd-carbon alkanes, long-chain monoketones, β-diketones, β-hydroxyketones, and secondary alcohols. Waxes are important components of the waxy CUTICLE covering the stems, leaves, flowers, and fruits of most plants. They are manufactured as oily droplets in epidermal cells, from which they migrate to the outer surface of the plant via tiny canaliculi in the cell wall, and crystallize as rods and platelets. Their pattern of deposition is sometimes used as a micromorphological taxonomic character below the genus level.

The function of waxes is not fully understood although it seems likely that they are involved in water balance. Plant waxes are obtained on a commercial scale from the leaves of the carnauba palm (*Copernica prunifera*) and from the stems of *Euphorbia antisyphilitica*.

WCMC see WORLD CONSERVATION MONITORING CENTRE.

WCU see WORLD CONSERVATION UNION.

weathering the physical and chemical breakdown of rock and the minerals in it. It occurs in exposed rocks and also at the underground boundary between rocks and SOIL or between rocks and water or ice covering them. See also SOIL STRUCTURE.

weed any plant growing where it is not wanted. Many weeds are adapted to exploit disturbed areas of land and, unless measures are taken to prevent it, weeds are the first plants to establish themselves on cleared patches of ground. Many weeds are EPHEMERAL plants that produce large quantities of seed and often pass through several generations in a single year. An example of such a weed is shepherd's purse (*Capsella bursa-pastoris*). Weeds, especially those closely related to crop plants, may also harbour various pests and diseases. For example, the roots of the black nightshade (*Solanum nigrum*) are often infected by the potato cyst eelworm. See also ROGUE.

weedkiller see HERBICIDE.

Weichselian see DEVENSIAN.

weighting in TAXONOMY, the assigning of greater or lesser importance to one character, as compared to another character, according to its known or assumed value. The initial choice of which characters to use in a classification is in itself a positive weighting process, termed selection weighting. Characters considered unreliable and consequently rejected, perhaps because they show too much environmental variation, are said to be given residual or rejection weighting. Once the characters have been selected some may be ascribed additional importance if, for example, they are known to be good diagnostically in other groups. When such weighting is applied before the classification is drawn

up it is termed *a priori* weighting. An example would be the placing together of two taxa that share the one character chromosome number but differ in a number of other characters. Alternatively a classification may be constructed in which all the chosen characters are considered of equal importance; this is the normal procedure in NUMERICAL TAXONOMY. In such a classification it will then be apparent which characters correlate well with other characters. Such characters are given *a posteriori* weighting. This second procedure provides a method for obtaining unbiased correlations of visible morphological characters with other characters, e.g. chemical, cytological, or genetic characters, that are not immediately apparent.

Weismannism the theory of the continuity of the germ plasm proposed by A. Weismann in 1886. It opposed the idea that acquired characteristics could be inherited. Weismann distinguished the body of the organism (the SOMA) from the reproductive cells (the *germ plasm*) and stated that it was the germ cells alone that affected inheritance and not the soma. He suggested that the germ plasm was set aside during early development and was not affected by subsequent changes in the soma. Weismann also formulated a theory of inheritance based on the behaviour of chromosomes. His ideas led to a rejection of LAMARCKISM.

Welwitschiales see GNETOPHYTA.

Western blotting a technique similar to SOUTHERN BLOTTING, but which is used to separate and identify proteins rather than nucleic acids. The mixture of proteins is separated by ELECTROPHORESIS and blotted onto a nitrocellulose filter. The proteins are identified by the use of antibodies which bind to specific proteins. Radioactively labelled antibodies specific to these antibodies then allow the proteins to be located by autoradiography.

wetland an area of land that is subject to seasonal or permanent inundation; examples of wetlands are MARSHES, SWAMPS, BOGS, and FENS. Such areas support ecosystems that are particularly vulnerable to destruction through drainage and cultivation of the land. See also

CONVENTION ON WETLANDS OF
INTERNATIONAL IMPORTANCE.

wet rot a plant disease in which there is
disintegration of tissues and release of cell
fluids. Brown rot of stored fruits, such as
plums and apples, is caused by fungi of the
genus *Sclerotinia*. Wet rot of structural
timber is caused by various fungi
including *Coniophora puteana* and *Poria
vaillantii*.

wetting agent a chemical, such as detergent
or soap, that lowers the surface tension of
water. Wetting agents are added to
fungicide, herbicide, and insecticide sprays
to increase the area of the spray droplets in
contact with the plant surface or insect
body.

whiplash flagellum (acronematic flagellum)
an UNDULIPODIUM that has a smooth
surface, i.e. it lacks the hairs that cover
mastigonemate undulipodia (tinsel flagella).

whisk ferns SEE PSILOPHYTA.

white alkali soil SEE SOLONCHAK.

whorled describing a form of leaf
arrangement in which three or more leaves
arise at each node, as in the bedstraws
(family Rubiaceae).

wide distribution (polychore distribution)
the situation where a taxon has a very
extensive range, spanning several
biogeographical or floralistic regions.

wilderness any area that has not been
subjected to significant human occupation
or interference, so that it remains in its
natural or near-natural state. In the United
States certain areas are specifically
designated wilderness areas, and no
economic activity or traffic is allowed
except by presidential decree, and the
number of visitors is strictly controlled. In
other countries, wilderness areas are often
zones of restricted public access within
national parks or reserves.

wildlife any undomesticated organisms,
especially animals.

wild-type the common form of a gene or
organism in natural (wild) populations.
Wild-type alleles are typically dominant.
They are usually designated '+'.

wilt a plant disease characterized by
WILTING. Wilting often occurs in the
advanced stages of root diseases when water
uptake becomes inadequate. However the

term 'wilt' is usually applied to diseases in
which wilting occurs in the absence of
marked root damage. Such wilting may
occur either because the vascular tissues are
blocked or because water is being
withdrawn by parasitic plants, such as
witchweeds (*Striga*) and broomrapes
(*Orobranche*). Blockage of vascular tissues is
caused by various fungi and bacteria. It may
be due either to the physical presence of
vast numbers of microorganisms or to
substances, such as gums and tyloses, that
the host forms in response to invasion.
Examples of wilt diseases are Dutch elm
disease, caused by the fungus *Ceratocystis
ulmi*, and wilts of potato, tobacco, and
banana, caused by the bacterium
Pseudomonas solanacearum.

wilting the state in which plants lose turgor
and the tissues become limp due to lack of
water. This may occur when the rate of
transpiration exceeds the rate at which
water is drawn in through the roots, or it
may be due to lack of water in the soil. See
also WILT.

wilting coefficient SEE PERMANENT WILTING
POINT.

wilting point SEE PERMANENT WILTING
POINT.

wind pollination SEE ANEMOPHILY.

wind-pruning damage to plants from coastal
winds carrying sand that abrades leaf
tissues, resulting in leaf drop and shoot
dieback. This leads to trees with a
characteristic shape in which their crowns
are shorter and lower on the side facing the
prevailing wind. They may also become
twisted and stunted.

wing **1** either of the two narrow lateral petals
of a 'pea' flower. See also KEEL, STANDARD.
2 the membranous outgrowth of certain
fruits, e.g. the samara.
3 a flange running down a stem or stalk as,
for example, seen along the stems of the
hairy vetchling (*Lathyrus hirsutus*).

Winkler method a method of determining
the dissolved oxygen content of a sample of
water using an iodine titration.

Wisconsin school a group of ecologists led
by J. T. Curtis and his colleagues who in the
1950s devised a range of ordination
methods which they applied to studying the
vegetation of Wisconsin. These included

the use of DENSITY–FREQUENCY–DOMINANCE (DFD) values for each species, then deriving *importance values* to determine the leading dominants. Stands with the same leading dominants were grouped together, then the importance values calculated for all the species in these new groups. This eventually produced a continuum index (see CONTINUUM). These methods were later modified to be more objective. See RELATIVE IMPORTANCE VALUE.

witches' broom a dense mass of deformed twigs, often resembling a bird's nest, caused by the host response to infection by certain insects, mites, viruses, fungi, or parasitic plants. Witches' broom of cacao is caused by the fungus *Marasmius perniciosus* and can result in almost total loss of yield as pods are also infected. Witches' broom of birch (*Betula*) is caused by the fungus *Taphrina betulina*.

wobble hypothesis a hypothesis proposed by Francis Crick in 1966 to explain the degeneracy of the GENETIC CODE. He suggested that during TRANSCRIPTION the interaction between the codon on the mRNA and the anticodon in the tRNA needs to be exact in only two of the three nucleotide positions, but it need not match in the third position and non-standard base-pairing may occur here. The degenerate base is said to be in the wobble position. This also explains why the number of different tRNAs does not exactly equal the number of amino acids coded for.

wood The SECONDARY XYLEM of dicotyledons and gymnosperms, which during secondary growth forms a dense, lignified tissue conferring great mechanical strength, allowing trees to reach considerable heights. See XYLEM.

woodland 1 a plant community similar to a FOREST, but which lacks a closed canopy because the large trees are more widely spaced (typically with 40% closure of the canopy or less). The understorey layer, too, is less dense and well-defined than in a forest. Consequently, the ground flora may consist of heath and scrub as well as grass and herbs.
2 in British colloquial use, the terms woodland and forest are used interchangeably, and the term woodland

may be used to describe any landscape containing broad-leaved woodlands.

wood sugar see XYLOSE.

woody perennial see PERENNIAL.

World Conservation Monitoring Centre (WCMC) an information service established by the WORLD CONSERVATION UNION, the WORLDWIDE FUND FOR NATURE, and the United Nations Environment Programme (UNEP). It compiles information on conservation and the sustainable use of the world's natural resources and assists in the development of other information sources and the training of staff to support them. Topics covered in its reports include data on species, threatened habitats, NATIONAL PARKS, and NATURE RESERVES, international agreements, and conservation and environment programmes. See also RED DATA BOOK.

World Conservation Union (IUCN, International Union for the Conservation of Nature, WCU) a union of governments, government agencies, and nongovernmental organizations, together with scientists and other experts, that works in the field and in the development of policies to protect nature. Its mission is to influence, encourage, and assist societies throughout the world to conserve the integrity and diversity of nature and to ensure that any use of natural resources is equitable and ecologically sustainable. One of the world's oldest international conservation bodies, it has offices throughout the world. It is a leader in the field of conservation and offers expertise in ecology and conservation, environmental law, protected area management, and environmental education. It helps governments and local conservation organizations to share information worldwide and provides a source of stimulation and expertise for putting policies into practice. See also CONVENTION ON BIOLOGICAL DIVERSITY, CONVENTION ON INTERNATIONAL TRADE IN ENDANGERED SPECIES OF WILD FAUNA AND FLORA, RED DATA BOOK.

Worldwide Fund for Nature (WWF, World Wildlife Fund) the largest independent CONSERVATION organization in the world, with over 4.7 million supporters and a global network of 24 National Organizations, 5 Associates, and 26 Programme Offices. WWF provides conservation services based on scientific information, fieldwork, and global policy. It has a network of local fieldwork groups and seeks to advise national and local governments and to raise public understanding of conservation issues. This advice and cooperation extend to large influential international bodies, such as the European Union, the WORLD CONSERVATION UNION (WCN), the United Nations Environment Programme, UNICEF, and the World Bank, and many international nongovernmental organizations. In 1998 it had an institutional presence in 53 countries.

World Wildlife Fund see WORLDWIDE FUND FOR NATURE.

wound cork a layer of PHELLEM that forms over a damaged part of the plant. It prevents desiccation of the underlying tissues and the entry of pathogens. See also CALLUS.

wound hormone any substance produced by damaged cells that diffuses into nearby undamaged cells and there stimulates meristematic activity, resulting in the formation of a protective CALLUS. Auxins, gibberellins, and certain products of wounding have been implicated in the wounding reaction. See also TRAUMATIC ACID.

wound wood see CALLUS.

wracks see FUCALES.

Würm see DEVENSIAN.

WWF see WORLDWIDE FUND FOR NATURE.

xanthin any of a group of oxidized isoprenoid compounds, usually coloured yellow or orange. See also ISOPRENE SUBUNIT.

xanthophyll any of a class of oxygenated hydrocarbons derived from the carotenes. Xanthophylls function mainly as photosynthetic ACCESSORY PIGMENTS, and are found in many different plant species. However the xanthophylls FUCOXANTHIN and peridinin of the brown algae (see PHAEOPHYTA) are of especial interest, because they are the primary light-absorbing pigments for these organisms. They absorb light at frequencies where chlorophyll has only poor absorption and transfer the absorbed energy to chlorophyll. See also LUTEIN.

Xanthophyta (yellow-green algae) a phylum of the PROTOCTISTA consisting mainly of freshwater and terrestrial algae (some are found in marine environments), including unicellular, colonial (palmelloid), filamentous, and siphonaceous forms. In some classification schemes they are placed in the division CHROMOPHYTA. They are characterized by having disclike yellow-green plastids (due to the large amounts of CAROTENOIDS present) and by accumulating oil and CHRYSOLAMINARIN rather than starch. Their chlorophylls a, c_1, c_2 and e, together with xanthins and beta carotene. The cellulose cell walls are rich in PECTIC SUBSTANCES and are made up of overlapping discontinuous parts. In filamentous forms disruption results in a number of H-shaped fragments. Many species are covered in scales. The motile stages have two unequal undulipodia (one hairy and one smooth), hence their older name of Heterokontae. Some species form dormant cysts whose walls are reinforced with iron or silica. See also TRIBOPHYCEAE.

xenia the modification of the form of a fruit or seed by the pollen due to its effect, through DOUBLE FERTILIZATION, on the nature of the endosperm. For example, in maize the endosperm may show a variety of colours depending on the origin of the pollen.

xenogamy fertilization that occurs as a result of the pollination of one flower by pollen from a different, genetically distinct plant (cross-pollination), i.e. the male and female gametes have different genetic makeups. Compare GEITONOGAMY.

xerophile see XEROPHYTE.

xerophyte a plant that is adapted to living in dry conditions caused either by lack of soil water or by heat or wind bringing about excessive transpiration. Many xerophytes are found in deserts, on sand dunes, and on exposed moors and heaths. Some xerophytes, e.g. the SUCCULENT desert cacti, store water in swollen stems and leaves. Many species reduce the rate of transpiration by having permanently rolled leaves, e.g. cross-leaved heath (*Erica tetralix*) or by having leaves that are rolled in dry weather, e.g. marram grass (*Ammophila arenaria*). Some have hairy leaves to trap moist air, e.g. great mullein (*Verbascum thapsus*) while others have stomata sunken into grooves producing pockets of moist air, e.g. *Pinus*. The leaves of many species are leathery with a thick CUTICLE and epidermis to reduce cuticular transpiration. Some species have stomata that close during the day and open at night (see CRASSULACEAN ACID METABOLISM). Compare HYDROPHYTE, MESOPHYTE. See also XEROSERE.

xerosere a pioneer plant community that develops in a dry region. A *lithosere* develops on bare rock, beginning with simple plant forms, such as lichens. Soil building and accumulation of organic matter continue so that xerophytic herbs develop followed by MESOPHYTES, such as hardy herbaceous plants and even larger woody plants. A *psammosere* develops on bare SAND DUNES and a HALOSERE on coastal mud flats and salt marshes. See also SERE.

X-ray diffraction analysis a technique used to determine the three-dimensional pattern of large molecules such as proteins or DNA. A beam of X-rays is fired at a crystal and the resulting diffraction pattern is recorded on a photographic plate behind the crystal. The crystal is rotated slightly and is subjected to more X-rays. By repeating this, the three-dimensional pattern can be determined. It was work of this nature that showed DNA to be a regular double helix.

xylan a polysaccharide in which the major monosaccharide subunit is XYLOSE. Xylans are abundant components of the HEMICELLULOSES. In monocotyledonous plants the dominant hemicellulose is an arabinoxylan, in which ARABINOSE side chains are attached to a backbone of xylose residues.

Xylariales (Sphaeriales) an order of fungi of the class ASCOMYCETES whose members produce well-developed stromata (see STROMA), which may be crustlike (e.g. *Ustulina*), hemispherical (e.g. *Hypoxylon*), club-shaped (e.g. *Xylaria polymorpha*), or antler-shaped (e.g. *Xylaria hypoxylon*). They typically have hard dark PERITHECIA, as, for example, in *Daldinia concentrica*. There are some 795 species in about 92 genera, most of them saprobic. Some, however, cause serious damage as parasites of economically important plants, including *Xylaria polymorpha* and *Ustulina deusta*, which infect beech, and *X. hypoxylon*, which causes root rot of apple.

xylary fibre see FIBRE.

xylem (wood) vascular tissue whose principal function is the upward translocation of water and solutes. It is composed mainly of vessels, tracheids, fibre-tracheids, fibres, and parenchyma cells. It should be noted, however, that all these cell types may not be present in any one wood sample. Wood anatomy is often very important taxonomically, the presence or absence of the various cell types and their distribution within the xylem being important diagnostic characters. The growth of large-diameter xylem vessels in spring followed by progressively smaller vessels through summer and autumn leads to the characteristic annual GROWTH RINGS in the wood of temperate trees. The xylem occurs in association with, and usually internal to, the phloem. See also PRIMARY XYLEM, SECONDARY XYLEM.

xylophilous describing an organism that thrives best growing on wood.

xylose (wood sugar) an aldopentose sugar (*see illustration at* aldose) widely found in plants, particularly in woody tissues. It occurs mainly in its polymerized form XYLAN. It is also a constituent of the rare disaccharide primeverose.

Yy

YAC see YEAST ARTIFICIAL CHROMOSOME.

yeast artificial chromosome (YAC) a vector used in GENETIC ENGINEERING to transfer and clone large pieces of DNA (up to 500 000 nucleotide bases long). YACs are PLASMIDS combined with chromosome segments derived mainly from yeast. The necessary elements for the YAC to function as a plasmid in *Escherichia coli* are incorporated, as well as centromere and telomere regions and autonomous replication sequences that will serve as replication origins. The YAC also carries sites where RESTRICTION ENZYMES can act, which allows for the insertion of foreign DNA for GENE CLONING.

yeasts see BLASTOMYCETES, SACCHAROMYCETALES.

yellow-green algae see XANTHOPHYTA.

yellowing see CHLOROSIS.

yellows **1** a plant disease, caused by mineral deficiency or a virus or MYCOPLASMA, in which there is yellowing of the foliage. Yellowing is a common symptom when there is a deficiency of elements important in chlorophyll production, namely, iron, magnesium, manganese, nitrogen, or sulphur. Barley yellow dwarf and beet yellows are caused by viruses and coconut lethal yellowing is caused by a mycoplasma. See also DEFICIENCY DISEASE, VIRUS. **2** a disease of cabbage caused by *Fusarium conglutinans*.

Zz

zeatin (6-(4-hydroxy-3-methyl but-2-enyl) aminopurine) the first plant hormone of the CYTOKININ type to be isolated from plant tissues. It was obtained from the kernels of sweet corn (*Zea mays*) hence its name. Subsequently other cytokinins similar in structure to zeatin (i.e. an adenine molecule substituted with an isoprenoid derivative) have been isolated. Some of these are found as minor bases in transfer RNA molecules (see IPA).

zein a low-molecular-weight simple protein found as a major storage protein in maize grains. See PROLAMINE.

zinc symbol: Zn. A metal element, atomic number 30, atomic weight 65.38, needed in trace amounts for successful plant growth, but toxic in large concentrations. Zinc ions are required as cofactors by certain enzymes, e.g. carboxypeptidase, carbonic anhydrase, and alcohol dehydrogenase. It has been suggested that zinc may play a role in the mechanism of action of the hormone ETHANE (ethylene) A common symptom of deficiency is leaf mottling.

Zingiberidae a subclass of monocotyledons containing mainly tropical and warm temperate herbaceous species, including many epiphytes and xerophytes as well as rainforest plants. The inflorescences are terminal, unisexual or bisexual, and usually borne in the axils of bracts. The ovary is made of three fused carpels, with three locules, and the fruit is a berry or capsule. The Zingiberidae includes two orders: the Bromeliales, which contains a single family, the Bromeliaceae (bromeliads, pineapple, Spanish moss); and the Zingiberales, which includes the Musaceae (bananas, hemp, bird of paradise flower, traveller's palm, heliconias), Lowiaceae, Zingiberaceae (ginger, cardamom, turmeric), Cannaceae (cannas, Queensland arrowroot), and Marantaceae (arrowroot).

zoidogamy the form of fertilization in which undulipodiate male gametes (antherozoids) are released from the pollen tube and swim to the egg apparatus. It is found in algae, mosses and liverworts, ferns, and some gymnosperms. Compare SIPHONOGAMY.

zonal soil see SOIL.

zoo- (zo-) suffix denoting animal-like or motile.

zoochlorella a symbiotic green alga (see CHLOROPHYTA) found in the cells of certain invertebrates and protoctists.

zoochory disperal of seeds or spoes by animals. In *endozoochory* the seeds or spores are dispersed after they have passed through the animal's gut. In *exozoochory* they are carried on the surface of the animal.

zoogeographical region a distinct region with its own unique collection of animal species. Several classifications exist, but most commonly, six regions are recognized: Australasian (Australia, New Zealand, and the surrounding regions), Neotropical (South America), Neartic (most of North America), Palaeartic (Europe, Northern Asia, North Africa), Oriental (Southern Asia), and Ethiopian (subSaharan Africa). Phytogeographical regions do not necessarily coincide as factors affecting growth and distribution and also physical barriers are sometimes different for plants and animals. See FLORISTIC REGION, PLANT GEOGRAPHY.

zoogloea see CAPSULE.

zoophilous describing flowers that are pollinated by animals.

zoosporangium a SPORANGIUM that produces ZOOSPORES.

zoospore a motile usually naked spore with one or more undulipodia. It may be produced either by a zygote or a zoosporangium and is dependent on water for dispersal. In some species, e.g. the green alga *Ulothrix*, two types of zoospore may be developed. In *Ulothrix* large macrozoospores with four undulipodia and small microzoospores with two undulipodia are seen. Zoospores may encyst under adverse conditions. Compare APLANOSPORE.

zooxanthella a symbiotic dinoflagellate (see DINOMASTIGOTA) found in the cells of some invertebrates, particularly cnidarians.

Zurich–Montpellier school of phytosociology (Montpellier school of phytosociology) a group of botanists led by J. Braun-Blanquet and his colleagues, working at Zurich and Montpellier from 1927 on, who devised a framework for the classification of vegetation based on its floristic composition. It emphasized characteristic species rather than dominant species or life-forms. In any plant community some species are better indicators of that community than others on account of their ecological relationships. These are the diagnostic species (see FAITHFUL SPECIES) for the ASSOCIATION, which forms the basic unit of this classification hierarchy. Originally intended as a tool for world vegetation classification, the method is most often used in regional and national surveys, especially in Europe. It requires detailed surveying in the field to identify plant associations, which are then grouped into a hierarchical classification resembling that used to classify species, comprising alliances, orders, classes, etc. Each rank has its own characteristic suffix. See also FIDELITY, PHYTOSOCIOLOGY.

zwitterion a dipolar ion, i.e. one that carries both a positive and a negative charge. At their isoelectric point (the pH at which they are electrically neutral) amino acids can form zwitterions, in which the amino group

becomes ionised to $-NH_3^+$ and the acid (carbonyl) group is ionized to $-COO^-$.

Zygnematales (Zygnemaphyceae) see GAMOPHYTA.

zygomorphy see BILATERAL SYMMETRY.

Zygomycetes a class of the ZYGOMYCOTA containing fungi with predominantly chitin or chitosan walls and typically producing spores in sporangia.

Zygomycota a phylum of the FUNGI (in some classifications a subdivision, Zygomycotina, of the EUMYCOTA) containing fungi with a mycelial thallus of aseptate hyphae. The cell walls contain chitin or chitosan. Zygomycotes do not produce motile cells, and asexual reproduction is by nonmotile ENDOSPORES produced in sporangia, or by conidia. The phylum includes both saprobic fungi and many that are parasitic, especially on arthropods. There are some 1056 species in about 173 genera. Some authorities recognize two classes, the ZYGOMYCETES and the TRICHOMYCETES. Others recognize seven, which in some classifications are considered to be orders: Dimargaritales, Endogonales, ENTOMOPHTHORALES, Glomales (which are important symbionts associated with plants in endomycorrhizas), Kickxellales, MUCORALES, and Zoopagales.

zygophore in fungi of the order Mucorales, a specialized hypha that bears a zygospore.

zygospore a thick-walled zygote that is formed by the fusion of isogamous gametes (see ISOGAMY). It is characteristic of organisms that reproduce by conjugation, e.g. fungi of the MUCORALES and algae of the GAMOPHYTA. The thickened walls enable it to survive adverse environmental conditions. Compare OOSPORE.

zygote the product of the fusion of two gametes, before it has undergone mitosis or meiosis. In lower algae and fungi it may retain the motile nature of the gametes. More usually the zygote is immobile and may develop a thickened resistant wall to form a ZYGOSPORE. In higher plants the zygote is typically protected by maternal tissue, and divides immediately after fertilization forming a PROEMBRYO. See also SEXUAL REPRODUCTION.

zygotene see PROPHASE.

zygotic meiosis meiosis that occurs in a zygote immediately after it is formed.

zymase the mixture of enzymes, isolated from yeasts, that brings about ALCOHOLIC FERMENTATION. It includes pyruvate decarboxylase, which catalyses the formation of acetaldehyde from pyruvate, and alcohol dehydrogenase, which catalyses the reduction of acetaldehyde to ethanol. It also includes the enzymes of the glycolytic pathway (see GLYCOLYSIS).

zymogen an inactive precursor of an enzyme, which requires a biochemical change such as a hydrolysis reaction that reveals the ACTIVE SITE, in order to become an active enzyme.

zymogenous describing an organism that is present temporarily in large numbers in a habitat usually following an increase in the available nutrients, for example after the addition of readily decomposable organic matter.

Appendix

Table 1 The meanings of some common specific epithets

Epithet	Meaning	Epithet	Meaning
acetosus	having an acid taste	comosus, comatus	tufted
		concolor	uniform in colour
aestivus, aestivalis	of summer	corniculatus, cornutus	having a horn-like appendage
agrestis	of fields or cultivated land	costatus, costatalis	ribbed
alatus	winged	crassus	thick, fleshy
albus	white	crispus, crispatus	finely waved
alpestris, alpinus	of the Alps or high mountains	cruentus	blood red
		cyaneus	dark blue
altissimus	very tall	demersus	growing in water
altus	tall, high	demissus	lowly, humble
amabilis	pleasing, lovely	discolor	not uniform in colour
angustatus	narrow, slender		
arborescens	treelike	dulcis	sweet
arenarius	of sandy places	dumosus	bushy
argenteus	silvery	echinatus	spiny
arundinaceus	reedlike	edulis	used for food
arvensis	of fields, especially ploughed fields	effusus	spread out thinly
		elatus	tall
aureus	golden yellow	elodes	of marshes
australis	southern	ensatus	swordlike
autumnalis	of autumn	esculentus	edible
borealis	northern	ferrugineus	rust coloured
caeruleus	sky blue	fistulosus	hollow and tubular
caesius	blue grey		
calcaratus	spurred	flabellatus	fanlike
campanulatus	bell-like	flavus	pale yellow
campestris	of fields	flexuosus	bending alternately in opposite directions
candidus	shining white		
canescens	hoary		
capreolatus	having tendrils		
carinatus	keeled	floribundus	flowering profusely
carneus	flesh coloured		
carnosus	fleshy, succulent	fluitans	floating
castaneus	chestnut coloured	fluviatilis	of rivers
caudatus	tailed	foetidus	foul smelling
cerasiferus	having cherry-like fruits	fontinalis, fontanus	of springs
		fulgens, fulgidus	bright, shining
ceriferus	producing wax	fulvus	yellow brown
cernuus	nodding, drooping	furcatus	forked
		gelidus	of cold regions
cinereus	ash grey	glutinosus	sticky
cirrhosus, cirrhatus	having tendrils	gracilis	slender
clavatus	club shaped	graveolens	strong smelling
coccineus	scarlet	griseus	pearl grey
collinus	of hills	hederaceus	ivy-like
communis	common	hepaticus	liver coloured

Epithet	Meaning	Epithet	Meaning
hibernus, hiemalis	of winter	nudus	naked
hirsutus	hairy	nutans	nodding, hanging
horridus	very bristly	occidentalis	western, American
hortensis	of a garden		
humilis	dwarf	officinalis	having medical use
hystrix	bristly		
incanus	grey, hoary	oleraceus	vegetable crop
incarnatus	flesh coloured	oratensis	of meadows
impudicus	shameless, immodest	orientalis	eastern, Asian
		paludosus, palustris	of bogs, marshes, or swamps
indicus	of India		
infundibuliformis	funnel shaped	pannosus	densely hairy
insignis	outstanding	parvus	small
integrifolius	having entire leaves	patens	spreading
		petraeus	growing among rocks
italicus	of Italy		
junceus	rushlike	plenus, pleniformis	full, double
lacustris	of lakes or ponds	praecox	developing early
laevigatus	smooth, polished	prasinus	bright green
lanatus, lanosus	woolly	pulcher	beautiful
latifolius	having broad leaves	pumilus	dwarf
		pusillus	very small; weak
leucanthus	having white flowers	ramosus	branched
		reniformis	kidney shaped
limosus	of muddy places	repens, reptans	creeping, prostrate
lineatus	marked with parallel lines		
		riparius	growing by rivers or streams
littoralis	of the seashore		
lividus	lead coloured	rivularis, rivalis	growing by streams or brooks
lunatus	half-moon shaped		
luridus	dull yellow		
luteus	deep yellow	roseus	pink
maculatus	spotted, blotched	rostratus	beaked
meridionalis	southern; flowers opening around midday	rubellus, rubens, rufus	reddish
		ruber	red
mollis	softly hairy	ruderalis	growing in rubbish
montanus, monticolus	of the mountains		
		rupestris	growing on rocks
moschatus	musk smelling	sativus	cultivated
muralis	of walls	saxatilis	growing among rocks
nanus	dwarf		
natans	floating on or under water	scaber	rough
		scandens	climbing
nemoralis, nemorosus	of shade or woodlands	sempervirens	evergreen
		sericeus	silky
nervosus	having conspicuous veins	serotinus	late
		setaceus, setasus	bristly
niger	black	sinensis	of China
nivalis, niveus	snow white	somniferus	sleep inducing

Epithet	Meaning	Epithet	Meaning
speciosus	good looking	tinctorius	used for dyeing
spectabilis	showy	umbrosus	of shade
squamatus	scaly	uncinatus	hooked
squarrosus	having overlapping leaves with outward-projecting tips	usitassimus, utilis	useful
		velutinus	velvety
		ventricosus	inflated, especially unevenly so
sylvaticus, sylvestris	of woods	vernalis, vernus	of spring
		versicolor	variously coloured
tenellus	delicate	vescus	small; edible
tenuis	slender	virens, viridus	green
terrestris	of dry ground	vulgaris	common

Table 2 The systematic names of some organic compounds

Common name	Systematic name	Common name	Systematic name
acetaldehyde	ethanal	α-ketoglutaric acid	1-oxybutanedioic acid
acetic acid	ethanoic acid	lactic acid	2-hydroxypropanoic acid
acetone	propanone		
alanine	2-aminopropanoic acid	lauric acid	dodecanoic acid
aspartic acid	aminobutanedioic acid	maleic acid	cis-butenedioic acid
catechol	benzene-1,2-diol	malic acid	2-hydroxybutanedioic acid
cinnamic acid	3-phenylpropenoic acid		
		malonic acid	propanedioic acid
citric acid	2-hydroxypropane-1,2,3-tricarboxylic acid	mercaptans	thiols
		myristic acid	tetradecanoic acid
		oleic acid	cis-octadec-9-enoic acid
ethylene	ethene		
ethylene glycol	ethane-1,2-diol	oxalic acid	ethanedioic acid
fatty acids	carboxylic acids	oxaloacetic acid	2-oxybutanedioic acid
formaldehyde	methanal		
fumaric acid	trans-butenedioic acid	oxalosuccinic acid	1-oxypropane-1,2,3-tricarboxylic acid
glutamic acid	2-aminopentane-dioic acid		
		palmitic acid	hexadecanoic acid
glycerol	propane-1,2,3-triol	phloroglucinol	benzene-1,3,5-triol
		pyruvic acid	2-oxypropanoic acid
glycine	aminoethanoic acid		
		stearic acid	octadecanoic acid
glycols	diols	succinic acid	butanedioic acid
hydroquinone	benzene-1,4-diol	toluene	methylbenzene
isoprene	methylbuta-1,3-diene	o-xylene	1,2-dimethylbenzene

Table 3 Base SI units

Physical quantity	Name	Symbol
length	meter	m
mass	kilogram(me)	kg
time	second	s
electric current	ampere	A
thermodynamic temperature	kelvin	K
luminous intensity	candela	cd
amount of substance	mole	mol

Table 4 Some derived SI units with special names

Physical quantity	Name	Symbol
absorbed dose	gray	Gy
activity	becquerel	Bq
catalytic activity	katal	kat
Celsius temperature	degree Celsius	°C
dose equivalent	sievert	Sv
electric conductance	siemens	S
electric capacitance	farad	F
electric charge	coulomb	C
electric potential difference	volt	V
electric resistance	ohm	Ω
energy	joule	J
force	newton	N
frequency	hertz	Hz
illuminance (illumination)	lux	lx
inductance	henry	H
luminous flux	lumen	lm
magnetic flux	weber	Wb
magnetic flux density	tesla	T
plane angle	radian	rad
power	watt	W
pressure	pascal	Pa
solid angle	steradian	sr

Until 1995, the radian and the steradian were regarded as supplementary units.

Table 5 Decimal multiples and submultiples used with SI units

Submultiple	Prefix	Symbol	Multiple	Prefix	Symbol
$\mathrm{S}10^{-1}$	deci-	d	10^1	deca-	da
10^{-2}	centi-	c	10^2	hecto-	h
10^{-3}	milli-	m	10^3	kilo-	k
10^{-6}	micro-	m	10^6	mega-	M
10^{-9}	nano-	n	10^9	giga-	G
10^{-12}	pico-	p	10^{12}	tera-	T
10^{-15}	femto-	f	10^{15}	peta-	P
10^{-18}	atto-	a	10^{18}	exa-	E
10^{-21}	zepto-	z	10^{21}	zetta-	Z
10^{-24}	yocto-	y	10^{24}	yotta-	Y

Internet links

There are many sites on the Internet giving information about botany and life science. A selection of useful sites is given below. In general, they are sites of universities or official organizations rather than web sites run by individuals.

General sites for biology resources

The US Geological Survey	http://biology.usgs.gov
National Biological Information Infrastructure	www.nbii.gov

Plant science

American Society of Plant Biologists	www.aspb.org/education
Botanical Society of America	www.botany.org
European Initiative for Biotechnology Education	www.eibe.info
International Plant Names Index	www.ipni.org/index.html
Internet Directory for Botany	www.botany.net/IDB/info.html
Millennium Seed Bank Project	www.pbgkew.org/msbp/index.html
Missouri Botanical Garden	www.mobot.org
National Museum of Natural History	www.si.edu/departments/ botany.html
New York Botanical Garden	www.nybg.org
Photosynthesis and the Web	photoscience.la.asu.edu/photosyn /photoweb/default.html
Plantlife International	www.plantlife.org
Royal Botanic Gardens, Kew	www.rbgkew.org.uk
Southwest School of Botanical Medicine	www.swsbm.com
University of California Botanical Garden	www.mip.berkeley.edu/garden

Museum sites

American Museum of Natural History	www.amnh.org
Canadian Museum of Nature	www.nature.ca
Harvard Museum of Natural History	www.hmnh.harvard.edu
Natural History Museum of Los Angeles County, California	www.nhm.org
Smithsonian Institution National Museum of Natural History	www.mnh.si.edu
The Natural History Museum, London	www.nhm.ac.uk

| University of Michigan Museum of Zoology | www.ummz.lsa.umich.edu |
| Yale Peabody Museum of Natural History | www.peabody.yale.edu |

Biochemistry and cell and molecular biology

European Bioinformatics Centre	http://ebi.ac.uk
Harvard University Department of Molecular and Cellular Biology	http://mcb.harvard.edu/BioLinks.html
Human Genome Project	www.ornl.gov/hgmis
International Union of Biochemistry and Molecular Biology (Nomenclature)	www.chem.qmul.ac.uk/iubmb
Leeds University Bioinformatics Links	www.bioinf.leeds.ac.uk/ bioinformatics.html
Oxford University Bioinformatics Centre	www.molbiol.ox.ac.uk
Princeton University	www-hhmi.princeton.edu/hhmi/ index.html
Southwest Biotechnology and Informatics Center	www.swbic.org
UK Human Genome Mapping Project	www.hgmp.mrc.ac.uk
University of Arizona, Biochemistry	www.biology.arizona.edu/biochemistry/ biochemistry.html
University of Arizona, Molecular Biology	www.biology.arizona.edu/ molecular_bio/molecular_bio.html
University of Southern Maine, Biochemistry Resources	www.usm.maine.edu/~rhodes/ Biochem

Environment

The US Environmental Protection Agency	www.epa.gov/enviroed
Environmental Sciences Division, Oak Ridge National Laboratory	www.esd.ornl.gov
IUCN Red List	www.redlist.org
Smithsonian Environmental Research Center	www.serc.si.edu

Biographies

| Nobel Prizes | http://nobelprize.org |
| University of Calfornia, Berkeley | www.ucmp.berkeley.edu/help/topic/ history.html |